D1675309

Handbook of Metrology

Edited by
Michael Gläser and
Manfred Kochsiek

Related Titles

Stock, R. (ed.)

Encyclopedia of Applied Nuclear Physics

ISBN: 978-3-527-40742-2

Stock, R. (ed.)

Encyclopedia of Applied High Energy and Particle Physics

2009
ISBN: 978-3-527-40691-3

Wilkening, G., Koenders, L.

Nanoscale Calibration Standards and Methods

Dimensional and Related Measurements in the Micro- and Nanometer Range

2005
ISBN: 978-3-527-40502-2

Riehle, F.

Frequency Standards

Basics and Applications

2004
ISBN: 978-3-527-40230-4

Trigg, G. L. (ed.)

Encyclopedia of Applied Physics

The Classic Softcover Edition

2004
ISBN: 978-3-527-40478-0

Brown, T. G., Creath, K., Kogelnik, H., Kriss, M. A., Schmit, J., Weber, M. J. (eds.)

The Optics Encyclopedia

Basic Foundations and Practical Applications. 5 Volumes

2004
ISBN: 978-3-527-40320-2

Gåsvik, K. J.

Optical Metrology

2002
ISBN: 978-0-470-84300-0

Keithley, J. F.

The Story of Electrical and Magnetic Measurements

From 500 BC to the 1940s

2001
ISBN: 978-0-7803-1193-0

Handbook of Metrology

Edited by
Michael Gläser and Manfred Kochsiek

Volume I

WILEY-VCH Verlag GmbH & Co. KGaA

The Editors

Dr. Michael Gläser
Physikalisch-Technische
Bundesanstalt
Bundesallee 100
38116 Braunschweig
Germany

Prof. Dr.-Ing. Manfred Kochsiek
Physikalisch-Technische
Bundesanstalt
Bundesallee 100
38116 Braunschweig
Germany

Library of Congress Card No.:
applied for

British Library Cataloguing-in-Publication Data
A catalogue record for this book is available from the British Library.

Bibliographic information published by the Deutsche Nationalbibliothek
The Deutsche Nationalbibliothek lists this publication in the Deutsche Nationalbibliografie; detailed bibliographic data are available on the Internet at ⟨http://dnb.d-nb.de⟩.

Composition Laserwords Private Ltd., Chennai, India
Printing and Binding T.J. International Ltd., Padstow, Cornwall
Cover Design Schulz Grafik-Design, Fußgönheim

Printed in the Federal Republic of Germany
Printed on acid-free paper

ISBN: 978-3-527-40666-1

Contents

Handbook of Metrology. Edited by Michael Gläser and Manfred Kochsiek

ISBN: 978-3-527-40666-1

List of Contributors

Felicitas Arias
Bureau International des
Poids et Mesures
Pavillon de Bretevil
92312 Sèvres cedex
France

Andreas Bauch
Physikalisch-Technische Bundesanstalt
Bundesallee 100
38116 Braunschweig
Germany

John H. Cantrell
University of Tennessee
NASA Langley Research Center
3 East Taylor Street
Hampton, VA 23681-2199
USA

Simona M. Cristescu
Radboud University
Heijendaalseweg 135
6525AJ Nijmegen
The Netherlands

Horst Czichos
University of Applied Sciences
BHT, Berlin
Luxemburger Straße 10
13353 Berlin
Germany

Richard S. Davis
Bureau International des
Poids et Mesures
Pavillon de Bretevil
92312 Sèvres cedex
France

Christian Fabjan
Institute of High Energy Physics of the
Austrian Academy of Science and
University of Technology
HEPHY
Nikolsdorfer Gasse 18
1050 Vienna
Austria

Joachim Fischer
Physikalisch-Technische Bundesanstalt
Bundesallee 100
38116 Braunschweig
Germany

J. L. Flowers
National Physical Laboratory
Teddington Middlesex
TW11 OLW
UK

Handbook of Metrology. Edited by Michael Gläser and Manfred Kochsiek

ISBN: 978-3-527-40666-1

Eckhart Förster
X-ray Optics Group
Institute of Optics and Quantum Electronics
Friedrich Schiller University Jena
Max-Wien-Platz 1
07743 Jena
Germany

Michael Gläser
Physikalisch-Technische Bundesanstalt
Bundesallee 100
38116 Braunschweig
Germany

Claus Grupen
Department of Physics
Siegen University
Walter-Flex-Str. 3
57072 Siegen
Germany

James E. Hardy
Engineering Science and Technology Division
Managed by UT-Battelle
LLC for the U.S. Department of Energy under contract DE-AC05-00OR22725
P.O. Box 2008
Oak Ridge, TN 37831-6003
USA

Frans J. M. Harren
Radboud University
Heijendaalseweg 135
6525AJ Nijmegen
The Netherlands

Massimo Inguscio
LENS-European Laboratory for Non-Linear Spectroscopy and Dipartimento di Fisica
Universita di Firenze
via N. Carrara 1
I-50019 Sesto Fiorentino-Firenze
Italy

Bryan P. Kibble
National Physical Laboratory
10 Warwick Close
Middlesex
Hampton TW12 2TY
UK

Manfred Kochsiek
Physikalisch-Technische Bundesanstalt
Bundesallee 100
38116 Braunschweig
Germany

Rainer Macdonald
Physikalisch-Technische Bundesanstalt
Bundesallee 100
38116 Braunschweig
Germany

Thomas H. Markert
Center for Space Research
Massachusetts Institute of Technology
Cambridge, MA 02139
USA

Stephan Mieke
Physikalisch-Technische Bundesanstalt
Bundesallee 100
38116 Braunschweig
Germany

Stefan Persijn
National Metrology Institute
Van Swinden Lab
Thijsseweg 11
2629 JA Delft
The Netherlands

B. W. Petley
National Physical Laboratory
Teddington Middlesex
TW11 OLW
UK

François Piquemal
Laboratoire National de Métrologie
et d'Essais (LNE)
29, rue Roger Hennequin
Trappes
France

Kenneth A. Rubinson
The Five Oaks Research Institute
Bethesda, MD 20817
USA

and

Department of Biochemistry and
Molecular Biology
Wright State University
Dayton, OH 45435
USA

Francis R. Ruppel
Alstom Power Inc.
1409 Centerpoint Boulevard
Knoxville, TN 37932-1962
USA

Giulia Rusciano
Dipartimento di Scienze Fisiche
Università di Napoli "Federico II"
Complesso Universitario
Monte Sant'Angelo
Via Cintia-80126
Napoli
Italy

Antonio Sasso
Dipartimento di Scienze Fisiche
Università di Napoli "Federico II"
Complesso Universitario
Monte Sant'Angelo
Via Cintia-80126
Napoli
Italy

Bernd R. L. Siebert
Physikalisch-Technische
Bundesanstalt
Bundesallee 100
38116 Braunschweig
Germany

Klaus-Dieter Sommer
Physikalisch-Technische
Bundesanstalt
Braunschweig und Berlin
Germany

Jörn Stenger
Physikalisch-Technische
Bundesanstalt
Bundesallee 100
38116 Braunschweig
Germany

1
Introduction

Michael Gläser and Manfred Kochsiek

Handbook of Metrology. Edited by Michael Gläser and Manfred Kochsiek

ISBN: 978-3-527-40666-1

Metrology is the science and technology of measurement. It is as old as human culture, because measurements were necessary even in ancient times for manufacturing tools, even if they were simple like axes, lances, and plows, or for construction of houses. Weighing of goods was common – at least since 5000 years – and this has been corroborated by the discovery of an old Egyptian balance beam. Uniform standards were first necessary only for particular manufacturing projects. A stone axe, for example, was probably made first by processing the stone and drilling the hole in the shaft. The shape of the hole was the standard for carving the shaft. The measurement was made by comparing the size of the shaft with that of the hole of the stone. Even though this is a very simple procedure, it shows the principle of measurement: comparing a measure of an object with that of a standard. An advanced measuring procedure is the comparison of multiples or submultiples of the standard with the object to be measured, as it is necessary for the construction of houses, temples, pyramids, or other buildings. In a larger community, uniform standards were necessary in commerce or technology, when goods were sold by weight or parts of a construction were premanufactured at a distant place. Ancient cultures with central rulers often kept national standards of weight, length, and volume in a temple of the capital. Even before the creation of the meter convention in 1875, cities kept their binding standards in the town hall or somewhere near the market place. Measurement of time has been important for agriculture since early times. The success and survival of a community sometimes depended on the time of seeding and harvesting. Ancient calendars of the Babylonian or the Maya cultures are well known. In Europe, the Julian calendar since Caesar's time and the Gregorian calendar since the sixteenth century successively optimized the measure of the length of the year by algorithms based on the day as a counting digit. The Babylonian divided the day into 12 hours at daylight and 12 hours at night, a division that is still common today. The only change now is that all the 24 hours of the day have the same length. Measurement and units of other quantities like density, force, velocity, acceleration temperature, electric quantities, frequency, amount of substance, energy, or power came in use at the end of the middle ages, when scientific research was progressively modernized. During the French revolution, efforts were made to unify the units of measurement not only in France but also on an international basis. However, it was only in 1875 that an internationally accepted system of units, the International System of Units

Handbook of Metrology. Edited by Michael Gläser and Manfred Kochsiek

ISBN: 978-3-527-40666-1

(SI) was accepted by 17 countries in what was called the *Meter Convention*. There are 52 members and 26 associated countries today in this convention. The SI comprises seven base units: meter, kilogram, second, ampere, kelvin, mole, and candela. Besides the kilogram, which is still defined by the mass of a material object, the base units are defined such that they can be realized at any place and at any time. The other units of measurement are derived from the base units as products and quotients. Two "dimensionless" units, the radian and steradian, are also part of the SI.

Metrology comprises (i) The calibration of a measurement standard or a measuring instrument from the realization of the definition of the unit, usually through a chain of several intermediate standards; (ii) the development and specification of reliable measuring methods and instruments for particular measurement quantities and measurement ranges, according to modern requirements in science, technology, commerce, environment, or health; (iii) the evaluation of measurement data, including reasonable corrections, to obtain the required measurement result; (iv) the evaluation of the measurement uncertainty by taking into account statistical rules and all influencing parameters; and (v) finally, from the knowledge of recent physical and technical research, projecting a new definition of a unit, based on a better constant of nature and developing a corresponding design for realizing such a new unit. Legal metrology is a particular branch of metrology. It deals with national regulations for the use of units, standards, and measurement procedures as well as institutions like National Metrology Institutes and verification offices. Calibration laboratories are working under the auspices of national accreditation bodies and mostly according to international rules, which, among others, require the traceability of their standards to national standards.

Measurement means comparison of a quantity between an object and a standard. The standard can be a ruler, a weight, a zener diode, or an atomic clock, for example. For the comparison, an instrument, often named *comparator*, is required. The measuring method of such an instrument comprises the conversion of the quantity to be measured, preferably in a linear function, to a visible or recordable, quantitative signal. Today, the majority of such signals make use of electrical quantities like voltages, currents, or resistances that are digitized and indicated as a decimal number and a unit on a display. There is always a sophisticated sensor that makes the conversion between the physical quantity and the corresponding electrical signal and provides such values. The quality of the sensor is crucial for the reliability and accuracy of the indicated value. Normally, a manufacturer of such instruments allows only as many digits to be indicated as the accuracy of measurement allows. Some scientific instruments can be adjusted for the zero point and the slope of the indication for allowing the user to calibrate the instrument – at least for its linear slope. Electronic balances of high accuracy, for example, are calibrated by the user for the zero point and the slope by using an internal reference weight.

The result of a measurement is given by a number, a unit, and the associated uncertainty. The uncertainty may be given as a number with a unit or as a relative value. For the evaluation of the uncertainty, the instrument's specification or class, the ambient conditions like temperature, humidity, or pressure, the number of repeated measurements and other parameters that influence the measured data, for example, the position of the sensor and its adaption to the measured object, are taken into account. A report or certificate of the measurement result proves the traceability of the reference standards to the national or international standards.

This handbook contains articles that are also published in the *Encyclopedia of Applied Physics* (Wiley-VCH) and it is divided into two parts. It contains articles dealing with general topics of measurement and articles on particular subjects in mechanics and acoustics, electricity, optics, temperature, time and frequency, chemistry, medicine, and particles. The contributions of the first part are summarized as follows.

"Units" presents not only the SI but also units outside the SI, writing rules, and international topics of this field.

"Fundamental constants" first points out their role and importance in physics, particularly for the Josephson and quantum Hall effects in the quantum electrodynamics, QED. The method by which empirical values of the constants are obtained, the relationships between the constants, and the least squares procedure for obtaining their values recommended by CODATA are presented. The role of fundamental constants in replacing the present definition of the kilogram and, finally, possible time variations of the constants are discussed.

"Fundamentals of material measurement and testing" describes the types of materials and the methods to characterize their properties, for example, their elemental and chemical composition, their microstructure, and essential physical properties. Deterioration analysis as well as nondestructive evaluations of the materials performance are presented.

"Measurement of mass and density" is devoted to two basic quantities in mechanics and their measurements, which are not independent from each other, to some extent. Mass as a physical quantity and the SI unit kilogram are presented in the first part. The method of mass determination using various types of balances and the need for and procedure of a correction for air buoyancy requiring density information are described. Methods for assessing astronomical and subatomic masses are mentioned as well. Different approaches for measuring densities, based on buoyancy or on mass and volume, and also density references are the themes of the second part.

"Measurement and instrumentation of flow" first starts with Bernoulli's differential flow equation, and then presents the various principles and instruments of flow measurement and discusses their advantages and disadvantages. Sections on measurement uncertainties and calibration methods complete this contribution.

"Ultrasonics" first develops the theories at conditions in boundless and bounded media and those in solids and liquids. Attenuation and wave dispersion models are discussed next. Finally the methods of ultrasonic generation, measurement techniques, and applications of ultrasound are dealt with.

" Measurement of basic electromagnetic quantities" first discusses about realizing the units farad, ohm, volt, and watt and continues with presenting working standards of voltage and resistance, for example, those that make use of the conventionally adopted constants R_k and K_J. Other primary electrical standards of National Measurement Institutes for DC voltage, DC resistance, capacitance, inductance, electric field and magnetic quantities, and AC current and voltage follow.

"Quantum electrical standards" deals with the presentation of the quantized Hall resistance standard and the Josephson voltage standard and an outlook for a new quantum standard for current, based on Coulomb blockade. Descriptions follow on the corresponding quantum effects, the determination of the involved constants, K_J and R_K, on the realization of the standards, and on measurement methods and devices.

"Metrology of time and frequency" first discusses the principles of characterizing clocks and oscillators, and then describes quartz crystal and atomic frequency standards, measurement of time and frequency, and comparisons between remote sites. The evaluation of the International Atomic Time (TAI) and the Coordinated Universal Time (UTC) is explained, followed by procedures of time dissemination and applications.

"Temperature measurement" first introduces the concept of temperature from basic laws of thermodynamics, and provides the definition of the kelvin and the envisaged future redefinition based on a fixed value of the Boltzmann constant. Various kinds of physical thermometry concepts follow, such as primary gas thermometry, thermometry based on noise, Doppler broadening, total radiation, and spectral radiation. The international temperature scale of 1990, the provisional low-temperature scale PLTS-2000 and candidates for high-temperature fixed points of metal–carbon eutectic phase transitions are described, as well as industrial resistance thermometers, thermocouples, and radiation thermometers.

"Metrology in medicine" is characterized by procedures that measure physical or chemical quantities tracing them to standards known not only from the other fields of metrology but also by other reference measurement procedures that refer to well-accepted physiological test signals in medical diagnostics. Examples for instruments of physical and physiological measurements like crank ergometers, spirometers, and air puff eye-tonometers are given. Chemical and biological measurands are discussed, together with metrological controls in laboratory medicine, quality assurance in hematology, primary methods in clinical chemistry, and biomeasurements for molecular medicine.

We hope the readers of the Handbook of Metrology will find the information contained herein new and useful.

2
Units

Bernd R. L. Siebert and Jörn Stenger

Handbook of Metrology. Edited by Michael Gläser and Manfred Kochsiek

ISBN: 978-3-527-40666-1

2.1 Introduction

The use of units is omnipresent in our daily life. Everywhere in trade, manufacturing, and science, measurements are quantified or compared with each other by the use of units.

Already James Clerk Maxwell (1873) stated that every expression of a quantity consists of two factors or components. In today's notation, this is expressed by the relation

$$Q = \{Q\} \times [Q] \tag{2.1}$$

where $\{Q\}$ denotes the numerical value and $[Q]$ is the unit of a physical quantity. The unit is a reference magnitude.

In other words, any quantitative measurement can be imagined as a comparison of a property of an object such as its mass, length, or temperature, with the same property of an object that represents the unit.

To ensure the correctness of this reference, it must be compared with a "higher order" reference. This is called *calibration*. Primary standards are the highest order references. They cannot be calibrated. Comparisons of different primary standards for the same unit, in many cases operated in different countries, are used to achieve highest confidence in the comparability of measurements.

When a measurement is referenced to a primary standard, directly or via an unbroken chain of references, it is called *traceable*. The trustworthiness of a measurement needs to be quantified by stating the uncertainty associated with measured value.

The choice of a unit is, in principle, quite arbitrary. Historically, units have been represented by artifacts. The magnitude of the quantity had been chosen for the sake of convenience or, in case of length, had been set by the length of cells or feet of pharaohs, kings or in later times by the average obtained from a group of people. This led to a variety of units for one and the same quantity that made comparison of measurement results across borders difficult and acted as a barrier for trade and later also for science.

The Convention du Mètre, signed by 17 states in Paris in 1875 was the first international agreement on metrology with the aim to use the same units worldwide. This furthers international trade and cooperation in science and is the basis for international standardization. Presently, there is a multitude of quantities that needs to be measured in the context of trade, manufacturing, the protection of life and environment, and science. Fortunately, virtually all of these quantities can be expressed as a product of powers of few basic quantities, forming a "system of base quantities." The units used to express

Handbook of Metrology. Edited by Michael Gläser and Manfred Kochsiek

ISBN: 978-3-527-40666-1

numerical values of base quantities are then termed *base units.*

The choice of the base quantities is to a certain degree arbitrary. The present worldwide convention is the International System of Quantities (ISQ). It is based on the seven base quantities of length, mass, time, electric current, thermodynamic temperature, amount of substance, and luminous intensity (BIPM, IEC, IFCC, ILAC, ISO, IUPAC, IUPAP, and OIML, 2008). The symbols for the dimensions of the base quantities are **L** for length, **M** for mass, **T** for time, **I** for electric current, **Θ** for thermodynamic temperature, **N** for amount of substance, and **J** for luminous intensity. Thus, the dimension of a quantity Q is denoted by

$$\mathrm{Dim}Q = \mathbf{L}^{\alpha}\mathbf{M}^{\beta}\mathbf{T}^{\gamma}\mathbf{I}^{\delta}\boldsymbol{\Theta}^{\varepsilon}\mathbf{N}^{\zeta}\mathbf{J}^{\eta} \tag{2.2}$$

where the exponents, named *dimensional exponents,* are positive, negative, or zero.

Requirements that must be met in order to make the system acceptable for all users are as follows:

1. Consistency with the laws of physics: a measurement of pressure, for example, through a simultaneous measurement of force and area must not lead to different results when obtained by comparison with a realization of the unit of pressure.
2. Flexibility for modifications due to new needs, while being reliable and stable.
3. Practicability in all intended uses and simplicity.

The International System of Units (SI) represents such a system (BIPM, 2006). It is described in the following sections.

2.2 The International System of Units (SI)

In 1948, the Ninth Conférence Générale des Poids et Mesures (GCPM) adopted a draft for a system of units, which was originally based on six and meanwhile on seven base units. All other units are related with these base units just through multiplication and division. In 1960, the GCPM named the new system as *Système International d'Unités,* abbreviated as SI, and it decided that this abbreviation was to be used in all languages. All member states of the Meter Convention have adopted the metric system and have taken steps to eliminate most uses of traditional measurements.

The Bureau International des Poids et Mesures (BIPM) was entrusted to ensure worldwide uniformity of measurements and their traceability to the SI and publish a Brochure about the SI. The latest edition of the Brochure was published in 2006.

2.2.1 SI Base Units

The base units of the International System are listed in Table 2.1, which relates the base quantity to the unit name and unit symbol for each of the seven base units.

Sections 2.2.1.1–2.2.1.7 present the presently adopted definitions in italics followed by additional clarifications provided in the SI-brochure. Only brief remarks on history and realization are given as details can be found on the BIPM homepage (see Section 2.4).

For practical and economic reasons, the units are in many cases realized by reference standards optimized to the practical needs. In order to harmonize these realizations of units, the Comité international des poids et mesures (CIPM) drew up recommendations, which are generally referred to as the *mise en pratique* for the definition. The first *mise en pratique* was given in 1983 at the time of the adoption of the present definition of the

Tab. 2.1 SI base units.

Base quantity		**SI base unit**	
Name	Symbol	Name	Symbol
length	$l, x, r, \ldots$	metre	m
mass	m	kilogram	kg
time, duration	t	second	s
electric current	I, i	ampere	A
thermodynamic temperature	T	kelvin	K
amount of substance	n	mole	mol
luminous intensity	I_v	candela	cd

meter. In 1992, the CIPM acting on the advice of its Consultative Committee for the Definition of the Metre (CCDM), approved a revised *mise en pratique* for the definition of the meter (Quinn, 1994). Thereafter, it was decided to publish *mises en pratiques* on the homepage of the BIPM. The page is regularly updated if new significant progress has been achieved. Presently, *mises en pratiques* are available for the definition of the meter and the definition of the kelvin; see *http://www.bipm.org/en/publications.*

2.2.1.1 The Meter

The meter, unit of length, was redefined in 1983 by the 17th Conférence Générale des Poids et Pesures (CGPM) that specified the current definition, as follows: *The metre is the length of the path travelled by light in vacuum during a time interval of* 1/299 792 458 *of a second.*

It follows that the speed of light in vacuum is exactly 299 792 458 metres per second, $c_0 = 299\,792\,458$ m s.[1]

The first definition of the meter by the first CGPM in 1889 was based on the international prototype of platinum–iridium. This prototype is still kept at the BIPM under the conditions specified in 1889. The 11th CGPM replaced this first definition in 1960 and selected the wavelength of a well-defined transition in krypton 86 as a basis for defining the meter as a multiple of this wavelength. This was an important step toward defining units via fundamental constants.

The definition of the meter allows to use either the relation $l = c_0 t$, where l is the length of the path traveled by a plane electromagnetic wave in vacuum and t is the measured time of flight, or the relation $\lambda = c_0/f$, where λ is the wavelength and f the frequency of electromagnetic radiation. Both relations reflect that the unit of length depends on the unit of time.

The first of these relations is well suited for large distances, whereas the second relation is practical for small distances. In practice, laser radiation is used for realization of the unit of length.

For example, the distance between the earth and the moon is measured by directing a short pulse of a high-power laser beam at the moon where it is reflected from a mirror. The distance can be calculated from the

1) This and the following definitions are cited from the 8th edition of the SI Brochure, BIPM 2006. With regard to the spelling of English-language words beside the quotations, we refer to the *United States Government Printing office Style Manual*, Which follows *Webster's Third New International Dictionary.* Form example, "meter" "liter," and "deka" are used instead of "metre," "litre," and "deca"

time which the laser pulse needs to travel from the earth to the moon and back to the earth. The Global Positioning System (GPS), too, is based on time-of-flight measurements of electromagnetic radiation. The relative uncertainties associated with measured values of length using $l = c_0 t$ are determined by the uncertainty of the associated time measurement.

The use of the relation $\lambda = c_0/f$ requires the establishment of frequency standards representing reciprocal time realizations.

2.2.1.2 The Kilogram

The kilogram, unit of mass, is the only unit that is still explicitly based on an artifact, the international prototype of the kilogram that is made of platinum–iridium and kept at the BIPM since it was sanctioned by the first CGPM in 1889. The third CGPM confirmed this in 1903 by the definition: *The kilogram is the unit of mass; it is equal to the mass of the international prototype of the kilogram.*

It follows that the mass of the international prototype of the kilogram is always 1 kilogram exactly, $m(\mathcal{K}) = 1\,\text{kg}$.

The masses of 1 kg secondary standards of the same alloy as the international prototype are compared in air with the mass of the international prototype by means of balances with a relative uncertainty approaching 1 part in 10^9.

2.2.1.3 The Second

The second, unit of time, was redefined in 1967/1968 by the 13th CGPM that specified the current definition, as follows: *The second is the duration of* 9 192 631 770 *periods of the radiation corresponding to the transition between the two hyperfine levels of the ground state of the cesium* 133 *atom.*

It follows that the hyperfine splitting in the ground state of the cesium 133 atom is exactly 9 192 631 770 hertz, $\nu(\text{hfs Cs}) = 9\,192\,631\,770\,\text{Hz}$.

As affirmed by the CIPM at its 1997 meeting, this definition refers to a cesium atom at rest at a temperature of 0 K.

This note clarifies that the definition of the SI second is based on a cesium atom unperturbed by black body radiation, that is, in an environment whose thermodynamic temperature is 0 K. The frequencies of all primary frequency standards should therefore be corrected for the shift due to ambient radiation, as stated at the meeting of the Consultative Committee for Time and Frequency (CCTF) in 1999.

Presently, the best primary standard produced the SI second with a relative standard uncertainty of some parts in 10^{16}.

2.2.1.4 The Ampere

The ampere, unit of electric current, was proposed by the CIPM in 1946 and, after replacing the expression "MKS unit of force" by "newton," adopted by the ninth CGPM in 1948:
The ampere is that constant current which, if maintained in two straight parallel conductors of infinite length, of negligible circular cross-section, and placed 1 metre *apart in vacuum, would produce between these conductors a force equal to* 2×10^{-7} newton per metre *of length.*

It follows that the magnetic constant, μ_0, also known as the *permeability of free space*, is exactly $4\pi \times 10^{-7}$ henries per metre, $\mu_0 = 4\pi \times 10^{-7}\,\text{H/m}$.

Note that this definition implies the value of the electric constant, ε_0 (permittivity of vacuum), is also fixed, since $\varepsilon_0 = (\mu_0 c_0^2)^{-1}$.

2.2.1.5 **The Kelvin**

The kelvin, unit of thermodynamic temperature, was adopted in its present form by the 13th CGPM in 1968:

The kelvin, unit of thermodynamic temperature, is the fraction 1/273.16 *of the thermodynamic temperature of the triple point of water.*

It follows that the thermodynamic temperature of the triple point of water is exactly 273.16 kelvins, $T_{tpw} = 273.16$ K.

As affirmed by the CIPM at its 2005 meeting, this definition refers to water having the isotopic composition defined exactly by the following amount of substance ratios: 0.00015576 mol of ^2H per mol of ^1H, 0.000 379 9 mol of ^{17}O per mol of ^{16}O, and 0.002 005 2 mol of ^{18}O per mol of ^{16}O.

Already the 10th CGPM selected the triple point of water in 1954 as the fundamental fixed point and assigned to it the temperature of 273.16 K, so defining the unit.

Because of the manner in which temperature scales used to be defined, it remains common practice to express a thermodynamic temperature, symbol T, in terms of its difference from the reference temperature $T_0 = 273.15$ K, the ice point. This difference is called the *Celsius temperature*, symbol t, which is defined by the quantity equation: $t = T - T_0$.

The unit of Celsius temperature is the degree Celsius, symbol °C, which is by definition equal in magnitude to the kelvin. A difference or interval of temperature may be expressed in kelvin or degree Celsius, the numerical value of the temperature difference being the same.

The kelvin and the degree Celsius are also units of the International Temperature Scale of 1990 (ITS-90).

The *mise en pratique* for the definition of the kelvin (last update in 2006) is available on *http://www.bipm.org/en/publications/*.

2.2.1.6 **The Mole**

The mole, the unit for the amount of substance was adopted in its present form by the 14th CGPM in 1971:

1. *The mole is the amount of substance of a system which contains as many elementary entities as there are atoms in* 0.012 kg *of carbon 12; its symbol is "mol."*
2. *When the mole is used, the elementary entities must be specified and may be atoms, molecules, ions, electrons, other particles, or specified groups of such particles.*

It follows that the molar mass of carbon 12 is exactly 12 grams per mole, $M(^{12}C) = 12$ g/mol.

This definition refers to unbound atoms of ^{12}C at rest and in their ground state.

The definition of the mole also determines the value of the universal constant that relates the number of entities to amount of substance for any sample. This constant is called the *Avogadro constant*, symbol N_A or L. If $N(X)$ denotes the number of entities X in a specified sample, and if $n(X)$ denotes the amount of substance of entities X in the same sample, the relation is $n(X) = N(X)/N_A$.

Note that since $N(X)$ is dimensionless, and $n(X)$ has the SI unit mole, the Avogadro constant has the coherent SI unit reciprocal mole.

2.2.1.7 **The Candela**

The candela unit of luminous intensity was adopted in its present form by the 16th CGPM in 1979.

The candela is the luminous intensity, in a given direction, of a source that emits monochromatic radiation of frequency 540×10^{12} hertz *and that has a radiant intensity in that direction of* 1/683 Watt per steradian.

It follows that the spectral luminous efficacy for monochromatic radiation of frequency of 540×10^{12} hertz is exactly 683 lumens per Watt, $K = 683\ \text{lm/W} = 683\ cdsr/W$.

The units of luminous intensity based on flame or incandescent filament standards in use in various countries before 1948 were replaced initially by the "new candle" based on the luminance of a Planck radiator (a black body) at the temperature of freezing platinum. This modification was then ratified in 1948 by the ninth CGPM, which adopted a new international name for this unit, the *candela*. In 1979, the present definition was given because of the difficulties in realizing a Planck radiator at high temperatures, and the new possibilities offered by radiometry, that is, the measurement of optical radiation power.

2.2.2 SI Derived Units

The unit of any quantity can be expressed by the relation

$$[Q] = \text{m}^{\alpha}\ \text{kg}^{\beta}\ \text{s}^{\gamma}\ \text{A}^{\delta}\ \text{K}^{\varepsilon}\ \text{mol}^{\zeta}\ \text{cd}^{\eta}.$$

Units obtained by this relation are coherent derived SI units. The base and coherent derived units of the SI form a coherent set, designated the set of *coherent SI units*.

For convenience, certain coherent derived units have been given special names and symbols. There are 22 such units (see Table 2.2).

These units and symbols can be used in forming other coherent derived units. As an example, the unit of dynamic viscosity is expressed in terms of SI base units $\text{m}^{-1}\,\text{kg}\,\text{s}^{-1}$, it can also be expressed in terms of coherent derived SI units pascal and the base unit second, it has then the unit symbol Pa. This convention allows to express the physical meaning of a given quantity.

2.2.3 SI Prefixes

The CGPM adopted, in 1960, a series of prefix names and prefix symbols to form the names and symbols of the decimal multiples and submultiples of SI units, ranging from 10^{12} to 10^{-12}. This range was increased in several steps. The presently approved prefix names and prefix symbols range from 10^{24} to 10^{-24}, and they are listed in Table 2.3. The prefixes are used to keep the numbers giving the values of a quantity in a practical order of magnitude. Names and symbols for decimal multiples and submultiples of the unit are generally formed by attaching prefix names to the unit name. However, one important exception is the base unit of mass, the kilogram, whose name and symbol, for historical reasons, already includes a prefix. For the unit of mass, decimal multiples and submultiples are formed by attaching a prefix to the gram, *g*.

The following points are noted:

- Units with a prefix are not coherent.
- SI prefixes refer strictly to powers of 10, and cannot be used for powers of 2. For example, 1 kilobit should not be used to represent 1024 bits (2^{10} bits), which instead is 1 kibibit (BIPM, IEC, IFCC, ILAC, ISO, IUPAC, IUPAP, and OIML, 2008). When SI prefixes are used, attention should be paid to the rules listed in Section 2.4.

2.2.4 SI on the BIPM-homepage

The BIPM issues the SI-brochure, which can be downloaded from its homepage *http://www.bipm.org*. The official version is the French text. The site allows both, online browsing and downloading of the entire brochure.

Tab. 2.2 Coherent derived units in the SI with special names and symbols.

	SI coherent derived unit			
Derived quantity	**Name**	**Symbol**	**Expressed in SI units**	**Expressed in SI base units**
plane angle	radian[a]	rad	1[a]	m/m
solid angle	steradian[a]	sr[a]	1[a]	m^2/m^2
frequency	hertz[b]	Hz	–	s^{-1}
force	newton	N	–	$m\ kg\ s^{-2}$
pressure, stress	pascal	Pa	N/m^2	$m^{-1}\ kg\ s^{-2}$
energy, work, amount of heat	joule	J	N m	$m^2\ kg\ s^{-2}$
power, radiant flux	watt	W	J/s	$m^2\ kg\ s^{-3}$
electric charge, amount of electricity	coulomb	C	–	s A
electric potential difference, electromotive force	volt	V	W/A	$m^2\ kg\ s^{-3}\ A^{-1}$
capacitance	farad	F	C/V	$m^{-2}\ kg^{-1}\ s^4\ A^2$
electric resistance	ohm	Ω	V/A	$m^2\ kg\ s^{-3}\ A^{-2}$
electric conductance	siemens	S	A/V	$m^{-2}\ kg^{-1}\ s^3\ A^2$
magnetic flux	weber	Wb	V s	$m^2\ kg\ s^{-2}\ A^{-1}$
magnetic flux density	tesla	T	Wb/m^2	$kg\ s^{-2}\ A^{-1}$
inductance	henry	H	Wb/A	$m^2\ kg\ s^{-2}\ A^{-2}$
Celsius temperature	degree Celsius	°C		K
luminous flux	lumen	lm	cd sr[a]	cd
illuminance	lux	lx	lm/m^2	m^{-2} cd
activity referred to a radionuclide[(f)]	becquerel[b]	Bq		s^{-1}
absorbed dose[d], specific energy (imparted), kerma	gray	Gy	J/kg	$m^2\ s^{-2}$
dose equivalent[d], ambient dose equivalent, directional dose equivalent, personal dose equivalent	sievert	Sv	J/kg	$m^2\ s^{-2}$
catalytic activity	katal	kat	–	s^{-1} mol

[a] The *radian* and *steradian* are special names for the number one that may be used to convey information about the quantity concerned. In practice, the symbols rad and sr are used where appropriate, but the symbol for the derived unit one is generally omitted in specifying the values of dimensionless quantities. However, in photometry, the name steradian and the symbol sr are usually retained in expressions for units.

[b] The *hertz* is used only for periodic phenomena, and the *becquerel* is used only for stochastic processes in activity referred to a radionuclide.

[c] Activity of a radionuclide is sometimes incorrectly called *radioactivity*.

[d] For details on quantities and units for ionizing radiation, see ICRU 1998a, 1998b.

Tab. 2.3 SI prefixes.

Factor	Prefix	Symbol	Factor	Prefix	Symbol
10^1	deca	da	10^{-1}	deci	d
10^2	hecto	h	10^{-2}	centi	c
10^3	kilo	k	10^{-3}	milli	m
10^6	mega	M	10^{-6}	micro	μ
10^9	giga	G	10^{-9}	nano	n
10^{12}	tera	T	10^{-12}	pico	p
10^{15}	peta	P	10^{-15}	femto	f
10^{18}	exa	E	10^{-18}	atto	a
10^{21}	zetta	Z	10^{-21}	zepto	z
10^{24}	yotta	Y	10^{-24}	yocto	y

Note that Appendix 2 of the brochure, titled "Practical realizations of the definitions of some important units" is published in electronic form only, as this allows updating in the course of advances in experimental realizations of units. This decision underlines the strict distinction between the definition and the realization of a unit.

Furthermore, this webpage features links to a "Brief history of the SI," a concise visualization of "Linking SI base units to fundamental and atomic constants" and to "Unit conversions" (see also Appendix).

2.3 Units Outside the SI

The consequent use of SI units and the few non-SI units that are accepted for use with the SI promotes communication in pure and applied science.

However, in nearly all countries traditional units are still in use, for example, units for time, such as minute, hour, and day, and there are also some fields of science that use non-SI units. The exclusive use of SI units cannot always be enforced; rather it is a process that is accelerated by evidencing its advantages. Along this line, the policy of the Consultative Committee for Units is to advocate the use of the SI instead of deprecating units that are still widely used. The eighth edition of the brochure promotes this process by providing listings of units accepted for use with the SI and tables for converting some units that are still widely used but not accepted to SI units and by referring to complete conversion table on the BIPM homepage.

2.3.1 Non-SI Units Accepted for Use with the SI

It is distinguished between

- non-SI units accepted for use with the SI (Table 2.4(a) and notes below);
- other non-SI units (Table 2.4(b));
- non-SI units whose values in SI units must be obtained experimentally (Table 2.5); and
- non-SI units associated with the CGS and the CGS-Gaussian system of units (Table 2.6).

Notes on Table 2.4(a):

- ISO 31 recommends that the *degree* (angle) be divided decimally rather than using the minute and the second. For navigation and surveying, however, the minute has the advantage that 1 minute of latitude on the surface of the Earth

Tab. 2.4a Non-SI units accepted for use with the International System of Units.

Quantity	Name of unit	Symbol for unit	Value in SI units
time	minute	min	1 min = 60 s
	hour	h	1 h = 60 min = 3600 s
	day	d	1 d = 24 h = 86 400 s
plane angle	degree	°	$1^\circ = (\pi/180)$ rad
	minute	′	$1' = (1/60)^\circ = (\pi/10\,800)$ rad
	second	″	$1'' = (1/60)' = (\pi/648\,000)$ rad
area	hectare	ha	$1\ \text{ha} = 1\ \text{hm}^2 = 10^4\ \text{m}^2$
volume	liter	L, l	$1\ \text{L} = 1\ \text{l} = 1\ \text{dm}^3 = 10^3\ \text{cm}^3 = 10^{-3}\ \text{m}^3$
mass	tonne[a]	t	$1\ \text{t} = 10^3\ \text{kg}$

[a] In English-speaking countries, the unit tonne is usually called *metric ton*.

corresponds (approximately) to one nautical mile.

- The *gon* (or grad, where grad is an alternative name for the gon) is an alternative unit of plane angle to the degree, defined as $(\pi/200)$ rad. Thus there are 100 gons in a right angle. The potential value of the gon in navigation is that because the distance from the pole to the equator of the Earth is approximately 10 000 km, 1 km on the surface of the Earth subtends an angle of one centigon at the center of the Earth. However, the gon is rarely used.
- For applications in *astronomy*, small angles are measured in *arcseconds* (i.e., seconds of plane angle), denoted as milliarcseconds, microarcseconds, and picoarcseconds, denoted mas, µas, and pas, respectively, where arcsecond is an alternative name for second of plane angle.
- The unit *hectare*, and its symbol ha, were adopted by the CIPM in 1879. The hectare is used to express land area.
- The *liter*, and the symbol lowercase l, were adopted by the CIPM in 1879. The alternative symbol, capital L, was adopted by the 16th CGPM (1979, Resolutions 6; CR, 101) in order to avoid the risk of confusion between the letter l (el) and the numeral 1 (one).

2.3.2 Other Non-SI Units Not Recommended for Use

There are many more non-SI units. Some are of historical interest and others are still used for specific applications such as the barrel of oil, or used in particular countries, for example, inch, gallon, and pound.

Although the advantages of a consequent use of SI units appear obvious, it is expected that some of these non-SI units will be used in future. The CIPM has therefore decided to compile a list of the conversion factors to the SI for such units and to make this available on the BIPM home page.

For the convenience of the reader, some more often needed conversion factors are provided in the Appendix.

2.4 Rules for Expressing Quantities

The compliance with rules and conventions enhances the readability of technical and scientific papers.

Tab. 2.4b Other non-SI units.

Quantity	Name of unit	Symbol for unit	Value in SI units
pressure[a]	bar	bar	1 bar = 0.1 MPa = 100 kPa $= 10^5$ Pa
	millimeter of mercury	mmHg	1 mmHg = 133.322 Pa
length	ångström[b]	Å	1 Å = 0.1 nm = 100 pm $= 10^{-10}$ m
distance	nautical mile[c]	M	1 M = 1852 m
area	barn[d]	b	1 b = 100 $\mathrm{fm}^2 = 100 \times (10^{-15}\ \mathrm{m})^2 = 10^{-28}\ \mathrm{m}^2$
speed	knot[c]	kn	1 kn = (1852/3600) m/s
logarithmic ratio quantities	neper[e,g]	Np	[see footnote[h] regarding the numerical value of the neper, the bel, and the decibel]
	bel[f,g]	B	
	decibel[f,g]	dB	

[a] Since 1982 1 bar has been used as the standard pressure for tabulating all thermodynamic data. It replaced the *standard atmosphere*, equal to 1.013 25 bar, or 101 325 Pa. The *millimeter of mercury* is a legal unit for the measurement of blood pressure in some countries.

[b] The *ångström* is widely used by x-ray crystallographers and structural chemists because all chemical bonds lie in the range 1–3 ångströms. However, it has no official sanction from the CIPM or the CGPM.

[c] The *nautical mile* is employed for marine and aerial navigation to express distance. As yet there is no internationally agreed symbol, but the symbols M, NM, Nm, and nmi are all used. One nautical mile on the surface of the Earth subtends approximately 1 minute of angle at the center of the Earth. The *knot* is defined as one nautical mile per hour. There is no internationally agreed symbol, but the symbol kn is commonly used.

[d] The *barn* is a unit of area employed to express cross sections in nuclear physics.

[e] The statement $L_A = n$ Np (where n is a number) is interpreted to mean that $\ln(A_2/A_1) = n$. Thus when $L_A = 1$ Np, $A_2/A_1 = \mathrm{e}$. The symbol A is used here to denote the amplitude of a sinusoidal signal, and L_A is then called the *neperian logarithmic amplitude ratio*, or the *neperian amplitude level difference*.

[f] The statement $L_X = m$ dB $= (m/10)$ B (where m is a number) is interpreted to mean that $\lg(X/X_0) = m/10$. Thus when $L_X = 1$ B, $X/X_0 = 10$, and when $L_X = 1$ dB, $X/X_0 = 10^{1/10}$. If X denotes a mean square signal or power-like quantity, L_X is called a *power level* referred to X_0.

[g] In using these units it is important that the nature of the quantity be specified, and that any reference value used be specified. These units are not SI units, but they have been accepted by the CIPM for use with the SI.

[h] The numerical values of the neper, bel, and decibel (and hence the relation of the bel and the decibel to the neper) are rarely required. They depend on the way in which the logarithmic quantities are defined.

Tab. 2.5 Non-SI units whose values in SI units must be obtained experimentally.

Quantity	Name of unit	Symbol for unit	Value in SI units[a]
Units accepted for use with the SI			
energy	electronvolt[b]	eV	$1\,\text{eV} = 1.602\,176\,53\,(14) \times 10^{-19}$ J
mass	dalton[c]	Da	$1\,\text{Da} = 1.660\,538\,86\,(28) \times 10^{-27}$ kg
	unified atomic mass unit	u	$1\,\text{u} = 1.660\,538\,86\,(28) \times 10^{-27}$ kg
length	astronomical unit[d]	ua	$1\,\text{ua} = 1.495\,978\,706\,91\,(6) \times 10^{11}$ m
Natural units (n.u.)			
speed	n.u. of speed (speed of light in vacuum)	c_0	299 792 458 m s^{-1}(exact)
action	n.u. of action (reduced Planck constant)	$\hbar$	$1.054\,571\,68\,(18) \times 10^{-34}$ J s
mass	n.u. of mass (electron mass)	m_e	$9.109\,382\,6\,(16) \times 10^{-31}$ kg
time	n.u. of time	$\hbar/(m_e c_0^2)$	$1.288\,088\,667\,7\,(86) \times 10^{-21}$ s
Atomic units (a.u.)			
charge	a.u. of charge, (elementary charge)	e	$1.602\,176\,53\,(14) \times 10^{-19}$ C
mass	a.u. of mass, (electron mass)	m_e	$9.109\,382\,6\,(16) \times 10^{-31}$ kg
action	a.u. of action, (reduced Planck constant)	$\hbar$	$1.054\,571\,68\,(18) \times 10^{-34}$ J s
length	a.u. of length, bohr (Bohr radius)	a_0	$0.529\,177\,210\,8\,(18) \times 10^{-10}$ m
energy	a.u. of energy, hartree (Hartree energy)	$E_\hbar$	$4.359\,744\,17\,(75) \times 10^{-18}$ J
time	a.u. of time	$\hbar/E_h$	$2.418\,884\,326\,505\,(16) \times 10^{-17}$ s

[a] The values in SI units of all units in this table, except the astronomical unit, are taken (Mohr, P.J. and Taylor, B.N. (2005) *Rev. Mod. Phys.*, 77, 1–107). The standard uncertainty in the last two digits is given in parenthesis (see Section 2.4, below).

[b] The *electronvolt* is the kinetic energy acquired by an electron in passing through a potential difference of 1 V in vacuum. The electronvolt is often combined with the SI prefixes.

[c] The *dalton* and the *unified atomic mass unit* are alternative names (and symbols) for the same unit, equal to 1/12 times the mass of a free carbon 12 atom, at rest and in its ground state. The *dalton* is often used with SI prefixes to express the masses of large molecules or small mass differences of atoms or molecules.

[d] The astronomical unit is approximately equal to the mean Earth–Sun distance. It is the radius of an unperturbed circular Newtonian orbit about the Sun of a particle having infinitesimal mass, moving with a mean motion of 0.017 202 098 95 radians per day (known as the *Gaussian constant*). The value given for the astronomical unit is quoted from the International Earth Rotation and Reference Systems Service (IERS) Conventions 2003 (McCarthy, D.D. and Petit, G. (eds) (2004) *IERS Technical Note* 32, Verlag des Bundesamts für Kartographie und Geodäsie, Frankfurt am Main, p. 12). The value of the astronomical unit in meters comes from the Jet Propulsion Laboratory (JPL) ephemerides DE403 (Standish, E.M. (1995) Report of the IAU WGAS Sub-Group on Numerical Standards, in *Highlights of Astronomy* (ed. Appenzeller), Kluwer Academic Publishers, Dordrecht, pp. 180–184).

The rules for quantity calculus are independent from the language used; however, the usage of names for quantities, dimensions, and units in a text does depend on the language used. For the rules discussed below, numbers in brackets refer to the examples given in Table 2.7.

Names for quantities and units are ordinary nouns. Therefore, in English, they are declined and start with a lowercase letter, even units named after person [1]. The plural is formed regularly, for example, "lengths" and "henries" are the plurals of "length" and "henry." However for lux, hertz, and siemens the plural is identical with the singular. The names of multiples or submultiples of a unit are written without a hyphen or a space [2]. Derived units that are products of units are written with a hyphen or a space as "multiplication sign" [3], those representing ratios are written using the word "per" to indicate the division [4]. Modifiers such as "squared" or "cubed" are placed after the unit name. However, for area and volume the words "square" and "cubic" may be used and are then placed before the corresponding unit [5].

Symbols for quantities, dimensions, and units are mathematical entities. The normal rules of algebra apply and one must neither use the plural nor mix symbols and names within one expression [6].

Symbols for quantities are generally single letters of the Latin or Greek alphabets, printed in an *italic* font. Symbols for quantities are *recommendations* [7].

Symbols for units are printed in roman (upright) type regardless of the type used in the surrounding text. Symbols for units are mandatory [8]. Unless they are named after a person, unit symbols begin with lowercase letter [9]. However, for the accepted non-SI unit liter the symbol L or l is allowed; L should be used to avoid that l is interpreted as the Arabic numerical for one [10].

The *symbols* for derived units reflect a product of base units with integer exponents, that are negative, zero, or positive, see Section 2.2. Base units with negative exponents are divisors, those with zero as exponents are omitted and those with positive exponents are multipliers. Multiplication must be indicated a half space () or a half-high centered dot (·), but never by a cross (×) [11] and division is indicated by a horizontal line (—), a solidus (/), or by the negative exponent [12].

SI *prefixes* and the unit again form a unit; therefore, they are also printed in roman (upright) type without a multiplication sign between prefix symbol and unit symbol and their names are also written as one word [13]. They can be raised to a positive or negative power [14]. Compound prefixes, that is, prefixes formed by juxtaposition of two or more SI prefixes are not permitted [15] as prefixes are not stand-alone abbreviations for $10^{\pm n}$, n being an integer. Names of decimal multiples or submultiples of the unit of mass are formed by attaching prefixes to the name "gram" and the symbols are formed by attaching prefixes to the unit g [16]. SI prefixes should only be used in stating results and not in calculations in order to preserve coherence.

When multiplying or dividing the values of quantities, brackets should be used to exclude any ambiguities [17]. The terms *ppm*, *ppb*, or *ppt* should not be used to express relative values in the SI. In most English-speaking countries, 10^6, 10^9, 10^{12}, and 10^{15} are named *million, billion trillion*, and *quadrillion*, respectively. In French, German, and other languages, however, they are named *million, milliard, billion*, and *billiard*. On the other hand, the internationally recognized symbol % (percent) may be used with the SI to

Tab. 2.6 Non-SI units associated with the CGS and the CGS-Gaussian system of units.

Quantity	Name of unit	Symbol for unit	Value in SI units
energy	erg[a]	erg	$1\ \text{erg} = 1\ \text{g cm}^2\ \text{s}^{-2} = 10^{-7}\ \text{J}$
force	dyne[a]	dyn	$1\ \text{dyn} = 1\ \text{g cm s}^{-2} = 10^{-5}\ \text{N}$
dynamic viscosity	poise[a]	P	$1\ \text{P} = 1\ \text{g cm s}^{-1} = 0.1\ \text{Pa s}$
kinematic viscosity	stokes	St	$1\ \text{St} = 1\ \text{cm}^2\ \text{s}^{-1} = 10^{-4}\ \text{m}^2\ \text{s}^{-1}$
luminance	stilb[a]	sb	$1\ \text{sb} = 1\ \text{cd cm}^{-2} = 10^4\ \text{cd m}^{-2}$
illuminance	phot	ph	$1\ \text{ph} = 1\ \text{cd sr cm}^{-2} = 10^4\ \text{lx}$
acceleration	gal[b]	Gal	$1\ \text{Gal} = 1\ \text{cm s}^{-2} = 10^{-2}\ \text{m s}^{-2}$
magnetic flux	maxwell[c]	Mx	$1\ \text{Mx} = 10^{-4}\ \text{cm}^2\ \text{T} = 10^{-8}\ \text{Wb}$
magnetic flux density	gauss[c]	G	$1\ \text{G} = 1\ \text{cm}^{-2}\ \text{Wb} = 10^{-4}\ \text{T}$
magnetic field	œrsted[c]	Oe	$1\ \text{Oe} \mathrel{\hat{=}} (10^3/4\pi)\ \text{A m}^{-1} \cong 7.957\,747 \times 10\ \text{A m}^{-1}$

[a] This unit and its symbol were included in Resolution 7 of the 9th CGPM (1948; CR, 70).
[b] The gal is a special unit of acceleration employed in geodesy and geophysics to express acceleration due to gravity.
[c] These units are part of the so-called electromagnetic three-dimensional CGS system based on unrationalized quantity equations, and must be compared with care to the corresponding unit of the International System, which is based on rationalized equations involving four dimensions and four quantities for electromagnetic theory. The magnetic flux, Φ, and the magnetic flux density, B, are defined by similar equations in the CGS system and the SI, so that the corresponding units can be related as in the table. However, the unrationalized magnetic field, $H(\text{unrationalized}) = 4\pi \times H(\text{rationalized})$. The equivalence symbol $\mathrel{\hat{=}}$ is used to indicate that when $H(\text{unrationalized}) = 1\ \text{Oe}$, $H(\text{rationalized}) = (10^3/4\pi)\ \text{A m}^{-1}$.

represent the number 0.01, and a half space separates number and % [18].

Multiplication or division of several quantities again leads to a quantity; however, addition or subtraction is only physically meaningful for quantities with the same dimension and equivalent units. Equations involving physical quantities must have the same dimensions on both sides, and the dimensions must be the correct ones for the quantity calculated.

For a complete statement of the result of a measurement, it is required

- to indicate the measurand;
- to provide the numerical value and the unit; and
- to state a measure for the trustworthiness of the result.

The measurand should be defined in sufficient detail. For instance, the quantity "temperature at the triple point of water" is not well defined, unless the isotopic composition of water is given (cf. Section 2.2.1.5).

Even if in a given field, one and only one unit is used, it must be given explicitly and a prefix should be used, such that the leading digits range from 0.001 to 1000. If a result has many digits, it enhances the legibility to group them in blocks of three, separated by a half space [19]. However, within a column of a table, the same unit should be used. In stating the values of quantities of dimension one, powers of 10 are used.

The most informative measure for the trustworthiness of a numerical value is the expanded uncertainty.

The Guide to the expression of uncertainty in measurement (GUM) ISO

Tab. 2.7 Examples for demonstrating rules.

1	Several kelvins, *not* Kelvins or Kelvin
2	milligram, *neither* milli-gram *nor* milli gram
3	newton meter or newton-meter *not* newtonmeter
4	meter per second, *not* divided by second
5	meter per second squared, *not* square second, *but* kilogram per meter cubed *or* cubic meter
6	Potential difference is measured in volts, *not* in V; its value is 1000 V *not* 1000 volts
7	There are *several* symbols in use for length quantities, for example, l, x or r
8	*Only* the symbol m is allowed for meter
9	Ω is the symbol for the unit ohm, *not* ω lx is the symbol for the unit lux, *not* Lx
10	1 L but *neither* 1 l *nor* 1 l
11	Multiplication: *either* N m *or* N·m, but *neither* Nm *nor* N × m
12	Division: $\frac{\text{m}}{\text{s}}$, m/s, m s^{-1} or m·s^{-1}
13	The symbol for millimeter is mm, *neither* the symbol for milli meter is m m *nor* the symbol for milli-meter is m·m the symbol for the unit millisecond is ms, the symbol for the unit of speed is m s^{-1}
14	$1\,\text{cm}^3 = (10^{-2}\,\text{m})^3 = 10^{-6}\,\text{m}^3$ or $1\text{ns}^{-1} = (10^{-9}\,\text{s})^{-1} = 10^9\,\text{s}^{-1}$
15	10^5 Pa = 0.1 MPa or 100 kPa *but not* 10^5 Pa = 1 hkPa "(hectokilo)"
16	10^{-6} kg = 1 mg *not* 10^{-6} kg = 1 µkg
17	$(0.5\,\text{m s}^{-2})(2.0\,\text{s}) = (0.5\,\text{m s}^{-2} \times (2.0\,\text{s}) = 1\,\text{m}^{-1})$ in some countries × is used for expressing areas $a = 5\,\text{m} \times 5\,\text{m} = 25\,\text{m}^2$ or volumes $(a/b)/c \neq a(b/c)$, unless $c = 1$; $a/b/c$ is undefined
18	$w\ (x) = 0.032\%$, and $w(x) = 0.032\%$
19	$l = 1000$ mm is appropriate, $l = 1\,000000$ µm is *less* appropriate, $l = 987.654\ 32$ mm *or* $l = 0.987\ 654\ 32$ m, but *neither* $l = 0.000\ 987\ 654\ 32$ km *nor* $l = 987\ 654.32$ µm are easy to read

(1993) denotes the standard uncertainty associated with the best estimate x of the value of a quantity X by $u(x)$. The GUM provides rules for the evaluation of the uncertainty and for the determination of the so-called coverage factor. The product of the coverage factor k and the uncertainty is called *expanded uncertainty* for which the symbol U is used. For example, the following statement would be complete:

The measured value of the nominal 50-mm *gauge block is* (49.999 926 ± 0.000 073) mm. *The reported expanded uncertainty of measurement is stated as the standard uncertainty of measurement multiplied by the coverage factor* $k = 2$, *which corresponds to a coverage probability of approximately* 95% *for a normal distribution.*

In some literature, the uncertainty is often given by adding two digits in parenthesis. For the result given above, it is 49.999 926 (73) mm. Here, too, the used coverage factor is to be stated.

Finally, there are no rules on choosing a unit for a given nonbase quantity. However, if possible, the unit should reflect the physics of the addressed quantity, for example, for torque, one would use newton and not joule.

2.5 International Metrology

2.5.1 Convention du Mètre

International metrology aims at global comparability of measurements. The first international agreement on metrology is the Convention du Mètre, signed by 17 states in Paris in 1875. It is still effective and currently has (2008) 51 signatory states and 27 associated states and economies. This treaty gives authority to the

- CGPM;
- the CIPM; and the
- BIPM.

to act in matters of world metrology, particularly concerning the demand for measurement standards of ever-increasing accuracy, range, and diversity, and the need to demonstrate equivalence between national measurement standards.

Table 2.8 visualizes the structure of international metrology under the authority of the Convention du Mètre.

The CGPM is the highest authority. Representatives of all member states and observers from the associate states meet every four years, last in 2007. The CGPM is responsible for

- discussing and instigating the arrangements required to ensure the propagation and improvement of the SI;
- confirming the results of new fundamental metrological determinations and the various scientific resolutions of international scope, mainly prepared by the CIPM; and
- adopting the important decisions concerning the organization and development of the BIPM.

Principal task of the CIPM is to promote worldwide uniformity in units of measurement and this is done by direct action or by submitting draft resolutions to the CGPM. The CIPM meets annually at the BIPM. It has 18 members from different member states of the Meter Convention and the director of the BIPM is a member *ex officio*.

The BIPM was established by the Convention du Mètre. It operates under the exclusive supervision of the CIPM and has the task to ensure worldwide uniformity of measurements and their traceability to the SI. The BIPM carries out measurement-related research, takes part in and organizes international comparisons of national measurement standards, and carries out calibrations for National Metrology Institutes (NMIs).

In view of the extension of the work entrusted to the BIPM, the CIPM has set up Comités Consultatifs (CC) since 1927. These CCs are responsible for coordinating the international work carried out in their respective fields and they advise on matters of the system of units. At present, 10 CCs exist (see Table 2.8).

The BIPM also forms Joint Committees with other international organizations, for particular tasks of common interest (see Table 2.8).

2.5.2 Mutual Recognition Arrangement

The Comité International des Poids et Mesures Mutual Recognition Arrangement (CIPM-MRA) (CIPM, 1999), is an agreement of NMIs with the following objectives:

- to establish the degree of equivalence of national measurement standards;

Tab. 2.8 International Metrology.

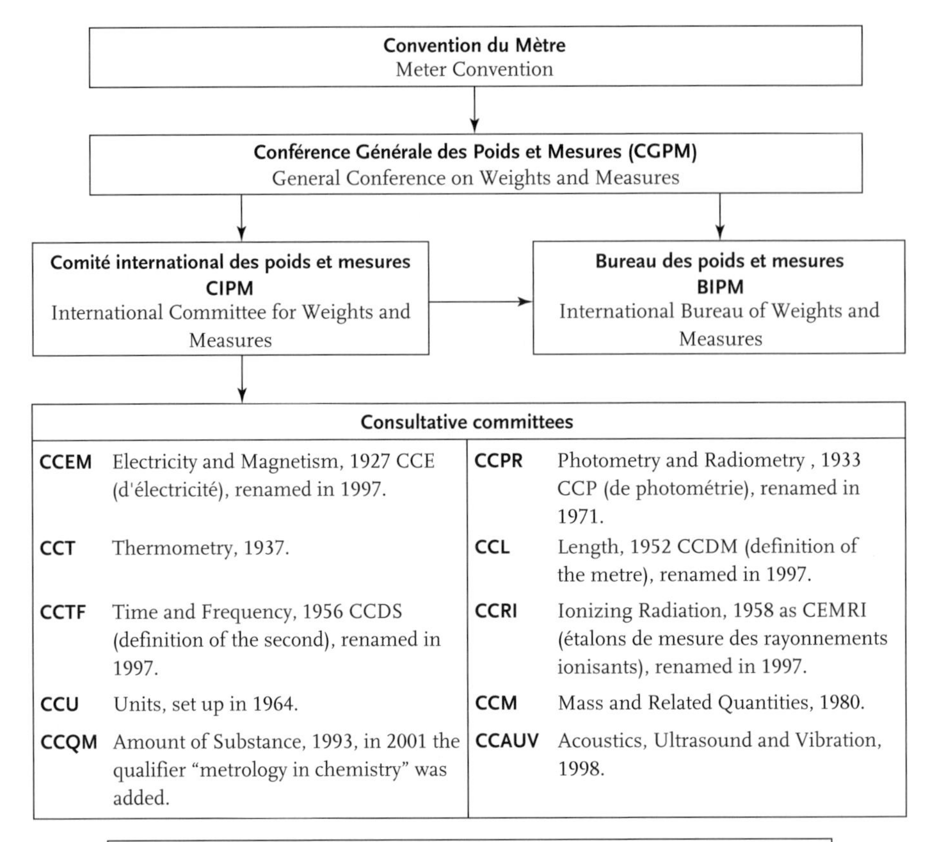

Consultative committees			
CCEM	Electricity and Magnetism, 1927 CCE (d'électricité), renamed in 1997.	**CCPR**	Photometry and Radiometry , 1933 CCP (de photométrie), renamed in 1971.
CCT	Thermometry, 1937.	**CCL**	Length, 1952 CCDM (definition of the metre), renamed in 1997.
CCTF	Time and Frequency, 1956 CCDS (definition of the second), renamed in 1997.	**CCRI**	Ionizing Radiation, 1958 as CEMRI (étalons de mesure des rayonnements ionisants), renamed in 1997.
CCU	Units, set up in 1964.	**CCM**	Mass and Related Quantities, 1980.
CCQM	Amount of Substance, 1993, in 2001 the qualifier "metrology in chemistry" was added.	**CCAUV**	Acoustics, Ultrasound and Vibration, 1998.

Joint Committees of the BIPM and other international organizations

JCDCMAS:	Joint committee for the coordination of technical assistance to developing countries in metrology, accreditation and standardization.
JCGM:	Joint committee for guides in metrology. It has two working groups (WG). WG 1 promotes the use of the guide to the expression of uncertainty in measurement (GUM), prepares supplemental guides for its broad application. WG 2 revises and promotes the use of the International vocabulary of metrology (VIM), the acronym stands for it French version "Vocabulaire international de métrologie"
JCRB:	Joint committee of the regional metrology organizations and the BIPM. Its most important tasks is the coordination of the activities among the RMOs in establishing confidence for the recognition of calibration and measurement certificates, according to the terms of the CIPM MRA. It reports annually to the CIPM and to the signatories of the CIPM MRA.
JCTLM:	Joint committee for traceability in laboratory medicine. The goal of the JCTLM is to provide a worldwide platform to promote and give guidance on internationally recognized and accepted equivalence of measurements in laboratory medicine and traceability to appropriate measurement standards.

For more details see http://www.bipm.org/en/committees/

- to provide for the mutual recognition of calibration and measurement certificates; and
- thereby to provide governments and other parties with the metrological basis for wider agreements negotiated for international trade, commerce, and regulatory affairs.

The NMIs, one per country, are signatories of the CIPM-MRA. However, in many countries also, other institutes hold national standards. These institutes can be designated by the signatory NMI and participate in the CIPM-MRA. These institutes are referred to as designated institutes (DIs).

The CIPM-MRA was introduced in the year 1999 at the 21st General Conference on Weights and Measures (CGPM). Presently (2008), it has been signed by the representatives of 74 institutes – from 45 member states, 27 associates of the CGPM, and 2 international organizations – and covers a further 121 DIs.

The objectives of the CIPM-MRA are reached by the following key elements:

- peer review of the declared Calibration and Measurement Capabilities (CMCs);
- international comparisons of measurement standards (key comparisons or supplementary comparisons); and
- peer review of the quality systems.

International comparisons take many forms. Some are carried out by the BIPM; others are organized by a Consultative Committee and may or may not include the BIPM; many more are organized and carried out, usually as part of the activities of a regional metrology organization, by the NMI and other institutes. Most BIPM comparisons are among the key comparisons chosen by the Consultative Committees to establish the *degree of equivalence* between national measurement standards.

The *degree of equivalence* of a measurement standard is a quantified expression of the consistency of the value of a measurement standard with the *key comparison reference value*. This is expressed quantitatively by the deviation from the key comparison reference value and the uncertainty associated with that value. The degree of equivalence between two measurement standards is expressed as the difference between their respective deviations from the key comparison reference value and the uncertainty associated with the value of this difference.

A *key comparison* is a set of comparisons selected by a Consultative Committee to test the principal techniques and methods for measuring quantities they are concerned with. Since only a limited number of participants is technically feasible in a given key comparison, key comparisons are also carried out by Regional Metrology Organisations (RMOs):

- APMP: Asia Pacific Metrology Programme;
- COOMET: Euro–Asian Cooperation of National Metrological Institutions;
- EURAMET: European Association of National Metrology Institutes;
- SADCMet: Southern African Development Community Cooperation in Measurement Traceability;
- SIM: Sistema Interamericano de Metrologia.

The results of key comparisons are accessible from a database maintained by the BIPM. Only key comparisons carried out by a Consultative Committee or the BIPM lead to a key comparison reference value. For a key comparison carried out by a regional metrology organization the link to the key comparison reference value is obtained by reference to the results from those institutes that have also taken part in the CIPM key comparison. For procedures to evaluate key comparisons see, for instance, Cox [2002a, 2002b].

2.6 Further Development of the SI

The SI is, and must be, a stable and reliable system because it affects almost all aspects of human life. However, the specific definitions of the units eventually need to be revised, since the levels of uncertainties, which can be achieved with the primary standards built to realize the units for practical use, depend on the specific definition of the units.

In fact, a quantum-based definition of the volt and ohm through the Josephson and the Quantum-Hall effect enables the realization of these units with smaller uncertainties than those possible through the classical SI definition. This led to a parallel use of these quantum-based units since 1990. In addition, an unsatisfactory relative drift of national kilogram prototypes versus each other and versus the international prototype kept at BIPM, was observed.

These and other arguments motivated the CIPM at its 94th meeting in 2005 to ask the metrology community to take preparative steps toward a redefinition of the kilogram, ampere, kelvin, and mole so that these four base units are linked to fixed values of fundamental constants.

It was advised that a possible redefinition should not introduce a discontinuity in the value of the unit. Furthermore, a significant benefit for one or more units must be achieved. Indeed, even an increased uncertainty in one specific unit may be tolerated, depending on the practical demands, if it is obviously outweighed by a related benefit of reduced uncertainty of another unit or of the system as a whole.

In the following, options that are currently being discussed but have not been decided yet are described. Some papers concerned with that discussion are cited below (see Further Reading) and briefly summarized there. The selection mirrors the main ideas and propositions, but not claimed to be exhaustive.

The possible redefinitions are such that one or more of the following fundamental constants are fixed: the elementary charge e, Avogadro's number N_A, Boltzmann's constant k_B, and Planck's constant h.

For instance, the definition of the ampere could refer to the number of electrons per second carrying the elementary charge. Approaches to realize the ampere according to such a definition are the single-electron transport (SET) using a chain of tunnel contacts, quantum dots, or nanostructured semiconductor channels.

The definition of the mole could refer to the amount of substance that contains N_A elementary entities.

The kelvin could refer to a change of temperature such that the thermodynamic temperature and the thermal energy E expressed in Joule are linked via $E = k_B T$.

For the definition of the kilogram, essentially two options are discussed: the first is the so-called Watt-balance experiment, in which the weight of a test body (which then represents a reference mass) is balanced by an electromagnetic force, thus effectively relating the mass to a measurement of a current and a voltage. If the voltage is measured via Josephson effect with the Josephson constant $K_J = 2e/h$, the current is measured via Quantum-Hall effect with the von Klitzing constant $R_K = h/e$, and applying

Ohm's law, the kilogram would be linked to Planck's constant h. As an alternative, the kilogram could refer to the atomic mass unit m_e or to Avogadro's constant N_A, as realized in the so-called Avogadro experiment.

Any definition of a system of units is based on the definition of a certain set of values of constants or material parameters. In the present SI, exact values are defined for the speed of light in vacuum c_0, the mass of the international prototype of the kilogram $m(K$, the triple-point temperature of water T_{tpw}, and the magnetic constant μ_0. All other constants, if not automatically exact by definition such as the electric constant ε_0 via $\varepsilon_0 = 1/(c_0^2\mu_0)$, are quantities that need to be measured with a particular uncertainty given by the experiment.

The possible redefinitions as outlined before that are based on defining exact values of e, N_A, k_B, h (and c_0), would require to measure the values of $m(\mathcal{K})$, $M(^{12}C)$, T_{tpw}, and μ_0 and nonvanishing uncertainties would inevitably be associated with those values. As one of the most prominent benefits, however, the quantum-based definitions of volt and ohm through Josephson constant K_J and the von Klitzing constant R_K would become part of SI. In addition, the definition of the kelvin would become independent of possible drifts in water triple point cells and the kilogram would become independent of drifts of the mass of the international kilogram prototype.

Some fundamental constants such as e, N_A, k_B, and h are linked by fundamental laws of nature. Thus, one is not free to choose their values. Any redefinition of units that defines an exact value of one or more of these fundamental constants requires the consistency of the current experiments on them. This is not yet satisfactorily the case for the Watt-balance and Avogadro experiments. Values for h can be deduced from both experiments, which differ significantly. This disagreement needs to be solved before the CGPM may adopt redefinitions as outlined earlier.

Acknowledgment

We wish to thank Professor A. J. Wallard, Directeur du Bureau International des Poids et Mesures. With his kind permission, passages and tables from the brochure "Le Systeme International d'Unites (SI)," 8th edition 2006, have been used. We are also grateful to the BIPM for their generous access to publications on its home page.

Appendix: Conversion Factors for Some Non-SI Units

This appendix provides a list of some more frequently used non-SI units (Table 2.9). For a complete list, the reader is referred to *http://www.bipm.org/en/si/si_constants.html* and to Appendix B in the National Institute of Standards and Technology (NIST) Special Publication 811, (Thompson and Taylor, 2008). Bold printing indicates *exactly defined* conversion factors.

Glossary

The glossary lists scientific terms of relevance to units. Main sources are the Guide to the Expression of Uncertainty in Measurement (GUM) (ISO, 1993), the Mutual Recognition Arrangement (MRA) (CIPM, 1999) and the Vocabulaire International de Métrologie (VIM) (JCGM, 2008).

Base Quantity: Quantity in a conventionally chosen subset of a given system of

Tab. 2.9

Unit name (symbol, when defined)	SI equivalent	Notes
Length		
1 inch (in)	**0.0254 m**	12 in = 1 ft
1 foot (ft)	**0.3048 m**	36 in = 3 ft = 1 yd
1 yard (yd)	**0.9144 m**	1760 yd = 1 mi
1 mile	**1.609 344 km**	
1 parsec	$3.085\,678 \times 10^{16}$ m	Used in astronomy
1 light year (l.y.)	$9.460\,73 \times 10^{15}$ m	Based on 1 d = 86 400 s, and 1 Julian century = 36 525 d
Volume		
1 barrel (bbl)	0.158 987 3 m^3	For trading crude oil = 42 US gallons
1 bushel (bu)	$3.523\,907 \times 10^{-2}$ m^3	2150.42 in^3, for trading grain
1 imperial gallon (gal)	**4.546 09 L**	Used in Canada and formally in the United Kingdom
1 gallon (US) (gal)	3.785 412 L	
1 quart (US liquid) (liq qt)	9.463 529 dL	4 liq qt = 1 gal, for example, beer and juice
1 pint (liquid) (US) (liq pt)	4.73 1765 dL	8 liq pt = 1 gal, for example, beer and juice
1 cup (US)	2.365 882 dL	2 cup = 1 liq pt, used in cookbooks
1 tablespoon (tbsp)	1.478 676 cL	16 tbsp = 1 cup, used in cookbooks
1 teaspoon (tsp)	4.928 922 mL	3 tsp = 1 tbsp, used in cookbooks
Velocity or speed		
1 knot (nautical mile per hour)	0.514 444 4 m/s	Navigation and aviation
1 mile per hour (mi/h)	**1.609 344 km/h**[1]	
1 revolution per minute (rpm) (r min)	0.104 719 8 rad/s	
Mass		
1 ounce (avoirdupois) (oz)	$2.834\,952 \times 10^{-2}$ kg	16 oz = 1 lb
1 pound (avoirdupois) (lb)	**0.453 592 37 kg**	Usually, only 0.453 592 4 is used
Force		
1 kilogram-force (kgf)	**9.806 65 N**	for g_n= **9.806 65 m/s²** (standard value)
1 poundal	1.383 550	1 poundal =1 lb ft s^{-2}
1 pound-force (lbf)	4.448 222 N	Rounded values for poundal and pound-force

Tab. 2.9 (*continued*)

Unit name (symbol, when defined)	SI equivalent	Notes
Pressure or stress		
1 atmosphere, standard (atm)	**$1.013\,25 \times 10^5$ Pa**	
1 torr (Torr)	$1.333\,224 \times 10^2$ Pa	760 torr = 1 atm
1 inch of mercury, conventional (in Hg)	$3.386\,389 \times 10^3$ Pa	
Energy (includes work)		
1 British thermal unit$_{IT}$ (Btu$_{IT}$)	$1.055\,056 \times 10^3$ J	
1 calorie$_{IT}$ (cal$_{IT}$)	**4.1868 J**	
Temperature		
Degree Fahrenheit (temperature) (°F) to kelvin (K):		$T(\mathrm{K}) = (t/°\mathrm{F} + 459.67)/1.8$
Degree Fahrenheit (temperature) (°F) to degree Celsius (°C):		$t(°\mathrm{C}) = (t/°\mathrm{F} - 32)/1.8$
Power		
1 horsepower (550 ft-lbf s^{-1}) (hp)	$7.456\,999 \times 10^2$ W	
Radiology		
Curie (Ci)	**3.7×10^{10} Bq**	roughly the activity of 1 g of 226Ra
Röntgen (R)	**2.58×10^{-4} C kg^{-1}**	measure of exposure to X or γ radiation
Rad (rad)	**10^{-2} Gy**	1 rad = 10^2 ergs/g (CGS-unit)
Rem (rem)	**10^{-2} Sv**	absorbed dose × radiation quality

Motor vehicle efficiency
Is expressed in the US by miles per gallon (mi gal^{-1}) and in metric countries by liter per 100 km. Upon introducing x mi gal^{-1} and y L/(10^2 km), one obtains $y = (2.352\,145 \times 10^2)/x$ or $x = (4.251\,437 \times 10^{-3})/y$.

quantities, where no subset quantity can be expressed in terms of the others. The following amy be noted:

- The subset mentioned in the definition is termed the *Set of Base Quantities.*
- Base quantities are referred to as *being mutually independent* since a base quantity cannot be expressed as a product of powers of the other base quantities.
- *Number of entities'* can be regarded as a base quantity in any system of quantities.

Base Unit: Measurement unit that is adopted by convention for a base quantity.

Coherent Derived Unit: Unit that is defined as product of the number 1 and powers of the base units with at least two nonvanishing exponents. For example, kg/m is a coherent derived unit for the quantity mass density.

Degree of Equivalence of a Measurement Standard: The degree to which the value of a measurement standard is consistent with the key comparison reference value. This is expressed quantitatively by the deviation from the key comparison reference value and the uncertainty of this deviation. The degree of equivalence between two measurement standards is expressed as the difference between their respective deviations from the key comparison reference value and the uncertainty of this difference.

Derived Unit: Unit that is defined as product of a number and powers of the base units with at least two nonvanishing exponents. For example, g/cm is a derived unit for the quantity mass density, but it is not coherent.

Dimension: See quantity dimension.

International System of Units (SI): System of units, based on the International System of Quantities, their names and symbols, including a series of prefixes and their names and symbols, together with rules for their use, adopted by the General Conference on Weights and Measures (CGPM), see Sections 2.2 and 2.4 and Table 2.8.

Key Comparison: One of the set of comparisons selected by a Consultative Committee to test the principal techniques and methods in the field (note that key comparisons may include comparisons of representations of multiples and submultiples of SI base and derived units and comparisons of artifacts).

Key Comparison Database: The database maintained by the BIPM, which contains Appendices A, B, C, and D of the Mutual Recognition Arrangement (CIPM, 1999).

Key Comparison Reference Value: The reference value accompanied by its uncertainty resulting from a CIPM key comparison (Source: BIPM).

Measurement Unit: Real scalar quantity, defined and adopted by convention, with which any other quantity of the same *kind* can be compared to express the ratio of the two quantities as a number. Also called *unit of measurement or unit.*

Multiple of a Unit: Unit that is formed from a given unit according to a scaling convention and is larger than that unit; see also prefix.

Numerical Value of a Quantity: Ratio of that quantity to the unit used.

Prefix: SI-prefixes (see Table 2.3) are used to form a multiple or submultiples of a unit.

Quantity: Property of a phenomenon, body, or substance, where the property has a magnitude that can be expressed as a number and a reference.

Quantity Dimension: Expression of the dependence of a quantity on the base quantities of a system of quantities as a product of powers of factors corresponding to the base quantities, omitting any numerical factor.

Set of Base Quantities: Mutually independent quantities such as, for instance, the International System of Quantities (ISQ).

Submultiples of a Unit: Unit that is formed from a given unit according to a scaling convention and is smaller than that unit; see also prefix.

Symbol of a Unit: Conventional sign designating a unit.

System of Quantities: Set of quantities together with a set of noncontradictory equations relating those quantities.

Uncertainty (of Measurement): Parameter, associated with the result of a measurement, that characterizes the dispersion of the values that could reasonably be attributed to the measurand (ISO, 1993).

Unit: See measurement unit.

Value of a Quantity: Number and reference together expressing magnitude of a quantity; see also numerical value of a quantity.

References

BIPM (**2006**) *Le système International d' Unités* (SI), 8th édn, Bureau International des Poids et Mesures, Sèvres.

BIPM, IEC, IFCC, ILAC, ISO, IUPAC, IUPAP, and OIML (**2008**) International Vocabulary of Metrology – Basic and General Concepts and Associated Terms(VIM), *http://www.bipm.org/en/publications/guides/vim.html.*

CGPM (**1948**) Resolution 7 of the 9th CGPM.

CIPM (**1999**) Mutual Recognition of National Measurement Standards and of Calibration and Measurement Certificates Issued by National Metrology Institutes, International Committee of Weights and Measures, Issued by the CIPM on 14 October 1999, and revised in October 2003: *http://www.bipm.org/utils/en/pdf/mra_2003.pdf*.

Cox, M.G. (**2002a**) Evaluation of key comparison data. An introduction. *Metrologia*, **39**, 587–588.

Cox, M.G. (**2002b**) Evaluation of key comparison data. *Metrologia*, **39**, 589–595.

ICRU (**1998a**) Conversion Coefficients for use in Radiological Protection Against External Radiation, Report No. 57.

ICRU (**1998b**) Fundamental Quantities and Units for Ionizing Radiation, Report No. 60.

ISO (**1993**) Guide to the Expression of Uncertainty in Measurement, International Standards Organisation, Geneva.

JCGM (**2008**) Evaluation of Measurement Data – Supplement 1 to the "Guide to the Expression of Uncertainty in Measurement" – Propagation of Distributions Using a Monte Carlo Method, *http://www.bipm.org/en/publications/guides/gum.html.*

Mohr, P.J. and Taylor, B.N. (**2005**) CODATA recommended values of the fundamental physical constants. *Rev. Mod. Phys.*, **77**, 1–107.

Quinn, T.J. (**1994**) Mise en Pratique of the Definition of the Metre (1992). *Metrologia*, **30**, 523–541.

Maxwell, J.C. (**1873**) *Treatise on Electricity and Magnetism*, Oxford University Press, London.

Thompson, A. and Taylor, B.N. (**2008**) Guide for the Use of the International System of Units (SI), NIST Special Publication 811 2008 Edition, *http://physics.nist.gov/cuu/pdf/sp811.pdf*.

Further Reading

History and Quantity Calculus

de Boer J. (**1995**) On the history of quantity calculus and the International System. *Metrologia*, **31**, 405–429.

The author discusses the foundation of quantity calculus and its algebraic structure as well as the interrelation between quantity systems and unit systems. Furthermore he analyzes the evolvement of the International System.

Selected Papers on the Envisioned Redefinition of Some SI Base Units

Kose, V., Siebert, B.R.L., and Wöger, W. (**2003**) General principles for the definition of the base units in the SI. *Metrologia*, **40**, 146–153.

The authors formulate general principles that should guide the redefinition of the SI base units and derive requirement from the demands and the need of economy, science and society. To meet this requirements, they propose to principally separate the definition and the realisation of the SI units.

Mills, I.M., Mohr, P.J., Quinn, T.J., Taylor, B.N., and Williams, E.R. (**2006**) Redefinition of the kilogram, ampere, kelvin and mole: a proposed approach to implementing CIPM recommendation 1 (CI-2005). *Metrologia*, **43**, 227–246.

The authors propose that the four base units kilogram, ampere, kelvin and mole should be given new definitions linking them to exactly defined values of the Planck constant h, elementary charge e, Boltzmann constant k and Avogadro constant N_A, respectively. They present the background and discuss the merits of these proposed changes, and also present possible wordings for the four new definitions.

Becker, P., De Bièvre, P., Fujii, K., Glaeser, M., Inglis, B., Luebbig, H., and Mana, G. (**2007**) Considerations on future redefinitions of the kilogram, the mole and of other units. *Metrologia*, **44**, 1–14.

The authors discuss alternatives for redefining the kilogram and the mole. They find good reasons for fixing a number as the ratio between the kilogram and the mass of an atomic particle, and they propose to convert the Avogadro constant to the 'Avogadro number', and to link the mole to this number.

Milton, M.J.T., Williams, J.M., and Bennett, S.J. (**2007**) Modernizing the SI: toward an improved, accessible and enduring system. *Metrologia*, **44**, 356–364.

The authors conclude that, while the proposed changes to the definitions of the mole and kelvin appear to have some advantages, the case for changes in the definitions of the ampere and the kilogram is less well made at the present stage of development of the key experiments involved.

3
Fundamental Constants

J. L. Flowers and B. W. Petley

Handbook of Metrology. Edited by Michael Gläser and Manfred Kochsiek

ISBN: 978-3-527-40666-1

3.1
Introduction

This chapter discusses the fundamental physical constants of nature such as the speed of light in vacuum c, the mass of the electron m_e, the Newtonian gravitational constant G, the elementary charge e, the Planck constant h, the Avogadro constant N_A, Faraday constant F, the Rydberg constant (for infinite mass) R_∞, the fine-structure constant α, the Josephson constant K_J, and the von Klitzing constant R_K. Such constants are nature's true invariants in contrast to, say, the freezing point of gallium, or the time required for the Earth to revolve once about its axis, which can be changed–for example, the former by impurities and the latter by seasonal effects. It does not necessarily mean that there might not exist a deeper set of constants in terms of which the above fundamental constants may ultimately be expressed, as indeed some of them, such as the above Faraday, fine-structure constant, the Rydberg constant, and others may be already be expressed in terms of e, h, μ_0, and c. Thus, they have a unique place in science and technology and have even acquired a certain mystique. Determination of their values spurred a considerable amount of activity that peaked between about 1970 and 1985. There are many important combinations of the fundamental constants that are widely used in science and technology. The 2006 CODATA (Committee on Data for Science and Technology) evaluation of the "best" values for the fundamental constants contains some 300 recommended values for various combinations of the constants. It is convenient to discuss progress in this field of endeavor with respect to the recommended values referred to the chapter by Barry Taylor in the earlier edition of the *Encyclopedia of Applied Physics* on this topic that we update here.

In all of our fundamental theories, there appear a certain few parameters that characterize the fundamental particles and the interactions that we find in nature. These are termed the *fundamental physical constants*.The precision determination of the numerical values of these constants, in terms of International System of Units (SI) units when they are dimensioned constants, has long been and still remains one of the principal objectives of experimental science. This is not because there is any intrinsic virtue in accumulating lists of numbers with ever more decimal places, but because the fundamental constants are the quantitative links between our most basic theories and the physical reality of our universe that we wish them to describe. Our theories must stand

Handbook of Metrology. Edited by Michael Gläser and Manfred Kochsiek

ISBN: 978-3-527-40666-1

or fall according to their ability to make quantitative predictions that agree with experimental observations to the maximum achievable accuracy. The progress of our understanding of the physical world is therefore very much interwoven with the advance of the art of precision measurement and its application to the determination of the fundamental physical constants.

This chapter provides an illustration of the *raison d'etre* for striving for increased measurement accuracy that Maxwell and many other scientists have long recognized:

> *... the opinion seems to have got abroad that in a few years all of the great physical constants will have been approximately measured and the only occupation which will be left to men of science will be to carry on these measurements to another place of decimals.... But the history of science shows that even during that phase of her progress in which she devotes herself to the improving the accuracy of quantities with which she has long been familiar, she is preparing materials for the subjugation of new regions which would have remained unknown if she had been content with the rough methods of her early pioneers.* (Maxwell, 1871)

Advances in accuracy continue to be made. Over the last 80 years, measurement accuracy has advanced about 10-fold every 15 years, a process that may well continue until natural limits are reached. Not only has this topic had its share of Nobel Prize winners but, in recent years, there also have been notable advances in both theory and experiment relevant to the fundamental constants with applications moving out into the market place. New technologies have made possible new levels of precision: Computers have made possible more complex calculations. The advent of increasingly stable lasers has led to new levels of accuracy being achieved in many different experiments, for example, optical spectroscopy in the visible region with Hertz-level, that is fractional 10^{-16} accuracy (Coddington, Swan, and Newbury, 2008; Ludlow *et al.*, 2008), and measurements of the Rydberg constant with fractional 10^{-12} accuracy level (Fischer *et al.*, 2004; Hänsch *et al.*, 2005). The discovery of the Josephson effects (JEs) in superconductors and the quantum Hall effect (QHE) in semiconductors have provided new ways to determine several important fundamental constants. Ion traps and atom traps of single particles have increased measurement times. There has been steady progress in precision measurements of the properties of fundamental particles and simple atomic systems.

Current theoretical and experimental efforts in this field span almost the whole of physics, from elementary particle theory to experimental solid-state physics. The intercomparison of measurements at the highest accuracy via very different methods in different parts of science permits detailed examination to be made of the coherence of science as a single entity. In an increasing number of cases, the experimental accuracy has reached an artificial limit set by our ability to realize the appropriate combination of the practical units of measurement, the SI. A consequence of this has been the increasing application of methods of measuring fundamental constants either to maintain more stable SI units, or to incorporate them into the definitions of the unit(s) concerned–as happened in 1983 when the value of the speed of light was adopted as the present definition of the meter. This is approaching the point where most of the SI base units are definable in terms of fixed values for fundamental constants.

There are also interrelationships between the measurable combination of fundamental constants and related quantities, for example, the Rydberg and fine-structure constants, $2e/h$ via the JE, realizations of the absolute ohm and ampere, the quantized Hall resistance, the electron and muon anomalous magnetic moments, the proton-to-electron mass ratio, the proton gyromagnetic ratio, Faraday, and Avogadro constants. Consequently it is appropriate, indeed even necessary, for the solid-state theorist to be aware of the work of the electrical metrologist; the atomic mass spectroscopist of the work of the precision electrochemist; and the quantum electrodynamics (QEDs) theorist of the work of the experimental semiconductor physicist. The fundamental constants of nature and closely related precision measurements are truly the common meeting ground for many of the disciplines of science and technology.

It is clear from the earlier part of this introduction, that the field of fundamental constants too broad for any one does more than touch upon a few key selected topics. However, these should provide the reader with a reasonable picture of the field, especially the very important interplay between theory and experiment that is the key to scientific progress. As elsewhere in science, much of the beauty of the subject lies in the detail, including how the measurements were made in the laboratory or the detailed calculations of the theoretician were brought to the meeting ground between theory and experiment at the fractional part in a billion (1×10^{-9}) level of accuracy to $<1 \times 10^{-12}$ and beyond. Both the measurement and the calculation may well have taken 10–15 years but the work is distilled into a four-page article!

The earlier reviews of the recommended values of the fundamental constants proceed at appropriate intervals of about once every 10 years. The date attributed to each review refers to the latest date for which published data was included. Thus the 2006 evaluation had a closure date of 31 December 2006 and was published in 2008. The values in this review replaced those in the 2002 review. It is important for the coherence of science that as far as possible scientists use the same values for the fundamental constants. Birge pioneered the subject and published the first major review of the values in 1929 (Birge, 1929) and most subjects were happy to adopt his recommended values. Since then the reviews have usually been mainly made by two or three people. However, since the 1970s, reflecting that the subject has become increasingly international, the reviewers have been supported by an international committee known as the *CODATA Task Group on Fundamental Constants.*Recently, the CODATA Task Group reviews have taken place every four years. From time to time others have been motivated to perform their own reviews, for example, the Taylor, Parker, and Langenberg review in 1969 (Taylor, Parker, and Langenberg, 1969) and the 1975 review by Tuninskii and Kholin (1975) and Tuninskii (1986). These reviews have been welcomed and, in several cases, have helped clarify discrepancies and have thereby served to progress the subject.

The main body of the chapter (Section 3.2) begins with a brief discussion of the "origins" of the fundamental constants, that is, how they might be usefully categorized; and a lengthier discussion of "the challenge and romance of the next decimal place," their importance and why their determination to ever greater levels of accuracy can have a profound effect on physics as a coherent whole by allowing any small differences between constants measured by very different measurements to be thoroughly investigated.

Besides the earlier example of the Millikan oil-drop determination of the elementary charge, a later illustration, which is featured here, comes from work in the late 1960s through the determination of the quotient $2e/h$ via the JE in superconductors (a topic in which two of the present authors participated). The derivation of a value of the fine-structure constant (α) from it, when combined with other constants, enabled the use of the resulting value of α to test the QED theory of the ground-state hyperfine splitting of the hydrogen atom to well beyond the sixth decimal place. A similar example that is still under active investigation is also presented in which a value of α obtained from the QHE is used to test the QED theory of the electron magnetic moment anomaly a_e.

In Section 3.3, we discuss how sets of the "best values" constituting the recommended values of the fundamental constants are derived, particularly concentrating on the 2006 CODATA least-squares adjustment of the constants, which is the most recent internationally accepted evaluation currently available. In this section, we include some of these recommended values for a comprehensive group of constants, which are taken from the 2006 CODATA set.

The realization of the definitions of the SI units is very time consuming. Modern science and technology relies on making measurements involving time and frequency and electrical quantities, and there is a universal need throughout the world to maintain a continuous representation of the SI units. In Section 3.4, we discuss the importance of the fundamental constants to technology by reviewing the 1 January 1990 introduction of new practical representations of the volt and ohm as defined in the SI. These representations, based on the JEs and quantized Hall effects, and the consequent internationally adopted values of the Josephson and von Klitzing constants K_J and R_K, are now being used worldwide in electrical metrology to derive and maintain representations of the volt and ohm to fractional accuracies of 10^{-9} or better.

In Section 3.5, we discuss the principal advances made in the fundamental constants field since the 1986 CODATA adjustment was completed and discuss their implications for the least-squares adjustment of 2006. In Section 3.6, we further consider the importance of the fundamental constants to modern technology, and illustrate these by exploring the likely future role of the constants in replacing the definition of the kilogram, the only SI base unit that is still defined in terms of a prototype material artifact, and to similarly replace the definitions of other SI units. In Section 3.7, we briefly discuss the question of why the constants have the particular values that they do, and their possible variation with time. Finally in Section 3.8, which concludes the main body of the chapter, we look ahead in an attempt to see where the fundamental constants field might be heading over the next decade.

It is useful at this point to mention briefly some aspects of the history of the measurements of the fundamental constants. Modern measurements of the fundamental constants really began at the start of the twentieth century modern physics, with the advent of quantum theory, relativity, and atomic and nuclear physics. The two notable exceptions were c and

G, which were first measured in the eighteenth century.

Extraordinary advances in determining the values of the constants were made immediately after 1945 as a result of the significant advances made in electronics, microwaves, and other technologies during World War II. A second period starting in about 1970 has also been especially productive because of the wide introduction of highly sophisticated digital electronic instruments, lasers, computers, and cryogenic techniques. To a considerable extent, the accuracy with which the fundamental constants have been known at any given epoch has depended upon the level of sophistication of the measuring instruments and apparatus available at the time. However, it is not only improvements in measurement technology that have led to more accurate values of the constants. The discovery of new phenomena, such as the JEs and the quantized Hall resistance effects, that provide new and better methods for determining certain constants have also led to significant advances in the field.

Measurements of the constants and the frontiers of measurement science or metrology, especially the development of new reference standards for the units of measurement, have always been closely linked, but in recent years the distinctions between them have all but disappeared. Perhaps the most striking example of this is the determination of *c* in the early 1980s based on extremely difficult measurements of the frequencies and wavelengths of several different laser radiations. As a direct consequence of this work, the definition of the SI unit of length, the meter, which, since 1960s was based on a particular optical transition in the ^{86}Kr atom, was replaced in 1983 with one that was based on an adopted value for the speed of light (Petley, 1983; Giacomo, 1984; Hudson, 1984).

As discussed in Section 3.6, if current efforts to determine h or N_A with a relative uncertainty of about 1×10^{-8} are successful, it is very probable that this scenario will be repeated with regard to the kilogram in the second or third decade of this century. There may be consequent changes to our definitions of the electrical units, while a fixed value for the Boltzmann constant may well be incorporated into a revised definition of the kelvin. This will be an important topic for the next meeting of the General Conference on Weights and Measures (CGPM) scheduled for 2011. Any changes to the definitions of the SI units are framed so as to ensure continuity in the sizes of the units concerned while ensuring that they anticipate or keep pace with the improvements in measurement accuracy enabled by, and demanded by, future progress in science and technology.

3.2 Conceptual Origins and Importance of the Constants

3.2.1 Conceptual Origins of the Constants

Three distinct "sources" of the fundamental constants may be identified. The first is the physical theories that have been developed to describe nature's basic laws, for example, Maxwell's theory of electromagnetism, Einstein's theories of relativity, quantum mechanics, and QED, and their application as appropriate to atomic systems. c, G, h, α, and R_∞, fall within this category. These particular constants (see Section 3.1) are of particular importance in science because c is the velocity of all electromagnetic radiation or photons (in vacuum) and the limiting velocity of all energy and matter; G determines the strength of the gravitational force between bodies; h is the fundamental constant of quantum mechanics and determines the energy of a quantum of radiation of a given frequency; R_∞ sets the scale of the energy levels of atoms; and α determines the strength of the electromagnetic force that governs how electrically charged elementary particles and electromagnetic radiation interact.

The second source is the fundamental particles such as the electron, proton, and neutron. The elementary charge e, the rest mass of the electron m_e, the rest masses of the proton and neutron m_p and m_n, the magnetic moment of the electron in units of the Bohr magneton, namely μ_e/μ_B, where $\mu_B = eh/4\pi m_e$, and the magnetic moments of the proton and neutron in units of the nuclear magneton, μ_p/μ_N and μ_n/μ_N, respectively, where $\mu_N = eh/2m_p$, respectively. These are all examples of quantities that characterize a basic property of the elementary building blocks of nature.

Conversion factors are the third source. Although not true fundamental constants like those in the first two categories, these quantities nevertheless play a critical role in the measurement of fundamental constants in terms of the SI units. Examples include the local value of the gravitational acceleration g, which relates mass to force; the ratio of the old X-ray unit of length or kilo-x-unit to the nanometer; and also the ratio of various so-called as-maintained electrical units in a particular national standard's laboratory to their SI definitions, for example, Ω_{NPL} where Ω_{NPL} is the National Physical Laboratory (NPL) as-maintained ohm. Prior to 1 January 1990 the laboratories local ohm standards were maintained as the mean resistance of a particular group of wire-wound standard resistors that were kept in an oil bath at 25°C.

It should be noted that $\alpha = e^2/(4\pi\varepsilon_0\hbar c)$ and $R_\infty = m_e c\alpha^2/(2h)$ are examples of constants that are actually combinations of other quantities, but are considered fundamental constants in their own right since the combination always appears in theoretical equations in the same way. (Here the magnetic constant, $\mu_0 = 4\pi \times 10^{-7}$ N A^{-2}, exactly, is the permeability of vacuum, and the electric constant $\varepsilon_0 = c^2/\mu_0$.). Many other similar examples are found in Table 3.1 of this chapter.

3.2.2 Importance of the Constants

There are at least four reasons why the fundamental physical constants play a critical role in science and technology and hence should be known as accurately as possible.

1. Accurate values of the constants are required for the critical comparison of theory with experiment, and such comparisons that advances in our understanding of the physical world. A closely related outcome is that by comparing the numerical values of the same fundamental constants obtained from experiments in the different fields of physics, the self-consistency of the basic theories of physics themselves can be tested with increasingly high accuracy (see Section 3.1).
2. The determination of the values of the fundamental constants more accurately fosters the development of new state-of-the-art measurement methods that may have a wider application. Determining the next decimal place is never trivial and usually requires an entirely new measurement technology that, as mentioned elsewhere, can take

Tab. 3.1 Recommended values of some selected physical constants.

Quantity	Symbol	Value	Units	Relative uncertainty
Speed of light in vacuum	C	299 792 458	$\mathrm{m\ s^{-1}}$	Exact
Permeability of vacuum	μ_0	$4\pi \times 10^{-7}$	$\mathrm{N\ A^{-2}}$	(Exact)
Permittivity of vacuum, $1/\mu_0 c^2$	ε_0	$8.854187817\ldots \times 10^{-12}$	$\mathrm{F\ m^{-1}}$	(Exact)
Newtonian constant of gravitation	G	$6.67428(67) \times 10^{-11}$	$\mathrm{m^3\ kg^{-1} s^{-2}}$	1.0×10^{-4}
Planck constant	H	$6.62606896(33) \times 10^{-34}$	J s	5.0×10^{-8}
$h/2\pi$	$\hbar$	$1.054571628(53) \times 10^{-34}$	J s	5.0×10^{-8}
Elementary charge	E	$1.602176487(40) \times 10^{-19}$	C	2.5×10^{-8}
magnetic flux quantum, $h/2e$	Φ_0	$2.067833667(52) \times 10^{-15}$	Wb	2.5×10^{-8}
Conductance quantum,	G_0	$7.7480917004(53) \times 10^{-5}$	S	6.8×10^{-8}
Electron mass	m_e	$9.10938215(45) \times 10^{-31}$	kg	5.0×10^{-8}
Proton mass	m_p	$1.672621637(83) \times 10^{-27}$	kg	5.0×10^{-8}
Proton-electron mass	m_p/m_e	1836.15267247(80)		4.3×10^{-10}
Fine-structure constant, $e^2/4\pi_o\hbar\, c$	α	$7.2973525376(50) \times 10^{-3}$		6.8×10^{-10}
Inverse fine-structure constant,	α^{-1}	137.035999679(94)		6.8×10^{-10}
Rydberg constant	R_∞	10 973 731.568527(73)	$\mathrm{m^{-1}}$	6.6×10^{-12}
Avogadro constant	N_A, L	$6.02214179(30) \times 10^{23}$	$\mathrm{mol^{-1}}$	5.0×10^{-8}
Faraday constant	F	96 485.3399(24)	$\mathrm{C\ mol^{-1}}$	2.5×10^{-8}
Molar gas constant	R	8.314472(15)	$\mathrm{J\ mol^{-1}\ K^{-1}}$	1.7×10^{-6}
Boltzmann constant	K	$1.3806504(24) \times 10^{-23}$	$\mathrm{J\ K^{-1}}$	1.7×10^{-6}
Stefan–Boltzmann constant $(\pi^2/60)k^4/\hbar^3 c^2$	σ	$5.670400(40) \times 10^{-8}$	$\mathrm{W\ m^{-2}\ K^{-4}}$	7.0×10^{-8}
		Non-SI units accepted for use with the SI		
Electron volt	eV	$1.602176487(40) \times 10^{-19}$	J	2.5×10^{-8}
(Unified) atomic mass unit	U	$1.660538782(83) \times 10^{-27}$	kg	5.0×10^{-8}

The digits in parentheses represent the one-standard-deviation uncertainty in the last digit(s) of the given value.
Taken from the 2006 CODATA least-squares adjustment by Mohr, Taylor, and Newell (2008).

10–40 years to bring from conception to fruition. An example is the determination of the Avogadro constant N_A, which necessitated the development of techniques to measure the lattice spacing of pure, single crystals of silicon directly in terms of meter with a relative uncertainty of 1/10th of a part per million, all uncertainties are one-standard-deviation estimates.) The result has been the extension of our ability to make accurate measurements of length in SI units to the picometer or 10^{-12}-m range and the demise of the x-unit used for measurements of X-ray and γ-ray wavelengths. Work toward an improved Avogadro constant continues using isotopically pure silicon-28.

3. Measurements of the fundamental constants also provide the key to the development of a reproducible and invariant system of measurement units, which is also a major global goal of measurement science or metrology. This marks the culmination of a long path from the use of properties of seeds and the human body, and later via properties of water and properties of the Earth, to the units used at present based on fundamental constants. Notable early examples of the latter were the involvement of the Faraday constant to maintain the International ampere at the end of the nineteenth century and the more recent involvement of the speed of light in the definition of meter resulting from the improved measurement accuracy of high frequency radiation and other technology associated with the development of radar, which were later allied with the development of highly monochromatic and stable lasers through to the visible region and beyond, and the later satellite Global Positioning System (GPS).

From the time of Maxwell, Johnstone-Stoney, Planck, Dirac, and countless others, the development of SI units that are really constant in space-time has progressed to become a reality today. The future role of artifacts such as the platinum–iridium cylinder kept in a vault at the International Bureau of Weights and Measures (BIPM), that at present defines the SI kilogram, will be to act as a flywheel to maintain sufficiently constant representations of a mass unit defined in terms of a fundamental constant. The development of ever more accurate atomic clocks for terrestrial time in applications have led to astrophysical investigations well beyond the Solar System such as monitoring the stability of the millisecond pulsars or the comparison of different types of atomic clock enabling investigations pertaining to the invariance of dimensionless constants such as α and m_p/m_e over the age of the universe.

4. Finally, values of the fundamental constants are required for computations and measurements throughout science and technology–for example, the calculation of the excited states of atoms and molecules that play an important role in air pollution monitoring, nuclear energy, and nuclear fusion.

3.2.3 QED and the Josephson Effect Determination of $2e/h$

As an example of the significant role improved measurements of fundamental constants can play in increasing our understanding of physical theory, consider the accurate determination of the quotient $2e/h$ using the JE in weakly coupled superconductors by a method that was first reported in the late 1960s (Parker, Taylor, and Langenberg, 1967). This also provides an illustration of the way in which limitations in the accuracy of the realization of the electrical units in various countries impacted in a dramatic way on our knowledge of the fundamental physical constants.

When a Josephson tunnel junction–for example, two thin films of lead separated by a 1-nm-thick thermally grown oxide layer–is irradiated with microwave radiation of frequency f, its current versus voltage curve exhibits current steps at highly precise quantized Josephson voltages U_J (Barone and Paterno, 1982; Gallop, 1991). The voltage of the nth step, $U_J(n)$, where n is an integer, is related to f by

$$U_J(n) = nf/K_J \tag{3.1}$$

where K_J is known as the Josephson constant ($= 2e/h$). Note that it is the inverse of the superconducting magnetic flux quantum $\Phi_0 (= h/2e)$.

The theory of the JEs predicts that

$$K_J = 2e/h \tag{3.2}$$

and, to date, the experimentally observed universality of Equation 3.1 is entirely consistent with the prediction. The exactness of this equation when subject to changes in the operating conditions (frequency, temperature, material, different apparatus, etc.) has been verified with fractional uncertainties ranging from parts in 10^9 to 10^{16}. It follows from Equation 3.1 that a measurement of f and U_J for the nth step will yield a value of $2e/h$. The result reported in 1969 (Parker *et al.*, 1969) using 8–12 GHz microwaves was

$$K_J = 2e/h = 4.835\,976(12) \times 10^{14}\,\mathrm{Hz/V_{NIST-69}} \tag{3.3}$$

where $\mathrm{V_{NIST-69}}$ is the unit of voltage maintained at the US National Institute of Standards and Technology (NIST) in 1969[1]. We mention in passing that symbols for units are in upright Roman type, and in an italic font for quantities.

Their value was rapidly confirmed by the 36-GHz measurements of Petley and Morris (1970) at the NPL after due allowance was made for the different

1) A note on uncertainties. In Equation 3.3, the number in parentheses represents the one-standard-deviation uncertainty in the last digits of the quoted value. It should also be noted that the uncertainty is a quasi-statistical combination of essentially two types of uncertainty, essentially one that represents the fluctuations of the measured value during the measurement and the second from quantities, which may be in error but have not varied during the measurement. Currently, these are characterized by the terms Type A and Type B, respectively, and have built on the earlier concept of random and systematic uncertainties (fluctuations in the measured value may well not be truly random and part of the art of this type of measurement lies in identifying as many as possible of these causes (e.g., someone using a hammer, switching off equipment, or thermostats, etc.). International agreement has been made on how to evaluate such uncertainties, see the document known as the

GUM : Guide to the expression of the Uncertainty of Measurement JCGM (1995). ISO/IEC Guide 98:1995 Guide to the expression of uncertainty in measurement (GUM). Geneva, Switzerland, International Organization for Standardisation (International Organization for Standardisation) (see also, e.g., *http://www.bipm.org/en/committees/jc/jcgm/wg1_bibliography.html* and *http://physics.nist.gov/Pubs/guidelines/TN1297/tn1297s.pdf*. Interestingly, uncertainty estimates are made by theoreticians as well as experimenters and it has also become essential to give an uncertainty for the numerical results of theoretical calculations.

voltage units maintained in the two countries. It was swiftly confirmed by other national standards laboratories, notably in Australia and Germany. At that time the maintained volt was based on the mean electromotive force (emf) of a group of electrochemical standard cells. These had the disadvantage that their emf depends on temperature and other factors. The JEs enabled provided a known constant voltage and the existence of this enabled the technology of standard cell manufacture to be improved rapidly. In 1972, the Consultative Committee on Electricity (CCE) recommended a value for K_{J-72} for volt maintenance purposes. This value has since been revised as K_{J-90} and provides an internationally agreed representation of the SI volt.

The difference between K_{J-72} when expressed in SI units (hertz per volt) and the previous best value, namely, that resulting from the 1963 least-squares adjustment of the fundamental constants (Cohen and DuMond, 1965), was (35 ± 10) parts per million. As discussed later, the cause of this rather severe discrepancy was the use in the 1963 adjustment of a value of α derived from the early 1950s measurement of the fine-structure splitting of the deuterium atom (energy difference between the $2^2\mathrm{P}_{3/2}$ and $2^2\mathrm{P}_{1/2}$ atomic energy levels), which subsequently turned out to be incorrect.

Because QED, the relativistic quantum field theory that describes the interaction of charged particles and electromagnetic radiation or photons, is one of the most precise and important theories of modern physics (Kinoshita, 1990a), and involves knowledge of the fine-structure constant, it is essential to see how well the topic can withstand stringent experimental tests. The Josephson $2e/h$ determination is therefore significant for QED mainly because a reliable indirect value of QED-independent value of α can be derived from it, and this value can in turn be used to compare QED and experiment critically and unambiguously. This is in marked contrast to the situation that existed prior to 1967 when no such value was available and tests of QED were mainly checks of internal consistency.

The equation relating α to $2e/h$ may be written as

$$\alpha = \left[\frac{4R_\infty(\Omega_{\mathrm{LAB}}/\Omega)\gamma'_p(lo)_{\mathrm{LAB}}}{c(\mu'_p/\mu_{\mathrm{B}})(2e/h)_{\mathrm{LAB}}} \right]^{1/2} \tag{3.4}$$

where, as before, $\Omega_{\mathrm{LAB}}/\Omega$ is the ratio of the laboratory as-maintained ohm to the SI ohm; μ'_p/μ_{B} is the magnetic moment of the proton in units of the Bohr magneton (throughout, the prime means that the protons are not free but are in a spherical sample of pure water at 25°C); and $\gamma'_p(\mathrm{lo})_{\mathrm{LAB}}$ is the gyromagnetic ratio of the proton obtained by the so-called low (magnetic) field method and is measured in terms of the laboratory as-maintained unit of current, $A_{\mathrm{LAB}} = V_{\mathrm{LAB}}/\Omega_{\mathrm{LAB}}$. (Note that the more general subscript LAB (for laboratory) may be replaced by the subscript NIST, say, for measurements made at the US national standards' laboratory); and that $\gamma'_p = \omega'_p/B$, where ω'_p is the proton nuclear magnetic resonance or spin-flip angular frequency in the magnetic field (magnetic flux density) B. In the low-field method, B is established via a precision solenoid of known dimensions carrying a current known in terms of A_{LAB}.) The value of α derived from Equation 3.3 in 1969 (Taylor, Parker, and Langenberg, 1969) using the best data available was

$$\alpha^{-1} = 137.036\,08(26) \tag{3.5}$$

Included among the quantities that require an accurate value of α for comparing theory and experiment are the g-factors of the electron and muon; the energy levels in hydrogen-like atoms, especially the $n = 2$ Lamb shift ($2^2S_{1/2} - 2^2P_{1/2}$ splitting); and the ground-state hyperfine splitting in hydrogen, muonium (an electron bound to a positive muon or μ^+e^- atom), and positronium (a bound electron–positron pair or e^+e^- atom).

Of particular interest in the late 1960s was the hydrogen hyperfine splitting ν_{Hhfs}. This transition frequency, which corresponds to essentially the energy difference between a hydrogen atom in which the electron and proton spins are aligned and one in which they are in opposite directions, has been determined experimentally with the extraordinary accuracy of 1 part in 10^{12} using the hydrogen maser. The theoretical QED equation for ν_{Hhfs} is

$$\nu_{\mathrm{Hhfs}}(\text{theor}) = (16/3)\, cR_\infty \left(\mu_p/\mu_B\right) \times \left(1 + m_e/m_p\right)^{-3} \times \alpha^2 \left(1 + \text{h.o.c.}\right) \qquad (3.6)$$

where h.o.c. are higher order corrections.

In contrast to the 1×10^{-12} relative uncertainty of $\nu_{\mathrm{Hhfs}}(\text{expt})$, Equation 3.6 is limited to an accuracy of a few parts in a million because of the difficulty in calculating some of the higher order corrections. (Even in the late 1960s, c, R_∞, μ_p/μ_B, and m_e/m_p were sufficiently well known for their total contribution to the fractional uncertainty of $\nu_{Hhfs}(\text{theor})$ to be considerably less than 10^{-6}.)

The most uncertain higher order correction to Equation 3.6 was $\delta_N^{(2)}$, the proton polarizability contribution, which arises from the various excited states or internal structure of the proton. In the late 1960s, theoretical calculations predicted $\delta_N^{(2)} = (0 \pm 5) \times 10^{-6}$. This was in conflict with what was implied by the value of α accepted at that time, which, as noted earlier, was derived from a measurement of the fine-structure splitting in deuterium. The deuterium value of α was used to calculate $\nu_{Hhfs}(\text{theor})$ from Equation 3.6, when this result was compared to $\nu_{Hhfs}(\text{expt})$ as obtained from the hydrogen maser and, when their difference was assumed to arise solely from the existence of a polarizability term, it was found that $\delta_N^{(2)} = (43 \pm 9) \times 10^{-6}$. This meant that the probability for $\delta_N^{(2)}$ to be as small as predicted by direct calculation was only 1 in 20 000, an obvious inconsistency. In contrast, when the value of α derived from the JE measurement of $2e/h$, Equation 3.5, was used in place of the deuterium value, it was found that $\delta_N^{(2)} = (2.5 \pm 9) \times 10^{-6}$, in keeping with the theoretical calculations.

Hence the JE value of α removed a discrepancy that during the 1960s was termed one of the major unsolved problems of QED. This example further illustrates the overall unity of science–a low-temperature, solid-state physics experiment has provided information about the excited states of the proton–as well as how precise measurements of the fundamental constants can illuminate apparent inconsistencies in our physical description of nature. It also illustrates why one of the most accurately measured quantities, the hydrogen maser frequency, does not contribute to our knowledge of fundamental constants other than the polarizability of the proton.

The other aspect of Equation 3.4 lies in the two methods of measuring the gyromagnetic ratio of the proton that are characterized by the terms $\gamma_p'\left(\text{lo}\right)_{\text{LAB}}$ and $\gamma_p'\left(\text{hi}\right)_{\text{LAB}}$, respectively.

In the low-field determination, the spin precession frequency is measured in a weak magnetic field, such as is produced by passing a current $i(\text{lo})_{\text{LAB}}$ through a precisely wound helical solenoid and in the high-field method the flux density is "measured" by weighing the force mg on a precisely wound rectangular that is suspended between the poles of an electromagnet, carrying a current $i(\text{hi})_{\text{LAB}}$, where m is the mass required to balance the magnetic force when the local gravitational acceleration is g. Taking both currents as having measured in terms of the same laboratory units one may combine the resulting two equations in two ways, one to eliminate the currents to obtain γ_p' and the other to eliminate γ_p' and hence obtain a realization of the SI definition of the ampere, also remembering that

$$\gamma_p' B(\text{lo}) = \gamma_p'(\text{lo})_{\text{LAB}} B(\text{lo})_{\text{LAB}} \quad \text{and}$$
$$\gamma_p'(\text{hi}) B(\text{hi})_{\text{LAB}} = \gamma_p B(\text{hi}) \tag{3.7}$$

Much of the difficulty of either method came from the need to measure the coil dimensions very accurately, yet the dimensions of γ_p' are coulombs per kilogram. This led Kibble (1975) to wonder whether there might be a way of adapting and combining the two methods into a realization of both quantities (Flowers and Petley, 2001). This led him to propose the method of realization of the electrical watt, which is the method behind the present best measure of determining h and the proposed new definition of the kilogram that is discussed in Section 3.5.

3.2.4
QED and the Quantum Hall Effect Determination of h/e^2

Another example of a critical test of QED, is one that is still under active experimental and theoretical investigation at present, this involves the QHE and the magnetic moment anomaly of the electron $a_e = g_e/2 - 1 = \mu_e/\mu_B - 1$, where g_e is the g-factor of the electron.

The current best experimental value for $a_e = (g_e-2)/2$, is that by Gabrielse *et al.* ([2006, 2007]) at Harvard using a single electron stored in a Penning-type ion trap. This work achieved a formidable fractional accuracy of 6.6 parts in 10^{10} and this is about six times smaller than an earlier Penning trap measurement at the University of Washington by Van Dyck, Schwinberg, and Dehmelt (1987).

$$a_e = 1.159\,652\,85(76) \times 10^{-3} \tag{3.8}$$

The QED theoretical expression for a_e may be written as (Kinoshita and Lindquist, 1990b)

$$\begin{aligned} a_e(theor) &= A_1^{(2)}(\alpha/\pi) + A_1^{(4)}(\alpha/\pi)^2 \\ &\quad + A_1^{(6)}(\alpha/\pi)^3 + A_1^{(8)}(\alpha/\pi)^4 \\ &\quad + \ldots + \delta a_e A_1^{(2)} = 1/2 \\ A_1^{(4)} &= -0.328478965579\ldots \\ A_1^{(6)} &= 1.181241456 \\ A_1^{(8)} &= -1.7283(35) \end{aligned} \tag{3.9}$$

The total uncertainty in a_e(theor) assuming α is exactly known is now $2.4 \times 10^{-10}\ a_e$, with the dominant uncertainty being that ascribed to the tenth order term, $A_1^{(10)} = 0.0(3.7)$. A test of QED, by comparing a_e(theor) with a_e(expt), to the full accuracies of a_e(theor) and a_e(expt) therefore requires a measured value of α having an uncertainty at or significantly less than $2.4 \times 10^{-10}\ a_e$. Such a value is not yet available, but one close to this has been obtained from the Harvard (Gabrielse *et al.*, 2006) and Washington (Van Dyck, Schwinberg, and Dehmelt, 1987) values for the fine-structure constant that are

$$\alpha^{-1} = 137.035999711(96) \quad \text{and}$$
$$\alpha^{-1} = 137.03599883(50) \qquad (3.10)$$

respectively.
The fractional uncertainty contribution from the QED theory to the conversion from a_e to α is estimated to be about 2.4×10^{-10}–a triumph of achievement for both theory and experiment.

The theory and measurement of the muon anomaly at Brookhaven, (Brown *et al.*, 2001) has similarly led to a measurement of a value of α^{-1}, denoted α_{μ}^{-1} of

$$\alpha_{\mu}^{-1} = 137.03567(26) \qquad (3.11)$$

which provides valuable confirmation of our understanding of the muon as an elementary particle. The hyperfine splitting in muonium has also been measured, and this can lead to a value for α^{-1}, of 137.036 0017(80).

Other tests of QED have been made by comparing the spin precession and cyclotron frequencies of hydrogen-like heavier atoms (highly ionized atoms with one electron remaining). Measurements of the g-factor ratio of hydrogen-like carbon and oxygen have been made at GSI Darmstadt and the University of Mainz (Werth *et al.*, 2002) by comparing the ratios of the spin precession and cyclotron frequencies in a Penning trap an experimental value of

$$\frac{g_g - (^{12}\mathrm{C}^{5+})}{g_e - (^{12}\mathrm{O}^{7+})} = 1.000497273218(41) \qquad (3.12)$$

and similarly a double Penning trap measurement of the ratio of the spin precession to the cyclotron frequency of hydrogen-like $^{16}\mathrm{O}^{7+}$ (that is oxygen from which all but one of its electrons have been removed) gave

$$\frac{f_s(^{16}\mathrm{O}^{7+})}{f_c(^{16}\mathrm{O}^{7+})} = 1.000497273218(41) \qquad (3.13)$$

Both of these are also in good agreement with some formidable QED calculations.

The QHE is a behavior characteristic of a completely quantized two-dimensional electron gas (Prange and Girvin, 1990; Yoshioka, 2002). Such a gas may be realized in a high-mobility semiconductor device such as a silicon metal-oxide-semiconductor field-effect transistor (MOSFET), or a device made from gallium–aluminum–arsenide as a heterojunction of standard Hall bar geometry in an applied magnetic flux density of the order of 10 T and cooled to about 1 K. Hence, in common with the JE, the QHE is a low-temperature, solid-state physics phenomenon.

It is found that for a fixed current I (typically 10–50 μA) through the device, there are regions in the curve of Hall voltage U_H, versus gate voltage for a MOSFET, or of U_H versus B, for a heterojunction, where U_H remains constant as either the gate voltage or B is varied. These regions of constant U_H are termed *quantum Hall plateaus.*Inthe limit of zero dissipation (zero voltage drop) in the direction of current flow, the Hall voltage-to-current quotient $U_H(\mathrm{i})/I$ or Hall resistance $R_H(\mathrm{i})$ of the ith plateau, where i is an integer (we consider only the integral QHE), is quantized and given by

$$R_H(i) = U_H(i)/I = R_K/i \qquad (3.14)$$

where R_K is the von Klitzing constant (named in honor of Klaus von Klitzing, who discovered the QHE in 1980 and for which he was awarded a Nobel Prize in 1985).

The theory of his effect predicts that

$$R_K = \frac{h}{e^2} = \frac{\mu_0 c}{2\alpha} \tag{3.15}$$

Measurements show consistency with Equation 3.9 within a few parts in 10^{10}.

Since in the present SI, $\mu_0 = 4\pi \times 10^{-7}$ N A^{-2} exactly, and $c = 299792458\,\text{m s}^{-1}$ exactly as a result of the 1983 redefinition of the meter in terms of the speed of light, it is apparent from (Equation 3.10) that a measurement of R_K in SI units (i.e., ohms) with a given uncertainty will yield a value of the fine-structure constant α with the same uncertainty.

In practice, R_K is measured in terms of Ω_{LAB} (indicated as $(R_K)_{LAB}$) so that the conversion factor Ω_{LAB}/Ω must be determined in a separate experiment notably by using an apparatus known as a *calculable cross-capacitor* in which the unknown resistance of a reference resistor is compared with the known impedance of the capacitor. Nevertheless, the QHE has already yielded a value of α with an uncertainty of 2.4 parts in 10^8 (Shields, Dziuba, and Layer, 1989). When used to compare a_e(theor) with a_e(expt), it gives a difference of a_e(expt)$-a_e$(theor) $= a_e =$ $(4.1 \pm 2.5) \times 10^{-8}$. Since the 4.1×10^{-8} difference is only 1.7 times the 2.5×10^{-8} standard deviation of the difference, the difference between theory and experiment a_e is within statistically acceptable limits. Since QED actually predicts the value of the entire electron g-factor, then a consequence of the near equality of $g_e(\text{expt}) = 2(1 + a_e(\text{expt}))$ and $g_e(\text{theor}) = 2(1 + a_e(\text{theor}))$, is that this agreement confirms the validity of the renormalization processes inherent in QED theory to the hitherto unprecedented level of four parts in 10^{11}.

3.2.5 The Quantum Metrology Triangle

The JE provides a measurement of $2e/h$ in terms of voltage, V, and similarly the quantized Hall resistance a measurement of h/e^2 in terms of resistance. There is an interesting third measurement that can be obtained from single electron tunneling (SET), which provides a measurement of the elementary charge e in terms of a current, I, and a driving frequency ν giving three relations with two unknowns.

The measurement of e by SET has been implemented on the charging of an isolated capacitor by the transfer of single electrons by the application of an alternating voltage at close to the maximum that a stochastic frequency can be counted; if this is 10 GHz, then one has a charging current of about 1.6 nA on a 10^{-16} F capacitor–a rather small current to measure with a fractional accuracy of about 10^{-6} or better. However, Mark Keller and his colleagues have achieved this accuracy at NIST in the United States (Keller, Zimmerman, and Eichenberger, 2007). This has excited considerable interest throughout the world and work is presently in progress in at least four national institutes: in France, Germany, United Kingdom, and United States. To date, this leg of the triangle has been verified to about a fractional part in 10^6 and is hoped to improve this accuracy by a further two orders of magnitude.

The closure provided by combining the three types of measurement has become known as closing the quantum metrology triangle. That is, if sufficient accuracy can be achieved with the SET work, the three measurements should provide an internally consistent set and thereby provide a valuable check of the underlying theory.

The 2006 evaluation showed that the JE $2e/h$ leg could be verified by internal consistency involving the moving coil watt determinations to a fractional uncertainty of about $\varepsilon_J = (-77 \pm 80) \times 10^{-9}$. There is some doubt over the reliability of this value due to the small amount of data and it is suggested that the agreement may be partly fortuitous. Similarly, the QHE leg could be verified to an uncertainty $\varepsilon_K = (20 \pm 18) \times 10^{-9}$ by both external and internal consistency. The SET method at present limits the accuracy and, to date, the triangle has been shown to provide closure with a fractional uncertainty at the 10^{-6} level (Gallop, 2005; Keller, 2008).

3.3 Obtaining "Best Values" of the Constants

As science and technology advances, so too our ability to measure particular combinations of the fundamental constants improve in accuracy. The particular combinations of the constants that can be best measured also change.

3.3.1 Relationships among the Constants

Equations 3.4 and 3.7 together with Equations 3.8 and 3.10 can provide values of α. However, because, in practice, a value of $\text{unit}_{\text{LAB}}/\text{unit}_{\text{SI}}$ is required to obtain α from Equation 3.10 as well as from Equation 3.4, the two values will not be completely independent. Moreover, $2e/h$ (in SI units) can be obtained from the relation

$$\frac{2e}{h} = \left[\frac{16R_\infty\left(m_p/m_e\right)N_\text{A}}{\mu_0 c^2 M_p \alpha}\right] \tag{3.16}$$

where M_p is the molar mass (atomic weight in units of kilograms per mole) of the proton and $\gamma_p'(\text{lo})$ from the equation

$$\gamma_p'\left(\text{lo}\right)_\text{LAB} = \left(\frac{F_\text{LAB} K_\text{A,LAB}^2 \left(\mu_p'/\mu_B\right)\left(m_p/m_e\right)}{M_p}\right) \tag{3.17}$$

where F_LAB (Faraday constant) is measured in terms of the as-maintained ampere A_LAB in a particular laboratory, and

$$\begin{aligned} K_\text{A,LAB} &= A_\text{LAB}/\text{A} = (V_\text{LAB}/\Omega_\text{LAB})/(\text{V}/\Omega) \\ &= (V_\text{LAB}/\text{V})/(\Omega_\text{LAB}/\Omega) \\ &= K_{\text{V}-\text{LAB}}/K_{\Omega-\text{LAB}} \end{aligned} \tag{3.18}$$

Further, the Faraday constant is related to the Avogadro constant by $F = N_\text{A}e$, and $K_\text{A,LAB}$ may be obtained from direct measurements using a so-called ampere force balance as well as from the equation

$$K_\text{A,LAB}^2 = \frac{\gamma_p'\left(\text{lo}\right)_\text{LAB}}{\gamma_p'\left(\text{hi}\right)_\text{LAB}} \tag{3.19}$$

where $\gamma_p'\left(\text{hi}\right)_\text{LAB}$ is the gyromagnetic ratio of the proton as obtained by the so-called high (magnetic) field method. (The magnetic flux density B is determined by measuring the force it exerts on a conductor of known dimensions carrying a current known in terms of A_LAB.)

It is apparent that complex relationships can exist among groups of fundamental constants and conversion factors, and that a particular constant may be obtained either by direct measurement or indirectly by appropriately combining other directly measured constants. If the direct and indirect values have similar accuracy, both must be considered in order to arrive at a best value for the quantity. However, each of the various routes that can be followed to a particular constant, both direct and indirect, provides a somewhat different numerical value. The best way to treat this is by using the mathematical technique of least-squares, a technique that is backed by the rigor of statistics.

Least-Squares Adjustments of the Constants
The least-squares method furnishes a well-defined procedure for calculating best "compromise" values of the constants from all of the available data. It takes into account all possible routes to a particular constant and yields a single value for each constant by weighting the different routes according to their relative uncertainties. The weights themselves are obtained from the *a priori* uncertainties assigned to the individual direct measurements that constitute the original set of data.

The least-squares study of the constants was pioneered in the late 1920s by Birge (1929) and continued by others, notably DuMond and Cohen (1953) and Cohen and DuMond (1965). More recent reviews were those of Taylor, Parker, and Langenberg (1969), based on their JEs determination of $2e/h$; and of Cohen and Taylor ([1973, 1987]); and more recently by Mohr and Taylor ([1999, 2005], and Mohr, Taylor, and Newell (2008). Since 1973, reviews have usually been undertaken under the auspices of CODATA. The 2006 adjustment is the most recent one currently available, and the recommended set of constants resulting from it, are those officially adopted by CODATA for international use. Subsequent evaluations are scheduled to take place at four-year intervals and the values will be posted on the NIST Web Site at *http://physics.nist.gov/cuu/Constants/*.

3.3.2
The 2006 CODATA Adjustment

An important guiding principle for the wide acceptance of the evaluations for scientists is

1. to be clear how they are performed;
2. what data is included; and
3. the weight given to a particular input datum.

The standard statistical method gives weights to the input of the data according to the inverse square of its associated uncertainty. Common sense, backed by the history of the subject, supports the reality that there may be some unknown science that is as likely to be affecting the most accurate value as those of lesser accuracy. Indeed, history suggests that it would have been more realistic to double the uncertainty associated with the output values. This option has so far been rejected on the grounds that it might slow the pace of the subject because it could mask the recognition that there were unanticipated problems or unknown science associated with particular measurements.

An inherent consequence of applying the least-squares method is that a determination with an uncertainty that is three or more times greater than that of the smallest, has less than one-ninth of the impact on the weighted mean value than does the most accurate result. This puts very strong pressure on the ability of scientists to estimate their uncertainties. Therefore, it is usual to check the effect on the evaluated values of omitting the result with the smallest uncertainty. Similarly, one can test for the effect of all data pertaining to the QHE (say) or the JE and thereby verify the extent to which these phenomena do represent a measurement of the particular combination of the fundamental constants

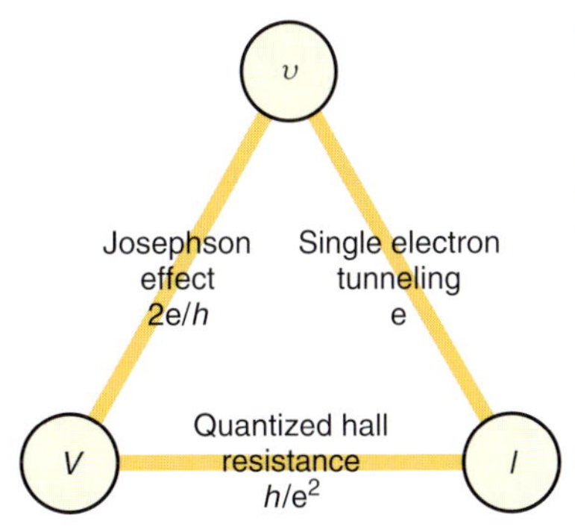

Fig. 3.1 The quantum metrology triangle links the Josephson effect, the quantum Hall effect and single electron tunneling, the sides of the triangle, to the electrical units and frequency, represented by the vertices, by using Ohm's law and tests the self-consistency of the expressions that are assumed in Equations 3.2 and 3.10.

concerned. The preliminary assessment of the input data is the most critical part of an evaluation because the computing thereafter is largely automatic–so the prudent approach to the rejection of "rogue" data suggested by the computer adage "garbage in: garbage out" can only be followed by subjective decisions by the reviewers backed by advice from the CODATA Task Group and other expert scientists.

If sufficient data on a particular topic are available, one may use a statistical quantity termed the *Birge ratio*,which amounts to comparing the value of χ^2 with the number of degrees of freedom for the group and expanding the assigned uncertainty to make the two values equal. Thus, in the 2006 evaluation, the determinations of G were analyzed in this way and the output uncertainty expanded by a factor of 4.26, and this is illustrated in Figure 3.2.

The present leading experimental values for G are also given in Table 3.2. It will be noted that the uncertainty in the CODATA recommended value was conservatively increased above that obtainable by statistical weighting of the results because it was judged that there were significant discrepancies among the published measurements.

Unfortunately, one of the major problems for the measurements of G result from the small size of the gravitational force between objects in the presence of the 10^9 times greater gravitational attraction of the Earth, or spurious electromagnetic fields. This places formidable requirement for the suspension, which must have a small and stable torsional constant combined with adequate strength and very good shielding from outside effects (including sidereal heating of the laboratory, or gravitational gradients). Consequently, the achievable experimental accuracy has not improved as rapidly as that of other constants and has only improved about a 100-fold over two centuries. Thus, the measurements of Cavendish (1798) led to $G = 6.75(5) \times 10^{-11}\ \mathrm{m^2\ s^{-2}\ kg^{-1}}$ and this may be compared with the CODATA 2006 recommended values of $6.67428(67) \times 10^{-11}\ \mathrm{m^2\ s^{-2}\ kg^{-1}}$.

In earlier evaluations, the data entering a least-squares adjustment of the constants were divided into two groups:

1. the more precise data or auxiliary constants and
2. the stochastic input data.

The auxiliary constants used in the 1986 adjustment, none of which had an assigned fractional uncertainty greater than 2×10^{-8}, included μ_0, c, R_∞, M_p, m_p/m_e, g_e, μ_e/μ_p, and the g-factor of the muon:

$$g_\mu = 2(1 + a_\mu) = 2\mu_\mu/(eh/4\pi m_\mu) \quad (3.20)$$

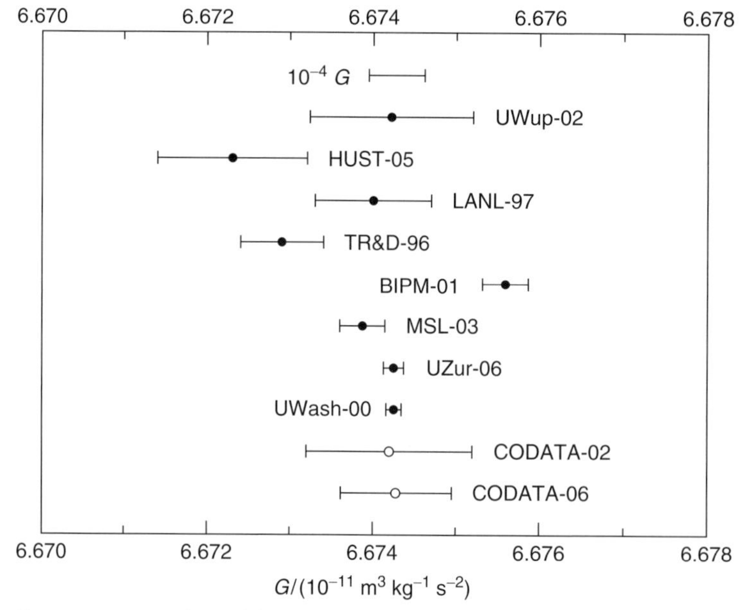

Fig. 3.2 Recent values of the gravitational constant. (Taken from CODATA 2006, see Mohr, Taylor, and Newell (2008) for details and designations.)

where a_μ is the magnetic moment anomaly of the muon and μ_μ is its magnetic moment.

Also included in the auxiliary constant category in 1986 were the representations of the volt, ohm, and ampere, expressed in terms of the common set of electrical units known as the *BIPM volt, ampere, ohm,* and the 1972 internationally agreed value for the Josephson constant $K_{J\text{-}72}$. In 1990, at the suggestion of the CCE, the latter was revised to the present internationally agreed value of K_J and assigned the designation $K_{J\text{-}90}$, together with a value for the von Klitzing constant $R_{K\text{-}90}$. The best values for K_J and R_K are obtained from the evaluations.

In the 1986 evaluation, the stochastic data were grouped into sets providing information on five constants. Since then, recognizing the greatly improved sophistication of calculations provided by modern computers that have progressively replaced the slide-rules and hand-cranked mechanical calculators of the Birge era, the distinction between auxiliary constants and stochastic input data has been dropped and all the relevant data has been included in the evaluation. This advance has now allowed any correlations between the input data to be taken into account.

Since the Rydberg constant, for example, is now measured to 12 decimal places, then, for the 2006 evaluation, the computer had to be able to perform a least-squares multivariate analysis comprising some 79 unknowns from some accepted 135 input data (the latter from an initial set of 150 input data), with the resulting large number of associated matrices, evaluated to a large number of decimal places. Birge could have scarcely imagined what scientific and technological progress would

Tab. 3.2 Data contributing to the CODATA 2006 value of the Newtonian gravitational constant G.

Source (see Mohr, Taylor, and Newell (2008), and references therein)	CODATA 2006 Identification	Method	Value 10^{11} $G\text{m}^{-3}$ kg^{-1} s^{-2}	Relative standard uncertainty u_r
2002 CODATA Adjustment	CODATA-02	–	6.6742(10)	1.5×10^{-4}
Karagioz and Izmailov (1996)	TR&D-96	Fiber torsion balance, dynamic mode	6.672 9(5)	7.5×10^{-5}
Bagley and Luther (1997)	LANL-97	Fiber torsion balance, dynamic mode	6.674 0(7)	1.0×10^{-4}
Gundlach and Merkowitz (2000), Gundlach and Merkowitz (2002)	UWash-00	Fiber torsion balance, dynamic compensation	6.674 255(92)	1.4×10^{-5}
Quinn *et al.* (2001)	BIPM-01	Strip torsion balance, compensation mode, static deflection	6.675 59(27)	4.0×10^{-5}
Kleinevos (2002), Kleinvos *et al.* (2002)	UWup-02	Suspended body, displacement	6.674 22(98)	1.5×10^{-4}
Armstrong and Fitzgerald (2003)	MSL-03	Strip torsion balance, compensation mode	6.673 87(27)	4.0×10^{-5}
Hu *et al.* (2005)	HUST-05	Fiber torsion balance, dynamic mode	6.672 3(9)	1.3×10^{-4}
Schlamminger *et al.* (2006)	UZur-06	Stationary body, weight change	6.674 25(12)	1.9×10^{-5}
2006 CODATA Adjustment	CODATA-06		6.674 28(67)	1.0×10^{-4}

Taken from Table 27 of CODATA 2006, see Mohr, Taylor, and Newell (2008) for details.

accomplish within 70 years of his seminal work–but he would surely have approved.

Although the output values have associated uncertainties, the least-squares analysis results on correlations between the various output values and, for some purposes, the user may need to use the associated variance and covariance matrices that are also available from the NIST web site. Thus, the recommended uncertainty of the quotient e/h may differ from root sum of the squares of the uncertainties assigned to e or h.

An equation that relates a stochastic input datum to the M adjustable constants (with the aid of the auxiliary constants if necessary) is known as an *observational equation*.Thus, if there are n_1 items of stochastic input data, there are n_1 observational equations. However, the observational equations for data items of the same type are of the same form. The weight w_i of each stochastic datum and hence of each observational equation is related to the datum's *a priori* assigned uncertainty s_i by $w_i = 1/s_i^2$.

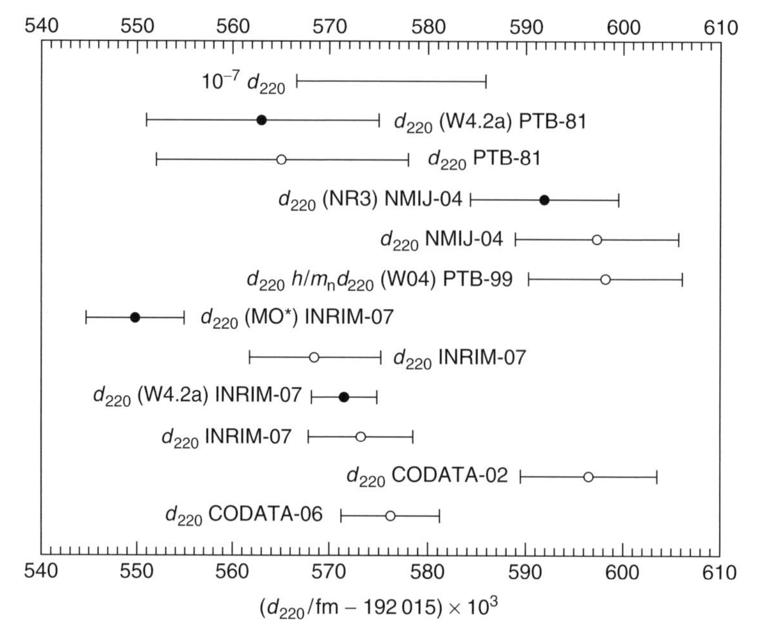

Fig. 3.3 Values of d_{220} inferred from various measurements. (Taken from CODATA 2006, see Mohr, Taylor, and Newell (2008) for details and designations.)

Consequently, if one value of a datum for a particular quantity has half the uncertainty of another, it carries four times as much weight in the adjustment. Taking into account the weight assigned to each observational equation, and with the aid of a present-day computer, the least-squares adjusted values of the M unknowns may be readily obtained. "Best values" in the least-squares sense for these M adjustable constants, with their variances and covariances, are thus the immediate output of the adjustment. The optimal values in the least-squares sense for most constants of interest not directly subject to adjustment may then be calculated from the M adjustable constants. In contrast to the auxiliary constants of the 1986 adjustment, for which no uncertainty exceeded 2×10^8, the fractional uncertainties assigned to the 138 items of stochastic input data initially considered in the 2006 adjustment were broadly in the range 10^{-12}–10^{-6}.

Because the d_{220} data were in such disagreement, as seen in Figure 3.3, a separate working group has been set up to coordinate the work in several countries in order to produce recommended value(s) for d_{220} and $V_m(\text{Si})$. The work in progress includes determination(s) of N_A resulting from measurements made with a sample of isotopically pure silicon produced in Russia (Becker, 2003; Becker *et al.*, 2006).

Because new results that can influence a least-squares adjustment of the constants are reported continually, it is always difficult to choose an optimal time at which to carry out a new adjustment and to revise the recommended values of the constants. In the case of the 2006 adjustment, all data available up to 1 January 2006 were considered for inclusion.

3.3.2.1 Uncertainty Assignment and Discrepant Data

Because the weight of each stochastic datum included in a least-squares adjustment of the constants is taken as $w_i = 1/s_i^2$, critical analysis of the input data and deciding what *a priori* uncertainty s_i should be assigned to each measurement is the main problem in successfully carrying out an adjustment. Modern data acquisition techniques mean that instead of making tens of observations one has thousands, but without sophisticated analysis and intelligent investigation of possible variables (e.g., by taking advantage of the sophistication provided by factorial replicate and other statistical analysis techniques), the apparently improved reduction in the uncertainty estimate may prove highly illusory. The uncertainty problem is made especially difficult because in most experiments sufficient data are taken to reduce the Type A uncertainties (formerly termed *random*effects) to negligible amounts, and the dominant uncertainty assigned to the experiment is primarily from estimates of the Type B uncertainties associated with the effects that remain substantially constant during the period of the measurement. Because such effects frequently arise from a phenomenon that the experimenter is not so familiar with, the quantitative estimation of the uncertainties is subjective and is necessarily based on "educated" guesses.

As shown above with the measurements of G, discussed above, another problem area is that of deciding how to handle "discrepant" data. These comprise measurements that differ from each other by statistically significant amounts in comparison with their assigned uncertainties s_i. Such data cannot be included in an adjustment uncritically because the inconsistencies imply either incorrect uncertainty assignments or the presence of unknown measurement errors. When confronted with such a situation, the reviewer can in general either include the inconsistent data, but only after expanding (increasing) their s_i so that they are no longer discrepant, or decide, on as sound a theoretical and experimental basis as possible, which of the inconsistent data are least reliable, and discard them. These two approaches clearly imply that different reviewers might treat the same data differently, thereby obtaining somewhat different sets of best values.

Motivated by the all-too-frequent occurrence of large changes in the recommended values of many constants from one adjustment to the next, new algorithms for handling discrepant data in least-squares adjustments of the constants were developed during the years between the 1973 and 1986 adjustments. These added increased objectivity to expanding *a priori* uncertainties and discarding inconsistent data were used extensively by Cohen and Taylor in carrying out the 1986 adjustment and, with their aid, the initial group of 38 items of stochastic input data was reduced to 22 items by deleting those that were either highly inconsistent with the remaining data or had assigned uncertainties so large that they carried negligible weight.

3.3.2.2 Recommended Values of the Constants

Table 3.3 gives the 2006 CODATA recommended values and the associated one-standard-deviation uncertainties ($\sim$65% probability) of a number of fundamental physical constants. As noted in Section 3.3, the direct output of a least-squares adjustment provides the best values (in the least-squares sense) for the adjustable constants, or unknowns, and their variances and covariances. For the 1986 adjustment, the five unknowns chosen for evaluation were α^{-1}, K_V,

Tab. 3.3 Comparison of the 1973, 1986, and 2006 CODATA recommended values and the assigned uncertainties of a selected group of constants implied by the data currently available.

Quantity	1973 fractional uncertainty σ_{73}	1986 fractional uncertainty σ_{86}	2006 fractional uncertainty σ_{2006}	Ratio $\sigma_{86}/\sigma_{2006}$	Fractional change δ of 2006 value from 1986 value	Change δ/σ_{86}
α^{-1}	8×10^{-7}	4.5×10^{-8}	6.8×10^{-10}	66.2	7.4×10^{-8}	1.73
E	2.9×10^{-6}	3.0×10^{-7}	2.5×10^{-8}	12	-5.3×10^{-7}	−1.75
H	5.4×10^{-6}	6.0×10^{-7}	5.0×10^{-8}	12	-9.9×10^{-7}	−1.65
m_e	5.1×10^{-6}	5.9×10^{-7}	5.0×10^{-8}	11.8	-8.3×10^{-7}	−1.40
N_A	5.1×10^{-6}	5.9×10^{-7}	5.0×10^{-8}	11.8	8.4×10^{-7}	1.43
m_p/m_e	3.8×10^{-7}	2.0×10^{-8}	4.3×10^{-10}	46.5	-1.6×10^{-8}	−0.79
F	2.8×10^{-6}	3.0×10^{-7}	2.5×10^{-8}	12.0	3.2×10^{-7}	1.07
$2e/h$, K_J	2.6×10^{-7}	3.0×10^{-7}	2.5×10^{-8}	12.0	4.6×10^{-7}	2.0
R	31×10^{-7}	8.4×10^{-6}	1.7×10^{-6}	4.9	-4.5×10^{-6}	−0.54
K	32×10^{-6}	8.5×10^{-6}	1.7×10^{-6}	5.0	-5.5×10^{-6}	−0.65
σ	125×10^{-6}	34×10^{-6}	7.0×10^{-6}	4.9	1.9×10^{-5}	−0.57
G	6.15×10^{-4}	3.4×10^{-4}	1.0×10^{-4}	3.4	2.5×10^{-4}	1.98
R_∞	7.6×10^{-8}	1.3×10^{-9}	6.6×10^{-12}	197	3.2×10^{-9}	0.70

K_Ω, d_{220}(Si), and μ_μ/μ_p. The number of "unknowns" was increased dramatically in the 2006 adjustment, However, the uncertainties of many of the output values were still largely determined by a comparatively few different types of measurements.

3.3.2.3 Comparison of the 1986 and 2006 Adjustments

The 1986 CODATA adjustment represented a major advance over its 1973 counterpart; the uncertainties of many recommended values were reduced by roughly an order of magnitude, and further progress has been made through advances made throughout the precision measurement-fundamental constants field between 1986 and 2006. This is apparent from column 4 of Table 3.3, which compares the recommended values and uncertainties of a selected group of constants as obtained from the two adjustments and from Figure 3.4 a plot of the changes in uncertainty tabulated in columns 1–3. It is also clear from the last column of Table 3.3 that unexpectedly large changes have occurred in the 1973 recommended values of several of these constants (there are changes that are large relative to the uncertainty assigned to their 1973 value). Many of these changes are a direct consequence of the 7.8×10^{-6} fractional decrease in the quantity K_V from 1973 to 1986, and the very high correlation between K_V and the calculated values of e, h, m_e, N_A, and F. Since $2e/h = K_{J-72}/K_V$, the 1986 value of K_V also implies that the value of $2e/h$ that was adopted by the CCE in 1972, (which was believed to be consistent with the SI value, and which most national standards laboratories adopted to define and maintain their national representation of the volt), was actually 7.59 μV smaller than the correctly realized SI value. As discussed

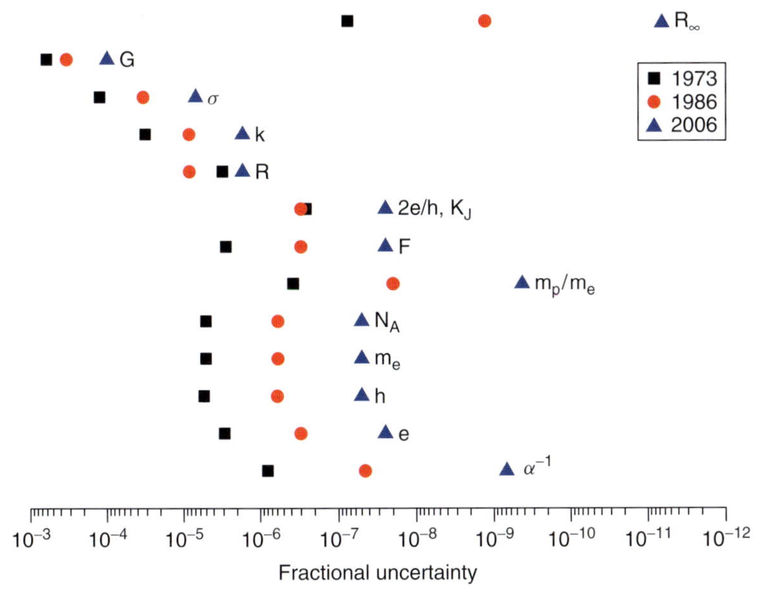

Fig. 3.4 Improvements in the fractional uncertainty of some selected constants from 1973 to 2006.

in Section 3.4, this unsatisfactory situation was rectified adopting the revised value for $2e/h$ with effect from 1 January 1990.

The large change in K_V between 1973 and 1986 would have been avoided if two determinations of F that seemed to be discrepant with the remaining data had not been deleted in the 1973 adjustment. In retrospect, the disagreement was comparatively mild. Fortunately, there are fewer similar disagreements in the 2006 adjustment; the measurements that were deleted were so discrepant, or of such low weight that if retained they would have a negligible effect on the adjusted values. Thus, it is unlikely that any reasonable alternative evaluation of the data considered in the 2006 least-squares adjustment could lead to significant changes in the 2006 recommended values. It is interesting that the spectacular improvement in the experimental and theoretical aspects relating to the accuracy of the Rydberg constant, largely as a result of the development of the ability to make direct frequency multiplication measurements through to the visible region and beyond (de Beauvoir *et al.*, 2000; Hänsch *et al.*, 2005), was not significantly changed when expressed in terms of the fractional uncertainty assigned to the 1986 value. None of the 1986 recommended values in Table 3.3 has been changed by more than two standard deviations.

3.4
Josephson and Quantum Hall Effect Representations of the Volt and Ohm

As pointed out in Section 3.2.2, the fundamental constants have a key role to play in the establishment of reproducible and invariant units of measurement. Exemplifying this role was the worldwide introduction on 1 January 1990 of new practical units of voltage and resistance based on standardized values for the

Josephson constant and quantum Hall constant, in the sections that follow. We consider first some aspects of the SI.

3.4.1
The SI and Representations of the Volt and Ohm

The SI is the system of units used throughout the world for expressing the results of physical and chemical measurements. The three SI base mechanical units are the meter (m), kilogram (kg), and second (s) (Taylor, 1991b). The derived SI unit of force, the newton (N), is then defined in terms of these three base mechanical units by $1N = 1\,\mathrm{m\,kg\,s^{-2}}$. The SI base unit of current, the ampere (A), is defined as that constant current which, if maintained in two straight parallel conductors of infinite length, of negligible circular cross section, and placed 1 m apart in vacuum, would produce between these conductors a force equal to 2×10^{-7} N per meter of length. This definition leads to $\mu_0 = 4\pi \times 10^{-7}\,\mathrm{N\,A^{-2}}$ exactly.

The SI derived unit of electric potential difference, the volt (V), and the SI derived unit of electric resistance, the ohm (Ω), are then defined in terms of the three base mechanical units and the base electric unit, the ampere. Formally, $1V = 1\,\mathrm{m\,kg\,s^{-3}\,A^{-1}}$ and $1\,\Omega = 1\,\mathrm{m^2\,kg\,s^{-3}\,A^{-2}}$.

In practice, the volt may be realized using a force balance in which an electrostatic or electromagnetic force is compared with a mechanical force; and the ohm may be realized by comparing the impedance of a calculable capacitor or inductor with that of a resistor. However, because both experiments are exceedingly difficult and time consuming to carry out and are of limited accuracy, it has become necessary to establish practical, or as-maintained, units of voltage and resistance, that is, representations of the volt and ohm that provide better long-term repeatability and constancy than the direct realizations of the SI units themselves.

Historically, practical representations of the volt were based on the mean emf of large groups of electrochemical standard cells. However, in 1972 (Terrien, 1973; Taylor, 1987), the CCE suggested that the national standards laboratories base their national representations of the volt on the JE in order to avoid the variation with time and severe temperature dependence of the emf of such cells. In particular, the CCE proposed, and the Conference International des Poids et Mesures (CIPM) subsequently approved the proposal, that the national standards laboratories adopt $K_{J\text{-}72} = 483\,594$ GHz $\mathrm{V^{-1}}$, exactly, as the conventional value of the Josephson constant for use in realizing and maintaining an accurate and stable national representation of the volt by means of the JE (Taylor and Witt, 1989). However, while most national laboratories accepted this value, three did not. Moreover, as pointed out in Section 3.3.3, the 1986 CODATA adjustment showed that $K_{J\text{-}72}$ was nearly 8×10^{-6} smaller than the correct SI value, implying that the national representations of the volt of those countries that used it should be changed accordingly.

During the late 1970s and early 1980s, the CCE also became increasingly concerned that, because most national standards laboratories based their representation of the ohm on the mean resistance of a particular group of precision wire-wound standard resistors, and because these artifact standards age, the various national representations of the ohm differed significantly from each other, and from the ohm of the SI definition, and some groups of resistors were aging excessively. Although, in principle, a calculable cross-capacitor can now be used to realize the ohm with an uncertainty of $<10^{-7}$, it is a difficult experiment to perform routinely. Only one laboratory in the world has had such an apparatus in continuous operation since the method was first developed in the early 1960s. Consequently, electrical metrologists throughout the world enthusiastically welcomed von Klitzing's 1980 discovery of the QHE since it promised to provide a method for basing a representation of the ohm on fundamental constants in much the same manner as the JE, discovered 19 years earlier in 1961, provided a method for basing a representation of the volt on fundamental constants.

3.4.2 The Conventional Values K_{J-90} and R_{K-90}

To address the problems associated with national representations of the volt and ohm discussed earlier, the CCE established the CCE Working Group on the JE and another working group on the QHE. Their proposals were subsequently approved by the CIPM and by the 18th CGPM (Giacomo, 1987; Quinn, 1989) and these proposals went into effect worldwide from 1 January 1990. The values are

$$K_{J-90} = 483\,587.9\,\text{GHz/V exactly}$$
$$R_{K-90} = 25\,812.807\,\Omega \quad \text{exactly} \qquad (3.21)$$

The respective CCE Working Groups estimated that $K_{J\text{-}90}$ was consistent with the SI value of K_J to within a conservatively assigned one-standard-deviation uncertainty of 0.2 GHz V^{-1} or relative uncertainty of 4×10^{-7}, and that $R_{K\text{-}90}$ is consistent with the SI value of R_K to within a conservatively assigned one-standard-deviation uncertainty of 0.005 Ω or a relative uncertainty of 2×10^{-7}. This implies that an ideal representation of the volt based on the JE and $K_{J\text{-}90}$ should be consistent with the SI volt to within 0.4 μV, and that an ideal representation of the ohm based on the QHE and $R_{K\text{-}90}$ should be consistent with the SI ohm to within 0.2 μΩ. Reassuringly, the 2006 evaluation improved on this uncertainty and suggested that the maintained units were actually 0.019(25) μV lower and 0.0212 μΩ higher, respectively, and that the corresponding maintained ampere A_{90} differed from that of the SI definition by −0.003(25) μA. Hopefully, these small differences will be confirmed by measurements currently in progress.

Thus, measurements of the fundamental constants of nature have led to a very practical and significant improvement of the international uniformity of electrical measurements and their consistency with the SI. This will no doubt be of major benefit to science, commerce, and industry throughout the world for many years to come; not least because all measurements are a comparison of one quantity relative to another like quantity, and one is never otherwise sure whether the problem lies with a measurement or the reference quantity.

3.5 Implications of Recent Advances in the Fundamental Constants Field

As clearly discussed in Section 3.4, many advances have been made in the fundamental constants field since the completion of the 1986 CODATA least-squares adjustment. The new results lead to values of the constants that in general have uncertainties five to seven times smaller than the uncertainties assigned to the 1986 values. However, the changes in the values themselves are less than twice the 1986 one-standard-deviation uncertainties and thus are not highly significant. Although much new data has become available since 1986, three new results dominate the analysis: a value of h obtained from a realization of the watt; a value of α^{-1} obtained from a_e; and a value of the molar gas constant R obtained from a new measurement of the speed of sound in argon.

Surprisingly, the measurement of $K_W = W_{90}/W$ is actually a measurement of the Planck constant. This follows from the fact that $W_{90} = V_{90}^2/\Omega_{90}$, $W = V/\Omega$, so that $K_W = K_V^2/K_\Omega$. Also $K_V = K_{J\text{-}90}/K_J$, $K_\Omega = R_K/R_{K-90}$, $K_J = 2e/h$, and $R_K = h/e^2$. These relations lead to $h = 4/K_J^2 R_K$ and hence

$$h = \frac{4K_W}{K_{J-90}^2 R_{K-90}} \tag{3.22}$$

Since $K_{J\text{-}90}$ and $R_{K\text{-}90}$ are defined constants (with no uncertainty), a measurement of K_W (i.e., a measurement of W_{90} in watts or realization of the SI watt) with a given relative uncertainty yields a value of h with the same relative uncertainty.

Since 1986, the value of the molar gas constant R from which the Boltzmann constant k, Stefan–Boltzmann constant σ, and other thermophysical constants were derived has been replaced by a new value of R with a relative uncertainty of 1.7×10^{-6}, nearly five times smaller than the 8.4×10^{-6} relative uncertainty of the 1986 value.

3.6 Possible Role of the Fundamental Constants in Replacing the Kilogram

As indicated in Section 3.4, this section further demonstrates the important role that the fundamental constants can play in the practical metrology.

3.6.1 Problems with Present Kilogram

The unit of mass in the SI, the kilogram, is the only SI base unit still defined in terms of a material artifact; it is equal to the mass of the international prototype of the kilogram, a 90% platinum, 10% iridium (by weight) right circular cylinder about 39 mm in height and diameter. Established by the CGPM in 1889, it is rarely used for fear of damage. In fact, it has only been used on three occasions to compare against copies of the prototype since it was made.

The problems with the present material artifact kilogram may be summarized as follows (Quinn, 1991; Taylor, 1991):

1. it can be damaged, or even destroyed;
2. it cannot be used routinely for fear of wear;
3. it is available in only one laboratory;
4. its value may have changed in an unknown manner, perhaps by as much as 75 μg in the last 100 years (7.5 parts in 10^8); and
5. its mass is not well defined because, despite its being enclosed by three concentric bell jars, it accumulates foreign material and the latter are difficult to remove reproducibly and an empirical, "lavage" process has been specified.

Using the best available balances and cleaning the masses under the proper experimental conditions, Pt–Ir kilogram standards can be intercompared with an uncertainty of about 0.001 μg (or a fractional amount of 1×10^{-9}) or even less. However, it is generally agreed that a new definition of the kilogram based on a constant of nature, such as the mass of a stable elementary particle or atom, that could be realized, in practice, with a relative uncertainty of the order of 1×10^{-8} at the 1-kg level is all that is required to replace the current definition. The reason is that the uncertainties associated with the cleaning and aging of the international prototype, the air buoyancy correction required when masses of different densities are compared, and the scaling of masses up and down from the 1-kg level limit practical mass measurements to a relative uncertainty of a few parts in 10^8 at best. There are also losses of accuracy in moving to other in-house mass balances and to other national standards laboratories and to the ultimate users.

3.6.2 An Alternative Based on Fundamental Constants

There are at present three alternative approaches that might be developed to have the required precision.

3.6.2.1 The Measurement of *h*

The most promising method for redefining the kilogram, to date, arose from the result of the measurement of $\gamma_p'(\text{hi})$ by Kibble and Hunt, as mentioned in Section 3.2.3. The method proposed by Kibble (1975) effectively combined aspects of the measurements of $\gamma_p'(\text{hi})$ and $\gamma_p'(\text{lo})$ into one measurement process that is performed on the same coil of n turns and horizontal width w in the same magnetic flux density B in the presence of a vertical flux gradient $\partial\Phi/\partial z$. The two-aspect measurement process involves

1. measuring the force mg, required to balance the coil when a current I was passed through it and
2. measuring the induced emf, E, when the coil was moved vertically in a flux rate of change of $\partial\Phi/\partial t\,(= u\partial\Phi/\partial z)$, at constant velocity, u.

This resulted in two equations from which $\partial\Phi/\partial z$ could be eliminated to give the electrical power VI in absolute mechanical units. If the voltage is measured in terms of a Josephson voltage $V_J = hf_J$ and the current $i = \left(f_J' n\right) e/2$ is measured by combining the JE ($v = hf/2e$) and QHE ($R = h/ne^2$), (where n is a small integer, and f and f' are small frequencies). Thus, since $(h/2e)^2 \big/ (e^2/2h)) = h/8$, the Planck constant h can be measured directly in terms of its SI units, that is, in Joule-second. The experimental determination requires knowledge of the mass m, the vertical coil velocity u (which is measured by optical fringe-counting techniques), and the gravitational acceleration g local to the apparatus (again involving optical fringe-counting of a cube corner reflector catapulted vertically in a commercially available vacuum apparatus). The remaining parameters involve frequency measurements that can readily be measured with the required accuracy. To date, there have been two groups making determinations, Williams *et al.* at NIST, and Kibble *et al.* at NPL. The latest results from the two groups are

Source	Value for *h*
Steiner *et al.* (NIST) (Steiner *et al.*, 2007)	6.626 068 91(24) $\times 10^{-34}$ J.s
Robinson and Kibble (NPL) (Robinson and Kibble, 2007)	6.626 070 95(44) $\times 10^{-34}$ J.s

These two-watt balance measurements differ at present by 308(48) nW/W, with the

Robinson and Kibble value being closer to that obtained from the Avogadro constant as shown in Figure 3.5.

It seems likely at present that this route will prove the most capable of achieving the desired 10^{-8} level of accuracy. That being so, the definition of the kilogram might be defined via the Einstein mass–energy equation ($E = mc^2$) in terms of the mass equivalence $h\nu_{def}/c^2$ (where ν_{def} is the frequency defined, e.g., in the definition below) and read, for example, Mills *et al.* (2005):

> *the kilogram, unit of mass is the mass of a body whose equivalent energy is equal to that of a number of photons whose frequencies sum to exactly [(299 792 458)^2/662 606 93] × 10^{41} hertz.*

It has to be admitted that this definition is likely to be difficult for many unsophisticated users to comprehend, despite representing the amount of mass converted into energy in a very powerful hydrogen bomb. Also, $h\nu_{def}$ is largely independent of h, rather depending on $\alpha^4 m_e$ instead (Flowers and Petley, 2005). The latter is a consequence of there being no single fundamental constant whose dimensions have solely the dimension of time, aside from the Planck time $t_P\left(=\left(\hbar G/c^5\right)^{1/2}\right)$, unless one considers, for example, such quantities as the Compton frequencies of particles, or the Rydberg constant as being more fundamental than those of the conventional basic set comprising h, e, c, G, k, and m_e. Much will depend on the degree of agreement achieved between the various methods, together with the combined views of users throughout the world, by the time that the decision is discussed by the CGPM at its next meeting in 2011.

3.6.2.2 Methods Involving the Avogadro Constant, N_A

An alternative approach to eliminating the artifact kilogram as the definition of the SI unit of mass is to replace it with a definition based on the mass of an atom, for example, carbon-12 (^{12}C). A possible definition might read as follows: the kilogram is the unit of mass; it is equal to the mass of 5.018×10^{25} free ^{12}C atoms at rest. To avoid introducing a discontinuity in the SI mass unit, the actual number $n(^{12}\mathrm{C})$ given in the definition would, to the fullest possible extent, be taken equal to the number of free ^{12}C atoms at rest whose total mass equals the mass of the international kilogram prototype. Thus we have

$$n\left(^{12}\mathrm{C}\right) = \frac{1kg}{m\left(^{12}\mathrm{C}\right)} = 1kg\frac{N_{\mathrm{A}}}{M\left(^{12}\mathrm{C}\right)} \qquad (3.23)$$

where $m(^{12}\mathrm{C})$ is the mass in kg of a free ^{12}C atom at rest; N_A is the Avogadro constant in units of mol^{-1}; and also $M(^{12}\mathrm{C})$ is the molar mass of ^{12}C in units of kilograms per mole, and from the definition of the base unit mole in the SI is exactly 0.012 kg mol^{-1}. It follows from Equation 3.15 that once the new definition is implemented (i.e., once an exact value for $\mathrm{n}(^{12}\mathrm{C})$ is adopted), N_A and the atomic mass unit $u = m_u = m(^{12}\mathrm{C})/12$ will become exactly defined constants.

Mass–ratio measurements of single charged particles and ions cooled and trapped in a Penning trap have reached the point where the molar masses of most particles or atoms of interest, such as the electron, proton, or ^{1}H, are now known or soon will be known with a relative uncertainty of $(1–3) \times 10^{-9}$. Similarly, the 20×10^{-9} relative uncertainty of the reported value for the proton–electron mass ratio, $m_p/m_e = M_p/M_e$, should soon

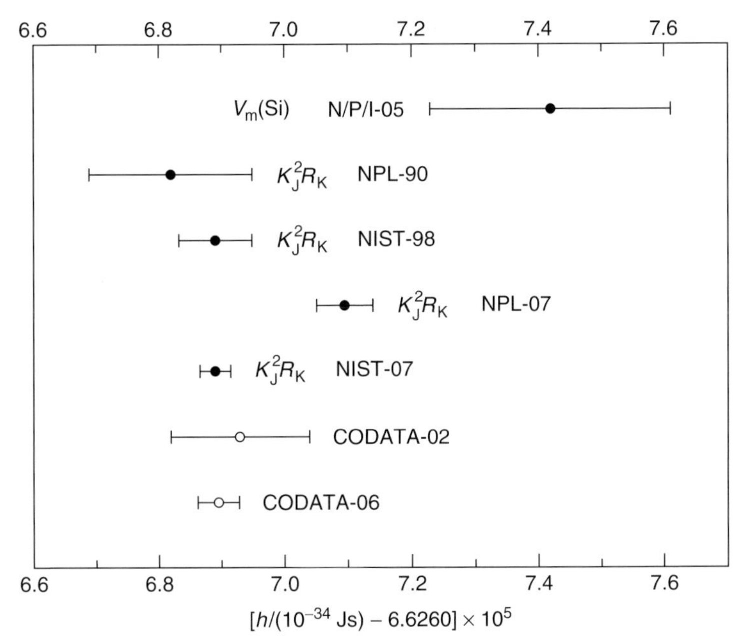

Fig. 3.5 Recent values of the Plank constant based on watt balance ($K_J^2 R_K$) and Avogadro V_m(Si) methods. (Taken from CODATA 2006, see Mohr, Taylor, and Newell (2008) for details).

be reduced by a factor of 10 or so. These small uncertainties imply that if the mass m of any elementary particle or atom can be obtained in kilograms with a relative uncertainty of 1×10^{-8}, then $m(^{12}C)$ and hence $n(^{12}C)$ will be known with essentially the same uncertainty. Further, Equation 3.16 implies that determining N_A with a relative uncertainty of 1×10^{-8} will lead to a value of $n(^{12}C)$ with the same uncertainty. Consequently, one needs to only consider the determination of N_A and, since the electron is the particle whose mass can be obtained most accurately at present, the determination of m_e.

Determining N_A and m_e As discussed earlier, values for constants may be obtained by direct measurement or indirectly from appropriate combinations of other directly measured quantities. Hence, as explained in Section 3.3, N_A may be determined

1. directly by the X-ray crystal density (XRCD) method in which one measures the molar mass, density, and unit-cell volume of a material such as silicon that can be prepared in the form of extremely pure and nearly perfect single crystals. For silicon, a cubic crystal with eight atoms per unit cell of edge length $a = d_{220}\sqrt{8}$ and volume $v = a^3$, the basic relation is $N_A = 8M(Si)/\rho(Si) \times \left(d_{220}\sqrt{8}\right)^3$.
2. N_A may also be obtained indirectly by a number of different routes but, since $m_e = M_e/N_A = M_p/[N_A(m_p/m_e)]$, any indirect measurement of N_A may be viewed as an indirect measurement of m_e.

Since m_e cannot yet be measured directly (one cannot yet "weigh" electrons), only indirect measurements (other than via N_A) need be considered. On the basis of the equations $R_\infty = m_e c\alpha^2/2h$, $\alpha = \mu_0 ce^2/2h$, and relations previously discussed, one can show that

$$m_e = \frac{8R_\infty \alpha^{-2} K_W}{cR_{K-90}K_{J-90}^2} = \frac{16R_\infty \alpha^{-1} K_V^2}{\mu_0 c^2 K_{J-90}^2} = \frac{4\mu_0 R_\infty \alpha^{-3} K_A^2}{R_{K-90}^2 K_{J-90}^2} \quad (3.24)$$

where K_W, K_V, and K_A have the same meaning as in Section 3.5, namely, $K_W = W_{90}/W$, $K_V = V_{90}/V$, and $K_A = A_{90}/A$.

Now, in the SI, μ_0, c, R_{K-90}, and K_{J-90} are all exact constants; R_∞ is known with a relative uncertainty that is currently limited by the practical realization of the meter at visible wavelengths or about 6.6×10^{-12} and α^{-1} as obtained from a_e has a relative uncertainty of 6.9×10^{-10} and it should eventually be possible to reduce this to $(1 - 2) \times 10^{-10}$. Consequently, realization of the watt, volt, or ampere (i.e., determination of K_W, K_V, or K_A) with a relative uncertainty of about 1×10^{-8} could lead to a new definition of the SI unit of mass.

One can also show that

$$m_e = \frac{2\left(\mu'_p/\mu_B\right)}{C_{K-90}K_{J-90}\gamma'_p(hi)_{90}} = \frac{2M_p\alpha^{-2}K_w}{R_{K-90}K_{J-90}(m_p/m_e)F_{90}} \quad (3.25)$$

where again the various quantities entering these relations have the same meaning as in the previous sections. The current relative uncertainty of μ_p/μ_B is 11×10^{-9} and can likely be significantly reduced if need be. The current relative uncertainty of m_p/m_e is 20×10^{-9}, and, as noted in Section 3.6.2, this will eventually be reduced by a factor of 10 or more. Consequently, a determination of either $\gamma_p(\mathrm{hi})_{90}$ or F_{90} with an uncertainty of about 10^{-8} can also lead to a value of m_e, and thus $m(^{12}\mathrm{C})$ and $n(^{12}\mathrm{C})$, having essentially the same uncertainty and therefore possibly lead to a new definition of the kilogram.

It should be recognized that Equations 3.17 and 3.18 are potentially useful as a basis for replacing the kilogram only because it is possible to calibrate laboratory voltage standards in terms of V_{90} with a relative uncertainty of about 1×10^{-9} or better using a Josephson-array voltage standard, and laboratory resistance standards in terms of Ω_{90} with a similar uncertainty using the QHE and all-cryogenic techniques.

The data pertaining measurements of the Avogadro constant provided an important input to the evaluations. This is a modernized version of the method from the 1930s era of Bragg and others, whereby the Avogadro constant is determined from a measurement of the lattice spacing d_{220}, and molar density $M(\mathrm{Si})$ of a pure silicon crystal.

There is considerable interest in this work for technological reasons because it is one of the two types of project providing the basis for decisions of whether and how to redefine the definition of the kilogram. This is a truly international project and technological input and measurements of various aspects are being made in USA, Japan, Belgium, Italy, Germany, Russia, and Australia.

A quantity d_{220} was introduced as an unknown in the 1986 and later adjustments. The two quantities d_{220} and $V_m(\mathrm{Si}) = M(\mathrm{Si})/(\mathrm{Si})$ were treated as separate items of stochastic input data, because the two available experimental values of d_{220} were in gross disagreement. If they had not been discrepant, then their weighted mean could have been used in conjunction with the single available value of $V_m(\text{Si})$ to obtain a single value of N_A through the relation $N_A = 8V_m(\text{Si}) / \left(d_{220}\sqrt{8}\right)^3$. This result could then have been included in the adjustment directly and there would have been no need to introduce d_{220} as an unknown. For the 2006 evaluation, four measurements of d_{220} were considered. One measurement differed markedly from the other three, but was included because it was supported by the value deduced from measurements of the Compton wavelength of the neutron, expressed as a measurement of $h/(m_n d_{220})$, where m_n is the neutron mass.

Work on resolving the present discrepancies continues. Silicon is not a single isotope element, and the relative isotopic abundances vary from one sample to another, or even throughout the volume of a particular sample. Consequently, it is very difficult to produce silicon crystals that are completely homogeneous. Moreover, there can be small faults and possibly impurities affecting both the density and lattice parameters. The vacuum zone refining method of purification produces gradients in the silicon isotopic abundances and, with this in mind, considerable efforts are being made to produce single isotope samples of very pure silicon and results are expected shortly.

3.6.3 Practical Mass Metrology Making Use of a New Definition of the Kilogram

Suppose that one of the above approaches eventually yields a value of m_e or h with the requisite relative uncertainty of about 1×10^{-8} and the kilogram is redefined in terms of either $n(^{12}\text{C})$ or $h\nu_{\text{def}}/c^2$, how would the newly defined kilogram be used in practice?

First, it should be recognized that either present-day platinum–iridium national kilogram standards (particularly those also made along with the prototype), or the prototype kilogram itself, would not be discarded. Rather, they would serve to act as "flywheels," that is, as a means of preserving the unit of mass between realizations based on the new definition. Their periodic calibration would be performed by carrying out, for example, a watt-realization experiment, or an Avogadro-constant determination, in terms of the mass standard to be calibrated, taking as its value that is currently assigned to it. Any deviation of the measured value of h, or m_e and thus h, or $n(^{12}\text{C})$ from their adopted value would/could be attributed to a shift in the mass standard (always assuming that there was not a really major problem with the data being input to the "best values" available by 2011). A new value would then be assigned to the standard, thereby recalibrating it.

For many years, only a few laboratories in the world would need to have the apparatus necessary to carry out such experiments, because the results obtained could be readily transferred between laboratories with sufficient accuracy by intercomparing mass standards. If more accurate values of relevant fundamental constants (e.g., R_∞, α^{-1}) became available, the adopted value of $n(^{12}\text{C})$ or $h\nu_{\text{def}}/c^2$ would not be changed. Rather, one would take into account any change in $n(^{12}\text{C})$, or $h\nu_{\text{def}}/c^2$, implied by the improved values of the constants by assigning new values to mass standards. Of course, before any new kilogram definition is adopted, the international prototype would be measured periodically (i.e., monitored) in terms of the proposed new definition to minimize introducing a discontinuity in the SI unit of mass.

A second major advantage of the change would be the elimination of a common, correlated, uncertainty from the values of many of the dimensioned fundamental constants, with the choice in terms of h having the added advantage of improved accuracy by removing the uncertainty in m_p/m_e from many constants.

The new kilogram definition would be realized for a kilogram that was in vacuum, but the transfer to a kilogram that is in air can already be made with the requisite accuracy, see Robinson and Kibble (2007), Robinson (1990), and Davidson, Brown, and Berry (2004). It remains to be seen whether future primary kilogram standards will be sustained in air or vacuum, or both.

3.7 Numerology and Time Variations of the Constants

Any chapter on the fundamental physical constants would not be complete without at least a brief discussion of why the constants have the values they do, and whether they are in fact really constant. For example, why does the ratio of the rest mass of the proton to that of the electron happen to be 1836.15267247(80)? Why does α^{-1} happen to equal 137.035999679(94)? Why not other, perhaps simpler, values?[2)]

2) Viewed in the round, there are probably, to some extent, relationships between the digits in any group of numbers, with the frequencies of particular integers of the first and successive digits depending on the empirical Benford's Law. This was proposed in 1938 by the late Dr Frank Benford, and extended a discovery reported in 1881, by the astronomer/mathematician Simon Newcomb. If absolute certainty is defined as 1 and absolute impossibility as 0, then the probability of any number "d" from 1 to 9 being the first significant digit is given by

$$P\{d\} = \frac{\ln\left(1+\frac{1}{d}\right)}{\ln(10)} = \log_{10}\left(1+\frac{1}{d}\right) \tag{3.26}$$

and for the nth decimal place it is

$$\begin{aligned} P_n\{d\} &= \frac{1}{\ln(10)} \sum_{k=10^{n-1}}^{10^n-1} \ln\left(1+\frac{1}{10k+d}\right) \\ &= \sum_{k=10^{n-1}}^{10^n-1} \log_{10}\left(1+\frac{1}{10k+d}\right) \end{aligned} \tag{3.27}$$

(For numbers to base B replace the number 10 by B) Benford found that this formula also predicts the frequencies of numbers found in many categories of statistics. Although after a few decimal places all integers are equally probable, it is interesting that this law also applies to the fundamental constants (see, e.g., *http://www.mathpages.com/home/kmath302/kmath302.htm* and Torres *et al.*, 2007) and their determination. Thus, for the first digit is 1 for about 30% of the constants and 7, 8, or 9, for about 5% of the constants, rather than an equal spread of 10% for the digits 1–9 that one might have otherwise expected.

Three rather different points of view have evolved about the constants, especially their dimensionless combinations:

1. The numbers are not at all arbitrary, but can be calculated from some as yet unknown basic theory (or theories) in much the same way that the g-factors of the electron and muon can be calculated from QED, or even in the way that the ratio of the circumference of a circle to its diameter need not be experimentally determined, but can be calculated to arbitrary accuracy from mathematics. For example, in the late 1960s, an expression was derived for α and m_p/m_e involving only simple integers and, based on the volumes of

certain bounded spaces associated with the invariance group of a relativistic quantum wave equation, and giving a value for them which were in excellent agreement with the recommended value at the time (Wyler, 1969), although not in accordance with present-day values. However, the physical basis for his derivation was not at all clear, and it remains safe to say that there does not yet exist a physically meaningful "derivation" of an accurate value for a fundamental constant, or dimensionless ratio or combination of constants. When one considers that evaluation of the tenth order QED term of the calculation of the electron $g_e - 2$ requires the evaluation of some 12 672 Feynman diagrams, one can imagine that this task remains a formidable challenge for the future. It would, of course, be nice if some development led to the problems associated with renormalization in QED going the way of the epicycles of pre-Keplerian astronomy.

2. The second viewpoint is not aimed at explaining the values of the physical constants, but simply notes that if they were terribly different from their currently observed values, we would not be here to measure them–life as we know it on earth, and perhaps the existence of the universe itself, depends on complex physical processes that require the constants to have their observed values. This is the idea behind the so-called weak anthropic principle (or WAP) (Barrow and Tipler, 1988). The strong anthropic principle (or SAP) goes even further by speculating that the fundamental constants and laws of physics must be such that life, and the universe as we know it, can exist–which represents a considerable extension of the "Cogito ergo sum," of Descartes of nearly six centuries ago.
3. An extreme statement of the third point of view is that the fundamental constants change with time as the universe evolves and therefore the currently observed values are not particularly significant. This idea had its origins in 1937 when Dirac noted that the ratio of the electric to gravitational forces between the electron and proton is of the order 10^{40}, very nearly equal to the age of the universe expressed in an appropriate atomic time unit, for example, the time required for light to travel a distance equal to the classical electron radius. He speculated that either the equality is accidental or it indicates a causal relationship among electromagnetism, gravitation, and cosmology. If such a relationship is in fact true, then since the universe is continually aging, the equality would remain unchanged only if one or more of the fundamental constants entering the two ratios changed–for example, the Newtonian gravitational constant G, or the elementary charge e. Grand unification theories (GUTs) imply that the fundamental constants were very different during the first second of the "ig bang" when there was a unification among the strong, weak, and electromagnetic forces, thereby providing a basis for some of Dirac's ideas on the topic.

These speculations have stimulated a considerable amount of theoretical and experimental work over the last 50 years, the latter quite often involving astronomical or geophysical observations. As a result, stringent limits (albeit frequently model dependent) have been placed on the possible time dependence of a number of physical constants. For example, it has been claimed that h changes less than 2 parts in 10^{12} per year; that G changes less than 3 parts in 10^{11} per year (these are essentially the variations of α and $\alpha_G = Gm_p^2 / \hbar c$, respectively).

We have already entered the situation where the accuracy of laboratory determinations gives measures of the constancy of certain of the constants. This particularly applies to measures of the stabilities of atomic clocks that rely on energy transitions in different atoms.

The frequencies of different types of atomic clock using transitions in different atoms can also be intercompared and also provide information about the stability of α over laboratory timescales. Thus, by comparing transitions, frequencies in Al^+ and Hg^+ single-ion optical atomic clocks Rosenband *et al.* at NIST (Rosenband *et al.*, 2008) measured

$$\nu_{Al}^{+}/\nu_{Hg}^{+} = 1.052871833148990438(55) \tag{3.28}$$

and, by comparing measurements taken at various times over a period of about a year, they deduced that α changes by less than $(-1.6 \pm 2.3) \times 10^{-17}$ per year.

One can only infer the constancy of dimensionless quantities because one cannot usually assume that the reference quantity remains constant. In the above example, other dimensionless quantities are also involved in the theoretical expressions for the transition frequencies.

In the near future, the changes in the values of the fundamental constants caused by mankind's present limitations in understanding and implementing the science behind measuring them, will dominate any changes that might be caused by the inconstancy of Nature's laws. The underlying assumptions behind some of the above can be questioned. For example, there has been at least one example of a cosmological "measurement" of the constancy of h where a null result was guaranteed through measuring a quantity by a method that predetermined a null result.

3.8 Future Trends and Advances

The fundamental-constants field is ever evolving; new experiments are undertaken with each major new advance in measurement science. Indeed, roughly speaking, our knowledge of the values of many of the fundamental constants improved by an order of magnitude each decade during the last half of the twentieth century. However, the question naturally arises as to whether this sort of improvement can continue indefinitely. Will phenomena and measuring techniques yet to be discovered lead to continued reductions in uncertainty of this magnitude into the twenty-first century?

We have almost arrived at the era predicted by Maxwell (1870) when he forecast, at a time when the atomic nature of our universe was increasingly emerging, that

> *If we wish to obtain standards of mass of length, mass or time that shall be absolutely permanent, we must seek them not in the dimensions, or motion, or the mass of our planet, but on the wavelength, the period of vibration and mass of those imperishable and unalterable and perfectly similar molecules.*

The projected change to the international measurement system (the basis for the SI units) means that theoretical and experimental science will at last be performed in terms of the same sets of quantities with an exactness that our forbears could only dream of. An important consequence will be that it will be possible for experiments relating to the measurement of dimensioned fundamental constants to revert from the national standard's laboratories to the universities and elsewhere.

We are discussing changing the definition of the SI unit of mass at a time when the discovery of the Higgs boson seems imminent. Why is the Higgs so important? Because it is thought to pervade the universal vacuum; not, as with the old ether, to provide a material substrate for the propagation of electromagnetic waves, but rather to interact with particles and confer mass upon them. The Higgs ministrations are usually hidden away in the vacuum, but if enough energy is brought to bear in a tiny volume of space, at the point where two energetic particles collide, then the Higgs can be turned into an actual particle whose existence can be detected. Theoretical calculations made using the standard model of particle physics combined with previous experiments serve to limit the possible range of masses for the Higgs particle. Right now that mass is thought to lie between 114 and about 190 GeV. However, it is difficult to imagine that an increased understanding of the nature of mass will immediately impact on how we choose to define the mass unit of our very practical SI unit system, but in another 50 years metrology may be heading along a very different direction.

The American Physicist Richtmyer (1932) echoing Maxwell of 50 years earlier, again asked and answered the question: "*Why should one wish to make measurements with ever increasing precision?* He too responded: *Because the whole history of physics proves that a new discovery is quite likely to be found lurking in the next decimal place.*" The last 50 years have amply demonstrated the wisdom of their response. A reply that will undoubtedly continue to be demonstrated in the future.

Some day we must reach the ultimate limits. However, we can be sure that scientists will continue their efforts to advance our knowledge of the fundamental constants unabated as long as they believe there is even a slim chance of success.

Glossary

a_e: Electron magnetic moment anomaly; $a_e = g_e/2 - 1 = \mu_e/\mu_B - 1$, where g_e is the g-factor of the electron, μ_e is the magnetic moment of the electron, and μ_B is the Bohr magneton: $\mu_B = eh/2\pi m_e$, where e is the elementary charge, h is the Planck constant, and m_e is the electron mass.

α: Fine-structure constant, approximately 1/137, is the dimensionless coupling constant or the measure of the strength of the electromagnetic force that governs how electrically charged elementary particles (e.g., electron, muon) and light (photons) interact.

α^{-1} Inverse fine-structure constant; $\alpha^{-1} \approx 137$.

BIPM: Bureau International des Poids et Mesures (International Bureau of Weights and Measures), located in Sèvres, France, a suburb of Paris. BIPM is the international standards laboratory.

CCE: Comité Consultatif d'Électricité (Consultative Committee for Electricity), one of the eight advisory committees to the CIPM.

CGPM: Conférence Général des Poids et Mesures (General Conference on Weights and Measures), the highest international metrological authority established by the Convention du Mètre (Treaty of the Meter) signed in 1875; it is a diplomatic organization.

CIPM: Comité International des Poids and Mesures (International Committee for Weights and Measures), one of the organizations established by the Convention du Mètre (Treaty of the Meter) signed in 1875. The CIPM supervises the BIPM and is under the authority of the CGPM.

CODATA: Committee on Data for Science and Technology, an interdisciplinary committee of the International Council of Scientific Unions. It seeks to improve the compilation, critical evaluation, storage, and retrieval of data of importance to science and technology.

JE: Abbreviation for Josephson effect, a solid-state, low-temperature physics phenomenon characteristic of weakly coupled superconductors cooled below their transition temperatures. An example is two thin films of superconducting lead separated by a thermally grown oxide layer about 1-nm thick.

K_J**:** Josephson constant: characteristic of the Josephson effect and assumed equal to $2\,e/h$, where e is the elementary charge and h is the Planck constant.

K_J**−90:** Conventional value of the Josephson constant adopted by the CIPM for international use in basing a representation of the volt on the Josephson effect starting 1 January 1990; $K_{J-90} = 483\,597.9\,\text{GHz V}^{-1}$ exactly.

Muonium: Hydrogen-like atom composed of a positive muon and an electron (μ^+e^-atom).

NIST: National Institute of Standards and Technology, the national standards laboratory of the United States (called the National Bureau of Standards or NBS prior to 23 August 1988).

NPL: National Physical Laboratory, the national standards laboratory of the United Kingdom.

ppm: Symbol often used to denote an uncertainty of a fractional part in 10^6. Being an English language, specific symbol it is not part of the internationally supported set of quantities in the International Standards Organization Standard: ISO 31 and, because it is not used in a consistent way, is not an internationally recommended symbol, and that consequently, where it is used by authors, its meaning must always be defined in the text.

QED: Quantum electrodynamics, the relativistic quantum field theory of the interaction of charged particles and electromagnetic radiation (photons). QED is an amalgamation of Maxwell's theory of electromagnetism, quantum mechanics, and Einstein's special theory of relativity.

QHE: Abbreviation for quantum Hall effect, a solid-state, low-temperature physics phenomenon characteristic of a two-dimensional electron gas. The latter may be realized in a high-mobility semiconductor device such as a silicon metal-oxide-semiconductor field-effect transistor (MOSFET) or GaAs-$Al_xGa_{1-x}As$ heterojunction cooled to about 1 K. The applied magnetic field (magnetic flux density) B must be of order 10 T.

R_K**:** von Klitzing constant characteristic of the quantum Hall effect and assumed equal to h/e^2, where h is the Planck constant and e is the elementary charge.

R_K**−90:** Conventional value of the von Klitzing constant adopted by the CIPM for international use in basing a representation of the ohm on the quantum Hall effect starting 1 January 1990; $R_{K-90} = 25812.807\Omega$ exactly.

SI: Système International d'Unités (International System of Units), the modernized metric system with the seven base units meter (m), kilogram (kg), second (s), ampere (A), Kelvin (K), mole (mol), and candela (cd); and numerous derived units such as the Hertz (Hz), Joule (J), Newton (N), watt (W), volt (V), and ohm (Ω).

References

Armstrong, T.R. and Fitzgerald, M.P. (**2003**) *New Measurements of G using the Measurement Standards Laboratory Torsion Balance*, Phys. Rev. Lett., **91**, 201101.

Bagley, C.H. and Luther, G.G. (**1997**) *Preliminary results of a determination of the Newtonian constant of gravitation: a test of the Kuroda hypothesis*, Phys. Rev. Lett., **78** (16), 3047–3051.

Barone, G. and Paterno, A. (**1982**) *Physics and Applications of the Josephson Effect*, John Wiley & Sons, Inc., New York.

Barrow, J.D. and Tipler, F.J. (**1988**) *The Anthropic Cosmological Principle*, Oxford University Press.

de Beauvoir, B., Schwob, C. *et al.* (**2000**) Metrology of the hydrogen and deuterium atoms: determination of the Rydberg constant and Lamb shifts. *Eur. Phys. J. D*, **12**, 61–93.

Becker, P. (**2003**) Tracing the definition of the kilogram to the Avogadro constant using a silicon single crystal. *Metrologia*, **40** (6), 366.

Becker, P., Schiel, D. *et al.* (**2006**) Large-scale production of highly enriched ^{28}Si for the precise determination of the Avogadro constant. *Meas. Sci. Technol.*, **17** (7), 1854.

Birge, R.T. (**1929**) Probable values of the general physical constants (as of January 1, 1929). *Phys. Rev. Suppl.*, **1** (1), 1–73.

Brown, H.N., Bunce, G. *et al.* (**2001**) Precise measurement of the positive muon anomalous magnetic moment. *Phys. Rev. Lett.*, **86** (11), 2227–2231.

Cavendish, H. (**1798**) Experiments to determine the density of the earth. *Philos. Trans. R. Soc. Lond. (1776–1886)*, **88**, 469–526.

Coddington, I., Swann, W.C., and Newbury, N.R. (**2008**) Coherent multiheterodyne spectroscopy using stabilized optical frequency combs. *Phys. Rev. Lett.*, **100** (1), 013902.

Cohen, E.R. and DuMond, J.W.M. (**1965**) Our knowledge of the fundamental constants of physics and chemistry in 1965. *Rev. Mod. Phys.*, **37** (4), 537–594.

Cohen, E.R. and Taylor, B.N. (**1973**) The 1973 least-squares adjustment of the fundamental constants. *J. Phys. Chem. Ref. Data*, **2** (4), 663–734.

Cohen, E.R. and Taylor, B.N. (**1987**) The 1986 adjustment of the fundamental physical constants. *Rev. Mod. Phys.*, **59** (4), 1121–1148.

Davidson, S., Brown, S., and Berry, J. (**2004**) A Report on the Potential Reduction in Uncertainty from Traceable Comparisons of Platinum– Iridium and Stainless Steel Kilogram Mass Standards in Vacuum, NPL. CMAM 88.

DuMond, J.W.M. and Cohen, E.R. (**1953**) Least-squares adjustment of the atomic constants, 1952. *Rev. Mod. Phys.*, **52** (3), 691–708.

Fischer, M., Kolachevsky, N. *et al.* (**2004**) New limits on the drift of fundamental constants from laboratory measurements. *Phys. Rev. Lett.*, **92** (23), 230802–230804.

Flowers, J.L. and Petley, B.W. (**2001**) Progress in our knowledge of the fundamental constants of physics. *Rep. Prog. Phys.*, **64** (10), 1191–1246.

Flowers, J.L. and Petley, B.W. (**2005**) The kilogram redefinition-an interim solution. *Metrologia*, **42** (5), L31.

Gabrielse, G., Hanneke, D. *et al.* (**2006**) New determination of the fine structure constant from the electron g value and QED. *Phys. Rev. Lett.*, **97** (3), 030802.

Gabrielse, G., Hanneke, D. *et al.* (**2007**) Erratum: new determination of the fine structure constant from the electron g value and QED [Phys. Rev. Lett. 97, 030802 (2006)]. *Phys. Rev. Lett.*, **99** (3), 039902.

Gallop, J. (**2005**) The quantum electrical triangle. *Philos. Trans. R. Soc. A*, **363**, 2221–2247.

Gallop, J.C. (**1991**) *SQUIDs, the Josephson Effects and Superconducting Electronics*, CRC Press.

Giacomo, P. (**1984**) News from the BIPM. *Metrologia*, **20** (1), 25–30.

Giacomo, P. (**1987**) News from the BIPM. *Metrologia*, **24** (1), 45–51.

Gundlach, J.H. Merkowitz, S.M. (**2000**) Measurement of newton's constant using a torsion balance with angular acceleration feedback. *Phys. Rev. Lett.*, **85** (14), 2869–2872.

Hänsch, T., Alnis, J. *et al.* (**2005**) Precision spectroscopy of hydrogen and femtosecond laser frequency combs. *Philos. Trans. R. Soc. A: Math. Phys. Eng. Sci.*, **363** (1834), 2155–2163.

Hu, Z.-K., Guo, J.-Q., and Luo, J. (**2005**) Correction of source mass effects in the HUST-99 measurement of G.. *Phys. Rev. D*, **71** (12), 127505.

Hudson, R.P. (**1984**) Documents concerning the new definition of the metre. *Metrologia*, **19** (4), 163–177.

JCGM (**1995**) ISO/IEC Guide 98:1995. International Organization for Standardisation Guide to the Expression of Uncertainty in Measurement (GUM), Geneva, Switzerland.

Karagioz, O.V. and Izmailov, V.P. (**1996**) Measurement of the gravitational constant with a torison balance. *Izmer. Tekh.*, **39** (10), 3–9.

Keller, M.W. (**2008**) Current status of the quantum metrology triangle. *Metrologia*, **45** (1), 102–109.

Keller, M.W., Zimmerman, N.M., and Eichenberger, A.L. (**2007**) Uncertainty budget for the NIST electron counting capacitance standard, ECCS-1. *Metrologia*, **44** (6), 505.

Kibble, B.P. (**1975**) A measurement of the gyromagnetic ratio of the proton by the strong field method, in *Atomic Masses and Fundamental Constants*, vol. **5** (eds J.H. Sanders and A.H. Wapstra), Plenum Press, London and New York, pp. 545–551.

Kinoshita, T. (ed.) (**1990a**) *Quantum Electrodynamics*, Advanced Series on Directions in High Energy Physics, vol. 7, World Scientific.

Kinoshita, T. and Lindquist, W.B. (**1990b**) Eighth-order magnetic moment of the electron. V. Diagrams containing no vacuum-polarization loop. *Phys. Rev. D*, **42** (2), 636.

Kleinevoß, U. (**2002**) Bestimmung der Newtonschen Gravitationskonstanten G, *University of Wuppertal*. Ph.D. Thesis.

Ludlow, A.D., Zelevinsky, T. *et al.* (**2008**) Sr lattice clock at 1 × 10-16 fractional uncertainty by remote optical evaluation with a Ca clock. *Science*, **319** (5871), 1805–1808.

Maxwell, J.C. (**1870**) Address to the mathematical and physical sections of the British Association, liverpool, Reproduced in *The Scientific Papers of James Clerk Maxwell*, vol. **2** (ed. W.D. Niven), Cambridge University Press, Cambridge, 1980, p. 245 [From the British Association Report, Vol. XL].

Maxwell, J.C. (**1871**) Inaugural address as Cavendish Professor, Reproduced in *The Scientific Papers of James Clerk Maxwell*, vol. **2** (ed. W.D. Niven), Cambridge University Press, Cambridge, 1980, p. 244.

Mills, I.M., Mohr, P.J. *et al.* (**2005**) Redefinition of the kilogram: a decision whose time has come. *Metrologia*, **42** (2), 71–80.

Mohr, P.J. and Taylor, B.N. (**1999**) CODATA recommended values of the fundamental physical constants: 1998. *J. Phys. Chem. Ref. Data*, **28** (6), 1713–1852.

Mohr, P.J. and Taylor, B.N. (**2005**) CODATA recommended values of the fundamental physical constants: 2002. *Rev. Mod. Phys.*, **77** (1), 1–107.

Mohr, P.J., Taylor, B.N., and Newell, D.B. (**2008**) CODATA recommended values of the fundamental physical constants: 2006. *Rev. Mod. Phys.*, **80** (2), 633–698.

Parker, D.H., Taylor, B.N., and Langenberg, D.N. (**1967**) Measurement of $2e/h$ using the ac Josephson effect and its implications for quantum electrodynamics. *Phys. Rev. Lett.*, **18** (8), 287–291.

Parker, W.H., Langenberg, D.N., Experiment, I. *et al.* (**1969**) Determination of e/h, using macroscopic quantum phase coherence in superconductors. *Phys. Rev.*, **177** (2), 639–664.

Petley, B.W. (**1983**) New definition of the metre. *Nature*, **303** (5916), 373–376.

Petley, B.W. and Morris, K. (**1970**) A measurement of $2e/h$ by the ac Josephson effect. *Metrologia*, **6** (2), 46.

Prange, R.E. and Girvin, S.M. (eds) (**1990**) *The Quantum Hall Effect*, Springer-Verlag.

Quinn, T.J. (**1989**) News from the BIPM. *Metrologia*, **26** (1), 69–74.

Quinn, T.J. (**1991**) The kilogram: the present state of our knowledge. *IEEE Trans. Instrum. Meas.*, **40** (2), 81–85.

Quinn, T.J., Speake, C.C., Richman, S.J. *et al.* (**2001**) A new determination of G using two methods. *Phys. Rev. Lett.*, **87**, 111101.

Richtmyer, F.K. (**1932**) The romance of the next decimal place. *Science*, **75**, 1–5.

Robertson, B. (**1971**) Wyler's Expression for the Fine-Structure Constant α. *Phys. Rev. Lett.*, **27** (22), 1545.

Robinson, I.A. (**1990**) Comparing in-air and in-vacuum mass standards without buoyancy corrections via in-vacuum weighing. *Metrologia*, **27** (3), 159.

Robinson, I.A. and Kibble, B.P. (**2007**) An initial measurement of Planck's constant using the NPL Mark II watt balance. *Metrologia*, **44** (6), 427–440.

Rosenband, T., Hume, D.B. *et al.* (**2008**) Frequency ratio of Al+ and Hg+ single-ion optical clocks; metrology at the 17th decimal place. *Science*, **319** (5871), 1808–1812.

Schlamminger, S., Holzschuh, E., Kundig, W. *et al.* (**2006**) Measurement of Newton's gravitational constant. *Physical Review D*, **74** (8), 082001–25.

Shields, J.Q., Dziuba, R.F., and Layer, H.P. (**1989**) New realization of the ohm and farad using the NBS calculable capacitor. *Instrum. Meas., IEEE Trans.*, **38** (2), 249–251.

Steiner, R.L., Williams, E.R. *et al.* (**2007**) Uncertainty improvements of the NIST electronic kilogram. *IEEE Trans. Instrum. Meas.*, **56** (2), 592–596.

Taylor, B.N. (**1987**) History of the present value of 2e/h commonly used for defining national units of voltage and possible changes in national units of voltage and resistance. *IEEE Trans. Instrum. Meas.*, **IM-36**, 659–664.

Taylor, B.N. (**1991**) The possible role of the fundamental constants in replacing the kilogram. *IEEE Trans. Instrum. Meas.*, **40** (2), 86–91.

Taylor, B.N., Parker, W.H., and Langenberg, D.N. (**1969**) *The Fundamental Constants and Quantum Electrodynamics*, Academic Press, New York and London. Reprint of *Reviews of Modern Physics*, 41 (3), 375 (1969).

Taylor, B.N. and Witt, T.J. (**1989**) New international electrical reference standards based on the Josephson and quantum Hall effect. *Metrologia*, **26**, 47–62.

Terrien, J. (**1973**) News from the Bureau International des Poids et Mesures. *Metrologia*, **9** (1), 40.

Tuninskii, V.S. (**1986**) Adjusting correlated values of fundamental constants. *Meas. Tech.*, **29** (8), 702–706.

Tuninskii, V.S. and Kholin, S.V. (**1975**) Concerning changes in the methods for adjusting the physical constants. *Metrologia*, **8**, 3 (in Russian).

Van Dyck, R.S.Jr , Schwinberg, P.B., and Dehmelt, H.G. (**1987**) New high precision comparison of electron and positron g-factors. *Phys. Rev. Lett.*, **59** (1), 26–29.

Werth, G., Beier, T. *et al.* (**2002**) A new value for the mass of the electron from an experiment on the g factor in 12C5+ and 16O7+. *Can. J. Phys.*, **80** (11), 1241–1247.

Wyler, A. (**1969**) *C. R. Acad. Sci. Ser. A*, **269**, 743.

Yoshioka, D. (**2002**) *The Quantum Hall Effect*, Springer.

Further Reading

Flowers, J. (**2004**) The route to atomic and quantum standards. *Science*, **306** (5700), 1324–1330.

Mills, I.M., Mohr, P.J. *et al.* (**2006**) Redefinition of the kilogram, ampere, kelvin and mole: a proposed approach to implementing CIPM recommendation 1 (CI-2005). *Metrologia*, **43** (3), 227–246.

Mohr, P.J. and Taylor, B.N. (**2000**) CODATA recommended values of the fundamental physical constants: 1998. *Rev. Mod. Phys.*, **72** (2), 351–495.

Petley, B.W. (**1985**) *The Fundamental Physical Constants and the Frontier of Measurement*, Adam Hilger, Bristol.

Quinn, T. and Burnett, K. (**2005**) The fundamental constants of physics, precision measurements and the base units of the SI. *Philos. Trans. R. Soc. A*, **363** (1834), 2097–2327.
This special issue contains many chapters on the status of aspects of fundamental constants and measurement.

Uzan, J.-P. (**2003**) The fundamental constants and their variation: observational status and theoretical motivations. *Rev. Mod. Phys.*, **75**, 403.

Useful Web Sites

SI brochure
http://www.bipm.org/en/si/si_brochure/
NIST fundamental constants
http://physics.nist.gov/cuu/Constants/
BIPM web site *http://www.bipm.org/*

4
Fundamentals of Materials Measurement and Testing

Horst Czichos

Handbook of Metrology. Edited by Michael Gläser and Manfred Kochsiek

ISBN: 978-3-527-40666-1

4.1
Introduction

In its most general context, the term *materials measurement and testing* denotes principles, techniques, and operations to distinguish qualitatively and to determine quantitatively characteristics of materials. As materials comprise all natural and synthetic substances and constitute the physical matter of engineered products and systems, materials characterization methods have a wide scope and impact for science, technology, economy, and society. This Handbook Contribution gives a concise overview on the essential features of materials measurement and testing. Details of principles, equipment, and application operations – as well as international standards of materials measurement and testing – are comprehensively compiled in the Springer Handbook of Materials Measurement Methods (Czichos, Saito, and Smith, 2006a).

4.1.1
Scope of Materials Measurement and Testing

Materials measurements are aiming to characterize the features of materials quantitatively; this is often closely related with analysis, modeling and simulation, and the qualitative characterization of materials through testing (see Figure 4.1).

Generally speaking, measurement begins with a definition of the *measurand*, the quantity that is to be measured, and it always involves a comparison of the *measurand* with a known quantity of the same kind. However in *metrology, the science of measurement*, measurands are based on and have to be traced back to the well-defined SI units; for materials there is a broad spectrum of "materials measurands." This is due to the broad variety of metallic, inorganic, organic, and composite materials, their different chemical and physical nature, and the manifold of attributes, which are related to materials with respect to composition, microstructure, scale, synthesis, properties, and applications. Some of these attributes can be expressed in a metrological sense as numbers, such as density; some are Boolean, such as the ability to be recycled; some, such as resistance to corrosion, may be expressed as a ranking (for instance, poor, adequate, and good) and some can only be captured in text and images (Ashby *et al.*, 2004). As background for the materials measurement system and the classification of materials characterization methods, the essential features of materials are reviewed in brief.

Handbook of Metrology. Edited by Michael Gläser and Manfred Kochsiek

ISBN: 978-3-527-40666-1

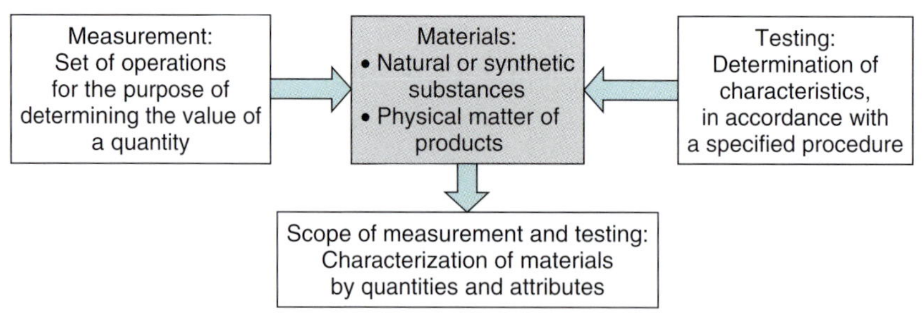

Fig. 4.1 The materials measurement and testing system.

4.1.2 Overview on Materials and Their Characteristics

Materials can be of natural (biological) origin or synthetically processed and manufactured. According to their chemical nature, they are broadly grouped traditionally into inorganic and organic materials. Their physical structure can be crystalline, amorphous, or mixtures of both structures. Composites are combinations of materials assembled together to obtain properties superior to those of their single constituents. Composites are classified according to the nature of their matrix: metal, ceramic, or polymer matrix composites, often designated as metal matrix composites (MMCs), ceramic matrix composites (CMCs) and polymer matrix composites (PMCs), respectively. Figure 4.2 illustrates with characteristic examples the spectrum of materials between the categories *natural*, *synthetic*, *inorganic*, and *organic*.

4.1.2.1 Nature of Materials

From the view of *materials science*, the fundamental features of a solid material are as follows:

- **Materials atomic nature:** The atomic elements of the periodic table, which constitute the chemical composition of a material.
- **Materials atomic bonding:** The type of cohesive electronic interactions between the atoms (or molecules) in a material, empirically categorized into the following basic classes:
  - Ionic bonds form between chemical elements with very different electron-negativity (tendency to gain electrons), resulting in electron transfer and the formation of anions and cations. Bonding occurs through electrostatic forces between the ions.
  - Covalent bonds form between elements that have similar electron-negativities; the electrons are localized and shared equally between the atoms, leading to spatially directed angular bonds.
  - Metallic bonds occur between elements with low electron-negativities, so that the electrons are only loosely attracted to the ionic nuclei. A metal is thought of as a set of positively charged ions embedded in an electron sea.
  - van der Waals bonds are due to the different internal electronic polarities between adjacent atoms or molecules leading to

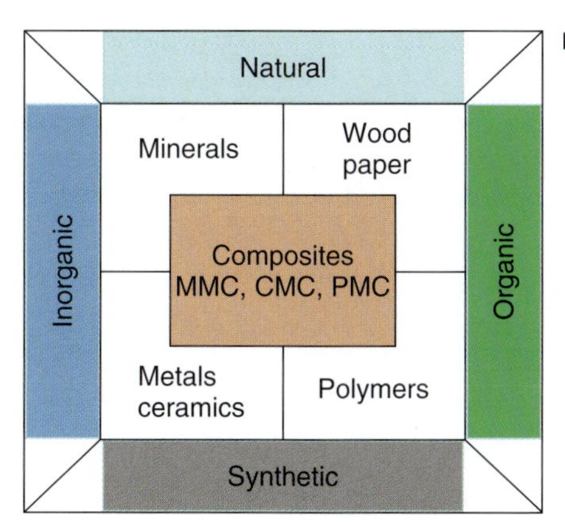

Fig. 4.2 Classification of materials.

weak (secondary) electrostatic dipole bonding forces.

- **Materials spatial atomic structure:** The amorphous or crystalline arrangement of atoms (or molecules), resulting from "long-range" or "short-range" bonding forces. In crystalline structures, it is characterized by unit cells that are the fundamental building blocks or modules repeated many times in space within a crystal.
- **Grains:** Crystallites made up of identical unit cells repeated in space, separated by grain boundaries.
- **Phases:** Homogeneous aggregations of matter with respect to chemical composition and uniform crystal structure: grains composed of the same unit cells are in the same phase.
- **Lattice defects:** Deviations of an ideal crystal structure:
  - **Point defects or missing atoms:** vacancies, interstitial, or substituted atoms;
  - **line defects or rows of missing atoms:** dislocations;
  - **area defects:** grain boundaries, phase boundaries, and twins; and
  - **volume defects:** cavities and precipitates.
- **Microstructure:** The microscopic collection of grains, phases, and lattice defects.

Together with *bulk materials characteristics*, surface and interface phenomena have to be considered.

4.1.2.2 Types of Materials

It has been estimated that there are between 40 000 and 80 000 materials that are used or can be used in today's technology (Ashby *et al.*, 2004). Figure 4.3 lists the main conventional *families* of materials together with examples of *classes, members*, and *attributes*. For the examples of attributes, necessary characterization methods are named.

- **Metallic materials – alloys:** In metals, the grains as the buildings blocks are held together by the *electron gas*. The free valence electrons of the electron gas account for the high electrical and thermal conductivity, and the optical gloss of metals. The metallic bonding – seen as interaction between the total of atomic nuclei and the electron gas – is not significantly influenced by a displacement of atoms. This is the reason for the good ductility and formability of metals. Metals and metallic alloys are the most important

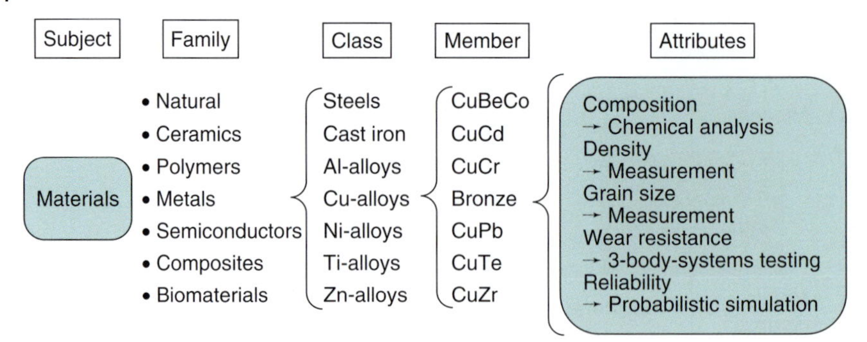

Fig. 4.3 Hierarchy of materials and examples of attributes and necessary characterization methods.

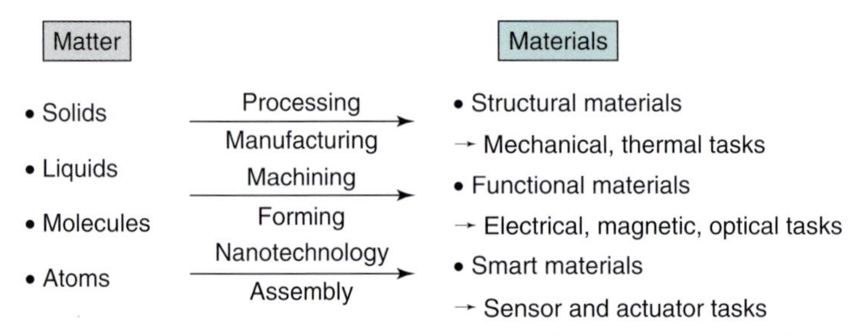

Fig. 4.4 Materials and their characteristics result from the processing of matter.

group of the so-called *structural materials* (see below) whose special features for engineering applications are their mechanical properties, for example, strength and toughness.

- **Semiconductors:** Semiconductors have an intermediate position between metals and inorganic nonmetallic materials. Their most important representatives are the elements silicon and germanium, possessing covalent bonding and diamond structure and the similarly structured III–V compounds, such as gallium arsenide (GaAs). Being electric nonconductors at absolute zero temperature, semiconductors can be made conductive through thermal energy input or atomic doping, which leads to the creation of free electrons contributing to electrical conductivity. Semiconductors are important *functional materials* (see below) for electronic components and applications.
- **Inorganic nonmetallic materials – ceramics:** Their atoms are held together by covalent and ionic bonding. As covalent and ionic bonding energies are much higher than metallic bonds, inorganic nonmetallic materials, such as ceramics have high hardness and high melting temperatures. These materials are basically brittle and not ductile: In contrast to the metallic bond model, a displacement of atomistic dimensions theoretically already breaks localized covalent bonds or transforms anion – cation attractions into anion–anion or cation–cation repulsions. Because of missing free valence electrons, inorganic nonmetallic materials are poor conductors for electricity and heat, and this qualifies

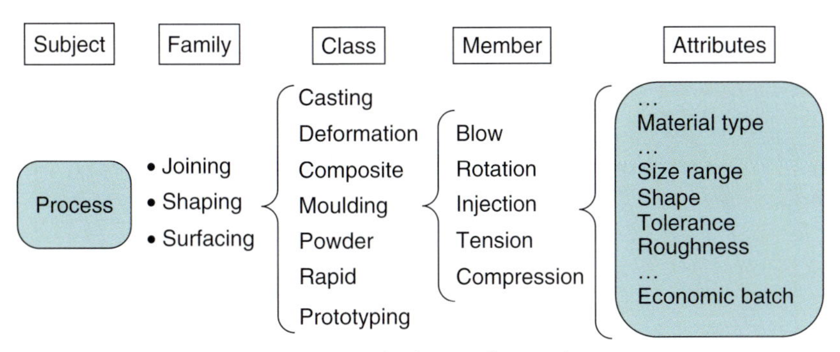

Fig. 4.5 Hierarchy of the processing technologies of materials.

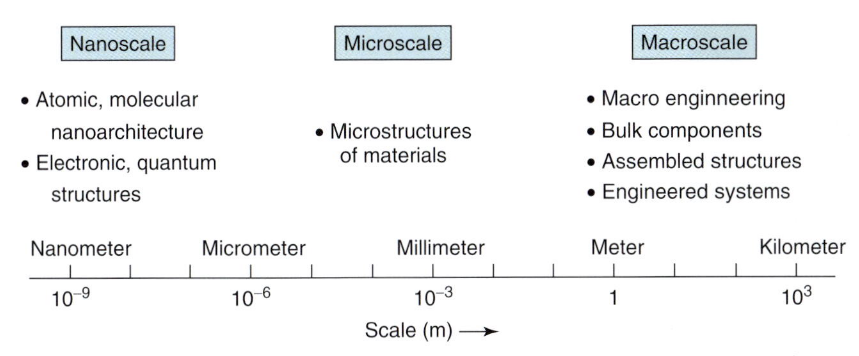

Fig. 4.6 Scale of material dimensions to be recognized in materials measurement and testing.

them as good insulators in engineering applications.

- **Organic materials – polymers and blends:** Organic materials, whose technologically most important representatives are the polymers, consist of macromolecules containing carbon (C) covalently bonded with itself and with elements of low atom numbers (e.g., H, N, O, and S). Intimate mechanical mixtures of several polymers are called *blends*. In thermoplastic materials, the molecular chains have long linear structures and are held together by (weak) intermolecular (van der Waals) bonds, leading to low melting temperatures. In thermosetting materials, the chains are connected in a network structure and do not melt. Amorphous polymer structures (e.g., polystyrene) are transparent, whereas the crystalline polymers are translucent to opaque. The low density of polymers gives them a good strength-to-weight ratio and makes them competitive with metals in structural engineering applications.
- **Composites:** Generally speaking, composites are hybrid creations made of two or more materials that maintain their identities when combined. The materials are chosen so that the properties of one constituent enhance the deficient properties of the other. Usually, a given property of a composite lies between the values for each constituent, but not always. Sometimes, the property of a composite is clearly superior to those of either of the constituents. The potential for such a

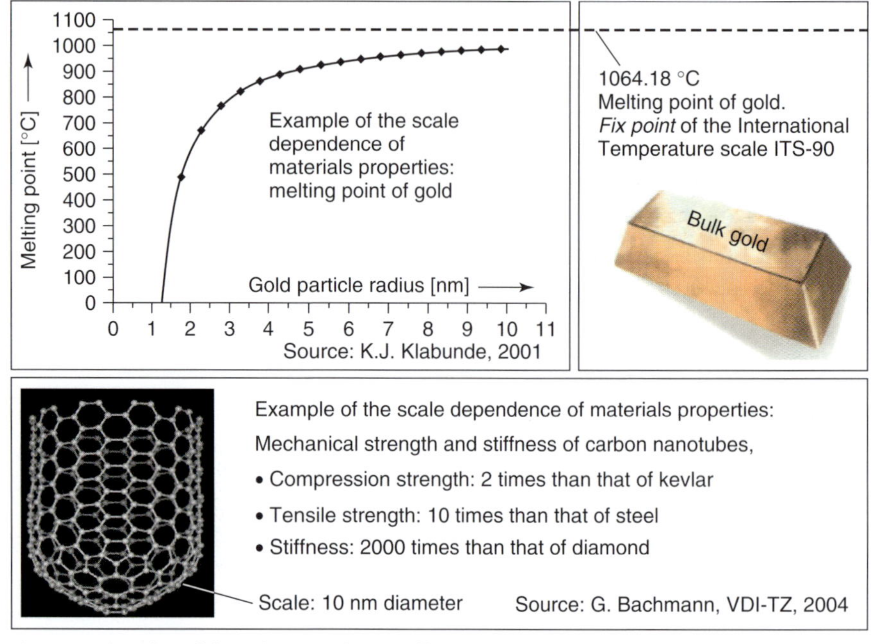

Fig. 4.7 Examples of the influence of scale effects on materials properties.

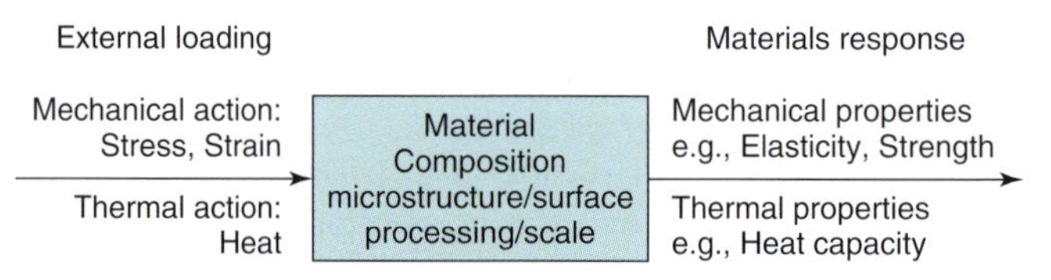

Fig. 4.8 Classification of *structural materials.*

synergy is one reason for the interest in composites for high-performance applications. However, because manufacturing of composites involves many steps and is labor intensive, composites may be too expensive to compete with metals and polymers, even if their properties are superior. In high-tech applications of advanced composites, it should also be borne in mind that they are usually difficult to recycle.

- **Natural materials:** Natural materials used in engineering applications are classified into natural materials of mineral origin, for example, marble, granite, sandstone, mica, sapphire, ruby, diamond, and those of organic origin, for example, timber, India rubber, natural fibers, such as cotton and wool. The properties of natural materials of mineral origin, as for example high hardness and good chemical durability, are determined by strong covalent and ionic bonds between their atomic or molecular constituents and stable crystal structures. Natural materials of organic origin often possess complex structures with direction-dependent properties. Advantageous application aspects of natural materials are recycling and sustainability.

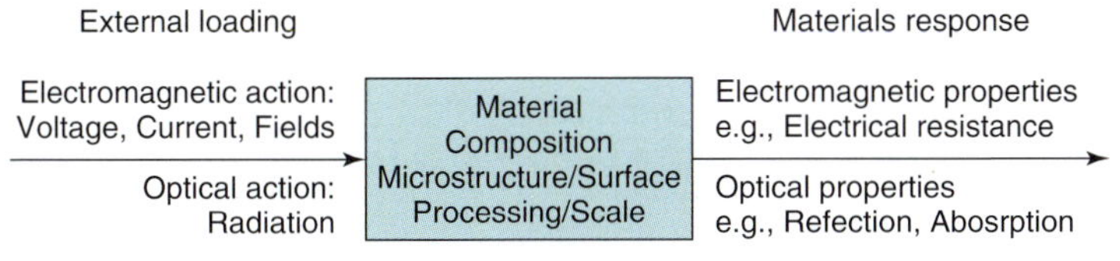

Fig. 4.9 Classification of *functional materials*.

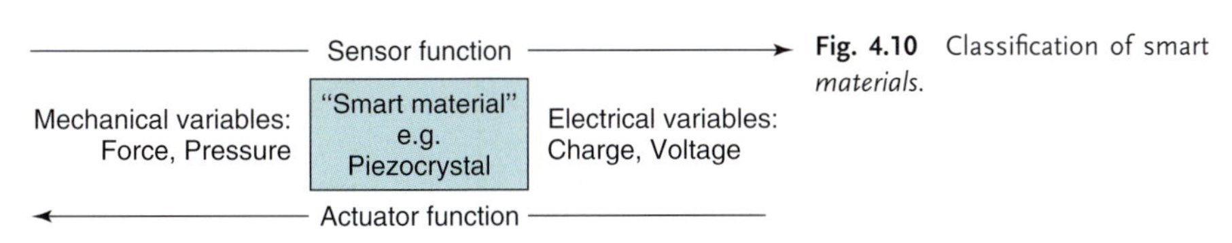

Fig. 4.10 Classification of smart *materials*.

- **Biomaterials:** *Biomaterials* can be broadly defined as the class of materials suitable for biomedical applications. They may be synthetically derived from nonbiological or even inorganic materials or they may originate in living tissues. The products that incorporate biomaterials are extremely varied and include artificial organs; biochemical sensors; disposable materials and commodities; drug-delivery systems; dental, plastic surgery, ear and ophthalmological devices; orthopedic replacements; wound management aids; and packaging materials for biomedical and hygienic uses. For the application of biomaterials, the understanding of the interactions between synthetic substrates and biological tissues is of crucial importance to meet the needs of clinical requirements.

4.1.2.3 **Processing of Materials**

For their applications, materials have to be *engineered* by processing and manufacture in order to fulfill their purpose as physical base of products designed for the needs of economy and society. There are the following main technologies to transform matter into engineered materials (see Figure 4.4):

- net forming of suitable matter, for example, liquids and molds;
- machining of solids, that is, shaping, cutting, drilling, and so on; and
- nanotechnological assembly of atoms or molecules.

It has been estimated that there are at least 1000 different ways to produce materials (Ashby *et al.*, 2004). Figure 4.5 lists some of the *families* of processing materials together with examples of *classes I members*, and *attributes*.

4.1.2.4 **Scale of Materials**

The geometric length scale of materials has more than 12 orders of magnitude. The scale ranges from the nanoscopic materials architecture to kilometer-long structures of bridges for public transport and pipelines and oil-drilling platforms for the energy supply of society. Figure 4.6 illustrates the dimensional scale relevant for today's materials science and technology.

Material specimens of different geometric dimensions have different bulk/surface ratios and may also have different bulk/surface-microstructures. This can heavily influence the properties of materials, as illustrated in Figure 4.7, for example, thermal and mechanical properties. Thus, *scale effects* have to be meticulously

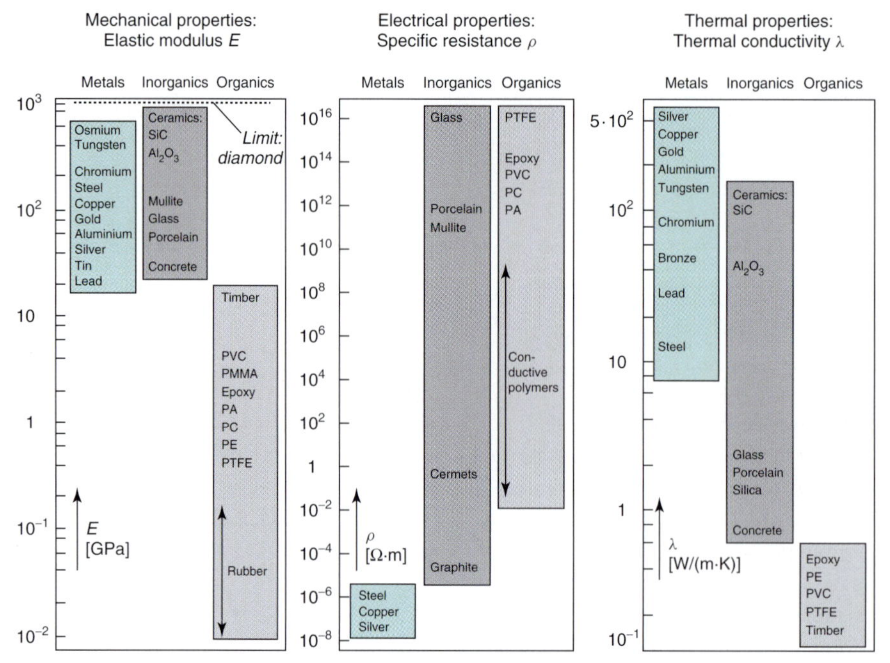

Fig. 4.11 Overview on data of materials properties for the basic types of materials.

considered in material measurement and testing.

4.1.2.5 **Properties of Materials**

The properties of materials – which are of fundamental importance for their technical applications – can be categorized in three basic groups, broadly classified as in Figures 4.8–4.10:

Structural materials: engineered materials, which have specific mechanical or thermal properties in responding to an external loading by a mechanical or thermal action.

Functional materials: engineered materials, which have specific electrical, magnetic, or optical properties in responding to an external loading by an electromagnetic or an optical action.

Smart materials: engineered materials with intrinsic or embedded *sensor* and *actuator functions*, which are able to accommodate materials in response to external loading, aiming at optimizing material behavior according to given requirements for materials performance.

The numerical values of the various materials properties can vary over several orders of magnitude for the materials of the different material types described in Section 4.1.2.2. An overview on the broad numerical spectra of some mechanical, electrical, and thermal properties of metals, inorganics, and organics gives Figure 4.11 (Czichos *et al.*, 2007).

It must be emphasized that the numerical ranking of materials in Figure 4.11 is based only on "rough average values." Precise data of materials properties require the specification of various influencing factors described above and symbolically expressed as follows:

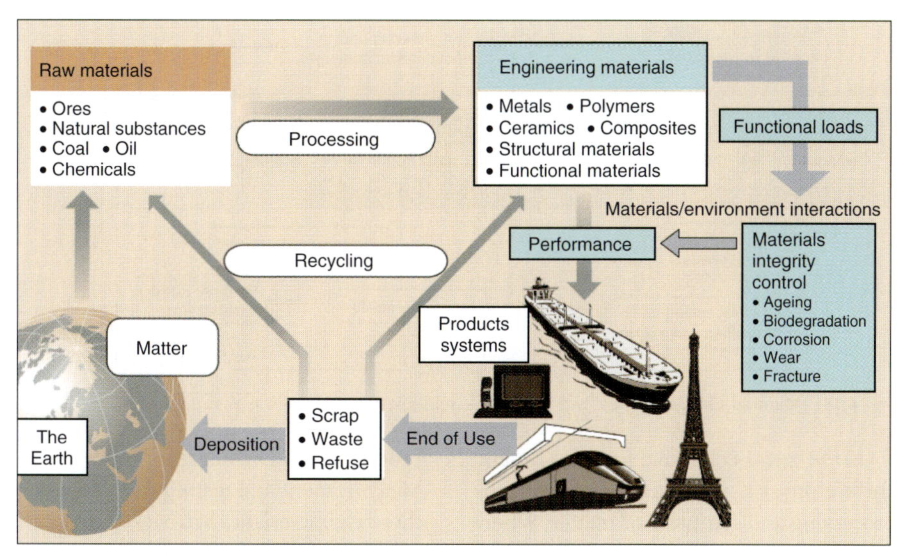

Fig. 4.12 The *materials cycle* of all products and technical systems.

Materials properties data = f (composition microstructure scale; external loading).

4.1.2.6 Application of Materials

For the application of materials – as constituents of products or of engineered components and systems – *performance characteristics,* such as *quality, safety,* and *reliability,* are of special importance. This adds performance control and material failure analysis to the tasks of application-oriented materials measurement and testing. As there is no usage of materials without interaction with the environment, detrimental influences on the integrity of materials must also be considered. Figure 4.12 gives an overview on the manifold of aspects to be recognized in the characterization of materials performance.

Figure 4.12 shows the so-called *materials cycle,* which applies to all man-made technical products in all branches of technology and economy. It illustrates that materials (accompanied by the necessary flow of energy and information) move in "cycles" through the techno-economic system: from raw materials to engineering materials and technical products and finally, after the termination of their task and performance, to deposition or recycling. For the proper performance of engineered materials, materials deterioration processes and potential failures – such as materials aging, biodegradation, corrosion, wear, and fracture – must be controlled.

Concluding this overview on the materials measurement and testing system, from Figure 4.12 it is obvious that for the different stages of the materials cycle, various measurement and testing techniques are needed, which are categorized in the following section.

4.1.3 Categories of Materials Characterization Methods

From a realization concerning the application of all material, a classification of materials characterization methods can be outlined, as shown in Figure 4.13.

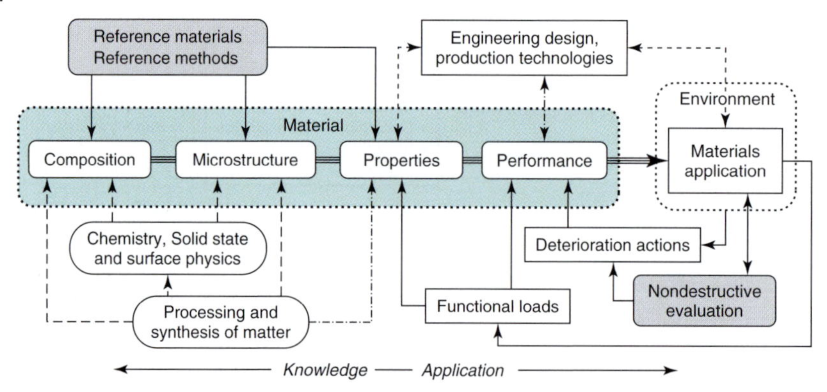

Fig. 4.13 Material characteristics and influences relevant for materials measuring and testing.

On the basis of chemistry, solid state, and surface physics, materials result from the processing and synthesis of matter. Whenever a material is being created, developed, or produced, the properties or phenomena the material exhibits are of central concern. Experience shows that the *properties* and *performance* associated with a material are intimately related to its *composition* and *structure* at all scale levels and are also influenced by the engineering component design and production technologies. The final material – as constituent of an engineered component – must perform a given task and must do so in an economical and societal acceptable manner.

The fundamental features of materials named in the middle part of Figure 4.13, namely,

- composition;
- microstructure;
- properties; and
- performance

and the interrelationship among them define the main categories of materials characteristics. They can be broadly classified as follows:

- **Intrinsic characteristics:** are materials composition and materials microstructure. They result from the processing and synthesis of matter. Measurement and testing to determine these characteristics are backed up by suitable reference materials (RMs) and reference methods, if available.
- **Extrinsic characteristics:** are materials properties and materials performance. They describe the response of materials and engineered components (with given composition and microstructure) to functional loads and environmental deterioration actions on materials integrity. Measurement and testing to determine these characteristics are backed up by suitable reference methods and nondestructive evaluation (NDE).

4.2 Characterization of Materials Composition

Analytical chemical measurements determine attributes and quantities of bulk materials and of materials surfaces with respect to their elemental and molecular chemical nature.

4.2.1 Elemental Chemical Analysis

Classical inorganic chemical analysis – *gravimetry, titrimetry, and coulometry* – are frequently used in assays of primary standard reagents that are used as calibrants in instrumental techniques. The methods are capable of very high precision and low relative uncertainties compared with other techniques of chemical analysis (Czichos, Saito, and Smith, 2006b).

- **Gravimetry:** Gravimetry is the determination of an analyte (element or species) by measuring the mass of a definite, well-characterized product of a stoichiometric chemical reaction (or reactions) involving that analyte. The product is usually an insoluble solid, although it may be an evolved gas. The solid is generally precipitated from solution and isolated by filtration. Preliminary chemical separation from the sample matrix by ion-exchange chromatography or other methods is often used. The mass fraction of the analyte in a gravimetric determination is measured by weighing a sample and the separated compound of known stoichiometry from that sample on a balance that is traceable to the kilogram. Appropriate ratios of atomic weights (gravimetric factors) are applied to convert the compound mass to the mass of the analyte or species of interest. Gravimetry is an absolute method that does not require reference standards. Thus, it is considered a direct primary reference measurement procedure.
- **Titrimetry:** The fundamental basis of titrimetry is the stoichiometry of the chemical reaction that forms the basis for the given titration. The analyte reacts with the titrant (typically added as a solution of the given reagent) according to the stoichiometric ratio defined by the corresponding chemical equation. The equivalence point corresponds to the point at which the ratio of titrant added to the analyte originally present (each expressed as an amount of substance) equals the stoichiometric ratio of the titrant to the analyte defined by the chemical equation. The practical determination of the equivalence point (endpoint) is obtained using visual indicators or instrumental techniques. Visual indicators or indicator – analyte compounds react with the added titrant at the endpoint, yielding a product of a different color. Titrimetry is considered a primary ratio measurement, since the measurement itself yields a ratio (of concentrations or amount-of-substance contents). The result is obtained from this ratio by reference to a standard of the same kind.
- **Coulometry:** This technique is based on Faraday's laws of electrolysis, which relate the charge passed through an electrode to the amount of analyte that has reacted. A coulometric cell has two main compartments. The sample is introduced into the sample compartment, which contains the working (generator) electrode. The other main compartment contains the counter-electrode. These main compartments are connected via one or more intermediate compartment(s) in series, providing an electrolytic link between the main compartments. Controlled-current coulometry is essentially titrimetry with electrochemical generation of the titrant. The amount-of-substance content of the analyte is calculated directly from the current I, passing through the coulometric cell, the time t, the stoichiometric ratio of electrons to analyte, the Faraday constant F, and the mass of sample. Since I and t can be measured very accurately, coulometry is (100% current efficiency assumed) capable of the smallest uncertainty and highest precision of all chemical analyses.

- **Spectroscopy:** In addition to the classical methods of inorganic chemical analysis described above, *spectroscopy* is broadly used in physical and analytical chemistry for the identification of substances through the spectrum emitted from or absorbed by them. In analytical chemistry, it is used for inorganic as well as for organic materials. Intensities of characteristic parts of spectra are used as basic information to carry out quantitative analyses. An overview on the principles and the key features of the main spectrometric methods for elemental chemical analysis are given in Table 4.1.

4.2.2 Compound and Molecular Specific Analysis

Molecular systems can be identified by their characteristic molecular spectra, obtained in the absorption or emission mode from samples in the gaseous, liquid, or solid state. Upon interaction with the appropriate type of electromagnetic radiation, characteristic electronic, vibrational, and rotational energy term schemes can be induced in the sample. These excited states usually decay to their ground states within 10^{-2} seconds, either by emitting the previously absorbed radiation in all directions with the same or lower frequency or by radiationless relaxation, thus providing spectral information for chemical analysis. Basic features of instrumental analytical methods are summarized in the following overview tables (compiled by Peter Reich, BAM, Berlin, 2004) (Table 4.2).

Further complementary structural information about molecular systems may be obtained by investigating the nuclear magnetic resonance (NMR) spectroscopy of a sample being irradiated with radio frequency (rf) in a magnetic field (Table 4.3).

Structural information can also be determined by analyzing the intensity distribution mass fragments of a sample bombarded with free electrons, photons, or ions in the *analytical mass spectroscopy* (*MS*) (Table 4.4).

Additional information on the near neighbor order in the solid state are provided in particular by the methods *infrared* (*IR*) and *Raman spectroscopy*, *EPR*, and *Moessbauer spectroscopy*. These techniques provide images of the interactions mentioned above and contain analytical information about the sample (Table 4.5).

4.2.3 Surface Chemical Analysis

Besides the bulk materials chemistry, surface characteristics of materials are of great significance. They are responsible for the appearances of materials and surface phenomena, and they have a crucial influence on the interactions of materials with gases or fluids (as in corrosion for example), contacting solids (as in friction and wear), or biospecies and materials – environment interactions. The main techniques of surface chemical analysis are listed as follows (Czichos, Saito, and Smith, 2006c):

- auger electron spectroscopy (AES);
- X-ray photoelectron spectroscopy (XPS); and
- Secondary ion mass spectrometry (SIMS).

The basic features of these techniques are compiled in Table 4.6. AES is excellent for elemental analysis (exception H, He) at spatial resolutions down to 10 nm, and XPS can define chemical states down to 10 μm. Both analyze the outermost atom layers and, with sputter depth profiling, layers up to 1–μm thick. Dynamic SIMS

Tab. 4.1 Spectrometric methods for elemental chemical analysis (compiled by Ralf Matschat, BAM, Berlin, 2008).

Analytical method	Measurement principle	Sample volume/mass	Limit of detection	Output, results	Applications
Inorganic mass spectroscopy *TIMS* *ICP-MS* *GD-MS* *SS-MS*	Measurement of the mass spectra of ions generated due to the ionization in an ionization source (thermal-ionization TI, inductively coupled plasma (ICP), glow discharge GD or electrical spark source SS; ICP-MS, sometimes combined with laser ablation). A mass spectrum consists of peaks of ions of a definite ratio of their atomic masses divided by their charge number (m/z). It results from a stream of gaseous ions with different values of m/z. The intensity of the spectral mass peaks corresponds with the concentration of the element in the plasma	TIMS: wide range depending on sample preparation, (final subsample 1–10 μl) ICP-MS: 1–10 ml GD-MS and SS-MS (direct solid sampling methods): ablated sample mass small, depending on instrument and parameters used: about 1–20 mg	TIMS: picograms nanograms absolute (upper picograms per grams to lower nanograms per grams range relative) ICP-MS: 0.001–0.1 ng ml^{-1} (using high-resolution spectrometers and in aquaous solution up to three orders of magnitude lower) GD-MS, SS-MS: 0.1–10 μg kg^{-1}	Simultaneous and sequential multielement analyses TIMS: High precision, high accuracy, highest metrological level with isotope dilution technique (ID-MS), but raw analyte isolation needed. Advantages of ICP-MS: Simultaneous multielement capability, very low limits of detection, very high dynamic range. Combined with laser ablation for solid samples; combined with isotope dilution technique (ID-ICP-MS): high metrological level results of high accuracy. GD-MS: Simultaneous multielement capability, direct sampling without chemical sample handling, very low detection limits, very high dynamic range. SS-MS: Not yet widely used	Qualitative and quantitative elemental and isotopic analysis, for example, semiconductors, ceramics, pure metals, environmental and biological materials, high-purity reagents, nuclear materials, geological samples; Speciation analysis when ICP-MS is combined with chromatographic methods. ID-MS: Merologically high-level technique for reference values of international comparisons in CCQM and certification of reference materials

(*continued overleaf*)

Tab. 4.1 *(continued)*

Analytical	Measurement principle	Sample volume/mass	Limit of detection	Output, results	Applications
Optical emission spectroscopy (atomic emission spectroscopy) *OES (AES)* *Flame-OES* *ICP-OES* *Arc-OES* *Spark OES* *Glow-discharge-OES* *(GD-OES)*	Measurement of the optical emission spectra of atoms or ions due to the excitation with flame, arc, spark, glow discharge, inductively coupled plasma (ICP-OES, sometimes combined with laser ablation or electrothermal vaporization ETV). An emission spectrum consists of lines produced by radiative de-excitation from excited levels. The intensity of the spectral lines corresponds with the concentration of the element in the plasma	Flame-OES: 5–10 ml ICP-OES: 1–10 ml Arc-, spark-, and glow-discharge-OES are direct solid sampling methods Arc-OES: 1–50 mg Spark-OES: 10–30 mg GD-OES: ~10 mg (mainly surface/layer analysis)	Flame-OES: 1–1000 ng ml^{-1} ICP-OES: 0.1–50 ng ml^{-1} Spark-OES, GD-OES: 1–100 mg kg^{-1} Arc-OES, ETV-ICP OES: 0.01–10 mg kg^{-1}	Qualitative and quantitative analysis. Determination of traces to main constituents Sequential and simultaneous multielement analysis possible	Qualitative and quantitative elemental analysis of metals and alloys (directly especially by spark-OES), technical materials, environmental and biological samples

Atomic Absorption Spectrometry *AAS* *F AAS* *GF AAS* *(ET AAS)* *HG AAS* *SS ET AAS*	Measurement of the optical absorption spectra of atoms based on the absorption of radiation (emitted by a primary radiation source) by the atoms in the ground state in the gaseous volume of the atomized sample. The strength of the line absorption corresponds with the concentration of the element in the absorbing gaseous volume	Flame AAS (FAAS): 1 mg–10 ml, graphite furnace (GFAAS) (= electro thermal atomization AAS, ETAAS): 0.01–0.1 ml, hydride generation AAS (HG AAS): 0.5–5 ml, solid sampling (SSETAAS): 0.1–100 mg	F AAS: 1–1000 ng ml^{-1}, GF AAS: 0.01–1 ng ml^{-1}, HG AAS: 0.01–1 ng ml^{-1}, SS ET AAS: 0.01–100 ng g^{-1}	Quantitative analysis, sequential multielement analysis possible, with some spectrometers multielement analysis possible by new techniques, especially by high-resolution continuum source AAS	Quantitative analysis of elements in the trace region. Extremely low detection limits by ET AAS. Environmental samples and samples from technical materials. Combination with flow injection analysis (FIA) and solid sampling possible. AAS was in the near past certainly the most commonly used method of elemental analysis

(*continued overleaf*)

Tab. 4.1 *(continued)*

Analytical	Measurement principle	Sample volume/mass	Limit of detection	Output, results	Applications
X-ray Fluorescence Spectroscopy *XRF* *TXRF*	The sample is irradiated with X-rays. The atoms in the solid (or liquid) sample are excited and emit characteristic fluorescence X-rays. The energies (or wavelengths) of these characteristic X-rays are different for each element. Basis for quantitative analysis is the proportionality of the intensity of characteristic X-rays of a certain element and its concentration, but this relation is strongly influenced by the other constituents of the sample. The total reflection XRF (TXRF) is based on the effect of total reflection to achieve extremely high sensitivity	XRF: Direct analysis of polished metal disks, pellets of pressed powders, thin films, liquids, particulate material on a filter. Especially relevant in metrological exercises list the use of fused borate samples in combination with the so-called "reconstitution technique" TXRF: direct analysis of flat (< 10 μm) microsamples or of thin deposits	Strongly dependent from element and matrix as well as from spectrometer (wave length dispersive or energy dispersive). 1–100 mg kg^{-1} for elements with medium atomic number, up to 1–5% for the lightest measurable elements (B, Be). Advantages of XRF are high precision and wide dynamic range. TXRF: down to nanograms per milliliters and below for dried droplets of aquaous solutions	Qualitative and quantitative analysis. Semiquantitative results are easily obtained; truly quantitative results require careful sample preparation and calibration of the instrument	Qualitative and quantitative analysis of metals, alloys, cement, slags, rocks inorganic air pollutions, minerals, sediments, freeze-dried biological materials TXRF: all kinds of flat microsamples and residues of solutions

Activation Analysis *NAA* *PAA*	Activation Analysis is based on the measurement of the radioactivity of indicator radionuclides formed from stable isotopes of the analyte elements by nuclear reactions induced in an irradiation of the samples with suitable particles (neutrons: neutron activation NAA, high-energy photons: photon activation PAA)	Solid or liquid samples: ~50 mg to ~2 g, in some systems up to 1 kg	Trace and ultratrace region, strongly depending on the element, sample composition and the parameters of irradiation	Qualitative and quantitative analysis. Easy calibration procedure (low matrix influences), high sensitivity and freedom of reagent blanks together make the NAA a unique method regarding achievable limits of detection and accuracy. Homogenity of elemental distribution can be checked by using small subsamples. Important supplement to other methods, for example, to check losses or contamination from their wet-chemical sample preparation	Simultaneous trace and ultatrace multielement determination. Destructiveless analysis for ~70 elements Possible: Precision analysis of high element contents

Tab. 4.2 Instrumental methods for compound and molecular specific analysis.

Analytical method	Measurement principle	Specimen type	Sample amount	Output results	Applications
UV/VIS spectoscopy	Measurement of absorption of emission in the UV/VIS region due to changes of electronic transitions in π-system	Organic compounds, inorganic complexes, ions of transition elements	Solution in transmission, powder in reflexion	Relations between structure and color. Photochemical processes	Qualitative and quantitative analysis of aromatic and olefinic compounds. Detector for chromatographic methods
Fluorescence spectroscopy	Measurement of fluorescence emission in relation to the wavelength of the excited radiation	Fluorescent organic compounds, dyes, inorganic complex compounds	Solid, liquid or in solution, $\sim$50 µl	Structure/ fluorescence-relations; most sensitive opto-analytical method	Qualitative/ quantitative analysis of aromatic compounds with low-energy $\pi - \pi^*$ transitions (conjugated chromophors)

incorporates depth profiling and can detect atomic compositions significantly below 1 ppm. Static SIMS retains this high sensitivity for the surface atomic or molecular layer but provides chemistry-related details not available with AES or XPS.

4.2.4 Metrology in Elemental Analysis – An Example to Establish Traceability

Chemical measurements are measurements of quantities of substances, where the "chemical identity of the substance" has to be defined. For this purpose, a *metrological traceability system for measurement results of inorganic chemical analysis* was set up in Germany in cooperation between the National Metrology and Materials Research Institutes PTB and BAM (Kipphardt *et al.*, 2006). Currently, the system comprises primary elemental standards for Cu, Fe, Bi, Ga, Si, Na, K, Sn, W, and Pb. In this system, core components are as follows:

- pure substances (National Standards) characterized at the highest metrological level;
- primary solutions prepared from these pure substances; and
- secondary solutions deduced from the primary solutions intended for technical applications.

For certifying a sample representing one chemical element in the *System of National Standards*, all relevant trace elements of the periodic table have to be metrologically considered and their mass fractions have to be subtracted from 100% mass fraction (= the ideal mass fraction of the

Tab. 4.3 Instrumental methods for compound and molecular specific analysis.

Analytical method	Measurement principle	Specimen type	Sample amount	Output results	Applications
Nuclear magnetic resonance spectroscopy	Determination of chemical shift and coupling constants due to magnetic field excitation and analysis of rff–emission	Inorganic, organic and organometallic compounds: gaseous, liquid, in solution, solid	Samples in all aggregation states ~100 μg	Contribution to the molecular structure: bond length, bond angle, interactions about several bonds	Identification of substances in combination with other techniques (MS, IR, and Raman spectroscopy, chromatography)

Tab. 4.4 Instrumental methods for compound and molecular specific analysis.

Analytical method	Measurement principle	Specimen type	Sample amount	Output results	Applications
Analytical mass spectroscopy	Generation of gaseous ions from analyte molecules, subsequent separation of these ions according to their mass-to-charge (m/z) ratio, and detection of these ions. Mass spectrum is a plot of the (relatice) abundance of the ions produced as a function of the m/z ratio	Organic and organometallic compounds: quantitative mixture analysis	Samples in all aggregation states; MS is the most sensitive spectroscopic technique for molecular analysis	Determination of the molecular mass and elemental composition; contribution to the structure elucidation in combination with NMR, IR, and Raman spectroscopy	Structure elucidation of unknown compounds; MS as reference method and in the quantitation of drugs; detector for gas chromatographic (GC) and liquid chromatographic (LC) methods

Tab. 4.5 Instrumental methods for compound and molecular specific analysis.

Analytical method	Measurement principle	Specimen type	Sample amount	Output results	Applications
Infrared spectroscopy	Measurement of absorption (extinction) of radiation in the infrared region due to the modulation of molecular dipole moment	Inorganic, organic, and organometallic compounds: adsorbed molecules	Samples in all aggregation states; in solution, embedded or in matrix isolation, ~100 µmol	Contribution to the molecular structure: bond length, bond angle, force constants; identification of characteristic groups and compounds by means of data bases	Identification of substances in combination with other techniques (IR spectroscopy, Raman, MS, and chromatographhy); quantitative mixture analysis; surface analysis of adsorbed molecules; detector for chromatographic methods
Raman spectroscopy	Measurement of Raman scattering (in the UV–VIS–NIR● region) due to the modulation of molecular polarizability	Inorganic, organic, and organometallic compounds, surfaces and coatings	Samples in all aggregation states, in solution, and in matrix isolation: ~50 µl, ~1 µg	Contribution to the molecular structure; identification of characteristic groups and compounds in combination with IR data	Identification of substances; surface and phase analysis; detector for thin layer chromatographic methods
Electron paramagnetic resonance spectroscopy	Selective absorption of electromagnetic microwaves due to reorientation of magnetic moments of single electrons in a magnetic field	Organic radicals, reactive intermediates, internal defects in solids in biomatrices	*In situ* investigation of organic radicals, oriented paramagnetic single crystals, crystal powders	*In situ* detection of organic radicals and their reaction kinetics (time-resolved EPR). Paramagnetic centers of crystals	Studies of photochemical and photophysical processes (requirement: 10^{11} single electrons); semiconductor studies, trace detection of 3D elements in biomaterials
Moessbauer spectroscopy	Measurement of the isomeric shift, line width, and line intensity due to the recoil free gamma quantum absorption	Inorganic compounds and phases, organometallic compounds	Samples in the solid state, or as freezing solutions, several milligrams	Determination of the oxidation number and the spin state of the Moessbauer nuclei	Phase analysis including amorphous phase on glasses, ceramics, and catalysts

Tab. 4.6 Methodology selection table for surface analysis of materials.

Method	AES	XPS	Dynamic SIMS	Depth profiling SIMS	Static SIMS
Best spatial resolution	5 nm	20 μm	50 nm	50 μm	80 nm
Best depth resolution (nm)	0.3	0.3	3	0.3	0.3
Approximate sensitivity	0.5%	1%	0.01 ppm	0.01 ppm	0.01%
Typical information depth (nm)	10	10	3	3	0.5
Information					
Elemental	x	x	x	x	x
Isotopic			x	x	x
Chemical		x			x
Molecular		x			x
Material					
Metal	x	x	x	x	x
Semiconductor	x	x	x	x	x
Insulator or ceramic		x	x	x	x
Polymer, organic or bio		x			x

investigated element) to establish the real mass fraction of the main component with an uncertainty below 0.01%. This value of the uncertainty is one order of magnitude lower than the uncertainties achieved with direct measurements of the mass fraction of the main component using best metrological methods of elemental analysis, such as ID-MS. Also for ID-MS measurements, the National Standards of Elemental Analysis are intended to be used in the future as instruments of metrological traceability, namely, by using them as "natural backspikes for ID-MS" of known mass fraction of the main component. To determine all trace elements in the pure materials, different methods of elemental analysis have to be applied. About 70 metallic impurities can be determined using inductively coupled plasma with high-resolution mass spectrometry (ICP-HRMS). For supplementation and validation, inductively coupled optical emission spectroscopy (ICP-OES) and atomic absorption spectrometry (AAS) are used. Classical spectrophotometry is applied for the determination of phosphorous, sulfur, and fluorine. Classical carrier gas hot extraction (CGHE) is used to determine oxygen and nitrogen and combustion analysis is used for carbon and sulfur. In addition to C, O, and N, chlorine, bromine, and iodine are determined using photon activation analysis (PAA). Hydrogen is measured using nuclear reaction analysis (RNA). If possible, also direct methods, typically, electrogravimetry (e.g., for copper) or coulometry (see Section 4.3.1) are applied in order to determine the mass fraction of the matrix directly, but, of course, with a higher uncertainty.

For those high-purity materials, which are most convenient to handle in the form of salts (alkali metals and alkaline earth metals), additionally to the measurements of the metallic and nonmetallic impurities, the anionic impurities are also determined. These measurement results are consistent with those of the other element-specific methods for I, Br, S, P, or N. For the radioactive elements, Tc, Pm, Po, At, Rn, Fr, Ra, Ac, and Pa upper

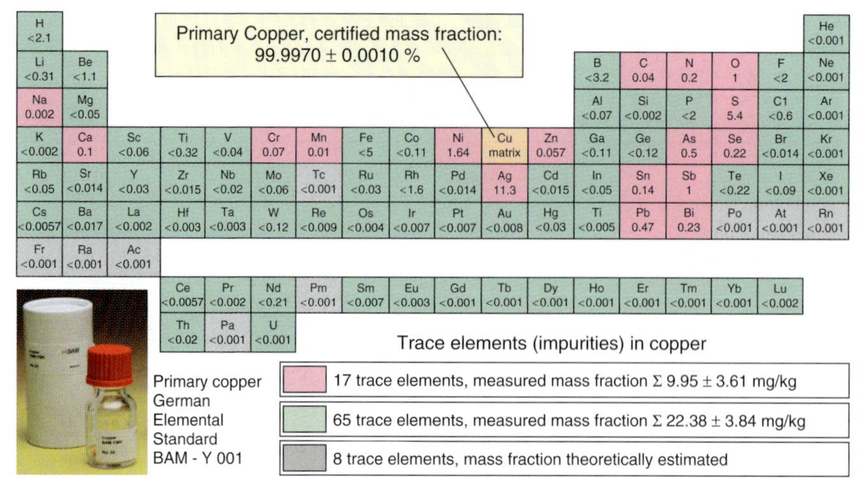

Fig. 4.14 Metrology in elemental analysis: the example of copper as primary standard; mass fractions of trace elements in milligrams per kilograms.

limits are estimated from theoretical considerations. For noble gases, either measurements are carried out by static mass spectroscopy or limits are also estimated from theoretical considerations. All uncertainties are calculated according to the *ISO Guide to the Expression of Uncertainty in Measurement* (*GUM*).

An example of this metrology-based system for elemental materials characterization is given in Figure 4.14, showing the certified mass fraction data of *primary copper* together with the data of all measured trace elements of the periodic table of chemical elements (Kipphardt *et al.*, 2006). The certified mass fraction of primary copper results from subtracting the mass fractions of all impurities from 100%.

4.3 Characterization of Materials Microstructure

The term *materials microstructure* denotes the crystalline or amorphous arrangements of materials chemical constituents as well as the collection of deviations from the ideal bulk or surface structures (Czichos, Saito, and Smith, 2006d). The microstructure is closely connected with materials composition and is fundamentally influencing the properties and performance of materials treated in Chapters 4 and 5.

4.3.1 Microstructural Analysis

As already mentioned in Section 4.2.4, the elementary physical structure of a material can be crystalline, amorphous, or a mixture, resulting from *short-range* or *long-range* bonding actions between the elementary chemical materials constituents described in Section 7.1.2.1. An overview on the different types of materials microstructures and the analytical techniques for their study is given by considering as examples the classes of polymeric and metallic materials.

Different from the more or less crystalline structural order of metals and ceramics, a polymeric material has no fixed configuration but takes a variety of forms. Owing to the flexibility of polymeric chains

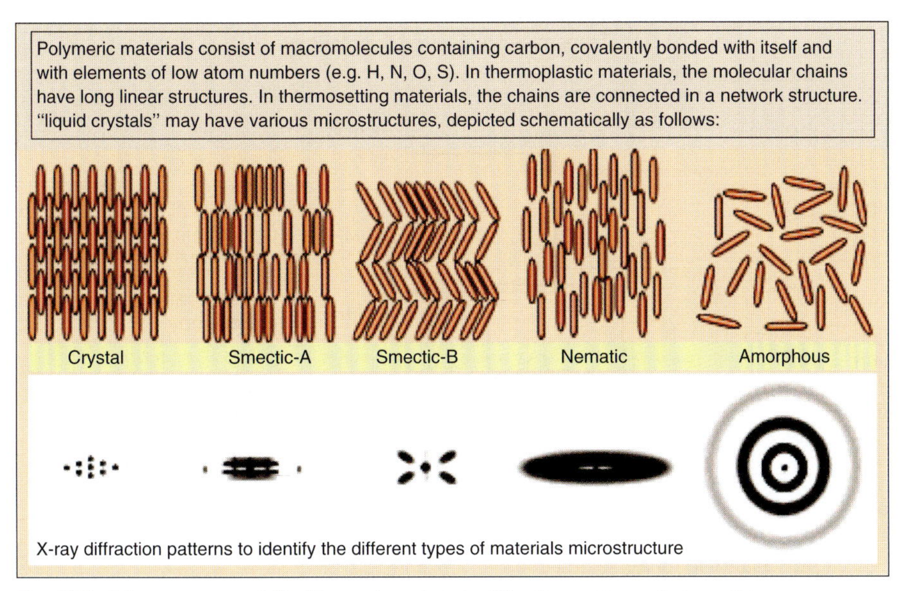

Fig. 4.15 Microstructure of *liquid crystals* and their diffraction patterns (schematic).

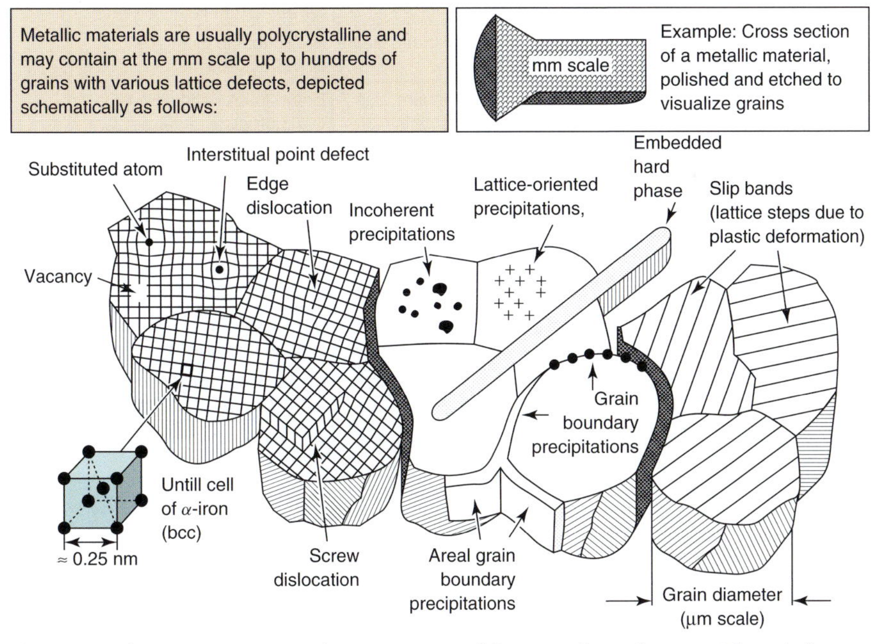

Fig. 4.16 Schematic overview on the microstructural features of metallic materials and alloys.

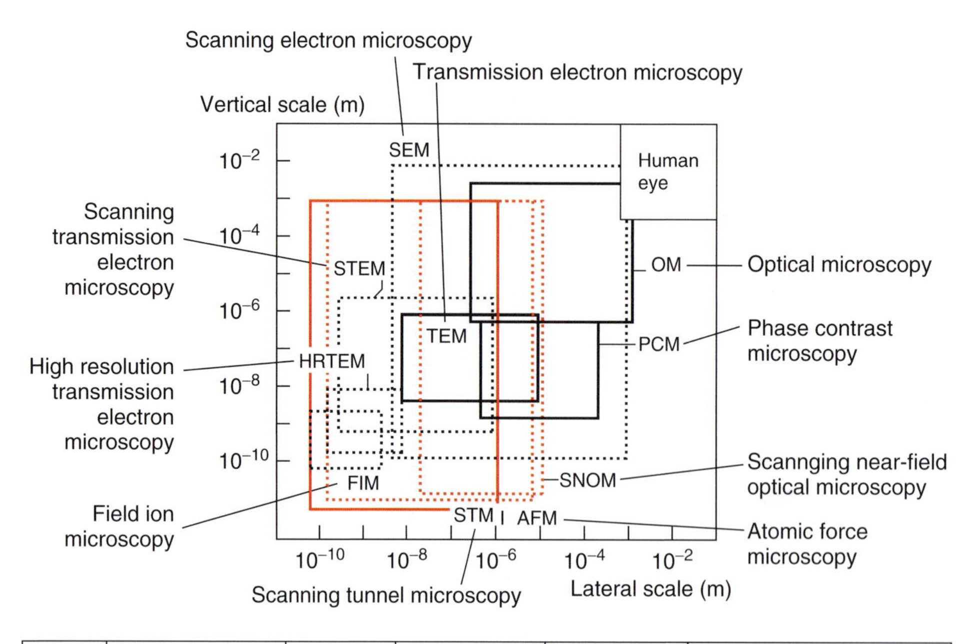

Method	Object	Resolution	Sample requirement	Test environment	Specification
OM	Surface morphology	$< 1\ \mu m$	Transparent or reflective	Ambient, liquid	Easy operation, wide survey range, dynamic observation
PCM	Organic microstructures	$< 1\ \mu m$	Transparent	Ambient, liquid	Applicable to biological cell structures
SEM	Surface morphology	< 1 nm	Electrically conductive	Vacuum, gas	Broad lateral and vertical scale range
TEM	Dislocations, planar defects	< 100 nm	Thickness $< 1\ \mu m$	Vacuum	Crystallographic analysis, dynamic observation
STEM	Extended defects, impurities	< 1 nm	Thickness < 100 nm	Vacuum	Insensitivity to focusing and sample thickness
HRTEM	Extended defects, layer stacking	≈ 0.2 nm	Thickness < 100 nm	Vacuum	Atomic-level resolution, dynamic observation
FIM	Surface atoms	< 0.1 nm	Conductive thin needles	UHV	Atomic resolution, difficult sample preparation
STM	Surface states, subsurface defects	< 0.1 nm	Conductive clean surface	Ambient, UHV, liquid	Affinity with various micro-spectroscopic techniques
AFM	Surface atoms, surface morphology	< 0.1 nm	Roughness $< 1\ \mu m$	Ambient, UHV, liquid	Applicable to nonconductive samples
SNOM	Surface atoms, surface morphology	< 100 nm	Optically active	Ambient, UHV, liquid	Flexible applicability to non- and conductive samples

Fig. 4.17 Microscopic techniques for materials microstructural analyses: scale range and basic technical data.

and the tendency for mutual entanglement, polymers are likely to solidify in an amorphous or glassy form and are difficult to crystallize perfectly. Thus, it is common that the ordered structure, if present, extends only to a limited range. Between completely three-dimensional crystalline and glassy arrangements, some molecular solids have structures in which the crystalline order is lost in some dimensions. For example, in *liquid crystals* the molecules are commonly elongated rigid rods with a typical length of flattened cross section of 0.6–0.4 nm. Figure 4.15 shows various forms of liquid crystals in different degrees of order; the lower drawings illustrate the schematic X-ray diffraction patterns of these phases (Czichos, Saito, and Smith, 2006d).

Apart from ideal theoretical crystalline and amorphous arrangements, the "real microstructure" of materials is of foremost importance for the properties and performance of all types of materials. As also already mentioned in Section 7.1.2.1, the deviations from "ideal solid matter configurations" are categorized as follows:

- **Zero-dimensional lattice deviations or point defects:** vacancies, interstitials, or atomic substitutes:
  - influencing, for example, the conductance of semiconductors.
- **One-dimensional lattice deviations or line defects:** dislocations in the form of missing or additional rows of atoms:
  - responsible for the elementary steps of plastic deformation.
- **Two-dimensional lattice deviations or area defects:** grain and phase boundaries:
  - influencing, for example, the mechanical strength of materials.
- **Three-dimensional lattice deviations or volume defects:** cavities and precipitates:
  - influencing, for example, thermal properties of materials.

In Figure 4.16 an overview of the microstructural features of materials is depicted schematically.

To characterize the microstructural features of materials with lateral or spatial resolution, various microscopic techniques are applied. Figure 4.17 gives an overview of the lateral and vertical scales covered by the different methods together with basic technical data of these techniques (Czichos, Saito, and Smith, 2006d).

4.3.2 Surface Topography Analysis

Surface topography is – besides surface chemistry, discussed in Section 7.2.3 – of paramount importance for the functional behavior of a materials surface, strongly interplaying with material properties and operating conditions. The fundamental topographical features of materials surfaces, which have to be characterized by appropriate measuring and testing techniques, are *surface roughness* and subsurface *microstructure* (Czichos, Saito, and Smith, 2006c), as illustrated schematically in Figure 4.18.

- **Surface roughness** is the general term to describe the microgeometrical features of a materials surface. The international standard International Organization for Standardization (ISO) 4287 defines three series of *surface two-dimensional parameters*: (i) P-parameters for the unfiltered profile, (ii) R-parameters for the roughness profile, and (iii) W-parameters for the waviness profile. The *bearing curve* is a graph of the percentage of cross-sectional horizontal peak lines from the highest surface-roughness-peak to the deepest surface roughness valley.

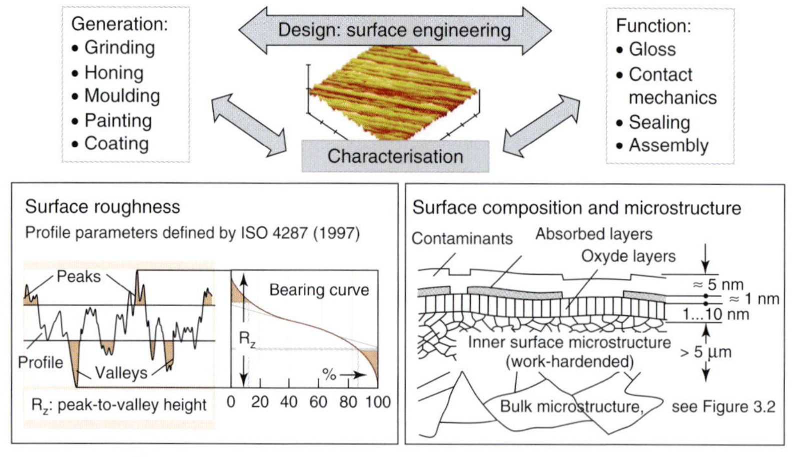

Fig. 4.18 Surface topography characterization links design, generation, and function.

- **Surface composition and microstructure** may significantly differ from materials bulk characteristics, depending on the surface generation process and physicochemical interactions with the environmental atmosphere. As illustrated schematically in the lower right part of Figure 4.18, materials are usually covered by contaminants and physically adsorbed or chemically bonded layers of nanometer thickness; metallic materials have also oxyde layers as well. The inner surface microstructure, which adjoins the materials bulk volume, is generated by the materials' shaping process. It may consist of "work-hardened" layers of several 10 µm, differing considerably from the bulk microstructure, with respect to grain size and ("frozen-in") residual strains, that is, distortions of crystallinity.

Specific methods for surface topography measurements are *stylus profilometry*, *optical interferometry*, *scanning electron microscopy* (*SEM*), and *atomic force microscopy* (*AFM*); Figure 4.19 gives an overview on data and experimental principles.

- **Stylus profilometry:** The pick-up draws a stylus (diamond tip, 60 or 90°, radius 1–10 µm) over the surface, and an electrical signal analog to the surface profile (see Figure 4.18) is produced by a piezoelectric, inductive or laser interferometric transducer. The spatial resolution achieved by this method, generally in the range 2–20 µm, is limited by tip geometry and local plastic deformation.
- **Optical interferometry:** Interference microscopy combines an optical microscope and an interferometer objective into a single instrument. These optical methods allow fast noncontacting measurements on essentially flat surfaces. Interferometric methods offer a subnanometer vertical resolution, being employed for surfaces with average roughnesses down to 0.1 nm and peak-to-valley heights up to several millimeters.
- **Scanning electron microscopy:** SEM allows an excellent visualization of surface topographies achieved through the very high depth of focus of this

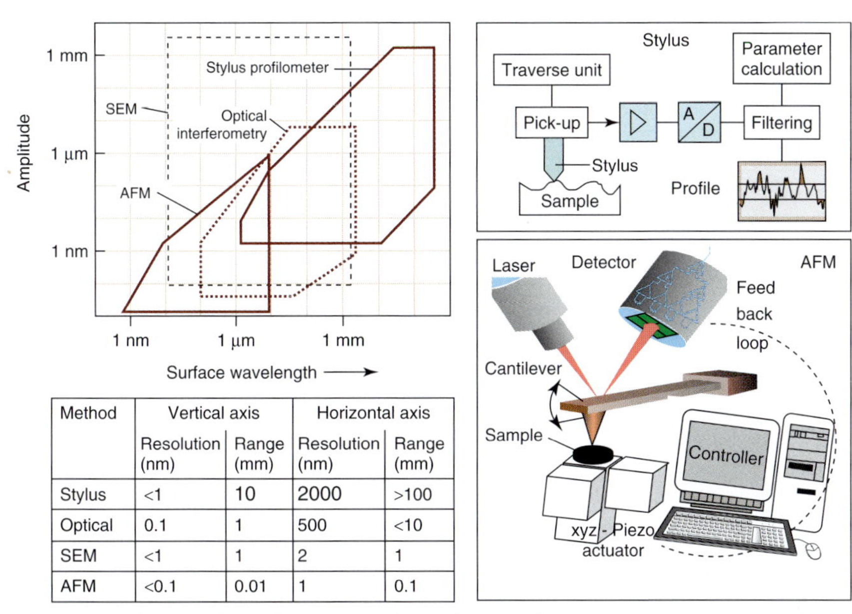

Method	Vertical axis		Horizontal axis	
	Resolution (nm)	Range (mm)	Resolution (nm)	Range (mm)
Stylus	<1	10	2000	>100
Optical	0.1	1	500	<10
SEM	<1	1	2	1
AFM	<0.1	0.01	1	0.1

Fig. 4.19 Techniques for surface topography analyses: scale ranges and basic technical data (left) and principles of stylus profilometry and atomic force microscopy (right).

technique. SEM measurements require conductive sample materials, or sample preparation through deposition of a gold layer on the surface. Possible magnification levels from less than 100 up to 100 000 either be on the order of 1 mm^2 or just 1 µm^2. There is no other microscopy method as flexible as SEM in terms of the range of scalability.

- **Atomic force microscopy:** In an AFM, as illustrated in the bottom right part of Figure 4.19, a sharp tip with a radius of approximately 5–20 nm, mounted on a very soft cantilever with a spring constant in the range of 1 N m^{-1}, is scanned over the surface by a *xyz* actuator with a resolution of much less than a nanometer and a dynamic range on the order of 10 µm in the *z* direction and up to 100 µm in the *x* and *y* directions. At the start of a measurement, the cantilever is positioned toward the sample. When the tip touches the sample, the cantilever begins to bend proportionally to the force exerted by the tip on the surface. The height of the probe, detected with a laser-detector unit as function of the *x* and *y* position in a feed back loop, is recorded as an image and the topography of the surface is built up into a three-dimensional image.

4.4 Characterization of Materials Properties

The properties of materials must be obtained by experiment methods. Materials are too complicated and theories of solids insufficiently sophisticated to obtain accurate theoretic determinations. In Chapter 1 it has been pointed out that materials properties depend on composition, microstructure, scale, and so on. Materials properties

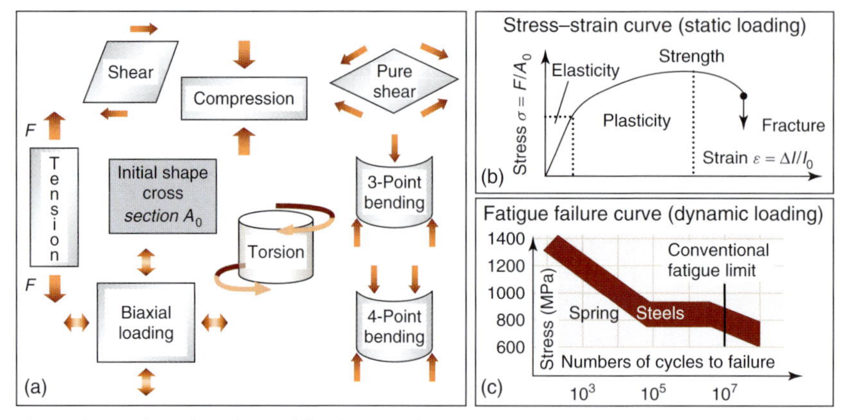

Fig. 4.20 Modes of mechanical loading used to measure mechanical properties of materials (a) and the typical stress–strain curve of ductile materials for quasistatic loading (b) and the fatigue failure curve (S–N curve or Wohler curve) for dynamic loading (c).

may also be dependent on temperature and time ("materials aging"). In this chapter, the fundamentals of measurement and testing to determine basic physical properties of materials – mechanical, thermal, electrical, magnetic, and optical – are compiled (Czichos, Saito, and Smith, 2006e).

4.4.1 Mechanical Properties

The mechanical properties of materials characterize the response of a material to an external loading. The mechanical loading modes are categorized as *tension, compression, bending, shear, and torsion* (see Figure 4.20).

As illustrated in the Figure 4.20b, *stress*, $\sigma = F/A_0$, gives the intensity of a mechanical force F that passes through the body's cross-sectional area A_0. *Strain* $\varepsilon = \Delta l/l_0$ gives the relative displacement of points within the body. Stress – strain curves depend basically on materials composition and microstructure and are influenced by the loading manner, strain rate, temperature, and chemical environment. They show typically three different regimes:

- First, when the applied load is small, deformation is reversible, that is, elastic deformation occurs. Stresses are proportional to elastic strains (Hooke's law: $\sigma = E \cdot \varepsilon$). The slope between tensile stress and tensile elastic strain is called the *Elastic (or Young's) modulus E*. A material with high E-modulus is strong or rigid to elastic deformation.
- Second, when the applied stress becomes higher, materials usually show plastic deformation as result of submicroscopic dislocation movements (see Figure 4.16). The stress at the onset of plastic flow is called the *yield strength*, which has no direct relation with Young's modulus. The yield strength indicates resistance against plastic flow, and the ultimate tensile strength is an important mechanical property.
- Third, materials show fracture under applied stress. There are many fracture mechanisms, and the fracture strength is different from the yield strength.

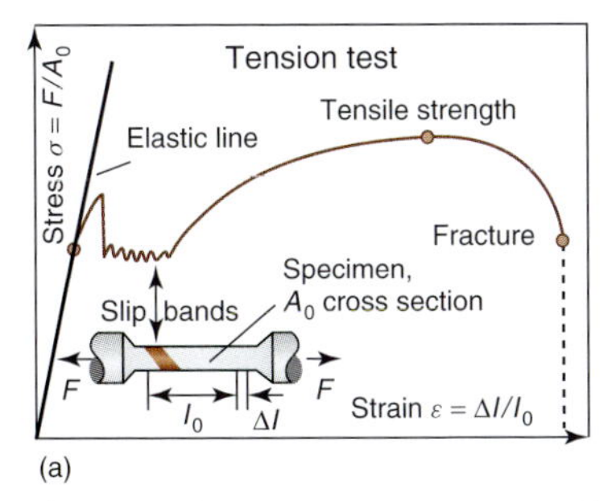

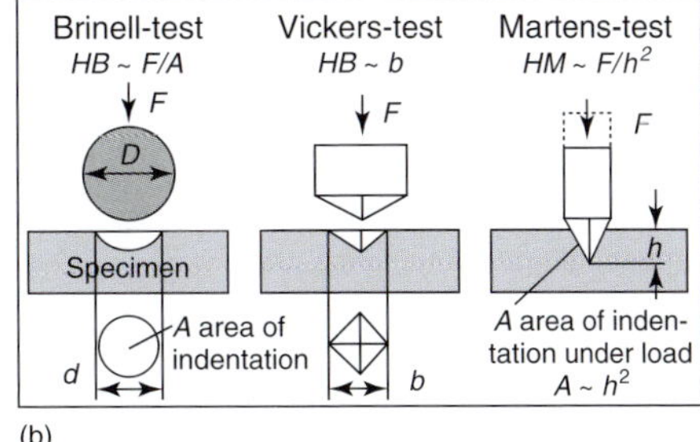

(a) (b)

Fig. 4.21 (a) Tension test: stress–strain curve of a typical engineering material (e.g., mild steel) showing discontinuous yielding (due to grain-size-dependent slip bands). (b) Hardness tests.

Under dynamic loading, the "strength" of materials decreases with the number of load cycles (see Figure 4.20c). At cyclic loading with small stresses below the yield strength, slip occurs locally and repeatedly (see Figure 4.16), resulting in "intrusions," "surface dents" at which stress is concentrated. Plastic flow occurs locally at the tip of then formed "microcracks," which may gradually grow. When the crack length has increased enough to satisfy the fracture condition given by the fracture mechanics approach (see Section 7.5.2.5), final fracture takes place. The stress at 10^7 cycles is called the *fatigue strength* or *endurance limit* σ_W. At stresses higher than σ_W, the number determines the time to fracture and is called *fatigue life*. Thus, the fatigue strength and fatigue life are utilized for the design of machines or structures.

Elementary mechanical properties, such as elastic modulus, yield strength, tensile strength, elongation, can be obtained from the results of *tension tests* (see Figure 4.21). They are usually performed at room temperature, with a strain rate of 10^{-4} – 10^{-1} per second. The testing procedure and the shape and dimensions of a tensile specimen should be taken from the ISO or other related standards. The loading speed, or extending speed, should be controlled within the strain-rate range using a gear-driven-type or servo-hydrostatic-type tester. During testing, the applied load and displacement of a gauge length of a tensile specimen should be recorded. Then, the load – displacement diagram can be obtained. The curve is converted to a nominal stress – strain curve.

Hardness is defined as the resistance of a material to permanent penetration by another material. Hardness tests are categorized as follows (see Figure 4.21b):

- **Brinell hardness test:** An indenter (hardmetal ball with diameter D) is forced into the surface of a test piece and the diameter of the indentation d left in the surface after removal of the force F is measured. The Brinell hardness (HB) is proportional to the quotient obtained by dividing the test force by the surface area of the indentation.
- **Vickers hardness test:** A diamond pyramid indenter with a square base and with a specified angle between opposite faces at the vertex is forced into the surface of a test piece, followed by the measurement of the diagonal length of the indentation left in the surface after removal of the test force F. An experimental relationship has been found to connect the Vickers hardness

number (HV) to the yield or tensile strength (Y) of materials: $HV = cY$, where the coefficient c is 2.9–3.0. Therefore, the Vickers hardness test is not a precise, but a convenient way to check approximately the strength of materials, particularly for engineering products.

- **Martens hardness test:** Martens hardness is measured under an applied test force. Martens hardness is determined from the values given by the force/indentation depth curve when the test force increases, preferably after reaching the specified test force. Martens hardness includes the plastic and elastic deformation; thus, this hardness value can be calculated for all materials.
- **Mechanical properties:** These – especially the combination of elasticity, strength, and weight – are of paramount importance in the selection of structural materials for engineering applications. Figure 4.22 shows elasticity/weight and strength/weight maps for the main classes of engineering materials (Czichos *et al.*, 2007).

4.4.2 Thermal Properties

Thermal properties are associated with a material-dependent response when heat is supplied to a solid body. This response might be a temperature increase, a phase transition, a change of length or volume, an initiation of a chemical reaction, or the change of some other physical or chemical quantity. Basically, almost all of the other materials properties, namely mechanical, electrical, magnetic, and optical properties, are temperature dependent. (For example, temperature influences mechanical hardness, electrical resistance, magnetism, or optical emissivity.) There are two main thermophysical properties associated with heat transfer in materials:

- **Heat capacity:** If a specific amount of heat ΔQ is supplied to a thermally isolated specimen of mass m, the relationship between heat and temperature increase ΔT is given by $\Delta Q = mc_p\Delta T$. Consequently, the ability of a material to store heat is characterized by the specific heat capacity at constant pressure c_p. If two equal volumes of different materials are compared to each other, the ability to store heat is described by the product of density ρ and specific heat capacity c_p. This product ρc_p is used if simultaneous heat conduction and storage processes are investigated by means of the transient heat conduction equation. In many cases, the density of a material or the mass of a specimen is well known or can be easily determined with sufficient accuracy. Therefore, the problem is reduced to the determination of the specific heat capacity.
- **Thermal conductivity:** λ is the material property associated with heat conduction and is defined by Fourier's law $q = -\lambda\Delta T/d = \Delta T/R_{\text{th}}$. Here, q is the heat flux, the heat conducted during a unit time through a unit area, driven by a temperature gradient. In practice, for thermal conductivity measurements, the temperature difference ΔT between two opposite surfaces of a sample with a separation of d is determined. The quotient of distance and thermal conductivity is the thermal resistance R_{th}.

On the basis of the fundamental laws of heat conduction and storage, different approaches to measure thermophysical properties of materials can be distinguished (Czichos, Saito, and Smith, 2006f):

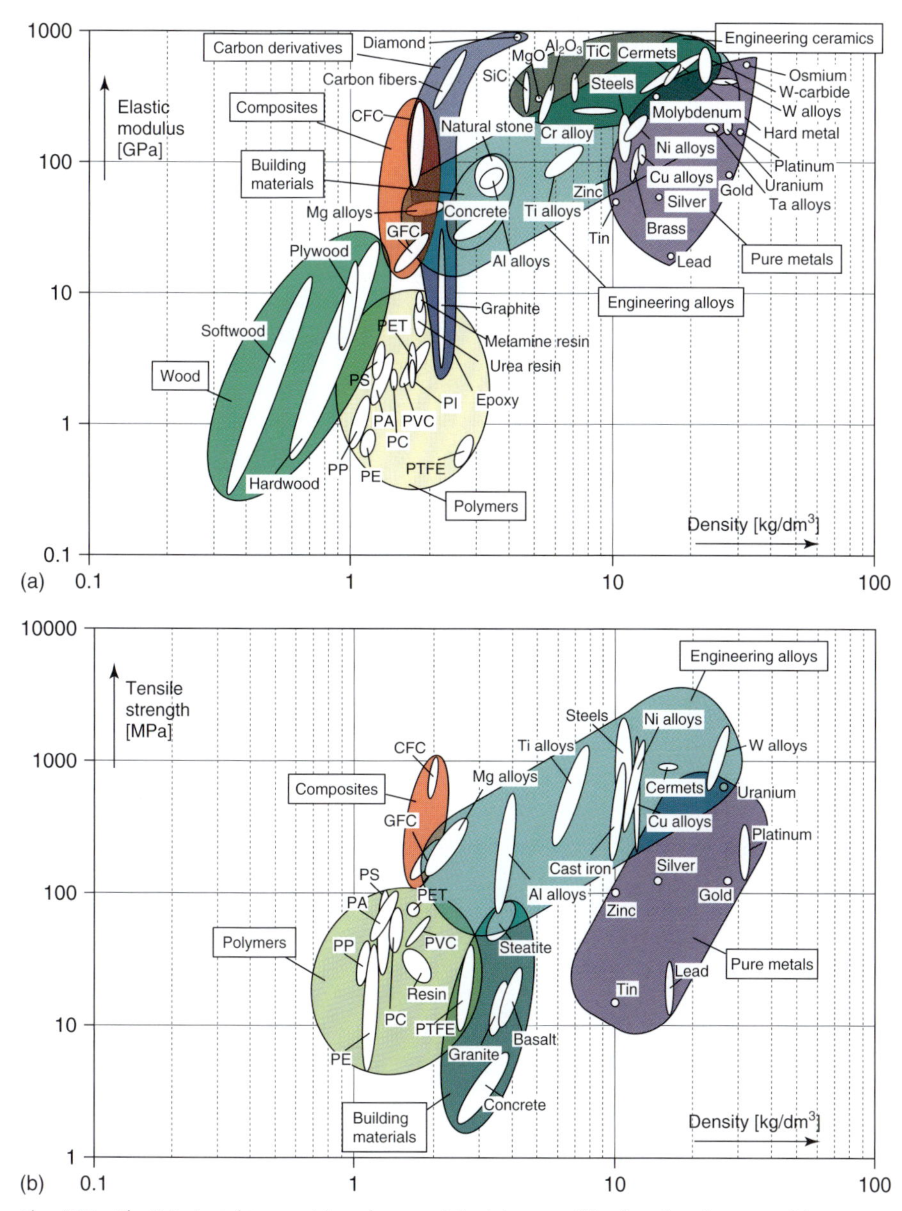

Fig. 4.22 Elasticity/weight map (a) and strength/weight map (b) of engineering materials.

Tab. 4.7 Measurement methods for the determination of thermal conductivity and thermal diffusivity.

Method	Temperature range	Uncertainity (%)	Materials	Merit	Demerit
Guarded hot plate	80–800 K	2	Insulators, glasses, polymers	High accuracy	Long measurement time, large specimen size, low conductivity material
Cylinder	4–1000 K	2	Metals	Temperature range, simultaneous determination of electrical conductivity possible	Long measurement time
Heat flow meter	−100 to 200°C	3–10	Insulators, glasses, polymers	Simple construction and operation	Measurement uncertainity, relative measurement
Comparative	2–1300 °C	10–20	Metals, ceramics, polymers	Simple construction and operation	Measurement uncertainty, relative measurement
Direct heating (Kohlrausch)	400–3000 K	2–10	Metals	Simple, fast measurements, simultaneous determination of electrical conductivity	–
Pipe method	20–2500 °C	3–20	Solids	Temperature range	Specimen preparation, long measurement time
Hot wire hot strip	20–2000 °C	1–10	Low conductivity solids	Temperature range, fastness, accuracy	Limited to low conductivity materials
Laser flash	−100 to 3000°C	3–5	Solids, liquids	Temperature range, small specimen, fastness, accuracy	Expensive, not for insulation materials

- **Steady-state methods:** using $q = -\lambda \Delta T/d = \Delta T/R_{th}$ in order to determine the thermal conductivity. Steady-state conditions mean that the temperature at each point of the sample is constant, that is, not a function of time. The determination of the thermal conductivity λ is based on the measurement of a heat flux q and a temperature gradient ΔT, that is, mostly a temperature difference between opposite surfaces of a sample.
- **Calorimetric methods:** based on $\Delta Q = mc_p\Delta T$, if the specific heat capacity c_p is the property of interest. In this case, heat ΔQ is supplied to a sample that is isolated from the surroundings, and the change of the (mean) sample temperature ΔT is measured.
- **Transient techniques:** to determine simultaneously both thermal conductivity λ and capacity c_p. For that purpose, numerous solutions of the transient heat conduction equation based on one-, two-, or three-dimensional geometries have been derived.

The main experimental problem for all methods for the determination of thermal properties is that ideal thermal conductors or insulators do not exist. In comparison to electrical transport properties (see Section 4.5.3), the ratio of thermal conductivities of the best conductors and the best insulators is many orders of magnitudes smaller (see Figure 4.8). Therefore, instruments for the determination of thermal properties are often optimized or restricted for a specific class of materials or temperature range. Table 4.7 gives an overview of the most important methods used for the measurement of thermal conductivity and thermal diffusivity (Czichos, Saito, and Smith, 2006f).

4.4.3 Electrical Properties

The theoretical basis for the characterization of the electrical properties of materials is the *electron band model*, illustrated in a simplified manner in Figure 4.23.

According to Bohr's atom model, electrons of isolated atoms (e.g., in a gas) can be considered to orbit at various distances about their nuclei, illustrated schematically by different energy levels (see Figure 4.23a). These distinct energy levels, which are characteristic for isolated atoms, widen into energy bands when atoms approach each other and form a solid. Quantum

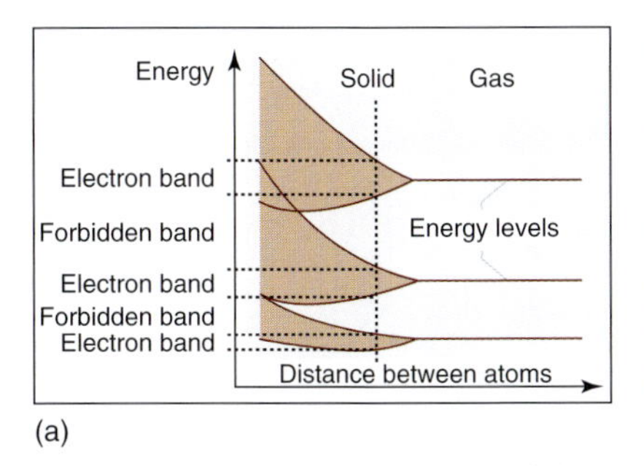

(a)

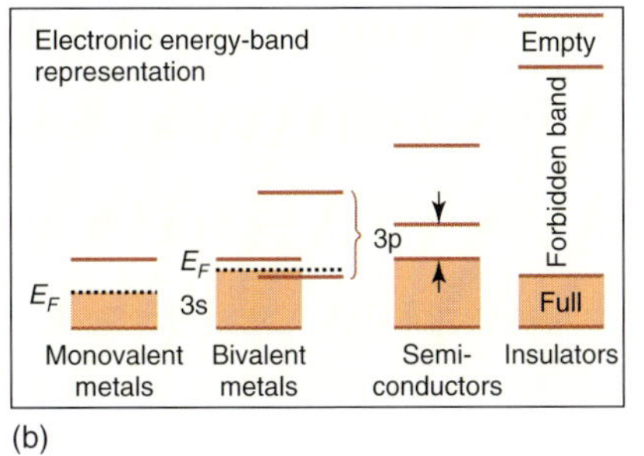

(b)

Fig. 4.23 Schematic representation of electron energy levels (a) and the electronic energy-band representation of materials (b).

mechanics postulates that electrons can only reside within these bands, but not in the areas outside of them. The highest level of electron filling within a band is called the *Fermi energy* E_F. As shown in Figure 4.23b, the electron energy-band representation classifies the electrical properties of materials as follows:

- Monovalent metals (such as copper, silver, and gold) have partially filled bands. Their electron population density near the Fermi energy is high, which results in a large conductivity.
- Bivalent metals are distinguished by overlapping upper bands and by a small electron concentration near the bottom of the valence band. As a consequence, the electron population near the Fermi energy is small, which leads to a comparatively low conductivity. For alloys, the residual resistivity increases with increasing amount of solute content.
- Semiconductors have completely filled electron bands, but at elevated temperatures, the thermal energy causes some electrons to be excited from the valence band into the conduction band. They provide there some conduction. The electron holes, which have been left behind in the valence band, cause a hole current, which is directed in the opposite direction compared to the electron current. The number of electrons in the conduction band can be considerably increased (and thus the conductivity "tailored") by adding, for example, to silicon small amounts of Group-V elements called *donor atoms*.
- Insulators have completely filled and empty electron bands, which result in a virtually zero population density. Thus, the conductivity in insulators is virtually zero.

Measurement principles for the determination of electrical properties of materials may be categorized with respect to two basic processes: electrical energy conduction (and dissipation) and electrical energy storage (Czichos, Saito, and Smith, 2006e) (see Figure 4.24).

Electrical conductivity measurements describe the ability of a material to transport charge through the process of conduction, normalized by geometry. Electrical dissipation comes as the result of charge transport or conduction. Dissipation or energy loss results from the conversion of electrical energy to thermal energy (Joule heating) through momentum transfer during collisions as the charges move. According to their electrical conductivity σ – and explained by the electron band model of Figure 4.23 – materials are divided into

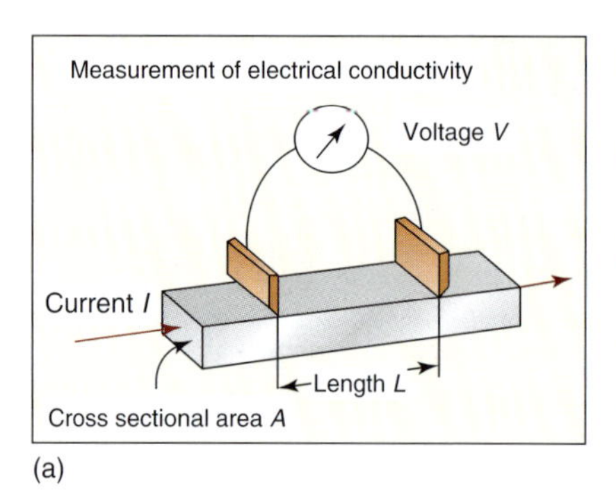

(a)

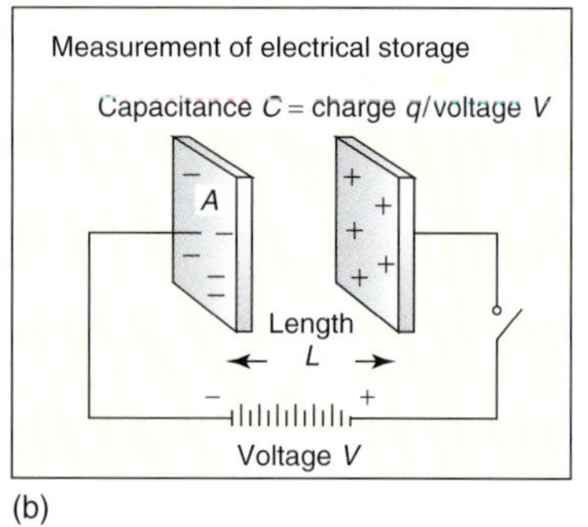

(b)

Fig. 4.24 Principles of the measurement of electrical conductivity (a) and electrical storage (b).

conductors, semiconductors, and insulators (dielectrics). The inverse of the conductivity is called *resistivity*, that is, $\rho = 1/\sigma$. The resistance R of a piece of conducting material is proportional to its resistivity and to its length L and is inversely proportional to its cross-sectional area $A : R = L\rho/A$. To measure the electrical resistance, a direct current is applied to a slab of the material. The current I through the sample (in ampères), as well as the voltage drop V on two potential probes (in volts), is recorded as depicted in Figure 4.24a. The resistance (in ohms) can then be calculated by using Ohm's law $V = RI$. Another form of Ohm's law $j = \sigma E$ links current density $j = I/A$, that is, the current per unit area (amperes per square centimeters), with the conductivity (Ω^{-1} cm^{-1} or siemens per centimeter) and the electric field strength $E = V/L$ (volts per centimeter).

Electrical storage measurements characterize the result of charge storing energy. This process is dielectric polarization, normalized by geometry to be the material property called *dielectric permittivity*. As polarization occurs and causes charges to move, the charge motion is also dissipative. When a voltage is momentarily applied to two parallel metal plates that are separated by a distance L, as shown in Figure 4.24b, the resulting electric charge essentially remains on these plates even after the voltage has been removed (at least as long as the air is dry). This ability to store an electric charge is called the *capacitance* C, which is defined to be the charge q per applied voltage V, that is, $C = q/V$, where C is given in coulombs per volt or farad. The higher the capacitance, the larger the area A of the plates and the smaller the distance L between them. Further, the capacitance depends on the material that may be inserted between the plates. The experimental observations lead to $C = \varepsilon\varepsilon_0(A/L)$, where $\varepsilon = C/C_{vac}$ determines the magnitude of the added storage capability. It is called the *dielectric constant*. ε_0 is a universal constant having the value of 8.85×10^{-12} farad per meter (F m^{-1}) or As V^{-1} m^{-1} and is known by the name permittivity of vacuum.

The most characteristic electrical property, namely, the electrical conductivity of materials at room temperature, spans more than 25 orders of magnitude as depicted in Figure 4.8. Moreover, if one takes the conductivity of superconductors, measured at low temperatures, into consideration, this span extends to 40 orders of magnitude. This is the largest known variation in a physical property.

4.4.4 Magnetic Properties

Magnetic materials are classified according to their response to an applied magnetic field, that is, the magnetization induced by the external field. A more fundamental understanding results from considering the microscopic mechanisms that determine the behavior of materials in the magnetic field.

- **Diamagnetism:** is a property of all materials. It results from an additional orbital angular momentum of electrons induced in a magnetic field and manifests itself as a magnetization oriented in the opposite direction to the external field.
- **Paramagnetism:** means a positive magnetic susceptibility, that is, the induced magnetization is along the direction of the external magnetic field. It is observed in materials that contain atoms with nonzero angular momentum, for example, transition metals. The susceptibility scales with $1/T$ (Curie's law). Paramagnetism is

also found in certain metals (e.g., aluminum) as a consequence of the spin of conduction electrons (Pauli spin susceptibility). The related susceptibility is essentially independent of temperature. The paramagnetic susceptibility is generally several orders of magnitude larger than the diamagnetic component; therefore, diamagnetism is not observable in the presence of paramagnetism.

- **Ferromagnetism:** describes the fact that certain solid materials exhibit a spontaneous magnetization, even in the absence of an external magnetic field. It is a collective phenomenon resulting from the spontaneous ordering of the atomic magnetic moments due to the exchange interaction among the electron spins. In general, the magnetization of a ferromagnetic object is preferentially oriented along certain axes that correspond to an energy minimum, which are known as *easy axes of magnetization*. The directions of the respective energy maxima are called *hard axes* or *hard directions*. This magnetic anisotropy is the key to many technical applications of ferromagnetic materials; for example, a binary memory element requires a uniaxial anisotropy characterized by two stable states. The strength of the anisotropy is expressed by anisotropy constants; it can be measured by the external field required to rotate the magnetization from an easy direction to a hard one. Materials with small/large anisotropy constants are called *magnetically soft/hard*.

The basic quantities to characterize the magnetic properties of materials are as follows:

- *Magnetic field H* and *flux density B* (which are equivalent in free space) are measured with Hall probes, inductive probes (e.g., flux-gate magnetometers for very low fields), or NMR probes for very high accuracy.
- *Magnetic moments* are measured (i) via the force exerted by a magnetic field gradient (e.g., Faraday) or (ii) via the voltage induced in a pick-up coil by the motion of the sample (e.g., the vibrating-sample magnetometer).
- *Magnetization* of a magnetic object (average) is usually determined by dividing the magnetic moment by the volume of the sample. Care must also be taken concerning the vector component of the magnetic moment that is detected.
- *Magnetic moments* related to different chemical elements in a specimen can be measured selectively by means of magnetic X-ray circular dichroism (MXCD) using circularly polarized X-rays from a suitable synchrotron radiation source. In principle, this method allows the separation of orbital and spin magnetic moments.

An important instrument for the characterization of bulk permanent magnets is the hysteresisgraph (Czichos, Saito, and Smith, 2006e) (Figure 4.25). It measures the magnetic field H and the magnetic polarization J and is mainly used to record the second-quadrant hysteresis loop. From this, material parameters such as the remanence B_r, the maximum energy product $(BH)_{\max}$, and the coercive field strengths H_{cJ} and H_{cB}, can be determined.

4.4.5 Optical Properties

Optical properties characterize a material with respect to its interaction with optical radiation. Materials are optical transparent

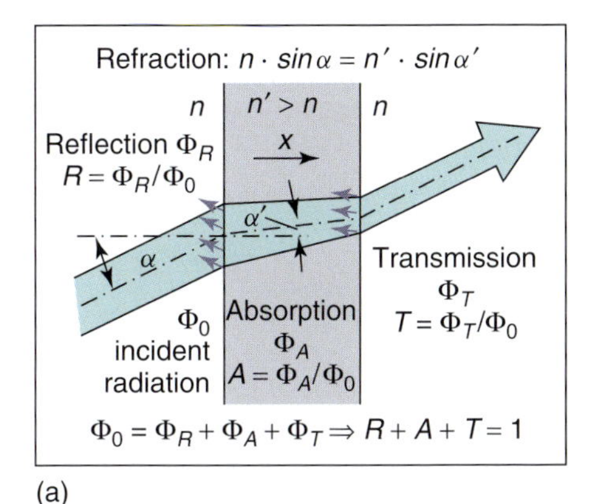

(a)

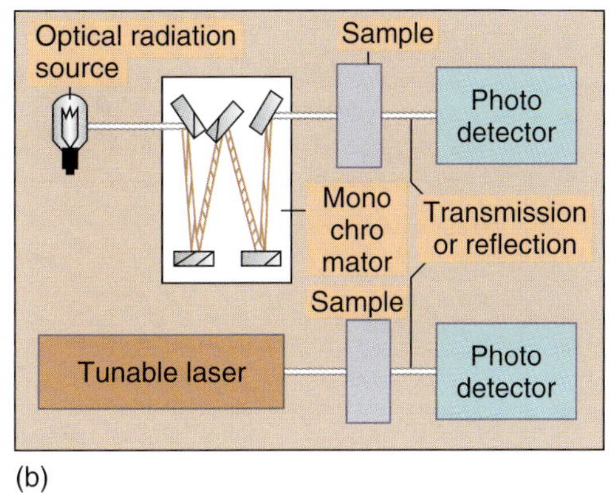

(b)

Fig. 4.25 Basic optical characteristics of transparent materials (a) and the principle for reflection or transmission measurement (b).

if no internal photon absorption occurs, for example, glasses or insulators with ionic or atomic bonding. If certain wavelengths λ of the radiation are absorbed, the material appears colored. In metals, electrons are excited by the incident optical radiation and the largest part of the optical radiation is reflected when they fall back to their initial electron energy levels (see Figure 4.23). The amount of incident light that is reflected at the specular reflectance angle γ of the mean of that surface characterizes the *materials surface gloss*. It can be measured with a specular glossmeter (ISO Standard 2813). Materials with smooth surfaces appear glossy if their peak-to-valley height roughness R_z (see Figure 4.18) is smaller than the wavelength λ of the radiation: $R_z < \lambda/(8\cos\gamma)$ (Rayleigh criterion). Rough surfaces reflect no specular light and therefore appear matt.

The basic optical characteristics of transparent materials are related to *refraction, reflection, scattering, absorption*, and *transmission*, as illustrated in Figure 4.26a.

For a light beam (power Φ_0) incident on a material surface in the x direction, the electric field is described by $E(x, t) = E_0 \exp[i(kx - \omega t)]$, where k is the wavevector of the light and ω is an angular frequency. The *refractive index n of a transparent material* is defined as the ratio of the phase velocity c of light in vacuum to the phase velocity v_p in the material: $n = c/v_p$. In a transparent material with refractive index n, k and the wavelength λ in vacuum are related to each other through $k = 2\pi n/\lambda = n\omega/c$.

- **Refraction:** As shown in Figure 4.26, the incident light beam is refracted according to Snell' law of refraction: $n \cdot \sin\alpha = n' \cdot \sin\alpha'$ and propagates through the material.
- **Reflection:** The reflectivity R, the ratio of reflected to incident power, can be expressed as $R = [(n-1)^2/(n+1)^2]$ at perpendicular incidence. For example, typical glass has a refractive index of 1.5, which means that light travels at $1/1.5 = 0.67$ times the speed in air or vacuum and the reflectivity is 4% at each air/glass interface.
- **Scattering:** During the propagation in the material, the light may be attenuated due to scattering. *Elastic scattering* occurs due to variations of the refractive index of the material, but the frequency remains unchanged (Rayleigh scattering

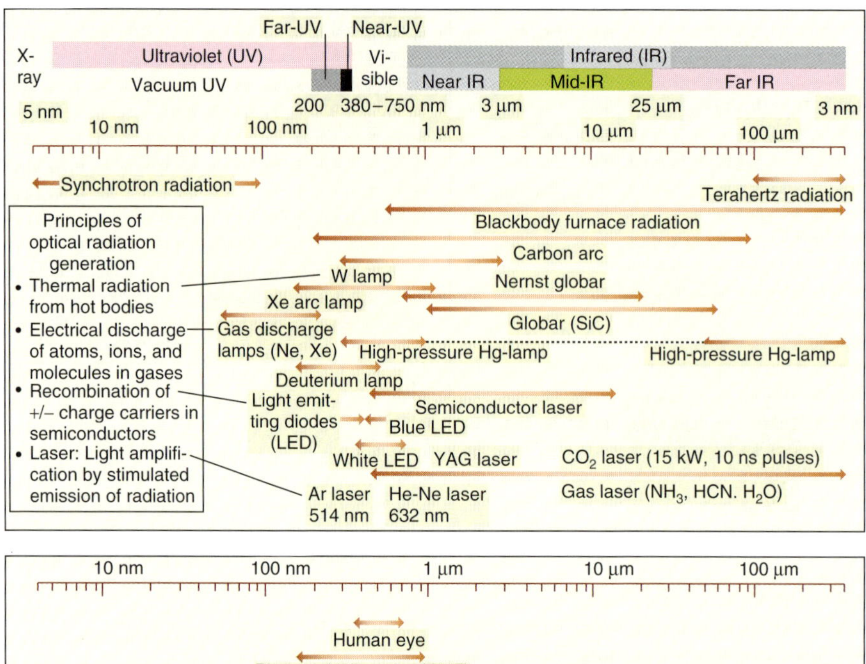

Fig. 4.26 Overview on optical light sources (top) and photodetectors (bottom) (Czichos, Saito, and Smith, 2006g).

and Mie scattering). *Inelastic scattering* occurs due to acoustic phonon emission/absorption and the frequency shift can be used to determine the energy of phonons or other elementary excitations (Raman spectroscopy, see Section 4.3.2)).

- **Absorption:** If the beam propagates in the x direction, the change in intensity $I(x)$ from the location $x = 0$ at the surface is $\Delta I(x) = I(x + dx) - I(x) = \beta I(x)dx$. It follows that $I(x) = I_0 \exp(-\beta x)$, where I_0 is the intensity of the incident light and β is the absorption coefficient (Beer's law). For example, typical glass has absorption of $A \approx 0.005$ for a sample thickness of 10 mm.
- **Transmission:** If the scattering is negligible, the transmittance T is described by $T = (1 - R_1) \exp(-\beta d)(1 - R_2)$, where R_1 and R_2 are the reflectivities of the front and back surfaces, respectively, and d is the sample thickness.

The principle for measuring the basic optical properties of reflectivity R or

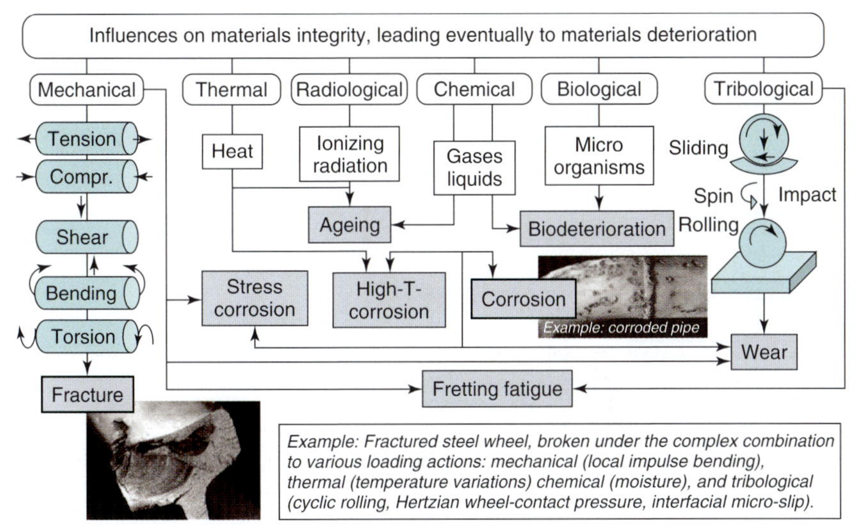

Fig. 4.27 Fundamentals of materials performance characterization: influencing phenomena.

transmittance T is shown in Figure 4.26b. A tunable light source, a combination of a white light and a monochromator, or a tunable laser is used. Changing the wavelength of the tunable light source, one can obtain a reflectivity spectrum $R(\lambda)$ or a transmittance spectrum $T(\lambda)$.

An overview on typical light sources and photdetectors is given in Figure 4.27 (Czichos, Saito, and Smith, 2006g).

4.5
Characterization of Materials Performance

In the preceding chapters, the fundamentals of materials measurement and testing have been considered with respect to the *composition, microstructure,* and *properties* of materials. In addition, depending on their technical applications, there are various other factors, which influence the integrity of materials – leading eventually to materials deterioration – which must be recognized for materials performance characterization (Czichos, Saito, and Smith, 2006h).

4.5.1
Influences on Materials Integrity

The operating conditions and influencing factors for the performance of a material in a given application stem from its functional tasks. In their applications, materials have to fulfill technical functions as constituents of engineered products or parts of technical systems. They have to bear mechanical stresses and are in contact with other solid bodies, aggressive gases, liquids, or biological species. As there is no usage of materials without interaction with the environment, these aspects have to be recognized in order to characterize materials performance. Figure 4.28 gives an overview of influences on materials integrity and possible failure modes.

Figure 4.28 illustrates in a generalized simplified manner that the influences on the integrity of materials, which are essential for their performance, can be categorized in mechanical, thermal, radiological, chemical, biological, and tribological terms. The basic materials deterioration mechanisms are *aging, biodegradation,*

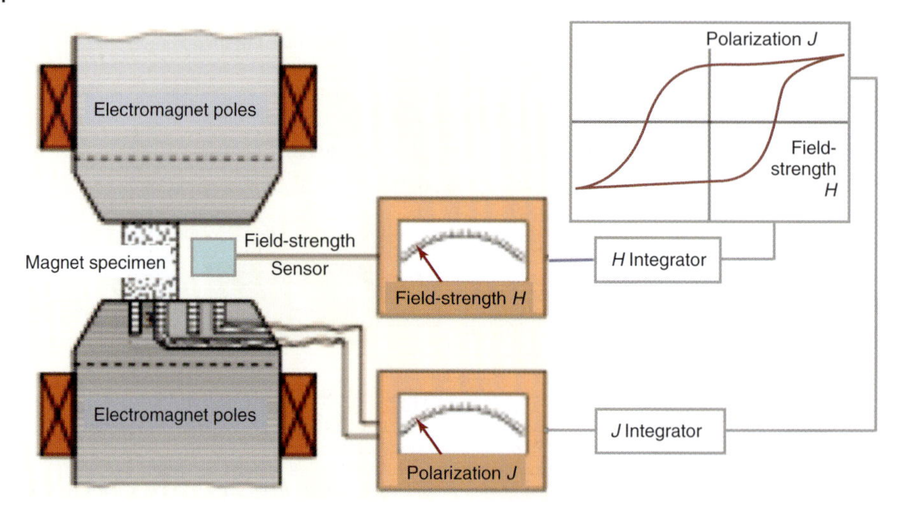

Fig. 4.28 Principle of the hysteresisgraph for the measurement of magnetic properties.

corrosion, *wear*, and *fracture* (Czichos, Saito, and Smith, 2006h).

4.5.2 Materials Deterioration Analysis

The deterioration and failure modes illustrated in Figure 4.28 are of different relevance for the two elementary classes of materials, namely, *organic* and *inorganic materials* (see Figure 4.2). Although *aging* and *biodegradation* are main deterioration mechanisms for organic materials such as polymers, the various types of *corrosion* are prevailing failure modes of metallic materials. *Wear* and *fracture* are relevant as materials deterioration and failure mechanisms for all types of materials.

4.5.2.1 Aging

Aging results from all the irreversible physical and chemical processes that occur in a material during its service life. Thermodynamically, aging is an inevitable process; however, its rate ranges widely as a result of the different kinetics of the single reaction steps involved. Aging of polymeric materials by the impact of *weathering* is mainly initiated by the action of solar radiation. However, other climatic quantities such as heat, moisture, wetting, and ingredients of the atmosphere influence the photochemically induced materials deterioration processes. Aging becomes noticeable through changes of different properties, varying from slight loss in appearance properties to total mechanical – technical failure. Relevant environmental factors for the aging of polymers are as follows:

- **Mechanical stress:** In the case of mechanical exposure, free radicals can be formed by rupturing chemical bonds, which is used as the initiation step in stress chemiluminescence. Ultrasonic degradation can be used to produce polymers with a definite molecular size or to degrade polymeric waste products. Frequent freezing/thawing cycles may also lead to degradation of dissolved polymers.
- **Heat:** Because of their low bond energies, technical polymers can only be used in a limited temperature range. For example, polyvinylchloride (PVC) is only

suited for permanent use up to 70° C and polycarbonate (PC) up to 115°C.

- **Ionizing radiation:** The impact of high-energy radiation (γ-, X-ray) leads to degradation and/or crosslinking. In the presence of oxygen, degradation predominates.
- **Chemical environment:** Exposed to water some polymers, such as polyamides (PAs), polyester, and polyurethane, can be hydrolized, the mechanism for which may depend on acidity. Other polymers show a high sensitivity to atmospheric pollution, for example, polyamides and polyurethanes even at room temperature are attacked by SO_2 and NO_2. To estimate the service life of products made from polymers or for the development of more-stable polymers, the conditions of outdoor exposure are simulated in accelerated artificial-weathering devices. Most of these tests procedures are based on the following ISO standards:
  - **Natural exposure of polymeric materials:** *ISO 877 plastics – methods of exposure to direct weathering*. Weathering by using glass-filtered daylight and intensified weathering by daylight using Fresnel mirrors.
  - **Artificial exposure of polymeric materials:** *ISO 4892 plastics – methods of exposure to laboratory light sources*, applying xenon-arc sources or fluorescent UV lamps. Filtered xenon-arc radiation is applied as artificial radiation for coatings.

Generally, weathering-induced degradation of a material first affects its surface on a molecular level before it accumulates and spreads over the bulk of the material. This brings up two consequences for the determination of property changes: (i) especially in the early stages of an environmental impact a heterogeneous distribution of property changes between the surface and bulk will result and (ii) the early effects of weathering will be able to be detected by surface-emphasizing detection techniques (see Sections 7.2.3 and 4.4.2).

4.5.2.2 Biogenic Impact

Detrimental changes of materials and their properties may be caused by microorganisms (bacteria, algae, higher fungi, and basidiomycetes) as well as by insects (termites, coleoptera, and lepidoptera) and other animals. In the marine environment, molluscs and crustaceans are usually considered the main deteriorating organisms. In the course of the proliferation of microorganisms, a slimy matrix is produced by microbial communities at the interface with a material called a *biofilm*. In technical systems "biofouling" occurs. Natural materials, such as timber, pulp, paper, leather, and textiles, are particularly susceptible to deterioration by biological systems. However, also many contemporary materials such as paints, adhesives, plastics, plasters, lubricating materials and fuels, technical liquids, waxes, and so on can support microbial growth. Even the properties of inorganic products, such as concrete, glass, minerals, and metals may suffer from biological attack. Biocorrosion of metals and metal alloys is also known to occur under a wide variety of conditions. Figure 4.29 shows examples of biodeterioration of an inorganic material and biodegradation of an organic material and explains elementary biodegradation steps of polymers (Czichos, Saito, and Smith, 2006i).

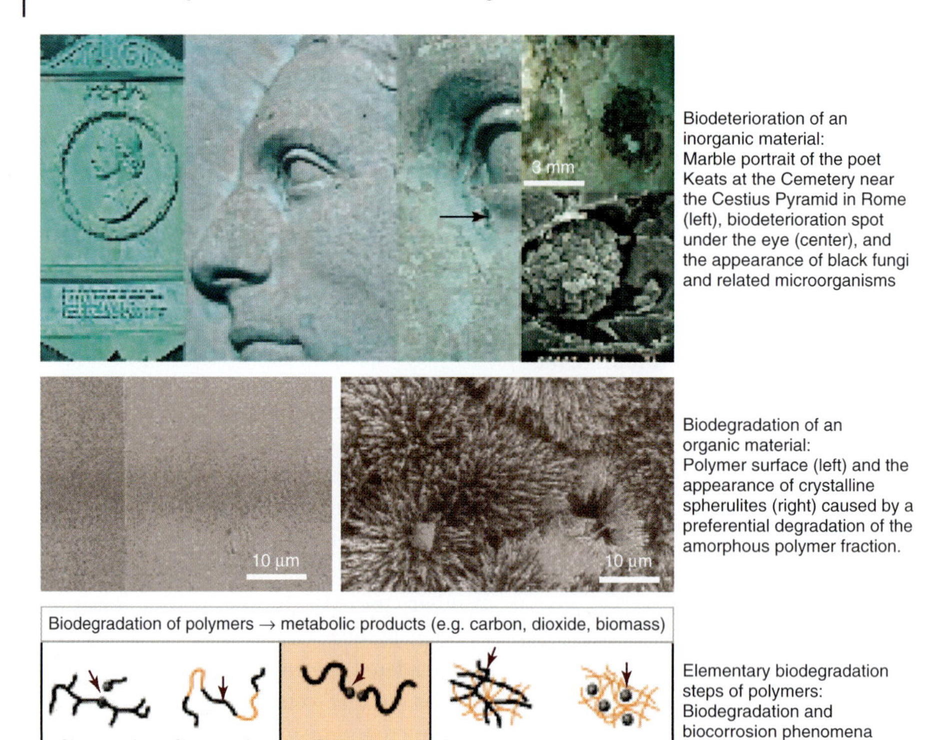

Fig. 4.29 Biodeterioration and biodegradation examples of inorganic and organic materials.

For the experimental determination of materials performance under biogenic attack, various ISO standards have been worked out, for example, ISO 846: plastics – determination of behavior under the action of fungi and bacteria, examination of properties.

4.5.2.3 Corrosion

Corrosion is defined as an interaction between a metal and its environment that results in changes in the properties of the metal (ISO Standard 8044). In most cases, the interaction between the metal and the environment is an electrochemical reaction where thermodynamic and kinetic considerations apply. From a thermodynamic point of view, the driving force, as in any electrochemical reaction, is a potential difference between anodes and cathodes in a short-circuited cell. The result of corrosion is a corrosion effect, which is generally detrimental and may lead to material loss, environment contamination with corrosion products, or impairment of a technical system. There are generally several forms of corrosion illustrated in Figure 4.30: (i) general (uniform) corrosion, (ii) localized corrosion, (iii) galvanic corrosion, (iv) cracking phenomena, (v) intergranular corrosion, (vi) dealloying, and (vii) high-temperature corrosion.

With respect to measurement and testing, the various forms of corrosion can be divided into three categories:

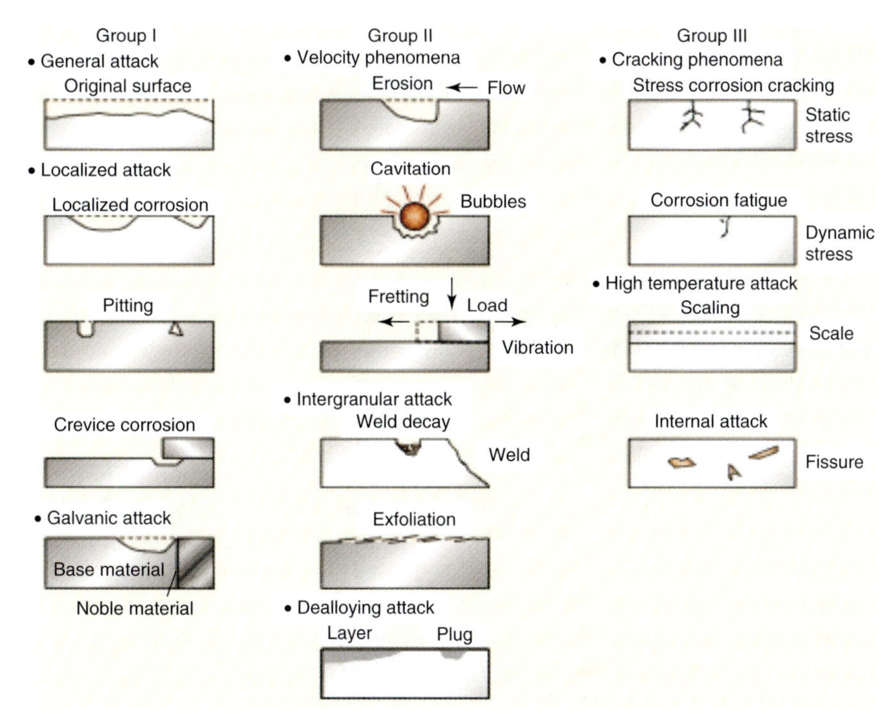

Fig. 4.30 Overview on the types of corrosion.

- **Group I:** Corrosion forms readily identified by visual examination (forms 1, 2, and 3).
- **Group II:** Corrosion forms that may require supplementary examination (forms 5, 6, and 7).
- **Group III:** Corrosion forms that usually should be verified by microscopy, optical, or scanning electron microscope, although they are sometimes apparent to the naked eye.

According to the ISO Standard ISO 8044, a corrosion investigation includes corrosion tests and their evaluation and is directed toward the following objectives: (i) the explanation of corrosion reactions, (ii) obtaining knowledge on corrosion behavior of materials under corrosion load, and (iii) selecting measures for corrosion protection.

A corrosion test is a special case of a corrosion investigation; the corrosion load and the assessment of the results being prescribed by regulations (standards and test sheet). The significance of the corrosion tests selected for an investigation differs. Although corrosion tests under service load conditions give the best estimation of the real corrosion behavior, accelerated corrosion tests are less suitable to give a reliable prognosis of behavior in practice, as the corrosion mechanism could differ from that under practical conditions. Nevertheless, accelerated corrosion tests are more suitable for determining certain material properties (e.g., resistance to intergranular corrosion or stress corrosion cracking), although they can be reliably used only if sufficient experience exists for the transfer of the test results to practical situations.

4.5.2.4 **Wear**

Tribology is the science and technology of interacting surfaces in relative motion (Czichos, 2001). Friction and wear represent two key aspects of the tribological behavior of materials. Like other measurements of material performance, such as corrosion or biodegradation, both friction and wear result from exposure of the material to a particular set of conditions. They do not, therefore, represent intrinsic material properties, but are "system-dependent attributes" and must be measured under well-defined test conditions. Materials in contact are subjected to relative motion in many different applications, for example, in rotating plain bearings, pistons sliding in cylinders, automotive brake disks interacting with brake pads, or the processing of material by machining, forging, or extrusion. Figure 4.31 shows examples of typical tribological machine elements subject to wear and the characteristic tribological model system with the basic structural elements and operating parameters, which must be recognized – and appropriately simulated – in wear testing and wear measurements.

An overview on the basic features of tribological systems is given in Figure 4.32 (Czichos, 2001). The *friction force* is defined as the force acting tangentially to the interface resisting motion, when, under the action of an external force, one body moves or tends to move relative to another. The *coefficient of friction f* is a dimensionless number, defined as the ratio between the friction force F_F and the normal force F_N acting to press the two bodies together. The Stribeck curve shown in the upper part of Figure 4.32 illustrates the different friction regimes of sliding motion.

Wear can be defined as damage to a solid surface, generally involving progressive loss of material, due to relative motion between that surface and a contacting

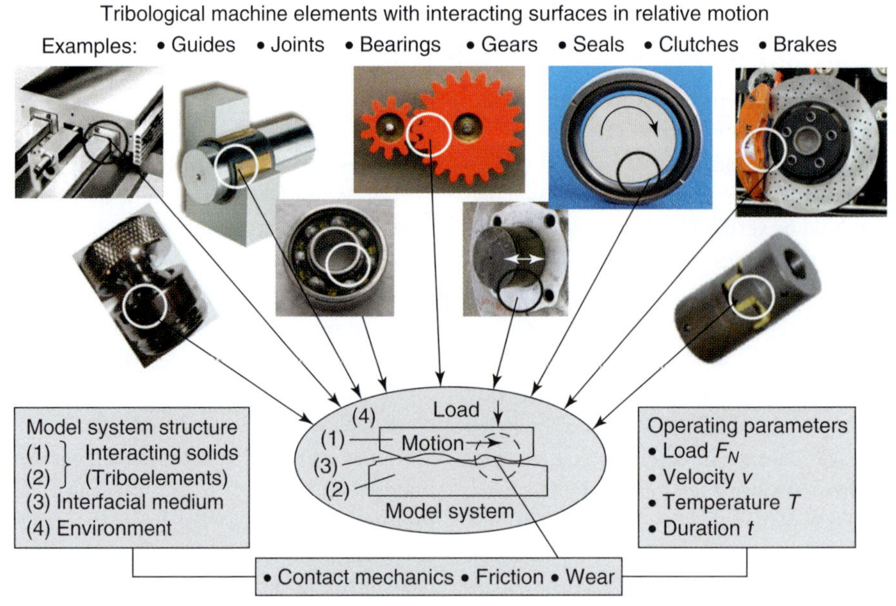

Fig. 4.31 Machine elements subject to friction and wear (examples) and the tribological model system.

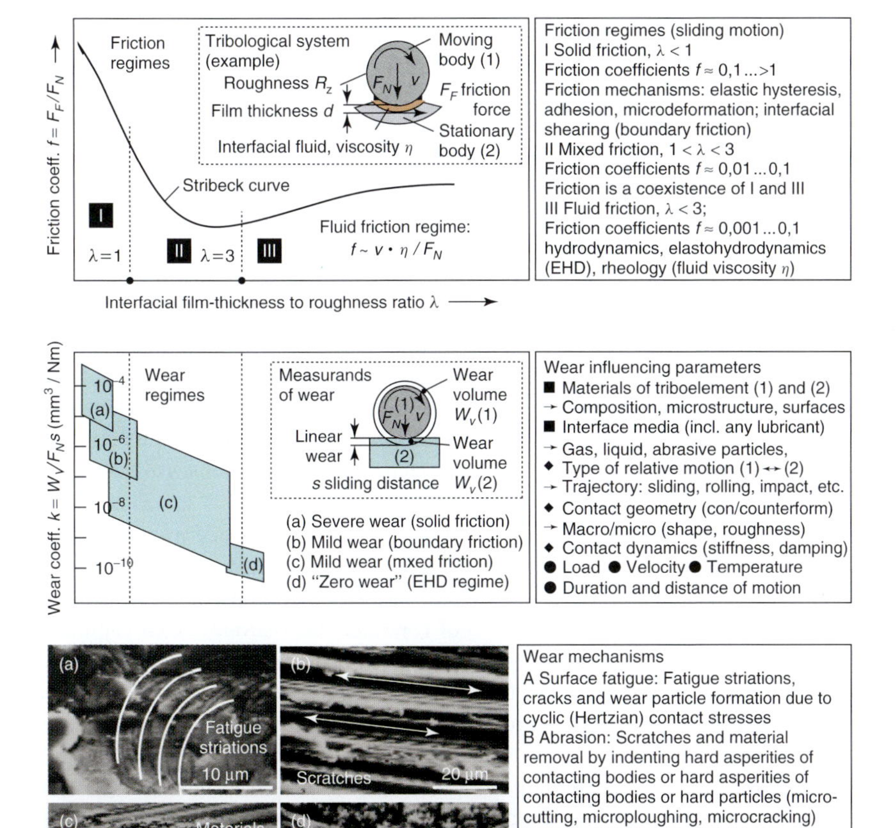

Fig. 4.32 Overview on the basic characteristics of tribological systems.

substance or substances. The middle part of Figure 4.32 illustrates the regimes of wear and lists the main influencing parameters. Wear tests – with well-defined triboelements (1) and (2) and a specified interfacial medium – are designed to produce specific types of relative motion, which can often lead to different mechanisms of wear in different materials or at different loads or speeds. The basic wear mechanisms are explained and illustrated in the lower part of Figure 4.32.

For the testing of wear, contact mechanics and motion dynamics must be well defined. For example, sliding and rolling motions may be either continuous or interrupted and may occur along the same track on a rotating counterbody or on a continuously fresh track. They may involve constant relative velocity (continuous sliding) or varying velocity (such as reciprocating motion). When analyzing the motion in a tribological contact, the nature of the contact must also be examined. For the measurement of wear, it is necessary to consider how the contact region moves with respect to the surfaces of the bodies in contact. In the simple example of a rotating shaft (1) sliding on a stationary counterbody (2) shown in Figure 4.32, the contact region is fixed relative to the block (2), which is in continuous contact. The circumferential elements of the rotating shaft (1) experience contact only for the time when they pass through the contact zone. Thus, the tribological impact on the interacting bodies is different and may cause different wear processes on triboelement (1) and triboelement (2):

- **Moving triboelement (1):** Intermittent contact of circumferential areas, cyclic local contact stresses on surface elements passing through the contact zone, intermittent frictional heating, cyclic noncontact exposure to the environmental atmosphere (possibility of the formation and subsequent rupture of surface layers), continuous scratching attack by hard asperities of counterbody (2) or abrasive particles embedded in (2).
- **Stationary triboelement (2):** Permanent (Hertzian) contact stresses, permanent frictional heating, no direct interaction with the environmental atmosphere, intermittent scratching attack by hard asperities of moving counterbody (1) or abrasive particles embedded in (1).

The removal of material from (1) and (2) is usually quantified in terms of the masses or the volumes of material removed, W_{V1} and W_{V2}. Volume changes in more complex geometries may be derived computationally from the difference between profile traces recorded by profilometry (see Section 4.4.2) before and after wear. Experimentally it has been observed that wear often increases with load F_N and sliding distance s. An empirical wear coefficient $k = W_V/F_N s$ has been defined to distinguish between the various regimes of wear. Figure 4.32 (middle) shows, very approximately, the range of values of k seen in various types of wear. Under dry sliding conditions, k can be as high as 10^{-2} mm^3 N m^{-1}, although it can also be as low as 10^{-6} mm^3 N m^{-1}. Often two distinct regimes of sliding wear are distinguished, termed *severe* and *mild*. These not only correspond to quite different wear rates but also involve significantly different mechanisms of material loss. In metals, "severe" sliding wear is associated with relatively large particles of metallic debris, whereas in "mild" wear the debris is finer and formed of oxide particles. In the case of ceramics, the "severe" wear regime is associated with brittle fracture, whereas "mild" wear results from the removal of reacted (often hydrated) surface material.

It must be emphasized that wear testing is ambiguous when the test conditions are not exactly specified: If in a tribological test system, a test specimen (to be well-defined with respect to composition, microstructure, shape, and intrinsic properties) is tested either as moving triboelement (1) or as stationary element (2), *different numerical wear values* W_{V1} *or* W_{V2} *may be measured for one and the same material*, because the nature of the contact, the tribological actions, and the wear surface patterns are generally different for (1) and (2). In summary, wear is a complex performance characteristic of *interacting materials*. It depends on the structure (see Figure 4.31) and the wear influencing parameters (see Figure 4.32) of the pertinent tribological system: Wear $= f$ (systems structure; operating parameters).

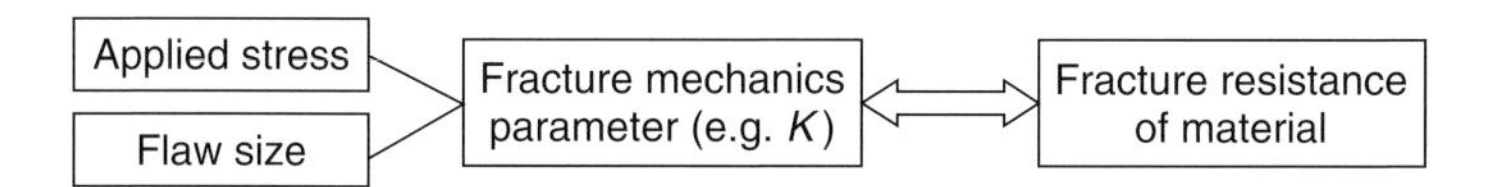

4.5.2.5 Fracture

Fracture – the separation of a solid body into pieces under the action of stress – is the ultimate failure mode. It destroys the integrity and the structural functionality of materials and engineered components and must be avoided and prevented. In materials that are components of engineering structures, the presence of fabrication flaws and their development during operational service should be anticipated. Generally, the strength of a structural component that contains a crack or cracklike flaw decreases with increasing crack size and cannot be deduced from the mechanical properties obtained in conventional tensile tests (see Figure 4.21) using smooth specimens.

The fracture mechanics approach – initially developed for brittle materials and later extended – states that the stress and strain fields ahead of a sharp crack can be characterized by the *stress intensity factor* K, that is, a function of the applied stress, crack length, and geometrical boundary conditions. For an infinite plate subjected to uniform tensile stress σ containing a through-thickness crack of length $2a$, the stress intensity factor is $K = \sigma(\pi a)^{1/2}$. As basis for fracture mechanics measurement and testing, an elastic body of arbitrary configuration with an ideally sharp crack, subjected to arbitrary external forces, is considered. Figure 4.33a illustrates the three fundamental types of relative movements of the two crack surfaces. Figure 4.33b–d shows typical specimens for fracture mechanics testing.

The **fracture mechanics approach** has three variables for the evaluation of strength, characterized as shown in the table above (Czichos, Saito, and Smith, 2006j):

In linear elastic fracture mechanics, it is postulated that fracture occurs when the stress intensity factor K reaches a critical value K_C.

From microscopic examination of fractured surfaces, important information can be obtained. This study is usually called *fractography*. The SEM (see Figure 4.17) is commonly employed in fractography because of its large depth of focus. Fractography is often applied in order to find the cause of accidental failures. Three of the most common fracture mechanisms, which are caused by monotonic tension in metallic engineering materials, are explained and illustrated in Figure 4.34 (Czichos, Saito, and Smith, 2006j).

4.5.3 Nondestructive Evaluation

NDE and nondestructive testing (NDT) are indispensable for the analysis of materials deterioration and the control of materials integrity. As described in the preceding section, flaws and especially cracks in structural materials can be crucially detrimental to the functional performance of whole engineering systems. This can be detected by *NDE sensors* on the basis of acoustical, magnetic, electrical, or radiological principles. Applications of NDE in

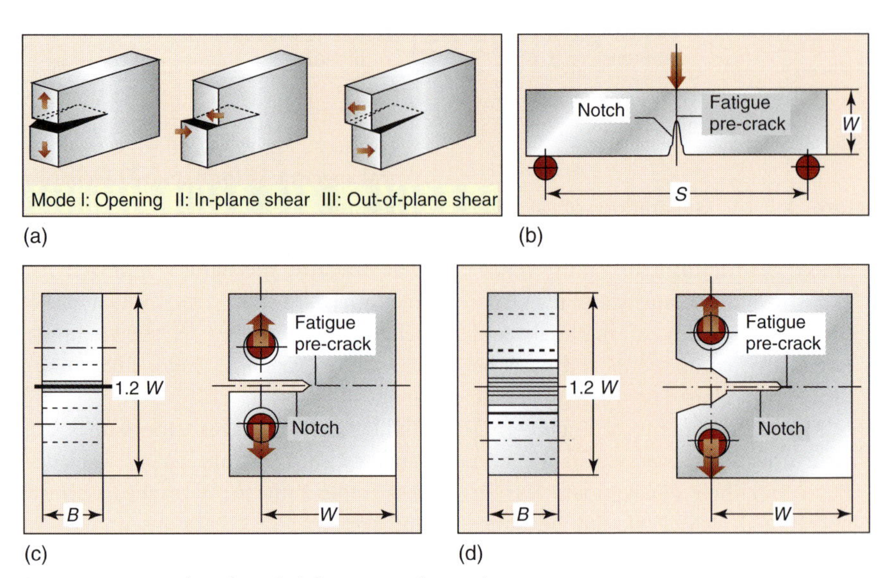

Fig. 4.33 (a) Modes of crack deformation, (b) single edge notched specimen (thickness $= W/2$), and (c and d) compact test specimen (recommended notch width: 0.45–0.55 W).

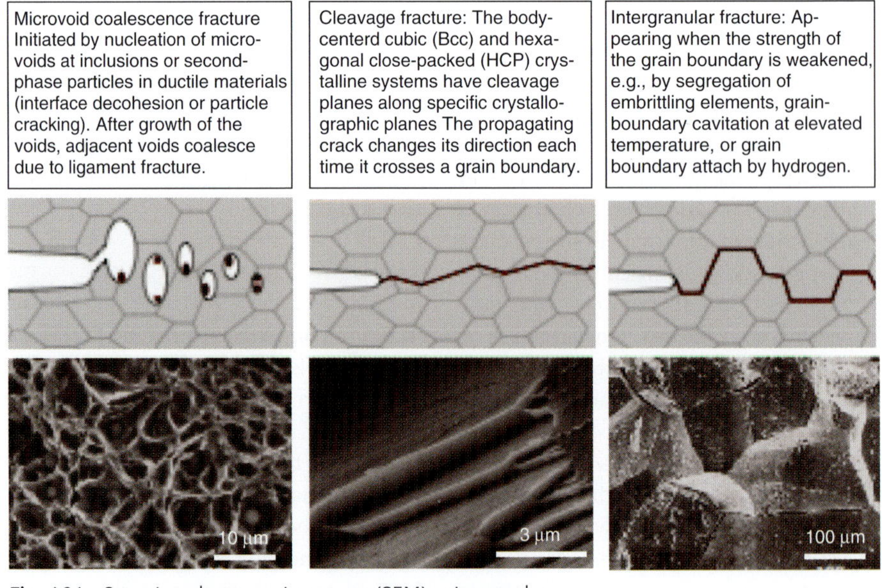

Fig. 4.34 Scanning electron microscope (SEM) micrographs for most common fracture appearances.

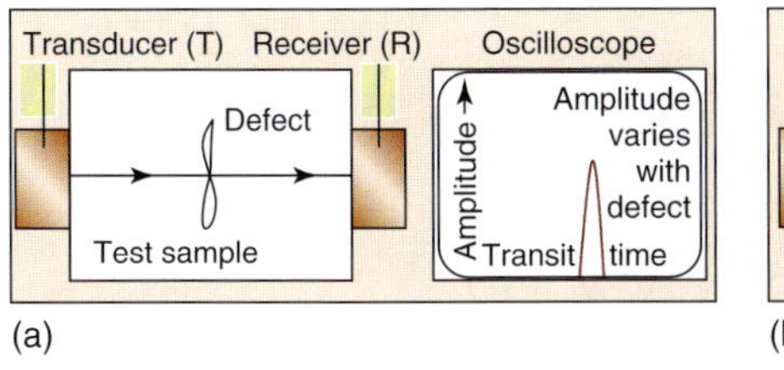

(a)

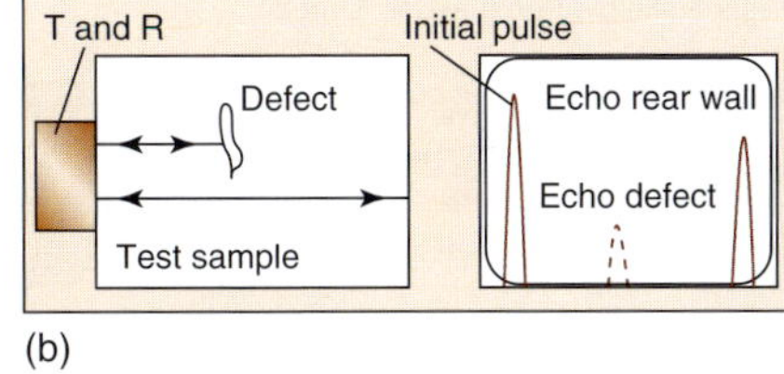

(b)

Fig. 4.35 Ultrasonic NDT techniques: (a) through-transmission technique and (b) pulse-echo technique.

industry include mechanical engineering, aerospace, civil engineering, oil industry, electric power industry, and so on (Czichos, Saito, and Smith, 2006k). In the public technological infrastructure, NDE also supports condition monitoring for the availability and proper functioning of the daily use of electricity, gas, or liquids, where the correct operation of components (e.g., pressure vessels, pipes) under applied stress plays an important role for safety and reliability.

4.5.3.1 **Ultrasonic Evaluation**

Sound travels in solids with a velocity depending on the mechanical properties of the material. Imperfections such as cracks, pores, or inclusions cause sound – wave interactions, which result in reflection, scattering, diffraction, and general dampening of the sound wave. NDT is carried out using ultrasonic waves at high frequencies above the audible range, higher than approximately 20 kHz and up to the range of some hundred megahertzes. The active sound-field generation tool for NDT is in most cases a piezomaterial. If a piezomaterial is deformed by an external mechanical pressure, electric charges are produced on its surface (sensor effect). The reverse phenomenon, that such a material, if placed between two electrodes, changes its form when an electric potential is applied, is called the *inverse piezoelectric effect (actuator effect)*. In NDT, the actuator effect is used in a transducer to apply mechanical pressures, deformations, and oscillations to a test sample. Defects in materials cause scattering and reflection of the ultrasonic wave and the detection of the reflected or transmitted waves with an ultrasonic receiver allows to determine the location of the defect. A general illustration of the through-transmission and pulse-echo methods is shown in Figure 4.35.

The through-transmission technique (i) normally requires access to both sides of a test sample and is applied if the back-reflection method cannot provide sufficient information about a defect. The pulse-echo technique (ii) requires access to only one side of the component. Another advantage of the pulse-echo method is given due to the time-of-flight measurement because the defect location (distance from the coupling surface) can be estimated from the time required for the pulse to travel from the probe to the discontinuity and back and can be displayed on an oscilloscope. This method is extensively used in practice to evaluate materials integrity.

4.5.3.2 **Magnetic and Electrical Evaluation**

Surface defects in magnetized ferromagnetic materials cause that some of the lines of magnetic force depart from the surface. The defects create a magnetic leakage field

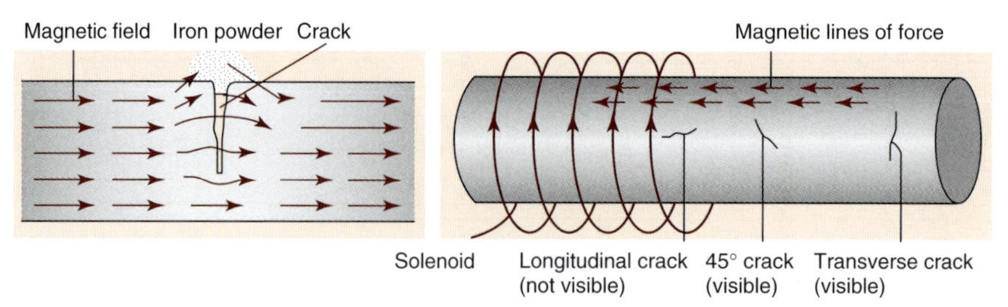

Fig. 4.36 Principle of magnetic nondestructive inspection of materials.

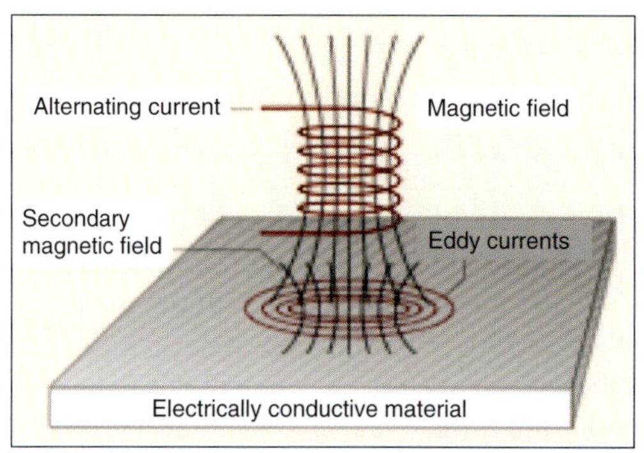

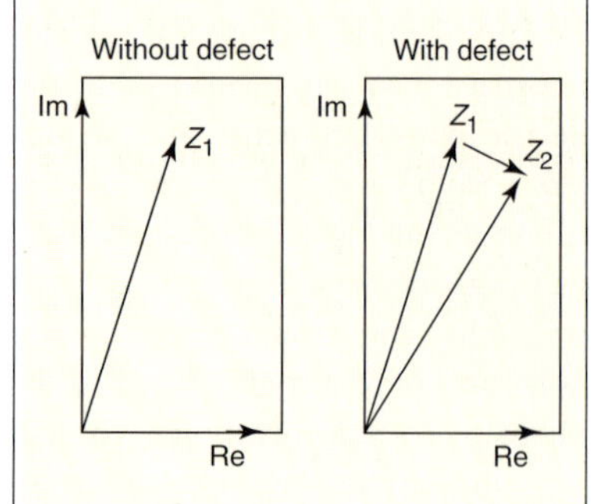

Fig. 4.37 (a) Principle of eddy-current NDT. (b) Influence of materials defects on the electric coil impedance Z, represented as a complex number change in the complex Im–Re plane.

and can be visualized by iron (powder) particles applied to the surface of the specimen (see Figure 4.36a). The influence of the crack orientation on crack detectability is illustrated in Figure 4.36b. A cylindrical test piece was magnetized with a solenoid coil that provides longitudinal magnetization, suitable for the examination of transverse cracks. Longitudinal cracks are not visible using this magnetization procedure. Thus, magnetic NDE of materials should always be carried out with the magnetic field in two different directions, preferable in two orthogonal directions.

Electrical NDE of materials uses the eddy-current effect. An alternating electric current flowing in a coil (eddy-current probe) generates a varying magnetic field. In a specimen with electric conductivity in the proximity of this coil, eddy currents flowing in a closed loop in the surface layer of the specimen are produced (see Figure 4.37a). The opposite of the eddy-current probe can be detected by the change of the electric coil impedance, which will be, for example, displayed for evaluation by the test instrument connected to eddy-current probe. Cracks and other surface inhomogeneities or discontinuities modify the eddy currents generated in the specimen. The coil impedance is changed in relation to a surface without defects (Figure 4.37b). Practical application using eddy-current techniques is only possible if sensitive electronic devices are available that are able to measure the very small changes of the magnetic field caused by a defect. The application of eddy-current

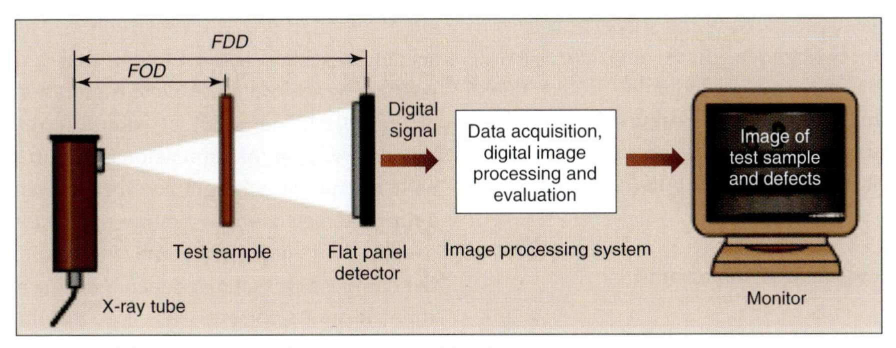

Fig. 4.38 Schematic set up of a digital industrial radiology system.

techniques embraces a wide industrial field – from the aircraft industry and the railway industry to the power-generation industry – to control materials integrity of engineered components in technical systems.

4.5.3.3 **Radiological Evaluation**

Radiology is typically applied for the volumetric materials integrity inspection of industrial products and installations. The basic set up consists of a radiation source in front of the object to be inspected and an area detector behind the object. The classical detector is an X-ray film. New electronic area detectors are gradually substituting film. The radiation source can be an X-ray generator, a gamma source, or a particle accelerator. Objects of all possible materials and thicknesses can be inspected, provided the right radiation source and energy is selected. There exist practical limitations to the upper material thickness, for example, 50-cm penetration length in steel or 2 m in concrete (at a radiation energy of 12 MeV). For portable radiation sources, the limits are much lower.

Fast digital X-ray inspection systems are used in the serial examination of industrial products since this technique is capable of detecting flaws rather different in their nature such as cracks, inclusions, or shrinkage. They enable a flexible adjustment of the beam direction and of the inspection perspective as well as online viewing of the radioscopic image to cover a broad range of different flaws. This economic and reliable technique is of essential significance for different applications. The configuration of such systems is schematically represented in Figure 4.38.

The test sample, irradiated from one side by X-rays, causes a radioscopic transmission image in the detector plane via central projection. The relation between the focus – detector distance (*FDD*) and the focus – object distance (*FOD*) determines the geometrical magnification of the image. An image converter, such as an X-ray image intensifier, a fluorescence screen, or a digital detector array also called a *flat-panel detector*, converts the X-ray image to a digital image. This testing technique of materials performance characteristics can be applied for industrial components, which have to undergo a 100% X-ray inspection for safety. For example, light alloy castings are widely used, especially in automotive manufacturing, in safety-relevant applications, such as steering gears, wheels, and wheel suspension components. Owing to imperfections of the casting process, these components are prone to material defects (e.g., shrinkage cavities and inclusions). A fully automated X-ray inspection system for unattended inspection can provide objective and reproducible defect detection. These NDT systems for materials integrity evaluations are known as automated defect-recognition (ADR) units.

4.5.4 Computerized Tomography

An important NDT technique for materials integrity evaluations is tomography with X-rays supported by computers – in short computerized tomography (CT) (Czichos, Saito, and Smith, 2006l). This technique is able to identify 0.1% differences in the density of materials. Thus, it can

- measure the dimensions of components (dimensional metrology);
- detect internal geometric defects of materials;
- perceive the evolution of mechanical, physical, and chemical damage;
- display homogeneous variation related to processing; and
- observe differences in texture and chemical concentration.

The tomographic system measures the attenuation of intensity in a materials test object from different angles to determine cross-sectional configuration with the aid of a computerized reconstruction. The section-by-section reconstruction makes it possible to achieve a complete exploration of the object. The study of volume properties as well as of dimensional features with computed tomography requires an optimized selection of source – detector combination depending on the material composition (energy-dependent linear attenuation coefficient), the size of the samples, and the maximum thickness of material that has to be irradiated. Additionally, the manipulator system and the mounting need to suffice the required accuracy. A CT system consists in principle of the following basic components (Figure 4.39): an X-ray source, a collimator, a testpiece turntable, a detector array, and a computer system with data-storage media and image processing.

The application of CT-NDE to organic and inorganic materials requires different CT techniques. Organic materials with low density ($<4\,g\,cm^{-3}$) can often be evaluated with medical scanners. For example, fiber-reinforced helicopter blades have been investigated, using commercial medical scanners. Inorganic materials require radiation sources with higher energy and appropriate detector systems. For example, steel samples of thickness

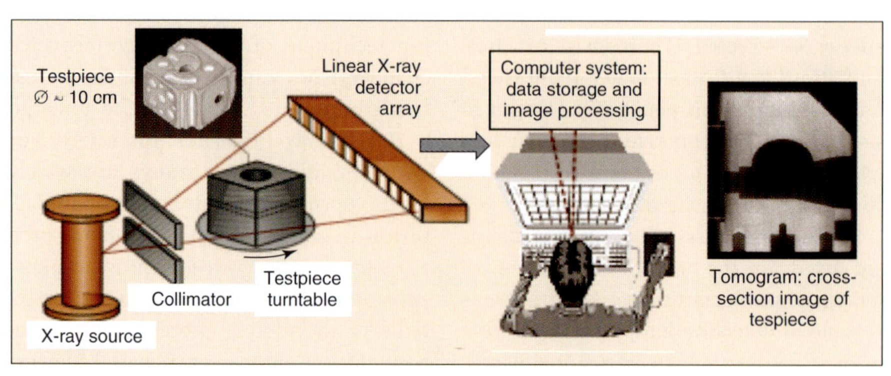

Fig. 4.39 Schematic representation of a computerized tomography system.

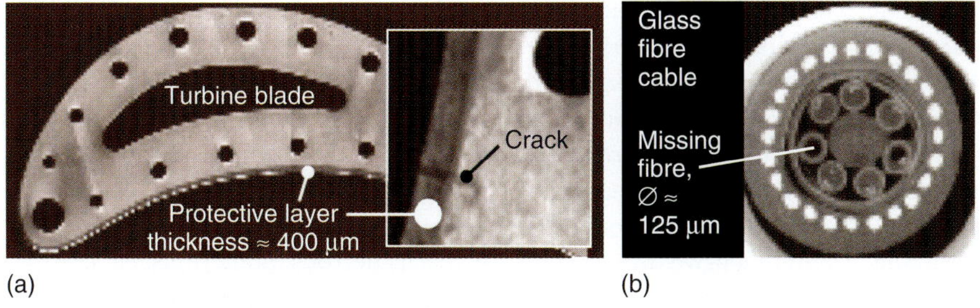

(a) (b)

Fig. 4.40 (a) Materials integrity analysis of a turbine blade. (b) Materials integrity analysis of a telecommunication glass-fiber cable.

20–30 mm can be inspected with a 320-kV microfocus X-ray tube and the optimum spatial resolution is approximately 35 μm. A spatial resolution of about $5 \times 5 \times 5\,\mu m^3$ detector voxel (volumetric pixel) can be achieved with synchrotron radiation (e.g., BAM*line* at BESSY, Berlin). The application of CT to evaluate materials integrity of engineered industrial components is illustrated in Figure 4.40 (Czichos, Saito, and Smith, 2006l):

- In engine components, materials are loaded by heavy mechanical and thermal stresses. Figure 4.40a shows the investigation of flaws of coated turbine blades. High-resolution CT could analyze the crack configuration. This information is important for the design of practice-oriented fracture mechanics tests (see Section 7.5.2.5).
- In telecommunication equipment, flaws in glass-fiber cables can disturb the transmission characteristics and may lead to failure of IT systems. Figure 4.40b shows that CT could identify nondestructively a missing fiber in one covering as well as glass-fiber flaws with a resolution in the 10–μm range. This fault analysis can help to optimize production processes in order to improve materials performance.

4.6 Reference Materials and Reference Procedures

Introducing the materials measurement and testing system, it had been stated that

1. measurement of characteristics of materials requires a precise definition of the *measurand* and the meticulous analysis of influencing parameters and
2. measurement always involves a comparison of the *measurand* with a known quantity of the same kind.

Although the basic categories of (1) have been treated in Chapters 2–5 in terms of (i) composition, (ii) microstructure, (iii) properties, (iv) erformance, important tools for task (2) are (i) RMs and (ii) reference procedures (RPs), they are fundamental for *measurement strategy* and *quality* (Czichos, Saito, and Smith, 2006m).

RMs are materials or substances whose property values or quantities are well established to be used for

a. the calibration of an apparatus;
b. the assessment of a measurement method; or
c. for assigning values to materials.

Certified reference materials (*CRMs*) are RMs whose property values are certified by a procedure, which establishes traceability to an accurate realization of the unit in which the property values are expressed. Every certified value must be accompanied by an *uncertainty parameter* that characterizes the dispersion of the values that could reasonably be attributed to the measurand. Comprehensive information on certified RMs are available from the international database for certified reference materials (COMAR) that contains information on more than 10 000 *CRMs* from about 200 producers in 27 countries (*www.comar.bam.de*). The following keywords give an overview on functions and applications of CRMs:

Function of RMs (*examples*)

- Definition of measuring scales;
- identification and qualitative analysis;
- assessment of analytical methods;
- calibration of apparatus and measurement systems;
- testing of measurement devices;
- interlaboratory comparisons; and
- education and training.

Application fields of RMs (*examples*)

- Industrial raw materials and products (fuels, glass, cement, etc.);
- chemicals (gas, solvents, paints, etc.);
- pure materials (chromatography, isotopes, etc.);
- metals (ferrous, nonferrous, etc.);
- physical properties of materials (optical, electrical, etc.);
- biological and clinical (blood, urine, etc.);
- food and agriculture (meat, fish, vegetable, etc.); and
- environment (mater, soil, sediment, etc.).

An example of a CRM for dimensional metrology is shown and explained in Figure 4.41 (Senoner *et al.*, 2004). The certified RM is an embedded cross section of epitaxially grown surface layers on a GaAs substrate. The surface of the sample provides a flat pattern with strip widths

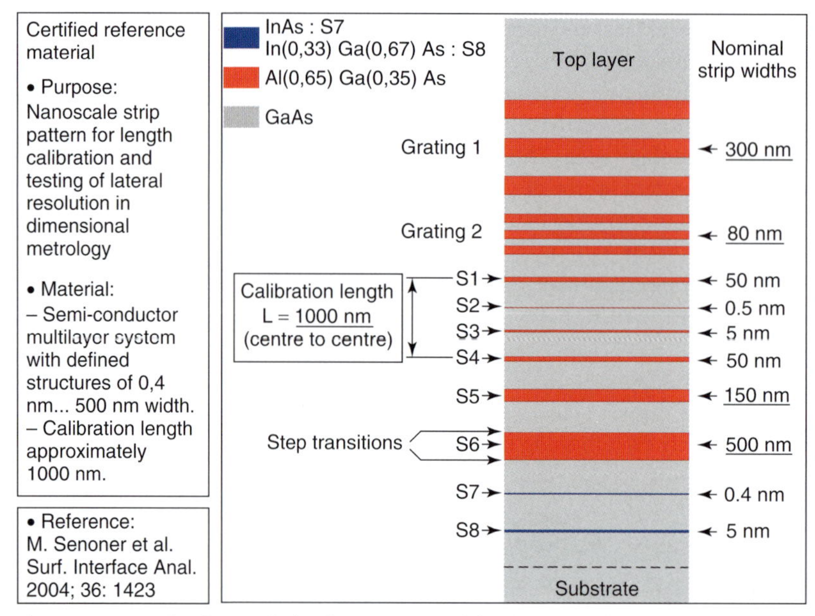

Fig. 4.41 Example of a certified reference material for dimensional metrology.

Tab. 4.8 Examples of reference procedures in materials measurement and testing.

Key words	Composition topic	Reference range	Uncertainty range	Methodology and instrumentation	Application examples
Characterization of materials composition					
Inorganic materials: analysis of chromium in steel					
Highly alloyed steel, alloys, titimetry	Chromium content in steels and alloys	5–500 g kg^{-1}	0.02–2 g kg^{-1}	Titration with combined Pt-electrode/ reference electrode; pure potassium titration solution	Certification of reference materials, standards for the iron and steel industry
Organic materials: analysis of pesticides in food					
Food, DDT, pesticides, GC-IDMS, chromatography	Chloropesticides in edible oils and fats or vegetables	Amount of single chloropesticide: 0.1 μg to 10 μg g^{-1}	1.5–3%	Sample preparation and clean up: GPC; measurement: GC-IDMS; calibration: gravimetry, certified purity; ^{13}C isotopes	Food control: analysis of pesticides (e.g., in oil or vegetables), environmental control samples

(*continued overleaf*)

Tab. 4.8 (*continued*)

Key words	**Microstructure topic**	**Reference range**	**Uncertainty range**	**Methodology and instrumentation**	**Application examples**
Characterization of materials microstructure					
Microstructure analysis: morphology analysis of inorganic and organic compounds					
Polymorphy, X-ray analysis, powder diffractometry	3D structure; configuration, bond length/angles, intermolecular interactions	Crystalline mixtures consisting of several polymorphs	1–5%, depending on number and crystallinity of polymorphs	X-ray structure and phase analysis: single crystal diffractometer, CCD• detector, powder diffractometer; order–dis-order theory	Validation of NMR. IR and Raman analysis (see Section 4.2.2), characterization of reference materials, importance for pharmacy developments

Key words	**Surface topic**	**Reference range**	**Uncertainty range**	**Methodology and instrumentation**	**Application examples**
Surface analysis: measurement of nanolayer thickness of oxidized silicon					
Nanoscience, silicon, ESCA	Oxid overlayer thickness of oxidized silicon	1–10 nm	0.3–3 nm	ESCA measurement of Si 2p photoelectron intensities in the substrate and in the overlayer	Characterization of materials used in the silicon technology and nanotechnology

Key words	Property topic	Reference range	Uncertainty range	Methodology and instrumentation	Application examples
Characterization of materials properties					
Mechanical properties: determination of hardness and other plastic–elastic characteristics					
Hardness, elasticity indentation test	Indentation depth Martens hardness HM	HM: $2–10^4$ N mm^{-2}	1–1.5%	Instrumented indentation test DIN EN ISO 14577 (Vickers pyramid)	Certified reference materials, round robin test, traceability of calibration
Optical properties: determination of optical and dielectric material constants					
Materials surfaces, coatings, spectroscopic ellipsometry	Optical constants (n, k), dielectric constants, layer thickness h	Thickness $h = 1$ nm to 10μm	0.5 nm to 1%	Reflection with polarized light, light sources: white light, HeNe laser, synchrotron radiation (BESSY Berlin)	Development and certification of reference materials, standards for surface technology

Key words	Performance topic	Defect dimensions	Uncertainty range	Methodology and instrumentation	Application examples
Characterization of materials performance					
Nondestructive evaluation: reference defects for nondestructive testing (NDT)					
Safety technology, fault analysis, NDT	Reference defects: length l, width w, depth d	$l > 100\,\mu m$ $w > 20\,\mu m$ $d <$ 2 mm	$\Delta l: 10–100\,\mu m$ $\Delta w: 5–20\,\mu m$ Δd: $10–100\,\mu m$	Spark-erosive preparation of defects and optical measurement of dimensions	Safety-relevant components: pressure vessels, engines, machine parts, gears

of 0.4–500 nm. The certified RM can be used for the calibration of a length scale (e.g., for a SEM), the determination of instrument parameters, the optimization of instrument settings, and the quality control of measurement methods for surface analysis, such as AES, XPS, and SIMS (see Section 4.4.2).

To achieve globally accepted *measurement and calibration standards and certificates* issued by National Metrology Institutes (NMI), the member states of the *Meter Convention* signed in 1999 the Mutual Recognition Arrangement (MRA). The database on the calibration and measurements capabilities (CMC) of the NMIs is hosted by the Bureau International des Poids et Mesures (BIPM) (*www.bipm.org*).

Besides RMs, RPs are instrumental for materials measurement and testing; they can be defined as follows:

- **RP**: procedure of testing, measurement, or analysis, thoroughly characterized and proven to be under control, intended for
  - quality assessment of other procedures for comparable tasks;
  - characterization of RMs including reference objects; or
  - determination of reference values.

The uncertainty of the results of an RP must be adequately estimated and appropriate for the intended use.

A system of RPs has been established by BAM, the German Federal Institute for Materials Research and Testing (*www.bam.de*). It comprises about 70 well-defined, quality-controlled RPs for (i) inorganic analysis, (ii) organic analysis, (iii) gas analysis and gas measurement, (iv) microprobing and microstructural analysis, (v) testing of mechanical – technological properties, (vi) testing of optical and electrical properties, (vii) nondestructive testing, and (viii) testing of surface and layer properties. Examples of RPs for the fundamental areas of materials measurement and testing are given in Table 4.8

In conclusion, RPs apply to materials measurement, testing, and analysis, that is, all procedures to characterize materials and engineered components. They are used for determining reference values of materials characteristics, to validate products and procedures, and to control material-related aspects of quality, reliability, and safety of products and technical systems.

Acknowledgments

Thanks are due to the author-team-colleagues of (Czichos, Saito, and Smith, 2006a) who created the scientific-technological base for this article as well as to Dr Birgit Skrotzki and Dr Heinrich Heidt. BAM Berlin for their helpful comments.

References

Ashby, M.F., Brechet, Y.J.M., Cebon, D., and Salvo, L. (**2004**) Selection strategies for materials and processes. *Mater. Des.*, **25**, 51–67.

Czichos, H. (**2001**) Tribology and its many facets: from macroscopic to microscopic and nano-scale phenomena. *Meccanica*, **36**, 605–615.

Czichos, H., Saito, T., and Smith, L. (**2006a**) *Springer Handbook of Materials Measurement Methods*, Springer, Heidelberg.

Czichos, H., Saito, T., and Smith, L. (**2006b**) Chemical composition: classical chemical methods, *Springer Handbook of Materials Measurement Methods*, Springer, Heidelberg, pp. 132–138.

Czichos, H., Saito, T., and Smith, L. (**2006c**) Surface and interface characterization, *Springer Handbook of Materials Measurement Methods*, Springer, Heidelberg, pp. 229–280.

Czichos, H., Saito, T., and Smith, L. (**2006d**) Nanoscopic architecture and microstructure, *Springer Handbook of Materials Measurement Methods*, Springer, Heidelberg, pp. 153–227.

Czichos, H., Saito, T., and Smith, L. (**2006e**) Measurement methods for materials properties, *Springer Handbook of Materials Measurement Methods*, Springer, Heidelberg, pp. 281–607.

Czichos, H., Saito, T., and Smith, L. (**2006f**) Thermal properties, *Springer Handbook of Materials Measurement Methods*, Springer, Heidelberg, pp. 399–429.

Czichos, H., Saito, T., and Smith, L. (**2006g**) Fundamentals of optical spectroscopy, *Springer Handbook of Materials Measurement Methods*, Springer, Heidelberg, pp. 532–549.

Czichos, H., Saito, T., and Smith, L. (**2006h**) Measurement methods for materials performance, *Springer Handbook of Materials Measurement Methods*, Springer, Heidelberg, pp. 609–829.

Czichos, H., Saito, T., and Smith, L. (**2006i**) Biogenic impact on materials, *Springer Handbook of Materials Measurement Methods*, Springer, Heidelberg, pp. 711–787.

Czichos, H., Saito, T., and Smith, L. (**2006j**) Fracture mechanics, *Springer Handbook of Materials Measurement Methods*, Springer, Heidelberg, pp. 353–371.

Czichos, H., Saito, T., and Smith, L. (**2006k**) Performance control and condition monitoring, *Springer Handbook of Materials Measurement Methods*, Springer, Heidelberg, pp. 831–912.

Czichos, H., Saito, T., and Smith, L. (**2006l**) Computerized tomography – application to inorganic materials, *Springer Handbook of Materials Measurement Methods*, Springer, Heidelberg, pp. 864–870.

Czichos, H., Saito, T., and Smith, L. (**2006m**) Measurement strategy and quality, *Springer Handbook of Materials Measurement Methods*, Springer, Heidelberg, pp. 17–94.

Czichos, H., Skrotzki, B., and Simon, F.-G. (**2007**) Werkstoffe, in *HÜTTE Das Ingenieur-wissen* (H. Czichos and M. Hennecke), Springer, Heidelberg.

Kipphardt, H., Matschat, R., Rienitz, O., Schiel, D., Gernand, W., and Oeter, D. (**2006**) Tra-ceability system for elemental analysis. *Accred. Qual. Assur.*, **10**, 633–639.

Senoner, M., Wirth, Th., Unger, W., Österle, W., Kaiander, I., Sellin, R.L., and Bimberg, D. (**2004**) BAM-L002 – a new type of certified reference material for length calibration and testing of lateral resolution in the nanometre range. *Surf. Interface Anal.*, **36**, 1423–1426.

5
Measurement of Mass and Density

Richard S. Davis

Handbook of Metrology. Edited by Michael Gläser and Manfred Kochsiek

ISBN: 978-3-527-40666-1

5.1 Introduction

Measurements of mass and density have had commercial importance since antiquity. With the rise of modern physics, beginning with Newton, mass (or density) became one of the fundamental quantities used to describe the natural world. In the eighteenth century, Lomonosov in Russia and Lavoisier in France demonstrated that mass is conserved in chemical reactions. Analytical balances soon became the one piece of equipment common to all chemistry laboratories. Balances have remained essential tools of physics and chemistry even though our understanding of mass has changed greatly since the time of Lavoisier.

The first section of this chapter is devoted to mass measurements. After some basic principles, the unit of mass in the International System of Units (SI) is defined, and research toward a redefinition in terms of physical constants is mentioned. Practical considerations in the specification of secondary mass standards are introduced. Balances, the essential tools that allow comparison of one mass with another, are then discussed. We conclude the section with the link between the laboratory standard of mass and the mass of astronomical bodies or subatomic particles.

The second section, on density measurements, reviews the most widely used techniques for the accurate determination of this quantity.

5.2 Mass

5.2.1 A Few General Principles

A parameter called *mass* appears in fundamental equations of physics. Instead of defining this quantity, we take the view that the equations themselves show how the mass ratio of any two entities can be measured.

Newton recognized that the quantity now called *mass* fulfilled two roles in his physics: as the proportionality constant between force and acceleration (inertial mass) and as a parameter in describing the force of gravitation (gravitational mass). We may further distinguish two types of gravitational mass: active mass, responsible for generating the gravitational field, and passive mass, influenced by the field. In Newtonian physics, the two types of gravitational mass are identical, but

Handbook of Metrology. Edited by Michael Gläser and Manfred Kochsiek

ISBN: 978-3-527-40666-1

their experimentally observed equivalence to inertial mass appears fortuitous.

The postulate that the ratio of inertial mass to passive gravitational mass is the same for all objects is now known as the *weak equivalence principle or the universality of free fall* (see, e.g., Misner, Thorne, and Wheeler, 1973). The principle may safely be assumed for the practical measurements described later.

According to special relativity, the rest mass of an object is not simply the total rest mass of all its parts. The rest mass also includes a contribution E/c^2, where E is the sum of all internal potential and kinetic energies and c is the speed of light. Thus, for example, the Coulomb energy within an atomic nucleus contributes to its rest mass (e.g., about 0.4% in the case of gold). It is the total mass energy that is conserved, and there is no mass measurement that can distinguish between the individual components of rest masses and energies.

In general relativity, mass plays the role of generating the curvature of space-time.

5.2.2
SI Unit, Relation to Atomic Mass

The SI provides a practical framework for the precise measurement of physical quantities. The kilogram is among the base units of the SI and is now unique in that its definition remains tied to an artifact (BIPM, 2006):

> *The kilogram is the unit of mass: it is equal to the mass of the international prototype of the kilogram.* 3rd General Conference of Weights and Measures (CGPM), 1901.

The international prototype, the artifact referred to, is a cylinder made of an alloy of 90% platinum and 10% iridium. Since its fabrication in the 1880s, it has been stored and used at the International Bureau of Weights and Measures (BIPM) in Sèvres, France, according to provisions of the Convention du Mètre (BIPM, 2006). The national laboratories of most industrialized countries are equipped with 1-kg prototypes that are copies of the defining artifact. The mass in kilograms of these national prototypes is traceable through calibration at the BIPM to the international prototype. It is then the task of the national laboratories to calibrate standards whose masses are multiples or submultiples of 1 kg for the requirements of science, industry, and commerce.

In essence, the mass m (in kilograms) of any object is the ratio m/m_0, where m_0 is the mass of the international prototype. Operationally, the desired ratio is achieved by a chain of calibrations:

$$\left(\frac{m}{m_n}\right)\left(\frac{m_n}{m_{n-1}}\right)\cdots\left(\frac{m_i}{m_j}\right)\cdots\left(\frac{m_1}{m_0}\right) \tag{5.1}$$

where intermediate ratios involve secondary standards. Every intermediate ratio m_i/m_j in the chain represents a measurement (or series of measurements) and thus contributes to the uncertainty of the final result. Intermediate ratios involving artifacts whose mass ratio is not nominally unity or whose densities are not nominally equal pose particular problems, as discussed later (Sections 5.2.3 and 5.2.4).

In the case of other base units of the SI, such as the meter, it became clear long ago that the original artifact standard was noticeably unstable when compared with a "natural" standard, such as the wavelength of light emitted in certain atomic transitions. An analogous situation has yet to be confirmed as regards the kilogram. However, the precision with which 1-kg artifacts can be compared

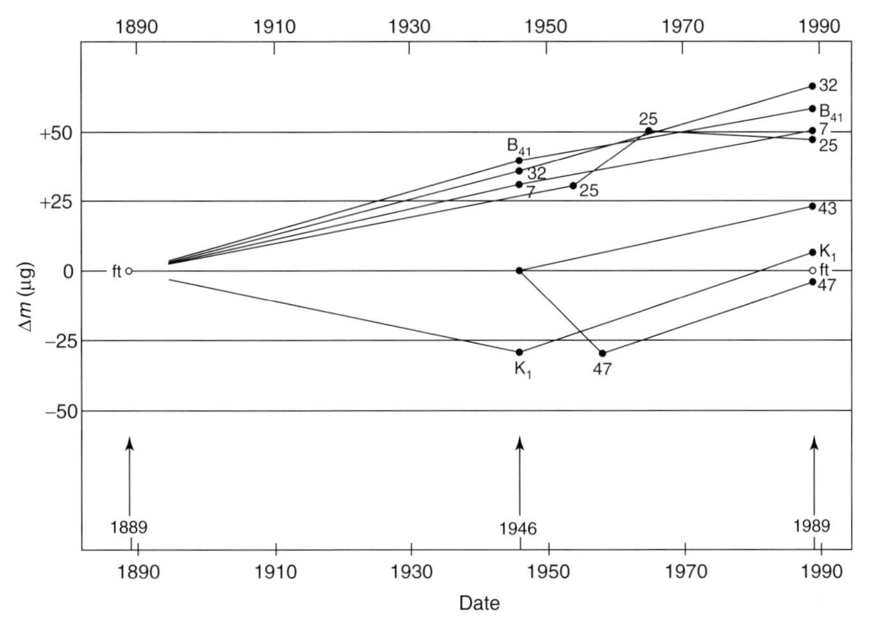

Fig. 5.1 The change in mass of seven similar artifacts since their first calibration, always with respect to the mass of the international prototype of the kilogram. The mass of the latter is defined to be 1 kg. The artifacts numbered 43 and 47 were first measured in 1946; the others in 1889. Courtesy of the Bureau International des Poids et Mesures.

is about 1 μg (1×10^{-9}), and one can easily see that a group of Pt/Ir kilogram artifacts made at the turn of the nineteenth century is slowly drifting with respect to the international prototype (Figure 5.1; Girard, 1994). It is conceivable, though unsubstantiated, that the entire group is drifting with respect to the constants of nature (Quinn, 1991). Several schemes are being pursued with a goal of checking the stability of the international prototype with respect to physical constants (Mohr, Taylor, and Newell, 2008). As yet, none of these schemes has achieved a precision sufficient to interpret the drifts shown in Figure 5.1. However, redefinition of the kilogram in terms of a physical constant, such as Planck constant, *h*, would already benefit electrical metrology and some areas of physics (Mills *et al.*, 2006). Redefinition of the kilogram, if it occurs, would maintain historical continuity.

Atomic masses can be measured to high precision relative to the mass of carbon-12 (Audi, Wapstra, and Thibault, 2003). The SI mole is the unit of "amount of substance" linking macroscopic and atomic mass: 1 mol of unbound atoms of carbon-12 at rest in their ground state has by definition a mass of exactly 0.012 kg. Each such atom has a mass of exactly 12 atomic mass units (u). The number of entities in 1 mol of any specified substance is the Avogadro constant $N_A(6.022\,141\,79(30) \times 10^{23}\ \mathrm{mol}^{-1}$ (Mohr, Taylor, and Newell, 2008)), numerically equivalent to the number of atomic mass units in 0.001 kg. Hence, a measurement of the Avogadro constant to sufficient accuracy (and repeated at regular intervals) would provide a practical means of realizing the kilogram in terms of the mass of a single atom of carbon-12. Since the value of the molar Planck constant, $N_A h$, is already well known, redefinition of the kilogram could also be made in terms of h (Mills *et al.* 2006; Mohr, Taylor, and Newell, 2008).

5.2.3 Mass Balances (General)

The mass ratios in Equation (5.1) may be determined to high accuracy by the use of analytical balances. In principle, these devices are transducers whose output can be related to the difference in mass between two objects. For the smallest uncertainty, the two objects are nominally equal and the range of the transducer (or on-scale range of the balance, to use more traditional terminology) is a small fraction of the mass of each object.

The balance capacity puts an upper limit on the nominal mass of each of the objects no matter how small the difference between them. The use of such a transducer requires a companion set of mass standards that can be assembled to approximate the mass of the unknown.

Modern analytical balances have on-scale ranges equal to their capacity. This is a major convenience. If the scale is linear, it can be calibrated with a single mass standard. However, demonstrating linearity requires additional standards, so that even these balances achieve lowest uncertainty when used to compare the mass of nominally equal objects.

The existence of narrow-range mass transducers is sufficient to create standards whose mass is any multiple or submultiple of 1 kg. For instance, it is possible, using a sensitive balance, to demonstrate that 10 artifacts are approximately equal in mass and that the 10 taken as a group have a mass approximately equal to 1 kg. Thus, the mass of each of the 10 artifacts is nominally 100 g. Corrections to the nominal mass can be calculated by taking the balance readings into account. The uncertainty in each correction is ultimately determined by the precision of the balances that were used (see, e.g., Prowse and Anderson, 1974).

5.2.4 Need for a Buoyancy Correction

Weighings carried out in air (or any other fluid) will require correction for buoyancy. If comparison weighing is done in air of density ρ_a, then the unknown object acts as if its effective weight were $m(\text{eff})g$:

$$m(\text{eff}) = m - \rho_a V \tag{5.2}$$

$$= m(1 - \rho_a/\rho) \tag{5.3}$$

where V is the volume of the object, ρ is its density, and g is the gravitational acceleration in the laboratory. For simplicity, assume the weighing is done on a linear balance whose scale range equals its capacity. The balance output is routinely calibrated weighing a standard of known mass m_s, with nominal value near the capacity. The calibration procedure adjusts the scale output to read (m_s), where $\{\}$ indicates the numerical value of the quantity within the brackets. However, the balance has actually responded not to m_s but to the weight $m_s(\text{eff})g$

$$m_s(\text{eff}) = m_s(1 - \rho_a/\rho_s) \tag{5.4}$$

where g is the local gravitational acceleration and ρ_s is the density of the standard. Consequently, m is related to its balance reading $\{y\}$ by

$$m = \{y\}\,\frac{(1 - \rho_a/\rho_s)}{(1 - \rho_a/\rho)} \tag{5.5}$$

The quantity $\{y\}$ is often referred to as the *apparent mass of the unknown object*. The local gravitational acceleration is the same for the standard and unknown and thus does not appear in Equation (5.5) as long as the centers of mass of the two objects are at the same elevation (the relative gradient of g is about -3×10^{-9} cm^{-1} at the surface of the earth). The buoyancy terms are more problematic, especially when one is weighing materials much less dense than the standard. Since ρ_s is typically 8000 kg m^{-3} and ρ_a is about 1.2 kg m^{-3} in laboratories at sea level, the correction for air buoyancy is, for example, roughly 0.1% for water.

The air density in a given laboratory whose temperature is controlled to within a few degrees will be mainly a function of ambient air pressure and thus will not vary beyond 5% of its average value. Air pressure is a function of elevation so that the mean air density in laboratories at 1600 m above sea level is about 1.0 kg m^{-3}. For lower uncertainty, the density of air can be computed from an equation of state whose input parameters are temperature, barometric pressure, relative humidity (or dew point), and atomic fraction of carbon dioxide (Picard *et al.*, 2008). A simplified version of this equation (OIML, 2004a) is adequate to achieve an uncertainty within 0.01% in air density:

$$\rho_a = \frac{0.34848p - 0.009(hr) \times \exp(0.061t)}{273.15 + t} \quad (5.6)$$

with ρ_a in kilograms per cubic meter, and where the air in the balance case has the following measured parameters: barometric pressure p (hectopascal or millibar), temperature t (degree Celsius), and relative humidity hr (percent).

5.2.5
Specification of Secondary Standards

The SI unit of mass is disseminated by means of secondary standards. Analytical mass standards are readily available commercially in denominations from 1 mg to 50 kg. Summations up to 400 tons of much larger artifacts may be needed to calibrate force transducers such as strain-gauge load cells. Because of the difficulty in their manipulation, large masses are most easily calibrated at the point of use (Bray, Barbato, and Levi, 1990). In recent years, there has been a growing need for standards with mass smaller than 1 mg. Production of such standards, previously left to the ingenuity of the individual user, is now being taken up by a number of national metrology institutes.

Mass standards used in legal metrology, named *weights*, are manufactured to specifications set by each country, usually in accordance with Recommendation R111-1 of the International Organization of Legal Metrology (OIML, 2004a). Thus, the mass of artifacts of a particular class are specified to be within certain maximum permissible errors, as shown in Table 5.1. If the user needs to know the mass to better accuracy, then each artifact must undergo calibration so that individual deviations from the nominal mass can be known.

The need for buoyancy corrections to weighings in air has led to conventions for the construction of weights. OIML, specifies the following convention (OIML, 2004b).

1. An artifact of mass m_s and density ρ_s has a conventional mass m_c defined as

$$m_c = m_s \frac{(1 - 1.2/\rho_s)}{(1 - 1.2/8000)} \quad (5.7)$$

It is the conventional mass that is referred to in Table 5.1 and in typical

Tab. 5.1 Maximum permissible errors for three best accuracy classes of weights, as recommended by the OIML.

Nominal value	Class E_1	Class E_2	Class F_1
50 kg	25 mg	80 mg	250 mg
20 kg	10	30	100
10 kg	5.0	16	50
5 kg	2.5	8.0	25
2 kg	1.0	3.0	10
1 kg	0.5	1.6	5
500 g	0.25	0.8	2.5
200 g	0.10	0.3	1.0
100 g	0.05	0.16	0.5
50 g	0.03	0.10	0.3
20 g	0.025	0.08	0.25
10 g	0.020	0.06	0.20
5 g	0.016	0.05	0.16
2 g	0.012	0.04	0.12
1 g	0.010	0.03	0.10
500 mg	0.008	0.025	0.08
200 mg	0.006	0.020	0.06
100 mg	0.005	0.016	0.05
50 mg	0.004	0.012	0.04
20 mg	0.003	0.010	0.03
10 mg	0.003	0.008	0.025
5 mg	0.003	0.006	0.020
2 mg	0.003	0.006	0.020
1 mg	0.003	0.006	0.020

reports of calibration, although the mass m_s may be given as well. To limit confusion, m_s is sometimes be referred to as the *true mass* or *vacuum mass* of the standard if it is reported along with the conventional mass. The user is normally expected to treat the artifact standard as if its mass were m_c and its density were 8000 kg m^{-3}. From Equations (5.4), (5.5), and (5.8), this fiction does not propagate as error provided $\rho_a = 1.2\,\text{kg m}^{-3}$. In general, however, the use of conventional mass does introduce an error e given as

$$e \approx m_s(1.2 - \rho_a)\left(\frac{1}{8000} - \frac{1}{\rho_s}\right) \tag{5.8}$$

Recommendation R111-1, therefore, contains a second specification.

2. The density ρ of an artifact standard must be such that a deviation in ρ_a of 10% from the assumed value of 1.2 kg m^{-3} will lead to an error that does not exceed one-fourth the absolute value of the maximum permissible error given in Table 5.1.

The introduction of conventional mass and density is intended to simplify the data analysis of routine weighing near sea level. The convention must, of course, be abandoned if its use will lead to serious error, for instance, if mass standards are used in vacuum.

The magnetic properties of secondary mass standards are sometimes of interest to experimentalists. Austenitic alloys of stainless steel and plated brass (where it meets the density requirements) are the most often used materials. The best grades of stainless steel have relative permeabilities less than 1.004. Common grades of brass and 18/8 stainless steel are much worse.

5.2.6 Analytical Balances

5.2.6.1 Design Considerations

The design of analytical balances has evolved significantly due to the perfection of electromagnetic servocontrol. The basic mechanical configuration is, however, traceable to ancient origins: two pans suspended symmetrically from the ends of a horizontal beam. The operating principle, shown schematically in Figure 5.2, is that a mass imbalance between the two pans creates a net torque on the beam so that its equilibrium angle is no longer zero. The sensitivity S is

$$s \equiv \frac{d\phi}{dm_2} = \frac{L}{m_1 l + (m_2 + m_3)h} \tag{5.9}$$

which will depend on the total load if the three pivots are not aligned. The most important design features of this type of balance are reviewed by Quinn (1992) and Speake (1987).

The so-called top-loading balances are designed with the pans above the beam. In order that an object produces the same torque no matter where it is placed, the pan must be guided so that it does not rotate with the beam. This can be accomplished by using twin beams in the form of a parallelogram. For large masses, the pan may be additionally equipped with a self-leveling mechanism. This design, though allowing unimpeded access to the pan, did not become practical for high-precision balances until the advent of servocontrol.

5.2.6.2 Servocontrol and Automation

The convenience and practical precision of balances are improved by automation and servocontrol. One can, in fact,

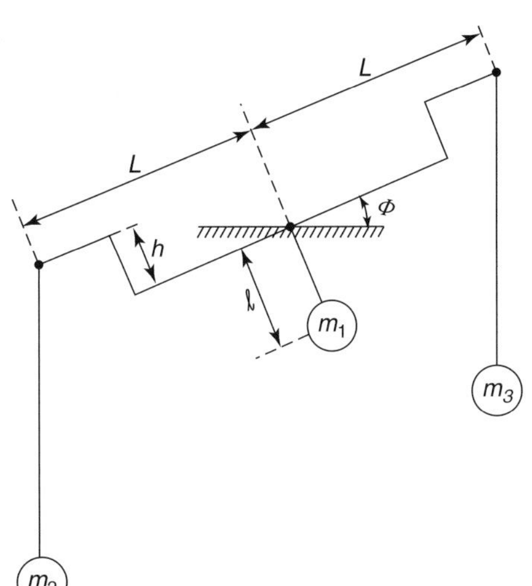

Fig. 5.2 Schematic of an equal-arm balance without servocontrol. The three pivot points are represented by solid circles. The mass of the beam is m_1 and is centered a distance ℓ below the central pivot. The loads on each pan are represented by m_2 and m_3.

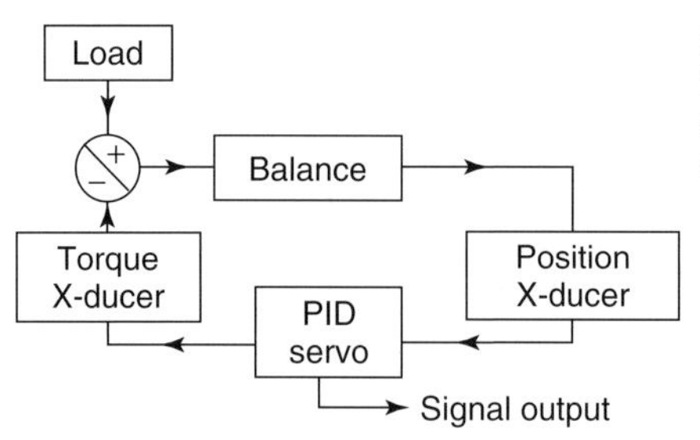

Fig. 5.3 Typical arrangement of a servocontrolled balance. The load, which would cause an unserved balance beam to rotate, is compensated by an electromagnetic torque transducer.

consider electronic balances as servocontrolled, modified versions of the balances presented above (Schoonover, 1982). Figure 5.3 shows a typical control loop where the balance is seen to be a transducer that converts a torque imbalance to an angular deflection that is itself converted to a voltage. An amplifier acting through an electromagnetic transducer supplies feedback to maintain the beam at constant angle. To optimize accuracy and transient response, the feedback signal is typically a linear combination of a force proportional to the angular deflection, its time integral, and its time differential (PID feedback).

There are several practical benefits of servocontrol. The rotation of the beam can be greatly restricted so that relatively robust leaf springs (also known as *flexure strips*) can replace delicate knives at the pivot points. The beam can be maintained precisely horizontal, thereby rendering the balance immune to spurious horizontal forces. The sensitivity of the unservoed balance can be adjusted until a tilt of the balance case causes the beam to tilt through exactly the same angle. A balance adjusted to this "autostatic" condition is relatively immune to common sources of mechanical noise (Speake, 1987). Finally, the on-scale, linear range of the balance can be greatly increased. Quinn (1992) describes a servocontrolled balance with a precision of order 0.05 μg for loads up to 2 kg. Speake (1987) derives fundamental theoretical limitations to balance precision. Commercial balances are now available that can compare 1-kg mass standards to a precision of 0.1 μg in air or vacuum.

Commercial balance manufacturers include a variety of convenience features in servocontrolled balances. The most common of these are automatic or semiautomatic scale calibration and suppression of the zero drift. Depending on the sophistication of the balance and its accessories, complete weighing operations can be fully automated.

5.2.6.3 **Force**

Analytical balances cannot distinguish the gravitational force on a mass from other vertical forces. In the case of air buoyancy, corrections can be applied and error avoided. Other forces must be made negligible. Of these, the principal concerns are moving air and electrostatic and magnetic interactions.

Thermal effects are minimized by keeping the objects to be weighed in good thermal equilibrium with the interior of the balance. Otherwise, a convection cell can be created that may result in serious systematic errors (Gläser, 1999).

Electrostatic problems often occur when one is weighing good insulators such as fused silica or poly(tetrafluoroethylene). Equipping the balance pan with a Faraday cage of stable mass can eliminate such problems, as can exposing the insulator to ionized air.

The weighing to small uncertainty of permanent magnets or even ferrous materials is often troublesome. To elucidate a problem, the object in question can be weighed in various orientations on the balance pan to see whether the results are independent of such maneuvers. If not, the object can be suspended well below the pan of the balance in hopes that the unwanted interaction is between the object and a part of the balance itself.

That analytical balances are blind to the origin of the force that they sense can be turned to advantage. Thus, laboratory balances are commonly used in measuring phenomena such as surface tension and magnetic susceptibility. Since the pan of a servocontrolled balance is maintained at fixed position with respect to the laboratory, measurement of forces that vary strongly with position is convenient with these balances.

5.2.7 Microbalances

5.2.7.1 Analytical Type, Adapted to Microweighing

Although modern analytical balances can have standard deviations of less than 1 μg, microbalances have traditionally been used for different types of measurements than those required of normal analytical balances. Typically, microbalances have relatively small capacity but precisions of 1 μg or better. They will operate in vacuum or controlled atmospheres and are often used to follow the evolution in mass of a sample that is undergoing some physical or chemical change.

5.2.7.2 Vibrating Transducers

Some commercial microbalances measure inertial mass. Typically this is accomplished by relating the change in the frequency of a mechanical resonance to the change in mass of the resonant system. In special cases, it is possible to design transducers that measure inertial mass to picogram sensitivity. The most widely known example is the quartz-crystal microbalance (Ward and Buttry, 1990; Lu and Czanderna, 1984). In this case, the piezoelectric effect is exploited in order to excite the shear resonance in a plate of AT-cut quartz. The resonant frequency f_0, usually about 10 MHz, decreases by an amount f that is proportional to the mass per unit area m/A loading the free surface. The mass sensitivity of such a device is approximated by a relation first derived by Sauerbray (1959):

$$\Delta m/A = -4.42 \times 10^6 \Delta f/f_0^2 \tag{5.10}$$

with $\Delta m/A$ in kilograms per meter square, for AT-cut quartz.

The field of MEMSs (micro-electromechanical systems) and NEMSs (nano-electromechanical systems) is burgeoning, and devices with attogram (10^{-18} g) precision have been reported.

5.2.8 Astronomical and Subatomic Mass

The product Gm_E, where m_E is the mass of the Earth, is approximately equal to gr^2, where r is the Earth's radius and G is the gravitational constant. Cavendish's measurement of G was popularly known as *weighing the Earth* in recognition of the role of G in relating planetary and other astronomical masses to laboratory standards. More generally, a small satellite in an elliptic orbit of semimajor axis a and angular frequency ω about a mass m will, in the appropriate limits, obey Newtonian mechanics:

$$m = \frac{a^3 \omega^2}{G} \tag{5.11}$$

Thus, for example, the mass in kilograms of the Sun can be known. The currently recommended value of G (Mohr, Taylor, and Newell, 2008) is $6.674\,28(67) \times 10^{-11}\ \mathrm{m^3\ kg^{-1}\ s^{-2}}$.

It is difficult to accurately determine the mass of celestial objects having no observable satellites. Consider the asteroids, for example. It is sometimes possible to infer the mass of the largest asteroids from observations of gravitationally perturbed neighbors. Otherwise, even more indirect methods must be used: the volume of the asteroid is estimated from its brightness (assuming a value for albedo), and the mass is then estimated by assuming a value for the asteroid's density.

The mass of subatomic particles may be found by applying the conservation laws of energy and momentum. The acceleration in electromagnetic fields of either the particle itself or its decay products may be measured, as well as the energy and momentum of emitted photons. By the nature of these measurements, the mass is most directly reported in electron volts (eV), an energy unit that is outside the SI. To convert electronvolts to kilograms, it is necessary to multiply by a factor equal to the numerical value of the elementary charge, e, divided by c^2 (Mohr, Taylor, and Newell, 2008).

5.3 Density

The density ρ of an infinitesimal volume dV of material of mass dm is

$$\rho = \frac{dm}{dV} \tag{5.12}$$

In the SI, density is a derived quantity with units of kilograms per cubic meter. Depending on the application, it may be convenient to measure the density of the unknown relative to a known density (most usually that of water). Alternatively, one may measure directly the mass and volume of the unknown. The latter technique is, of course, the only method of determining a reference density directly in terms of SI units.

5.3.1 Measurements Based on Buoyancy

Buoyancy methods rely on Archimedes' principle, just as do the corrections for air buoyancy described in Section 5.2.4. Immersion of the sample in a liquid results in a buoyancy "correction" so large that the density of the sample may be thus determined with respect to the density of the immersion liquid.

The special case of neutral buoyancy forms the basis of the flotation methods discussed in Sections 5.3.1.2 and 5.3.1.3.

5.3.1.1 Hydrostatic Weighing

Two methods of hydrostatic weighing are shown schematically in Figure 5.4. Method 1 is the classic technique capable of relative uncertainty below 10^{-6} in favorable cases. Method 2 suffers from a greater minimum uncertainty but is more convenient when one is using a servocontrolled top-loading balance.

Let x_1 be the ratio of the balance reading when the sample is placed on the pan in air to the reading when immersed in the liquid bath; ρ_w and ρ_a the density of the water bath and the ambient air; and ρ the unknown density to be determined. It

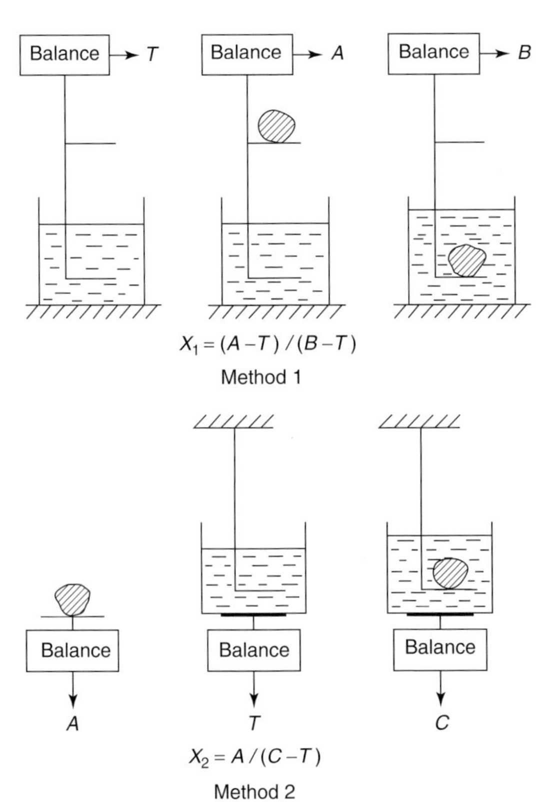

Fig. 5.4 Two methods for determining the density of a solid sample relative to an immersion liquid. The dimensionless quantities x_1 and x_2 are used in Equations (5.13) and (5.14).

is then easy to show that

$$\rho = \frac{x_1 \rho_w - \rho_a}{x_1 - 1} \text{ (Method 1)} \tag{5.13}$$

For Method 2, let x_2 be the ratio of the balance reading when the sample is placed on the pan in air to the difference in readings upon immersion. Then

$$\rho = x_2(\rho_w - \rho_a) + \rho_a \text{ (Method 2)} \tag{5.14}$$

Note that x_1 and x_2 are dimensionless ratios so that the balances used need not be calibrated to any particular unit of mass (although their linearity must be assured). This feature is often cited as an advantage to using the same balance for both the air and water weighings. The two methods will only be sensitive if the balance is precise compared to $\rho_w V$. In addition to the obvious requirements of knowing ρ_w and ρ_a, achieving accurate results depends on careful attention to the suspension wire and prevention of gas bubbles adhering to the sample. The report of Bowman and Schoonover (1967) still serves as a useful introduction to these subjects and a paper by Fujii (2006) shows the present state of advancement. The method can also be used to determine the density of the surrounding liquid if the density of the solid is known. Techniques to measure the density of mass standards of various nominal values are described in OIML (2004a).

The suspension wire must be eliminated when determining the density of fluids far from ambient conditions (under high pressure, for instance). In such cases, the solid can take the form of a magnet that is servocontrolled to a position within the liquid by means of external coils (see the review by Wagner and Kleinrahm, 2004 and included references). Analogs of both Methods 1 and 2 shown above are practical.

It may happen that water is not suitable as an immersion liquid for a given sample. In this case, a different liquid is used and its density at the time of the measurements determined from published tables, either by performance of an auxiliary experiment or by means of a solid density reference standard (Bowman *et al.*, 1974). There are obvious advantages if the reference standard has approximately the same mass and density as the unknown.

5.3.1.2 Flotation Methods

As density is an intensive quantity, it should be possible to determine very small samples to high accuracy. Flotation methods accomplish this feat. When a sample of volume V whose density ρ is to be determined is immersed in a liquid of density ρ_L, the vertical force on it will be $(\rho - \rho_L)Vg$. The sample will, in general, rise or sink depending on whether this quantity is negative or positive. If the value of ρ_L can be adjusted so that the sample does neither (neutral buoyancy), then $\rho = \rho_L$ and the problem then becomes that of determining the density of a relatively large volume of liquid.

The method is most easily realized with transparent liquids. Since the sample and liquid generally have different thermal expansion and compressibility, equilibrium may be achieved through temperature or pressure changes as well as by adjusting the composition of a binary liquid.

When the surrounding liquid approaches the density of the sample, it becomes increasingly difficult to determine whether the sample will ultimately rise or fall. Greater sensitivity can be achieved by the use of a centrifuge rather than gravity to produce the forces. Alternatively, one can create a uniform density gradient in a column of liquid. Samples introduced in the column will, unattended, eventually come to rest at their position of neutral buoyancy.

Flotation methods are useful in quantifying small differences in density between two objects of the same material. Objects of different sizes and shapes can be compared in this way (Bettin and Toth, 2006).

5.3.1.3 Hydrometry

Hydrometers have been used as a relatively convenient method to find the density of liquids, especially when only small differences among samples are expected. The hydrometer consists of a graduated stem with a weighted volume at the bottom. The hydrometer floats in the test liquid, with the base of its stem in the liquid and the top of the stem in the air. Depending on the desired measurand, the stem is calibrated in units of density, relative density, or a specialized quantity related to density.

The review paper by Lorefice and Malengo (2006) describes how these instruments are calibrated and includes an extensive bibliography.

5.3.2 Measurements Based on Mass and Volume

If a sample of unknown density ρ has the same volume as a known standard of density ρ_s, then the ratio of masses m/m_s of the unknown and standard must equal ρ/ρ_s, the ratio of their densities. Although the volume of the unknown and the standard must be the same, it need not be known. Thus, mass determinations constitute the only precise measurements needed. Special containers, known as *pycnometers (or pyknometers)*, are designed to hold a reproducible volume of liquid.

5.3.2.1 **Liquid Pycnometers**

A variety of liquid pycnometers have been designed over the years and marketed by laboratory supply companies. Although all these have design advantages specific to given applications, it may be sufficient to use a simple hypodermic syringe as a volume standard. In this case, the syringe is first weighed empty. If V_i is the internal volume of the syringe, V_e its external volume, m its mass, ρ_a the density of ambient air, and y_1 the balance reading, then

$$y_1 = \rho_a(V_i - V_e) + m \qquad (5.15)$$

The syringe is then filled with the standard liquid of density ρ_L, usually water, and weighed again:

$$y_2 = \rho_L V_i - p_a V_e + m \qquad (5.16)$$

Finally, it is filled with the sample of unknown density ρ and weighed for the third time:

$$y_3 = \rho V_i - \rho_a V_e + m \qquad (5.17)$$

The desired density is then determined from the known densities of the liquid standard and the ambient air:

$$\rho = (\rho_L - \rho_a)\frac{y_3 - y_1}{y_2 - y_1} + \rho_a \qquad (5.18)$$

This example suggests many of the potential problems of pycnometry: the interior volume must be reproducible to the needed precision. The chamber must be carefully cleaned between fillings. The pycnometer itself must be a suitable object for weighing. If all measurements cannot be carried out at the same temperature, then the thermal expansion of the pycnometer must be known. Liquids must be introduced without entraining air bubbles (ISO, 1999).

Some commercially available pycnometers are designed for density measurements of solids, often in the form of grains or powders. In this case, the first observation is the mass of the solid sample itself. The third observation is the mass of the pycnometer with its interior volume containing the solid sample plus enough of the standard liquid to fill the interior volume completely. The density of the solid is then determined. This technique will only work well if atmospheric gases occluded by the solid can be eliminated.

Pycnometers were traditionally designed for use with a chemist's analytical balance with a capacity of about 200 g. Thus, their interior volume is typically 25–100 ml.

The so-called gas pycnometers are available commercially for the determination of solid volumes. They are based on the fact that, at constant temperature, the pressure of a fixed number of atoms of a noble gas varies inversely with the volume to which the gas is confined. In a gas pycnometer, introduction of a sample decreases the original volume by that of the sample. The measurements can be automated to an extent. The method is especially suited to the determination of powders or samples with open pores, since noble gases penetrate better than liquids.

5.3.2.2 **Vibrating-Tube Densimeters**

Pycnometry, as developed by Gay-Lussac and refined by later investigators, has been superseded in many laboratories by the vibrating-tube densimeter. First developed by Kratky, Leopold, and Stabinger

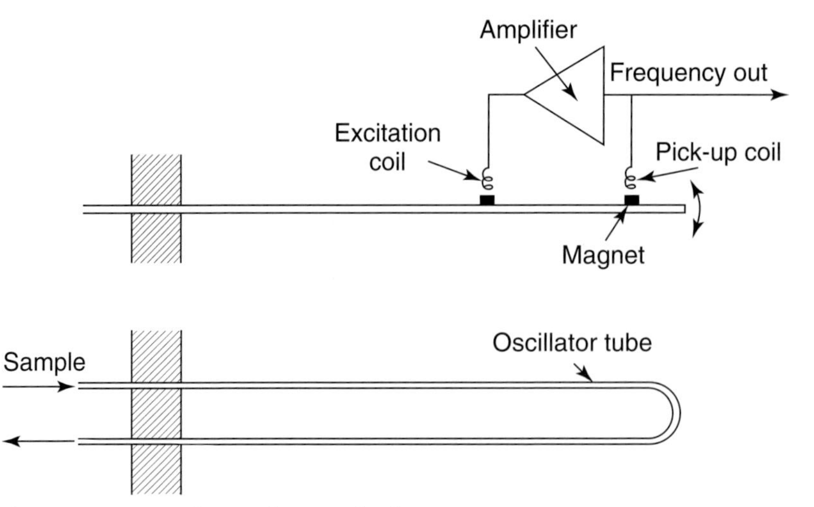

Fig. 5.5 Two views showing the principle of a vibrating-tube densimeter. The natural frequency of the hollow oscillator tube will depend on the density of the sample that fills it. Based on a sketch provided by H. Stabinger, used with permission.

(1969), this device differs from the traditional liquid pycnometer in only one fundamental respect, namely, that the inertial mass of the material within the volume chamber is sensed rather than the gravitational mass. This one conceptual change leads to many experimental advantages.

To realize the device, the liquid container takes the form of a hollow tube bent into the form of a "U" and anchored at each end (Figure 5.5). Like a tuning fork, the tube will have a mechanical resonance whose frequency depends on elastic properties, geometry, and mass. In particular, the natural period τ of the tube is proportional to $m^{1/2}$, the square root of its mass. The latter quantity is equal to the mass of the tube itself plus ρV_e, the mass of any fluid filling the vibrating region of the tube. In practice, V_e is about 1 ml, that is, much smaller than the volume of typical pycnometers. On the basis of this simple model,

$$\rho = A\tau^2 - B \tag{5.19}$$

The parameters A and B are constants, and thus the device must be calibrated at a minimum of two points over its range. Dry nitrogen and water are frequently used as standards.

The highest relative precision obtainable with such devices is of order 1×10^{-6}. Deviations from Equation (8.19) are noticeable in highly precise data, and so more careful calibration is required to achieve comparable accuracy. Devices based on the same principle of operation have been constructed for temperatures to 400°C and pressures to 40 MPa. Practical considerations are given in the ISO (1998) standard.

5.3.3
Miscellaneous Techniques

The techniques discussed above are those most generally used for density determinations. Nevertheless, myriad other methods have been developed to meet special needs.

Where volume or density appears in a physical equation, it has probably formed the basis for a determinative method. Numerous examples of traditional methods are reviewed by Davis and Koch (1992).

5.3.4 Reference Standards

Distilled or deionized tap water has a density that is known to a relative uncertainty of about 1×10^{-5} at 0.1 MPa and between 0 and 40°C and is the single most useful reference standard. Uncertainty may be improved by about a factor of 3 by correcting for or eliminating atmospheric gases dissolved in the water and by taking account of the isotopic variability of individual water samples.

A particular isotopic mix of water is designated Vienna Standard Mean Ocean Water (V-SMOW), small samples of which can be obtained from the International Atomic Energy Commission in Vienna for use in determining the relative isotopic abundances of other water samples of interest. A table of water of density suitable for typical metrological needs is given by Tanaka *et al.* (2001). A much more general formulation of the thermodynamic properties of V-SMOW has been published by Wagner and Pruß (2002).

Good use is sometimes made of the fact that the density of water has a maximum near 4°C. Problems that might be caused by convection are greatly reduced near this temperature since the density is insensitive to small changes in temperature.

The density of selected mercury samples was determined with respect to standards of mass and length by Cook (1961), who achieved a relative uncertainty of about 1×10^{-6} at 0.1 MPa and 20°C. In the succeeding years, these samples have played a key role in precision manometry, gas thermometry, and measurements of fundamental physical constants.

Glossary

Apparent Mass: The reading, uncorrected for air buoyancy, of a perfectly calibrated balance. The ratio of mass to apparent mass is a function of the ambient air density, the density of the object in question, and the density of the standards used to calibrate the balance (Section 5.2.4).

Atomic Mass Unit (Dalton): The unit mass, u, in a system where the mass of an atom of carbon-12 is taken to be exactly 12 u. One mole of carbon-12 atoms has a mass of exactly 0.012 kg.

Avogadro Constant: The number of entities in 1 mol of substance and hence the number of atomic mass units in 0.001 kg.

Balance: Originally, an ancient lever device used to demonstrate that two objects have the same (gravitational) mass. The term is now used more generally to designate almost any transducer configured to make mass measurements.

Buoyancy Correction: The correction applied to balance readings in order to eliminate the component due to air buoyancy.

Cook's Mercury (Informal): Mercury stock whose density was measured by A. H. Cook. The stock has been used to advantage in several areas of precision measurement.

Conventional Mass: At a reference temperature of 20°C, the mass of material of density 8000 kg m^{-3} that will balance an object of interest, usually a secondary standard of mass, in air of density 1.2 kg m^{-3} (Section 5.2.4).

Dalton: See atomic mass unit.

Effective Mass: The mass of an object less the product of its volume and the density of the ambient fluid, usually air (Section 5.2.4).

Gas Pycnometer: A device used to find the volume of porous solids and powders through application of the gas laws.

Gradient Column: A column of liquid whose density decreases uniformly from the bottom to the top. The densities of a sample or set of samples can be determined from their equilibrium heights in the column.

Gravitational Mass: The mass that appears in Newtonian equations for gravitational force.

Hydrostatic Weighing: A traditional method of measuring the density of a solid by weighing it both in air and submerged in a liquid, usually water (Section 5.3.1.1).

Inertial Mass: The mass that appears in Newton's second law and in special relativity.

International Prototype of the Kilogram: The artifact that defines the unit of mass in the SI.

Maximum Permissible Error: The maximum allowed absolute difference between the conventional mass of a weight (2) and its nominal mass.

Microbalance: A balance of microgram or better resolution.

Neutral Buoyancy: The condition in which an object at equilibrium neither rises nor sinks in a surrounding fluid. It is assumed that the system is subject to a uniform acceleration, usually gravitational but sometimes centripetal.

Nominal Mass: The intended mass of a standard weight (2).

Pycnometer (Pyknometer): A vessel with precisely reproducible interior volume, traditionally used to determine the relative densities of liquids.

Quartz-Crystal Microbalance: A device that measures mass per unit area in terms of the oscillation frequency of a quartz crystal (Section 5.2.4).

Relative Density: A dimensionless quantity representing the density ratio of a given sample to a standard of reference. Unless otherwise stated, the standard is assumed to be distilled water at 20°C. Relative density is the modern name for specific gravity.

True Mass: See vacuum mass.

Vacuum Mass: Another term for mass. The term vacuum mass or true mass is sometimes used to distinguish mass from conventional mass in reports of calibration.

Vibrating-Tube Densimeter: A device that measures the relative densities of fluids by a mechanical resonance technique (Section 5.3.1.2).

Vienna Standard Mean Ocean Water (V-SMOW): Water having a certain, specified isotopic composition. Distilled tap water generally has a density about $0.003\ \text{kg m}^{-3}$ less than V-SMOW.

Weak Equivalence Principle: A basic principle of physics whose consequence for mass measurements is as follows: equivalent results are obtained independent of whether the measuring device responds to the inertial or the gravitational properties of mass.

Weighing: The process of determining mass by means of a balance.

Weight (1): The product of a mass and the local gravitational acceleration. The term *standard weight* is used to signify weight at a conventional acceleration, usually $9.80665\ \text{m s}^{-2}$.

Weight (2): An artifact manufactured to embody a specified nominal mass.

References

Audi, G., Wapstra, A.H., and Thibault, C. (**2003**) *Nucl. Phys. A*, **729**, 129–336, 337–676.

Bettin, H. and Toth, H. (**2006**) *Meas. Sci. Technol.*, **17**, 2567–2573.

BIPM (**2006**) *The International System of Units (SI)*, 8th edn, Bureau International des Poids et Mesures, Sèvres, *http://www.bipm.org/en/si/*.

Bowman, H.A. and Schoonover, R.M. (**1967**) *J. Res. Natl. Bur. Stand.*, **71C**, 179–198.

Bray, A., Barbato, G., and Levi, R. (**1990**) *Theory and Practice of Force Measurement*, Academic, London.

Cook, A.H. (**1961**) *Philos. Trans. R. Soc. Lond. A*, **254**, 125–154.

Davis, R.S. and Koch, W.F. (**1992**) Chapter 1, in *Physical Methods of Chemistry*, 2nd edn, vol. **6** (B.W. Rossiter and R.C. Baetzold), John Wiley & Sons, Inc., New York.

Fujii, K. (**2006**) *Meas. Sci. Technol.*, **17**, 2551–2559.

Girard, G. (**1994**) *Metrologia*, **31**, 317–336.

Gläser, M. (**1999**) *Metrologia*, **36**, 183–197.

http://physics.nist.gov/cuu/Constants/RevModPhys_80_000633acc.pdf.

ISO (**1998**) *Oscillation-type density meters – Part 1*, 15212-1:1998, International Organization for Standardization.

ISO (**1999**) *Laboratory glassware – Pyknometers*, 3507:1999, International Organization for Standardization.

Kratky, O., Leopold, H., and Stabinger, H. (**1969**) *Z. Angew. Phys.*, **4**, 273–277.

Lorefice, S. and Malengo, A. (**2006**) *Meas. Sci. Technol.*, 2560–2566.

Lu, C. and Czanderna, A.W. (eds) (**1984**) *Applications of Piezoelectric Quartz Crystal Microbalances*, Elsevier, Amsterdam.

Mills, I.M., Mohr, P.J., Quinn, T.J., Taylor, B.N., and Williams, E.R. (**2006**) *Metrologia*, **43**, 227–246.

Misner, C.A., Thorne, K.S., and Wheeler, J.A. (**1973**) *Gravitation*, Freeman, San Francisco.

Mohr, P.J., Taylor, B.N., and Newell, D.B. (**2008**) *Rev. Mod. Phys.*, **80**, 633–730.

OIML (**2004a**) International Recommendation R111-1 (E): International Organization of Legal Metrology. *http://www.oiml.org/publications/*.

OIML (**2004b**) International Document D28 (E): International Organization of Legal Metrology. *http://www.oiml.org/publications/*.

Picard, A., Davis, R.S., Gläser, M., and Fujii, K. (**2008**) *Metrologia*, **45**, 149–155.

Prowse, D.B. and Anderson, A.R. (**1974**) *Metrologia*, **10**, 123–128.

Quinn, T.J. (**1991**) *IEEE Trans. Instrum. Meas.*, **40**, 81–85.

Quinn, T.J. (**1992**) *Meas. Sci. Technol.*, **3**, 141–159.

Sauerbray, G. (**1959**) *Z. Phys.*, **155**, 206–222.

Schoonover, R.M. (**1982**) *Anal. Chem.*, **54**, 973A–980A.

Speake, C.C. (**1987**) *Proc. R. Soc. Lond. A*, **414A**, 333–358.

Tanaka, M., Girard, G., Davis, R., Peuto, A., and Bignell, N. (**2001**) *Metrologia*, **38**, 301–309.

Wagner, W. and Kleinrahm, R. (**2004**) *Metrologia*, **41**, S24–S39.

Wagner, W. and Pruß, A. (**2002**) *J. Phys. Chem. Ref. Data*, **31**, 387–535.

Ward, M.D. and Buttry, D.A. (**1990**) *Science*, **249**, 1000–1007.

Further Reading

(a) Kochsiek, M. and Gläser, M. (eds) (**2000**) *Recent Monographs on Mass Metrology are: Comprehensive Mass Metrology*, Wiley-VCH Verlag GmbH; (b) Jones, F.E. and Schoonover, R.M. (eds) (**2002**) *Handbook of Mass Measurement*, CRC Press.

Experimental progress in linking the SI unit of mass to physical constants are reported at the Conference on Precision Electromagnetic Measurements, held biannually. Proceedings are published in a special issue of the IEEE Transactions on Instrumentation and Measurement.

A brief introduction to mass measurements below the microgram level can be found in the chapter by Davis, R.S. (**2007**) in *Metrology and Fundamental Constants* (eds T.W. Hänsch, S. Leschiutta, and A.J. Wallard), IOS Press, pp. 473–497.

6
Measurement and Instrumentation of Flow

Francis R. Ruppel and James E. Hardy

Handbook of Metrology. Edited by Michael Gläser and Manfred Kochsiek

ISBN: 978-3-527-40666-1

6.1 Introduction

The measurement of flow is one of the most common requirements for processes in research and industry. In the development of new processes or products, the flow rate of the various ingredients is needed to assess the impact of mixture ratios on product or process quality. In industry, once a particular process has been "set up," the recipe must be controllable and repeatable, which often requires the measurement and control of flow. In energy usage and custody transfer, the accurate measurement of flow is necessary to track usage trends and customer energy consumption for billing purposes. Instrumenting pipelines and vessels with flow sensors is often required to ensure process, equipment, or personnel safety or to monitor for leaks or infiltration.

Because flow is so commonly measured, a wide variety of flowmeter types based on a variety of measurement principles are available. A universal flowmeter does not exist. Each type has its own advantages and limitations. Careful consideration of the measurement criteria must be given to each application before deciding what type of flowmeter should be used.

As in all areas of instrumentation, advances in microelectronics have significantly improved flowmeter accuracy, ease of use, maintainability, and cost of ownership. "Smart" flow sensors are becoming more popular, extending the range and accuracy of flowmetering. The predominant use of flowmeters is in the measurement of single-phase fluids in closed systems (pipes or ducts) and in open channels. In this article, the use of flowmeters for these two areas is discussed. The measurement of the flow of solids or multiphase flow is not considered here. The basic principle of operation of various flowmeters is discussed along with some applications and major advantages and limitations. A few flow velocity sensors are also described. Finally, some parameters that affect flow measurement uncertainty as well as calibration techniques that may quantify some of the uncertainties are briefly outlined. This article is intended to give an overview of flow instrumentation but is not all inclusive. Most popular meters as well as a few recent innovations are discussed.

Handbook of Metrology. Edited by Michael Gläser and Manfred Kochsiek

ISBN: 978-3-527-40666-1

6.2
Classification, Principle of Operation, Applications, and Advantages and Disadvantages of Flowmeters

6.2.1
Differential Producing

Bernoulli's equation for a nonviscous fluid shows that the pressure P, fluid density ρ, elevation z, and velocity v at two points are related as follows, where g_c = gravitational constant:

$$\frac{v_1^2}{2} + g_c z_1 + \frac{P_1}{\rho_1} = \frac{v_2^2}{2} + g_c z_2 + \frac{P_2}{\rho_2} \tag{6.1}$$

This equation is the basis for all differential-producing flow meters. In different hardware configurations, a differential-producing flow meter uses available measurements of differential pressure at a minimum and may measure temperature and static pressure as well for compressible fluids, where the density is not constant.

The first hardware configuration to consider is the Pitot tube as shown in Figure 6.1. In this configuration, there are two pressure measurements, the impact pressure of the fluid stream (also referred to as *velocity pressure*), indicated as P_2. The second measurement is the static pressure at the back side of the insertion tube, indicated as P_1. When measuring gas flow rate, the pressure difference between the two measurement points for Pitot tubes is usually a draft-range difference, meaning only a few centimeters of water column. Therefore, the density at the two points can be considered to be the same. For liquid measurements, the fluid is considered to be incompressible, so the density at the two points is the same. Also, because the elevation at the two measurement points are the same, and the velocity at the impact point is zero, Equation (6.1) can be simplified and rearranged as shown in Equation (6.2)

$$v_1 = \sqrt{\frac{2(P_2 - P_1)}{\rho}} \tag{6.2}$$

The velocity of the fluid is related to the differential pressure between the pressure taps. The velocity can be converted to volumetric flow rate q_V by using the cross-sectional area A of the pipe.

$$q_V = vA \tag{6.3}$$

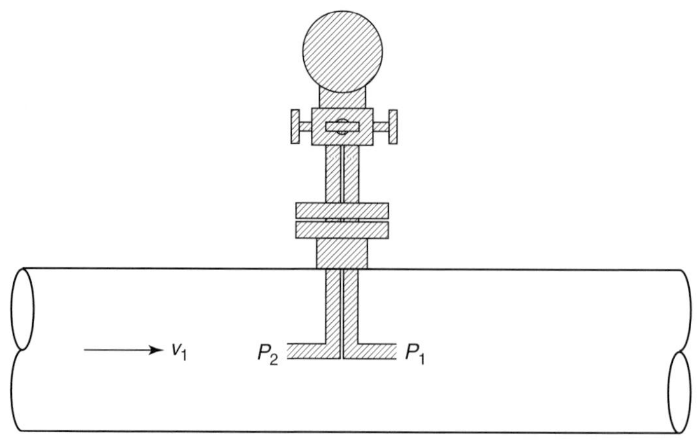

Fig. 6.1 Pitot tube flowmeter with fluid velocity v_1, impact pressure P_2, and static pressure P_1.

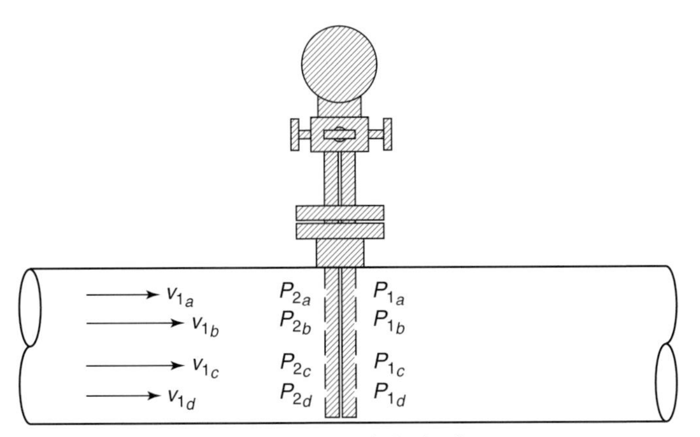

Fig. 6.2 Averaging Pitot flowmeter with fluid velocity v_1, impact pressure P_2, and static pressure P_1. Subscripts refer to points along the cross section of pipe.

Because the velocity pressure measurement is measured at a single point, it is sensitive to the velocity flow profile of the fluid. For example, the velocity at the middle of the pipe is not necessarily the same as the velocity at one-third of the distance into the pipe. Because of this, the inaccuracy is probably not better than ±5%. Another configuration of the Pitot tube is the averaging Pitot tube, as shown in Figure 6.2. This type of flow meter has several openings in each of the upstream and downstream tubes that measure velocity pressure and static pressure. The overall effect is that a single pressure measurement is taken in each tube, which represents the averaged pressure measurement. Some vendors with advanced, aerodynamic designs for the measuring tubes claim inaccuracies on the order of ±1% for averaging Pitot flow meters. Both designs should be used in applications with Reynolds number of 20 000 or more and have turndown ratios on the order of 4 : 1. A disadvantage of Pitot-type flowmeters versus other differential-producing flowmeters to be discussed in this section is that the developed differential pressure is a function only of the fluid parameters and diameter of the pipe line. For gas applications, it is common that very low differential pressure is produced, making for a challenging measurement. In contrast, for an orifice plate for example, it is always possible to reduce the diameter of the orifice to produce more differential pressure.

The second hardware configuration to consider is the orifice plate. A typical cross-sectional view of flow through an orifice and the corresponding pressure distribution are shown in Figure 6.3. The fluid stream contracts as it passes through a constriction. The velocity at each point is assumed to be inversely proportional to the cross-sectional area of flow at that point (not necessarily the same as the cross-sectional area of the pipe or orifice). The flow area contracts to a minimum at the vena contracta, the point of smallest flow cross section. The kinetic energy of the stream increases as it contracts, and the potential energy (pressure) decreases. Therefore, a pressure difference is created and can be measured. The relationship is more complicated with compressible flow because factors that affect the internal energy of the fluid, such as the density, compressibility, and

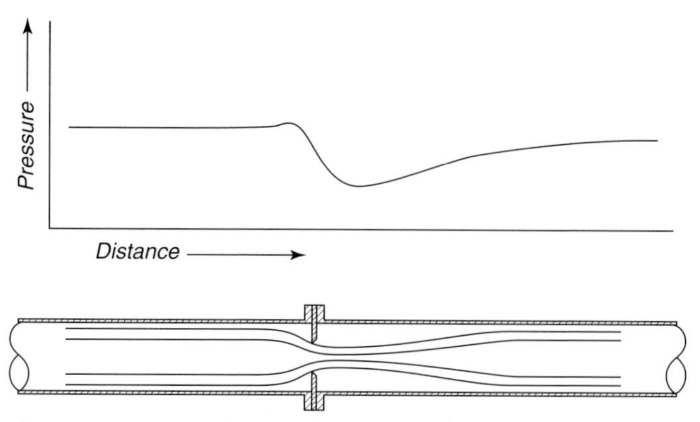

Fig. 6.3 Cross-sectional view of flow through an orifice.

isentropic exponent, have to be taken into account. However, the net result is the same: by measuring differential pressure across an engineered flow restriction, flow rate can be determined as a function of the square root of differential pressure when the cross-sectional flow area of each point is known.

The velocity at two points is related to the cross-sectional area as follows, where A_1 and A_2 are the cross-sectional areas at each point:

$$v_2 = v_1 \frac{A_1}{A_2} \tag{6.4}$$

By rearranging Equation (6.1) and using the relationship in Equation (6.4), the relationship of flow velocity for a constant density situation is given as follows:

$$v_1 = \sqrt{\frac{2}{1-\left(\frac{A_1}{A_2}\right)^2}}\sqrt{\frac{\Delta P}{\rho}} \tag{6.5}$$

By knowing the measurements of the pipe and orifice cross-sectional area and the density, the flow velocity is a function of the measured differential pressure across the orifice. For incompressible fluids, Equation (6.5) represents the actual flow fairly well, except that the cross-sectional area in the equation is of the fluid stream, not of the pipe and the orifice. Many empirical relations have been made to relate flow rate using dimensions of the pipe and orifice to actual flow rate. The relations are a function of where the pressure measuring taps are located. A term called the *discharge coefficient* C is added to the equation. The *discharge coefficient* is defined as the ratio of actual flow rate to theoretical flow rate. The discharge coefficient corrects the theoretical equation for velocity profile, actual energy loss between pressure taps for viscous fluids, and for the particular pressure tap location. ISO 5167 defines the discharge coefficient as a function of beta ratio β (the ratio of orifice diameter to pipe internal diameter D), Reynolds number Re, distance from upstream pressure tap to upstream face of the plate l_1, and distance from downstream pressure tap to downstream face of the plate l_2' as follows (ISO, 2003):

$$\begin{aligned} C = {} & 0.5959 + 0.0312\beta^{2.1} - 0.184\beta^8 \\ & + 0.0029\beta^{2.5}\left(\frac{10^6}{Re}\right)^{0.75} \\ & + 0.090\frac{l_1}{D}\frac{\beta^4}{(1-\beta^4)} - 0.033\frac{l_2'}{D}\beta^3 \end{aligned} \tag{6.6}$$

For compressible flow, Equation (6.5) has to be modified to account for the density at each measuring point. The density at the upstream measuring tap can be calculated from the measurements of pressure and temperature at that point. It is necessary to determine the density at the downstream measuring point.

The expansion of the gas is a constant entropy process. A gas expansion factor Y_1 can be derived to determine density at the downstream point, where x_1 is the differential pressure divided by the upstream pressure, and k is the isentropic exponent (ISO, 2003).

$$Y_1 = 1 - (0.41 + 0.35\beta^4)\frac{x_1}{k} \tag{6.7}$$

The flow equation with the discharge coefficient and gas expansion factor is then expressed as shown in the following equation, where d is the orifice diameter, ρ_{f1} is the upstream fluid density, and the mass flow rate q_m is used for convenience.

$$q_m = \frac{\pi}{4}\frac{CY_1 d^2}{\sqrt{1-\beta^4}}\sqrt{2\Delta P \rho_{f1}} \tag{6.8}$$

Therefore, it becomes necessary to iteratively solve Equation (6.8), because the downstream pressure and downstream density are related to each other, and the equation cannot be solved outright. If Equation (6.8) is rearranged to solve for the differential pressure, the following equation results.

$$\Delta P = \frac{8q_m^2(1-\beta^4)}{(\pi CY_1 d^2)^2 \rho_{f1}} \tag{6.9}$$

From Equation (6.9), an invariant is defined as follows:

$$Inv = \frac{8q_m^2(1-\beta^4)}{(\pi Cd^2)^2 \rho_{f1}} \tag{6.10}$$

The differential pressure is then as follows:

$$X = \Delta P = \frac{Inv}{Y_1^2} \tag{6.11}$$

A residual δ is defined in the following equation:

$$\delta = X - \frac{Inv}{Y_1^2} \tag{6.12}$$

The iteration calculation is as follows, where X_n is the value of the calculation at the current iteration, X_{n-1} is the value of the calculation at the last iteration, and X_{n-2} is the value of the calculation two iterations before:

$$X_n = X_{n-1} - \delta_{n-1}\frac{X_{n-1} - X_{n-2}}{\delta_{n-1} - \delta_{n-2}} \tag{6.13}$$

For the first calculation, it is necessary to "seed" the calculation with estimated values for the past calculations. A typical starting seed for the gas expansion factor is 1. For each iterative calculation, X_n, the residual is calculated. The iteration continues until the residual is extremely small. The iteration will need to be seeded for the first two iterations, because past values of the residual are not available. The coefficient is strongly dependent on the Reynolds number *Re* (see Section 6.14).

As we see in Section 6.3.3, the determination of flow rate by differential pressure measurement is heavily influenced by the flow profile. This is because skewed or swirling flow causes differential pressure errors because of the different velocity vector at the point of measurement as compared to the average velocity in the measurement plane. We have also seen that empirical equations are necessary for differential pressure flowmeters because the vena contracta of the fluid flow does not occur at a plane in the fluid flow that has a corresponding physical dimensional measurement. In other words, the vena contracta does not occur in the plane of the orifice.

A relatively new type of differential-producing flow meter reduces the effects of the two parameters that are discussed in the last paragraph by shaping the flow profile in a controllable way. The cone-type flow meter is shown in Figure 6.4. By shaping the flow profile around the cone, the downstream differential pressure measurement is made in a known annular cross-sectional area. This is not possible with an orifice plate, because it is not possible to place a pressure tap in the exact same plane of the orifice. Also, by "shaping" the flow by mixing the low and high velocities to the downstream portion of the cone, a more consistent flow profile is produced (Ifft and Mikkelsen, 1993). Similar to the effect in orifice plates, the consistency of the shaped flow profile is a function of the meter's beta ratio (see Section 6.3.3). Comparison testing against other conventional flowmeters has shown the superior accuracy of the cone-type meters (Lloyd, Guthrie, and Peters, 2002; Dahlstrom, 1994). Independent tests have also shown the meter's ability to deal with swirling flow (Shen, Bosio, and Larsen, 1995). The cone-type flowmeter has an advertised inaccuracy of ±0.5% of flow rate, a turndown ratio of 10 : 1, straight line requirements between 0–3 pipe diameters upstream and 0–1 pipe diameters downstream, and can measure fluids with a minimum Reynolds number of 10 000. Because the manufacturing tolerances are not as good as, say a simple orifice, it is advised to have each meter calibrated in the factory before being placed in service.

Perhaps, the most important hydraulic parameter to flow metrology is the dimensionless *Re*. It is the ratio of inertia forces to viscous forces in a flow system as described by

$$Re = \frac{\rho v D}{\mu} \tag{6.14}$$

where ρ = fluid density, v = fluid velocity, D = characteristic dimension of the flow (often called *hydraulic diameter*), μ = absolute viscosity of fluid.

Dissimilar flow systems can be compared by the use of the *Re*. The flow condition in a system can be described as laminar, transitional, or turbulent. The magnitude of the *Re* indicates the type of flow condition present. For pipeline flows, laminar flow exists if the *Re* is less than 2000. Transitional flow occurs for the *Re* range of 2000–10 000, and turbulent flow exists for *Re* over 10 000. The advantage of using the nondimensional *Re* is that the above ranges hold for any gas, vapor, or liquid. Most flowmeters function in a particular flow condition; thus, the

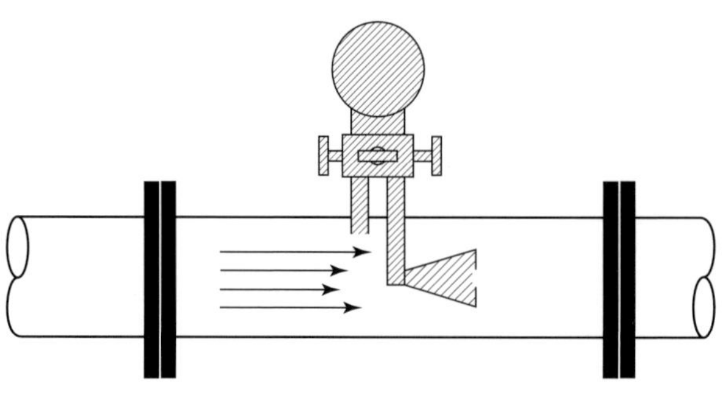

Fig. 6.4 Cone-type flowmeter.

Re should be calculated with the appropriate characteristic dimension for each case. When thermal properties are important, the Prandtl number is also a dimensionless number that can be very useful in characterizing dissimilar flow systems.

By far, the most predominant flowmeter in existence by number, the differential-producing-type flowmeter is popular because the technology of pressure and differential pressure sensing is quite advanced in comparison to other measurements and because there are many empirical relationships to account for the difference between theory and practice. The concentric orifice has been the workhorse in industrial applications. Other differential producers include the integral orifice, the venturi tube, and the simple pipe elbow with pressure taps upstream and downstream. The integral orifice meter combines in one assembly an orifice and pressure taps along with a honed pipe section in which the inner surface has been deburred and smoothed. It is also common to include the differential pressure transmitter with the assembly. The venturi-tube meter uses a gradual constriction and expansion of the flow stream instead of an orifice plate. Its advantages over the orifice plate are that the downstream pressure recovers closer to the upstream pressure, thereby causing less energy loss, and it is also less susceptible to erosion when used with solids-bearing fluids. The pipe elbow meter is very simple but not very accurate.

Highly sophisticated computer programs have been developed for the orifice meter to accurately determine flow on the basis of differential pressure, flowing temperature, absolute pressure, dimensions of the orifice and pipeline, thermal expansion factor of orifice material, tap location, and fluid viscosity. For the custody transfer of valuable fluids, these types of calculations are carried out to several decimal points – more resolution than can possibly be justified. However, on a contractual basis, these are the accepted calculations regardless of the inaccuracy. Many configurations exist for the orifice meter. Pressure tap locations vary in their distance upstream and downstream of the orifice plate but usually fall into one of several standard configurations. The most common orifice plate type is square-edged concentric, but quadrant- and conical-edged eccentric and segmental orifices exist as well.

A big disadvantage of differential-producing devices has been that because of the square-root relationship between flow and differential pressure, an error in differential pressure, especially at the low end of the range, becomes a magnified error in the flow reading. For example, in Figure 6.5, a 2% error in differential pressure at 60% of full-scale differential pressure causes a 1.3% of full-scale error in flow. However, the same differential pressure error at 3% of full-scale differential pressure causes a 5% of full-scale error in flow. The term *turndown ratio* (or *rangeability*) refers to the maximum feasible reading of a flow device divided by its minimum feasible reading. Differential producers, in the past, were typically limited to a turndown ratio of three. With improved accuracy and repeatability of modern pressure measurement devices, and on-line determination of the Reynolds number and gas expansion factor as discussed in Section 6.3.2, the turndown ratio is claimed to have been increased to 10. Inaccuracies vary from 0.25 to 2% of full scale. Flow ranges vary from very high for industrial applications to somewhat low for research and development applications. However, the flow must remain turbulent (the Reynolds number $>10\,000$).

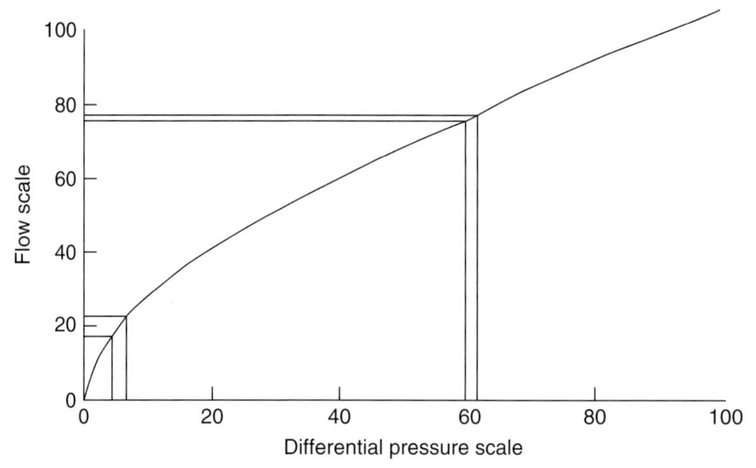

Fig. 6.5 Typical flow versus differential pressure curve for an orifice, arbitrary flow, and differential pressure units.

Straight-pipe requirements, which ensure a symmetric velocity profile through the meter (see Section 6.3.3), vary depending on the type of flow obstruction and the beta ratio. The range of straight-pipe requirements usually falls within 5–50 pipe diameters upstream and 2–4 pipe diameters downstream of a pitot tube or orifice, but much less for a cone-type meter.

6.2.2 Volumetric

For this class of meters, volumetric flow rate q_V is generally inferred by the fluid flow velocity v multiplied by the meter cross-sectional area A (see Equation (6.3)). Because the meter reading is actually based on fluid velocity, care must be taken to ensure that the entire meter cross-sectional area is filled with fluid; otherwise, errors will occur in the volumetric flow rate.

6.2.2.1 Vortex Shedding

The principle of the vortex-shedding flowmeter, as shown in Figure 6.6, is that a blunt, nonstreamlined object perpendicular to the direction of flow sheds vortices alternately from its sides when flow attains a high enough Reynolds number. The frequency that vortices are formed on alternate sides of a bluff body exposed to flow is proportional to the velocity of the flow around the bluff body. Vortices roll into

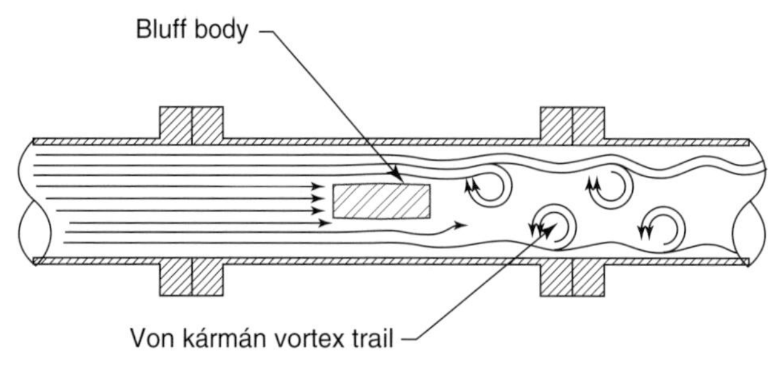

Fig. 6.6 Vortex-shedding flowmeter.

low-pressure areas behind the bluff body. As the vortex is shed on one side of the bluff body, fluid velocity increases, which causes a pressure decrease on that side. On the opposite side of the bluff body, the velocity decreases and the pressure increases – there is a net pressure change across the bluff body. The velocity and pressure change at the same frequency as the vortex-shedding frequency. The Strouhal number Sr is a dimensionless number that relates shedding frequency f, a characteristic dimension of the bluff body l, and the fluid velocity v,

$$Sr = \frac{fl}{v} \tag{6.15}$$

The vortices downstream of the bluff body are known as the *Von Kármán vortex trail.* The Strouhal number is essentially independent of the Reynolds number and process fluid for Reynolds numbers greater than 20 000. Below that, the relationship between frequency and flow velocity is nonlinear. At Reynolds numbers below 4000 or 5000, the stability of the vortex-shedding degrades. Vortex-shedding flowmeters are not generally applicable for Reynolds numbers less than 5000 (Mattar, 2004).

The relationship between shedding frequency and flow rate is a function of only the dimensions of the bluff body – Vortex shedding is not a function of fluid properties. Expansion of the bluff body dimensions can cause errors. Installation requirements to meet upstream straight-pipe requirements range from 10–20 diameters upstream and 5 diameters downstream to 25–45 diameters upstream and 8 diameters downstream. Fluids with viscosities lower than 8 cP do not work well. The flow signal will have a dropout below the minimum Reynolds number. Also, because the sensing methods are passive, a minimum amount of energy must also be available in the fluid. This is expressed as the minimum velocity constraint. Usually 30–60 cm s^{-1} for liquids, but greatly varies for gases (Spitzer, 2005). The combination of the minimum Reynolds number and the minimum velocity constraint causes the flow reading to fall to zero; therefore, low flows cannot be detected. However, this can also be an advantage as compared with the orifice meter. With the orifice meter, small differential pressure errors can falsely indicate flow at near zero differential pressure. In applications where flow is being totalized, this error can lead to large totalization errors over long durations of low or no flow when there is an error in the differential pressure reading. For the vortex-shedding flowmeter, the flow signal simply falls to zero, and no flow totalization occurs. If the meter is correctly sized for the minimum anticipated flow, having the flow signal drop to zero for flow rates below the minimum will cause no substantial errors in flow totalization. On the negative side, if the flow signal is used in a control loop, dropout can cause control loop instability. For example, consider the case where a control valve is being used to control flow rate. When the flow signal drops out, the controller interprets this as zero flow and opens the control valve farther. When it surpasses the dropout point, the flow meter will read flow again. However, at this time, the flow is higher than needed, so the controller will close the valve farther. The opening and closing of the control valve will cause a condition known as *limit cycling*. The meter can be sensitive to vibration. In the past, passive techniques with fixed range band pass filters have been used to overcome vibration effects. Recent efforts are using digital signal processing to dynamically adjust the band pass filters to overcome the vibration effects (Ostling and Oki, 2001).

For calibration, a disadvantage is that the meter cannot be verified in-line. For maintenance, the flowmeter is fixed in line and cannot be repaired without opening up the pipeline (as opposed to differential pressure meters that use transmitters that are mounted remotely from the line). However, some designs allow replacement of piezo sensors while the meter is in-line. On the plus side, a "sealed" meter can be an advantage to reduce fugitive emissions. There are redundant configurations available to increase the safety integrity level of the measurement (Mattar, 2004). This is important when the measurement is critical to the safety of personnel or the integrity of the process.

A typical vortex-shedding flowmeter consists of a pipe section, a bluff body in the center of the pipe section, a means to detect the vortex frequency, and power and signal conditioning. The measurement can be made in different ways depending on what is to be measured:

- an oscillating disc measured by a magnetic sensor;
- the pivot of the shedder measured by a piezo element;
- alternating pressure;
- fluid flow through a passage in the bluff body that alternates back and forth, as measured by a temperature sensor;
- the twist of a torque tube measured by a piezo element; and
- downstream vortices measured ultrasonically.

The meter has an inaccuracy of around 0.5–1% of flow rate for liquids, and 1.5–2% of flow rate for gases. Turndown ratios of up to 40 : 1 are claimed by manufacturers. A word of caution is in order for these claims, though – they are stated relative to the maximum flow possible for the size of pipe in which the flowmeter is installed. This is almost always higher than the maximum design flow expected in an application, because pipe sizing criteria is usually conservative with regard to the pipe diameter. The net effect is that turndown for an application, defined as maximum design flow rate divided by minimum flow rate readable by the flow meter is much lower than 40 : 1. A common practice to improve turndown has been to decrease the pipe diameter for the meter run only. New vortex flow meters are available in reduced body sizes, that is, they are connected to the large diameter pipe, but have decreased cross-sectional area within the meter to increase velocity and improve turndown. However, data on the line-size reduction and expansion at the meter coupled with not having straight runs before and after reduced bore meters is not mature yet. The meter can be used for liquids, vapors, and gases; however, to obtain the minimum Reynolds number, gas flows must be at relatively high pressure. It is quite common to use these meters to measure steam flow rate. Care must be taken to avoid cavitation of liquid flow by keeping the pressure at the vortices above the vapor pressure of the fluid. The performance can be affected by the thermal expansion of the bluff body. In applications that are susceptible to condensation, the bluff body should not be on the bottom of the line (Spitzer, 2005).

Some claims have been made that pressure and temperature compensation is not required for compressible flow. However, this would assume that an output in actual volumetric flow or velocity is desired. In practice, most users are interested in volumetric flow that is expressed in "standard" (US) or "normal" (Europe) flow units, which are actually mass flow rates. There are configurations available with pressure and temperature sensors that allow converting to these flow units (Mattar, 2004).

6.2.2.2 Magnetic

For conductive fluids, the magnetic flowmeter is a very reliable flow device. The meter is based upon Faraday's law. Voltage V will be induced in a conductor that is moving through a magnetic field proportional to the length of the conductor l (the diameter of the pipe in this case), the velocity of the conductor v, and the strength of the magnetic field B,

$$V = vBl \tag{6.16}$$

The induced voltage is perpendicular to both the fluid flow direction and the magnetic field. By knowing the area of the flowing cross section, a volumetric flow is calculated once the velocity is known as shown in Equation (6.3).

A typical magnetic flowmeter is shown in Figure 6.7. Electromagnets surrounding the flow tube generate a magnetic field within it. Process fluid passing through the magnetic field acts as a moving conductor, inducing voltage in the fluid. Flush-mounted electrodes inside the primary tube pick up the voltage. A nonconducting pipe-liner section is required. The liner is typically made from a polymer, rubber, or ceramic. The meter must be electrically bonded to earth ground to prevent stray currents from flowing through the meter and causing a zero shift. For nonconductive pipe, grounding rings are required to be installed upstream and downstream of the meter when the conductivity of the fluid is less than 100 µs cm^{-1}; otherwise, grounding electrodes can be used. The grounding rings may have to be made of exotic materials when exposed to acidic or corrosive fluids and can be expensive. Grounding straps are acceptable for conductive, unlined pipe. Buildups of conductive solids on the electrodes may

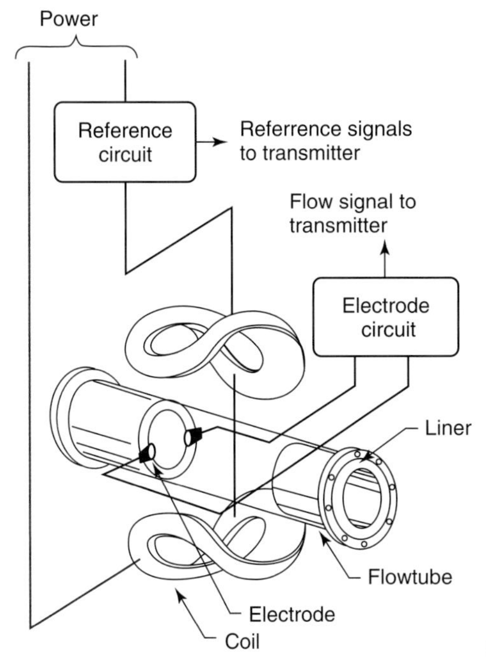

Fig. 6.7 Magnetic flowmeter.

form, which decrease the cross-sectional flow area, and therefore cause an error in the inferred volumetric flow rate. Almost all manufacturers of magnetic flow meters today incorporate means for optional electrode cleaning to overcome coating problems. The coating can be sensed by measuring the relaxation time of a pulse across the electrodes and ground (Walker, 2001).

Typically, the minimum conductivity of the fluid must be 1 μs cm^{-1}. The temperature range of the fluid should be taken into account because the fluid conductivity is a function of temperature. The measured flow signal is not a function of the conductivity so long as the conductivity is above the minimum threshold level. Because this type flowmeter is relatively insensitive to velocity-profile effects, only ~3 pipe diameters of straight piping is required before the meter (Rose and Vass, 1993). Inaccuracies vary from ~0.2 to 1% of flow rate. Turndown ratio varies from ~10 : 1 to 30 : 1. Pipe sizes from 0.1- to 244-cm diameter can be accommodated. Two types of magnetic-field excitation are commonly used: ac and pulsed dc. The ac excitation methods were very common in the past, but pulsed dc excitation methods are mostly used now. The trade-offs between the two methods are that ac methods have good signal-to-noise ratios and are more stable in two-phase applications such as liquids with either entrained gas or high suspended-solids content. The dc-pulsed methods have a stable zero, are not as susceptible to coatings on the electrodes, are smaller and lighter, and can be made to be intrinsically safe in areas that are exposed to explosive gases and dusts because of their lower power. Steady dc is not used to avoid polarization of ions in the fluid. It is typical for an ac meter to have a zero drift of as much as 1% per month. To correct the drift, it is necessary to fill the meter with fluid, stop flow, and zero the meter. The pulsed dc units have typical drifts of only 0.1% per month. Meters powered with ac have a signal that includes components of flow and noise. Dc-pulsed meters can measure the noise during periods when the coil is not energized and subtract this from the signal during energized pulses to obtain better accuracy. The ac meter would require a nonflow zero calibration to achieve the same effect. Pulsed dc meters do not have as good a time response as ac because of time required for the pulsed technique. Two-wire meters are available that use energy from the 4 to 20 mA loop. However, due to the limited amount of energy, low- flow measurement is less accurate. Also the pulsing frequency is low, so response time is correspondingly slow. Meters designed to measure pipelines that are only partially full are available with built-in level measurement. The flow tube is generally made of stainless steel that does not affect the magnetic field, as would carbon steel. Applications are limited in vacuum service due to the possibility of the liner separating. Magnetic flowmeters are good choices for slurries because of their obstructionless design. Electrodes should be oriented in a horizontal plane to avoid being uncovered by gas bubbles (Spitzer, 2005). A rule of thumb for flow velocity through the meter is 60–600 cm s^{-1} for normal service, 90–300 cm s^{-1} for abrasive slurry, and 150–450 cm s^{-1} for nonabrasive slurry. Flow can be laminar, transitional, or turbulent. This meter can even measure bidirectional flow. Noncontact electrodes are becoming popular. Meters with noncontact probes can measure fluids down to 0.05 μs cm^{-1} but are very expensive. The obstructionless design of the noncontact probes is good for slurries and corrosives. Organic liquids, in general, do not have high enough conductivity to work.

6.2.2.3 Turbine

A turbine flowmeter consists of a bladed rotor suspended axially in a pipe section in the direction of flow such that flow through the pipe causes the rotor to rotate. Speed of rotation of the rotor, which is correlated to flow velocity, is measured by a pickup outside the pipe section. For a turbine blade with angle θ with respect to the pipe axis, as shown in Figure 6.8, the ideal turbine rotational velocity ω is related to the flowing velocity v and the rotor radius r by resolving vectors as

$$\omega = \frac{v}{r} \tan\theta \tag{6.17}$$

Therefore, by combining Equation (6.3) with Equation (6.17), the volumetric flow rate q is given by

$$q = rA\frac{\omega}{\tan\theta} = K\omega \tag{6.18}$$

K is the called the *K factor*. Although the K factor is derived theoretically here, it is used to characterize the meter on the basis of meter calibration tests (Sydenham, 1999). It provides a flow output based on the measured rotational speed of the turbine.

The K-factor curve is valid only above a minimum threshold in flow. Then it typically shows a peak, as shown in Figure 6.9, before settling out at a fairly constant value. The K factor that is used in practice is usually a constant over the whole operating range of the meter, because the range of actual K-factor values from a high at the top of the peak to the low at higher flow rates does not vary from an average K value more than would cause the meter to exceed its tolerance specification (Lipták, 2003). However, with more and more microprocessor-based instruments being built and benefiting from smaller component size and surface-mount technology, it is now possible to store the K-factors in electrically erasable, programmable, read-only

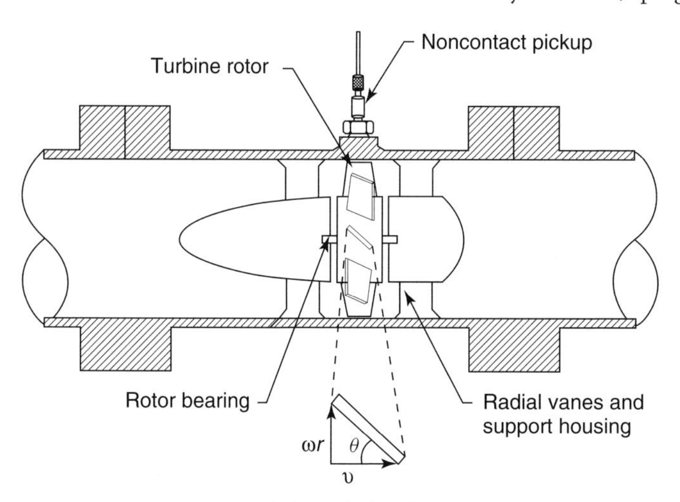

Fig. 6.8 Turbine flowmeter arrangement with turbine rotational velocity ω, rotor radius r, turbine blade with angle θ with respect to the pipe axis, and fluid velocity v.

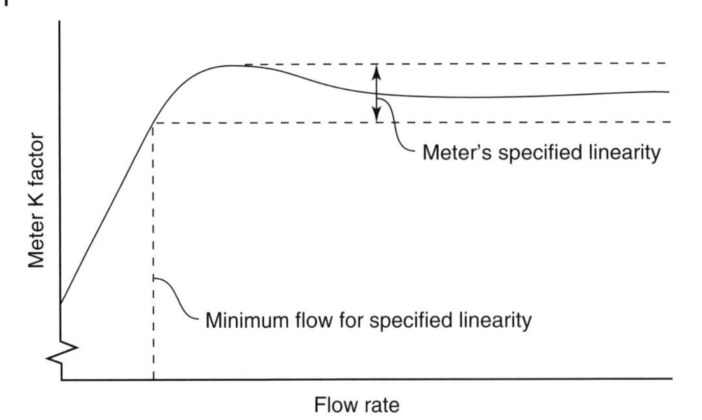

Fig. 6.9 Typical *K*-factor curve for a turbine flowmeter.

memory chips within the manufacturer's signal conditioning units (McCoy, 1992).

The relationship between fluid velocity and rotor speed is linear over a wide range of flow conditions. Retarding forces cause the rotor to spin at a speed less than theoretical. This effect is called *slip*. Factors influencing the *K* factor are changes in rotor-blade angle (due to collisions with solids), swirl and velocity-profile distortion, rotor-bearing friction, viscous drag, and fluid density. These factors can combine to cause errors from a few tenths of a percent to 15% (Miller, 1996).

Two types of noncontact pickups are used: reluctance and inductance. These types of meters can be used for liquids, vapors, and gases. Low-cost units have inaccuracies of ~1% of flow rate with a 10 : 1 turndown ratio; high-cost units have inaccuracies of 0.1–0.5% of flow rate with turndown of 10 : 1 to 50 : 1. The volumetric flow is measured at the operating temperature. It may be necessary to correct for temperature effects (Lipták, 2003).

Primary uses include flow totalization (by counting pulses from the pickups) and inventory control. Frequency-to-analog signal conditioning must be used to convert to standard analog signals. Sizes range from 1.25 to 30 cm. End connections include threaded, male flared tube, flanged, and plain for slip-on couplings (McMillan and Considine, 1999).

Changes in fluid viscosity affect the rotor speed-to-flow velocity relationship. The use of "universal viscosity curves" that plot flowmeter sensitivity versus Reynolds number allow the combined effects of velocity, density, and absolute viscosity to be taken into account (Miller, 1996). Turbine flowmeters can compensate for viscosity by knowing the fluid's viscosity versus temperature curve. The temperature of the fluid is measured and is fed into a microprocessor-based signal conditioner (Madison, 1995).

6.2.3
Positive Displacement

When all applications of flowmeters are considered, positive-displacement meters by far outnumber all others because of the almost exclusive use of positive-displacement meters for household gas and water metering and for the metering of motor fuels. The positive-displacement meter, in these applications, is not actually a flowmeter but a volume meter. The fundamental positive-displacement meter has no time base; but for industrial applications where a flow rate is desired to be monitored, the positive-displacement meter can be outfitted to provide flow readings. Otherwise, the receiving instrument can be made to compare volume readings over time to provide flow readings.

Positive-displacement meters work on the principle that successive chambers within the meter are filled with the metered fluid. As one chamber is emptied, the successive chamber is filled. Each chamber holds a fixed volume of fluid. The filling and emptying is translated into a rotary motion. The speed of rotation is converted to pulses. The pulses are sensed optically or magnetically. The pulses can be totalized, or the frequency can be measured for flow rate. Any gas will be treated as "volume" and thus cause errors in liquid flow rate indications that contain gas bubbles. Because positive-displacement meters measure volume, the pressure and temperature must be taken into account to arrive at a mass exchange value. There are no Reynolds numbers or straight-pipe constraints. Viscosity changes cause an error (which is function of temperature). There can be slippage at low viscosity. Liquids that do not lubricate the rotor bearings can cause errors at low flow rates. The following are the different types available (Spitzer, 2005):

- *Helical gear*: These are usually used for extremely viscous service, up to 300 000 cP. They have an inaccuracy of ~0.4% of rate. This style causes relatively little differential pressure for high viscosity fluids as compared to other types.
- *Nutating disc*: These are often used for water service where inaccuracy (~2% of rate) is not a primary concern. Their turndown ratio is on the order of 5 : 1 to 10 : 1.
- *Oscillating piston*: These are used for viscous liquid service where turndown is not a concern. They have 0.5% of rate inaccuracy, turndown of 3 : 1 to 4 : 1, and can work with a viscosity range of 0.2–10 000 cP.
- *Oval gear (rotor)*: These are used for viscous service where other types of flow meters have Reynolds number constraints. They are available with an inaccuracy of 0.25–1% of rate, turndown up to 20 : 1 for intermittent service, but as low as 3 : 1 for continuous service. The viscosity range is between 0.2 and 500 000 cP.
- *Piston*: These are used in low-flow viscous liquid service where other types of meters would have Reynolds number constraints. They have inaccuracy of 0.5–1% of rate. The turndown can be as high as 100 : 1, and the viscosity range is 0.5–10 000 cP.
- *Rotary*: These are used for liquid service where accuracy is important but turndown is not. They are available with 0.1–0.2% of flow rate inaccuracy, turndown of 5 : 1 to 10 : 1, and viscosity ranges of 1–25 000 cP.

6.2.4 Drag

The drag force F_D exerted on an object by a fluid is a function of the object's projected area A_p, the drag coefficient C_D, and the density of the fluid ρ:

$$F_D = \frac{1}{2} A_P C_D \rho v^2 \tag{6.19}$$

The drag coefficient is a function of the fluid's Reynolds number and a shape factor associated with the object. For disks with

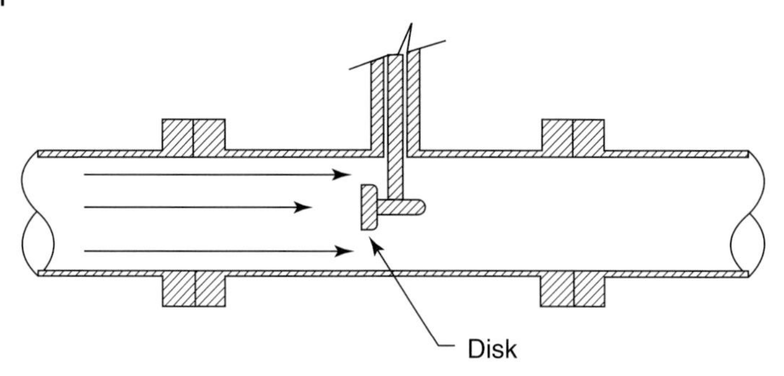

Fig. 6.10 Drag flowmeter.

the flat side perpendicular to flow, the drag coefficient is nearly a constant for Reynolds numbers higher than ~4000.

Flowmeters use the principle by exposing a disk to flow as shown in Figure 6.10. The force exerted on the disk is measured by force balance or strain gauge methods. By taking the square root of the force signal, a signal proportional to flow velocity is obtained. Volumetric flow rate is inferred from the velocity and the cross-sectional area of the pipe line as shown in Equation (6.3).

The use of a drag meter is good for liquids, gases, steam, and slurries. Inaccuracy usually lies between 0.5 and 2% of full scale. Absolute inaccuracy increases as flow decreases because of the square-root relationship (similar to the orifice meter). Turndown ratios of 10 : 1 to 15 : 1 can be obtained. The drag force depends on the geometry of the target. Erosion of or buildups on the disk will cause errors. The meter requires about the same straight-pipe lengths as orifices. The meter is usually used for reliability, speed of response, repeatability, and usable turndown ratio. It is not known for good accuracy. This design is economically attractive for large pipe sizes using insertion-style designs, because flanges or pipe wafers, that can be very expensive in large diameters, are not required. Pipe vibration can affect the measurement (Spitzer, 2005).

6.2.5 Variable Area

The variable-area flowmeter, or rotameter, is a device for measuring flow that maintains a relatively constant differential pressure across the meter by increasing the annular area around a float within the meter as the flow rate increases. As shown in Figure 6.11, the rotameter is oriented vertically. The float is situated inside a tapered-glass or -plastic tube. In the more common case, where the float is denser than the metered fluid, flow enters at the bottom of the tapered tube and exits at the top. It is common to have a needle valve in either the inlet or the outlet flow tube to control the flow rate. The buoyant forces, F_A, and upward hydraulic forces, F_D, of the fluid are balanced by the weight of the float, F_G.

$$F_G = F_A + F_D \tag{6.20}$$

The weight of the float is given by the following, where V_S is the volume of the float and ρ_S is the density of the float.

$$F_G = V_S \rho_S g_c \tag{6.21}$$

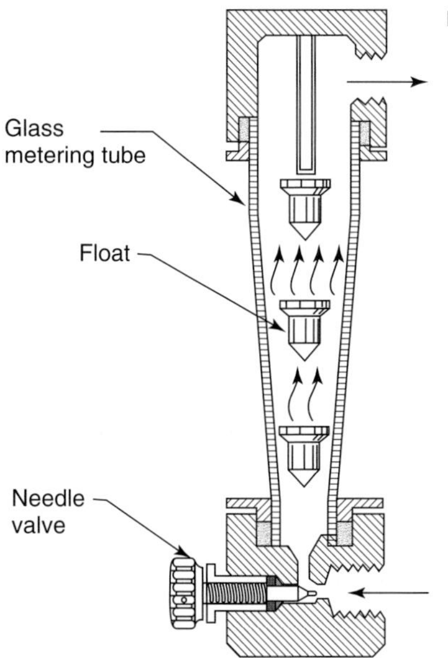

Fig. 6.11 Variable-area flowmeter.

The buoyancy of the float is given by the following, where ρ_m is the density of the fluid:

$$F_A = V_S \rho_m g_c \quad (6.22)$$

As we saw in the previous section, the upward hydraulic drag forces are represented in Equation (6.19). For these purposes, we will use ρ_m in the equation, for the density of the fluid. The volumetric flow rate through the meter is represented as in Equation (6.3), where A_r, the annular surface area between the float and the tube wall is substituted into the equation.

By combining Equations (6.3) and 9.19–22, we obtain a volumetric flow rate as follows:

$$q_V = A_r \sqrt{\frac{2g_c V_S(\rho_S - \rho_m)}{C_D A_P \rho_m}} \quad (6.23)$$

As the tube taper is a linear function of height, the annular area A_r is directly proportional to height. The other parameters are constant, and as we see from this equation, the volumetric flow rate is directly proportional to the height of the float.

The float is positioned in the tapered tube in proportion to fluid flow rate and annular area. As the float rises with increasing flow rate, the annular area around the float increases. As the annular area increases, the differential pressure decreases. The glass or plastic tube is marked with gradations for flow. When the rotameter is used properly, the flow rate is determined by noting the level of the float and the corresponding flow gradation on the tube.

The rotameter must be used with the design fluid and the operating parameters, or else a correction factor must be applied to the reading obtained. The variable-area flowmeter typically has no interface into data acquisition and control systems, although some manufacturers do offer an electrical interface, especially when metal flow tubes are used. In this case, the float position is measured magnetically or electrically. Output types are available with pneumatic, electric, or pulse output.

Variable-area flowmeters measure liquids with viscosities less than 3 cP, gases, and steam. Inaccuracies vary from 0.5 to 5% of flow rate, with a turndown ratio of ~10 : 1. The meter is usually limited to less than 8-cm pipe diameters, tends to be somewhat self-cleaning, and is relatively insensitive to piping configuration (no straight-pipe requirements).

Some rotameters use a spring that counteracts the fluid forces. It is not necessary for these to be oriented vertically. Ball floats are almost exclusively used for purge flow meters. For liquids, the minimum Reynolds number is 250–950 for low flows and 4400 for high flows. Gas flows have no minimum Reynolds number requirement, but must have high enough density to raise the float (Spitzer, 2005).

6.2.6 Ultrasonic

Ultrasonic flowmeters and flow sensors work on one of the two basic principles – Doppler shift or time of flight. The Doppler-shift method functions by transmitting an ultrasonic signal of known frequency into a flow stream (Figure 6.12). The ultrasonic beam reflects off particles or other discontinuities in the stream. The difference in frequency (Doppler shift) between the original beam and the reflected signal is related to the flow velocity of the reflecting medium. The argument made is that the small reflecting particles travel at the same velocity as the transporting fluid. Most commercially available Doppler-type flowmeters require particles or air bubbles with a minimum size of 25 – 30 μm and a concentration of at least 25 ppm. The ultrasonic transceivers (transmitter and receiver) can clamp onto the outside of a pipe; for better accuracy, they can be mounted through the pipe wall and flush with the inside wall (wetted transducer type). These devices can be used on liquids, aerated liquids, and slurries. The turndown ratio is at least 10 : 1 with an inaccuracy of ±1 – ±5% of full scale depending on whether clamp-on or wetted transducers are used and whether *in situ* calibration is possible. The time-of-flight or transit time meter operates by pulsing an ultrasonic beam across a pipe in the direction of flow and then in the reverse direction (Figure 6.13). The difference between the travel times in the two directions is related to the velocity of the flow. This relationship is described by the following equations:

$$V_{AB} = C_0 + V_m \cos\varphi \tag{6.24}$$

$$V_{BA} = C_0 - V_m \cos\varphi \tag{6.25}$$

$$t_{AB} = \frac{L}{C_0 + V_m \cos\varphi} \tag{6.26}$$

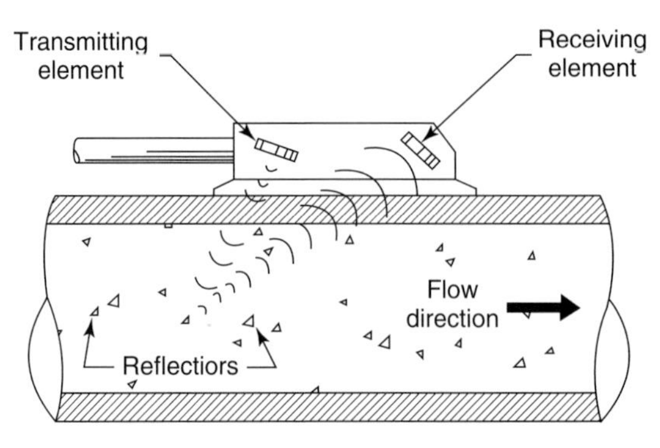

Fig. 6.12 Doppler-type ultrasonic flowmeter.

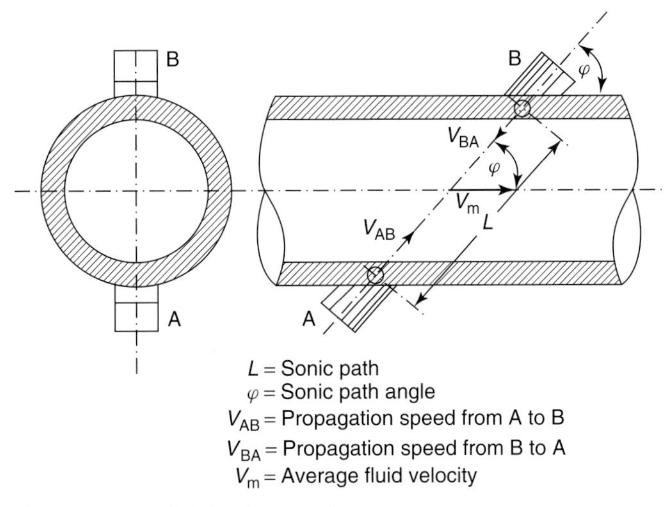

Fig. 6.13 Time-of-flight ultrasonic flowmeter.

$$t_{BA} = \frac{L}{C_0 - V_m \cos\varphi} \tag{6.27}$$

where L = sonic path length, φ = sonic path angle, C_0 = speed of sound for the fluid medium, V_m = average fluid velocity, V_{AB} = propagation speed from A to B, and V_{BA} = propagation speed from B to A. The parameters t_{AB} and t_{BA} are the measured ultrasonic wave flight times from points A and B, L and φ are known constants, and C_0 and the desired value V_m can be calculated. This method requires a clean fluid, and transducers can be of the wetted or clamp-on type. The turndown ratio is 40 : 1, and the meter can measure flow velocities down to $0.03\ \mathrm{m\,s^{-1}}$. Inaccuracies of $\pm 0.5 - \pm 2\%$ of flow are obtainable. Both ultrasonic methods can be nonintrusive, have no moving parts, and induce no pressure loss. Although these meters are typically used on liquids, they have been extended to gas flows as well as to solid-liquid and gas-liquid mixed flow (Rudroff, 2005; Magness, 2002; Soo, 1999; Witte, 2002).

An area of growth for ultrasonic flowmeters is in custody transfer measurements. To increase the accuracy (or lower the uncertainty of the ultrasonic flowmeter), multipath (multiple beams) systems have been developed. Typically, three to five beam paths are used at a single location in the transit time mode (Figure 6.14). The multiple paths help in reducing uncertainties that are caused by asymmetric velocity and temperature profiles, turbulence, and other flow disturbances. Uncertainties have been lowered to $\pm 0.1\%$ or less (Cousins, 2002; Basrawi, 2004).

A new ultrasonic method combines the Doppler and time-of-flight technique to expand the flow conditions over which the meter will operate accurately. An ultrasonic-Doppler velocity profile (UVP) technique (Mori et al., 2002) determines the velocity profile in small "channels" across a pipe section and then integrates those velocity profiles to yield the flow rate. The time-of-flight technique is used in its typical fashion. When one of the methods does not function well (too many bubbles or particles or the fluid is too

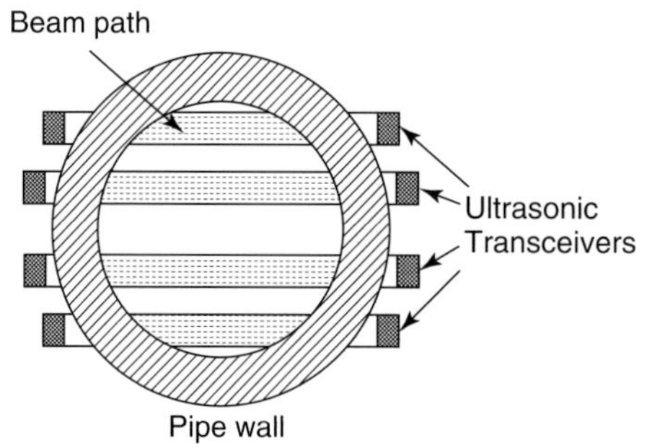

Fig. 6.14 Multibeam ultrasonic flow meter.

clean), the hybrid flow meter switches techniques to obtain a quality signal and flow measurement. The usable range and types of flow conditions that it can handle are substantially extended. (Yamamoto, Yao, and Kshiro, 2004).

A nice summary of the history of ultrasonic flowmeters is presented in a report (Lynnworth and Liu, 2006).

6.2.7 Mass Flow

True mass flowmeters are usually based on one of the two principles of operation. The Coriolis type is based on the Coriolis force that is generated when a section of pipe is vibrated while flow is passing through it (Figure 6.15). This measurement technique is essentially the application of Newton's second law of motion, where F = force, m = mass, and a = acceleration.

$$F = ma \tag{6.28}$$

The acceleration is known (from the tube vibrator), and the force is measured by the instrument itself. The amplitude of the generated force is a function of the acceleration and the fluid velocity and its mass, or the mass rate of flow. The major advantages of these meters are direct mass flow measurement, good turndown ratio (better than 10 : 1 and up to 100 : 1), small inaccuracy ($\pm 0.5\%$ of rate), independence of fluid properties, measurement of either gas or liquid, and a requirement for little straight piping. Disadvantages include sensitivity to equipment-induced pipe vibrations, pressure loss, and the expense of the meter. Recent advances include lower cost designs, straight piping for lower pressure loss, and digital electronic signal processing designs to lower the response time from seconds to tenths of a second (Henry, Clark, and Mattar, 2004). This lower response time allows Coriolis flowmeters to work with very short batches or with pulsating flow from pumps. New flowmeters have been developed to work in

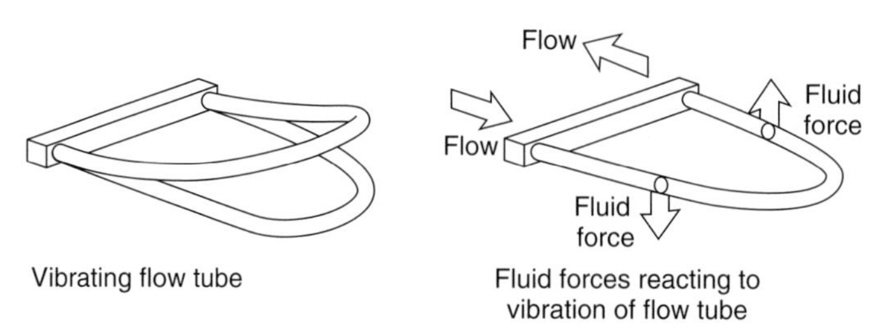

Fig. 6.15 Operating principle of a Coriolis force flowmeter.

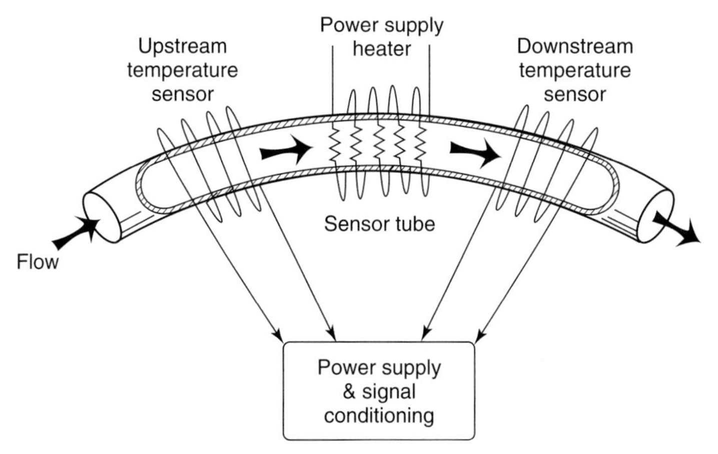

Fig. 6.16 Operating principle of a thermal mass flowmeter.

aerated liquids and continuous two-phase flow by using more than one flow tube driver and redesigning the flow tube geometry (Henry, Yeung, and Mattar, 2004). A Coriolis flowmeter was developed to handle higher flows, temperatures, and pressures. This design, an Omega Tube, combines a half circle measuring tube, torsion bar oscillation system, and a media feed tube section. This flowmeter is commercially available and is described in Hettrich and Daly (2004)).

The second mass flowmeter type is based on thermal mass effects (Figure 6.16). Typically, a temperature sensor is heated to a temperature above that of the fluid stream. The amount of cooling that the sensor undergoes is a function of the mass velocity of the fluid stream. Advantages are that the meter measures mass flow directly, has low pressure loss, has the ability to measure low velocities (Viswanathan et al., 2001), has a $>10:1$ turndown ratio, and has small inaccuracy (down to $\pm 0.5\%$ of full scale in some models). The disadvantages are sensitivity to fluid properties (especially thermal conductivity and capacity) and its potential for undetected calibration shifts caused by vapor condensation or coatings deposited on the sensing element.

Over the last 10 years, a significant amount of research has taken place to better understand the sensitivity of thermal mass flowmeters to fluid properties. This body of research has included modeling and improved calibration methods. Models were created to better understand the changes in linearity and accuracy (Hardy, 1999; Han, 2005; Kim, 2007) due to fluid property changes, geometry, and transients. Calibration techniques have also improved, which led to lower uncertainties (Hardy, 1999; Tison, 1996). The industry standard was to calibrate a thermal mass flowmeter on a benign gas and then use a simple thermal-physics model to "transfer" that calibration to a hazardous, corrosive, or toxic gas. This was essentially a linear transform. Rigorous testing in the early to mid-1990s showed that the correction factor should be a function (nonlinear) and errors using the factor could be as much as 70% (Hinkle, 1991). Gravimetric calibration devices were developed to handle toxic and corrosive gases; calibration functions were derived using experimental data and improved modeling to significantly reduce the "transfer" calibration errors (Remenyik, 1995).

6.2.8 Open Channel

Open channel refers to any conduit that allows a liquid to flow with a free surface such as a sewer line or a river. Open-channel flow measurement is often necessary at water and sewage treatment plants, waste water sites, and many other environmentally sensitive areas. This flow measurement is most often made by installing a specially designed device into the flow stream. The most commonly used insertion devices are weirs and flumes. Both devices are based on the principle that the flow in the open channel is a function of the water head (or level) that is created by the weir or flume. Weirs operate on the principle that a weir inserted into a stream will create a fluid head behind (upstream of) it. This head is a function of the flow velocity over the weir. Weirs have large turndown ratios, 20 : 1 to 60 : 1, and ±2 to ±3% full-scale inaccuracies, but they can trap entrained matter, which can lead to measurement inaccuracies. Flumes are specially designed structures that are similar to venturis in closed pipes. The flow through a flume can be related to the head at a specific location. Flumes have large turndown ratios, 40 : 1, have inaccuracies of ±2 to ±3% full scale, and are self-cleaning. Both devices use various methods to measure the head, including bubblers, differential pressure sensors, and ultrasonic instrumentation. Bubblers measure the pressure required to release a single air bubble from the end of a submerged tube. The required pressure is related to the head of liquid above the tube end. As the level of liquid changes above the tube, the required pressure changes and is measured and converted to a head or level measurement. A differential pressure sensor (submerged in the fluid) measures the pressure difference between the fluid surface and the bottom of the weir or flume. The pressure difference is directly related to the gravity head. By knowing the density of the fluid, then the fluid level can be determined. The ultrasonic head or level detector uses a time-of-flight technique to measure the distance from the ultrasonic sensor head to the fluid surface (the ultrasonic sensor is mounted above the water surface). As the level changes (the surface rises or falls), the ultrasonic sensor detects the change and converts it to a head measurement.

Open-channel flow can also be measured using an area-velocity method. This method is used when the channel geometry is known and a weir and flume is not practical. Using the channel geometry and typically a fluid level measurement, the flow area can be calculated. Multiplying the flow area by the average flow velocity results in the flow rate value. The average flow velocity is normally measured by one of the several techniques. These techniques include two ultrasonic methods (Doppler and Time-of-Flight), electromagnetic, radar, and pitot tubes. As described earlier, Doppler systems bounce acoustic signals off particle in the flow and the time-of-flight method send ultrasonic signals upstream and downstream into the flow. Electromagnetic flow measurement is based on Faraday's principle that a moving conductor (the fluid) passing through a magnetic field generates a current proportional to the velocity of the conductor. The fluid is the moving conductor in this case. Radar essentially uses a Doppler technique but bounces the signal off the moving fluid surface instead of submerged particles to determine the fluid velocity. Pitot tubes are differential pressure measuring devices that translate the pressure into a flow velocity. Typically, they are used as a point measurement; however, multiple port Pitot tube are available that measure an average pressure and thus, an average flow velocity. A fluid level or height measurement is also required (to determine the flow area) and the methods described in the previous paragraphs can be used with the average velocity techniques to derive the average flow rate. The area-velocity method works best in open pipe or conduit flow because the geometry is very well known. This method has been using in flowing streams with higher uncertainties but can be much more practical than a flume or weir in this application (see also: Gonzalez and Chanson, 2007; Sturm, 2001).

6.2.9 Anemometers

There are two major types of anemometers; one for measuring wind speed and one for measuring velocities at a point in a flow field.

The most widely used wind anemometer is the cup type. The cup type has three or four cups mounted on horizontal arms that are attached to a centered vertical shaft. The arms are equidistant apart, radially. The wind flows past the cups causing rotation. The rotation is proportional to the wind velocity. Other types of velocity anemometers include the windmill type where the axis of rotation is vertical and sonic anemometers. The windmill type usually includes a propeller and a tail on the same axis. These devices often measure the wind direction as well. The sonic instrument uses ultrasonic waves to measure wind speed and direction. These devices do not have any moving parts, which can result in higher reliability.

Various instruments have been developed to measure the velocity at a single point in the flow field. The two most commonly used in the research arena are the hot-wire or film anemometer and the laser Doppler velocimeter (LDV), sometimes called the *laser Doppler anemometer* (LDA). The hot-wire or film-sensor operation is based on the fact that the heat transfer from a heated probe depends on the properties of the flow past it. Two methods are used to determine the cooling effect. One method attempts to keep the sensor at a constant temperature by controlling the power to it. The power can then be related to the flow velocity. The second method is to keep the power to the hot film constant and monitor the probe's temperature changes, which can be related to flow velocity. The hot film and wire probes can be made very small to make flow measurements in boundary layers and small channels. Gases, liquids, and conducting metals are all acceptable media. Their time response is very fast. A schematic diagram of a hot-wire sensor is shown in Figure 6.17.

An LDV operates by splitting a laser beam into two equal-intensity parallel beams. A focusing lens crosses the two beams at its focal point, thus forming a measuring point or probe volume. At this location, the beam wavefronts interfere to form alternate regions of high- and low-intensity light "fringes." As a particle passes perpendicularly through the fringes, it results in variations in the intensity of the scattered light (Figure 6.18). The frequency of these variations can be measured, and by knowing the fringe spacing from geometry and lens characteristics, the particle velocity can be determined. The probe volume can be made very small for essentially point measurements. A

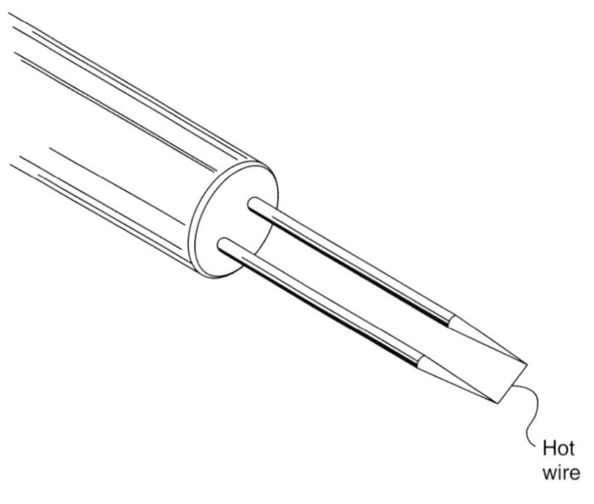

Fig. 6.17 Schematic diagram of a hot-wire anemometer.

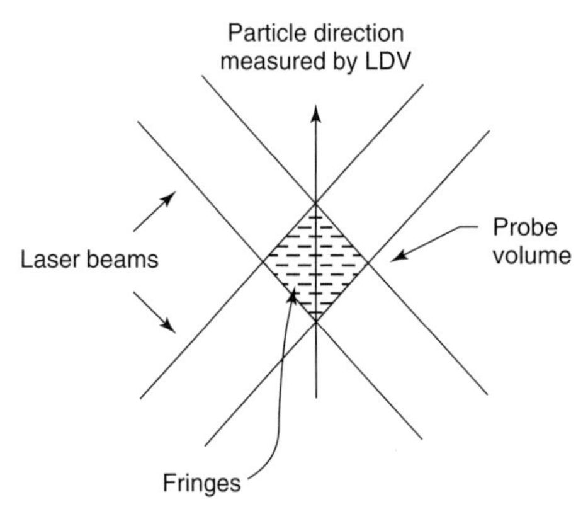

Fig. 6.18 Schematic diagram of a probe volume created by a laser Doppler velocimeter system.

major advantage of the LDV is that it is nonintrusive, but it does require particles to be present in the flow stream. A typical layout of a system is schematically shown in Figure 6.19. LDVs are available that can simultaneously measure the flow velocities in all the three dimensions. This option is often used in measuring the main velocity and orthogonal velocities to obtain an indication of turbulent intensities (see also: Durst et al., 1981; Zimmel et al., 2002).

6.2.10
MEMS Flowmeters

Over the last 30 years, MEMS (micro electro-mechanical systems) sensor technology has gone from a research interest to commercialized products. The drivers for these advancements have been the hope for very low-cost sensors when produced in large quantities and for the small footprint. An additional benefit is the low power consumption of MEMS sensors. To date, trade-offs have included moderate performance and relatively low temperature operation. Because of the large market potential, automotive applications have received the most attention, at least initially. The earliest flow sensors were designed and fabricated in miniature versions of thermal mass flowmeters. These MEMS devices demonstrated higher flow sensitivity, a faster time response, and lower power requirements than their larger counterparts. The largest application for

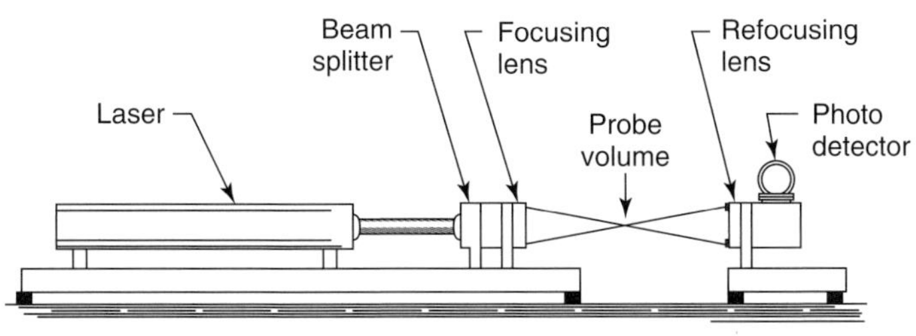

Fig. 6.19 Schematic diagram of a laser Doppler velocimeter system.

the thermal-based MEMS flow sensors is for heating, ventilation, and air conditioning systems. MEMS sensors integrated with signal conditioning electronics have a footprint of less than 25 mm × 25 mm. The thermal MEMS flow sensors typically include a heater and temperature sensing elements in a bridge circuit. The gas flows directly over the sensing elements and the heater. The electronics are designed to produce a voltage proportional to the flow rate. Some units have dual temperature elements on each side of the heater to measure flow direction. These sensors can be sensitive to flows as low as 0.1 sccm. Recent developments include a dust "proof" design for insertion in ventilation systems and inexpensive, disposable dust filters that mount immediately upstream of the flow meter.

A variety of MEMS flow sensors are becoming available and new designs are being developed. A MEMS-based vortex flow sensor (Pedersen and Andersen, 2003) that operates by measuring the vortex-shedding frequency behind a bluff body in a flow stream has been created. A MEMS-based pressure sensitive detector is located in the fluid stream downstream of a bluff body. Owing to the high sensitivity of the MEMS pressure membrane, very low flow rates can be sensed, lower than usually possible with vortex-shedding sensors (Reynolds number as low as 2000 instead of ~10 000). A coriolis mass flow sensor has also been developed with a sealed MEMS chip (Sparks et al., 2003). At the heart of the flow sensor is a U-shaped, vacuum-sealed resonating silicon microtube. The tube is driven into resonance and its motion is sensed capacitively using metal electrodes under the tube. In addition to measuring mass flow rate, chemical concentration, fluid density, and temperature can also be monitored.

As with the coriolis MEMS sensor, other multisensor MEMS chips are being developed (Xu et al., 2005). These chips can measure multiple parameters simultaneously. Parameters such as shear stress, temperature, and pressure have been combined into one chip. Flow rate measurements have been demonstrated using the pressure drop measurement and temperature values with thermal anemometry principles.

6.2.11 Summary of Flowmeters

Table 6.1 summarizes the major attributes of the flowmeters described in this article.

6.3 Flow Measurement Uncertainties

It is practically impossible to place a flowmeter in service and expect to achieve the manufacturer's performance

Tab. 6.1 Summary of flowmeter attributes (FS, full scale; D, pipe diameter).

Flowmeter	Turndown	Accuracy	Piping requirements	Reynolds number	Major advantage	Major disadvantage
Pitot tube	4 : 1	±5% FS, ±1% FS averaging Pitot design	10D up/5D down	>20 000	Low permanent pressure drop, low cost	Small differential pressure for gases, low accuracy, and turndown
Orifice	4 : 1	±1–2% FS	10D up/5D down	>10 000	Low cost, lots of empirical data	Low turndown
Cone type	10 : 1	±0.5% rate	0-3D up/0-1D down	>10 000	Minimal flow conditioning	Each meter has unique calibration
Vortex shedding	10 : 1 to 20 : 1	±1% rate	10D up/5D down	>10 000	Accuracy, no moving parts	Signal dropout at low flows, viscosity dependent
Magnetic	10 : 1 to 30 : 1	±0.5–1% rate	3D up/none down	No limit	Accuracy, turndown, and minimal flow conditioning	Cost, fluid must be conductive
Turbine	10 : 1	±1% rate	10D up/5D down	>4000	Accuracy	Bearings, viscosity dependent
Positive displacement	10 : 1	±0.5–1% FS	None	None	Rugged	Viscosity dependent
Variable area	10 : 1	±0.5–1% rate	None	Not highly viscous fluids	Low cost	More difficult to provide with electrical interface, vertically mounted
Drag	10 : 1 to 15 : 1	±0.5–2% FS	10D up/5D down	>4000	Handles dirty fluids	Erosion of drag target
Ultrasonic Doppler	10 : 1	±2% FS	10D up/5D down	No limit	Noninvasive	Needs particles in fluid
Ultrasonic transit time	40 : 1	±1% rate	10D up/5D down	No limit	Accuracy, turndown	Need clean fluid
Coriolis	>10 : 1	±0.2% rate	None	No limit	Accuracy	Cost, size
Thermal mass flow	10 : 1	±1% FS	None	No limit	Very low flows	Cost

specifications without paying attention to application-specific details. The end user faces the responsibilities of knowing the effects of flow measurement uncertainties for each application and ensuring that the flowmeter installation is engineered to mitigate or negate these uncertainties.

6.3.1 Installation Effects

The geometry of many flowmeter elements is critical to accurately measure the flow. Factors that could change the dimensions of the flowmeter element should be addressed before the meter is installed. Thermal expansion of the element and erosion/corrosion from solids-bearing flow or as a result of cavitation or liquid impingement should be addressed (Spitzer, 2005). The potential for coating an element should also be evaluated.

The flow transmitter electronics should also be taken into account when installation effects are addressed. Temperature and relative-humidity (RH) cycles of flow transmitters need to be considered for the actual installation location. The effect of exposing the flow transmitter to direct sunlight may, in some cases, exceed the transmitter's temperature specification. Many flow transmitter electronics are rated for noncondensing RH service. At least, one end-user company tests its flowmeters at 100% RH because, even if ambient conditions are not 100% RH, if the electronics are not sealed, leakage into the electronics during the day at warmer temperatures will cause condensation of moisture when the temperature drops below the dew point at night (DeBoom, 1992).

For volumetric flow measurements inferred from the inside cross-sectional area of the pipe, an accurate measurement of the inside diameter should be taken. For example, the nominal inside cross-sectional area of a Schedule 40, 6-in.-diameter (15.2-cm-diameter) pipe is allowed to vary as much as 4.5% according to American Society of Testing Materials guidelines (Kopp, 1989). The diameter error translates to a squared error in flow for differential pressure flow meters.

In the past, differential pressure flow meters used impulse tubes to transfer the pressure measurements from the flow element (e.g., orifice) to the flow transmitter. Recently, new flow transmitter and manifold designs have allowed direct mounting of the flow transmitter to the line, which eliminates the impulse tubing. Impulse tubing is susceptible to the following error sources (Strom and Livelli, 2002).

- *Plugging*: This can be caused by a precipitate left by steam flashing, a low spot in the line where condensation has occurred (gas flow), accumulation of solids, and freezing (liquid flow).
- *Leakage*: The number of potential leaks increases with increased number of connections (tube fittings, manifolds, seal pots, blowdown valve connection, and so on).
- *Hydrostatic head uncertainty*: The upper and lower side of dP measurement are not at same level, or not along the same axis of pipe line.

6.3.2 Fluid Properties

Several fluid properties may have significant effects on the performance of a flowmeter or a sensor. Fluid density, compressibility, vapor pressure, and viscosity are important to fluids in motion.

In addition, thermal properties such as gas constant, specific heat capacities, and thermal conductivity are important in heat transfer. The thermodynamic state of a fluid, determined by its pressure and temperature, affects most of the above-mentioned properties. Other parameters of importance, which are combinations of fluid properties, include the Prandtl number (involves viscosity, specific heat, and thermal conductivity), thermal diffusivity (involves thermal conductivity, density, and specific heat), and the Reynolds number (inertial to viscous force ratio and an indication of laminar or turbulent flow).

Because the fluid properties affect the flow instrumentation's performance and usually its calibration, one must be careful if the flowmeter experiences fluid properties that are different from the ones at calibration or if the fluid properties are not constant. Errors in the output of the flowmeter may occur, increasing the uncertainties obtained. In some cases, these errors can be very high (easily greater than 20%). Depending on the particular flowmeter's sensitivity to fluid properties, conditions that are known to be different from the calibration state or that are time-varying may require monitoring. Corrections to the output of the flowmeter can then be made, if possible, to reduce property-induced errors.

In the past, the most sophisticated method to correct for property-induced flow rate effects over just a plain differential pressure measurement was to measure temperature and pressure and to perform density compensation on the flow signal. As can be seen by Equation (6.8), there are other variables in the flow equation. Modern flow transmitters (smart transmitters) are able to continuously compensate for the following (Rowe, 2004; Strom and Livelli, 2002):

- changes in fluid density with specific algorithms optimized for fluid type (i.e., liquid, gas, or vapor);
- dimensional changes in primary device and adjacent piping due to pressure and/or temperature;
- changes of primary device discharge coefficient with Reynolds number;
- changes in saturation temperature for saturated vapor (e.g., steam, ethylene);
- velocity of approach factor; and
- gas expansion factor.

6.3.3 Velocity Profiles

Fully developed velocity profiles for laminar flow are parabolic and symmetric about the pipe axis. Fully developed velocity profiles for turbulent flow are symmetric about the pipe axis and are almost constant from the pipe axis outward, except near the pipe walls where a sharp gradient occurs. The velocity profile through a flowmeter should be taken into account when the meter is sensitive to this effect. Asymmetric flow profiles should be avoided. Swirling-type flow velocities are caused by upstream pipe elbows. Jets can be caused by high-velocity, smaller diameter streams that enter larger diameter streams upstream. Straight-pipe lengths upstream of the flowmeter are the best way to avoid asymmetric flow profiles. Flow-straightening vanes, of either the plate type or the honeycomb type, should be avoided because they inhibit lateral mixing of the fluid, which is required to reduce the asymmetric profile. Downstream straight-pipe lengths should be designed as well. Other considerations include pipe ovality, smoothness of the pipe wall, concentricity of the pipe and

Tab. 6.2 Minimum straight-pipe diameters for orifices, venturis, and nozzles with respect to diameter ratio [a].

Diameter ratio, beta	Single 90° elbow	Two 90° elbows, same plane	Two 90° elbows, out of plane	Globe valve fully open	Downstream
0.2	10	14	34	18	4
0.4	14	18	36	20	6
0.6	18	26	48	26	7
0.8	46	50	80	44	8

[a] Length for straight pipe diameters measured in same units as pipe diameter. Last column is required straight pipe downstream of flow element.

meter, and gaskets projecting into the fluid stream. Flangeless wafer-type meters may not provide adequate centering (Kopp, 1989).

Table 6.2 is an excerpt from ISO 5167-1 regarding the minimum pipe diameters of straight line that are required for different geometries for orifice plates, nozzles, and venturi tubes. Similar tables are available for other flow devices. Notice that the diameter ratio of the flow element has a large impact on the straight line required. This is because with small diameter ratios, the flow element itself plays a large role in shaping the velocity profile through the meter. At large diameter ratio diameters, the velocity profile is not affected much by the flow meter and is more susceptible to the effects of nonideal geometries (ISO, 2003).

6.4 Calibration Methods

Calibration with the meter in place is recommended over having it calibrated in a laboratory because installation effects are taken into account. By the same token, it is always better to calibrate with the same fluid that will be monitored by the meter. Pressure and temperature ranges also should be the same because they affect both the fluid and the meter (Sydenham, 1999). Calibration methods have evolved throughout the years, and this section provides an insight into the various methods.

6.4.1 Time-Collection Techniques

Time-collection techniques involve diverting a fluid in series with the meter being calibrated into either a volume- or weight-measuring device and recording time and either volume or weight data points. The techniques are probably among the oldest methods of performing a flowmeter calibration, but modern data acquisition equipment will greatly improve the results from these tests if it can be used. This calibration technique allows traceability to primary standards: mass or volume and time.

6.4.1.1 Monitored Volume Methods

Time versus volume techniques consist of measuring the volume of flow through a flowmeter over a given time duration. For example, in liquid flows, the flow could be diverted to a tank. The volume of the tank would be measured before and after the test. The volume change divided by the time duration provides the flow reference. For gases, a "bell prover" can be used. An inverted bell fills with gas during the calibration test. As the bell fills, it rises. The amount that the bell rises provides an indication of the volume of gas that has flowed through the flowmeter. Other gas calibration devices include the bubble meter, where a soap-film bubble rises in a graduated cylinder relative to flow; and the piston meter, where a sealed piston is used rather than a soap-film bubble (Wight, 1994).

The disadvantage of time versus volume methods is that the volume devices have not, in general, been outfitted for automatic data collection. Only beginning and ending data points are available, and flow was assumed to be steady during the entire test. The time versus volume method is acceptable if a volumetric flowmeter is being tested; but if a mass flowmeter is being tested, the density of the fluid must also be determined and used to calculate true mass flow.

6.4.1.2 **Monitored Weight Methods**

Time versus weight methods are similar to time versus volume methods, except that the reference variable is weight instead of volume. Many weigh-cell meters have the advantage that they have electronic interfaces for data output and respond fast enough to provide dynamic data during a calibration test. The advantage here is that the weight data can be examined for periods of unsteady flow during the test. The periods of unsteady flow can be dropped from the analysis when the calibration comparison is made.

Time versus weight methods are good if a mass flowmeter is being tested; but if a volumetric flowmeter is being tested, the density of the fluid must be determined and used to calculate true volumetric flow.

Gravimetric calibration devices are weight-method- based calibrators. In the mid-1990s, one such device was developed to handle toxic and corrosive gases for low-flow thermal mass flow meters. This gravimetric calibrator is schematically shown in Figure 6.20. The 20-l vessel was designed to be slightly heavier than neutral buoyancy so its mass is essentially tared out; thus, only the mass of the gas is weighed by a precision load cell. This technique allows the incoming gas mass to be accurately measured, even in small amounts (just a few grams of gas need to be collected). The vessel was passivated to handle corrosive gases and after completing a calibration run, the vessel was vented to a toxic gas incinerator. Flow rates to several hundred cubic centimeters a second could be run and accuracies to $\pm 0.1\%$ could be obtained in just a few minutes. A total calibration (10 data points) could be completed in a couple of hours. Standard gravimetric calibration techniques typically take one to two days to complete because large quantities of the gas must be collected to allow for an accurate measurement (the mass of the collection device is much larger than to the mass of the gas). To reduce the measurement uncertainty, the mass of the gas collected needs to be a reasonable percentage of the mass of the collection device, thus the long calibration times.

6.4.2 **Transfer Standards**

Generally, it is not practical or economical to use a primary standard for calibrating a flowmeter. The primary standard may be too expensive to build and maintain. Routine commercial practice is to have high-quality flowmeters calibrated against a primary standard (and have this done periodically like once in a year) and

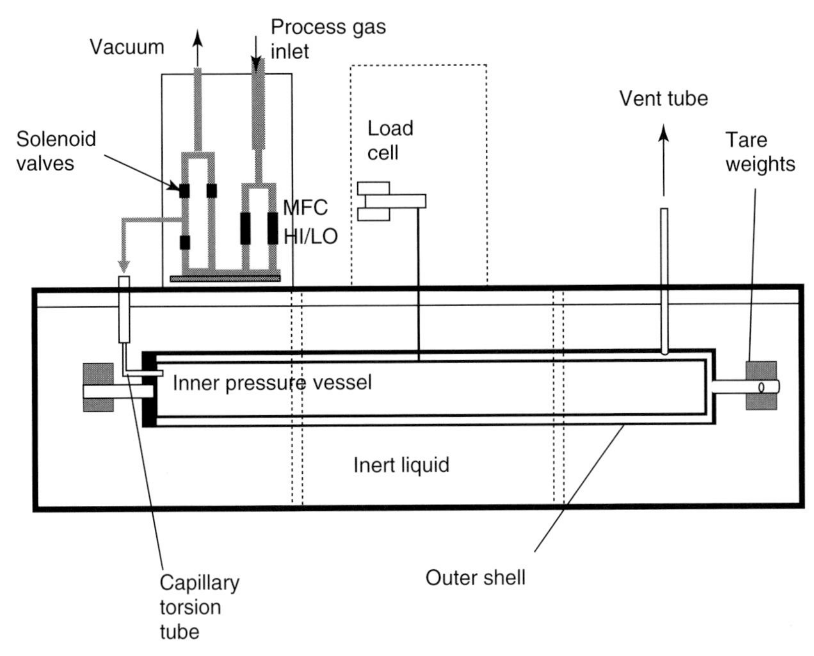

Fig. 6.20 Gravimetric calibrator schematic for mass flow controller (MFC).

then use this high-quality flowmeter as a secondary standard. The secondary or transfer standard is usually based on fundamental operating principles that conform to theoretical predictions very well. The secondary standard must be stable and predictable for long periods of time (a year or more). The transfer of the secondary standard to a flowmeter that needs calibrating is accomplished by installing the secondary standard and the device requiring calibration in series in a flow loop. The secondary reference method of calibration consists of allowing the fluid to flow through the flowmeter under calibration in series with a standard meter. The readings of the two meters are compared, and the meter under test is calibrated as necessary on the basis of the standard meter readings. Care must be taken not to allow flow obstructions caused by the upstream meter to affect the velocity profile entering the other meter. Temperature and pressure changes between the two meter locations must be accounted for as well. The most common secondary standard flowmeters are the venturi meter, critical nozzles, squared-edged orifices, and laminar flow elements.

6.4.3
Metered Volume Methods

Metered volume methods are the opposite of monitored volume methods. Metered volume methods of calibration consist of engineering a system that will meter a precise volume of fluid through a flowmeter in a measured time duration, rather than measuring the volume produced by another means. One system consists of a pipe displacer or "pig" that is driven through a honed section of pipe. The pig is timed from the time it passes a beginning point of the honed pipe until it passes the ending point. The pig travels between the points at a constant speed. If the diameter and length of the honed pipe and the transit duration are known, volumetric flow rate can be calculated and compared with a flowmeter reading in series with the honed section. The test duration of this type calibrator is very short. Another system uses a bidirectional U-tube system where a four-way valve switches flow at the ends of the U-tube to ensure a continuous metered flow. The flow in the U-tube switches direction when the pig reaches either of the end points. Detectors are mounted near the ends of the U-tube to sense the passing of the pig. The system can run indefinitely, as compared to the once-through pig method (Paton, 2005).

Glossary

Accuracy: The closeness of agreement between an observed value and the true value.

Boundary Layer: A very thin layer of fluid in the neighborhood of a body where friction plays an essential part in the hydrodynamics. Outside this thin layer, friction can be neglected. The boundary layer also controls the heat-transfer characteristics from a surface to the surrounding fluid.

Calibration: The procedure of comparing a particular instrument with either a primary standard or a secondary standard with a higher accuracy than the instrument to be calibrated. This process allows the inaccuracy of the instrument to be quantitized against a known, accepted reference.

Compressibility: The measure of the change of volume of a fluid under the action of external forces. The compressibility of liquids is typically very small, and, thus, liquids are considered to be incompressible. Gases do exhibit significant volume changes and are considered as compressible fluids.

Drift: An undesired change in the output–input relationship over a period of time.

Flow Totalization: The summation of the flow sensed by an instrument as a function of time. Flow totalization is used in product custody transfer and to determine home and business water and energy usage.

Flow Transmitter: A device that responds to a measured variable by means of a flow element, and converts it to a standardized transmission signal that is a function only of the measured variable.

***In situ* Calibration:** Calibration of an instrument carried out with the instrument located in its measurement or field position. The instrument is calibrated *in situ* – in place – rather than removing the device and calibrating it in a laboratory.

Invariant: A mathematical term that does not change under a set of transformations.

Isentropic: Taking place with no entropy production, that is, at constant entropy.

Laminar: The state of a fluid in motion where the velocity is free of macroscopic fluctuations or mixing.

Liquid Impingement: A process by which a material is eroded by liquid droplets, contained in a liquid/vapor stream, striking the material at high velocities.

Pressure Tap: The physical entry into a system from which the system pressure may be measured. The "tap" is usually a hole through the system boundary from which the system pressure may be sensed.

Repeatability: The closeness of agreement among a number of consecutive measurements of the output for the same value of the input under the same operating conditions, approaching from the same direction. It is usually measured as a non-repeatability in percent of span.

Residual: The difference between the current iteration value and the value calculated during the previous iteration step.

Signal-to-Noise Ratio: The ratio of the electrical magnitude of the desired measured quantity (signal) to the undesired component of that magnitude (noise). The undesired component comes from flow turbulence, environmental factors, and electrical interferences.

Smart Flow Sensor: Flow sensor that contains a microprocessor that allows the sensor to make corrections for changing environmental conditions or flow range or calibration changes. These changes may allow the flow sensor to improve its accuracy or extend its rangeability.

Turbulent: The state of a fluid in motion where the velocity is subject to macroscopic fluctuations or mixing. These velocity fluctuations are superimposed on the mean motion.

Uncertainty: The range within which the true value of the measured quantity is expected to be; an indication of the variability associated with a measured value.

Velocity Profile: The distribution of the magnitude of the velocity vector in a flow stream. In a closed system, the velocity is the highest at the centerline and drops to zero at the system's fixed boundary or wall.

Viscosity: The constant of proportionality between the applied shear stress and the rate of deformation or to the velocity gradient normal to the velocity of the fluid. It is the measure of the fluid's resistance to flow.

References

Basrawi, Y.F. (**2004**) Ultrasonic Flow Metering Devices and Their Applications to Royalty and Custody Transfer Measurements, ISA AUTOMATION WEST, 2004.

Cousins, T. (**2002**) Proving of Multi-Path Liquid Ultrasonic Flowmeters, 20th NSFMW.

Dahlstrom, M.J. (**1994**) V-cone meter-gas measurement for the real world. North Sea Flow Measurement Workshop, Peebles, Scotland.

DeBoom, R.J. (**1992**) Rating a Flowmeter's Field Performance, InTech, July.

Durst, F., Melling, A., and Whitelaw, J. (**1981**) *Principles and Practice of Laser-Doppler Anemometry*, Academic Press, London.

Gonzalez, C. and Chanson, H. (**2007**) Experimental measurements of velocity and pressure distribution on a large broad-crested weir. *Flow Meas. Instrum.*, **18** (3–4), 107–113.

Han, I.Y. (**2005**) Study of the transient characteristics of the sensor tube of a thermal flow meter. *Int. J. Heat Transfer*, **48** (13), 2583–2592.

Hardy, J.E. *et al.* (**1999**) *Flow Measurement Methods and Applications*, John Wiley & Sons, Inc., pp. 135–203.

Henry, M., Clark, C., and Mattar, W. (**2004**) The Dynamic Response of Coriolis Mass Flow Meters: Theory and Application, ISA, Houston, October 2004.

Henry, M., Yeung, H., and Mattar, W. (**2004**) How a Coriolis Mass Flow Meter Can Operate in Two-Phase Flow, ISA, Houston, October 2004.

Hettrich, U. and Daly, J. (**2004**) Omega Tube Design Extends the Usability of Coriolis Technology in Mass Flowmetering Applications, ISA, Houston, October 2004.

Hinkle, L. (**1991**) MFC Accuracy: is simple gas correction enough? Semiconductor International, March.

Ifft, S.A. and Mikkelsen, E.M. (**1993**) Pipe elbow effects on the V-Cone flowmeter. Presented at ASME Fluids Engineering Conference 1993, Washington, DC.

ISO (**2003**) 5167-1. *Measurement of Fluid Flow by Means of Pressure Differential Devices*, International Organization for Standards, Genève.

Kim, S.J. (**2007**) Study of the steady-state characteristics of the sensor tube of a thermal flow meter. *Int. J. Heat Transf.*, **50** (5–6), 1206–1211.

Kopp, J.G. (**1989**) How to Get the Accuracy You Expect from Flowmeters, I&CS, September.

Lipták, B.G. (**2003**) *Instrument Engineers' Handbook*, Process Measurement and Analysis, vol. **1**, CRC Press, Boca Raton.

Lloyd, K.E., Gutherie, B.D., and Peters, R.J.W. (**2002**) A flowmeter calibration facility developed at The University of Iowa to evaluate custody transfer steam flowmeters with a cone differential pressure meter used as the metering standard. ISA 2002 Technical Conference, October 21-24, 2002, Chicago.

Lynnworth, L.C. and Liu, Y. (**2006**) Ultrasonic flowmeters: Half-century progress report, 1955-2005. *Ultrasonics*, **44** (Suppl 1), 1371–1378.

Madison, R.M. (**1995**) Modern electronics meet turbine meters, Measurements & Control, April.

Magness, M. (**2002**) Ultrasonic Flow Measurement of Natural Gas Flow, ISA 2002, Chicago, October 2002.

Mattar, W. (**2004**) Vortex shedding flowmeters – a technology finally comes of age. Presented at the ISA 2004, Houston, Texas, October 5-7, 2004.

McCoy, M. (**1992**) Sensor data integration for improved flow data accuracy, Sensors, February.

McMillan, G.K. and Considine, D.M. (**1999**) *Process/Industrial Instruments and Controls Handbook*, 5th edn, McGraw-Hill, New York.

Miller, R.W. (**1996**) *Flow Measurement Engineering Handbook*, McGraw-Hill, New York.

Mori, M. *et al.* (**2002**) Development of a novel flow metering system using ultrasonic velocity profile measurement. *Exp. Fluids*, **32** (2), 153–160.

Ostling, H. and Oki, S. (**2001**) Spectral Signal Processing (SSP) enhances vibration immunity for vortex flowmeters. ISA TECH/EXPO Technology Update Conference Proceedings, Houston, Texas, 416, pp. 411–417.

Paton, R. (**2005**) Calibration and standards in flow measurement, in *Handbook of Measuring System Design* (P.H.Sydenham), John Wiley & Sons, Inc., New York.

Pedersen, N. and Andersen, P.E. (**2003**) A MEMS-based vortex flow sensor for aggressive media. *Proc. IEEE Sens.*, **2** (1), 320–325.

Remenyik, C.J. (**1995**) An instrument for gravimetric calibration of flow devices with corrosive gases. Proceedings of the 41st International Instrumentation Symposium, May 7-11, 1995, Aurora.

Rose, C. and Vass, G. (**1993**) Obstructionless magenetic flowmeters. Sensors, March.

Rowe, J.D. (**2004**) Advanced sensor technology key to new multivariable transmitter. Presented at ISA Expo 2004, October 5-7, 2004, Houston.

Rudroff, D. (**2005**) Ultrasonic meters ready for gas distribution market. *Pipeline Gas J.*, **232** (7), 24–26.

Shen, J.S., Bosio, J., and Larsen, S. (**1995**) A performance study of a V-Cone meter in swirling flow. 13th North Sea Flow Measurement Workshop, October 24-26, 1995, Lillehammcr.

Soo, S. (ed.) (**1999**) *Instrumentation of Fluid-Particle Flow*, Noyes Publications, pp. 172–196.

Sparks, D. *et al.* (**2003**) A portable MEMS coriolis mass flow sensor. *Proc. IEEE Sens.*, **2** (1), 337–339.

Spitzer, D.W. (**2005**) Industrial Flow Measurement, ISA, Research Triangle Park.

Strom, G.R. and Livelli, G. (**2002**) A New Generation of DP Measurement Instrumentation Eliminates Traditional Limitations, Intech, August.

Sturm, T. (**2001**) *Open Channel Hydraulics*, McGraw-Hill, Boston.

Sydenham, P.H. (**1999**) *Handbook of Measurement Science*, John Wiley & Sons, Inc., New York.

Tison, S.A. (**1996**) A critical evaluation of thermal mass flow meters. *J. Vac. Sci. Technol. A Vacuum, Surf. Films*, **14** (4), 2582–2591.

Viswanathan, M. *et al.* (**2001**) Development, modeling and certain investigations on thermal mass flow meters. *Flow Meas. Instrum.*, **12** (5–5), 353–360.

Walker, J.R. (**2001**) Diagnostics advance in electromagnetic flow metering. Presented at ISA 2001, September 10-12, 2001, Houston.

Wight, D.W. (**1994**) *Fundamentals of Air Sampling*, CRC Press, Boca Raton.

Witte, J.N. (**2002**) Ultrasonic gas meters from flow lab to field: a case study. American Gas Association Operations Conference Chicago, Illinois.

Yamamoto, T., Yao, H., and Kshiro, M. (**2004**) Advanced Hybrid Ultrasoinc Flow meter Utilizing Pulsed-Doppler Method and Transit Time Method, ISA, Houston, October 2004.

Xu, Y. *et al.* (**2005**) A MEMS multi-sensor chip for gas flow sensing. *Sens. Actuators A Phys.*, **121** (1), 253–261.

Zimmel, M. *et al.* (**2002**) Comparison of velocity measurements by high temperature anemometer and laser-Doppler anemometer with results of CFD-simulation. *Therm. Sci.*, **6** (1), 3–13.

Further Reading

Stauss, T. (ed.) (**2004**) *Flow Handbook*, Schaub Medien AG, Reinach.

Sydenham, P.H. (ed.) (**2005**) *Handbook of Measuring System Design*, John Wiley & Sons, Inc., New York.

Webster, W.C. (ed.) (**1999**) *Measurement, Instrumentation, and Sensors Handbook*, CRC Press, Boca Raton.

Yoder, J. (**2001**) Flow meter shootout part II, Control, February.

7
Ultrasonics

John H. Cantrell

Handbook of Metrology. Edited by Michael Gläser and Manfred Kochsiek

ISBN: 978-3-527-40666-1

7.1 Introduction

Generally, sound (sonics) refers to vibratory mechanical disturbances at all frequencies in all substances. Ultrasound (ultrasonics) refers to such disturbances in the range of frequencies above that of audible sound (approximately 20 kHz) up to pretersonic frequencies (10 GHz and higher). Recently, the term *hypersonics* has also come into use to describe sound at frequencies above 1 GHz. However, the term is somewhat ambiguous, because in aeroacoustics hypersonics refers to the study of phenomena associated with objects traveling faster than five times the velocity of sound in air.

The study and application of ultrasonics as a serious scientific discipline began about 1910 with the development of piezoelectric transducers. Early applications of ultrasound centered on the development of a primitive form of sonar, communications devices, and light modulators used in early experiments with television. Following improvement in high-frequency electronics during World War II, the pace of ultrasonics research and applications accelerated. The first acoustic delay lines were applied to radar systems to produce a moving target indicator and used as short term memories in early computers. Today, ultrasonic methods and techniques are widely used in such far-ranging fields of application as low temperature and solid state physics, signal processing and devices, materials processing and testing, nondestructive evaluation, medical diagnostics and imaging, Doppler sonar, microstructural materials characterization, acoustic microscopy, chemical kinetics, viscoelasticity, lubrication, fluid flow, ultrasonic motors, materials forming, and welding among others.

This article is divided into three broad categories: (i) physical and technical principles of ultrasound; (ii) ultrasonic generation, detection, and measurements; and (iii) applications. The first category is covered in four main sections (Sections 7.2–7.5), which establish the physical basis and phenomenological constructs of ultrasound. The development emphasizes the intrinsic connection to reversible and irreversible thermodynamics. The second category is covered in two main sections (Sections 7.6 and 7.7), which provide an overview of the most commonly used methods and equipment for the generation and detection of ultrasonic signals and techniques for measuring ultrasonic parameters. The article concludes in Section 7.8 with a survey of representative applications of ultrasound including some

Handbook of Metrology. Edited by Michael Gläser and Manfred Kochsiek

ISBN: 978-3-527-40666-1

recent applications in the final stage of development.

7.2 Ultrasonic Waves in Boundless Media

Many applications of ultrasound involve wave propagation that is minimally influenced by geometrical constraints or boundaries. Such conditions occur when the time duration of an ultrasonic wave, that is, an ultrasonic pulse or toneburst (gated continuous wave (cw)), is short compared to the propagation time through the medium. For sufficiently long tonebursts, the wave propagation may be modeled to a good approximation as a cw propagating in a boundless medium. The relative simplicity of such models emphasizes the basic physical properties of the propagation medium. We begin by considering the propagation of ultrasonic waves in thermodynamically reversible solids and fluids to explore the most basic physical features of these media that influence wave propagation.

7.2.1 Solids

We may consider a solid to be a material in which the thermodynamic state of equilibrium depends on the fixed mean configuration of the particles comprising the material and either the temperature T or the entropy S of the system. We denote the present configuration of the material particles by the set of particle vectors $\{\boldsymbol{x}\}$. The relevant state functions describing the system of particles are the Helmholtz free energy per unit mass $F(\boldsymbol{x}, T)$ and the internal energy per unit mass $U(\boldsymbol{x}, S)$.

7.2.1.1 Stress, Strain, and Elastic Constants

We assume that the present configuration of material particles of the solid is obtained by means of a small (infinitesimal) elastic deformation from some initial configuration of particles denoted by the set of particle vectors $\{\boldsymbol{a}\}$. We describe the elastic deformation by the set of transformation coefficients $\alpha_{ij} = \partial x_i/\partial a_j$ such that $x_i = \alpha_{ij} a_j$. The indices i and j take the values 1, 2, 3 representing three mutually orthogonal spatial reference axes. We define the resulting particle displacements u_i by $u_i = x_i - a_i$ and write $\alpha_{ij} = \delta_{ij} + u_{ij}$, where the displacement gradients $u_{ij} = \partial u_i/\partial a_j$, and δ_{ij} is the Kronecker delta ($\delta_{ij} = 1$ if $i = j$, $\delta_{ij} = 0$ if $i \neq j$). The deformation of the solid results in a change in the mass density of the material described by the relation (Wallace, 1970)

$$\frac{\rho_0}{\rho} = J = \det[\alpha_{ij}] \tag{7.1}$$

where $\rho_0 = \rho_0(\boldsymbol{a})$ is the mass density in the initial state and $\rho = \rho(\boldsymbol{x})$ is the mass density in the final state.

Let $\Delta\boldsymbol{a}$ be the vector between any two material particles in the initial state and $\Delta\boldsymbol{x}$ be the vector between the same two particles in the final state. If the deformation is constant over the region defined by $\Delta\boldsymbol{x}$, then the relation (Murnaghan, 1951) (repeated suffix denotes summation)

$$|\Delta\boldsymbol{x}|^2 - |\Delta\boldsymbol{a}|^2 = 2\eta_{ij}\Delta a_i \Delta a_j \tag{7.2}$$

holds where

$$\begin{aligned}\eta_{ij} &= \frac{1}{2}(\alpha_{ki}\alpha_{kj} - \delta_{ij}) \\ &= \frac{1}{2}(u_{ij} + u_{ji} + u_{ki}u_{kj})\end{aligned} \tag{7.3}$$

The symmetric parameters η_{ij} are called the *Lagrangian strains*. For sufficiently small strains, the terms $u_{ki}u_{kj}$ can be neglected in Equation (7.3) and the Lagrangian strains reduce to the stain measure $\varepsilon_{ij} = (1/2)(u_{ij} + u_{ji})$ conventionally defined in linear elasticity. The advantage of the Lagrangian strains is that the distances between all pairs of material particles can be completely specified by the initial configuration of particles $\{\boldsymbol{a}\}$ and the Lagrangian strains η_{ij} defined over any region for which the η_{ij} are constant. Because the Helmholtz free energy and the internal energy depend only on the relative positions of the particles, we may write that $F(\boldsymbol{x}, S) = F(\boldsymbol{a}, \eta_{ij}, S)$ and $U(\boldsymbol{x}, S) = U(\boldsymbol{a}, \eta_{ij}, S)$. For small deformations, we may expand $U(\boldsymbol{a}, \eta_{ij}, S)$ about $\boldsymbol{a}$ (i.e., about $\eta_{ij} = 0$) to first order in η_{ij} as (Wallace, 1970)

$$\rho_0 U(a, \eta_{ij}, S) = \rho_0 U(a, 0, S) + C_{ij}^S \eta_{ij} + \frac{1}{2} C_{ijkl}^S \eta_{ij} \eta_{kl} + \cdots \quad (7.4)$$

where

$$C_{ij}^S = \rho_0 \left(\frac{\partial U}{\partial \eta_{ij}} \right)_s, \quad C_{ijkl}^S = \rho_0 \left(\frac{\partial^2 U}{\partial \eta_{ij} \partial \eta_{ij}} \right)_s \quad (7.5)$$

A similar expansion may be obtained for the Helmholtz free energy with coefficients C_{ij}^T and C_{ijkl}^T defined in terms of appropriate strain derivatives of the Helmholtz free energy per unit mass.

The coefficients C_{ij}^S are the isentropic initial stresses in the solid $(\sigma_0)_{ij}$ (sometimes called the *thermodynamic tensions*) and the coefficients C_{ijkl}^S are the isentropic elastic constants (often called the *second-order Brugger elastic constants*) of the solid. The C_{ij}^T and C_{ijkl}^T are the corresponding isothermal coefficients. For sufficiently small strains such that the terms $u_{ki}u_{kj}$ can be neglected in Equation (7.3), we may take the derivative of Equation (7.4) with respect to the strains to obtain the isentropic equation of state

$$\sigma_{ij} = (\sigma_0)_{ij} + C_{ijkl}^S \varepsilon_{kl} + \cdots \quad (7.6)$$

where $\sigma_{ij} = \rho_o(\partial U/\partial \varepsilon_{ij})$ are the stress components in the final state, $(\sigma_0)_{ij}$ are the stress components in the initial state, and ε_{ij} are the conventional strains. A similar isothermal expression holds using the Helmholtz free energy. Equation (7.6), known as the *generalized Hooke's law*, clearly shows that the second-order elastic constants serve to linearly couple the stress and the strain in perfectly elastic bodies.

Both the isentropic and the isothermal C_{ijkl} are fourth rank tensors that generally have 81 components. However, the stress symmetry $(\sigma_{ij} = \sigma_{ji})$ requires $C_{ijkl} = C_{jikl}$. The strain symmetry $(\varepsilon_{kl} = \varepsilon_{lk})$ requires $C_{ijkl} = C_{jilk}$. These two conditions reduce the number of independent second-order elastic constants to 36. The quadratic dependence of the energy on the strain, shown in Equation (7.4), means that the energy must remain unchanged not only under an interchange of strain suffix pairs i, j and k, l, but also under an interchange of i, k and j, l. Thus, the relation $C_{ijkl} = C_{jikl} = C_{ijlk} = C_{klij}$ holds for the elastic constants. This relationship reduces the number of independent constants to 21.

The crystal symmetry of a material may also reduce the number of independent elastic constants, because the constants are derivatives of the thermodynamic state functions with respect to crystallographically equivalent strains. A detailed derivation of the number of independent elastic constants may be found in Nye (1960). The number of independent constants for the different crystal systems is summarized in Table 7.1. The point groups in Table 7.1

Tab. 7.1 Number of independent elastic constants for different crystal systems.

Crystal system	Laue group	Point groups	Elastic constants
Triclinic	N	$1, \bar{1}$	21
Monoclinic	M	m, 2, $\frac{2}{m}$	13
Orthorhombic	O	2mm, 222, $\frac{2}{m}\frac{2}{m}\frac{2}{m}$	9
Tetragonal	TII	$4, \bar{4}, \frac{4}{m}$	7
Tetragonal	TI	4mm, $\bar{4}$2m, 422, $\frac{4}{m}\frac{2}{m}\frac{2}{m}$	6
Rhombohedral	RII	$3, \bar{3}$	7
Rhombohedral	RI	3m, 32, $3\,\frac{2}{m}$	6
Hexagonal	HII	$\bar{6}, 6, \frac{6}{m}$	5
Hexagonal	HI	$\bar{6}$2m, 6mm, 622, $\frac{6}{m}\frac{2}{m}\frac{2}{m}$	5
Cubic	CII	23, $\frac{2}{m}\bar{3}$	3
Cubic	CI	$\bar{4}$ 3m, 432, $\frac{4}{m}\,\bar{3}\,\frac{2}{m}$	3
Isotropic	I		2

reflect all symmetry operations possible for a given crystal system. A symmetry operation for a boundless crystal is an operation that interchanges the positions of the various atoms of the crystal into equivalent positions such that the crystal appears exactly the same as before the operation. For example, crystals belonging to the monoclinic system have one of three possible point groups given as m, 2, or $2/m$ in the full international notation. The symbol m denotes an operation corresponding to the reflection of the crystal atoms across a mirror plane, 2 denotes an operation corresponding to 180° rotations (twofold rotations) around a symmetry axis, and $2/m$ denotes an operation corresponding to reflection across a mirror plane normal to a twofold rotation axis. The point groups are given in the full international notation in Table 7.1 with the full set of possible point groups listed for each crystal system.

The full tensor notation C_{ijkl} for the elastic constants is often replaced by the abbreviated notation (sometimes called the *matrix* or *Voigt notation*) $C_{\alpha\beta}$ according to the prescription:

ij or $kl =$	11	22	33	23, 32	13, 31	12, 21
α or $\beta =$	1	2	3	4	5	6

It is instructive to write the set of elastic constants in Voigt notation for triclinic, cubic, and isotropic symmetry classes in the half matrix forms (the full matrices are symmetrical about the diagonal):
Triclinic : 21 constants

$$\begin{matrix} c_{11} & c_{12} & c_{13} & c_{14} & c_{15} & c_{16} \\ & c_{22} & c_{23} & c_{24} & c_{25} & c_{26} \\ & & c_{33} & c_{34} & c_{35} & c_{36} \\ & & & c_{44} & c_{45} & c_{46} \\ & & & & c_{55} & c_{56} \\ & & & & & c_{66} \end{matrix}$$

Isotropic : 2 constants

$$\begin{matrix} c_{11} & c_{12} & c_{12} & 0 & 0 & 0 \\ & c_{11} & c_{12} & 0 & 0 & 0 \\ & & c_{11} & 0 & 0 & 0 \\ & & & c_{44} & 0 & 0 \\ & & & & c_{44} & 0 \\ & & & & & c_{44} \end{matrix}$$

Cubic (all classes) : 3 constants

$$\begin{matrix} c_{11} & c_{12} & c_{12} & 0 & 0 & 0 \\ & c_{11} & c_{12} & 0 & 0 & 0 \\ & & c_{11} & 0 & 0 & 0 \\ & & & \frac{1}{2}(c_{11}-c_{12}) & 0 & 0 \\ & & & & \frac{1}{2}(c_{11}-c_{12}) & 0 \\ & & & & & \frac{1}{2}(c_{11}-c_{12}) \end{matrix}$$

For the remaining symmetry classes, see Nye (1960).

For isotropic symmetry Hooke's law, Equation (7.6) may be written as (Pollard, 1977) $\sigma_{ij} = (C_{11} - C_{12})\varepsilon_{ij} + C_{12}\varphi\delta_{ij} = 2G\varepsilon_{ij} + \lambda\varphi\delta_{ij}$, where the dilatation $\varphi = \varepsilon_{11} + \varepsilon_{22} + \varepsilon_{33} = \text{div}\,\boldsymbol{u}$ ($\boldsymbol{u}$ is the displacement vector). The Brugger elastic constants in the last equality are replaced by the Lamé constants • and G, where $\lambda = C_{12}$ and $G = (1/2)(C_{11} - C_{12})$. The constant G is also called the *shear modulus* or *rigidity modulus*.

Another important modulus is the *bulk modulus B* defined as the ratio of the change in hydrostatic pressure, Δp, to the fractional change in the volume of the material, $\Delta V/V_0 = \Delta J$, produced when pressure is applied to an isotropic material (including fluids). It is related to the Brugger constants and the Lamé constants as $B = (1/3)(C_{11} + 2C_{12}) = \lambda + (2/3)G$.

A related constant is the compressibility $K = 1/B$.

7.2.1.2 **Modal Wave Propagation**

For a solid of arbitrary crystalline symmetry, we may substitute Equation (7.6) into Newton's Law, $\rho_0(\partial^2 u_i/\partial t^2) = (\partial\sigma_{ij}/\partial a_j)$, where t is time and body forces are neglected. We obtain the isentropic wave equation

$$\rho_0 \frac{\partial^2 u_i}{\partial t^2} = C^S_{ijkl} \frac{\partial^2 u_k}{\partial a_j \partial a_l} \tag{7.7}$$

It is important to note that for the case of small elastic deformations, the transformation coefficients $\alpha_{ij} \approx \delta_{ij}$ and the wave equation, Equation (7.7), may be written with the variables a_i (often called the *material or Lagrangian coordinates*) replaced by x_i (often called the *spatial or Eulerian coordinates*) (Thurston, 1964). Thus, in the "linear acoustic approximation," the coordinate variables in the wave equation may be either the set a_i or x_i. In this article, both sets are used depending on notational convenience.

An equation similar to Equation (7.7) holds for isothermal wave propagation. Whether isentropic or isothermal wave propagation occurs depends on the thermodynamic properties of the material (Pierce, 1989). Although isentropic (strictly adiabatic) conditions prevail in most cases, we shall drop the isentropic and isothermal designations except where it is appropriate to distinguish a specific condition.

We assume the propagation of plane harmonic waves of the form $u_i = w_i \exp[i(k_j a_j - \omega t)]$, where w_i are the Cartesian components of the normalized wave amplitude $\boldsymbol{w}(|\boldsymbol{w}| = 1)$, k_j are the components of the wave vector $\boldsymbol{k}$, and ω is the angular frequency of the wave. Substituting this form into Equation (7.7), defining $\boldsymbol{n} = \boldsymbol{k}/|\boldsymbol{k}|$ to be the unit vector along the wave propagation direction, and defining the wave phase velocity $c = \omega/|\boldsymbol{k}|$, we obtain the Christoffel equation

$$(\rho_0 c^2 \delta_{ik} - C_{ijkl} n_j n_l) w_k = 0 \qquad (7.8)$$

For a given direction of propagation $\boldsymbol{n}$, Equation (7.8) is a set of three homogeneous equations in three unknowns w_k, where $k = 1, 2, 3$. Nonzero solutions to Equation (7.8) exist if the determinant of the set of terms in parentheses (the secular determinant) is zero. Evaluation of the secular determinant leads to a cubic equation in c^2. The three roots of this equation correspond in general to three different propagation velocities. Substitution of each of these velocities back into Equation (7.8) yields a solution set of w_k, called the *wave polarization vector*, corresponding to that velocity. The polarization vectors are mutually orthogonal. If the polarization vector is parallel or nearly parallel to the propagation vector, the wave is called a *longitudinal or quasi-longitudinal wave.* If the polarization vector is perpendicular or nearly perpendicular to the propagation vector, the wave is called a *transverse (shear) or quasi-transverse (quasi-shear) wave.* In general, the one longitudinal wave and the two shear waves have different velocities.

From Equation (7.8), we find that the velocity of propagation of a wave of polarization $\boldsymbol{w}$ along the direction $\boldsymbol{n}$ is determined as $c = (C_{ijkl} n_j n_l w_i w_k / \rho_0)^{1/2}$. It is apparent from this expression that the crystalline symmetry influences the sound velocity through the elastic constants. For example, for plane wave propagation along the [100] crystalline axis of cubic solids, the longitudinal velocity $c_1 = (C_{11}/\rho_0)^{1/2}$ and the shear wave velocities $c_2 = c_3 = (C_{44}/\rho_0)^{1/2}$; along the [110] direction $c_1 = [(C_{11} + C_{12} + 2C_{44})/2\rho_0]^{1/2}$, $c_2 = (C_{44}/\rho_0)^{1/2}$, and $c_3 = [(C_{11} - C_{12})/2\rho_0]^{1/2}$; and along [111] $c_1 = [(C_{11} + 2C_{12} + 4C_{44})/3\rho_0]^{1/2}$ and $c_2 = c_3 = [(C_{11} - C_{12} + C_{44})/3\rho_0]^{1/2}$.

Along any wave propagation direction in materials having isotropic symmetry, the longitudinal velocity $c_1 = (C_{11}/\rho_0)^{1/2} = [(\lambda + 2G)/\rho_0]^{1/2} = [(B + (4/3)G)/\rho_0]^{1/2}$ and the shear wave velocities $c_2 = c_3 = [(C_{11} - C_{12})/2\rho_0]^{1/2} = (G/\rho_0)^{1/2}$. This result can also be obtained by substituting the isotropic Hooke's law, $\sigma_{ij} = 2G\varepsilon_{ij} + \lambda\varphi\delta_{ij}$, into Newton's law to obtain (after some manipulation) (Pollard, 1977)

$$\rho_0 \frac{\partial^2 \boldsymbol{u}}{\partial t^2} = (\lambda + 2G)\nabla(\nabla \cdot \boldsymbol{u}) - G\nabla \times (\nabla \times \boldsymbol{u}) \qquad (7.9)$$

where ∇ is the vector operator $\nabla = (\partial/\partial a_1)\boldsymbol{e}_1 + (\partial/\partial a_2)\boldsymbol{e}_2 + (\partial/\partial a_3)\boldsymbol{e}_3$ and $\boldsymbol{e}_i$ are the unit vectors along the Cartesian axes. For the case of pure dilatational (compressional or longitudinal) waves, $\nabla \bullet \boldsymbol{u} \neq 0$, $\nabla \times \boldsymbol{u} = 0$ and Equation (7.9) reduces to $(\partial^2 \boldsymbol{u}/\partial t^2) = c_l \nabla^2 \boldsymbol{u}$, where $c_1 = [(\lambda + 2G)/\rho_0]^{1/2}$ is the longitudinal wave velocity. For the case of pure rotational (shear or transverse) waves, $\nabla \bullet \boldsymbol{u} = 0$, $\nabla \times \boldsymbol{u} \neq 0$ and Equation (7.9) reduces to $(\partial^2 \boldsymbol{u}/\partial t^2) = c_t \nabla^2 \boldsymbol{u}$, where $c_2 = c_3 = (G/\rho_0)^{1/2}$ is the shear wave velocity.

7.2.2 Fluids

We define a fluid as a material that will yield under the action of a shear stress, no matter how small it is. A fluid is ordinarily understood to be isotropic. We define a perfect fluid to be an idealized material in which there never exists a shear stress. Hence, the shear modulus G must be zero. The wave equation, Equation (7.7) or Equation (7.9), also holds for fluids and the results obtained for isotropic solids can be directly used for fluids with appropriate consideration of the isotropic symmetry and the vanishing of the shear modulus. Thus, the transverse wave velocity $c_2 = c_3 = 0$ and the longitudinal (also called *compressional*) wave velocity $c_1 = [B/\rho_0]^{1/2}$, where B is the bulk modulus.

It is instructive to consider an alternative form of the wave equation often used for fluid media. For a frictionless or nonviscous perfect fluid, the stress σ_{ij} is the thermodynamic pressure p obtained as $\sigma_{ij} = -p\ \delta_{ij}$. For small variations in the mass density ρ about the initial mass density ρ_0, the equation of state, Equation (7.6), may be approximated using the isotropic symmetry properties of the elastic constants and Equation (7.1) as $\rho = \rho_0 + (\rho_0/B_S)(p - p_0) + \ldots$. Substituting this equation into the equation of continuity $(\partial\rho/\partial t) = \partial(\rho v_i)/\partial a_i \approx \rho_0(\partial v_i/\partial a_i)$, where $v_i = \partial u_i/\partial t$ is the particle velocity, we obtain $(\partial p/\partial t) = -B_S(\partial v_i/\partial a_i)$. Taking the derivative of this expression with respect to time and using Newton's Law $\rho_0(\partial^2 u_i/\partial t^2) = \rho_0(\partial v_i/\partial t) = (\partial\sigma_{ij}/\partial a_j) = -(\partial p/\partial a_i)$, we obtain the pressure wave equation

$$\frac{\partial^2 p}{\partial t^2} = \frac{B_S}{\rho_0}\frac{\partial^2 p}{\partial a_i \partial a_i} = \frac{B_S}{\rho_0}\nabla^2 p \qquad (7.10)$$

where ∇^2 is the Laplacian operator. It is apparent from Equation (7.10) that the compressional wave velocity $c_1 = (B_S/\rho_0)^{1/2}$.

The isentropic equation of state used in obtaining Equation (7.10) may be recast in the more traditional Hooke's Law form $p = p(\rho, S) = p_0 + (\partial p/\partial\rho)_S(\rho - \rho_o) + \ldots = p_0 + (B_S/\rho_0)(\rho - \rho_o) + \ldots$. This form reveals that the isentropic wave velocity for fluids, $c = (B_S/\rho_0)^{1/2}$, can be generally obtained from the equation of state as $c = [(\partial p/\partial\rho)\rho_0]^{1/2}$. For the special case of an ideal gas, the adiabatic condition is expressed as $p = p_0(\rho/\rho_0)^\gamma$, where γ is the ratio of isobaric to isochoric specific heats. Hence, the adiabatic sound velocity is obtained as $c = [\gamma(p_0/\rho_0)]^{1/2}$. The ideal gas equation of state involving temperature T may be written as $p = p(\rho, T) = \rho RT$, where $R = R_0/M$, R_0 is the universal gas constant, and M is the average molecular weight of the molecules comprising the gas. Thus, we find the temperature dependence of the adiabatic sound velocity to be $c = (\gamma RT)^{1/2}$.

7.3 Ultrasonic Waves in Bounded Media

The presence of surfaces and interfaces bounding two media having different acoustic properties gives rise to a variety of acoustic phenomena having important applications. We consider here conditions leading to wave reflection, refraction, acoustic mode conversion, inhomogeneous and evanescent waves, and guided waves.

7.3.1 Reflection and Refraction

We consider a plane harmonic pressure wave of unit amplitude of the form $p_{inc} = \exp[i(\boldsymbol{k}\bullet\boldsymbol{x} - \omega t)]$ incident on a plane boundary between two liquid or gas media as illustrated in Figure 7.1. We orient and center the coordinate axes such that the planar interface occurs at $z = 0$. We assume that the plane of the incident wave is coincident with the xz-plane such that $k_x = k\sin\theta$, $k_y = 0$, $k_z = k\cos\theta$, and $k = \omega/c$, where c is the sound velocity in the incident wave propagation medium (upper medium in figure) and θ is the angle of incidence formed by the normal to the wave front and the z axis. The density of

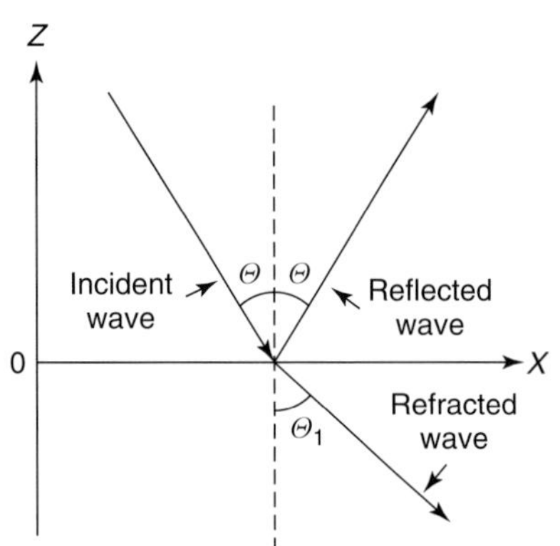

Fig. 7.1 Reflection and refraction of an incident sound wave.

the upper medium is ρ and the density and the sound velocity of the lower medium are ρ_1 and c_1, respectively.

In general, we expect that the incident wave of unit amplitude will be partially reflected from the interface with some amplitude R called the *stress (pressure) reflection coefficient*. The expressions for the incident and reflected waves are written respectively as $p_{inc} = \exp[ik(x \sin\theta - z\cos\theta)]\exp(-i\omega t)$ and $p_{refl} = R\exp[ik(x \sin\theta + z\cos\theta)]\exp(-i\omega t)$. Note that we assume specular (geometrical) reflection of the reflected wave such that the angle of incidence is equal to the angle of reflection. The total pressure field in the upper medium is then

$$p = p_{\text{inc}} + p_{\text{refl}} = (e^{-ikz\cos\theta} + Re^{ikz\cos\theta}) \times e^{i(kx\sin\theta - \omega t)} \tag{7.11}$$

The remaining energy of the incident wave will be transmitted into the lower medium with an amplitude W called the *stress (pressure) amplitude transmission coefficient*. The pressure field in the lower medium consists only of the transmitted wave, which we write as

$$p_1 = We^{ik_1(x\sin\theta_1 - z\cos\theta_1) - i\omega t} \tag{7.12}$$

where $k_1 = \omega/c_1$ is the wavenumber in the lower medium. The angle of refraction θ_1 does not generally equal the incident angle θ, indicating that the wave in the lower medium (the refracted wave) deviates from the incident wave direction.

The quantities R, W, and θ_1 are determined from the conditions that the pressure and the normal component of the particle velocity $v_z = \partial u_z/\partial t$ must be continuous across the boundary. In acoustics, often analogies between acoustic and electromagnetic quantities are used. The acoustic pressure can be identified with the electromotive force or voltage and the particle velocity with the electrical current (Pierce, 1989). Thus, in analogy with the concept of electrical impedance (ratio of voltage to current), we define the acoustic impedance Z normal to the surface as

$$Z = -\frac{p}{v_z} \tag{7.13}$$

The boundary conditions can now be expressed equivalently as the continuity of pressure and acoustic impedance at $z = 0$.

The continuity of pressure requires that $p = p_1$ at $z = 0$. From Equations (7.11) and (7.12), we find that $1 + R = W\exp[i(k_1\sin\theta_1 - k\sin\theta)x]$. Because the left side of this equation is independent of x, so must be the right side. We thus obtain

Snell's law

$$\frac{\sin\theta}{\sin\theta_1} = \frac{k_1}{k} = \frac{c}{c_1} = n \quad (7.14)$$

where n is the index of refraction. Thus, we have $1 + R = W$.

The continuity of particle velocity may be replaced by the continuity of acoustic impedance. In the lower medium, the impedance $Z_1 = -i\omega\rho_1 p_1(\partial p_1/\partial z)^{-1} = \rho_1 c_1/\cos\theta_1$; in the upper medium $Z = (\rho c/\cos\theta)\,[\exp(-ikz\cos\theta) + R\exp(ikz\cos\theta)]/[\exp(-ikz\cos\theta) - R\exp(ikz\cos\theta)]$. Because $Z = Z_1$ at $z = 0$, we obtain from the above equations that

$$R = \frac{Z_1 - Z_{inc}}{Z_1 + Z_{inc}},\ Z_{inc} = \frac{\rho c}{\cos\theta} \quad (7.15)$$

For fluid–solid and solid–solid interfaces, the number of reflected and refracted waves is dramatically increased, because both longitudinal and shear wave motion must be considered. Correspondingly, the reflection and transmission coefficients must include shear wave impedance terms as well as longitudinal impedance terms. For further information, see Brekhovskikh (1980).

7.3.2 Resonance Phenomena

Consider the one-dimensional case of a sample of length $L/2$ having flat and parallel opposing faces. We assume complete reflection of the sound waves normally incident on the surfaces. We assume that the face $x = 0$ is caused to vibrate with a time harmonic disturbance having particle velocity $v = \partial u/\partial t = \cos(\omega t)$. The disturbance propagates from the surface as a plane wave along the direction x described by the complex expression $v' = \exp[i(\omega t - kx) - \alpha x]$. The factor $\exp(-\alpha x)$ represents the decay of the waveform as the result of attenuation in the medium (see below); the parameter α is the attenuation coefficient. We assume that the wave is totally reflected at the opposing surface $x = L/2$ and the reflected wave after propagating back to the surface $x = 0$ is superimposed on the original driving vibration. The wave continues to reflect between the opposing surfaces giving rise to a superposition of waves of the form (Bolef and Miller, 1971)

$$\begin{aligned} v' &= \exp(i\omega t)(1 + \exp[-(ikL + \alpha L)] \\ &\quad + \exp[-2(ikL + \alpha L)] + \ldots) \\ &= \exp(i\omega t)/(1 + \exp[-(ikL + \alpha L)] \end{aligned} \quad (7.16)$$

Taking the real part of the above expression, we obtain

$$v = \operatorname{Re} v' = v_1\cos\omega t + v_2\sin\omega t \quad (7.17)$$

where

$$v_1 = \frac{e^{\alpha L} - \cos kL}{2(\cosh\alpha L - \cos kL)},\quad v_2 = \frac{\sin kL}{2(\cosh\alpha L - \cos kL)} \quad (7.18)$$

The term v_1 is in phase with the driving vibration and the term v_2 is 90° out of phase (in quadrature). The amplitude of v is

$$|v| = (v_1^2 + v_2^2)^{1/2} = \frac{e^{\alpha L/2}}{[2(\cosh\alpha L - \cos kL)]^{1/2}} \quad (7.19)$$

Plots of these terms as a function of $kL(= \omega L/c)$ are given in Figure 7.2. The curves show a resonance behavior at a value of kL corresponding to the condition that the sample length $L/2$ is equal to an integral number of half wavelengths (wavelength $\lambda = 2\pi c/\omega$). Such standing wave resonances occur at all frequencies

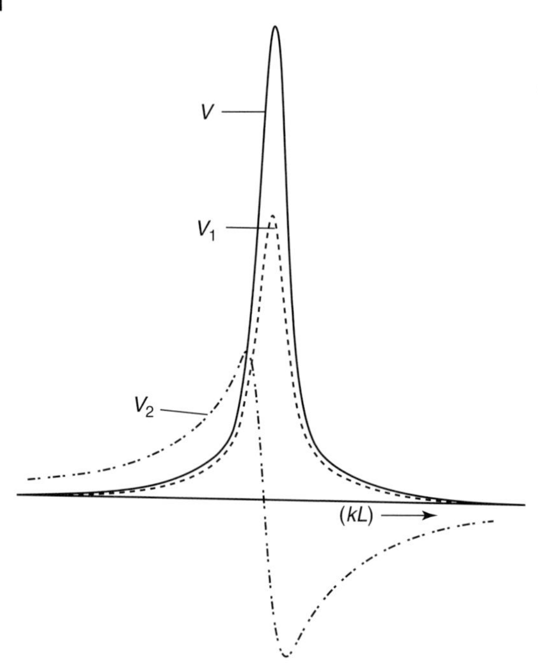

Fig. 7.2 Plots of ν (solid curve), ν_1 (dashed curve), and ν_2 (dashed-dotted curve) as functions of kL about a resonance value.

ω_m such that the resonance condition $L/2 = m(\lambda_{m/2})$ is satisfied, where m is an integer (harmonic number) and $\lambda_m = 2\pi c/\omega_m$. Using $k = \omega/c$, we obtain from the resonance condition that the mth resonance frequency is $\omega_m = 2\pi mc/L$ and that $\omega_m - \omega_{m+1} = 2\pi c/L$. The latter expression can be used to determine the ultrasonic velocity from the measurement of two successive resonance frequencies. The attenuation can be determined from the frequency width $\Delta\omega$ of the resonance at half power as $\alpha = \Delta\omega/(2c)$ (Bolef and Miller, 1971).

7.3.3
Surface and Interface Waves

We consider the situation at a fluid–fluid interface where the index of refraction $n < 1(c_1 > c)$ and the angle of incidence of a plane wave exceed the critical angle $\theta_0 = \arcsin n$. For $\theta > \theta_0$, we rewrite Equation (7.15) in the complex form $R = [(\rho_1/\rho) \cos\theta - i(\sin^2\theta - n^2)^{1/2}]/[(\rho_1/\rho)\cos\theta + i(\sin^2\theta - n^2)^{1/2}]$. Because $|R| = 1$, there is total reflection of the wave. Writing $R = \exp(i\phi)$, we find that the phase ϕ of the reflected wave varies, $\phi = -2\arctan[(\rho(\sin^2\theta - n^2)^{1/2}/\rho_1\cos\theta)]$. For the case of complete internal reflection, the field in the lower medium can be written as (Brekhovskikh, 1980)

$$z < 0,\ p_1 = (1 + R)e^{\delta z}e^{ikx\sin\theta - i\omega t},$$
$$\delta = k(\sin^2\theta - n^2)^{1/2} \tag{7.20}$$

The wave represented by Equation (7.20) is an example of an inhomogeneous wave. Such waves propagate along a given direction with an amplitude that decays exponentially in a direction perpendicular to the direction of propagation. The wave of Equation (7.20) propagates along the boundary (a surface wave) with exponential decay perpendicular to the boundary.

Inhomogeneous waves can propagate on a boundary in the absence of an incident wave. Consider the case of an isotropic elastic halfspace that forms a planar interface with a vacuum on the $z = 0$ plane (we assume same axes orientation and centering as in Figure 7.1). A monochromatic wave propagating along the solid–vacuum surface (x direction) must satisfy the wave equation in Equation (7.7). In analogy with Equation (7.20), we assume that the form of the surface wave is given by $u_i = \exp(\delta z)\exp[i(kx - \omega t)]$. From Equation (7.14) (Snell's law) and the relations $k_1 = \omega/c_1$, $k_{1x} = k_1 \sin\theta_1$, we may recast the δ given in Equation (7.20) in the form $\delta = [k_{1x}^2 - (\omega/c_1)^2]^{1/2}$. We assume for the present case that δ can be written in the analogous form $\delta = [k^2 - (\omega/c)^2]^{1/2}$ where the subscripts have been dropped.

Let us resolve the displacement $\boldsymbol{u}$ (or u_i) into a longitudinal displacement $\boldsymbol{u}_l$ (or u_{li}) and a transverse displacement $\boldsymbol{u}_t$ (or u_{ti}) as $\boldsymbol{u} = \boldsymbol{u}_l + \boldsymbol{u}_t$(or $u_i = u_{l,I} + u_{t,i}$). The transverse component must satisfy the condition $\nabla \bullet \boldsymbol{u}_t = (\partial u_{t,1}/\partial x) + (\partial u_{t,3}/\partial z) = 0$. We assume $u_{t,I} = \exp(\delta_t z)\exp[i(kx - \omega t)]$, where $\delta_t = [k^2 - (\omega/c_t)^2]^{1/2}$ and c_t is the shear wave velocity. We find from these relations that $u_{t,1} = \delta_t A_1 \exp(\delta_t z)\exp[i(kx - \omega t)]$ and $u_{t,3} = -ikA_t\exp(\delta_t z)\exp[i(kx - \omega t)]$, where A_t is a constant (Pollard, 1977).

The longitudinal part of the displacement satisfies the condition $\nabla \times \boldsymbol{u}_l = (\partial u_{l1}/\partial z) - (\partial u_{l,3}/\partial x) = 0$. Performing an analysis similar to that for the transverse components above, we find that $u_{l,1} = kA_l\exp(\delta_l z)\exp[i(kx - \omega t)]$ and $u_{l,3} = -i\delta_l A_l \exp(\delta_l z)\exp[i(kx - \omega t)]$, where A_l is a constant. The boundary conditions to be satisfied for a free surface are $\sigma_{ij}n_j = 0$, where n_j are the components of the unit vector $\boldsymbol{n}$ normal to the surface (parallel to z axis). Application of these boundary conditions provides the relationship $2ik\delta_l A_l + (k^2 + \delta_t^2)A_t = 0$. We obtain from these expressions that the component of the displacement parallel to the surface $u_1 = u_{l,1} + u_{t,1}$ and perpendicular to the surface $u_3 = u_{l,3} + u_{t,3}$ are (in real form) (Briggs, 1992)

$$u_1 = A_2 k \left(e^{-\delta_1 z} - \frac{2\delta_1 \delta_t}{\delta_t^2 + k^2} e^{-\delta_t z} \right) \sin(kx - \omega t)$$

$$u_3 = A_2 \delta_1 \left(e^{-\delta_1 z} - \frac{2k^2}{\delta_t^2 + k^2} e^{-\delta_t z} \right) \cos(kx - \omega t) \tag{7.21}$$

and that the wave velocity $c_R = \omega/k$ is given by

$$16G^2 \left(1 - \frac{c_R^2}{c_l^2}\right)\left(1 - \frac{c_R^2}{c_t^2}\right) = \left[\frac{4G}{3} - \left(\lambda + \frac{4G}{3}\right)\frac{c_R^2}{c_l^2}\right]^2 \left[2 - \frac{c_R^2}{c_l^2}\right]^2 \tag{7.22}$$

According to Equation (7.21), both the longitudinal and transverse components decay exponentially in the direction z perpendicular to the propagation direction x. The wave represented by Equation (7.21) is thus a type of inhomogeneous wave known as the *Rayleigh wave*. An approximate solution to Equation (7.22) yields the expression (Scruby, Jones, and Anatoniazzi, 1987) $c_R \approx c_t(1.14418 - 0.25771\mu_P + 0.12661\mu_P^2)^{-1}$ for the Rayleigh wave velocity, where $\mu_P = [1 - 2(c_t/c_l)^2]/\{2[1 - (c_t/c_l)^2]\}$ is Poisson's ratio for the material (see Section 7.3.4).

Figure 7.3 shows the particle displacement field given by Equation (7.21) for a Rayleigh surface wave. The motion is generally elliptical with the major axis perpendicular to the surface. The motion reverses the sense of rotation below a depth of approximately 0.2λ.

If the solid surface on which the Rayleigh wave propagates is in contact with a fluid medium, the velocity of the Rayleigh wave

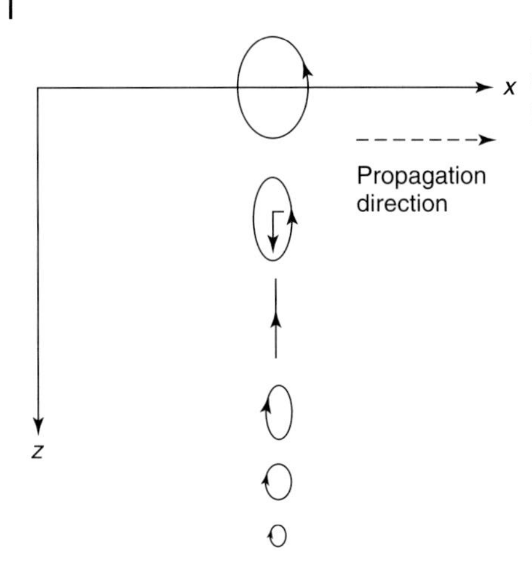

Fig. 7.3 Particle displacement field for a Rayleigh surface wave. The arrows indicate sense of particle motion. From Sabine and Cole, 1971.

will be slightly altered as a result of the reaction of the fluid medium. If the velocity of the Rayleigh wave is greater than the velocity of sound in the fluid medium, then the wave will be partially radiated into the fluid medium. The Rayleigh wave in this case belongs to a class of surface acoustic waves (SAWs) called "*leaky*" waves.

Many other types of surface waves can propagate in materials that depend on the properties of the materials forming the interface. The interface between two elastic media gives rise to an interface wave called the *Stonely wave*. The field distribution in each medium is that of two partial waves with particle displacements parallel and normal to the interface, the motion on either side of the interface being of the Rayleigh type. The Stonely wave velocity lies between that of the Rayleigh waves in each medium and the shear wave in the denser medium.

Other interface waves include the Love wave and the lateral wave. The Love wave involves solid–solid interfaces and is characterized by a transverse component parallel to the surface with particle displacement perpendicular to the direction of wave propagation. The lateral wave may occur at various interface combinations (solid–solid, fluid–solid, fluid–fluid) and is generally associated with the reflection of spherical waves. See Brekhovskikh (1980) for more information.

7.3.4 Guided Waves

Guided waves are characterized by energy flow primarily along the direction of some guiding configuration. Strictly, SAWs are guided waves, but we concern ourselves here with propagation in configurations such as cylinders and plates. These structures are capable of supporting an infinite number of waveguide modes. For simple configurations, the modes may be analyzed as composed of traveling waves along the waveguide axis and standing waves in the transverse direction (the transverse resonance principle). A general expression for the propagation of a sound wave along the direction x in a medium bounded in the (y, z) plane can written as an expansion in terms of the complete set of eigenfunctions $f_n(y, z)$ arising from the solution of the three-dimensional wave equation subject to transverse boundary conditions (i.e., boundary conditioned referred to the y and z dimensions) (Morse and Ingard, 1968). We thus write

$$v(x, y, z, t) = \sum_n b_n f_n(y, z) e^{-\alpha_n x} e^{i(\omega t - k_n x)} \quad (7.23)$$

where the constant $k_n = [(\omega/c)^2 - K_n^2]^{1/2}$ is the propagation constant for the nth mode and K_n is the eigenvalue corresponding to the eigenfunction $f_n(y, z)$, b_n is an expansion coefficient, and a_n is the attenuation coefficient for the nth mode.

If a solution corresponding to the eigenvalue $K_n = 0$ exists, it is called the *fundamental mode* and corresponds to one-dimensional wave propagation along the x direction. Higher order modes corresponding to $K_n^2 > 0$ reflect from the sides of the waveguide in a zigzag fashion but have a net propagation in the x direction. For a given frequency, eigensolutions exist only if $(\omega/c)^2 > K_n^2$. A given mode thus has a frequency cutoff below which the mode does not propagate. Such nonpropagating disturbances that are periodic only in time and decay exponentially in all spatial coordinates are called *evanescent waves.*

For a solid isotropic cylinder, the simplest modal solution may be obtained by considering the solid to be subjected to a stress σ_{11} along the cylindrical axis x and stress free conditions on the sides of the cylinder. The application of Hooke's law for an isotropic solid to this case yields the equations: $\sigma_{11} = 2G\varepsilon_{11} + \lambda\varphi; 0 = 2G\varepsilon_{22} + \lambda\varphi; 0 = 2G\varepsilon_{33} + \lambda\varphi$, where the dilatation $\varphi = \varepsilon_{11} + \varepsilon_{22} + \varepsilon_{33}$. From these equations, we define and obtain Young's modulus as $E = \sigma_{11}/\varepsilon_{11} = G(3\lambda + 2G)/(\lambda + G)$ and Poisson's ratio $\mu_p = -\varepsilon_{22}/\varepsilon_{11} = \lambda/[2(\lambda + G)]$. Substituting the defining relation for Young's modulus into Newton's law gives the wave equation $\rho_0(\partial^2 u_1/\partial t^2) = E(\partial^2 u_1/\partial a_1^2)$. We identify from this wave equation that $c_L = (E/\rho_0)^{1/2}$ is the velocity of propagation of bounded longitudinal waves in an isolated solid cylinder. This velocity is in general less than that of a plane longitudinal wave propagating in an unbounded solid. Rayleigh obtained a correction to c_L by accounting for "lateral inertia" and obtained the longitudinal phase velocity $c'_L = c_L[1 - \mu_p^2\pi^2(r_0/\lambda)^2]$, where r_0 is the radius of the cylinder. More general solutions require more powerful approaches such as that leading to Equation (7.23) or methods employing Green's functions. For more information, see Auld (1990).

For a free isotropic plate, horizontally polarized shear (SH) wave modes (characterized by having a polarization direction parallel to the surface) and Lamb wave modes (characterized by a coupling of vertically polarized shear waves and longitudinal waves) may propagate. In the limit $kL \to \infty$, where L is the thickness of the plate, the fundamental symmetrical and antisymmetrical Lamb wave modes degenerate into Rayleigh waves bound to the surfaces. An isotropic plate on an isotropic elastic substrate gives rise to generalized Lamb waves which yield tightly bound Stonely waves at the plate-substrate interface in the limit $kL \to \infty$. For further information on guided waves, see Brekhovskikh (1980) and Auld (1990).

7.4 Ultrasonic Attenuation

Attenuation refers to the total loss of amplitude (or energy) of a sound wave along its propagation path resulting from all mechanisms responsible for such losses, including but not limited to absorption, scattering, and diffraction. Absorption refers to the loss of amplitude (or energy) of an acoustic wave that results in an increase in temperature (however slight) in the propagation medium and is often associated with relaxation phenomena. Scattering results in a redirection of wave energy from the original propagation direction. Diffraction refers to scattering that occurs from the boundaries or edges of objects. Generally, attenuation produces an exponential decrease in the wave amplitude. For plane wave propagation along the x direction, we may write

$$u(x,t) = u_0 e^{-\alpha x} e^{i(kx-\omega t)} \tag{7.24}$$

where α is the attenuation coefficient. It is generally assumed that the contributions from all sources to the attenuation is additive such that $\alpha = \alpha_{\text{abs}} + \alpha_{\text{sc}} + \alpha_{\text{diff}} \ldots$ where the subscripts refer to the absorption, scattering, and diffraction contributions, respectively.

7.4.1 Scattering

Generally, scattering refers to the redirection of wave energy from the original propagation direction or equivalently to the deviation of a ray or pencil of ultrasound from a straight path. Scattering may be classified into geometric contributions and diffraction-based contributions (Überall, 1997). Geometric contributions include specular reflections, transmission, refraction, and internal reflections of ultrasonic waves. Diffraction-based contributions refer to phenomena whereby an inhomogeneity in the path of a sound wave produces secondary waves spreading out from the inhomogeneity in various directions. Generally, an inhomogeneity refers to any object or region in the wave propagation path that exhibits a difference in density or elastic properties from that of the incident wave propagation medium. The object could be, for example, a defect in a crystal, a region of atmospheric turbulence, a fish in the ocean, or a tumor in the brain. Geometric scattering was considered in Section 7.3.1. We consider here general diffraction-based scattering. Diffraction is treated in Section 7.4.2.

If the region of inhomogeneity (having characteristic dimensions of order d) is small compared to the wavelength λ of the incident wave (i.e., $kd \ll 1$), the scattering is called *Rayleigh scattering*. The general features of Rayleigh scattering may be obtained from the consideration of constant frequency plane waves propagating along the direction $\boldsymbol{e}_k = \boldsymbol{k}/|\boldsymbol{k}|$ ($\boldsymbol{k}$ is the wave number vector) that impinge on a rigid immovable object centered at the origin of the coordinate system. The total acoustic pressure is $p = B\ \exp(i\boldsymbol{k}\bullet\boldsymbol{x}) + p_{\text{sc}}(\boldsymbol{x})$, where B is the amplitude of the incident wave and $p_{\text{sc}}(\boldsymbol{x})$ is the complex amplitude of the scattered wave. A rigid scattering surface imposes the requirement that $\boldsymbol{\nabla} p\bullet\boldsymbol{n} = 0$ at the surface S of the scatterer, where $\boldsymbol{n}$ is the unit normal pointing outward from the surface (Pierce, 1989). Substituting the total acoustic pressure equation into the boundary condition gives $\boldsymbol{\nabla} p_{\text{sc}}\bullet\boldsymbol{n} = -iB[\exp(i\boldsymbol{k}\bullet\boldsymbol{x})](\boldsymbol{k}\bullet\boldsymbol{n})$. Thus, the scattered wave field may be viewed as that of a radiating body of the same size and shape as that of the scatterer with an amplitude that is the negative of that associated with the incident wave field.

For a small wavenumber, we may expand the exponential in the above equation in a power series to first order in k and integrate the results over the surface of the scatterer. The procedure yields the results (Pierce, 1989) $p_{\text{sc}} = p_{\text{sc}}$, mono $+ p_{\text{sc}}$, dipole. The term $p_{\text{sc,mono}} = -(k^2 BV/4\pi)[\exp(ikr)/r]$, where r is the radial coordinate and V is the scatterer volume of order d^3, is due to monopole radiation. The term $p_{sc,dipole} = (k^2 B/4\pi)\{e_r \bullet M \bullet e_k[(1+(i/kr)]\}[\exp(ikr)/r]$, is due to dipole radiation. The factor M is a tensor, which scales as d^3, that is dependent on the scatterer size and shape. The sum of the monopole and the dipole terms produces a far-field scattered wave amplitude that scales as $(k^2 d^3/r)$. Thus, both the monopole and the dipole radiations are dependent on the square of the wavenumber (or frequency). Because the intensity of the scattered sound wave is proportional to the square of the pressure, the intensity of the scattered wave field is then dependent on the fourth power of the sound frequency. This strong dependence on the frequency in Rayleigh scattering has an analog in optics and leads to an explanation of why the sky is blue: higher frequency blue light scatters more than lower frequency red light.

In the limit $kd \gg 1$ (high frequencies), geometric scattering dominates and the scattering is more appropriately calculated in terms of impedance ratios described in Section 7.3. The transition region between the low- and the high-frequency limits is not easily assessed in terms of simple expressions. For wavelengths of the same magnitude as the dimensions of the scatterer, the scattered waves form complex patterns that must be calculated numerically. Figure 7.4 shows the angular distribution of scattered energy for waves incident from the left onto a sphere of radius d for $kd \ll 1$ (Rayleigh scattering), $kd = 6$, and $kd \to \infty$.

Relative motion between the scattering object (viewed as a source of sound) and the sound receiver causes a shift in the frequency of sound received relative to that of the source. This change in frequency is called the *Doppler shift* and can be understood by reference to Figure 7.5. We consider a sound source fixed in reference frame 2 moving with a velocity V_{21} relative to a sound receiver fixed in reference frame 1. A point vector x_1 referred to frame 1 is related to a point vector x_2 referred to frame 2 by the Galilean transformation $x_1 = x_2 + (t - t_0)V_{21}$, where at $t = t_0, x_1 = x_2$. We describe the phases of the same sound disturbance in each of the two reference frames by $\phi_1(x_1, t)$ and $\phi_2(x_2, t)$. The wave vector k in either frame is defined by $k = -\nabla\phi$ and the angular frequency $\omega = \partial\phi/\partial t$. For the same sound disturbance, we must have $\phi_1(x_1, t) = \phi_1(x_2 + (t - t_0)V_{21}, t) = \phi_2(x_2, t)$. Differentiation of this expression with respect to the spatial and time coordinates, respectively, yields the relations $k_1 = k_2$ and $\omega_2 = \omega_1 - k_1 \bullet V_{21}$.

We first consider the case where reference frame 1 (with fixed receiver) is stationary relative to the sound propagation medium and frame 2 (with fixed source) is moving through the propagation medium. Assuming that the source generates sound at frequency ω_0, we may write $\omega_2 = \omega_0$. Because the receiver is stationary in the propagation medium, we may also write that $k_1 = (\omega/c)e_R$, where c is the sound velocity and e_R is the unit vector determined at the instant of sound emission along the direction from the sound source to the receiver. We thus find $\omega_2 = \omega_0 = \omega_1 - k_1 \bullet V_{21} = \omega - (\omega/c)e_R \bullet V_{21}$. Solving this equation for ω, we obtain $\omega = \omega_0(1 - c^{-1}e_R \bullet V_{21})^{-1}$.

We now consider the case where reference frame 1 (with fixed receiver) is moving in the propagation medium and frame 2 (fixed sound source) is stationary in the propagation medium. Then we may write $\omega_2 = \omega_0$ and $k_1 = k_2 = (\omega_0/c)e_R$. Thus $\omega_2 = \omega_0 = \omega - (\omega_0/c)e_R \bullet V_{21}$. Solving for ω, we obtain $\omega = \omega_0(1 + c^{-1}e_R \bullet V_{21})$. In each case, we find that the frequency

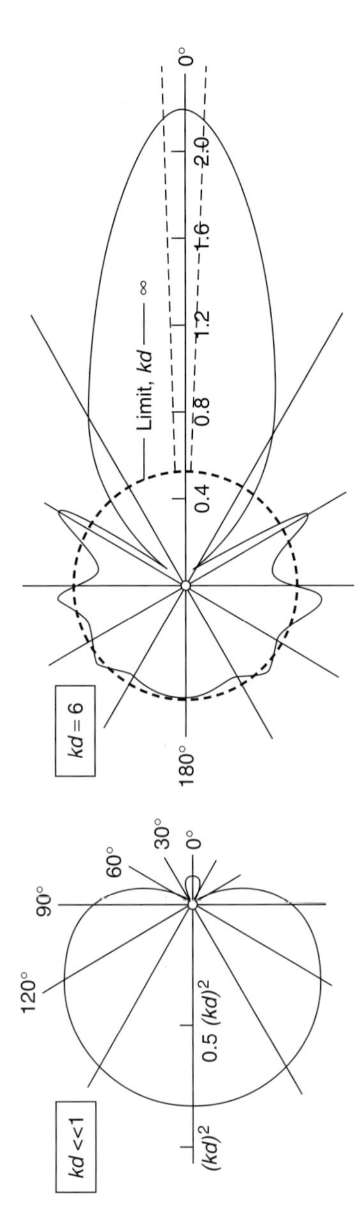

Fig. 7.4 Angular distribution of scattered wave energy for waves incident from the left onto a sphere of radius d for $kd \ll 1$, $kd = 6$, and $kd \to \infty$ (dotted curves). From Pierce, 1989.

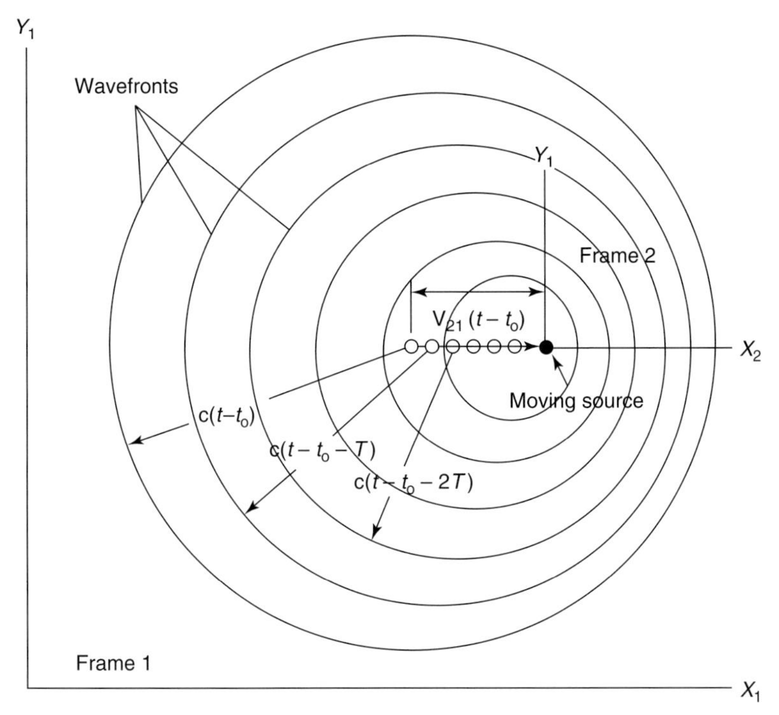

Fig. 7.5 Schematic of Doppler shifted crests of waveforms emitted with period T by source fixed in reference frame 2 moving at velocity V_{21} with respect to the propagation medium fixed in reference frame 1. Adapted from Pierce, 1989.

increases if the source and receiver are moving toward each other and decreases if the source and receiver are separating. The magnitude of frequency shift is determined by the component of the relative velocity along $\boldsymbol{e}_R$.

In many applications, a transmitting transducer fixed in a propagation medium (reference frame 1) is used to transmit sound to a scatterer fixed in a second propagation medium (reference frame 2) moving relative to reference frame 1 with velocity $\boldsymbol{V}_{21}$. A receiving transducer also fixed in frame 1 is then used to receive the scattered sound waves. The boundary between propagation media 1 and 2 is assumed to occupy well-defined instantaneous positions with respect to frame 1. With respect to the scatterer fixed in frame 2, the sound waves incident and scattered at the scatterer have the same frequency ω_2. However, with respect to frame 1, the transmitter frequency is $\omega_{\text{tr}} = \omega_2 + \boldsymbol{V}_{21} \bullet \boldsymbol{k}_{1\text{tr}}$ and the receiver frequency is $\omega_{\text{rec}} = \omega_2 + \boldsymbol{V}_{21} \bullet \boldsymbol{k}_{1\text{rec}}$, where $\boldsymbol{k}_{1tr}$ and $\boldsymbol{k}_{1\text{rec}}$ are the transmitted and received wave vectors referred to frame 1. Eliminating ω_2 from these two equations yields the frequency shift between the transmitted and received signals as $(\omega_{\text{rec}} - \omega_{\text{tr}}) = \boldsymbol{V}_{21}) \bullet \boldsymbol{k}_{1rec} - \boldsymbol{k}_{1\text{tr}})$. Typical applications of Doppler frequency shift measurements include the assessment of flow velocities and fluid volume flowing per unit time in blood vessels ducts and tubes, the remote sensing of tropospheric wind velocities, and the measurement of subsurface ocean current velocities.

7.4.2 Diffraction

Generally, diffraction refers to scattering that occurs at the "edges" of objects. It also includes the influence of boundaries on radiation sources such as piston transducers. Mathematically, diffraction phenomena can be described in terms of monopole and dipole radiation patterns as is done with diffraction-based scattering phenomena generally (see Section 7.4.1). One of the most important applications of diffraction theory in ultrasonics is to account for beam spreading from circular transducers, the most common shape for ultrasonic sources.

Because the transducer is of finite size, the acoustic wave generated in the sample spreads out into a diffraction field that leads to an attenuation of the wave along the original propagation direction and to errors in velocity and source amplitude measurements if not corrected. The effects of diffraction are usually treated for the case of circular, axially concentric transmitting and receiving transducers of the same radius r. The transmitting transducer is treated as a finite piston source in a semi-infinite medium. The acoustic field is found at each point in the propagation medium and an integration is performed over the area in the field presented by the receiving transducer. In one treatment (Benson and Kiyohara, 1974), the effects of diffraction can be represented for waves of wavelength λ propagating along the z direction in an isotropic medium by the expression $u(z,t) = 2(I_x^2 + I_y^2)^{1/2}\cos[kc_pt - kz - \phi]$, where $\phi = \tan(I_y/I_x)$ and the quantities Ix and Iy are integral functions of the dimensionless parameter $S = z\lambda/r^2$.

These equations show that the diffraction involves an intensity variation in the diffraction pattern and a distortion of the otherwise planar wave front. Phase variations in the surface plane of the receiving transducer give rise to interference effects and a loss in signal (attenuation) when integrated over a phase-sensitive transducer surface. The factor ϕ in the above equation produces a variation in the phase of the propagating wave that results in an error in the velocity measurements, if not properly corrected.

A useful aspect of diffraction occurs for the focusing of sound fields from curved transducers. The field of a focused sound beam has diffraction characteristics similar to that of a plane piston transducer. Along the axis of the curved transducer, the intensity of the sound field goes through a series of maxima and minima in accordance with the interference patterns produced by diffraction. The maximum intensity occurs at an interference maximum corresponding to the focal zone of the transducer.

7.4.3 Relaxation Processes

The propagation of sound waves based on Hooke's Law is such that the stress (or pressure) at any point in the body is a linear function only of the instantaneous strain (or mass density) at that point. Upon removal of the stress, the strain fully and instantly recovers. The reversibility of the elastic deformation means that no dissipation or loss of energy in the wave occurs and that the wave propagates indefinitely. However, wave propagation in real materials always exhibits an amplitude loss resulting from the conversion of acoustic energy into other forms. Such conversions may involve, for example, thermal motion, the vibration or rotation of atoms in a molecule, structural rearrangements (including chemical reactions and electrolytic processes), or the motion of lattice constituents such as dislocations, point defects, and so on. The energy dissipation destroys the isentropic character of the sound propagation (although it may still be adiabatic) and requires the formalism of irreversible thermodynamics for a proper description.

7.4.3.1 **Relaxation in Solids**

We consider a process described by the reaction (Beyer and Letcher, 1969) $n_1 + n_2 + \ldots + n_k \leftrightarrow n_{k+1} + n_{k+2} + \ldots + n_m$, where $n_i(i \leq k)$ is the number of moles of the ith reactant and $n_i(i > k)$ is the number of moles of the ith product involved in the reaction. The reaction process is driven by the acoustic wave and may describe, for example, the excitation of higher vibrational or rotational states of a gas or the jumping of interstitial atoms in a solid lattice to a new equilibrium site. We account for such reactions by modifying the internal and Helmholtz free energies per unit mass to include a dimensionless quantity ξ describing the extent to which the reaction has taken place. This variable, called the *degree of reaction*, ranges from a value of 0 when the state of the system is all reactant to a value of 1 when the state of the system is all product. In analogy to Equation (7.4), we may expand the modified internal energy per unit mass $U(\boldsymbol{a}, \eta_{ij}, S, \xi)$ to first order in all variables as

$$
\begin{aligned}
&U(a, \eta_{ij}, S, \xi) \\
&= U(a, 0, S_0, \xi_0) + \frac{1}{\rho_0} C_{ij}^S \eta_{ij} + \frac{1}{2\rho_0} C_{ijkl}^S \eta_{ij} \eta_{kl} \\
&\quad + T(S - S_0) + A(\xi - \xi_0) + \cdots \qquad (7.25)
\end{aligned}
$$

The constant $A = (\partial U/\partial \xi)_{\eta,S}$ is called the *affinity* for the reaction and is related to n_i as $A = \Sigma(\mu_{CP})_i n_i$, where $(\mu_{CP})_i$ is the chemical potential for the reaction leading to a change in n_i.

It can be established from thermodynamic considerations (Beyer and Letcher, 1969) that (i) if $Ad\xi > 0$, where $d\xi$ is a small variation in ξ, then the process naturally occurs and will take place; (ii) if $Ad\xi = 0$, the process is reversible; and (iii) if $Ad\xi < 0$, the process cannot occur by itself. Thus, the value $A = 0$ corresponds to equilibrium at some value ξ_0 of the degree of reaction. Acoustically, equilibrium corresponds to zero frequency. Displacements from equilibrium result in natural system forces reacting to restore equilibrium such that the displacement rate $(d\xi/dt) = bA$, where b is a constant. A constant ξ occurs when the acoustic frequency is so large that the system cannot follow the amplitude changes in the sound wave.

It is clear that irreversible thermodynamics requires the independent thermodynamic variable ξ in addition to the usual reversible thermodynamic variables. As in reversible thermodynamics, the irreversible thermodynamic "variables" σ_{ij}, η_{ij}, T, S, A, and ξ are not completely independent, but are related, for example, through Equation (7.25). Any of the irreversible thermodynamic "variables" may be expressed in terms of any three other irreversible thermodynamic "variables." Thus, the affinity A, although introduced as a constant in Equation (7.25) (as is the temperature T), may also be regarded as a variable in a different thermodynamic state equation. In particular, the affinity may be regarded as a function of σ_{ij}, S, and ξ, which can be expanded in a triple Taylor series about the equilibrium point $A[(\sigma_0)_{ij}, S_0, \xi_0] = 0$. Performing the expansion to first order, using standard relations among partial derivatives, and recombining terms, we obtain the relation (Bauer, 1965)

$$A(\sigma_{ij}, S, \xi) = \left(\frac{\partial A}{\partial \xi}\right)_{\sigma,S} [\xi - \xi'(\sigma_{ij}, S)] \tag{7.26}$$

$$\xi'(\sigma_{ij}, S) = \xi_0 + \left(\frac{\partial \xi}{\partial \sigma_{ij}}\right)_{A,S} [\sigma_{ij} - (\sigma_0)_{ij}] + \left(\frac{\partial \xi}{\partial S}\right)_{A,\sigma} (S - S_0) \tag{7.27}$$

where the subscripted zero denotes the initial value of the parameter. Substituting Equation (7.26) into the rate equation $(d\xi/dt) = bA$, holding the stress and entropy constant, and integrating, we obtain

$$\xi = \xi'(\sigma_{ij}, S) + const \cdot \exp(-t / \tau_{\sigma S})$$

$$\tau_{\sigma S} = -(1 / b)(\partial \xi / \partial A)_{\sigma S} \tag{7.28}$$

The parameter $\tau_{\sigma S}$ is the relaxation time for processes in which the stress and entropy remain constant. A corresponding relaxation time $\tau_{\eta S}$ may defined for processes in which the strain and entropy remain constant. The two time constants are related as $(\tau_{\sigma S}/\tau_{\eta S}) = (M_\infty/M_0) = (c_\infty/c_0)^2$, where M_∞ is the appropriate modulus (combination of elastic constants) and c_∞ is the sound speed at infinitely large frequency; M_0 is the modulus and c_0 is the sound speed in the limit of zero frequency.

Combining the above triple Taylor series expansion of $A(\sigma_{ij}, S, \xi)$ with a similar expansion of the strain $\eta_{ij}(\sigma_{ij}, S, \xi)$ and performing several transformations of partial derivatives, we obtain the equation of state for a given propagation mode in the form

$$\tau_{\sigma S}\frac{\partial \varepsilon}{\partial t} + \varepsilon = \frac{\tau_{\sigma S}}{M_\infty}\frac{\partial \sigma}{\partial t} + \frac{1}{M_0}(\sigma - \sigma_0) \tag{7.29}$$

where ε and σ denote (quasi)longitudinal or (quasi)shear strains and stresses, respectively, as appropriate. Combining Equation (7.29) with Newton's Law in the form $(\partial^2\sigma/\partial a^2) = \rho_0(\partial/\partial t)(\partial\varepsilon/\partial t) = \rho_0 (\partial^2\varepsilon/\partial t^2)$ and using the relation $(\tau_{\sigma S}/\tau_{\eta S}) = (M_\infty/M_0)$, we obtain the wave equation

$$\frac{\partial^2\psi}{\partial t^2} + \tau_{\sigma S}\frac{\partial^3\psi}{\partial t^3} = \frac{M_0}{\rho_0}\left(\frac{\partial^2\psi}{\partial a^2} + \tau_{\eta S}\frac{\partial^3\psi}{\partial a^2\partial t}\right) \tag{7.30}$$

where ψ denotes either σ or ε.

We assume harmonic waves of the form $\psi = \psi_0\exp[i(k'a - \omega t)$, where $k' = k + i\alpha = (\omega/c) + i\alpha$ is the attenuation coefficient, and $c = (\omega/k)$ is the phase velocity of the sound wave. Substituting this form into Equation (7.30) and defining $\tau = (\tau_{\sigma S}\tau_{\eta S})^{1/2}$, we obtain that the phase velocity is given by

$$c = \frac{\omega}{k} = \left[\frac{M_\infty}{\rho_0}\left(1 - \frac{\Delta M / M_\infty}{1 + \omega^2\tau^2}\right)\right]^{1/2} \tag{7.31}$$

where the relaxation strength ε_r is defined by

$$\varepsilon_r = \frac{\Delta M}{M_\infty} = \frac{M_\infty - M_0}{M_\infty} = \frac{c_\infty^2 - c_0^2}{c_\infty^2} \tag{7.32}$$

and the wave attenuation (absorption) coefficient α is given by

$$\alpha = \frac{\omega^2(\tau_{\eta S} - \tau_{\sigma S})}{2c(1 + \omega^2\tau^2)} \approx \frac{1}{2}\varepsilon_r\frac{\omega^2\tau}{1 + \omega^2\tau^2} \times \left(\frac{\rho_0}{M_\infty}\right)^{1/2} \tag{7.33}$$

Relaxation in solids can result from a number of different mechanisms. The diversity of mechanisms include atomic movement by diffusion, stress-induced ordering whereby point defects producing a localized strain will move from one state of equilibrium to another under the action of a sound wave, thermoelastic loss involving heat flow between hot and cold regions of the sound field, Bordoni peaks in metals involving the "jumping" of dislocation segments between equilibrium positions, and Akhieser loss in which time-varying strains continually change the equilibrium conditions for the thermal phonon distribution in the solid (Berry and Nowick, 1966).

7.4.3.2 **Relaxation in Fluids**

For a fluid medium, we again assume that the stress σ_{ij} is the thermodynamic pressure p obtained as $\sigma_{ij} = -p\delta_{ij}$. Assuming small variations in the mass density ρ about the initial mass density ρ_0 in Equation (7.1) and using the isotropic symmetry properties of the elastic constants, we may recast Equation (7.29) in the form

$$\tau_{\mathrm{pS}}\frac{\partial \rho}{\partial t} + \rho - \rho_0 = \frac{\tau_{\mathrm{pS}}}{c_\infty^2}\frac{\partial p}{\partial t} + \frac{1}{c_\infty^2}(p - p_0) \tag{7.34}$$

where we have used the relation $(t_{\mathrm{pS}}/\tau_{\rho S}) = (M_\infty/M_0) = (B_\infty/B_0) = (c_\infty/c_0)^2$, where B is the bulk modulus. Combining Equation (7.34) with Newton's law and assuming harmonic wave propagation as before, we obtain that the absorption coefficient at unit wavelength $\alpha\lambda$, where λ is the wavelength given as

$$\alpha\lambda = \left(\frac{c^2}{c_0^2}\right)\varepsilon_{\mathrm{r}}\frac{\pi\omega\tau_{\mathrm{pS}}}{1+\omega^2\tau_{\mathrm{pS}}^2} \tag{7.35}$$

where ε_r is the relaxation strength given by Equation (7.32). The phase velocity c is given by

$$\frac{c^2}{c_0^2} = \frac{1+\omega^2\tau_{\mathrm{pS}}^2}{1+\left(c_0^2\big/c_\infty^2\right)\omega^2\tau_{\mathrm{pS}}^2} \tag{7.36}$$

If we define $\tau = (\tau_{pS}\tau_{\rho S})^{1/2}$, then

$$\frac{c^2}{c_0^2} = 1 + \left(\frac{\varepsilon_{\mathrm{r}}}{1-\varepsilon_{\mathrm{r}}}\right)\left(\frac{\omega^2\tau^2}{1+\omega^2\tau^2}\right) \tag{7.37}$$

which can be recast in the form of Equation (7.31) using the definition of ε_r, Equation (7.32).

In the case of gases, the relaxation phenomena are often recast in terms of heat capacities. The ideal gas is assumed to consist of identical molecules all of which are capable of being excited into different energy states each characterized by specific energy differences per mole. The relaxation strength for ideal gases is often written as $\varepsilon_r = (C_p^0/C_p^\infty)[(C_\rho^\infty/C_\rho^0) - (C_p^\infty/C_p)]$, where the $C_p^j(j = 0, \infty)$ are the appropriately indicated heat capacities. An understanding of the physical phenomena involved for a given gas is obtained from an examination of the physical models used for the relaxation processes. It is generally assumed in gases that equilibrium involves collisions among pairs of molecules, so that the reaction rate (reciprocal of relaxation time) is directly proportional to the product of the probability that an excited molecule will lose its energy in a given collision and the average collision frequency. In an ideal gas, the collision frequency is a linearly proportional gas density or pressure, so the relaxation time $\tau \sim 1/p_0$. Thus, $\omega\tau \sim \omega/p_0$ and the absorption coefficient α and the phase velocity c are functions of the ratio of the sound frequency to gas pressure. Experimentally, this means that data can be taken equivalently as a function of sound frequency or gas pressure.

In liquids, the modeling is complicated by the fact that, unlike gases, the molecular collisions are not generally binary but involve both nearest neighbor and next nearest neighbor molecules. The collision rate in liquids is also usually much higher than in gases. This raises the relaxation frequency in most cases to ranges above that is experimentally attainable. Further, the number of vibrational states in liquids is considerable, complicating the relaxation processes. Nonetheless, the application of relaxation concepts has proven useful in connection with theories of the liquid state, and vibrational relaxations have been discovered in numerous liquids. The nature of the relaxation processes involved has been accurately established for liquids with rotational isomers. Generally, the attenuation in viscous fluids and in some biological tissues is proportional to the frequency or to some power of the frequency that is less than one. For more details, see Beyer and Letcher (1969) and Bauer (1965).

7.4.3.3 Viscoelasticity

Historically, viscoelasticity was introduced before the constructs of irreversible thermodynamics were established. The classical Maxwell (1867), Kelvin (1875) and Voigt (1892) models of viscoelasticity and the Zener (1948) anelastic solid model were based on various mathematical combinations of elastic and flow elements of the material. The elastic behavior is represented by springs and the viscous behavior is represented by a dashpot in these models. Although each of the classical models can be obtained from the more general relaxation model based on irreversible thermodynamics, it is instructive to consider each model independently.

Maxwell Material The Maxwell model (1867) was introduced to explain the behavior of fluids whereby the fluid responds to an applied shear stress with both elasticity and flow. Maxwell combined Hooke's law, $\sigma_{12} = G\varepsilon_{12}$, where σ_{12} is the shear stress, ε_{12} is the shear strain, and G is the shear modulus, with Newton's law for a viscous fluid, $\sigma_{12} = \mu(d\varepsilon_{12}/dt)$, where μ is the shear viscosity coefficient. He obtained the equation

$$\frac{d\varepsilon_{12}}{dt} = \frac{1}{G}\frac{d\sigma_{12}}{dt} + \frac{1}{\mu}\sigma_{12} \tag{7.38}$$

Using the relation $(\tau_{\sigma S}/\tau_{\eta S}) = (M_\infty/M_0) = (G_\infty/G_0)$ in Equation (7.29) and comparing the resulting expression with Equation (7.38) reveals that $G = G_\infty$ and that $\mu = \tau_{\eta S} G_\infty$. Assuming sinusoidal variation in σ_{12} and ε_{12} (i.e., sinusoidal wave propagation) and recalling the definition of the shear modulus (Hooke's law), we find from Equation (7.38) that

$$G = G_\infty \frac{\omega^2\tau_{\eta S}^2}{1+\omega^2\tau_{\eta S}^2} + iG_\infty \frac{\omega\tau_{\eta S}}{1+\omega^2\tau_{\eta S}^2} \tag{7.39}$$

Because the shear modulus G is complex, the sound velocity is also complex and gives rise to an attenuation of the sound wave along the propagation path. At very high frequencies, $G \to G_\infty$ and the fluid exhibits a shear rigidity that allows shear wave propagation. At very low frequencies, $G \to 0$ and the fluid loses its solidlike properties. Maxwell's model has been successfully applied to the characterization of the shear properties of such substances as pitch, simple liquids, and polymers.

Kelvin-Voigt Material Kelvin (1875) and Voigt (1892) independently suggested that

Hooke's law be modified to include a contribution from the strain rate as

$$\sigma_{ij} = C^S_{ijkl}\varepsilon_{kl} + \mu_{ijkl}\frac{\partial \varepsilon_{kl}}{\partial t} \tag{7.40}$$

where μ_{ijkl} is the viscosity tensor. Substituting Equation (7.40) into Newton's law and assuming isotropic symmetry, we obtain the following equation (using Lamé elastic constants) for longitudinal wave propagation:

$$\rho_0\frac{\partial^2 u_1}{\partial t^2} = (\lambda + 2G)\frac{\partial^2 u_1}{\partial a_1^2} + (\chi + 2\mu)\frac{\partial^3 u_1}{\partial a_1^2 \partial t} \tag{7.41}$$

where χ is the compressional viscosity coefficient.

We assume a plane harmonic wave disturbance of the form $u_1 = w_1\exp[i(k'a - \omega t)]$, where the wave number $k' = k + i\alpha = (\omega/c_1) + i\alpha$, α is the attenuation coefficient, and c_1 is the longitudinal velocity. Substituting this form into Equation (7.41) and solving for c_1 and α gives $c_1 = [(\lambda + 2G)/\rho_0]^{1/2}$ as before and

$$\alpha \approx \frac{\omega^2(\chi + 2\mu)}{2\rho_0 c_1^2} \tag{7.42}$$

A similar procedure for shear waves in isotropic materials yields the values $c_2 = (G/r_0)^{1/2}$ for the shear wave velocity and

$$\alpha \approx \frac{\omega^2 \mu}{2\rho_0 c_1^2} \tag{7.43}$$

for the shear wave attenuation coefficient.

7.5 Wave Dispersion

The dependence of sound velocity on wavelength, such as that arising from modal propagation in waveguides, for example, is known as *wave dispersion.* A wave disturbance of finite duration such as an ultrasonic pulse or toneburst (a gated cw) may be viewed as composed of an integrated sum of monochromatic waves where the integration is performed over a continuous range of infinitesimally differing frequencies of the waves. In a linear, dispersive, propagation medium, the acoustic waveform $u(x,t)$ at position x and at time t in the medium is given by the expression (Whitham, 1974)

$$u(x,t) = F(x,t)e^{i\Omega(x,t)} + c.c.$$
$$\Omega(x,t) = kx - \omega t \tag{7.44}$$

where $c.c.$ denotes the complex conjugate of the preceding term providing

$$\frac{d\omega}{dk} - \frac{x}{t} = 0$$
$$\frac{d^2\,\omega}{dk^2} \neq 0 \tag{7.45}$$

where ω is the angular frequency. The functional dependence of angular frequency on wave number k defines the dispersion relation $\omega = \omega(k)$. The condition $(d^2\omega/dk^2) \neq 0$ is the defining condition for a dispersive propagation medium and $(d\omega/dk) = c_g$ is defined to be the group velocity. For a given choice of x and t, Equation (7.45) may be solved for k if the dispersion function $\omega(k)$ is known. Thus, different parts of the waveform, corresponding to different given pairs (x, t), propagate with different group velocities $c_g(k)$ corresponding to specific wave numbers determined from Equation (7.45). The group velocity corresponds to the velocity of energy propagation of the wave (Whitham, 1974; Pierce, 1989).

Both the complex amplitude $F(x, t)$ and the phase function $\Omega(x,t)$ in Equation (7.44) are functionally dependent on x and t. Equation (7.44) has the form of an elementary waveform, but $F(x, t)$

is actually a position- and time-varying envelope of the waveform with the position- and time-varying phase $\Omega(x, t)$ defining nonuniformly spaced maxima and minima of the wave. Equation (7.44) thus defines a disturbance, which in general changes shape during its propagation through a dispersive medium.

The propagation of a given constant value of phase Ω_o defined by $\Omega(x, t) = \Omega_o$ moves with the phase velocity $c_p(x, t) = dx/dt = \omega/k$. The phase velocity $c_p(x, t)$ is generally different for different values of Ω_o and for different values of x and t. An observer following a particular phase (e.g., crest of a wave) moves with the local phase velocity having a locally varying frequency and wave number. Thus, the phase velocity of a wave of finite length propagating in a dispersive medium can only be defined locally (i.e., for given x and t) and for a given phase Ω_o. Hence, if a given pulse measurement technique (see Section 7.7) follows a given phase Ω_o of the waveform at different monotonically increasing or decreasing values of the space-time coordinates (x_i, t_i), then the local phase velocity at (x_α, t_α) for phase Ω_o can be obtained by plotting x_i as a function of ti and measuring the tangent to the curve at (x_α, t_α). For dispersive media, the curve is generally nonlinear, but for nondispersive media the curve is linear and coincides with the curve for the group velocity, $c_g = x/t$. But in general, the group velocity is related to the phase velocity as $c_g = c_p - \lambda(dc_p/d\lambda)$, where λ is the phase velocity wavelength (Pollard, 1977).

7.6 Ultrasonic Generation and Detection

The generation and detection of ultrasonic waves can be achieved using a variety of methods. One of the earliest and simplest methods for generating ultrasound was the Galton whistle, which produced sound waves in air by blowing high-pressure air against a sharp metal edge. Some methods require direct physical contact between the ultrasonic source and the propagation medium and are usually, but not always, associated with piezoelectric generation and detection of sound. Other methods, not requiring direct contact, are found to be useful in some applications.

7.6.1 Piezoelectric Transducers

When a mechanical stress σ_{ij} is applied to a piezoelectric crystal, an electric polarization P_i is produced according to the relation $P_i = d_{ijk}\sigma_{jk}$, where d_{ijk} are the piezoelectric coefficients. Similarly, when an electric field E_i is applied to a piezoelectric crystal, a strain ε_{jk} is produced in the crystal according to the relation $\varepsilon_{jk} = d_{ijk}E_i$. Thus, piezoelectric crystals can both generate and detect sound waves.

Although there are in general 27 piezoelectric coefficients, the crystal symmetry considerably reduces the number. Quartz, for example, has only five independent coefficients. If the quartz is cut such that a compressive stress σ_{11} occurs along the x (a_1) axis, then $P_1 = d_{111}\sigma_{11}$, $P_2 = P_3 = 0$. The electric polarization is also along the x axis. Conversely, if an electric filed is applied along the x axis, the resulting strains include a component $\varepsilon_{11} = d_{111}E_1$. Thus, the same piezoelectric crystal can be used to generate and detect compressional sound waves.

Generally, piezoelectric crystals can be cut in specific ways to generate and detect compressional waves or shear waves. The fundamental frequency of vibration f_1 for thickness longitudinal modes is

$f_1 = (1/2L_x)(c_{11}/\rho_0)^{1/2}$, where L_x is the thickness of the crystal along the x direction and c_{11} is the elastic coefficient. For shear modes, the fundamental vibration is determined from $f_1 = (1/2L_y)(c_{44}/\rho_0)^{1/2}$, where L_y is the thickness of the crystal along the y direction. Harmonics of the fundamental frequency in each case may be generated, but only for odd multiples of the fundamental. This occurs because the strain changes sign for each half wavelength and the polarization can be detected only when opposite charges appear at the opposing crystal surfaces.

Other single crystal materials used for piezoelectric transducers include ethylene diamine tartrate (EDT), sodium potassium tartrate (Rochelle salt), dipotassium tartrate (DKT), and ammonium dihydrogen phosphate (ADP).

Polycrystalline ceramic materials such as barium titanate, lead zirconate titanate (PZT), lead metaniobate, and sodium metaniobate have also been popular materials for ultrasonic transducers. In contrast to inherently piezoelectric materials such as quartz, these ferroelectric ceramics are initially isotropic. Because an essential feature of piezoelectricity is the absence of a center of crystallographic symmetry, the ceramics must be electrically polarized above the Curie temperature to induce the anisotropy in the materials responsible for the piezoelectric properties. A particular advantage of these materials is that the transducer can be made in a variety of shapes and focusing forms. Further, the resonances of the polycrystalline ceramic transducers (low Q value) are generally much less sharp than those of single crystal materials (high Q value). If sound absorbing material is added to the back of the ceramic material, the resonance will be flattened even more. This occurs because waves propagating in the backward direction will be eliminated from the superposition of resonance-forming waveforms (see Section 7.3.2). The result is a transducer having a broadband frequency capability for both generating and receiving sound.

High-frequency transducers made from thin film zinc oxide and cadmium sulfide deposited on a base material are capable of generating waves with frequencies in the pretersonic range of frequencies. Piezoelectric polymer film transducers such as polyvinylidene fluoride (PVDF) have a broad frequency bandwidth that is advantageous for use with short pulses. For more information on piezoelectricity, see Berlincourt, Curran, and Jaffe (1964) and O'Donnell, Busse, and Miller (1981).

7.6.2 Noncontacting Methods

In many applications, it is desirable to generate and detect sound by methods that are not in direct physical contact with the material to be measured. These applications may include, for example, materials for which the temperature is too high, materials that are too corrosive, or materials that are too inaccessible to use contacting transducers. A variety of noncontacting generation and detection techniques have been developed for such situations.

7.6.2.1 Capacitive Transducers

For solids, the capacitive transducer (also called the electrostatic acoustic transducer or ESAT) is a simple parallel plate arrangement in which a flat sample surface functions as the ground plate. The other plate is a flat electrode recessed (typically of the order 7 – 10 μm) from the sample surface. A dc bias V_b (typically 50–150 V) is applied through a large resistor to the electrode. An ultrasonic wave incident at the solid surface varies the gap spacing (hence capacitance) of the parallel plate arrangement. The measured output voltage V_{out} from this arrangement is related to the displacement amplitude u of the incident wave amplitude u as $V_{\text{out}} = V_{\text{b}}(2us^{-1})$, where s is the gap spacing between the plates in the absence of the sound wave. The capacitive transducer for solids has a flat frequency response from dc to well into the gigahertz range of frequencies.

The capacitive transducer may also be used as a sound generator by applying an ultrasonic drive voltage across the electrodes, which produces an electric field E at the sample surface. The resulting electrical force per unit area σ at this surface is $\sigma = (1/2)\varepsilon_0 E^2$, where ε_0 is the dielectric permittivity of air. For a sinusoidal drive voltage $E = E_0 \sin \omega t$, the driving stress $\sigma = (1/4)\varepsilon_0 E_0^2(1 - \cos 2\omega t)$. Thus, the driving stress and resulting ultrasonic wave have a frequency that is twice that of the driving voltage. Typically, in a capacitive driving transducer, a material with a high dielectric permittivity such as mica is sandwiched between the electrodes. The material allows higher drive fields to be obtained without electrical breakdown.

A capacitive transducer for immersion in liquids is generally constructed by stretching a conductive membrane over an electrically isolated, dc biased, cylindrical electrode again recessed approximately 10 μm from the membrane inner surface. Unlike the capacitive transducer for solids, the liquid immersible capacitive transducer does not have a flat frequency response. In the ultrasonic range of frequencies, the inertia of the membrane and the coupling to the fluid produces a frequency roll-off that varies as $[1 + (\sigma_A \omega / Z)]^{-1/2}$, where σ_A is the area density of the membrane and Z is the acoustic impedance of the immersion fluid (Cantrell and Yost, 1989).

7.6.2.2 Electromagnetic Acoustic Transducers (EMATs)

A useful noncontacting transducer for generating SAWs in electrically conductive materials is the electromagnetic acoustic transducer (EMAT). The transducers are based on the body forces produced by the presence of both static and time-varying magnetic fields. Figure 7.6 shows one configuration for generating surface waves. An alternating current is passed through a flat meander coil held near the surface of the conductor and produces an alternating magnetic field in the conductor. If a static magnetic field is also produced in the conductor having the field lines parallel to the surface, Lorentz forces on the electrons within the conductive material will couple to the material lattice. The lattice coupling will serve as a source for the generation of surface waves propagating along the static field lines. The elementary sources represented by each meander coil wire segment add coherently, if the periodic distance between the parallel wire segments equals one half of the acoustic wavelength at the frequency of operation. For more information on EMATs, see Thompson (1990) and Frost, Sethares, and Szabo (1977).

7.6.2.3 Laser Generation and Detection

The generation of sound waves by incident optical laser radiation occurs by a number of mechanisms including radiation pressure, material vaporization, thermal expansion, Brillouin scattering, and electrostriction (due to electric polarization-induced movement of atoms into regions of different optical intensity). In most cases, the thermal

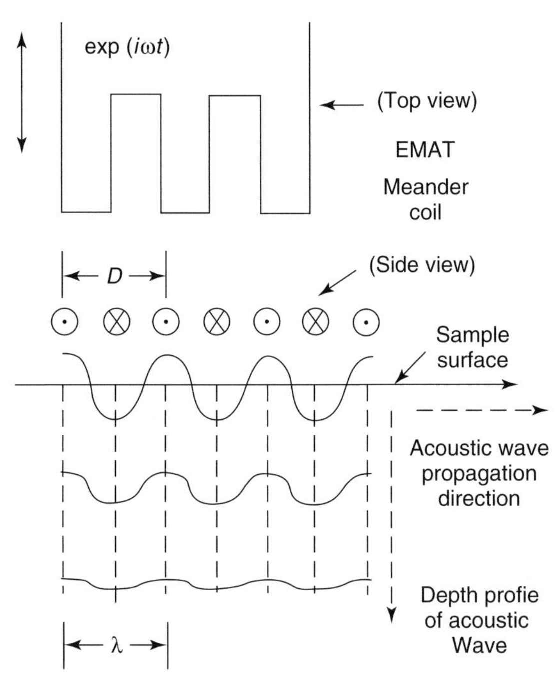

Fig. 7.6 EMAT meander coil configuration for generating surface acoustic waves. Adapted from Thompson, 1977.

expansion mechanism dominates. If the incident laser energy is a pulse having energy E_0 that passes through a length L of a medium having an absorption coefficient α, then it can be qualitatively shown (Hutchins, 1988) that the resulting acoustic pressure amplitude p in the material is $p = K\beta c E_0 \alpha / C_p$, where β is the volume expansion coefficient, C_p is the isobaric specific heat, c is the sound velocity, and K is a constant that depends on the configuration geometry and laser pulse shape. The details of the laser-induced acoustic pressure profile depends on whether the medium is solid or fluid, the shape of the acoustic source formed by the interaction of the laser light with the material, and the physical constraints placed on the propagation medium.

Laser detection of sound is commonly accomplished using an interferometric technique or a variation thereof. A variation useful for noncontact probing of pulsed SAWs is illustrated in Figure 7.7. Light from a laser is divided into two parallel and slightly separated collimated beams of equal intensity by a fixed beamsplitter. The beams are partially transmitted by a second beamsplitter and focused to points separated by a fixed distance on the surface of the specimen. Upon reflection, the beams are partially reflected by the second beamsplitter and superimposed to form a straight line interference pattern which is filtered with a spatial filter. This interferometer is sensitive to differential changes in the optical pathlengths of the two beams resulting from the passage of an SAW. The sound field causes a relative motion of the output fringe pattern with respect to the fixed spatial filter. For more information on optical detection of ultrasound, see Wagner (1990).

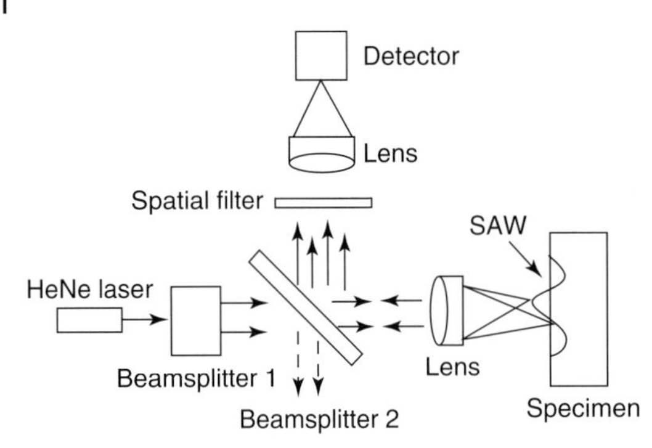

Fig. 7.7 Differential optical interferometric surface acoustic wave measuring system. From Claus and Cantrell, 1981.

7.6.2.4 Modulated Particle Beam Generation

Atomic particles incident on a material surface lose their energy in scattering or collision processes with target atoms as they penetrate into the material. Although the ions accelerated through an electric field have been used as particle beam sources, more typically accelerated electrons are used. Electron beams formed by accelerating electrons through a potential V (in kilovolts) will lose most of their energy due to atomic scattering in a roughly spherically shaped region (the interaction volume) of radius $d_s = V^{1.43}/(10\rho_0)$, d_s in microns, (Reimer, 1979). The electron scattering results in thermal generation as well as in momentum transfer. A modulated or chopped electron beam source generates both a thermal wave and acoustic waves simultaneously. A number of bulk and SAW modes are generated including modes resulting from thermal-to-acoustic wave mode conversion and from the driving source of electron in the interaction volume itself. Their amplitudes depend on the intensities of the driving source and the thermal waves and on their gradients at the material boundaries (Qian and Cantrell, 1989).

7.7 Measurement Techniques

Measurements of sound velocity and attenuation can provide considerable information about the physical properties of solids, liquids, and gases. Measurement of the variation of the sound velocity as a function of some intrinsic material variable, such as temperature or pressure, provides additional information about basic (often nonlinear) material properties. It is important to distinguish between group and phase velocities in the measurements of velocity. This distinction was addressed in Section 7.5 in the context of acoustic propagation in a dispersive medium. Similarly, it is important to distinguish between absorption and other attenuation mechanisms in acoustic measurements as addressed in Section 7.4.

Numerous techniques have been reported in the literature for making velocity and attenuation measurements. The most common techniques are based on optical, pulse, and cw methods, or on hybrids of these methods. The choice of technique is dictated in part by the availability of equipment and ease of setup for the particular application, the specific material and ultrasonic properties of the material to be measured, and the accuracy and precision desired in the measurement. Here, we briefly consider only representative techniques. For more information on measurement methods, see the review articles by McSkimin (1964), Papadakis (1968, 1976), Breazeale, Cantrell, and Heyman (1981), and Cantrell and Yost (1997).

7.7.1 Pulse Techniques

Pulse techniques utilize either broadband acoustic pulses or gated continuous acoustic waves (tonebursts) generated by an ultrasonic source.

7.7.1.1 Basic Pulse-Echo Method

An equipment arrangement for the basic pulse-echo system that allows for both velocity and attenuation measurements is shown in Figure 7.8. A pulsed ultrasonic signal is transmitted into the sample by means of a transducer attached to the sample surface. The ultrasonic pulse travels through the sample and reflects between the sample boundaries, eventually decaying away because of sample attenuation. Each time the ultrasonic pulse strikes the end of the sample coupled to the transducer, an electrical signal is generated, which is amplified and displayed on an oscilloscope. The received signals are commonly rectified and filtered (i.e., detected) before being displayed on the oscilloscope. The velocity of ultrasonic wave propagation is determined by measuring the transit time between the corresponding reference points of the reflected pulses and the corresponding pulse propagation distance in the sample. The ultrasonic attenuation can also be read directly from the calibrated attenuator by adjusting the attenuator to maintain a constant receiver signal level for each displayed echo. Accuracies of a few parts in 10^3 for absolute velocity measurements and five parts in 10^2 for attenuation measurements are typically achieved with the basic pulse-echo method. Numerous modifications and variations of the basic pulse-echo system have been developed that result in measurement accuracies of a few parts in 10^4 for absolute velocity measurements, one part in 10^7 for relative velocity measurements, and two parts in 10^3 for attenuation measurements. The reader is referred to McSkimin (1964), Papadakis, (1968, 1976), and Breazeale, Cantrell, and Heyman (1981) for details.

7.7.1.2 Pulse Superposition Method

The pulse superposition method uses a series of tonebursts introduced into the sample by a radiofrequency (rf) pulse generator. The repetition rate of these pulses is controlled by the frequency of a cw low-frequency oscillator. The repetition rate is adjusted to correspond to some multiples of an acoustic round-trip transit time in the sample corresponding to an "in-phase" superposition of pulses such that maximum amplitude is achieved. The measured time delay between the superimposed pulses is used to obtain the sound velocity. Accuracies of two parts in

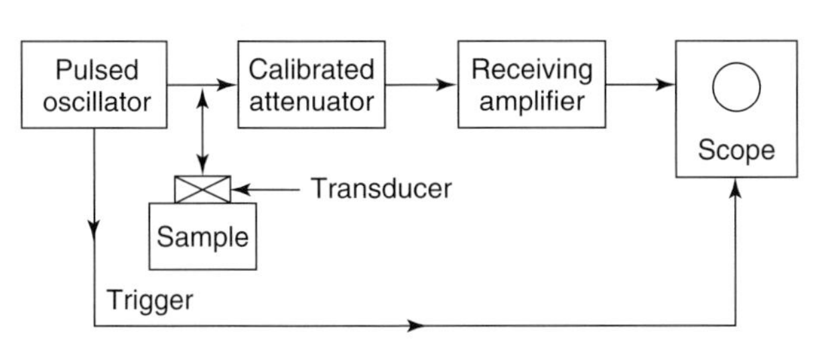

Fig. 7.8 Basic ultrasonic pulse-echo measurement system.

10^4 for absolute velocity measurements and one part in 10^7 for relative velocity measurements are possible with the pulse superposition method (see McSkimin, 1964, for more details).

7.7.1.3 Echo-Overlap Method

In the pulse-overlap method, a series of rf pulses from an rf-pulsed oscillator is transmitted into the sample. The trigger signal is derived by frequency dividing the output of a cw low-frequency oscillator (typically 100–1000 kHz) by a factor of 10^3. The trigger is used to control the pulse repetition rate and to trigger a double time delay (strobe) generator that actuates the z axis intensity gate on the oscilloscope. The delays are adjusted to intensify any chosen pair of displayed echoes. The cw oscillator frequency is adjusted to correspond to the reciprocal of the travel time between the echoes so that the echoes can be made to overlap cycle-for-cycle. The sound velocity is calculated from the measured travel between echoes after correcting for phase shifts due to reflections. Accuracies of a few parts in 10^4 for absolute velocity measurements are typical with the pulse superposition method (see Papadakis, 1976, for more details).

7.7.1.4 Pulse Interferometer Methods

In the pulse interferometer methods, an ultrasonic toneburst is transmitted into the sample by gating a cw reference oscillator. The received pulse (an echo in a single transducer arrangement) is combined with the cw reference signal in a phase-sensitive detector. If the relative phase between the received acoustic signal and the cw reference signal is then changed, the output of the phase-sensitive detector varies in response. The change in relative phase may result, for example, from a variation in the acoustic path or from a variation in the sound velocity as a function of temperature.

Two basic means are reported to measure the change in relative phase. One of the means employs a phase shift network that monitors the amount of phase added to the reference signal necessary to maintain a constant output from the phase-sensitive detector. The frequency of the cw reference signal is held fixed during this process. The second means uses a change in the frequency of the cw reference signal to maintain a constant detector output. In the pulse phase-locked loop system, a voltage controlled oscillator network replaces the phase shift network.

In both the constant frequency and variable frequency techniques, the total phase seen by the phase detector includes acoustic contributions related to the acoustic wavelength relative to sample length as well as collective nonacoustic contributions. The frequency-dependent nonacoustic contributions include those from electrical signal propagation paths and phase perturbations associated with various electrical components. In measurements of phase velocity in fluids, for example, the change in phase corresponding to a given change in path length may be measured. For the constant frequency technique, both the wavelength and the nonacoustic phase term are constants. The phase velocity is obtained from the measurement of the slope of the linear curve giving the change in phase as a function of the change in path length. For the variable frequency technique, the wavelength and the nonacoustic phase term are not fixed and the velocity can be calculated only after correcting for the variations in the nonacoustic phase term. The pulse interferometer methods have a sensitivity of parts in 10^8 for relative velocity measurements. The accuracy of absolute velocity measurements is comparable to that of the

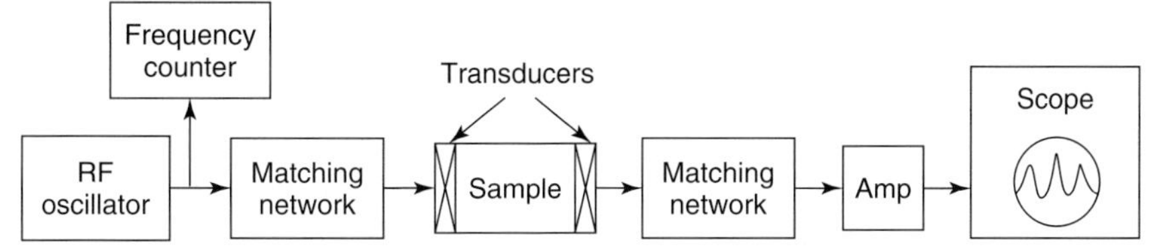

Fig. 7.9 Continuous wave ultrasonic interferometer.

pulse superposition and the echo-overlap methods. For more information on pulse interferometer methods, see Yost, Cantrell, and Kushnick (1992).

7.7.2 Continuous Wave Techniques

The cw techniques employ uninterrupted harmonic waves introduced into bounded material and are among the most expedient methods for measuring sound velocity and attenuation of materials with appropriate geometry. An advantage of cw methods is the increase in sensitivity to variations in velocity and attenuation resulting from the resonance conditions established by the material geometry and the physical properties. The effective gain in sensitivity can be measured in terms of "sensitivity enhancement factors."

Consider the one-dimensional resonances discussed in Section 7.3.2. For any observable A, the variation in amplitude $\Delta|A|$ may be expressed as $\Delta|A| = (\partial|A|/\partial\alpha)\Delta\alpha + (\partial|A|/\partial k)\Delta k$ for small variations in the attenuation α and wavenumber k. The partial derivatives are the sensitivity enhancement factors for attenuation and wavenumber, respectively. If the observable is the particle velocity $|v|$, the sensitivity enhancement factors are found from Equation (7.18) to be

$$\frac{\partial|v|}{\partial\alpha} = \frac{L(e^{-\alpha L/2} - e^{\alpha L/2})\cos kL}{8^{1/2}(\cosh\alpha L - \cos kL)^{3/2}} \tag{7.46}$$

and

$$\frac{\partial|v|}{\partial k} = \frac{Le^{\alpha L/2}\sin kL}{8^{1/2}(\cosh\alpha L - \cos kL)^{3/2}} \tag{7.47}$$

These factors together with similar factors for v_1 and v_2 can be used to separate the effects of dispersion and attenuation on the measurement parameters.

A simple transmission cw measurement system called the *continuous wave* ultrasonic interferometer is shown in Figure 7.9. In this technique, an oscillator is swept through a series of resonance peaks as observed on an oscilloscope screen. The frequencies at which the peak amplitudes occur are used to determine the ultrasonic velocity, and the frequency width $\Delta\omega$ at half power is used to calculate the attenuation in the manner described in Section 7.3.2. The addition of a mixer and 0 and 90° phase shifters to the system permits the assessment of the in-phase (v_1) and quadrature (v_2) resonance signals. Several modifications to the basic cw transmission spectrometer have been reported. The reader is referred to Breazeale, Cantrell, and Heyman (1981) for details.

For materials having a fixed length and sufficiently small dispersion, the phase velocity *cp* can be determined from the frequency difference ($f_n - f_{n-1}$) between two adjacent resonances in the manner discussed in Section 7.2.2. The attenuation α can be calculated from the determination of the frequency width at half power, $\Delta\omega$ as $\alpha = \Delta\omega/2c_p$. These results are based on the assumption that the sample responds as a simple resonator driven by noncontacting transducers. If contacting transducers are used, the sample and the transducers form a composite resonating system whose resonant frequencies are predicted by the roots of a transcendental equation. Approximate solutions to the transcendental equation have resulted in correction factors for the above formulas (Chern, Cantrell, and Heyman, 1981). However, errors in the velocity determination can exceed 10% under certain conditions.

For samples requiring analysis in a three-dimensional geometry, the method of resonant ultrasonic spectroscopy (Demarest, 1971; Ohno, 1976; Migliori *et al.*, 1993; Anderson, 1992; Visscher *et al.*, 1991) can be used. In this method, the modal frequencies for a sample of given three-dimensional geometry are compared to predictive analytical models of the sample modal frequencies. Analysis of the amplitude–frequency spectrum allows an assessment of the elastic constants as well as the internal friction (attenuation) of the sample.

7.7.3 Optical Techniques

Optical techniques can be used for sound velocity measurements in transparent media. Generally, the methods are based on the optical determination of acoustic wavelength either from diffraction or from scattering of light by the sound waves. The wavelength together with the acoustic frequency allows the determination of the phase velocity. For techniques based on diffraction, it is essential to know the type of diffraction experienced by the light.

7.7.3.1 Diffraction of Light by Sound

Consider an ultrasonic transducer generating compressional waves in a transparent propagation medium. If the propagating sound wave encounters light traversing the same medium at some angle incident to the sound wave, the light is diffracted. Assuming that the Fraunhofer conditions hold for the light diffraction (i.e., parallel light rays are brought to a focus), two distinct physical processes can produce the diffraction effects. One involves the formation of a corrugation in the phase fronts of the light due to the acoustically induced spatial variation of the index of refraction. This phenomenon is called *Raman–Nath diffraction*. The second process involves the reflection of light from the (evenly spaced) crests of the sound wave that occurs under conditions similar to that of X-ray diffraction. This phenomenon is called *Bragg diffraction*.

A dimensionless parameter has been introduced to determine which type of diffraction predominates for given experimental conditions. The parameter Q is defined by $Q = k^2L/\mu_0 k^*$, where k is the ultrasonic propagation constant, L is the width of the ultrasonic beam, μ_0 is the index of refraction, and k^* is the propagation constant of the light in vacuo (Klein, Cook, and Mayer, 1965). If $Q > 9$, Bragg diffraction is predominant. If $Q < 1$, Raman–Nath diffraction occurs. If $1 < Q < 9$, the diffraction is mixed. Both types of diffraction can be used to determine the sound velocity of the material.

For Raman–Nath diffraction with the light incident at an angle $(\pi/2 - \phi)$ to the sound beam direction, the light is diffracted into diffraction orders given by the expression $\sin(\theta_n + \phi) - \sin(\phi) = n\lambda^*/\lambda$, where θ is the angle of diffraction with respect to the incident beam, λ^* is the wavelength of light in the medium, λ is the ultrasonic wavelength, and n is an integer denoting the diffraction order. The ultrasonic wavelength can be determined from the above equation.

For Bragg diffraction, the angle of diffraction θ_n is set to be equal to twice the negative angle of incidence $(-\phi)$ such that $n\ \lambda^* = 2\lambda \sin\phi_B$, where ϕ_B is the Bragg angle. Bragg diffraction is used to measure wave velocities in the frequency range from approximately 100 MHz to several gigahertz. The accuracy of velocity measurements using diffraction of light by sound methods is estimated to be of the order of 0.01–0.1% for homogeneous samples.

7.7.3.2 Brillouin Scattering

Lattice vibrations are caused by thermal and other excitations of the crystal lattice. These vibrations form vibrational modes (normal modes) within the lattice and play a major role in determining the thermal properties (e.g., heat capacity) of the solid. The normal modes form two major branches. The acoustic branch includes normal mode frequencies, which vanish linearly as the propagation vector $\boldsymbol{k}$ approaches zero (long wavelength limit). The optical branch includes normal mode frequencies that do not vanish as the propagation vector approaches zero.

The quantum mechanical treatment of the lattice vibrations predicts that the normal modes gain and lose energy in quantized units just as in the case of the harmonic oscillator. Because of the similarity of the lattice vibrations to the electromagnetic radiation field in a cavity, a quantum of lattice vibrational energy called the *phonon* is defined, primarily for convenience, in analogy to the electromagnetic quantum of energy called the *photon*. As with photons, phonons are given the property that each phonon carries an energy $\hbar\omega$, where $\hbar$ is the Planck's constant divided by 2π, and under certain conditions behaves as if it carries a crystal momentum of value $\hbar k$.

Light interacting with the normal modes in the material can gain or lose energy and momentum. Brillouin scattering occurs when the modes of the acoustic branch interact with light. Consider the interaction of a photon with a material whose index of refraction is μ_o. Application of the laws of conservation of energy and momentum, respectively, to the scattering process gives $\hbar\boldsymbol{\omega}^{*'} = \hbar\boldsymbol{\omega}^* \pm \hbar\omega$ and $\hbar\mu_o\boldsymbol{k}^{*'} = \hbar\mu_o\boldsymbol{k}^* + \hbar\boldsymbol{k} + \hbar\boldsymbol{K}$, where $\omega^{*\prime}$ and $\boldsymbol{k}^{*\prime}$ are the frequency and propagation vector, respectively, of the incident photon, ω^* and $\boldsymbol{k}^*$ are the frequency and propagation vector of the scattered photon, ω and k are the frequency and the crystal momentum assigned to the phonon, and $\boldsymbol{K}$ is the reciprocal lattice vector (assumed to be zero in this case). Because the photon frequency shifts are small for this process, the propagation vectors are very nearly of the same length. The vector addition expressed by the above equations gives an isosceles triangle whose apex angle is θ, the complement of the scattering angle. The equations lead to the result (Ashcroft and Mermin, 1976)

$$\omega = \pm 2\,\omega^* \left(\frac{c}{C_1}\right) \sin\left(\frac{\theta}{2}\right) \tag{7.48}$$

where C_1 is the speed of the photon in the medium, c is the speed of the phonon, and θ is the scattering angle. The frequency ω is equal to the difference in frequencies between the incident and scattered photons.

The technique produces a scattered photon spectrum of three lines, one of which is unshifted. The two frequency-shifted lines are called the *Stokes line* (a phonon is created) and the *anti-Stokes line* (a phonon is annihilated). These lines can be separated and measured with optical devices such as the Fabry–Perot interferometer to obtain ω. If desirable, the Bragg condition can be used for constructive reinforcement by adjusting θ. Equation (7.45) can be solved for the velocity c in terms of the other quantities measured.

Brillouin scattering is useful in the investigation of the sound velocity of solids and liquids especially near a phase transition where attenuation can become very large. The technique is also advantageous in that it is one of the few techniques capable of assessing sound velocity at frequencies of several gigahertz. The technique has been used for sound velocity measurements for both longitudinal and mixed (longitudinal-transverse) modes.

7.7.4 Sources of Error

A number of sources of error arise in the measurements of velocity and attenuation. Errors arising from time or length measurements can be evaluated directly, but among the less easily evaluated errors are those arising from velocity dispersion, material inhomogeneity, phase cancellation effects, diffraction, and the effects of thickness of the transducer and transducer bond. Many aspects of the sources of error are best addressed in the context of the measurement technique used, but certain characteristics of the error sources have more generic implications.

For example, as indicated in Section 7.5, any technique employing a change in frequency ω to measure the sound velocity c_p must account for dispersion, because the velocity is a function of the frequency. Measurements using gated cws (tonebursts) of finite length contain a spectrum of frequencies around the frequency of the generating continuous waveform. Such measurements can be made to approximate narrowband cw conditions, hence minimizing the effects of dispersion, if the pulse lengths are sufficiently long. The appropriate pulse length L is dependent on the intrinsic error M of the pulse measurement method and is given by the expression $L \geq \pi(d^2\omega/dk^2)/Mc_p$, where k is the wave number (Yost, Cantrell, and Kushnick, 1992).

A variation in the phase of the wave over the area presented by the surface of a phase-sensitive transducer can give rise to phase cancellation of different portions of the received signal that sets limits on the measurement accuracy of both velocity and attenuation. Sources of such phase variation include diffraction, media inhomogeneity, temperature gradients, wedging, and nonparallelism of the sample surfaces. Nonparallelism of opposite surfaces of the sample by an angle θ is representative of the effects of phase cancellation. For example, in measurements employing acoustic pulses of wave number k reflecting from the ends of the sample, the amplitude of the pulse received after n double reflections is diminished by the factor $[J_1(2kan\theta)/kan\theta]$ due to nonparallelism Truell, Elbaum, and Chick, 1969). The use of phase-insensitive devices such as the "acoustoelectric transducer" can minimize phase cancellation errors (Heyman, 1978), although such devices are generally less sensitive than the commonly used phase-sensitive piezoelectric elements.

A major source of error in measurements of velocity results from the effects of the transducer and the bond that couples the transducer to the sample. Both the transducer and bond introduce additional phase into the measurement system that can lead to errors of the order of parts in 10^2 for the measurement of velocity. Methods for correcting such errors are well established and have been addressed as appropriate to the specific measurement techniques discussed (for details see the review articles by McSkimin (1964), Papadakis (1968, 1976), Breazeale, Cantrell, and Heyman (1981), and Cantrell and Yost (1997). An alternative to such corrections is the use of noncontacting transducers such as capacitive transducers (Cantrell and Breazeale, 1977), electromagnetic transducers (Meridith, Watts-Tobin, and Dobbs, 1969), and optical techniques.

7.8 Applications of Ultrasound

Ultrasonic methods have long been used to characterize various properties of materials ranging from the fundamental to the most practical applications. Ultrasonic measurements of fundamental material properties, for example, have provided some of the first experimental evidence of the double well interatomic potential in vitreous silica, have been used to quantify the superconducting state of materials, and have provided well-established methods of testing lattice dynamical theories of the solid state. More practical uses of ultrasound span a wide range of disciplines including medical and industrial applications, nondestructive evaluation and testing of materials, acoustic microscopy, and underwater acoustics.

7.8.1 Medical Applications

Many medical applications of ultrasound center on the scanning or imaging of organs and tissues. The simplest scanning technique is the basic pulse-echo technique, often called the *A-scan method*. In the A-scan, an ultrasonic pulse is transmitted into the body. Changes in the acoustic impedance along the propagation path at tissue interfaces result in the partial reflection and transmission of sound from the interfaces. The reflected pulses (echoes) are received by the transducer and the echoes are displayed on the x trace of an oscilloscope and recorded. The amplitude of the echo is proportional to the magnitude of the reflection coefficient at the interface involved. A measurement of the time delays, calibrated in terms of distance, of ultrasonic echoes reflected from various anatomical landmarks provides the basis for the measurements. The A-scan is useful, for example, in the assessment of the anterior and posterior chambers of the eye. A variation of the A-scan is the M-scan in which a time record of variations in the x trace echo pattern is serially recorded along the y trace of the oscilloscope. The M-scan has proven useful in heart monitoring.

A B-scan image is formed by linking the time-base direction in the oscilloscope with the direction of propagation of the ultrasonic beam. The image is built up by using the amplitude of the received echoes to modulate the intensity of the oscilloscope beam as the ultrasonic beam is scanned.

In the ultrasonic C-scan, the x and y positions of the oscilloscope beam are made to track the x and y positions of the transducer as the transducer is made to areally scan a portion of the body. The z direction is the direction of ultrasonic propagation. A gated time delay is used to set the depth (z direction) at which ultrasonic echo information is received. The brightness of a given image point is dependent upon the amplitude of the received ultrasonic signal. In the C-scan mode, the ultrasonic images most closely resemble X-ray images.

Various modifications of the basic A-, B-, and C-scan imaging techniques have been developed including transducer array methodologies with electronic beam steering and acoustic holographic techniques. Among the many applications of ultrasonic imaging in its various forms are fetal monitoring, heart and cardiac structure imaging, assessment of the depth and severity of skin burns, and more recently high-resolution ultrasonic breast examination.

Ultrasound is also used in numerous medical applications other than imaging. For example, the Doppler shift associated with sound scattering from moving targets is used to measure both blood flow velocity and flow volume in blood vessels. Lithotripsy, a nonsurgical procedure, uses high-intensity sound focused into a small volume and directed onto hard objects such as kidney stones in the body. The acoustic impedance mismatch between the kidney stone and the surrounding soft tissue produces a sufficiently large pressure on the stones to break them into smaller pieces which can be excreted. The absorption of ultrasound directed into a muscle is used for heat therapy and message. The ultrasonic syringe utilizes the ability of ultrasound to temporarily disrupt surface tissue organization to permit the injection of medicine through the skin into the body. More recent applications of ultrasound (currently undergoing clinical evaluation) include the nonsurgical assessment of decubitus ulcers (bedsores) using spectral analysis profiles of tissue reflected echoes (amplitude versus frequency curve) and the noninvasive monitoring of intracranial dynamics and pressure (ICP) using pulse phase-locked loop ultrasonic interferometry.

7.8.2
Industrial Applications

Among the most common uses of ultrasound in industry is that in the nondestructive evaluation and testing of materials. Basic ultrasonic pulse-echo techniques are widely used for the detection and locations of cracks and flaws during routine inspection of machine and structural components and fixtures. The propagation of cracks is also passively monitored using the acoustic signals emitted during microfracture (acoustic emission monitoring). Ultrasonic imaging techniques are widely used for the assessment of material integrity, particularly for the assessment of delaminations and debonds in advanced composite materials. Ultrasonic velocity and attenuation measurements are often employed to obtain mechanical property data, such as the elastic constants or parameters associated with relaxation processes (internal friction) of the material. Such data can also be used for the assessment of the purity or the processing conditions of the material. The data is also useful in the assessment of basic microstructure-property relationships for the development and manufacturing of materials with improved or modified properties.

Residual stresses formed during materials processing or alterations in the stresses during the fatigue of materials are often of concern in critical components subjected to high cycle stresses. Because the magnitude of the sound velocity is sensitive to the stresses in the material, ultrasonic techniques are often used for the evaluation of such stresses. The acoustoelastic birefringence method relies on a stress-induced anisotropy in the material, which splits an incident shear wave into two components having different measured sound velocities. The difference in the two sound velocities is proportional to the stress in the material. Another method of assessing the presence of stresses depends on the measurement of stress-induced changes in the longitudinal velocity using either ultrasonic interferometry or pulse time-of-flight measurement techniques. The latter technique is also used for the precise tensioning of threaded fasteners in critical applications such as the disconnect studs linking the external fuel tank to the space shuttle orbiter.

Among the most important uses of SAWs is that in electronic SAW signal processing devices. Generally, the SAW device consists of an interdigital transducer formed by depositing an array of interlacing, alternately, charged electrodes on the surface of a piezoelectric substrate. The interdigital transducer generates an SAW. The SAW propagates to an output interdigital transducer. The intervening path between the input and output transducer is designed to produce a desired change in the input signal that is captured at the output. The intervening path may provide, for example, a propagation time delay, a waveform distortion, or alterations in the spectral frequency components of the waveform. Among the many SAW devices are delay lines, mixers, filters, correlators, resonators, coders, oscillators, and convolvers. Because the sound velocity is dependent on temperature and pressure, measurements of the SAW velocity on calibrated standards are also used to provide a sensitive monitor for both temperature and pressure. SAWs have been used to sense the position of a pointer (stylus or human finger) on a computer-controlled graphical input screen. SAW filters are widely used in cellular telephones. For more information on SAW devices, see Kino (1987).

Optical interactions with SAWs are used to make surface wave optical modulators, deflectors, and filters. For example, SAWs generated onto a thin optical-guiding film causes periodic variations in the optical index of refraction of the material. These periodic variations serve as Bragg diffraction gratings for incident optical beams coupled into the film. The incident optical waves are deflected in accordance with the Bragg reflection conditions as they encounter the SAWs. Optical output couplers can then be placed to receive only optical signals having a desired angle of deflection. Ultrasonic-induced frequency shifts of several megahertz in the diffracted light of optical modulators are used for single sideband mixing or demodulation of laser Doppler devices.

Applications of ultrasound other than materials characterization or SAW devices are also widely used in industry. For example, acoustic Doppler shift measurements are often used for monitoring the flow velocity of gases and molten solids. Doppler techniques are also used for measuring the velocity of objects moving in a fluid and ultrasonic Doppler sensors are commonly used in entry alarm systems. High-frequency acoustic vibrations in crystals are used in devices requiring precise frequency or time control. Basic pulse time-of-flight techniques are used for measuring the distance between an object and the sound source (range-finding). Ultrasonic rangefinders are found on many autofocusing cameras. The pressure in resonating high-intensity sound fields can be used to support and manipulate objects in air without mechanical contact with the object (acoustic levitation). Such manipulation is useful for handling contact sensitive or corrosive materials. SAWs can be used to manipulate and position small particles on a material surface. High-intensity sound is used in ultrasonic cutting and drilling machines, ultrasonic cleaners, and ultrasonic humidifiers (atomizers). High-intensity ultrasound is also used for soldering, welding (both metals and plastics), stress relieving previously welded joints, and metal forming and drawing.

Ultrasonic motors have become an important development for self-focusing cameras and robotic devices. Although there are many variations, the ultrasonic motor basically consists of a vibration-driven stator frictionally coupled to a rotor. A piezoelectric element is used to excite a flexural wave in the stator. The stator is placed in direct contact with the rotor. The transverse undulations of the flexural wave produces points on the stator surface that are in direct contact with the rotor and other points that are not in contact. As the wave propagates, a given point on the stator surface moves in an elliptically shaped, retrograde pattern similar to that of the Rayleigh wave. The motion of the stator–rotor contact point corresponding to the maximum wave amplitude is opposite to that of the wave propagation direction. The frictional forces between the stator and the rotor cause the rotor to move in that direction (Inaba *et al.*, 1987).

Thermoacoustic heat pumps have been developed that utilize a heat source and heat sink configured to an acoustic resonator having a stacked heat exchanger. The action of acoustically driven thermodynamics processes in the resonator gases has allowed the development of reliable, inexpensive engines, and refrigerators with a significant Carnot efficiency (~40%) (Swift, 1995).

7.8.3 Acoustic Microscopy

The microstructural variation in the acoustic properties of materials (including biological material) has been the key to the development of a new class of microscopes that utilize acoustic waves in some fashion for obtaining microscopic images. The images so obtained are quite different from those obtained in conventional light or electron microscopes. Different types of acoustic microscopes use different energy sources such as electron beams or piezoelectric transducers for generating the acoustic images and are descriptively named, for example, the scanning acoustic microscope, the scanning electron acoustic microscope, the thermoacoustic microscope, and the scanning laser acoustic microscope (SLAM). Although the image contrast and spatial resolution attainable in such microscopes are in large measure dictated by the details of energy source-material interactions, the dependency on the local variations in the material elastic or thermoelastic properties, including residual stresses, is a necessary consideration in such interactions.

7.8.3.1 Scanning Acoustic and Laser Acoustic Microscopes

In scanning acoustic microscopy (SAM), plane acoustic waves are generated by a piezoelectric transducer attached to a cylindrical rod (typically made of sapphire) as shown in Figure 7.10. The opposite end of the rod is made into a concave spherical surface to form an acoustic lens. Coupling fluid (usually water) is placed between the rod surface and the specimen surface. Wave refraction at the acoustic lens-coupling fluid interface provides a focusing action for the plane waves incident from the transducer. The focused waves impinge on the specimen surface. In general, the waves incident on the specimen surface are specularly (geometrically) reflected back to the transducer. At certain critical angles of incidence (if the material properties permit), leaky surface

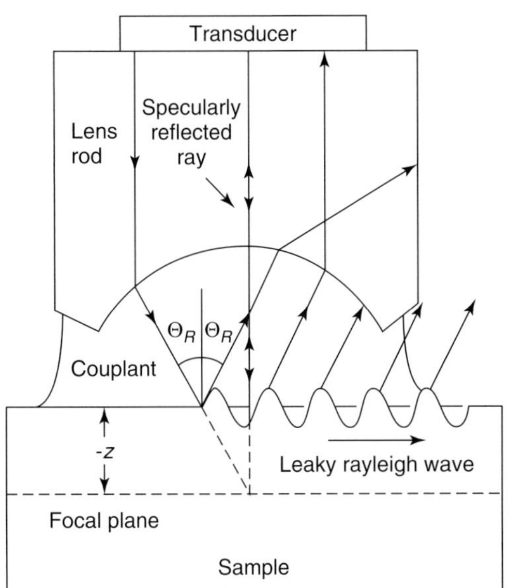

Fig. 7.10 Scanning acoustic microscope acoustic lens configuration for generating signals via the interference between "leaky" Rayleigh waves generated at the Rayleigh critical angle θ_R and specularly reflected waves (illustrated by the central specularly reflected wave). Adapted from Briggs, 1992.

waves (usually leaky Rayleigh waves) can also be generated. Under proper defocus conditions (lens-specimen distance), the leaky surface waves find their way back to the transducer. The total acoustic signal reaching the transducer in this mode of operation is composed of contributions from the geometrically reflected waves and the leaky surface waves. The interference of these two waves gives rise to the SAM output signal. If the acoustic lens–specimen distance is allowed to vary, the periodic variation in the signal amplitude as a function of this can be used to obtain the surface wave velocity of the material.

If the focused acoustic beam is rastered over the specimen surface, variations in the acoustic impedance of the specimen or in the surface wave velocity or geometrical irregularities in the specimen surface can change the magnitude and phase of the output signal. The variations in output signal at each point on the surface are recorded on a visual display to generate an SAM image. The images can be analyzed to obtain information related to the variations in the mechanical properties of the specimen over micron scale regions. Commercially available scanning acoustic microscopes use acoustic waves with frequencies as high as 2 GHz. For more information on SAM, see Briggs (1992).

The SLAM uses unfocused plane acoustic waves (usually in the range 100–500 MHz) to insonify the specimen. Internal elastic microstructures or other acoustic inhomogeneities in the specimen cause the propagating sound wave to locally scatter, distort, or attenuate. The sound field then impinges on and deforms a flat, polished area of the specimen surface producing a surface pattern reflective of the specimen inhomogeneity. A laser beam reflected from this surface is rastered to trace out an optical phase replica of the sound field at the surface. The laser signals are electronically processed and displayed on a television monitor to produce a visual image of the surface pattern. For more information on the SLAM, see Kessler and Yuhas (1979).

7.8.3.2 Scanning Electron Acoustic and Photoacoustic Microscopes

In scanning electron acoustic microscopy (SEAM), the signal generation source is derived by chopping or modulating the electron beam (most often in the range 0.1–1.0 MHz) in a scanning electron microscope (SEM) by means of electrostatic deflector plates inserted into the electron beam column of the SEM. The deflector plates are driven by a periodic signal (typically a square wave) of frequency f derived from a function generator. The chopped electron beam is focused on the specimen surface. The electrons lose their energies by means of atomic collision processes in a scattering interaction volume as discussed in Section 7.6.2.4, resulting in thermal and acoustic waves of the same frequency being generated simultaneously in the interaction volume. The acoustic waves are detected by a transducer (typically a piezoelectric element) attached to the specimen. The signals from the transducer are commonly processed by means of a lock-in amplifier that derives its reference signal from the modulation signal of the function generator. The signal from the lock-in amplifier is differentially amplified and used to drive the z axis of a cathode ray tube (CRT) display. As the chopped electron beam rasters the specimen surface, an electron-acoustic image is generated on the CRT display.

The acoustic SEAM signal output $V(f)$ may be approximated for isotropic materials as $V(f) = \alpha_T(3\lambda + 2G)R(f)/(\rho_0 C_H \kappa)^{1/2}$, where α_T is the thermal expansion coefficient, C_H is the specific heat, κ is the thermal conductivity, and ρ_0 is the mass density of the material (Qian and Cantrell, 1989). $R(f)$ is a function of the chopping frequency that collectively includes those factors not directly involving thermoelastic parameters. We see from this expression that the electro-acoustic signal is primarily governed by the thermoelastic properties of the material. The resulting SEAM images thus reflect variations in the thermoelastic parameters of the material and allow variations in material properties affecting the thermoelastic parameters to be imaged. Because the acoustic signal is derived from the electron interaction volume, both surface and subsurface features including residual stresses can be imaged with the SEAM, as illustrated in Figure 7.11. For further information on SEAM, see Cargill (1988).

The photoacoustic microscope uses intensity-modulated light rather than an electron beam to generate the acoustic signals. The acoustic signals are generally detected by a piezoelectric transducer bonded to the sample or by various optical methods. The original photoacoustic arrangement consists of a specimen enclosed in a gas-filled cell. Periodic heat flow and thermal expansion of the specimen leads to a corresponding pressure variation of the gas in the cell, which is detected by a microphone. The acoustic signals generated in the photoacoustic microscope are primarily governed by the thermoacoustic properties of the material in a manner similar to that of the SEAM. Measurements of the photoacoustic signal amplitude as a function of the light modulation frequency leads to a materials characterization modality known as *photoacoustic spectroscopy*. Among the uses of photoacoustic spectroscopy are the measurements of optical absorption and spectroscopic depth profile analyses of opaque materials. For more information on the phenomenological aspects of photoacoustic microscopy, see McDonald and Wetsel (1988) and for information on photoacoustic spectroscopy, see Sawada and Kitamori (1988).

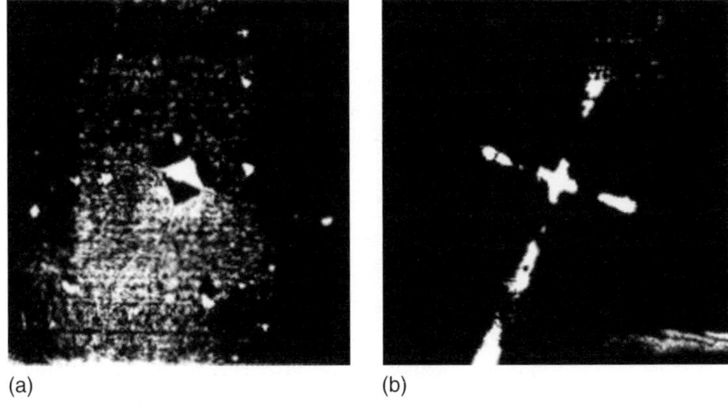

Fig. 7.11 Vickers indentation in reinforced alumina after a load of 15 kg has been applied and removed: (a) scanning electron microscope (SEM) secondary electron image showing surface cracks and deformation. (b) Scanning electron acoustic microscope (SEAM) image showing additionally the presence of residual stresses and subsurface cracks. From Cantrell *et al.*, 1990.

7.8.3.3 **Acoustic-Atomic Force Microscopes**

Acoustic-atomic force microscopy (A-AFM) is a dynamic implementation of atomic force microscopy (AFM) (Bennig, Quate, and Gerber, 1986) that combines the attributes of AFM with the bulk propagation properties of ultrasound to produce images of near surface and subsurface features within the material at the nanoscale. A-AFM includes such modalities as atomic force acoustic microscopy (AFAM) (Rabe *et al.*, 2002), force modulation microscopy (FMM) (Maivald *et al.*, 1991), ultrasonic force microscopy (UFM) (Kolosov and Yamanaka, 1993), heterodyne force microscopy (HFM) (Cuberes *et al.*, 2000), and resonant difference-frequency atomic force ultrasonic microscopy (RDF-AFUM) (Cantrell, Cantrell, and Lillehei, 2007). All modalities require that the AFM cantilever tip engage the sample surface, as indicated in Figure 7.12. Upon engagement, the cantilever tip encounters an interaction force that in general varies nonlinearly with the tip-surface separation distance. In hard tip-surface contact, the force-separation curve is linear to a good approximation. In soft tip-surface contact, the force-separation curve is extremely nonlinear at certain tip-separation distances.

Two basic equipment arrangements are used for A-AFUM depending on the modality. The simplest arrangement is that used for AFAM and FMM and is shown in Figure 7.12, where the indicated switches are in the open positions. AFAM and FMM utilize ultrasonic waves transmitted into the material by a transducer attached to the bottom of the sample. After propagating through the bulk of the material, the wave impinges on the sample top surface where it excites the engaged AFM cantilever. For AFAM and FMM, the cantilever tip is set to engage the sample surface in hard contact corresponding to the roughly linear interaction region of the force-separation curve. The response of the excited cantilever is monitored by a laser beam reflected from the cantilever

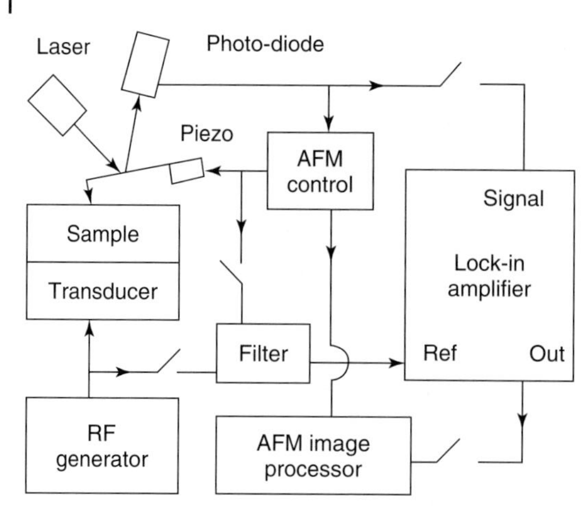

Fig. 7.12 Acoustic-atomic force microscope equipment configuration: switches open for AFAM, FMM, and UFM: switches closed for HFM and RDF-AFUM. Adapted from Cantrell, Cantrell, and Lillehei, 2007.

surface and detected by a photodiode. The signal from the photodiode is processed by the AFM image processor. The image processor and AFM controller are used to generate a raster image of subsurface, nanoscale, material features resulting from changes in the A-AFM signal produced by the features. The image may be an amplitude or phase-generated image depending on whether it is the change in the amplitude or the phase of the ultrasonic wave propagating through the material that is measured. The basic equipment arrangement used for UFM is the same as that for AFAM and FMM, except that the cantilever tip for UFM is set to engage the sample surface in soft contact, corresponding to the highly nonlinear interaction region of the force-separation curve. The UFM output response is a static or "dc" signal resulting from the interaction nonlinearity.

The equipment arrangement used for HFM and RDF-AFUM is shown in Figure 7.12, where the indicated switches are in the closed positions. In a manner similar to UFM, HFM and RDF-AFUM employ ultrasonic waves launched from the bottom of the sample, while the AFM cantilever tip engages the sample top surface in soft (nonlinear regime) contact. However, in contrast to UFM, the cantilever is also driven in the HFM and RDF-AFUM modalities. In RDF-AFUM, the cantilever drive frequency is set to differ from the ultrasonic frequency by one of the contact resonance frequencies of the cantilever. In HFM, the difference frequency is generally set to a value well below the lowest contact resonance of the cantilever. As shown in Figure 7.12, the cantilever drive and transducer drive signals are split and fed to a mixer. The mixer output signal, consisting of sum- and difference-frequency signals, is sent to the reference input of a lock-in amplifier that, because of its limited bandpass, filters out the sum frequency. The mixing of the oscillating cantilever and the ultrasonic wave in the nonlinear region defined by the cantilever tip-sample surface interaction force generates difference-frequency oscillations in the cantilever. The AFM photodiode signal, derived from the cantilever response at all frequencies, is sent to the signal input of the lock-in amplifier where all frequencies except the difference-frequency are filtered out. The lock-in amplifier measures both the amplitude and phase of the input difference-frequency signal. Variations in the amplitude and phase of the ultrasonic bulk wave due to the presence of subsurface nano/microstructures as well as variations in near surface material parameters affect the amplitude and phase of the difference-frequency signal. These variations are used to create spatial mappings generated by near surface and subsurface structures.

For a detailed mathematical analysis of signal generation, image contrast, and the relation of image generation to the physical properties of materials for each of the above-cited A-AFM modalities, see Cantrell and Cantrell (2008).

7.8.4 Underwater Acoustics

Because of the relative opacity of water to electromagnetic signals in the visible, infrared, radio, and radar ranges of frequencies, sound waves are used as the means of propagating information in underwater environments. The development of underwater sound systems has been among the most long-standing applications of ultrasound. Among the many developments are ultrasonic systems for both passive and active sound navigation ranging (sonar), depth sounding, mapping and profiling of the ocean floor, underwater geophysical prospecting, and fish detection.

Glossary

Acoustic Impedance: Ratio of the pressure to the particle velocity of the acoustic wave field.

Affinity: A measure of the ability of a reaction to occur and defined as the sum of the products of the chemical potential for a reaction and the number of moles of reactant–product involved.

Attenuation: The loss of acoustic amplitude (or energy) of a sound wave along its propagation path resulting from all mechanisms responsible for such losses, including but not limited to absorption, scattering, and diffraction.

Absorption: The loss of amplitude (or energy) of an acoustic wave that results in an increase in temperature (however slight) in the propagation medium.

Diffraction: Wave scattering occurring at the "edges" or boundaries of objects, including the influence of boundaries on radiation sources.

Dispersion: The variation of phase velocity with wavelength or wavenumber.

Elasticity: A material property characterized by a thermodynamically reversible relationship between stress and strain.

Evanescent Wave: A nonpropagating wavelike disturbance periodic only in time and decaying exponentially in all spatial directions.

Group Velocity: The velocity associated with the propagation of wave energy and defined as the derivative of the angular frequency with respect to the wavenumber.

Inhomogeneous Wave: A sound wave having an amplitude that decays exponentially in a direction perpendicular to the direction of propagation.

Phase Velocity: The sound velocity of a monochromatic wave defined as the ratio of the angular frequency to the wavenumber.

Refraction: The deviation in the propagation direction of a sound wave as the wave crosses the boundary at nonnormal incidence between two propagation media having different sound velocities.

Relaxation: A reaction process whereby the energy state of the material is caused to change from one state of equilibrium to another state of equilibrium by the driving action of an ultrasonic wave.

Relaxation Time: A time constant characterizing the time exponential approach to equilibrium following a sudden deviation from equilibrium of a relaxing medium.

Scattering: That phenomenon whereby an inhomogeneity in the path of a sound wave produces secondary waves spreading out from the inhomogeneity in various directions.

Viscoelastic Material: A material possessing both elastic and viscous properties.

Viscosity: A material property characterized by the dependence of the stress on the strain rate.

References

Anderson, O. (**1992**) *J. Acoust. Soc. Am*, **91**, 2245–2253.

Ashcroft, N.W. and Mermin, N.D. (**1976**) *Solid State Physics*, Saunders, Philadelphia.

Auld, B.A. (**1990**) *Acoustic Fields and Waves in Solids*, Kreiger Publishing Company, Malabar.

Bauer, H.J. (**1965**) in *Physical Acoustics*, vol. **IIA** (ed. W.P. Mason) Academic, New York, pp. 47–131.

Bennig, G., Quate, C.F., and Gerber, Ch. (**1986**) *Phys. Rev. Lett.*, **56**, 930–933.

Benson, G.C. and Kiyohara, O. (**1974**) *J. Acoust. Soc. Am.*, **55**, 184.

Berlincourt, D.A., Curran, D.R., and Jaffe, H. (**1964**) in *Physical Acoustics*, vol. **IA** (ed. W.P. Mason), Academic, New York, pp. 170–267.

Berry, B.S. and Nowick, A.S. (**1966**) in *Physical Acoustics*, vol. **IIIA** (ed. W.P. Mason), Academic, New York, pp. 1–42.

Beyer, R.T. and Letcher, S.V. (**1969**) *Physical Ultrasonics*, Academic, New York.

Bolef, D.I. and Miller, J.G. (**1971**) in *Physical Acoustics*, vol. **VIII** (eds. W.P. Mason and R.N. Thurston), Academic, New York, pp. 96–201.

Breazeale, M.A., Cantrell, J.H., and Heyman, J.S. (**1981**) in *Methods of Experimental Physics*, vol. **19** (ed. P.D. Edmonds), Academic, New York.

Brekhovskikh, L.M. (**1980**) *Waves in Layered Media*, Academic, New York.

Briggs, A. (**1992**) *Acoustic Microscopy*, Clarendon, Oxford.

Cantrell, J.H. and Breazeale, M.A. (**1977**) *J. Acoust. Soc. Am.*, **61**, 403–406.

Cantrell, J.H. and Cantrell, S.A. (**2008**) *Phys. Rev. B*, **77**, 165409.

Cantrell, S.A., Cantrell, J.H., and Lillehei, P.T. (**2007**) *J. Appl. Phys.*, **101**, 114324.

Cantrell, J.H., Qian, M., Ravichandran, M.V., and Knowles, K.M. (**1990**) *Appl. Phys. Lett.*, **57**, 1870–1872.

Cantrell, J.H. and Yost, W.T. (**1989**) *Rev. Sci. Instrum.*, **60**, 487–488.

Cantrell, J.H. and Yost, W.T. (**1997**) in *Encyclopedia of Acoustics*, vol. **2** (ed. M.J. Crocker), John Wiley & Sons, Inc., New York, pp. 629–637.

Cargill, G.S. (**1988**) in *Physical Acoustics*, vol. **XVIII** (eds. W.P. Mason and R.N. Thurston), Academic, New York, pp. 125–165.

Chern, E.J., Cantrell, J.H., and Heyman, J.S. (**1981**) *J. Appl. Phys.*, **52**, 3200–3204.

Claus, R.O. and Cantrell, J.H. (**1981**) *Acoust. Lett.*, **5**, 1–4.

Cuberes, M.T., Alexander, H.E., Briggs, G.A.D., and Kolosov, O.V. (**2000**) *J. Phys. D Appl. Phys.*, **33**, 2347–2355.

Demarest, H.H. Jr (**1971**) *J. Acoust. Soc. Am.*, **49**, 768.

Frost, H.M., Sethares, J.C., and Szabo, T.L. (**1977**) *J. Appl. Phys.*, **48**, 52.

Heyman, J.S. (**1978**) *J. Acoust. Soc. Am*, **54**, 243–249.

Hutchins, D.A. (**1988**) in *Physical Acoustics*, vol. **XVIII** (eds. W.P. Mason and R.N. Thurston), Academic, New York, pp. 21–118.

Inaba, R., Tokushima, A., Kawasaki, O., Ise, Y., and Yoneno, H. (**1987**) Proceedings of the IEEE Ultrasonics Symposium, Denver, Colorado, pp. 747–756.

Kelvin, L. (**1875**) *Encyclopedia Britannica*, 7th edition, p. 796.

Kessler, L.W. and Yuhas, D.E. (**1979**) *Proc. IEEE*, **67**, 526–536.

Kino, G.S. (**1987**) *Acoustic Waves*, Prentice-Hall, Englewood Cliffs.

Klein, W.R., Cook, B.D., and Mayer, W.S. (**1965**) *Acoustica*, **15**, 67–74.

Kolosov, O. and Yamanaka, K. (**1993**) *Jpn. J. Appl. Phys*, **32**, L1095–L1098.
Maivald, P., Butt, H.J., Gould, S.A., Prater, C.B., Drake, B., Gurley, J.A., Elings, V.B., and Hansma, P.K. (**1991**) *Nanotechnology*, **2**, 103–106.
Maxwell, J.C. (**1867**) *Philos. Trans. R. Soc. Lond.*, **157**, 49.
McDonald, F.A. and Wetsel, G.G.Jr (**1988**) in *Physical Acoustics*, vol. **XVIII** (eds. W.P. Mason and R.N. Thurston), Academic, New York, pp. 167–277.
McSkimin, H.J. (**1964**) in *Physical Acoustics*, vol. **IA** (ed. W.P. Mason), Academic, New York, pp 271–417.
Meridith, D.J., Watts-Tobin, R.J., and Dobbs, E.R. (**1969**) *J. Acoust. Soc. Am.*, **45**, 1393–1401.
Migliori, A., Sarrao, J.L., Visscher, W.M., Bell, T.M., Lei, M., and Fisk, Z. (**1993**) *Phys. B*, **183**, 1.
Morse, P.M. and Ingard, K.U. (**1968**) *Theoretical Acoustics*, McGraw-Hill, New York.
Murnaghan, F.D. (**1951**) *Finite Deformation of an Elastic Solid*, John Wiley & Sons, Inc., New York.
Nye, J.F. (**1960**) *Physical Properties of Crystals*, Oxford University Press, Oxford.
O'Donnell, M., Busse, L.J., and Miller, J.G. (**1981**) in *Methods of Experimental Physics*, vol. **19** (ed. P.D. Edmonds), Academics, New York, pp. 29–65.
Ohno, I. (**1976**) *J. Phys. Earth*, **24**, 355.
Papadakis, E.P. (**1968**) in *Physical Acoustics*, vol. **IVB** (ed. W.P. Mason), Academic, New York, pp. 269–328.
Papadakis, E.P. (**1976**) in *Physical Acoustics*, vol. **XII** (eds. W.P. Mason and R.N. Thurston), Academic, New York, pp. 277–374.
Pierce, A.D. (**1989**) *Acoustics*, Acoustical Society of America, Woodbury.
Pollard, H.F. (**1977**) *Sound Waves in Solids*, Pion, London.
Qian, M. and Cantrell, J.H. (**1989**) *Mater. Sci. Eng.*, **A122**, 57–64.
Rabe, U., Amelio, S., Kopychinska, M., Hirsekorn, S., Kempf, M., Goken, M., and Arnold, W. (**2002**) *Surf. Interface Anal.*, **33**, 65–70.
Reimer, L. (**1979**) *Scan. Electron Microsc.*, **2**, 111–124.
Sabine, H. and Cole, P.H. (**1971**) *Ultrasonics*, **9**, 103.
Sawada, T. and Kitamori, T. (**1988**) in *Physical Acoustics*, vol. **XVIII** (eds. W.P. Mason and R.N. Thurston), Academic, New York, pp. 347–401.
Scruby, C.B., Jones, K.R., and Anatoniazzi, L. (**1987**) *J. Nondestruct. Eval.*, **5**, 145–156.
Swift, G.W. (**1995**) *Phys. Today*, 22–28.
Thompson, R.B. (**1977**) Proceedings of the IEEE Ultrasonics Symposium, Phoenix, Arizona, p. 74.
Thompson, R.B. (**1990**) in *Physical Acoustics*, vol. **XIX** (eds. R.N. Thurston and A.D. Pierce), Academic, New York, pp. 157–200.
Thurston, R.N. (**1964**) in *Physical Acoustics*, vol. **IA** (ed. W.P. Mason), Academic, New York, pp. 2–109.
Truell, R., Elbaum, C., and Chick, B.B. (**1969**) *Ultrasonic Methods in Solid State Physics*, Academic, New York.
Überall, H. (**1997**) in *Encyclopedia of Acoustics*, vol. **1** (ed. M.J. Crocker), John Wiley & Sons, Inc., New York, pp. 55–67.
Voigt, W. (**1892**) *Ann. Phys.*, **47**, 671.
Wallace, D.C. (**1970**) in *Solid State Physics* (ed. H. Ehrenreich, F. Seitz, and D. Turnbull), Academic, New York, pp. 301–404.
Whitham, G.B. (**1974**) *Linear and Nonlinear Waves*, Wiley-Interscience, New York.
Yost, W.T., Cantrell, J.H., and Kushnick, P.W. (**1992**) *J. Acoust. Soc. Am.*, **91**, 1456–1468.
Visscher, W.M., Migliori, A., Bell, T.M., and Reinert, R.A. (**1991**) *J. Acoust. Soc. Am.*, **90**, 2154–2162.
Wagner, J.W. (**1990**) in *Physical Acoustics*, vol. **XIX** (eds. R.N. Thurston and A.D. Pierce), Academic, New York, pp. 201–266.
Zener, C. (**1948**) *Elasticity and Anelasticity of Metals*, University of Chicago Press, Chicago.

Further Reading

Bhatia, A.B. (**1967**) *Ultrasonic Absorption*, Clarendon, Oxford.
Crocker, M.J. (ed.) (**1997**) *Encyclopedia of Acoustics*, vols. **1–4**, John Wiley & Sons, Inc., New York.
Edmonds, P.D. (ed.) (**1981**) *Methods of Experimental Physics*, vol. **19**, Academic, New York.
Landau, L.D. and Lifshitz, E.M. (**1970**) *Theory of Elasticity*, Pergamon, Oxford.
Mason, W.P. and Thurston, R.N. (eds) (**1964–1988**) (eds. R.N. Thurston, A.D.

Pierce), (**1992–1995**), *Physical Acoustics*, Academic, New York. Each volume contains relevant information.

McDicken, N. (**1991**) *Diagnostic Ultrasonics*, Churchill Livingstone, New York.

Pierce, A.D. (**1989**) *Acoustics*, Acoustical Society of America, Woodbury.

Ziskin, M.C. and Lewin, P.A. (eds) (**1992**) *Ultrasonic Exposimetry* CRC Press, Boca Raton.

8
Measurement of Basic Electromagnetic Quantities

Bryan P. Kibble

Handbook of Metrology. Edited by Michael Gläser and Manfred Kochsiek

ISBN: 978-3-527-40666-1

8.1 Introduction

Electromagnetism deals with the relationships between two fields, electric and magnetic. Therefore, two independent electromagnetic units must be defined in terms of the meter, kilogram, and second to ensure that the electrical units are part of the International System of Units (SI) that are now used universally for practically all scientific and engineering measurements. The two necessary connections between the electrical and the mechanical units are made at present by

1. implicitly defining mechanical and electrical power to be equal by using the same unit (watt) for both; and
2. explicitly defining the ampere as "the constant current which, if maintained in two straight parallel conductors of infinite length and negligible circular section placed 1 m apart in vacuum, will produce a force equal to 2×10^{-7} N m^{-1} between these conductors." (This definition of the ampere is equivalent to defining the magnetic permeability of free space μ_0 as $4\pi \times 10^{-7}$ H m^{-1} exactly.)

These abstract definitions have to be put into practice (or realized, in the jargon of metrology). Realizations are carried out by more than one method and in more than one national measurement laboratory to secure confidence in their correctness. To realize an electrical unit, it is necessary to assume the complete accuracy of Maxwell's electromagnetic theory and to use an apparatus in a way that is, according to this theory, consistent with the definitions of the watt and the ampere. Up to about 40 years ago, the two apparatuses that realized the two independent electrical units were

1. some form of current balance in which the force between circular current-carrying coils was weighed as a direct extension of the ampere definition; and
2. a mutual inductor whose carefully measured dimensions established a link between the meter and the ohm.

Neither the current balance nor the mutual inductor is accurate enough for the present-day routine needs of science and commerce at the highest level of accuracy of a part per million (ppm) or better. They have been replaced by electrometers or watt balances that realize the volt or watt and the Thompson–Lampard calculable capacitor that realizes the ohm via the farad (Thompson and Lampard, 1956; see also Section 8.2.1). An alternative route for the ohm is via a measurement of the fine-structure constant and the quantum Hall effect (see Section 8.2.1.2).

Handbook of Metrology. Edited by Michael Gläser and Manfred Kochsiek

ISBN: 978-3-527-40666-1

The correctness of these realizations has been confirmed by examining the relationships between those fundamental physical constants whose measured values involve mass, length, and time as well as electrical quantities.

Once they have been realized by these difficult and time-consuming measurements (each of which may take many years), the electrical units are preserved for everyday use in national standardizing laboratories by ascribing values to the constants that govern the Josephson (1962) and quantum Hall effects (von Klitzing, Dorda, and Pepper, 1980). These precise effects of quantum physics, which occur at cryogenic temperatures, reproduce voltages and resistances very exactly (to 0.005 ppm at present).

Values of standards to represent other derived, electrical units must then be determined in terms of the ohm and volt. There is usually only one preferred method of establishing each derived unit. This method yields the most accurate measurement in volts and ohms of an artifact standard representing the derived unit.

At present, the stability and reproducibility of volt and ohm units preserved by means of the Josephson and quantum Hall effects greatly exceed the accuracy with which these units can be realized in terms of the meter, kilogram, and second. This situation suggests a possible future scenario: The present link to the meter, kilogram, and second could be replaced by a pair of basic electrical units (voltage defined in terms of the Josephson effect and impedance in terms of the quantum Hall effect) and then the kilogram could be derived from these. μ_0 would become an experimentally measurable physical constant, rather than a defined quantity, and the present prototype kilogram, which is the last unit based on a material artifact, would become a maintained standard whose value would have to be realized either from these new definitions of the volt and ohm via a watt balance or via a measurement of Avogadro's constant (see Section 8.2.2.3).

At the time of writing, the General Conference on Weights and Measures, which recommends changes to the SI for worldwide acceptance, is actively seeking to change the basis of the SI along these lines with a target date of 2011 (Mills *et al.*, 2006).

8.2
Realizing the Base Units

8.2.1
The Farad, Realized by a Calculable Capacitor

The Thompson–Lampard theorem (Thompson and Lampard, 1956) applied to a system of four electrodes of constant cross section in vacuum (Figure 8.1) states that the cross capacitances C_1 and C_2 per meter length are related by the expression

$$\exp(-\pi C_1/\varepsilon_0) + \exp(-\pi C_2/\varepsilon_0) = 1 \tag{8.1}$$

If C_1 is nearly equal to C_2(i.e., $C_1 - C_2 = \Delta C$), this simplifies to give the mean cross capacitance $C = (C_1 + C_2)/2$ as

$$\begin{aligned} C &= \left[\epsilon_0(ln2)/\pi\right]\left[1 + 0.086(\delta C/C)^2\right] F/m, \\ &= \left[(ln2)/\pi\mu_0 c^2\right]\left[1 + 0.086(\delta C/C)^2\right] \\ &= 1.953\,549\,043\{1 + 0.086(\delta C/C)^2\} pF/m \end{aligned} \tag{8.2}$$

The farad can therefore be realized from a single length measurement and the defined values of μ_0 and the speed of light in vacuum c.

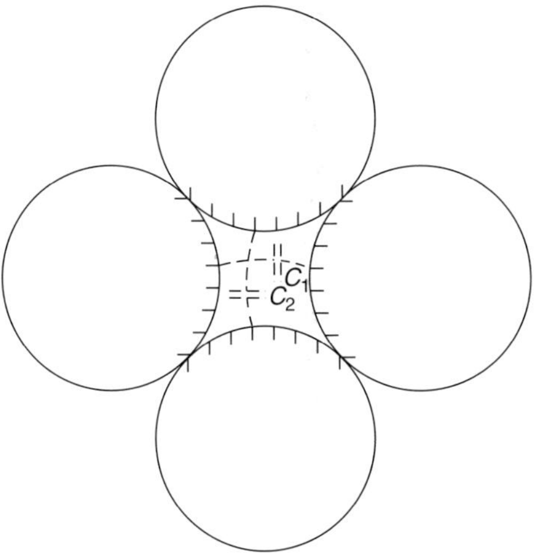

Fig. 8.1 Illustration of the Thompson–Lampard theorem, showing a cross section through four electrodes.

Usually, the cross capacitances are between the inward-facing surfaces of four cylindrical electrodes placed with their axes on the corners of a square (Figure 8.2). The small gaps between the electrodes cause negligible error. The whole structure is surrounded by a guard electrode. To avoid having to calculate the fringing fields at either end of the electrodes, guarding cylinders are inserted along the axis of the structure. The effect of the fringing fields between the ends of these guards and the cylinders is eliminated by observing the differences in the cross capacitances as the guards are moved a measured distance L toward each other. They must not approach each other enough to cause the fringing fields to overlap appreciably.

The accuracy of this realization of the farad is limited by the departure of real machined cylindrical electrodes from the constant cross section assumed by the Thompson–Lampard theorem. The fringing fields can be shaped by attaching extensions of lesser diameter to the ends of the inserted guards to reduce this limitation, but it is impossible to eliminate entirely the effect of changes in cross section over a length comparable with these extensions. The accuracy with which the cylinders can be made, therefore, sets a limit of the order of 0.01 ppm to the achievable accuracy in realizing the farad. A similar uncertainty arises from the small inductance of the electrodes, which causes the surfaces of the electrodes to be not quite equipotentials, because there are currents flowing along them.

8.2.1.1 The Ohm, Derived from the Farad

The ohm is obtained from the farad by scaling up the fraction of a picofarad of capacitance of a calculable capacitor to, usually, 1000 pF and then equating its impedance at a frequency of 10^4 rad s^{-1} to a resistance of 10^5 Ω, using ac bridges. This resistance is further scaled to yield the ac resistance of a 1000-Ω resistor so constructed (see Figure 8.3) that the difference in its resistance between ac and dc conditions is small and calculable.

Finally, the 1000-Ω dc SI resistance derived from a calculable capacitor by this long chain of measurements is scaled up, either by the reconfigurable resistor networks known as *Hamon resistors* (Hamon, 1954) or by a superconducting (cryogenic) current comparator (Harvey,

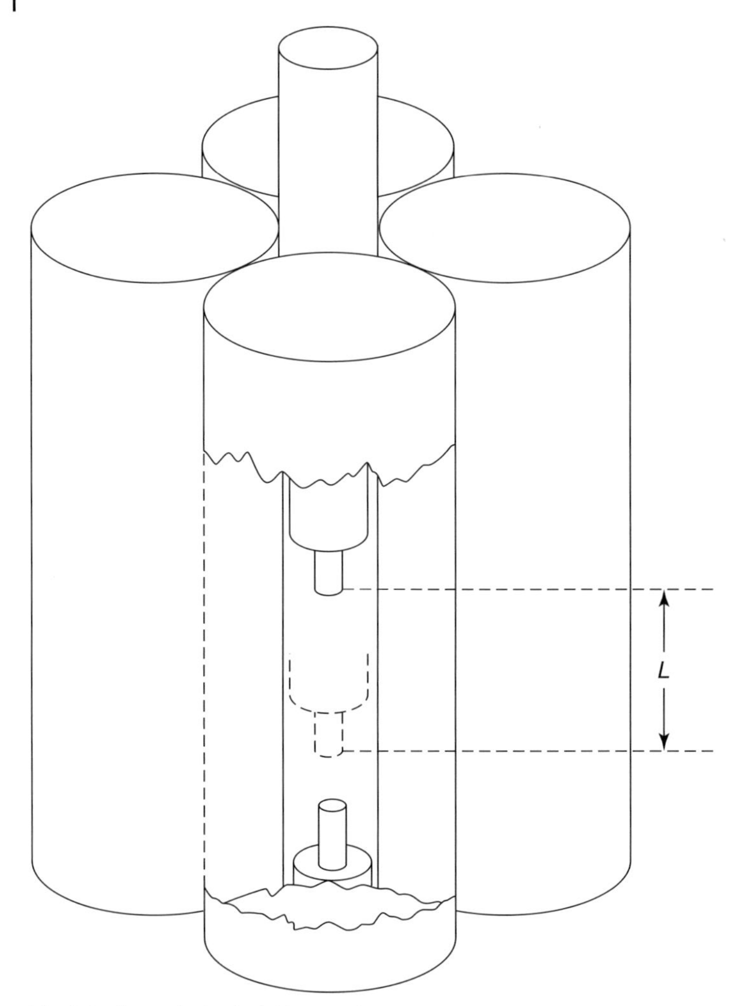

Fig. 8.2 A practical calculable capacitor.

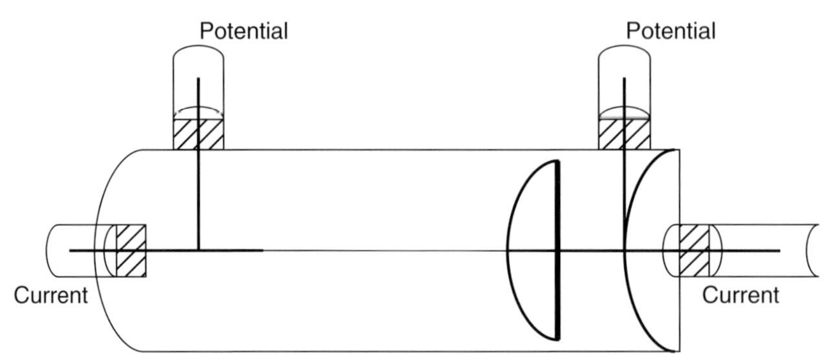

Fig. 8.3 A section through a coaxial resistor whose ac value is calculable from its geometry and dc value.

1972), to a value suitable for comparison with the quantum Hall effect.

A superconducting current comparator is also used to scale dc resistance ratios in major standardizing laboratories possessing liquid-helium cryogenic facilities to derive their maintained representation of the ohm from the quantum Hall effect. The construction and use of a superconducting current comparator is described in Section 8.3.2.

The parts-per-billion accuracy achieved in the AC measurements is only possible because the measuring conditions applying at the terminations of an ac standard are carefully defined. An example of these conditions is the four-terminal pair definition illustrated in Figure 8.4, where the *impedance of the standard* is defined as the ratio of the voltage appearing across the P_H coaxial connector when zero current flows through either inner or outer contact to the current flowing from the inner contact of the C_L coaxial connector. The measurement circuit must be so arranged that an equal and opposite current flows into the outer contact of the C_L coaxial connector, and that there is zero voltage between the inner and outer contacts of the associated P_L connector together with zero current flowing through either. The remaining C_H connector has a source connected between its inner and outer contacts, which is adjusted, together with details of the measurement circuitry, to set up these defining conditions. This complete and precise definition, while being somewhat strange at first sight, naturally leads to measuring networks ("bridges") of components joined by coaxial conductors. The definition can be readily understood by recognizing that the impedance Z in the inner circuit is a familiar four-terminal impedance having separate current and potential terminals and that the outer circuit is also a small four-terminal impedance z. The effect of the defining conditions is to make the total impedance of the device as the series combination, $Z + z$. Standard impedances defined in this way are compared in networks in which the current in the coaxial interconnecting conductor of each cable is arranged to be as nearly as possible equal and opposite to that in the outer conductor. Circuitry of this kind has the two important and related properties that the measurements are affected neither by the disposition of cables and components nor by externally generated electric and magnetic fields, including those at the measurement frequency. Further details and an introduction to the literature of this subject have been given by Kibble and Rayner (1984).

8.2.1.2 **Realizing the Ohm via Measurements of the Fine-Structure Constant**

If certain assumptions are made, the ohm may be derived directly from the quantum Hall effect. Under proper experimental

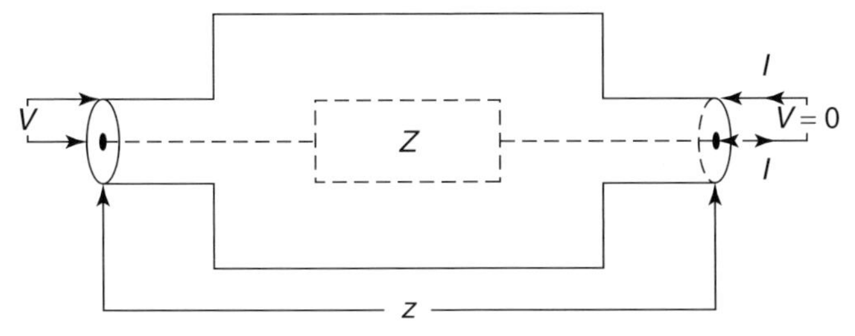

Fig. 8.4 Four-terminal pair definition of an impedance. $Z' = V/I = Z + z$.

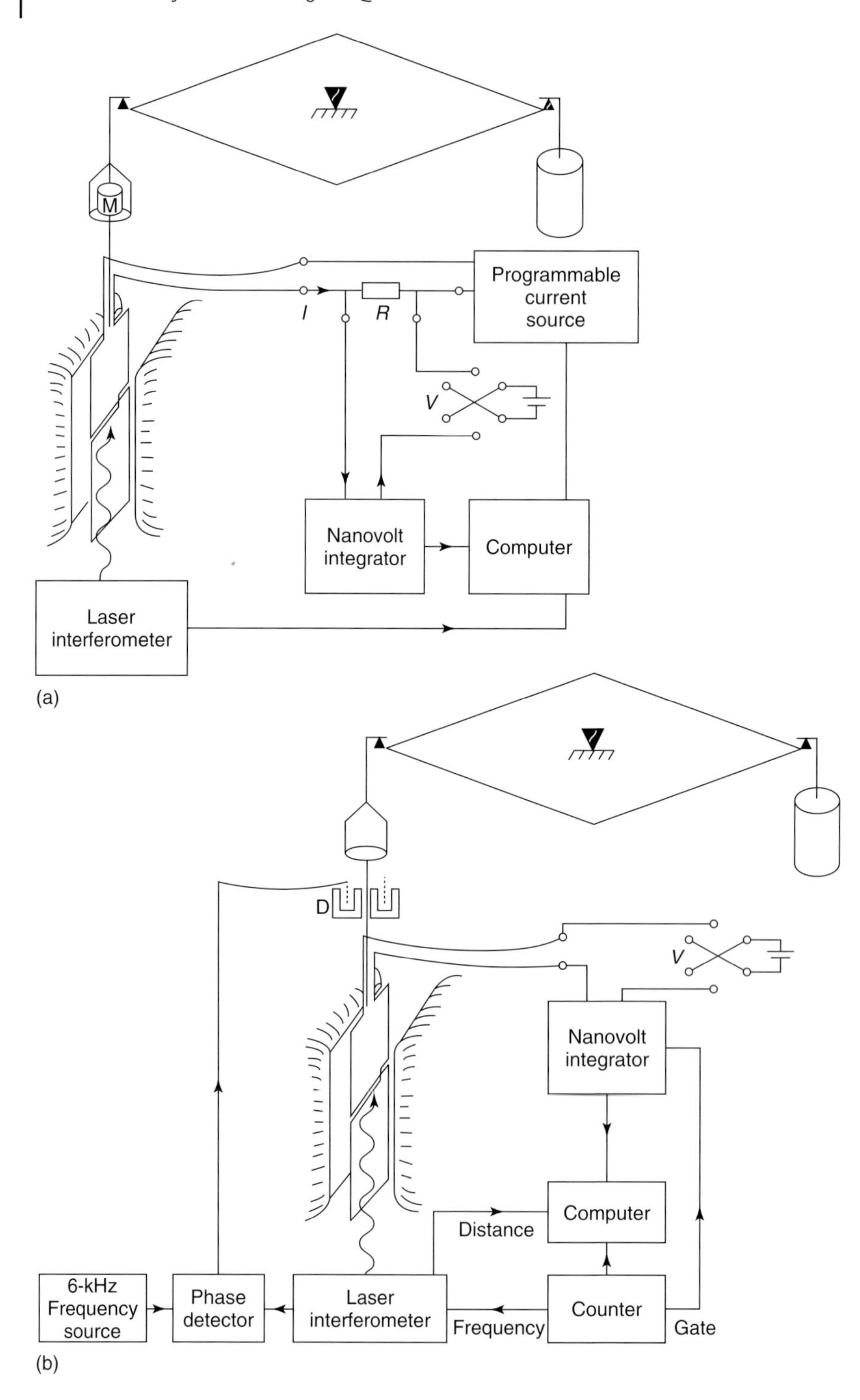

M
Programmable current source
I
R
V
Nanovolt integrator
Computer
Laser interferometer
(a)
D
V
Nanovolt integrator
Computer
Distance
6-kHz Frequency source
Phase detector
Laser interferometer
Frequency
Counter
Gate
(b)

◀ **Fig. 8.5** The moving-coil watt balance. (a) The force on a current from the programmable current source flowing in the coil is opposed by a mass M. Any difference between the voltage V of a Josephson voltage standard and the voltage across the resistor due to the current is measured by the nanovolt integrator. (b) The nanovolt integrator measures any small difference between the voltage V and the voltage induced in the coil when it is moved with constant velocity by the coil-and-magnet drive motor D.

conditions, the current-to-voltage ratio of a device exhibiting this effect is believed to be a multiple of e^2/h, where e is the elementary charge and h is the Planck constant. A measurement of the fine-structure constant, α, in SI units, $[\alpha = (\mu_0 c/2)(e^2/h)]$, would therefore constitute a realization of the SI ohm, because the values of μ_0 and c are defined. A very accurate value for α can be obtained from the measurements of the departure of the value of the magnetic moment of the electron from 2 exactly. This departure can be measured as the ratio of two frequencies and calculated (by a very long and involved calculation that requires many hundreds of hours on a powerful computer) (Kinoshita, 2006). The accuracy of the value of e^2/h, and therefore the ohm, obtained in this way is approaching 0.001 ppm, which is more accurate than that obtainable from calculable capacitors and associated bridges. However, this realization does assume that the theories of the quantum Hall effect and the "anomalous" magnetic moment of the electron are accurate to the uncertainty claimed.

8.2.2 The Volt or Watt

Realizing either of these units involves deriving a force from a mass M subject to gravitational acceleration g to balance against an electromagnetic force. g can be measured with a "free-fall" gravimeter (Niebauer *et al.*, 1995) to approximately 0.005 ppm. Mass was involved in realizations of the ampere using a current balance, and indeed this instrument inspired the definition of the ampere. However, because of experimental difficulties primarily associated with measuring the dimensions of the coils, the accuracy of a current balance was limited to about 5 ppm. The instruments described below have replaced it.

8.2.2.1 The Watt Balance

In Figure 8.5a, the force produced when a current I flows through a coil suspended from a balance so that the coil lies partly in the field of a large permanent magnet is made equal to Mg. In a separate measurement, the coil is moved vertically with velocity u and an electromotive force V is induced, as illustrated in Figure 8.5b. Making the measurements separately ensures that virtual, rather than real, power is involved and therefore that errors associated with real energy expenditure such as Joule heating in the coil are eliminated. Equality of mechanical and electrical power ensures that

$$IV = Mgu \tag{8.3}$$

If I is related to V via the voltage drop that it produces across a resistor R, then

$$V = \sqrt{MguR} \tag{8.4}$$

If the voltage is measured in terms of the Josephson effect voltage $\nu = hf/2e$ and the current I in terms of υ and the quantum Hall resistance $R = h/\,ne^2$ so that $I = f'\,ne/2$, where n is a small integer and f, f' are frequencies, the Planck constant

h can be measured (see Sections 8.3.1 and 8.3.2 below).

Neither the dimensions of the coil nor the density and distribution of the magnetic flux need to be measured. In principle, instantaneous values are required for u and V as the coil passes through the weighing position, but in practice the magnetic flux and coil can be so shaped that Vu^{-1} is almost constant and a value interpolated from measurements made over a finite distance is adequate. If the mean velocity is measured between two instants by counting optical fringes with a gated timer/counter and the mean voltage between the same two instants with a gated integrator, the correlated fluctuations in the actual velocity and voltage caused by vibration are removed from the data.

In the National Physical Laboratory (NPL) realization (Robinson and Kibble, 2007), a current of about 10 mA is controlled to keep the balance in equilibrium. The current is then reversed and a mass M of about 1 kg is placed on the same side of the balance as the coil. The knives and planes are not disturbed, so that the force caused by the current reversal is measured by substitution. Under good conditions, 0.003-ppm repeatability of the mean is obtained.

Then the coil is moved successively upward and downward some 15 times and the average value of uV^{-1} is measured with an accuracy of 0.015 ppm or better as the coil passes through the weighing position. Taking the mean for the two directions of travel eliminates extraneous voltages. The velocity and weighing parts of the realization each contribute to the total uncertainty, and the reproducibility of the result for the watt or for h is about 0.02 ppm.

Another version of this principle of watt realization is in progress at the National Institute of Standards (NIST) (Steiner *et al.*, 2007) and results accurate to 0.036 ppm have been obtained so far.

8.2.2.2 **Electrometers**

Another direct method of relating the mechanical and electrical SI units is to measure the force exerted on the electrode surface of a capacitor when a voltage V is applied to it. Clothier *et al.* (1989) measured the change in force per unit area as a hydrostatic pressure ρgh on a pool of mercury of density ρ when an electric field applied by a plate a distance d above the surface of the mercury was changed from V_1/d_1 to V_2/d_2 (see Figure 8.6a). The mercury pool is surrounded by a moat connected electrically to a guard electrode above it, to establish the unperturbed mercury level corresponding to $V = 0$ from which the height h of the surface of the mercury is raised and is measured. The density of mercury, ρ, was known from the measurements of Cook (1961) at NPL. No capacitance measurements were necessary, because the change in voltage may be calculated from the equation

$$V_2/\sqrt{h_2} - V_1/\sqrt{h_1} = \left(2\rho g/\varepsilon\right)\left(d_2 - d_1\right) \tag{8.5}$$

where $V_{1,2}$ and $d_{1,2}$ are so chosen that $h_1 \approx h_2$; ε is the relative permittivity of the residual vapor above the mercury surface. h was only about 0.6 mm (equivalent to 2000 optical fringes), but nevertheless this realization reached an accuracy of ± 0.32 ppm.

Other realizations of similar accuracy have involved an electrode suspended from a balance, as shown in Figure 8.6b. The force F_x and capacitance changes resulting from the displacement of a second electrode by a measured distance yield V in SI volts, because

$$V^2 = \frac{2F_x}{\partial C/\partial x} \tag{8.6}$$

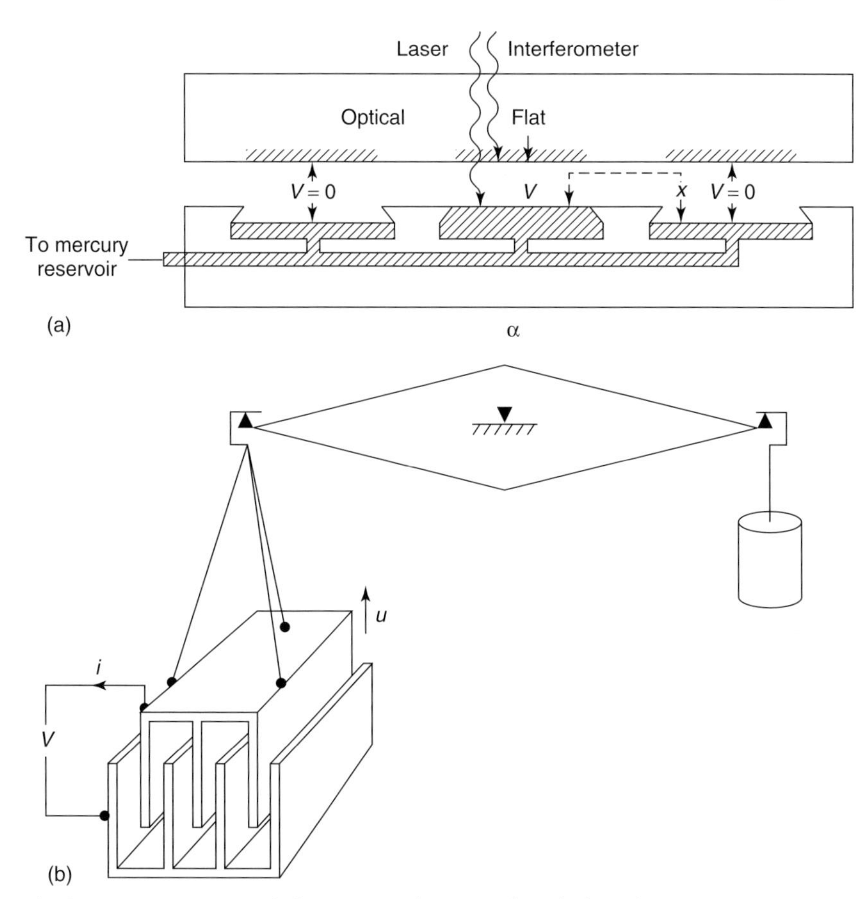

Fig. 8.6 (a) A mercury-pool electrometer. (b) An interleaved plate electrometer.

$\partial C/\partial x$ is obtained either from the dimensions of the apparatus by calculation (Elnekave, Fau, and Delahaye, 1976) or by measurement of capacitance (Bego, 1988; Funck and Sienknecht, 1988), in terms of a calculable capacitor. Because a watt balance seems to offer better accuracy, work on electrostatic balances has ceased for the present but it is also possible to proceed via the virtual work principle that lies behind the watt balance realization. The magnitudes of the quantities involved are less favorable than in a watt balance apparatus, and a measurement having useful accuracy has yet to be made by this technique.

8.2.2.3 **Defining the Kilogram via Avogadro's Constant**

A seemingly straightforward way to define the kilogram in terms of a fundamental constant would be to specify it as the mass of a (very large) number of atoms. Counting such a large number can be accomplished if the atoms are arranged in a crystal whose structure is as nearly perfect as possible (Becker, 2003). The refining techniques that have been developed for silicon make it the best candidate at present. The number of atoms in an almost perfectly spherical artifact of silicon can be calculated from the distance between atoms and the mean diameter of the sphere. The distance between atoms is measured by a device in which a laminar plate is moved a measured distance between two fixed plates while the passage of planes of atoms is counted by observing Moiré fringes in the passage of a collimated X-ray beam that passes perpendicularly through the plates. Obtaining and maintaining subatomic alignment of the planes while translating the middle plate through a few millimeters demands the ultimate precision from strip-hinge technology.

This atom-counting procedure constitutes a measurement of Avogadro's constant, N_A which is connected to Planck's constant, and therefore to the watt balance approach to the kilogram, by the expression

$$N_A = c_0 \alpha^2 A_r(\mathrm{e}) M_u / (2R_\infty h) \tag{8.7}$$

where the other constants, c_0, the velocity of light in free space, α, the fine-structure constant, $A_r(e)$, the relative atomic mass of the electron, M_u, the molar mass constant and R_∞, the Rydberg constant, have either defined or more accurately known values.

There are two major problems in this approach. First, the mass of a silicon atom is not unique because naturally occurring silicon consists of three isotopes in ratios of approximately 92 : 5 : 3. Determining the ratio of these amounts accurately enough is very difficult, but work is in progress to create artifacts from crystals consisting almost entirely of a single isotope. Second, the surface of a silicon sphere is covered with a layer a few atoms thick of silicon oxides and the thickness of this layer has to be accurately measured by ellipsometry to account for the mass discrepancy it produces.

At the time of writing (2007), there is up to one part-per-million discrepancy between the silicon crystal and watt balance approaches to redefining the kilogram, but it is hoped that this discrepancy will be resolved in the next few years.

8.2.2.4 Self-Consistency of the SI

The continuing progress toward basing the SI on the (presumed) unchanging values of fundamental physical constants has raised the question as to whether the von Klitzing constant is exactly h/e^2, the Josephson constant is exactly $2e/h$ and whether electron-counting experiments transport charge in elements of e exactly. Strong evidence for the validity of these assumptions would be obtained from experiments, which examine the closure of the so-called "metrological triangle" in which a current measured in counted electrons is passed through a resistance measured by the quantum Hall effect in terms of the von Klitzing constant and the resulting voltage drop measured against a Josephson effect voltage standard. That is, whether

$$(e) \times (2e/h) \times (h/e^2) = 2 \tag{8.8}$$

within experimental error. In principle, a single experimental arrangement in which electrons are counted, pass through a quantum Hall device and the resulting voltage is opposed to a Josephson junction voltage, will examine this. No other units are required provided that the radio frequencies needed to count out the electrons and generate the Josephson voltage are scaled from the same frequency source. At present, the 1-ppm accuracy with which electrons can be counted is a limitation, but this is likely to improve substantially in the future.

8.3
Working Standards of Voltage and Resistance

8.3.1
A Standard for the Volt

We can establish a representation of the SI volt from the ac Josephson effect in which microwaves of frequency f are shone on a junction formed from an insulating barrier a few nanometers thick between two superconductors (see Figure 8.7). A current of electron pairs tunneling through the barrier gives rise to a voltage $V = nf(h/2e) = nf/K_J$ across this insulated junction of superconductors, where n is a small integer, h is the Planck's constant, e is the electronic charge, and $K_J(= 2e/h)$ is the Josephson constant. A measurement of f and knowledge of n gained from the approximate value of the voltage enables V to be derived in terms of a conventional value assumed for $2e/h$. This value is so chosen that V is then expressed in terms of the SI volt as given by the best realizations available. If a single junction is used, V is only a few millivolts and has to be scaled up to 10 V for practical dissemination either by a potentiometric ratio method often involving Hamon resistors or by using a superconducting current comparator (described below in connection with the quantum Hall effect). The technology of connecting many Josephson junctions in a series array so that each is subjected to nearly the same incident microwave power has been developed (Niemeyer, Hinken, and Kautz, 1984; Hamilton *et al.*, 1985), and any multiple of 150 μV up to a few volts can be established. Hundred and fifty microvolts corresponds to the 72.5-GHz microwave radiation for which the array is designed. An array provides a greatly simplified and more accurate voltage standard than the use of a single junction. Modern arrays employ a superconductor fabrication technology, which leads to a nonhysteretic $I - V$ characteristic. The integer n, equal to the number of junctions, is stable and only

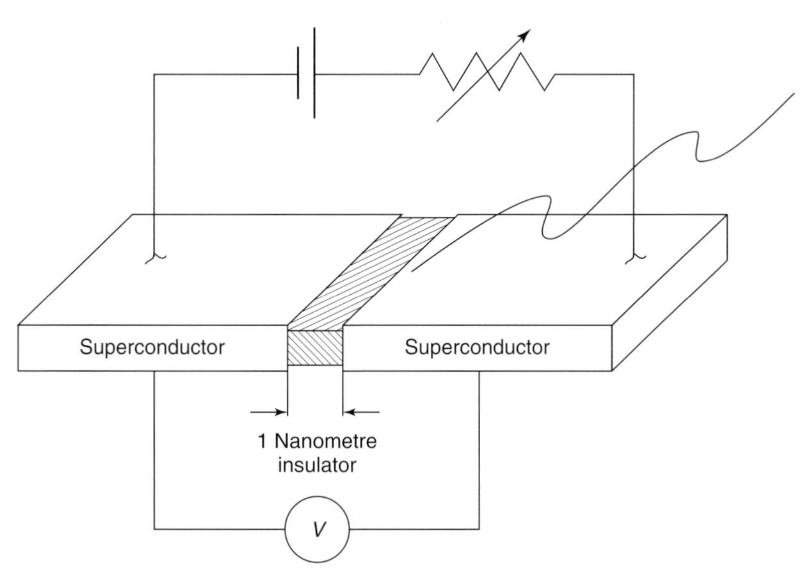

Fig. 8.7 A single Josephson junction.

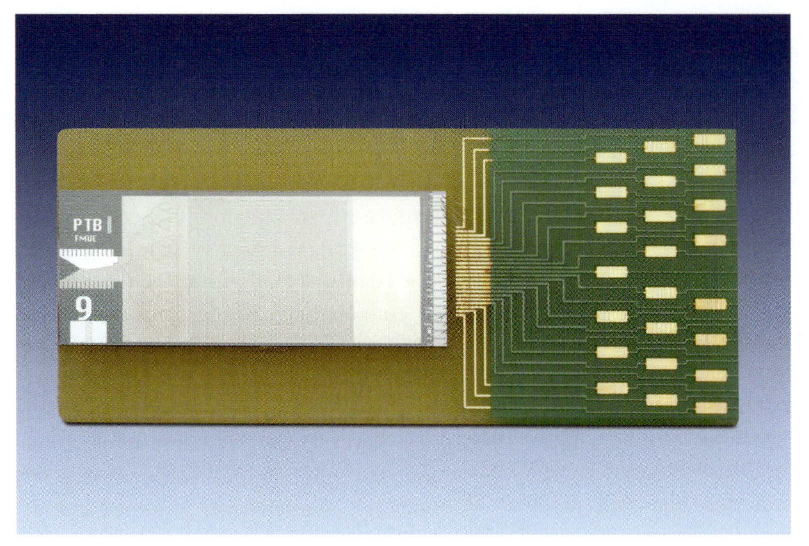

Fig. 8.8 A Josephson multijunction array.

needs to be determined once. Arrays consisting of subarrays with junction numbers forming a binary sequence can be used as programmable quantum digital-to-analog converters and provide a quantum standard for low-frequency ac voltages. (Hamilton *et al.*, 1989; Pöpel *et al.*, 1990) Figure 8.8 shows a modern 10-V array manufactured at the Physikalisch-Technische Bundesanstalt (PTB), Germany.

Using an array of Josephson junctions, calibrations reproducible to 0.01 ppm are readily achieved, the principle limitation being the variability of the thermal emfs arising from the thermal gradients along the conductors connecting the array, which is at a cryogenic temperature, to the room-temperature terminals. To eliminate thermal and other spurious emfs, both voltage sources being compared must be capable of reversal. Taking the mean of the comparison before and after reversal then eliminates all unwanted emfs, provided they do not vary significantly during this short time. A Josephson array can be made to give a voltage of either polarity by presenting its terminals with a voltage of the required magnitude and polarity while increasing the microwave power from zero.

The resolution with which voltage sources can be compared has recently been greatly increased by the availability of electronic null detectors. The noise these contribute to the measurement does not add significantly to the irreducible Johnson noise arising from source resistance. The semiconducting transistor and diode junctions in these electronic detectors are, however, prone to rectify any interference voltage appearing at their input terminals, leading to systematic errors.

8.3.2
A Standard for the Ohm

The quantum Hall effect affords precisely reproducible voltage-to-current ratios, and these constitute the best representation of the SI ohm now available. Both silicon MOSFETs and heterostructures of gallium arsenide-gallium aluminum arsenide at liquid helium temperatures provide the necessary conditions to establish a two-dimensional electron gas in a boundary layer of a few nanometers thick. When a strong ($1 - 14T$) magnetic field is applied in a direction perpendicular to this layer, the ratio of the voltage between the sides of the layer to the current flowing across the layer (i.e., the Hall "resistance" R_H) is found to take one of a number of discrete values depending on the strength of the applied magnetic field (see Figure 8.9). The peculiar quantum-mechanical nature of the device is such that any two-terminal measurement of the voltage-to-current ratio between any two points on the boundary of the layer reveals a true resistance equal to the step resistance plus that any contact resistances involved. By contrast, in a four-terminal measurement when the current and potential terminals alternate around the boundary, the voltage-to-current ratio equal to R_H is observed. The theory of the effect indicates that $R_H = h/ie^2$, where i is an integer, as the temperature approaches absolute zero for certain ranges of magnetic field values. Measurements made so far support this theory, at least to an accuracy of 0.01 ppm. Moreover, the values of R_H have been measured to be independent of the particular device used to generate them to within 0.00035 ppm.

Despite its purely superficial resemblance to the Josephson effect used as a voltage standard, the physical origins of the quantum Hall effect are quite different (Hartland, 1988), and in particular it is important to realize that the simple expression for R_H given above is exact only at the unattainable absolute zero of temperature, $T = 0$. However, fortunately, the approximation is sufficiently good for all practical purposes at temperatures between 0.3 and 4.2 K, depending on the particular device. A measure of the error involved is simply obtained by measuring the voltage drop along the direction of current flow, using extra connections made on the

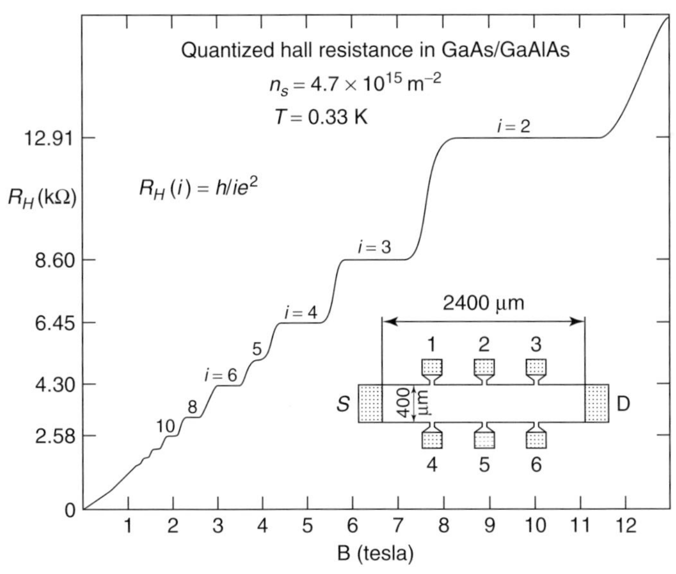

Fig. 8.9 A graph showing the variation of the voltage-to-current ratio as a function of magnetic field for the quantum Hall device shown in the inset.

same side of the layer between the current connections. This voltage drop is zero only at $T = 0$, but at finite temperatures its ratio to the perpendicular Hall voltage is of the same order of magnitude as the departure from the above expression for R_H. In practice, the temperature can usually be made low enough for the error to be less than 0.01 ppm.

The best resolution currently being obtained in relating R_H to the value of a conventional standard resistor at room temperature is a few times 0.001 ppm.

The measurement circuit used to attain the greatest accuracy is based on a superconducting current comparator (Harvey, 1972), a device whose accuracy depends on the Meissner effect of exclusion of magnetic flux from a superconductor. A long (in principle, infinite), superconducting tube is threaded by two windings of n_1 and n_2 turns carrying the two currents to be compared. The Meissner effect then ensures that any imbalance in their ampere-turns appears as a current flowing over the whole inside and outside surface of the tube so that the total current flowing through the inside of the tube is zero. The magnetic flux from the imbalance current $(n_1 I_1 - n_2 I_2)$ that flows uniformly back around the outside of the tube is independent of the positions of the windings within the tube and can be sensed with great sensitivity by a pickup coil coupled to a SQUID flux detector . A dc SQUID detector circuit contains a Josephson junction near which is a radio-frequency coil that can sense the changing impedance of the circuit when quanta of magnetic flux thread it. An ac SQUID circuit contains two Josephson junctions connected into a superconducting loop and the voltage across the loop is periodic in the number of flux quanta threading it. A dc SQUID is now more frequently employed, as it has greater sensitivity. By registering the state of the SQUID circuit when both the currents in n_1 and n_2 are zero, and maintaining this state with a feedback locking circuit as these currents are increased to their final value, flux balance from these currents is ensured so that $n_1 I_1 = n_2 I_2$. The need for an infinitely long superconducting tube is avoided by arranging its

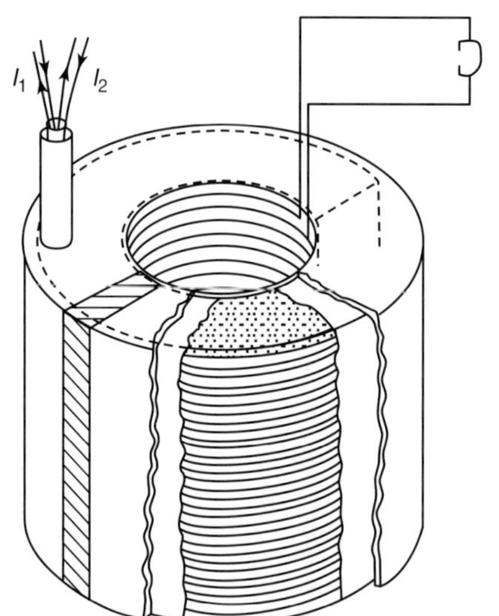

Fig. 8.10 A superconducting current comparator.

topology to be that of a snake swallowing its own tail for about three convolutions. The construction of a practical device is illustrated in Figure 8.10. The n_1 and n_2 turns carrying the currents I_1 and I_2 are wound on a bobbin, which is enclosed in a convoluted superconducting lead-and-insulation shield. The ends of the windings are taken out through a similarly constructed side tube whose layers are connected to the corresponding layers of the convoluted shield. The magnetic field caused by imbalance in the ampere-turns of the windings threads the center of the insulated bobbin where it is sensed by a coil connected to the SQUID.

The resistor and the quantum Hall device to be compared are connected to the windings of the current comparator as shown in Figure 8.11 with a direct connection between one set of potential terminals and a sensitive null voltage detector D between the others. When the voltage at the detector is nulled, either by a shim current in an auxiliary winding on the current comparator or by shunting one of the resistors with an adjustable high resistance, $I_1 R_H = I_2 R$ or $n_1/n_2 = R_H/R$.

As R_H can be equal to 25 812 Ω, 6453 Ω, and so on, 100 Ω is a good choice for the value of R_2. Hundred-ohm resistors are better than the 1-Ω resistors used hitherto as the starting point for comparing with the decade values of resistance in routine calibration work, both because of the relatively smaller effect that thermoelectric potentials have on the measurement circuit and because matching to the noise properties of null detectors is easier.

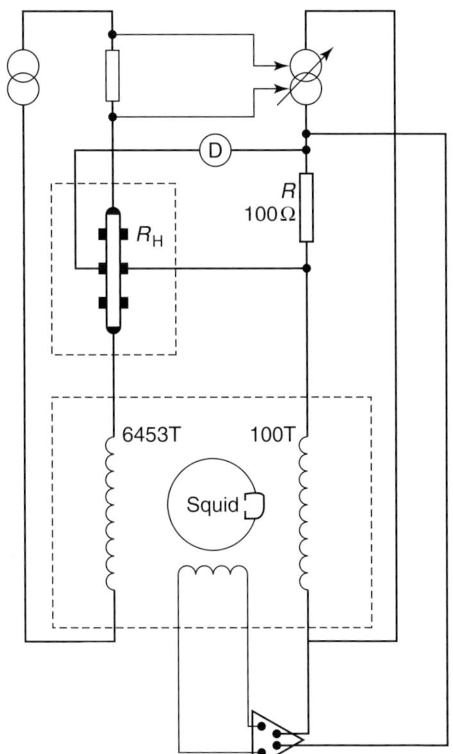

Fig. 8.11 A measurement circuit for comparing R_H with a resistor at room temperature.

8.3.3 Conventional Values Adopted for K_J R_K to Represent the SI Volt and Ohm

From 1 January 1990, the General Conference on Weights and Measures agreed to adopt exact conventional values $K_{J-90} = 483597.9\,\text{GHz}\,\text{V}^{-1}$ and $R_{K-90} = 25812.807\ \Omega$, respectively, for the Josephson and von Klitzing constants to represent the SI volt and Ohm for international conformity in ordinary measurements. It must be understood that if the actual value of K_J or R_K in SI units is required, for example, for the comparison of measured values of fundamental constants, an additional uncertainty must be ascribed to the measurements.

8.4 Other Primary Standards of National Measurement Institutes

8.4.1 dc Voltage

Other primary working voltage standards, such as Zener diodes and Weston standard cells, can be measured against an array of Josephson junctions. An array yields a representation of the SI volt by using the conventionally agreed value for the frequency-to-voltage conversion constant governing the Josephson effect. Higher and lower dc voltages are easily measured using the measured resistance ratios of resistive voltage dividers.

8.4.2 dc Resistance

Stable wire resistance standards of (usually) 100-Ω value can be calibrated directly in terms of the quantum Hall effect using the verifiably accurate current ratios created by a superconducting current comparator. Standards of other decade values can be derived either by a superconducting current comparator or by Hamon resistors (see Section 8.2.1.1), or simply by connecting 10 resistors in series to generate standards of 10 kΩ and greater. Typical accuracies of decade values derived in this way are given in Table 8.1.

8.4.2.1 The ac Behavior of Standard Resistors

An ideal resistive circuit component under ac conditions will have the potential across it exactly in phase with the current passing through it. Its resistance will be independent of the ac frequency. The value of a real resistor $R(f)$ will be its resistance at the frequency f, and will depend only slightly on f through causes such as the tendency of the current to be deflected outwards to a path of smaller cross section at the surface of the conductor by its associated magnetic field (the "skin effect") or the "proximity effect" caused by additional loss mechanisms associated with currents magnetically induced in neighboring conductors. The magnetic field of these induced currents couples back into the original conductor and generates an opposing in-phase emf.

A practical resistor will also have small self-generated modifying currents and

Tab. 8.1 Certified accuracies in parts per million of dc resistances.

Resistance (Ω)	10^{-3}	10^{-2}	10^{-1}	1	10	10^2	10^3	10^4	10^6	10^9	10^{12}
Accuracy	0.8	0.6	0.3	0.15	0.25	0.07	0.13	0.15	1.3	10	500

Tab. 8.2 Certified accuracies in parts per million of capacitances at 1 kHz.

Capacitance	1 pF	10 pF	1 nF	10 nF	1 μF	10 μF	1 mF	1 F
Accuracy	1	0.7	3	30	60	100	500	10 000

voltages at 90°-phase associated with its accompanying magnetic and electric fields. These effects are quantified as the self inductance, L, and self capacitance, C, of the resistor and their effects can, up to quite high frequencies, be described by a single parameter, the time constant τ of the resistor. The ac impedance of the resistor is

$$Z(f) = R(f) + j \times 2\pi f \tau \tag{8.9}$$

τ is equal to LR^{-1}, or $-RC$ if the self capacitance or inductance, respectively, is negligible compared to the other. τ can be as small as 10^{-9} s, particularly for resistors in the range of values from 10 to 1000 Ω when the inductive and capacitive effects tend to cancel. τ is measured in a bridge network by comparison or substitution with standards having known or negligible values of τ, measured in quadrature bridge networks from well-designed gas-filled capacitors having very small departures from 90°-phase between current and voltage.

8.4.3 Capacitance

Primary standards of capacitance are calibrated either via an artifact – a very stable fixed-value capacitor whose value is established from time to time directly in terms of the Thompson–Lampard calculable capacitor – or from a resistance measured in terms of the quantum Hall effect, using a resistance-capacitance ac bridge (Kibble and Rayner, 1984). The value of the best capacitance standards can be constant to within 0.1 ppm over a few years. Either route depends on coaxial ac bridge techniques, which imply complete electrical definition of impedance standards at kilohertz frequencies. Coaxial ac bridge techniques are also used to scale the value of a primary capacitance standard to values of other standards at decade intervals from 1 pF to 1 F with the accuracies given in Table 8.2.

8.4.4 Inductance

Inductance is derived from capacitance. Either the parallel-resonance technique illustrated in Figure 8.12a is employed, in which the reactances of a capacitor and an inductor are observed to be equal at a particular measured frequency, or the ratio of these reactances is measured using the Maxwell–Wien ac bridge illustrated in Figure 8.12b. An inductive ratio-arm ac bridge can scale inductance to other decade values with little further loss of accuracy, and the typical 100-ppm or better accuracies obtained are sufficient for all present scientific or commercial needs.

8.4.4.1 The Frequency Dependence of Standard Inductors

Skin and proximity effects, by excluding magnetic flux from the conductors and any constructional metalwork of an inductor, will alter its value slightly in a frequency-dependent manner. Its reactance will also be modified by the capacitance associated with its surrounding electric field. These effects are investigated in calibrating standard inductors by comparing them with specially constructed inductors wound with thin, spaced conductors

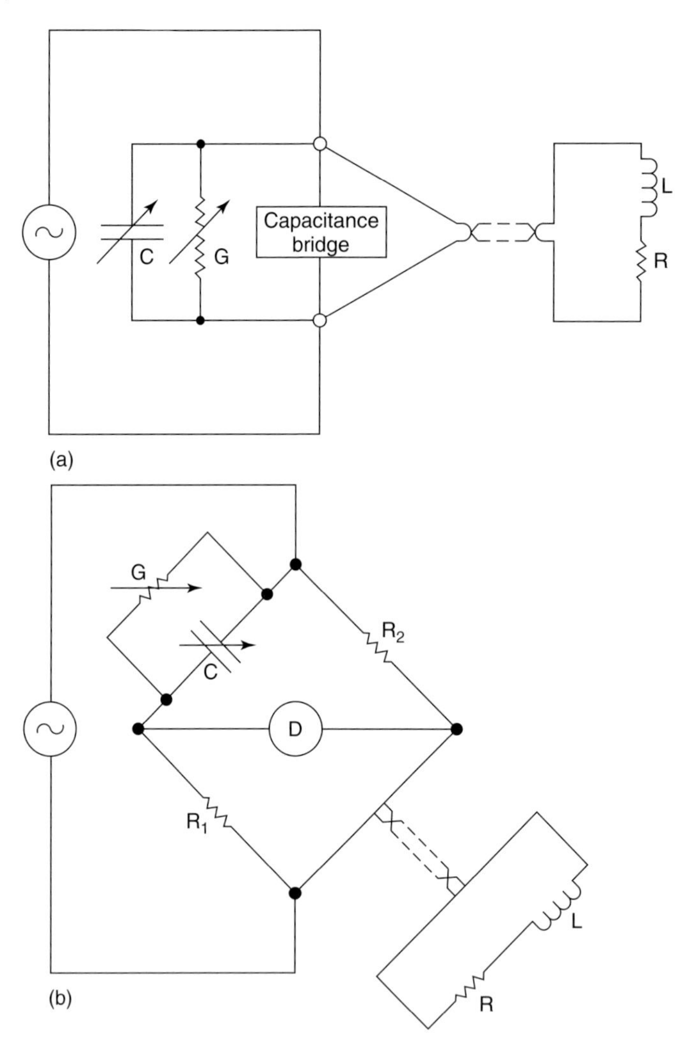

Fig. 8.12 (a) The parallel-resonance technique for establishing inductance L and associated series resistance R in terms of capacitance C and conductance G. (b) A Maxwell–Wien bridge. The screening that eliminates stray capacitances from the measurement has been omitted from the diagrams.

whose self capacitance has been measured by observing their self-resonant frequency when the terminals are shorted.

8.4.5
Electric Fields and Dielectric Constants

Electrical field strengths can be inferred directly in volts per meter from the voltages on conducting surfaces between which the electric field is generated, because conducting surfaces define the field-boundary conditions accurately.

The relative permittivity of materials is measurable in terms of the change of capacitance that a sample of the material produces when inserted in a suitable test capacitor. Permittivities are also needed at frequencies as high as in the microwave region, and are then measured by a waveguide insertion or an interferometric technique (Kemp *et al.*, 1996).

8.4.6
Magnetic Quantities

In contrast to electric flux and permittivity, magnetic flux cannot in general be deduced with any accuracy from the currents generating it if the boundary conditions at the surfaces of any magnetic media are involved (the superconductivity Meissner effect excluded), and the units and basic measurement of total flux and flux density of magnetic fields must be provided in other ways.

8.4.6.1 Total Magnetic Flux

The SI unit is the weber, and it is defined and realized in terms of the emf V produced around a circuit resulting from the rate of change of the total magnetic flux in webers threading it. That is,

$$V = -d\Phi/dt \qquad (8.10)$$

which may be integrated to yield

$$\Phi = \int V dt \qquad (8.11)$$

Therefore, the weber can be realized by directly measuring total flux by integrating the voltage induced in a search coil, which encompasses all the flux. The voltage integral is measured as the flux is established, or as the coil is moved into position from a region where the flux is negligible, with an electronic integrating circuit that is calibrated in terms of a known direct current charging a capacitor of known value. If the flux is caused by an alternating current of known wave form, the voltage integral can either be measured directly as a mean rectified value or be inferred from the wave form and a measurement of the peak or root mean square voltage.

8.4.6.2 Magnetic Flux Density

This can be measured using a similar search coil method in which the search coil is small enough to encompass only a region over which the flux is uniform. The flux density measurements made by the search coil can be calibrated either by measuring the voltage induced in it when it is inserted into, or its sense is reversed in, a known flux density (derived from a current-carrying coil of measured geometry).

A much more accurate indirect method is to observe the precessional frequency of protons in the flux (Williams *et al.*, 1985).

8.4.7
ac Current and Voltage

These are not "basic" quantities in that they do not require separate definition of their units, but relating them at the highest accuracy level to the corresponding dc

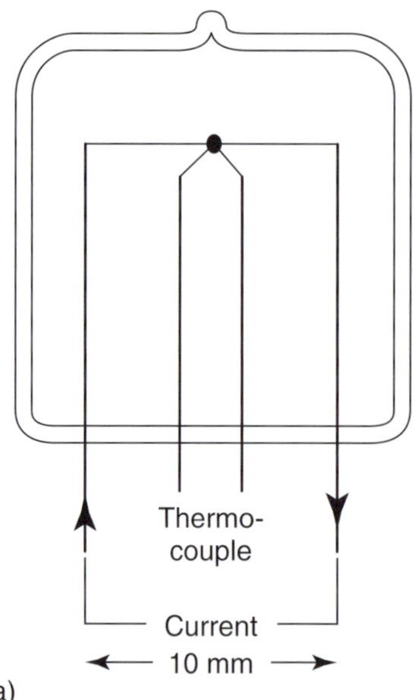

(a)

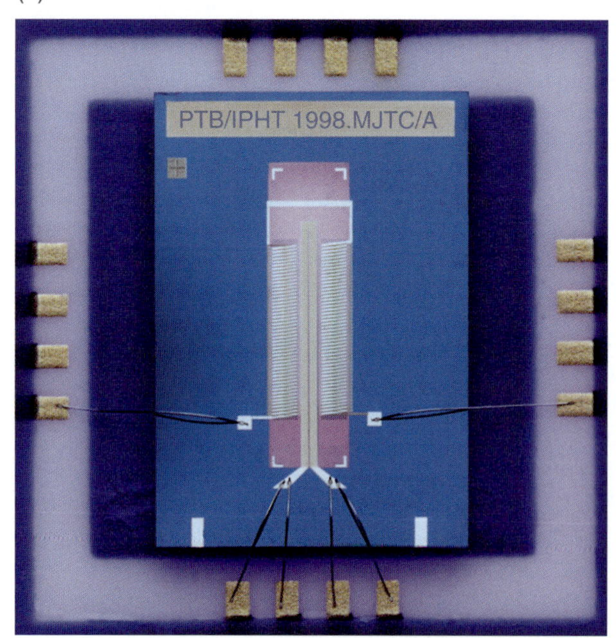

(b)

Fig. 8.13 A thermal convertor for relating ac and dc power: (a) design and (b) picture.

quantities does need a specialized measurement technique that is only to be found in national measurement laboratories.

The basic concept is to equate the heating of a dissipating element by a direct current with the same heating power generated by an alternating current. To ensure that the distribution of heat in the dissipating element is independent of frequency up to at least 100 kHz (i.e., eddy-current, capacitive, and inductive effects are negligible), the physical dimensions of the element are kept very small – usually by making it from a fine, straight length of resistance wire. The design of a typical single-element device is illustrated in Figure 8.13 where the temperature of a heater wire is sensed by a thermocouple attached to its center by an insulating glass bead. Sensitivity and accuracy can be somewhat enhanced, at the expense of high-frequency performance, by putting several thermocouple temperature sensors in series along a heater element (Dix and Bailey, 1975). Devices deposited on a substrate by integrated circuit techniques have now superseded these wire-and-bead convertors (Klonz and Weimann, 1989; Scarioni, Klonz, and Funck, 2005).

If the two currents are also compared via the voltages they produce in turn across a resistor whose value is independent of frequency, then the alternating and direct currents and voltages can be separately related.

Glossary

ac Josephson Effect: A phenomenon in which microwaves of frequency f are shone on a thin insulating barrier between two superconductors. A current of electron pairs tunneling through the barrier interacts with the nth harmonic of the microwave frequency, resulting in a precise voltage equal to $nf(h/2e)$ appearing across the barrier, where e is the electronic charge and h is the Planck constant.

Farad: The electrical capacitance between electrodes having a potential difference of 1 V when charged with 1 C.

Quantum Hall Effect: This effect gives rise to fixed ratios of transverse voltage to the current flowing through a semiconducting device in which the electrons are confined to two-dimensional motion at very low temperatures in a strong perpendicular magnetic field. These ratios are often termed as *quantized Hall resistances.*

Realize (a Unit): Use appropriate apparatus to derive a value for a standard consistent with the abstract definition of a unit. This standard can then be used to calibrate further standards and instruments.

References

Becker, P. (**2003**) Determination of the Avogadro constant via the silicon route. *Metrologia*, **40**, 366–375.

Bego, V. (**1988**) Determination of the volt by means of voltage balances. *Metrologia*, **25**, 127–133.

Clothier, W.K., Sloggett, G.J., Bairnsfather, H., Currey, M.F., and Benjamin, D.J. (**1989**) A determination of the volt with improved accuracy. *Metrologia*, **26**, 9–46.

Cook, A.H. (**1961**) A determination of the density of mercury. *Philos. Trans. R. Soc.*, **A254**, 125–154.

Dix, C.H. and Bailey, A.E. (**1975**) Electrical standards of measurement: 1. DC and low frequency standards. *Proc. IEEE*, **122**, 1018–1036.

Elnekave, N., Fau, A., and Delahaye, F. (**1976**) Evaluation provisoire du coefficient 2e/h à Partir de la dètermination absolute de volt. *BNM*, **7**, 3–8.

Funck, V.T. and Sienknecht, V. (**1988**) Darstellung der spannungseinheit volt mit der PTB-Spannungswaage. *Elecktrizitätswirtschaft*, **87**, 104–106.

Hamilton, C.A., Kautz, R.L., Steiner, R.L., and Lloyd, F.L. (**1985**) A practical Josephson voltage standard at 1 V. *IEEE Device Lett.*, **EDL-6**, 623.

Hamilton, C.A., Lloyd, F.L., Chieh, K., and Goeke, W.C. (**1989**) A 10-V Josephson voltage standard. *IEEE Trans. Instrum. Meas.*, **38** (2), 314–316.

Hamon, B.V. (**1954**) A 1–1000∼Ohm built-up resistor for the calibration of standard resistors. *J. Sci. Instrum.*, **31**, 450–453.

Hartland, A. (**1988**) Quantum standards for electrical units. *Contemp. Phys.*, **29**, 477–498.

Harvey, I.K. (**1972**) A precise low temperature DC ratio transformer. *Rev. Sci. Instrum.*, **43**, 1626–1629.

Josephson, B.D. (**1962**) Possible new effects in superconductive tunnelling. *Phys. Lett.*, **1**, 251–253.

Kemp, I. *et al.* (eds) (**1996**) Special issue on measurement techniques. IEEE Proceedings of the Seventh International Conference on Dielectric Materials, Measurements and Applications (Conf. Publ. No.430).

Kibble, B.P. and Rayner, G.H. (**1984**) *Coaxial AC Bridges*, Adam Hilger, Bristol.

Kinoshita, T. (**2006**) Improved α^4 term of the electron anomalous magnetic moment. *Phys. Rev. D*, **73**, 013003-1–013003-28.

von Klitzing, K., Dorda, G., and Pepper, M. (**1980**) New method for high accuracy determination based on quantized Hall resistance, *Phys. Rev. Lett.* **45**, 494–497.

Klonz, M. and Weimann, T. (**1989**) Accurate thin film multijunction thermal converter on a silicon chip [AC-DC standard]. *IEEE Trans. Instrum. Meas.*, **38** (2), 335–337.

Mills, I.A., Mohr, P.J., Quinn, T.J., Taylor, B.N., and Williams, E.R. (**2006**) Redefinition of the kilogram, ampere, Kelvin and mole: a proposed approach to implementing CIPM recommendation 1 (CI-2005). *Metrologia*, **43**, 227–246.

Niebauer, T.M., Sasagawa, G.S., Faller, J.E., Hilt, R., and Klopping, F. (**1995**) A new generation of absolute gravimeters. *Metrologia*, **32**, 159–180.

Niemeyer, J., Hinken, J.H., and Kautz, R.L. (**1984**) Microwave-induced constant voltage steps at one volt from a series array of Josephson junctions. *Appl. Phys. Lett.*, **45**, 478–480.

Pöpel, R., Niemeyer, J., Fromknecht, R., Meier, W., and Grimm, L. (**1990**) 1- and 10-V series array Josephson voltage standards in Nb/Al2O3/Nb technology. *J. Appl. Phys.*, **68**, 4294–4303.

Robinson, I.A. and Kibble, B.P. (**2007**) An initial measurement of Planck's constant using the NPL Mark II watt balance. *Metrologia*, **44**, 427–440.

Scarioni, L., Klonz, M., and Funck, T. (**2005**) Quartz planar multijunction thermal convertor as a new ac-dc current transfer standard up to 1 MHz. *IEEE Trans. Instrum. Meas.*, **54** (2), 799–802.

Steiner, R.L., Williams, E.R., Newell, D.B., and Liu, R. (**2007**) Uncertainty improvements of the NIST electronic kilogram, *IEEE Trans. Instrum. Meas.* **56** (2), 592–596.

Thompson, A.M. and Lampard, D.G. (**1956**) A new theorem in electrostatics and its application to calculable standards of capacitance. *Nature*, **177**, 888.

Williams, E.R., Jones, G.R., Song, J.S., Phillips, W.D., and Olsen, P.T. (**1985**) Report on the new NBS determination of the Proton Gyromagnetic Ratio. *IEEE Trans. Instrum. Meas.*, **IM-34**, 163–167.

Further Reading

Deslattes, R.D. and Kessler, E.G. (**1991**) Status of a silicon lattice measurement and dissemination exercise. *IEEE Trans. Instrum. Meas.*, **IM40**, 92–97.

Kibble, B.P. (**1981**) New amps for old. *Electron. Power*, **21**, 136–139.

Kibble, B.P. and Hunt, G.J. (**1979**) A measurement of the gyromagnetic ratio of the proton in a strong magnetic field. *Metrologia*, **15**, 5–30.

Petley, B.W. (**1985**) *The Fundamental Physical Constants and the Frontier of Measurement*, Adam Hilger, Bristol.

Samuel, M. (**1988**) Samuel Replies (He replies to a comment on his original article in PRL57). *Phys. Rev. Lett.*, **61**, 2899.

Seyfried, P. *et al.* (**1987**) The Avogadro Constant – recent results on the Molar Volume of Silicon. *IEEE Trans. Instrum. Meas.*, **IM-36**, 201–204.

9
Quantum Electrical Standards

François Piquemal

Handbook of Metrology. Edited by Michael Gläser and Manfred Kochsiek

ISBN: 978-3-527-40666-1

9.1 Introduction

Since nearly two decades, the SI units ohm and volt are maintained by two quantum phenomena of condensed matter physics: the quantum Hall effect (QHE) and the AC Josephson effect (JE). These phenomena have revolutionized electrical metrology, by linking resistance and electromotive force directly to the two fundamental constants, the von Klitzing constant R_K and the Josephson constant K_J, respectively. Theory predicts that $R_K = h/e^2$ and $K_J = 2e/h$, where h is the Planck constant and e the elementary charge.

If certain conditions are fulfilled, these two quantum effects allow the realization of quantum standards whose values are independent of space and time and set uniquely and simultaneously the representation of the ohm and the volt worldwide. Over the last two decades, their high reproducibility led to an improvement of the consistency and the maintenance of the electrical units (formerly based on wire wound resistors and saturated Weston cells) by a factor of 100 and gave rise to a better and simplified traceability of all resistance and voltage measurements to the SI. Moreover, the scalar property of the QHE and the JE allows the fabrication of resistance and voltage arrays composed of elements connected in large number on a same chip, which simplifies the realization of corresponding measurement scales as well.

In general, the great impact of the QHE and the JE in metrology and in physics lies also in their significant contribution to the quest for improving the knowledge of constants of nature. Indeed, through two routes involving electromechanical set-ups and measurement circuits referred to a quantum Hall resistance (QHR) standard or a Josephson voltage standard, the well-known fine structure constant $\alpha = \mu_0 c/(2h/e^2)$, where μ_0 is the magnetic constant of vacuum, and the Planck constant can be determined very accurately:

- The calculable capacitor, whose principle is based on a theorem in electrostatics (Thompson and Lampard, 1956), is implemented to realize the farad in the SI (linking the farad to the meter) (Figure 9.1). The comparison of this capacitance to the QHR by means of a quadrature bridge (introducing the second) results in an SI value of R_K and thus in an estimate of h/e^2 and α.
- The watt balance, relying on the equivalence of mechanical and electrical powers (Kibble, 1976), allows one to determine h through the measurement of the product $K_J^2 R_K$. Combining the two experiments constitutes both the SI

Handbook of Metrology. Edited by Michael Gläser and Manfred Kochsiek

ISBN: 978-3-527-40666-1

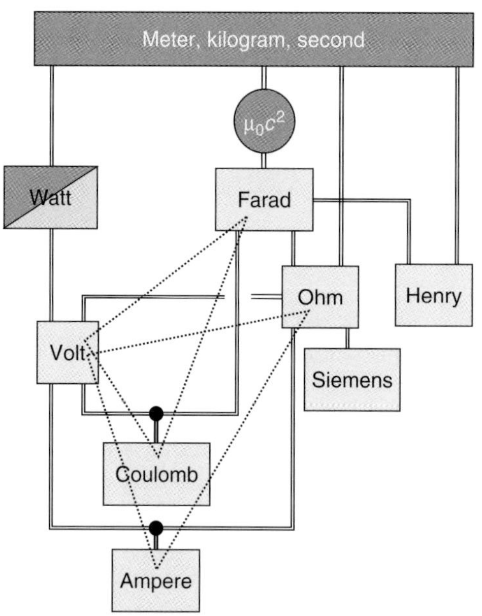

Fig. 9.1 Present chain of SI realizations of electrical units and metrological triangles (Piquemal *et al.*, 2007). According to the present definition of the ampere, the value of μ_0 is fixed: $\mu_0 = 4\pi \times 10^{-7}$ N A^{-2}. The value of the speed of light in vacuum being fixed for the definition of the meter, this leads to exact values for the permittivity of vacuum $\varepsilon_0 = 1/(\mu_0 c^2)$ and the free-space impedance $Z_0 = (\mu_0/\varepsilon_0)^{1/2}$. The farad undoubtedly occupies the first place in the hierarchy of electrical units. In theory, the link between electrical and mechanical units is made through an SI realization of the ampere and based on the equivalence principle between mechanical and electrical powers. However, in practice and for improved accuracy, it is more relevant nowadays to realize first the derivative electrical units, farad and ohm, on the one hand, and volt, on the other. This allows the determination of the ampere afterward with a higher accuracy.

realization of the volt and the determination of the Josephson constant K_J.

Consequently, these routes provide both a measurement of R_K and K_J, which have to be known with the highest accuracy for metrological purposes (consistency within the SI) and a validation test of theories (in solid state physics as well as in the other fields) by comparing their values with the ones obtained from different experiments in which α, h/e^2, and $2e/h$ (as well as other fundamental constants) are involved. Moreover, the watt balance is a very promising way to monitor the kilogram, which remains the sole material artifact in the SI defining both the unit of mass and the standard. In the frame of a revised SI, the unit of mass could indeed be redefined by fixing the value of the Planck constant. If we are convinced in the exactness of the relations $R_K = h/e^2$ and $K_J = 2e/h$, it would also make sense to redefine the ampere by fixing the elementary charge e, and consequently making μ_0 (and ε_0) a measurable constant with its value no longer fixed (Figure 9.1).

A quantum SI, that is, an SI fully based on quantum constants, is a genuine prospect (Mills *et al.*, 2005; Mills *et al.*, 2006). However, to become reality, some inconsistencies between values of fundamental constants have to be resolved (Mohr, Taylor, and Newell, 2008). For this issue, the single-electron tunneling (SET) devices and their main applications could play a significant role.

The exploitation of the underlying quantum phenomenon, the Coulomb blockade of electron tunneling, indeed makes the development of a quantum standard of current possible whose amplitude is directly linked to the elementary charge. The first important application of SET devices, which has been originally suggested by Likharev and Zorin (1985), is the quantum metrological triangle (QMT). This experiment enables a direct test at different uncertainty thresholds, of the coherence of the constants involved in the QHE, JE, and SET phenomena, which are strongly presumed to be given by the free-space values of h/e^2, $2e/h$, and e. It consists either in applying Ohm's law $U = RI$ or in following $Q = CU$ from the realization of an electron counting capacitance standard (ECCS) (Piquemal and Genevès, 2000; Keller, 2008). In the first way, the Josephson voltage is compared to the Hall voltage of a QHR standard crossed by a current delivered by a SET device. The second approach consists in transferring a well-known charge between the electrodes of a capacitor, measuring the voltage difference by means of a Josephson voltage standard and calibrating the capacitance against QHR standard via a quadrature bridge.

Moreover, the combination of the QMT with the watt balance and the calculable capacitor will lead to the determination of the elementary charge (Piquemal *et al.*, 2007; Keller *et al.*, 2008) and thus, will link up with the historical experiment of Millikan (in the early stage of the last century), demonstrating the quantization of the electric charge.

Although the ampere is at present indirectly represented with high accuracy through the QHR and Josephson voltage standards, the lower part of the current scale, below 1 nA, could be significantly improved by the direct use of the SET effect. The relevance of developing a quantum current standard based on SET for this purpose is even clearer in the frame of a redefinition of the ampere and the corresponding "mise en pratique" (put into practice), which could read, for example, as "The ampere is the electrical current equivalent to the flow of exactly $1/(1.602176487 \times 10^{-19})$ elementary charges per second". It follows that this definition fixes the elementary charge as exactly $1.602176487 \times 10^{-19}$ A s" (Mills *et al.*, 2006). The value of e taken here is the 2006 Committee on Data for Science and Technology (CODATA) recommended value.

This would drastically differ from the present definition of the ampere:

> *The ampere is that constant current which, if maintained in two straight parallel conductors of infinite length, of negligible circular cross section, and placed 1 m apart in vacuum, would produce between these conductors a force equal to $2\,10^{-7}$ N per meter of length.*

The chapter is structured not according to the present-day practical chain of electrical units (Figure 9.1) but according to the importance of the fundamental constants, which are given by the quantum phenomena used to realize the present and future standards, that is, the fine structure constant, the Planck constant, and the elementary charge. Thus, Sections 9.2

and 9.3 deal with the quantum standards of resistance and voltage, respectively. The SET standards are outlined in Section 9.4.

9.2 Quantized Hall Resistance Standard

9.2.1 Integer Quantum Hall Effect

The QHE (von Klitzing, Dorda, and Pepper, 2002) is observed at low temperature and under a high magnetic flux density in a two-dimensional electron gas (2DEG). The Hall resistance of the 2DEG exhibits a set of plateaux centered on quantized values that depend only on the fundamental constant h/e^2:

$$R_H(i) = h/ie^2 \tag{9.1}$$

where i is an integer. Each plateau of resistance is correlated with an abrupt drop of the longitudinal resistance R_{XX} to minimal values that can be less than 100 μΩ as observed for the $i = 2$ or 4 plateaux, usually used (Figure 9.2). The 2DEG is realized in the inversion layer either at the interface of a silicon MOSFET or at the junction of semiconductors with different energy band gaps. The GaAs/AlGaAs heterostructure is so far the prime system for metrological applications of the QHE. The Hall bar sample shown in Figure 9.2b is realized by lithographic techniques. It consists of an active area delimited by a 300-nm-thick mesa with AuGeNi contacts on each of the eight terminal pads.

Laughlin has given a general theoretical explanation of the QHE using a topological argument based on the concept of a mobility gap and gauge invariance (Laughlin, 1981). Büttiker has proposed a formulation based on edge states where the current is flowing only at the sample edges, emphasizing the role of the contacts acting as reservoirs (Büttiker, 1988). However, both these approaches predict the QHE properties at zero temperature and in the zero dissipation state. Actually, among the variety of the proposed models, very few of them at the present time are able to take into account first the real experimental conditions of the nonzero temperature, the finite magnetic field, and the high current value, which lead to a dissipative state, and

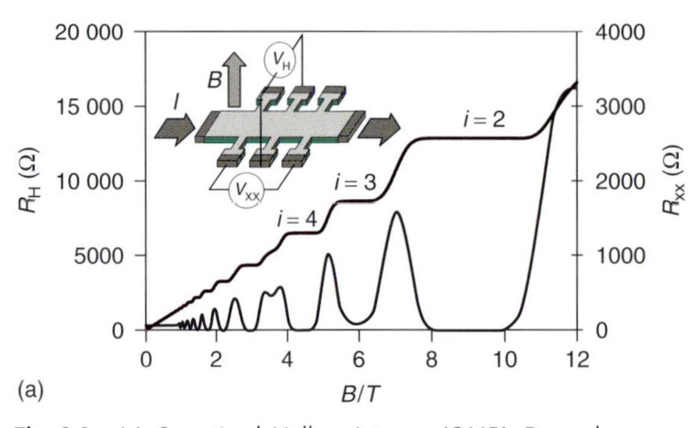

Fig. 9.2 (a) Quantized Hall resistance (QHR) R_H and longitudinal resistance R_{XX} as a function of magnetic flux density B for GaAs/AlGaAs heterostructure ($T = 180$ mK, $I = 8$ μA). (b) LEP514 Hall bar sample mounted on TO-8 holder (Piquemal *et al.*, 1993).

second the imperfect nature of the Hall sample and its contacts.

Nevertheless, this lack of models does not call in question again the universal character of the QHE (see Section 9.2.1.2). As pointed out by Laughlin and Pines, the QHE is a collective phenomenon and can, thus, not completely be explained by microscopic theories (Laughlin and Pines, 2000).

Detailed bibliographies on both theoretical and experimental works can be found in Prange and Girvin, 1987; Stone, 1992; Janssen *et al.*, 1994; Jeckelmann and Jeanneret, 2000; Yoshioka, 2002; and Poirier and Schopfer, 2009. Only a very brief description about the theory is given below.

9.2.1.1 **Basic Theoretical Principles**

At low temperature and for weak carrier density n_S, the conduction electrons confined in the potential well created at the interface of the device (Si-MOSFET or GaAs/AlGaAs) populate only the lowest subband energy E_0. Any motion perpendicular to the interface is forbidden. Under weak magnetic flux density B (perpendicular to the interface, $B = B_z$), the electrons of the 2DEG occupy a continuum of states up to the Fermi energy E_F. The expression for the classical Hall resistance gives

$$R_H = B/(n_S e) \tag{9.2}$$

Under strong magnetic field B, the electrons follow cyclotron orbits with an angular frequency $\omega_c = eB/m^*$, where m^* is the effective electron mass (e.g., $m^* = 0.068\ m_e$ in GaAs). The electrons occupy quantized states in Landau levels with energy:

$$E_n = E_0 + (n + 1/2)\hbar\omega_c \tag{9.3}$$

where n is the index of the Landau level and $\hbar = h/2\pi$ (neglecting the spin splitting of the Landau levels).

In the real case of a crystal with impurities and defects, the disorder and scattering make the energy spectrum no longer discrete. The density of states can be described (Figure 9.3) by a series of bands associated with E_n values whose central part is composed of extended states and whose tails correspond to localized states (i.e., to zero mobility gaps). The quantization of the energy will occur only if the energy bands do not overlap. Two conditions have to be fulfilled:

- The thermal activation energy of electrons does not exceed the energy separation between two Landau levels, $\hbar\omega_c \gg kT$, where k is the Boltzmann constant.
- The electrons have time to describe several cyclotron orbits before diffusion $\omega_c\tau \gg 1$, where τ is the diffusion time. In other terms, the electron mobility μ and the magnetic field must be high enough: $\mu B \gg 1$.

Within these conditions, it can be shown that the well-separated energy levels contain $N = eB/h$ states per unit surface area. By adjusting B such that the Fermi energy E_F lies in a region of localized states and corresponds to an integer number i of populated levels, the carrier density becomes

$$n_S = Ni = ieB/h \tag{9.4}$$

and, combining with the relation (9.2), it gives $R_H(i) = h/ie^2$.

The Fermi energy being in the localized states region, the longitudinal conductivity σ_{XX} is zero. Owing to the relations between resistivity and conductivity tensor components in a two-dimensional system, ρ_{XX}

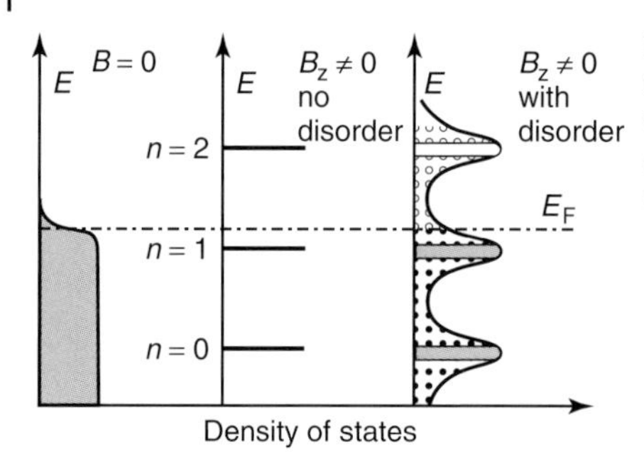

Fig. 9.3 Energy spectrum of a 2DEG at zero magnetic field (left), at magnetic field $B_z \neq 0$ without disorder and scattering (middle), at $B_z \neq 0$ in the case of disorder and scattering (right).

$= \sigma_{XX}/(\sigma_{XX}^2 + \sigma_{XY}^2)$ and $\rho_{XY} = -\sigma_{XY}/(\sigma_{XX}^2 + \sigma_{XY}^2)$, the longitudinal resistivity ρ_{XX} also vanishes. The Hall resistance, which depends on the total number of occupied extended states below the Fermi level, remains constant for small variations of E_F, or more rigorously for small variations of B, resulting in the observed plateaux. The transition between two plateaux occurs when the Fermi level lies in a region of extended states. The longitudinal resistivity is no more zero. For large variation of B, R_{XX} oscillates with a period $\propto 1/B$. This is the so-called Shubnikov de Haas oscillation characterizing the quantization of energy in Landau levels.

9.2.1.2 Universality and von Klitzing Constant R_K

The QHR, theoretically linked to h/e^2, may be used as a reference standard of resistance when the QHE sample respects the criteria given by CCEM technical guidelines (Delahaye and Jeckelmann, 2003). Under these conditions, QHE provides a highly reproducible reference standard. Indeed, measurements have shown values of the product $i \times R_H(i)$ independent of the sample properties (growth techniques, type of structure, and technological and geometrical parameters), of the plateau index ($i = 1, 2, 4, 6$, or 8), and of the experimental conditions (temperature and measuring current amplitude) with uncertainties as low as a few parts in 10^{10} (Hartland *et al.*, 1991; Jeckelmann, Inglis, and Jeanneret, 1995). Let us mention here two recent results:

1. A new method involving a Wheatstone bridge composed of four QHE devices and based on multiple connection techniques (see Section 9.2), which opens the way to reproducibility tests with an uncertainty reduced down to 10^{-11} (Schopfer and Poirier, 2007);
2. The unconventional QHE discovered in graphene, a monolayer of carbon atoms densely packed in a honeycomb crystal lattice, and in bilayer graphene (Novoselov *et al.*, 2005; Zhang *et al.*, 2005), for which the underlying physics drastically differs from QHE in semiconductors (in graphene, charge carriers behave as relativistic particles with zero rest mass) and promises novel QHR standards.

In addition, bilateral comparisons of complete QHE systems performed between Bureau International des Poids et Mesures (BIPM) and some National Metrological Institutes (NMIs) during the past decade (see data available on BIPM web site: *www.bipm.fr/* and, e.g., Delahaye *et al.*, 1995) have shown excellent agreement of a few parts in 10^9 (Figure 9.4). International comparisons involving 1 or 100 Ω traveling standards have also shown good results but with a less high accuracy, that is, no significant deviation at a uncertainty level of one part in 10^8, due to instabilities

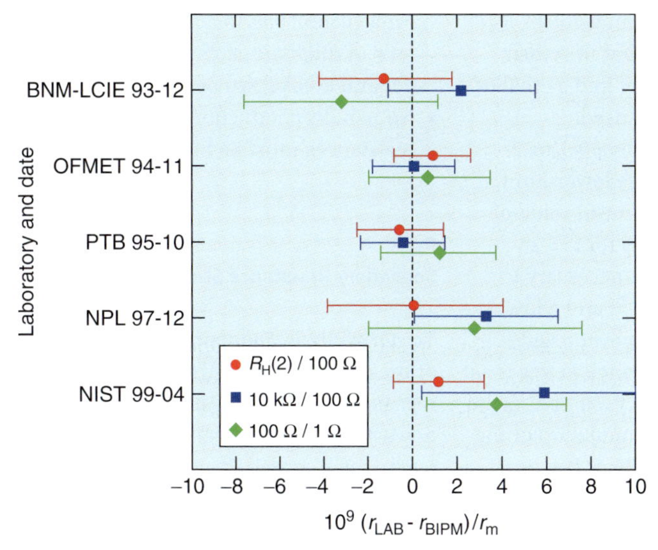

Fig. 9.4 Results of the bilateral comparisons between BIPM (with transportable QHE system) and LNE (formerly BNM-LCIE), METAS (formerly OFMET), PTB, NPL, and NIST. Error bars mean one standard deviation estimate of combined uncertainty.

of the traveling resistors (e.g., Schumacher, 2007).

All these results strongly support the universal aspect of the constant involved in the QHE and confirm the conclusiveness of the 1988 Comité International des Poids et Mesures (CIPM) recommendation for the use of the QHR as a resistance standard. In the framework of this recommendation, the von Klitzing constant R_K was defined as an estimate of h/e^2 and a conventionally true value $R_{K-90} = 25\,812.807\ \Omega$ was agreed upon for all calibration purposes (CIPM, 1988b). This unique representation of the ohm constitutes a watershed compared with the prior situation when national bases were constituted by wire-wound resistance standards drifting in time, depending on ambient conditions (temperature and pressure), and which materialized only a local value of the ohm.

It is noteworthy that the design requirements make the fabrication of QHE samples for metrological purposes difficult and reliable sources of samples are lacking. Specific projects for producing a large number of QHE samples have been undertaken by NMIs in order to partly solve this problem (Piquemal *et al.*, 1993; Poirier *et al.*, 1999; Inglis, 2004).

9.2.1.3 **QHE Samples of High Metrological Qualities**

The selection of QHE samples for getting a quantum standard reproducible at a level of one part in 10^9 is based on the following criteria (Delahaye and Jeckelmann, 2003) by taking into account the experimental conditions of nonzero temperature (typically, $300\ \text{mK} < T < 4.2\ \text{K}$) and of applied current high enough (typically, $10\ \mu\text{A} < I < 50\ \mu\text{A}$):

- A 2DEG with high carrier mobility: $10\ \text{T}^{-1} < \mu < 100\ \text{T}^{-1}$. Higher mobility yields a smaller minimum value of R_{XX} at a given temperature, but in return, the plateaux are narrower and a smaller driving current must be used.

- A carrier density n_s : 2.5×10^{15} m^{-2} < $n_s < 7.5 \times 10^{15}$ m^{-2}. The lower limit corresponds to the minimum value of 5 T magnetic flux density for $i = 2$ plateau (currently used), necessary to get $R_H(i)$ constant at 10^{-9}, and a low value of R_{XX} (< 100 μΩ) over a wide enough magnetic field range (a few tens of millitesla). The upper limit of n_s is the value above which the population of the second subband energy level becomes possible.
- A negligible dependence of the Hall resistance $R_H(i, T)$ corresponding to the i plateau and measured at a temperature T, on a nonzero value of $R_{XX}(i, T)$. The observed correlation between $R_H(i, T)$ and $R_{XX}(i, T)$ is in general linear and characterized by a slope depending on several parameters relative to the sample (nature, fabrication, and technology) and to the measurement (applied current and magnetic field polarity). This slope must not exceed a typical value of 10^{-8} mΩ^{-1}.
- Ohmic contacts (no diode like behavior) and contact resistance to the 2DEG as low as possible, preferably a few milliohms (but values of 1 Ω can be acceptable). In particular, the current contacts have to stay ohmic for currents up to 50 μA.
- Critical current density with value higher than 0.5 A m^{-1}, which corresponds to a critical value of $I_c = 200$ μA for usual sample with 400 − μm channel width. Below this value, the 2DEG remains in a dissipationless state. Once the current exceeds this limit, R_{XX} abruptly increases by several orders of magnitude for a current varying only over a few tens of microamperes and local or general breakdown of the QHE results.
- No parallel conduction. Leakage resistances must be higher than 10 TΩ.

9.2.2 Secondary Resistance Standard

To derive their maintained representation of the ohm from the QHE and to scale DC resistances from the QHR up to 1 PΩ and down to 1 nΩ, the NMIs have necessarily used a set of secondary resistance standards and appropriate instruments to carry out the decade resistance ratios.

In addition to the requirements that the material resistance standards must depend as little as possible on measurement conditions (temperature, pressure, humidity, and dissipated power), the first resistance standards in the metrological chain must be stable enough in time to avoid frequent setting up of the QHE (still time consuming and expensive in cryogenic fluid). Their nominal values are 1 Ω and 10 kΩ which are not so far from the QHR values corresponding $i = 2$ and $i = 4$ plateaux, that is, about 12 906.4 and 6453.2 Ω. However, the direct calibration against the QHE is not yet possible and requires the use of 100 Ω resistors as transfer standards. The typical relative uncertainty can be reduced down to one part in 10^8 and even less, by means of rather elaborate resistance bridges, such as those based on a cryogenic current comparator (CCC). Let us note that the requirements on time drift for the 100 Ω transfer standard can be less severe than for 1 Ω and 10 kΩ standards.

The choice of the alloy as material for the resistive element of these secondary standards is thus essential. It relies on four criteria: (i) a chemical stability with a well-defined physical state improved by annealing, (ii) a temperature coefficient less than 10^{-6} per °C at the working temperature (typically between 20 and 30°C), (iii) a thermoelectric power less than a few microvolts per degree Celcius relative to pure copper (taking into account the materials used for the terminals, the type of connections (soldering and clamping), and the temperature homogeneity of the overall system), and (iv) a high enough resistivity (40 – 130 μΩ cm).

The best resistance standards of 100 and 1 Ω (and for lower values down to 10 mΩ) consist generally of manganin (Cu 84%, Mn 12%, and Ni 4%) wire wound around an insulating cylindrical holder. For resistance values higher than 100 Ω, the manufacturers use alloys such as evanohm (Ni 75%, Cr 20%, Al 2.5%, and Cu 2.5%) or karma (Fe replacing Cu in evanohm composition), which show resistivity higher than for manganin (130 μΩ cm compared to 40 μΩ cm) and a very weak thermoelectric power (≈0.2 μV per°C). Moreover, a temperature coefficient at least five times less than for manganin can be obtained after the annealing process. Let us note that resistance standards of 1 Ω can also be made of evanohm wire, such as those fabricated by the National Metrological Institute of Australia (NMIA).

The fabrication of secondary resistance standards also needs some care on the structure of the standard itself to minimize undesired effects. The dissipated power due to the Joule effect has to be taken into account as well as the influences of variations in atmospheric pressure and relative humidity.

Aside from CCC resistance bridges (described below) allowing them to carry out the first resistance decades starting from the QHR with the highest accuracy (i.e., typically over the range 1 Ω to 10 kΩ, but extended to 1 MΩ or even 1 GΩ in a very few labs), the NMIs and other calibration laboratories use various kinds of instruments and artifacts for the rest of the scale. In general, room temperature current comparators are used for values less than 1 Ω, whereas in the upper part of the scale (>10 kΩ) calibration methods involve potentiometer bridges or modified Wheatstone bridges for resistances up to 100 GΩ then an integration bridge for the highest values (1 PΩ).

9.2.3 Resistance Bridge Based on Cryogenic Current Comparator

Most of the resistance bridges currently used in NMIs to calibrate resistances against QHR are based on a CCC. This is the instrument that has allowed to demonstrate the universality of R_K with the high accuracy mentioned above. Fitted with a very sensitive magnetic flux detector such as a superconducting quantum interference device (SQUID), the CCC presents two advantages compared to a conventional ferromagnetic-core-based comparator:

- the resolution is significantly improved by at least a factor of 100. The noise spectral density of a CCC, in terms of ampere-turn, is between 10^{-9} and 10^{-10} A t $\mathrm{Hz}^{-1/2}$ in the bandwidth and
- the uncertainty on current ratio does not exceed one part in 10^9 (and can be reduced down to one part in 10^{10} or less) compared to a few parts in 10^8 for a magnetic core comparator.

The CCC is briefly described below. More details on this instrument as well as on the SQUID can be found in the literature; see for example, Gallop and Piquemal (2006), Clarke and Braginski (2004), respectively.

The principle of the CCC, invented by Harvey (1972), rests on Ampère's law and the perfect diamagnetism of the superconductor in the Meissner state. Given two wires inserted in a superconducting tube (Figure 9.5), currents I_1 and I_2 circulating through these wires will induce a supercurrent I flowing up the inner surface of the tube and back down the outer surface so as to maintain a null magnetic flux density B inside the tube. Application of Ampère's law to a closed contour (a) in the bulk gives

$$\oint_a B \cdot dl = 0 = \mu_0 \cdot (I_1 + I_2 - I) \tag{9.5}$$

and leads to the equality of the currents:

$$I = I_1 + I_2 \tag{9.6}$$

If the tube contains N_1 and N_2 wires crossed, respectively, by currents I_1 and I_2, then Equation (9.6) becomes

$$I = N_1 I_1 + N_2 I_2 \tag{9.7}$$

These equalities are valid independent of the position of the wires inside the tube, which is the key reason for the high accuracy of the CCC. In practice, a CCC is made of two windings with N_1 and N_2 turns crossed by currents I_1 and I_2 circulating in opposite directions. These windings are enclosed in a superconducting torus (Sullivan and Dziuba, 1974), whose extremities overlap without being electrically connected on a length large enough to overcome the end effects, which distort the current equality in the real case of a finite length tube (Figure 9.5).

The outside magnetic flux, which results only from the supercurrent, is detected by a SQUID whose output voltage is converted into a current that feeds back to one of the two windings to null the magnetomotive forces:

$$N_1 I_1 - N_2 I_2 = 0 \tag{9.8}$$

The CCC is connected to a double constant current source (Delahaye, 1978), which feeds resistors R_1 and R_2 to be compared with primary and secondary currents I_1 and I_2, respectively, as shown in Figure 9.6. Here, the secondary current source is slaved on the primary current source in such a way that the current ratio can be reliably adjusted allowing the SQUID to be

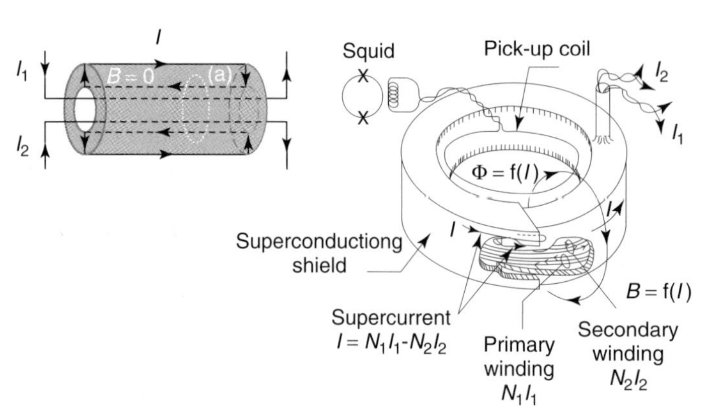

Fig. 9.5 Toroidal structure of a CCC and principle (in insert). The supercurrent flowing up the inner surface of the toroidal shield is given by $I = N_1 I_1 + N_2 I_2$.

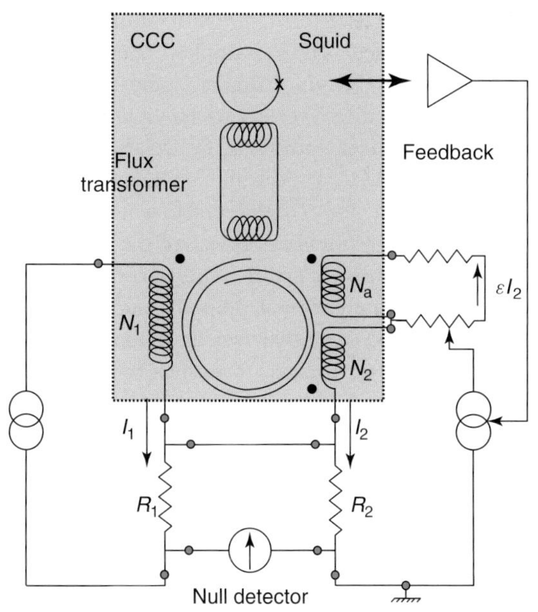

Fig. 9.6 Circuit diagram for a CCC-based resistance bridge. The fraction ε of the secondary current I_2 is deviated into an auxiliary winding by means of a resistive divider. The toroidal shield of the CCC is symbolized by a two-turn spiral. The three dots indicate the input of coils for applying currents, which induce magnetic fields of the same polarity.

properly and accurately flux locked (Gallop and Piquemal, 2006).

In the ideal case, the operation should consist of setting the current ratio to equal the voltage drops across the resistances, that is, to null the voltage drop through a detector connected to the low voltage terminals of the resistances: $R_1 I_1 = R_2 I_2$. Consequently, the resistance ratio R_1/R_2 will be equal to the winding ratio N_1/N_2.

In real cases, a fraction ε of the current I_2 has to be diverted to an auxiliary winding of N_a number of turns to balance the bridge both in voltage and in ampere turns:

$$N_1 I_1 - N_2 I_2 + \varepsilon N_a I_2 = 0 \tag{9.9}$$

$$R_1 I_1 - R_2 I_2 = 0 \tag{9.10}$$

Elimination of the currents from these two relations gives the resistance ratio:

$$R_1/R_2 = N_1/N_2 \cdot [1 + (N_a/N_2)\varepsilon] \tag{9.11}$$

In the bridge illustrated in Figure 9.6, the deviation ε is obtained by recording the output voltages of the null detector, which correspond to the two positions of the resistive divider $\varepsilon_\pm = \varepsilon \pm x$, where x is typically of the order of one part in 10^7. Thus, ε is given as

$$\varepsilon = (\varepsilon_- V_+ - \varepsilon_+ V_-)/(V_+ - V_-) \tag{9.12}$$

where the voltages V_+ and V_- of opposite signs correspond to ε_+ and ε_-, respectively (Piquemal, 1999). This deviation ε can also be obtained by using an amplifier at the output of the detector that generates a current through the auxiliary winding and a resistor placed in series. The value of ε is thus deduced by measuring the voltage drop across the resistor (Hartland, 1992). The advantage of this second method is the ability to fully automate the bridge.

Either way, the voltages are measured by periodically reversing the current polarity in order to compensate for the unwanted thermal electromotive forces. The typical working frequency is of the order of 0.1 Hz or less and might be in the range where the

SQUID generates $1/f$ noise. This flicker noise may be avoided by operating the bridge at 1 Hz (Delahaye, 1991). However, the current ratio has to be preadjusted both in phase and out of phase and the CCC needs a supplementary winding. Moreover, the dependence of the resistance standard on frequency has to be known. Working frequencies higher than 1 Hz might induce significant error ($>10^{-9}$) in the current ratio due to finite capacitive leakage and shunt between the CCC windings.

9.2.4
Quantized Hall Array Resistance Standard

The metrological applications of the QHE in a single Hall bar are only limited to resistance standards with a nominal value of $R_K/2$ and $R_K/4$, that is, around 10 kΩ. Plateaux corresponding to odd or high value of i are generally not well quantized. Moreover, the maximum current that can be supplied to single Hall bars does not exceed 100 μA typically. Special resistance bridges, such as those based on a CCC, are therefore required to calibrate material resistance standard with the QHE, and the use of these bridges is restricted to NMIs. Fortunately, by means of a multiple connection technique with redundant links between QHE samples connected in series or in parallel, other quantum resistance values can be obtained. This technique, proposed by Delahaye (1993), allows to cancel the contact resistance effect and consequently to define the four terminal resistance of the equivalent quantum resistors. This technique uses two fundamental properties of the QHE: the two-terminal resistance between any pair of probes and the four-terminal longitudinal resistance are ideally equal to R_H and zero, respectively. The equivalent electrical circuit, as shown in Figure 9.7, illustrates these properties.

Let us consider a device with eight probes, six for voltage and two for current, and assume a negligible longitudinal resistance R_{xx} and infinite leakage resistance. Each probe has a resistance:

$$r_j = r + r_{cj} + r_{wj} = r(1 + \varepsilon_j) \tag{9.13}$$

where $r = R_H(i)/2$, r_{cj} is the contact resistance to the 2DEG, r_{wj} is the resistance of the connection wire, and ε_j the relative deviation that results from these finite resistances. To take into account the nonzero value of R_{xx}, an internal resistance

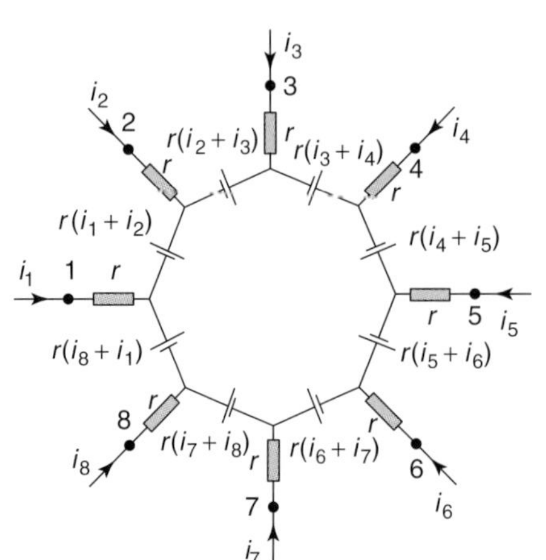

Fig. 9.7 Equivalent electrical circuit of a sample Hall bar fitted with eight probes.

r_{int} of the voltage generator $U_{j,j+1} = r(i_j + i_j + 1)$ can be introduced such that $r_{\text{int}} = R_{xx}/2$.

- In the usual case of measurement of the Hall voltage V_H (between points 2 and 8, 3 and 7, or 4 and 6) and of the longitudinal voltage V_{xx}, a direct current I crosses the points 1 and 5, $I = i_1 = -i_5$ and no current should cross the voltage probes. We then find again:

$$V_H = 2rI \quad \text{and} \quad V_{xx} = 0 \qquad (9.14)$$

- In the particular case of voltage $V_{j,k}$ measured between any two points j and k, one obtains

$$V_{j,k} = (r_j + r_k)I \approx 2rI \qquad (9.15)$$

The multiple connection technique consists in connecting two adjacent samples by means of both current and voltage probes and in increasing the number n of links (Figure 9.8). Consequently, the error due to the finite values of contact resistances R_C and of linking resistances R_L becomes negligible, the error being limited to ε^n, where $\varepsilon(\varepsilon \ll 1)$ corresponds typically to the ratio $(R_C + R_L)/R_H$.

The metrological proof of the efficiency of the multiple connection technique was first obtained for two Hall bars placed in parallel (Delahaye, 1993; Delahaye *et al.*, 1995), and then for Hall bars connected in autoseries (Piquemal *et al.*, 1999). After first realizing series arrays of 10 Hall bars, the Laboratoire National de métrologie et d'Essais (LNE) has developed quantum Hall array resistance standards (QHARSs) with nominal values in the wide range $R_K/200$ to 50 R_K (Poirier *et al.*, 2002; Poirier *et al.*, 2004; Bounouh *et al.*, 2003). The development of such a QHARS has also been undertaken by the Physikalisch Technische Bundesanstalt (PTB) (Hein, Schumacher, and Ahlers, 2004) and more recently by the National Metrology Institute of Japan (NMIJ) (Oeh *et al.*, 2008). In addition, the multiple connection technique makes the development of other interesting circuits possible, such as a quantum Wheatstone bridge (Schopfer and Poirier, 2007) or a voltage divider (Kaneko *et al.*, 2008) and allows highly accurate AC measurements of the QHR (see below).

As illustration, Figure 9.9 shows a QHARS, which consists of 100 multiply connected Hall bars in a parallel arrangement in such a way that the nominal value is equal to 129 Ω on the $i = 2$ plateau, and QHE curves obtained from this QHARS and from QHARS of 50 Hall bars in parallel. Their resistances have been found equal to their nominal values within an uncertainty of five parts in 10^9 with a measuring current as high as 4 mA and at a temperature of 1.3 K. The development of such

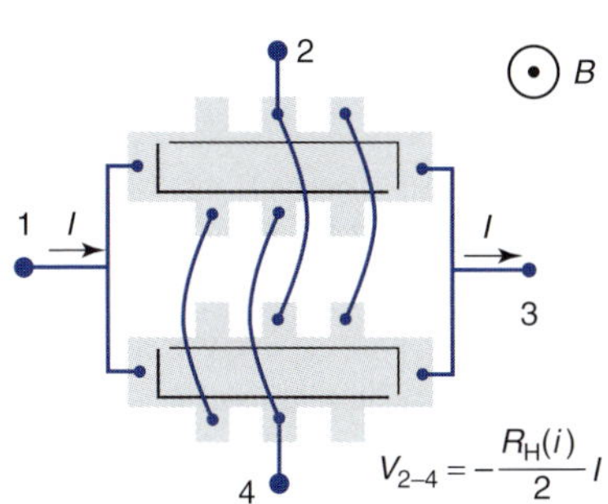

Fig. 9.8 Schematic of two Hall bars placed in parallel by means of a triple connection.

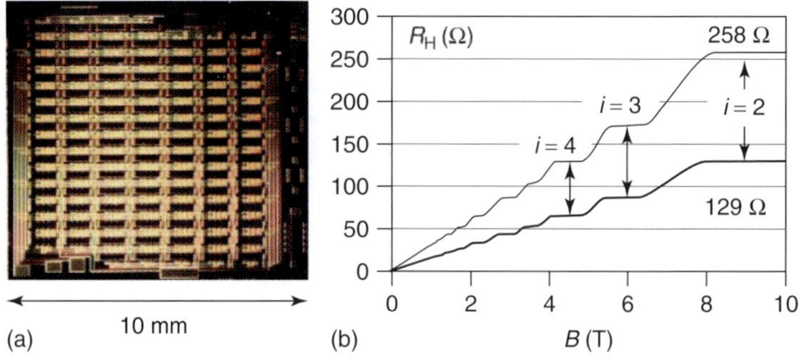

Fig. 9.9 (a) QHARS of 129 Ω nominal value (square of 10 mm of side) developed at LNE. The array is composed of 100 Hall bars (200 μm thick) placed in parallel by using the triple connection technique (Poirier *et al.*, 2004). (b) Dependence of R_H on B for two QHARS, with resistances of 129 and 258 Ω on $i = 2$ plateau (Poirier *et al.*, 2002). The 129 Ω QHARS is composed of 50 Hall bars connected in parallel by triple connections.

QHARS, which are real quantum standards, in view of the successful Josephson array voltage standards (JAVSs), needs to be pursued because they open new prospects on metrological applications of the QHE:

1. The calibration of resistances with nominal values up to 1 MΩ or down to 1 Ω is possible without using transfer standards of intermediate values, and consequently the uncertainties can be reduced by a factor of 10 or 100 (a few parts in 10^7 for 1 MΩ and one part in 10^9 for 1 Ω).
2. The consistency of the ohm representation in the NMIs might be checked with a very high accuracy. Using QHARS as traveling standards instead of resistors (1 or 100 Ω) or instead of moving a complete QHE system has obvious advantages and the uncertainties reached in international comparisons of resistances could be reduced at a level never reached before.
3. QHARS are compatible with commercially available bridges typically used in industrial calibration centers. For instance, resistance standards can be directly calibrated against QHARS of the parallel type, which might tolerate high current such as the one supplied by a room temperature current comparator.

9.2.5 AC QHR Standard

For more than one decade, NMIs and the BIPM have investigated the use of the QHR as an impedance standard in the audio frequency range (Delahaye, Kibble, and Zarka, 2000). The aim is to improve the metrological chain linking resistance and capacitance in the frame of the SI realization of the ohm (described in the following section) and vice versa in the reproduction of the farad from the ohm. The frequency correction of the artifact resistors used in the capacitance–resistance measurement chain needs to be determined. So far, this determination implies a comparison against an AC–DC resistor for which the frequency variation is calculable. Variations of less than one part in 10^9 from DC to 1 kHz reported for the best AC–DC resistors can be considered as a requirement for the use of an AC QHR standard in the frame of the determination of R_K.

The AC measurement techniques for the QHR have been significantly improved. In early work, the main difficulties arose from the fairly resistive cryogenic coaxial cables needed for connecting the QHE sample to the impedance bridge. The adoption of the multiple connections technique and the use of active current equalizers in place of passive ones have been found well suited to reduce the measurement uncertainties. As a consequence, some controversial results about the frequency effect on the QHR have been partly explained. However, the linear dependence of the QHR on frequency and current usually measured, which would be correlated to an AC loss mechanism in the QHE device, remains unclear. Indeed, one can reduce this linear variation by different methods by the use of either gated samples or specific sample holders (Delahaye, Kibble, and Zarka, 2000; Overney, Jeanneret, and Jeckelmann, 2003; Schurr *et al.*, 2006). A frequency coefficient of the QHR, more precisely of its real part $\mathrm{Re}[R_H]$, within a few parts in 10^8 per kHz has thus been obtained. In addition, it is noteworthy that the QHR measured at AC depends linearly on the longitudinal resistance R_{xx} as in DC but with a slope 10 times higher (Overney, Jeanneret, and Jeckelmann, 2003). Extrapolated to a null value of R_{xx}, $\mathrm{Re}[R_H]$ has been found in agreement with the DC value of R_H at an uncertainty less than one part in 10^8. From these results, it could be concluded that the calibration of capacitance in terms of the AC QHR is possible at this level of uncertainty and that the QHE provides a quantum impedance standard (Schurr, Bürkel, and Kibble, 2008).

9.2.6 Determination of R_K and the Fine Structure Constant α

The theory predicts that $R_K = h/e^2 = \mu_0 c/2\alpha$ where α is the famous fine structure constant, the dimensionless basic constant in atomic physics ($\alpha \approx 1/137$) characterizing the low-energy electromagnetic interaction. The direct measurements of R_K by realizing the ohm in the SI, thanks to the implementation of a calculable capacitor and an impedance measurement chain, constitute also a method to determine α, if one assumes the exactness of the relation. Conversely, the other determinations of α or h/e^2 improve the knowledge on R_K and the ongoing comparison of different experiments, leading on one side to R_K and to α or h/e^2 on the other side, establishes the exactness of the relation $R_K = h/e^2$ and thus is a test of validity of the theory.

9.2.6.1 Thompson Lampard Calculable Capacitor and SI Realization of the Ohm

The "mise en pratique" of the Thompson–Lampard theorem gives rise to a calculable capacitor, which makes the farad the electrical unit realized in SI units with the smallest uncertainty. This theorem stipulates that for a cylindrical system (Figure 9.10) composed of four isolated electrodes of infinite length and placed in vacuum, the direct capacitances per unit of length γ_{13} and γ_{24} of two pairs of electrodes obey to the relation:

$$\exp(-\pi\gamma_{13}/\varepsilon_0) + \exp(-\pi\gamma_{24}/\varepsilon_0) = 1 \tag{9.16}$$

where ε_0 is the permittivity of vacuum. Moreover, in the case of a perfect symmetry with identical capacitances per unit of length, it results

$$\gamma_{13} = \gamma_{24} = \gamma = \frac{(\varepsilon_0 \ln 2)}{\pi} = 1.953549043\ (\mathrm{pF\ m^{-1}}) \tag{9.17}$$

and then the value of the electrical capacitance can be directly linked to a length measurement.

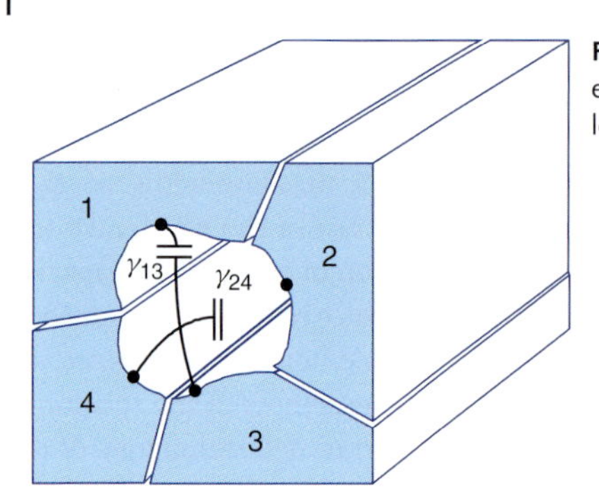

Fig. 9.10 Cross section of a structure with four electrodes. γ_{13} and γ_{24} are capacitances by unit of length.

From the outstanding work of Clothier at NMIA (Clothier, 1965), the most accurate implementation of the theorem is to assemble a system of four identical long parallel and slightly spaced cylindrical electrodes (bars), placed in vertical position at each corner of a square. A movable grounded bar is placed in the cross section of the four main bars. In practice, to avoid the end effects, the measurements are carried out by comparing a fixed capacitance to the capacitance variation of the calculable capacitor for two positions of the movable bar. The length of this displacement is measured by means of a laser interferometer allowing the link of the farad to the meter.

The theorem can be applied to a system composed of more than four electrodes. For example, the calculable capacitor formerly used at LNE consists of five electrodes in the horizontal position arranged at the vertices of a regular pentagon (Trapon *et al.*, 2003). This unique feature strongly differs from the calculable capacitors developed by the other NMIs. If one connects successive pairs of adjacent bars, a five-bar system is equivalent to five different four-bar systems in turn and the theorem can be applied to each of these five systems. In the ideal case of a perfect symmetry, the capacitance per unit of length between two nonadjacent bars is equal to 2γ such as

$$\exp(-2\pi\gamma/\varepsilon_0) + \exp(-\pi\gamma/\varepsilon_0) = 1 \tag{9.18}$$

leading to

$$\begin{aligned}\gamma &= (\varepsilon_0/\pi)\ln[2/(\sqrt{5}-1)](\text{F m}^{-1}) \\ &= 1.356235626\ (\text{pF m}^{-1})\end{aligned} \tag{9.19}$$

The main difficulties to overcome for elaborating any calculable capacitor lie in the alignment of all electrodes, the movable guard, and its trajectory. Moreover, the cross section of the cylindrical electrodes has to be as regular as possible. Fitting the end of the movable guard with a spike is the usual method for significantly reducing the residual cylindricality defect in the cross section of the capacitor (Clothier, 1965). Another error source arises from the frequency effect, which can be due, for example, to the self-inductance of the electrodes (Clothier, 1965) or to electrical connections (Trapon *et al.*, 2003). The most accurate calculable capacitors allow an SI value of the farad with an uncertainty of a few parts in 10^8 (Jeffery *et al.*, 1997).

By means of a complete measurement chain, the calculable capacitors have made possible SI realizations of the ohm and then of R_K. The keystone of the set up is a quadrature bridge, which allows one to compare, with a very high accuracy, impedances of two resistances R_1 and

R_2 against two capacitances C_1 and C_2 first linked to the calculable capacitor. The bridge is balanced when the equation $R_1 R_2 C_1 C_2 \omega^2 = 1$ is true, ω being the frequency of the applied voltages. In practice, C_1 and C_2 have nominal values ranging from 1 to 10 nF. They are compared to the calculable capacitor by successive measurements of capacitance ratios involving transfer standards of 1, 10, and 100 pF. The nominal values of R_1 and R_2 are contained between 10 and 100 kΩ and ω is about 10 000 rad s^{-1} typically. These resistances are then compared at DC to the QHR by means of a CCC resistance bridge. This yields an SI value of R_K, once the frequency variations of the resistances R_1 and R_2 are corrected. This correction is determined by comparing them to a calculable coaxial resistor whose frequency dependence can be calculated from its geometry and whose DC and AC values are nearly equal (Gibbings, 1963; Haddad, 1969).

9.2.6.2 Test of the Exactness of the von Klitzing Relation $R_K = h/e^2$

The best determinations of R_K have uncertainties between two and six parts in 10^8, whereas the most accurate atomic physics measurements lead to values of h/e^2 with uncertainties one or two orders better (Wicht *et al.*, 2002; Cladé *et al.*, 2006; Hanneke, Fogwell, and Gabrielse, 2008). Figure 9.11 shows the weighted mean values of α^{-1} determined through R_K and by indirect methods, which are related to the measurements of other constants and are listed below in order of decreasing uncertainties:

- the ground-state hyperfine transition frequency of muonium $\Delta\nu_{Mu}$;
- the shielded gyromagnetic ratio of proton $\Gamma'_{p-90}(\mathrm{lo})$ or helion $\Gamma'_{h-90}(\mathrm{lo})$ in low magnetic field;
- the quotient of the Planck constant and the neutron mass times the lattice spacing of a silicon crystal $h/m_n d_{220}$;
- the quotient of the Planck constant and the relative atomic mass of cesium or rubidium h/m;
- the anomalous magnetic moment of the electron a_e. The experimental value of a_e is equated to its theoretical value $a_e(\mathrm{th}) = a_e(\mathrm{QED}) + a_e(\mathrm{weak}) + a_e(\mathrm{strong})$, where the first term, given by quantum electrodynamics (QED), can be expressed as a power series in α, the other terms being related to weak and strong interactions. The expression of a_e (QED) has been calculated up to the fourth order (Mohr, Taylor, and Newell, 2008).

All measurements agree within a range of ±1 part in 10^7, confirming the former decision of CIPM in 2000 that it was appropriate to reduce the uncertainty of R_K from two parts to one part in 10^7 (CIPM, 2000) while keeping the conventionally true value R_{K-90} for metrological purposes. Let us note that the uncertainty could not be much more reduced unless relevant reasons were found to discard the $\Gamma'_{p-h-90}(\mathrm{lo})$ values that are inconsistent with the most accurate values of α (Figure 9.11).

There is no significant deviation between the weighted mean of the five R_K values of α on one hand, and, on the other hand, the most accurate values of α, that is, a_e value and the weighted mean of the two h/m values. Moreover, in the framework of the 2006 least-squares adjustment of fundamental constants performed by the CODATA group (Mohr, Taylor, and Newell, 2008), no significant deviation has been found in the relation $R_K = h/e^2$ at the uncertainty level of two parts in 10^8. Let us note that this 2006 CODATA adjustment leads to values of $\alpha^{-1} = 137.035999679$

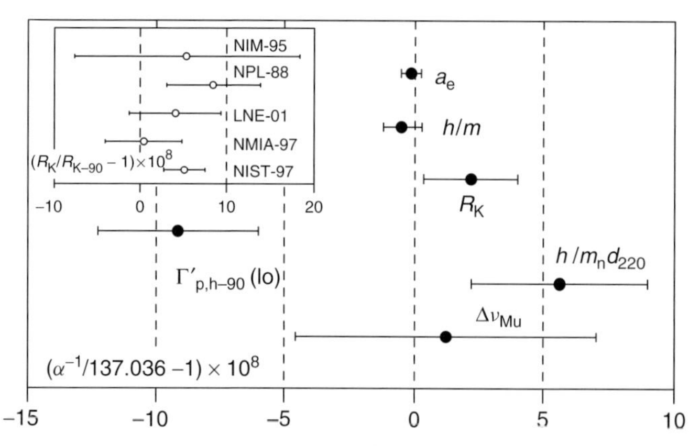

Fig. 9.11 Weighted mean values of α^{-1} obtained by SI determinations of R_K (in inset) and by other methods (Mohr, 2007).

and $R_K = R_{K-90}(1 + 2.16 \times 10^{-8})$ with the same standard deviation uncertainty of 6.8 parts in 10^{10}.

9.3
Josephson Voltage Standard

9.3.1
Josephson Effects and Universality of K_J

In 1962, Josephson discovered a remarkable consequence of the macroscopic coherence of the superconducting state (Josephson, 1962). The Josephson Effects (JE) take place at low temperature in a junction of two weakly coupled superconductors, for example, two superconducting electrodes separated by a thin insulating layer. In each superconductor, the electrons paired in Cooper pairs form a condensate, which is described by a unique wavefunction, $\Psi = Ae^{i\phi}$ with a phase ϕ coherently maintained over macroscopic distances. Josephson predicted two effects:

- Cooper pairs tunnel the junction even at zero voltage drop between the terminals, giving a DC supercurrent $I_J = I_C \sin\phi$, where I_C is a constant and $\phi = \phi_1 - \phi_2$ is the phase difference between the wavefunctions in the two superconductors.
- A constant voltage U across the junction induces an oscillation of the tunnel supercurrent at a frequency $f = (2e/h)U$ ($2e/h$ corresponding to the inverse of the flux quantum Φ_0). It leads to the time variation of the phase due to an energy change $2eU$ involved when a Cooper pair tunnels:

$$d\phi/dt = (2e/\hbar)U \tag{9.20}$$

This second effect, called the AC Josephson effect and first observed by Shapiro (1963), provides a perfectly reproducible and universal voltage standard, since the voltage is directly linked to the frequency.

The great interest of metrologists since the 1970s for the JE as a means of maintaining the volt results from both theoretical and experimental works, which tend to verify the validity of the relation $f = (2e/h)U$. From topological arguments, Bloch has shown that the constant involved in this relation is $2e/h$ exactly (Bloch, 1970). Numerous experiments have investigated

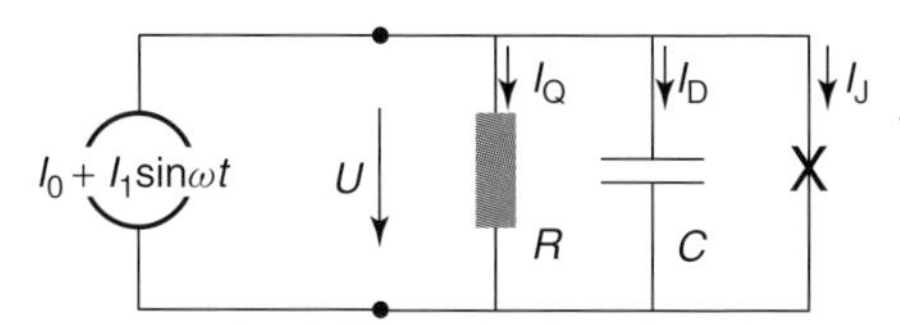

Fig. 9.12 Equivalent electrical circuit of a Josephson junction according to the model proposed by Stewart and Mc Cumber.

the dependence of the voltage to frequency ratio on materials (Pb, Sn, In, Nb, and more recently YbaCuO) or on junction types (microbridge, tunnel junction, or point contact). An upper limit of less than two parts in 10^{16} has been established so far (Tsai, Jain, and Lukens, 1983).

From early on, the JE was used for reproducing the volt. However, because of slightly different values of $2e/h$ obtained from SI realizations of the volt, no international consensus on a single valuc was possible. Fortunately, a better agreement was found between determinations of $2e/h$ performed in 1980s. This allowed the CIPM to recommend implementing the JE as a voltage standard, using the Josephson constant K_J as an estimate of $2e/h$, and for calibration purpose by assigning to it one single conventional value (CIPM, 1988a):

$$K_{J-90} = 483\,597.9 \text{ GHz V}^{-1} \text{ exactly} \tag{9.21}$$

In terms of SI units, the uncertainty on K_J is four parts in 10^7 until now.

9.3.2 Current–Voltage Characteristics of a Josephson Junction

The dynamic behavior of a real Josephson junction connected to an external circuit can be described by means of the resistively and capacitively shunted junction (RCSJ) model (Stewart, 1968; Mc Cumber, 1968). In this model, the equivalent circuit of the junction consists of three arms in parallel, respectively, crossed by

- the supercurrent $I_J = I_C \sin\phi$;
- the current carried by quasi-particles (due to broken Cooper pairs since $T \neq 0$), $I_Q = U/R$, where U is the voltage drop across junction terminals and R is approximately the tunnel resistance of the junction; and
- the displacement current $I_D = CdU/dt$ through the capacitance C between superconducting electrodes (Figure 9.12).

The sum of these currents has to be equal to the current $I(I = I_{dc} + I_1 \sin\omega t)$ supplied to the junction by the external source. Eliminating U given by relation (12.18) yields a second-order nonlinear differential equation:

$$I = (\hbar C/2e)d^2\phi/dt^2 + \hbar/(2eR)d\phi/dt + I_C \sin\phi \tag{9.22}$$

Introducing dimensionless quantities $i = I/I_C$ and $\tau = 2\pi f_P t$, where $f_P = (eI_C/(\pi hC))^{1/2}$ is the plasma frequency, Equation (9.19) becomes

$$i = d^2\phi/d\tau^2 + \beta_C^{-1/2} d\phi/d\tau + \sin\phi \tag{9.23}$$

where $\beta_C = (2\pi RCf_P)^2$ is the Mc Cumber parameter. In the general case of nonnegligible junction capacitance, these Equations (9.19) and (9.20) have to be solved numerically.

For overdamped junctions ($\beta_C \leq 1$) supplied with an alternating current at a microwave frequency f, it is found that the I–V characteristic can exhibit a series

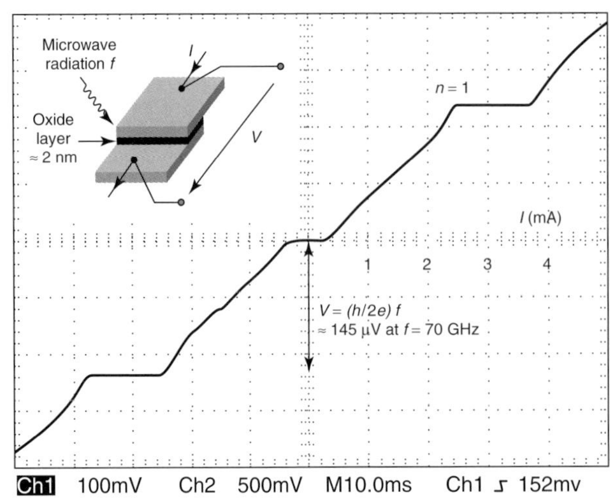

Fig. 9.13 First voltage steps from overdamped Josephson junctions irradiated at a frequency of 70 GHz. In inset, a drawing of Josephson junction based on a thin oxide layer separating two superconducting electrodes.

of stable voltage steps at constant values (Figure 9.13):

$$U_n = n(h/2e)f \tag{9.24}$$

with n being an integer.

9.3.3 Josephson Array Voltage Standard (JAVS)

The first Josephson standards were based on single overdamped junctions. They generated a voltage of only a few millivolts (typically operating on high-order steps, $n > 500$ at $f \approx 10$ GHz). The calibration of the former primary standards such as 1.018 V saturated Weston cells then implied the use of a voltage divider, and the final uncertainty was limited at a level of 100 nV. Reducing this uncertainty naturally motivated the increase in the Josephson voltage by placing in series a large number of junctions. Levinson *et al.* suggested the use of underdamped junctions ($\beta_C > 1$) that have the great advantage of delivering several voltage steps at zero bias current if the radiation frequency is much higher than the plasma frequency (Levinson *et al.*, 1977). Consequently, only a single current source is necessary to bias an array of N junctions placed in series. In the framework of a National Institute of Science and Technology (NIST)-PTB collaboration, the first 1-V voltage arrays made of 2000–3000 superconductor–insulator–superconductor (SIS) junctions were elaborated (Niemeyer, Hinken, and Kautz, 1984). 1-V and 10-V arrays, the latter being composed of more than 10 000 junctions, are now commercially available.

In addition to the design requirements (Hamilton *et al.*, 1989; Pöpel *et al.*, 1991) on the junction parameters (size, plasma frequency, and critical current), the condition of homogeneous irradiation over the array must be fulfilled. This requires a specific implementation of the junctions in an integrated circuit. Figure 9.14a shows a 10-V array developed by PTB (Pöpel *et al.*, 1991). The array is made of 13 924 Nb/AlO_X/Nb junctions placed along a microstrip line, which is divided in four arms of 3481 junctions in order to reduce the attenuation of the microwave radiation. A finline antenna allows to couple the microwave (80 GHz) to the transmission line. For such an array, each junction typically operates on the fourth

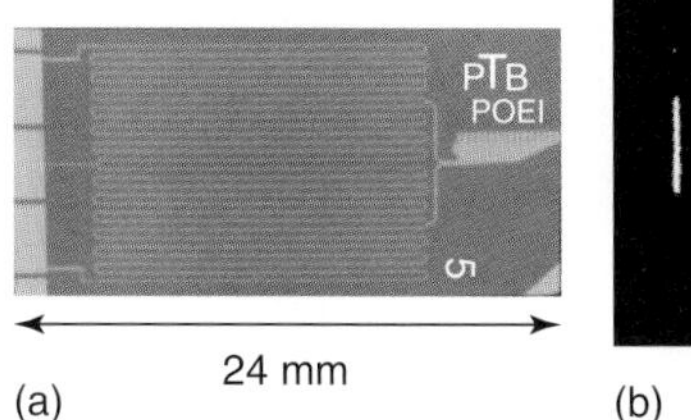

(a)

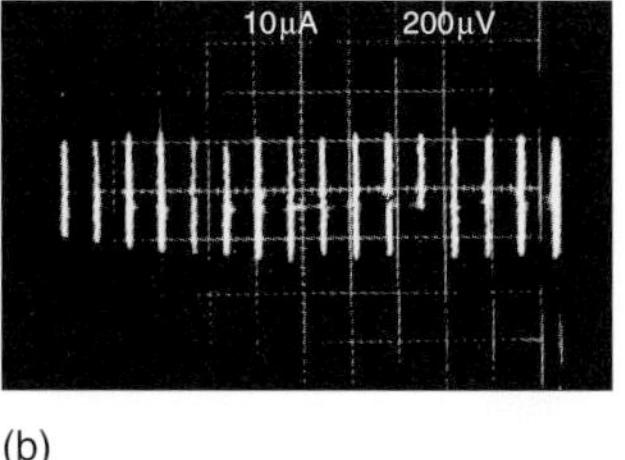

(b)

Fig. 9.14 (a) PTB 10-V array composed of 13 924 junctions ($18 \times 50\ \mu m^2$) in series along a transmission microstrip line. Photograph by courtesy of the PTB. (b) Voltage steps around 10 V observed on the I–V characteristic of a 10-V array with 80-GHz microwave irradiation.

or fifth step over the eight steps it can deliver and then the array provides a total voltage of 10 V, with steps equally spaced ($\Delta V = 165\ \mu V$ at 80 GHz), as shown in Figure 9.14b.

9.3.3.1 Josephson Calibration Bench

A typical measurement bench used for calibrating secondary standards such as 1.018-V Weston cells and 1- or 10-V Zener diode references (described in the following section) against JAVS is sketched in Figure 9.15. It is composed of four parts and can be fully computer controlled:

- *The cryogenic part*: The array is mounted in a cryoprobe fitted with a low loss wave-guide. Three pairs of wires are used for characterizing the voltage steps, biasing the array with DC current, and measuring the metrological voltage. The latter is filtered at the top against electromagnetic interference.
- *The microwave radiation source*: The circuit consists of a Gunn diode servo-controlled on a 10-MHz clock signal (rubidium crystal) by means of a phase-locked frequency counter and then referred to an atomic clock.
- *The electronics unit*: A DC current source is used to bias the array on the desired voltage step through an adjustable resistor placed in parallel.
- *The metrological circuit*: A digital nanovoltmeter coupled to a switch of low electromotive force measures the voltage difference between the JAVS and the device to be calibrated.

The typical uncertainties for a routine calibration of a Weston cell can be reduced down to a few tens of nanovolts (Hamilton and Tang, 1999). In the case of a Zener diode reference, the calibration uncertainties are much degraded mainly due to their $1/f$ noise (Witt, 2003).

9.3.3.2 Secondary Voltage Standards

Although the cadmium cell, invented by E. Weston in 1891, remains so far the electrochemical cell artifact standard of electromotive force with the highest stability with a time drift as low as $100\ nV\ V^{-1}$ per year, the Zener diode reference is taking its place as working standard directly calibrated against the JAVS. The latter is easier to use than the saturated Weston cell and significant improvements have been made concerning its stability, with an annual drift in the order of $1\ \mu V\ V^{-1}$, and its temperature dependence. The composition of these two types of standard is described below and their main characteristics are given.

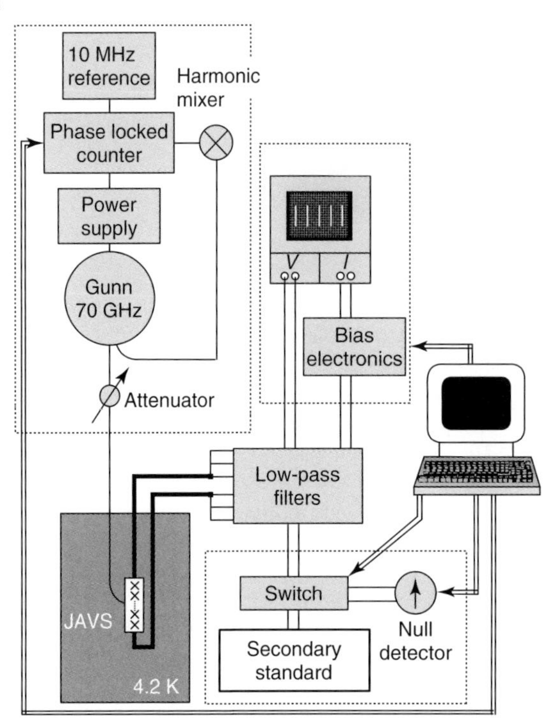

Fig. 9.15 Schematic of Josephson bench used for calibrating secondary voltage standards (1.018-V Weston cells and 1- or 10-V Zener diode references).

- *The Weston standard cell*: The electrolyte used is a solution of cadmium sulfate permanently saturated by hydrated crystals in excess ($CdSO_4, 8/3H_2O$). The positive electrode is composed of mercury surmounted by a mixture of cadmium and mercury sulfates ($CdSO_4 + Hg_2SO_4$), whereas the negative electrode is constituted of amalgam of mercury and cadmium (10–12.5% for the latter). The electromotive force (emf) delivered by the standard cell amounts to 1.01864 V at 20°C (with no load). Its strong dependence on temperature, about −40 μV per °C at 20°C, requires placing the cell in a thermoregulated enclosure, which guarantees a stability better than 0.01°C and a temperature homogeneity of the same order. The more difficult aspect about the use of the electrochemical cell is to absolutely avoid any current flow and vibration (neither shaken nor stirred). For example, let us consider a cell with internal resistance of 800 Ω. After a short circuit applied during 1 minute, the time needed for the cell to recover the initial emf at ±10 and at ±1 μV is typically about 80 minutes and more than five days, respectively.
- *The Zener diode reference*: The reverse current I_i of a p–n junction semiconductor diode abruptly increases once the reverse voltage exceeds an onset value V_Z. With suitable doping, reproducible values V_Z from 2 to 200 V can be obtained as well as a low dynamic resistance R_d characterizing the inverse of the variation slope (a few ohms up to a few tens of ohms). Moreover, the temperature coefficient of such diodes, so-called Zener diodes, is a monotonic function of V_Z and nulls at $V_Z \approx 6$ V.

The use of Zener diode as voltage standard comes from the possibility to compensate this temperature effect.

The Zener reference standards commercially available are temperature compensated by using a Zener diode of weakly negative temperature coefficient (V_Z between 6.5 and 7 V) placed in series with an n–p–n transistor of positive coefficient whose value depends on the collector current. The output voltage of the whole system (Zener-transistor) is then amplified to 10 V. In general, the reference standard is also fitted with another voltage output, delivering 1.018 V by means of a resistance divider. The temperature coefficient is in the order of a few 10^{-8} K^{-1}. Unlike the Weston cell, a short circuit creates no damage for the Zener reference. However, the latter generates a voltage with a $1/f$ noise up to a frequency of few tens of hertz or few hundreds of hertz, which limits the measurement time.

9.3.3.3 Josephson Voltage Comparison

In the framework of the natural exercise of NMIs to check the coherence of their own standards, a large number of international comparisons involving 1 and 10 V Josephson underdamped junction arrays have been performed (Reymann and Witt, 1993). The direct comparisons of JAVS are the most precise (in the indirect comparisons, the accuracy is mainly affected by the instabilities and noise of traveling secondary standards). In general, they show an agreement with an uncertainty of one or two parts in 10^{10} as shown in Figure 9.16 (Wood and Solve, 2008). Some comparisons with even smaller uncertainties have been reported (Reymann *et al.*, 1999). A very good agreement, a few parts in 10^{11}, has also been found in a recent comparison involving the new generation of JAVS designed to be programmable as described later (Djordjevic *et al.*, 2008).

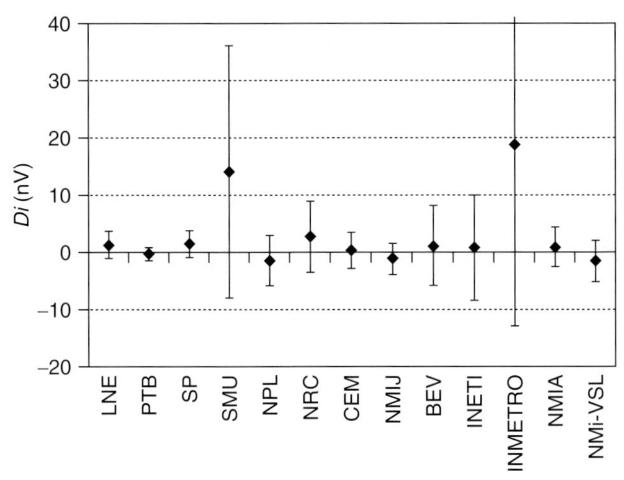

Fig. 9.16 Results of the 10 V direct on site Josephson voltage comparison obtained until November 2007 (BIPM.EM.K10.b). The data are the voltage differences from the BIPM voltage reference and given with uncertainties at two standard deviations (95% of confidence interval) (Wood and Solve, 2008).

9.3.4 Programmable JAVS

Although 1- and 10-V underdamped Josephson junction arrays provide NMIs with highly reproducible primary voltage standards, their working mode remains tricky and prevents other metrological applications. The voltage steps are not stable enough against environmental electromagnetic noise and they cannot be quickly and unambiguously settled. Consequently, a fully computer-controlled calibration of Weston cells (without any risk of current flow) or a calibration of digital voltmeters is a difficult task. Generating arbitrary waveform AC voltage at audio frequency from these arrays is also not possible. These applications and others such as fast-reversed DC voltage measurements for AC/DC testing of thermal converters (Burroughs *et al.*, 1999; Funck, Behr and Klonz, 2001), QMT (Steck *et al.*, 2008) and watt balance experiments (Steiner *et al.*, 2007), development of a quantum voltmeter (Behr *et al.*, 2001), a Josephson potentiometer (Behr *et al.*, 2003a), and last but not the least a quantum AC power standard (Burroughs *et al.*, 2007a; Palafox *et al.*, 2007), become possible by means of programmable arrays. These arrays, operating like a digital/analog (D/A) converter, are of two types: binary divided or pulse driven.

9.3.4.1 Binary Divided Josephson Junction Array

Invented by Hamilton, Burroughs, and Kautz (1995), the array is divided into segments of $M(= 2^N)$ overdamped Josephson junctions, and each segment can be independently set to the $n = 0, \pm 1$ steps by applying on it the corresponding bias current $I_B = 0, \pm I$, and the microwave irradiation f (Figure 9.17). The output voltage of the array is the sum of the voltages developed at each segment, $V_{seg} = 0, \pm Mf/K_J$, and hence this allows the generation of any voltage from $-M_{tot}f/K_J$ to $+M_{tot}f/K_J$ with an increment of $M_{min}f/K_J$, where M_{tot} is the total number of the junctions and M_{min} the number of junctions of the smallest segment.

In practice, the overdamped junctions are preferably made from SNS (Benz, 1995; Yamamori *et al.*, 2006) or SINIS (S superconductor, N normal, and I insulator) technology (Kohlmann *et al.*, 2000; Mueller *et al.*, 2007), but externally shunted SIS junctions (Hassel *et al.*, 2001) can be used as well. From these binary arrays, any arbitrary voltage can be adjusted without ambiguity and settled in a very short time, for example, less than 1 μs. Moreover, the large step width (1 mA for SNS or SINIS junctions compared with 100 μA at the most for SIS junctions) makes the voltage steps highly stable. However, because they occur at nonzero bias current,

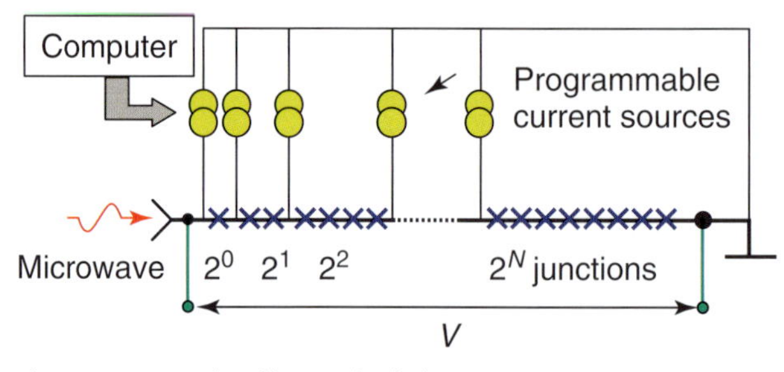

Fig. 9.17 Principle of binary-divided array.

the voltage steps may be sloped due to possible small series resistances within the array (Behr *et al.*, 2003b). The flatness has thus to be checked carefully (Behr *et al.*, 1999; Jeanneret, Rüfenacht, and Burroughs, 2001).

Numerous direct comparisons between the different kinds of binary arrays and the conventional SIS arrays have been reported in the literature. They have shown very good agreement within a few tenths of nanovolts at the 1-V level (Behr *et al.*, 1999; Behr *et al.*, 2003b; Jeanneret, Rüfenacht, and Burroughs, 2001; Lo-Hive *et al.*, 2003) as well as at the 10-V level (Djordjevic *et al.*, 2008). In addition, a direct comparison of the voltages delivered by two segments, each composed of 4086 junctions, of a same SINIS array, has shown no difference at a level of two parts in 10^{17} (Krasnopolin, Behr, and Niemeyer, 2002). All these results explain the trend that programmable JAVS based on binary arrays is replacing conventional JAVS based on SIS arrays.

In theory, the binary-divided array can also deliver AC voltages by selecting quickly and adequately voltage steps for each segment. The resulting signal is an approximate sinusoidal wave sampled by steps. However, during the period between two successive steps, the voltage is not quantized at all. This transient problem leads to errors on the root mean square (rms) value of the signal, which could be very important for metrological purposes (Hamilton, Burroughs, and Benz, 1997).

As already mentioned, the challenge is to extend the application domain of the JE to the field of electrical metrology at audio frequencies. In that field, the "core" standards are presently thermal voltage converters (an AC–DC transfer standard with a constant gain over a wide frequency range, from DC up to a few hundreds of MHz) and the starting point of any calibration is 1 V, then scaling down to a few millivolts and scaling up to few thousands of volts. A feature of such devices is a low input impedance, which consequently requires them to be provided with load current. Considering this further difficulty, investigations on the transients are being pursued by NMIs to clarify if with a binary-divided array the quantum calibration of these standards could be accurate enough, that is, with an uncertainty better than one part in 10^6 at 1 kHz (Behr *et al.*, 2006; Burroughs *et al.*, 2007b; Katkov, Behr, and Lee, 2008). Very recently, two new methods based on sampling techniques (differential or synchronous) that avoid the transient problem have been proposed (Behr *et al.*, 2007; Jeanneret *et al.*, 2008; Rüfenacht *et al.*, 2008).

9.3.4.2 **Pulse Driven Array**

In fact, an actual AC quantum voltage standard can be realized from a series array of overdamped Josephson junctions that, instead of being irradiated by sinusoidal microwaves, are driven with a train of short current pulses (Benz and Hamilton, 1996; Benz and Hamilton, 2004). For a fixed repetition frequency of the pulse f, the time-integrated value of the output voltage is $V = nNf/K_J$, where n is the step number that depends on the pulse height and N is the number of junctions.

In contrast to the case of sinusoidal excitation for which a frequency modulation induces a chaotic behavior of the junction, or strongly affects the step width, the voltage steps do not depend on the frequency f if the pulse width remains short enough: $\tau_p < 1/f$ with $\tau_p = 1/(2\pi I_C R K_J)$. Consequently, the output voltage of the array is easily adjustable by varying f.

Focusing on the main application of pulse-driven Josephson array in audio frequency electrical metrology (other applications exist such as the generation of waveforms with multiple harmonics, or pseudo-random noise for thermometry), the NIST, PTB, and NMI/VSL (Van Swinden Laboratorium) are developing pulse-driven AC Josephson voltage standards. The ultimate aim is to reach an rms voltage of 1 V (Benz *et al.*, 1999; van den Brom *et al.*, 2008). A digital code generator delivering pulses clocked at a high frequency (10 Gbit s^{-1}) in a predetermined sequence followed by a delta–sigma modulator used for D/A conversion allows to synthesize arbitrary waveforms. The rms voltage is calculable with a fundamental accuracy from the knowledge of the number of pulses and their positions in time. Promising results on synthesizing AC voltages with rms values of a few hundreds of millivolts have been reported by the two groups (Houtzager, 2008; Benz 2008), one of them claiming that a direct synthesis of accurate and low distortion waveforms of 1 V_{rms} is possible.

9.3.5 Determination of the Josephson and Planck Constants

The Josephson constant K_J can be directly determined from an SI realization of the volt. The principle consists in measuring the attraction force F between two electrodes, one fixed and the other movable, on which a voltage drop U is applied. The measurement of U in terms of the JE leads to an SI value of K_J:

$$K_J = nf(2F/(dC/dx))^{-1/2} \qquad (9.25)$$

where n and f refer to the step number and the microwave frequency and dC/dx is the variation of the capacitance between electrodes with their distance. dC/dx is measured in SI units using a Thompson–Lampard calculable capacitor. Two sets of apparatus have given rise to K_J values in 1980s, the liquid-mercury electrometer developed by NMIA (Clothier *et al.*, 1989) and the voltage balance at PTB (Funck and Sienknecht, 1991), with uncertainties around three parts in 10^7. Better uncertainties have been obtained by means of the moving coil watt balance.

9.3.5.1 Watt Balance Experiments

The experiment originally proposed by Kibble more than 30 years ago (Kibble, 1976) and described in more detail by Eichenberger, Jeckelmann, and Richard (2003) consists in two successive phases:

- weighing (with a mass m) the Laplace force created on a current (I) carrying coil in a magnetic flux and
- then measuring the electromotive force (U) created at the input of the same coil when it is moved in the same magnetic flux at constant speed (v). From these two phases, the following relation results:

$$UI = mgv \qquad (9.26)$$

which makes electrical and mechanical powers equivalent. The measurements of electrical quantities in terms of the JE and the QHE and of mechanical quantities in SI units lead to an SI determination of the product $K_J^2 R_K$. If R_K is measured in SI by an independent method, then K_J is given in the simplified form:

$$K_J = f/(mgvR_K)^{1/2} \qquad (9.27)$$

The ultimate uncertainty expected on K_J by this method is of the order of one part in 10^8.

The watt balance experiment combined with the SI realization of the ohm consequently leads to the SI determination of the volt as well as of the ampere. Moreover, if the relations $K_J = 2e/h$ and $R_K = h/e^2$ are assumed exact, then the experiment gives a value of the Planck constant:

$$h = 4/(K_J^2 R_K) = 4(mgv)/f^2 \quad (9.28)$$

9.3.5.2 Planck Constant and Validation of the Josephson Relation $K_J = 2e/h$

Figure 9.18 shows values of h determined until now by means of watt balances at National Physical Laboratory (NPL) and NIST and values obtained by other electrical methods (volt balance) and by indirect methods (involving the Faraday constant, the shielded gyromagnetic ratio of the proton in high field, and the molar volume of silicon). The relations between these different quantities are given in Mohr, Taylor, and Newell (2008). First, this figure indicates that the watt balance is the technique of highest accuracy so far (reaching so far 3.6 parts in 10^8). Second, all the electrical methods are in good agreement within three parts in 10^7. Let us note, however, that the two most accurate values of $K_J^2 R_K$ between NPL (Robinson and Kibble, 2007) and NIST (Steiner *et al.*, 2007) significantly disagree, by nearly three standard deviations. Third, the most difficult issue is the strong disagreement between the $K_J^2 R_K$ values of h and the $V_m(\mathrm{Si})$ value of h. This amounts to one part in 10^6 taking into account the weighted mean value of $K_J^2 R_K$.

In the framework of the 2006 least squares adjustment of fundamental constants and considering all these results except the last value of $K_J^2 R_K$ of NPL (not considered because it was published too late), the CODATA task group recommends a value $h = 6.62606896 \times 10^{-34}$ J s with a standard uncertainty of five parts in 10^8 (Mohr, Taylor, and Newell, 2008). The recommended value for K_J is very close to the 1990 CIPM recommended value, $K_J = K_{J-90}(1 - (1.9 \pm 2.5) \times 10^{-8})$, which mainly results from the value of NPL obtained at that time (Mohr, Taylor, and Newell, 2008).

The CODATA task group has also performed several least squares adjustments to test the validity of the Josephson relation and at the same time that of von Klitzing relation. These relations have been relaxed by introducing adjustable correction factors, using the same notations:

$$K_J = 2e/h(1 + \varepsilon_J) \text{ and } R_K = h/e^2(1 + \varepsilon_K) \quad (9.29)$$

Considering first the whole set of input data, the Josephson relation $K_J = 2e/h$ has been found valid, from a statistical point of view, with an uncertainty of eight parts in $10^8 (\varepsilon_J = (-77 \pm 80) \times 10^{-9})$. This could mean a convincing argument for CIPM to reduce by a factor of four the uncertainty on the recommended value K_{J-90} of K_J if expressed in terms of SI units.

However, by alternatively deleting the $V_m(\mathrm{Si})$ datum and the $\Gamma'_{p-90}(\mathrm{lo})$ and $\Gamma'_{h-90}(\mathrm{lo})$ data from the set of input data, opposite and huge significant deviations have been found on K_J:

$$\varepsilon_J = (-281 \pm 95) \times 10^{-9} \text{ with } V_m(\mathrm{Si}) \text{ datum deleted} \quad (9.30)$$

$$\varepsilon_J = (+407 \pm 143) \times 10^{-9} \text{ with } \Gamma'_{p-90}(\mathrm{lo}) \text{ and } \Gamma'_{h-90}(\mathrm{lo}) \text{ data deleted} \quad (9.31)$$

These deviations, which compensate each other in the first validation test, reflect the corresponding inconsistencies previously mentioned and are shown in Figures 9.11 and 9.18.

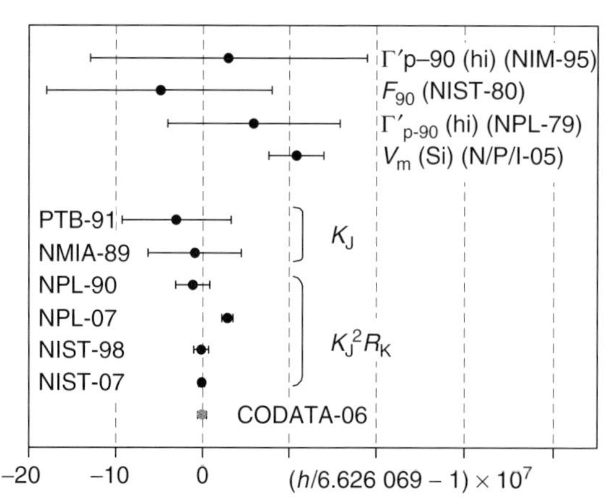

Fig. 9.18 Values of the Planck constant h obtained by direct measurements (volt and watt balances) and through indirect method involving measurements of the quantity F_{90} related to the Faraday constant, the shielded gyromagnetic ratio of proton in high field Γ'_{p-90}(hi), and the molar volume of silicon V_m(Si).

Finally, deleting all three discrepant data results in a value $\varepsilon_J = (+238 \pm 720) \times 10^{-9}$, which is consistent with zero but with an uncertainty around seven parts in 10^7. It must be noted that for each of these different least squares adjustments, the unknown correction factor ε_K has been found consistent with zero with an uncertainty of near two parts in 10^8.

NMIs are thus pursuing their efforts for solving the problem of the discrepancy between the two ongoing routes ($K_J^2 R_K$ and V_m(Si)) leading to h and consequently for testing the Josephson relation with a better accuracy. The targeted and reachable uncertainty is one part in 10^8. Once the discrepancy removed, monitoring the kilogram, for example, by means of a watt balance will become a relevant task and would allow, in time, to redefine the kilogram by fixing the Planck constant.[1] With this aim, NIST is improving its moving coil apparatus, whereas other institutes (METAS, LNE, and BIPM) are working on the development of new watt balances (Eichenberger, Jeckelmann, and Richard, 2003; Genevès *et al.*, 2005; Picard *et al.*, 2006).

1) Another competitive proposal is to fix Avogadro number N_A instead of h for a new definition of the unit of mass.

9.4 Toward New Quantum Standards Based on Coulomb Blockade

9.4.1 Representation of the Ampere

Although the ampere is the base electrical unit, its representation requires the volt and the ohm. This fact can be explained for a practical reason: the quantum-mechanical standards QHRS and JAVS enable one to achieve uncertainties much lower than any experiment or artifact for a representation of the ampere realized up to now. However, a quantum current standard is developed in the framework of the metrological triangle experiment. The simplest idea imagined by physicists is a current source controlled electron by electron. For 20 years, the development of the nanofabrication has made it possible to create sub-microndevices that allow the manipulation of individual electrons (Fulton and Dolan, 1987). These SET devices consist of tunnel junctions in series forming isolated conductors, essentially "metallic islands" whose charge state is controlled by means of gate electrodes. The Coulomb blockade is the physical phenomenon, which is briefly described in the following section. Detailed bibliographies can be found in Averin and Likharev, 1991; Grabert and Devoret, 1991; and Likharev, 1999. In theory, a SET device like an electron pump can transfer millions of single charges through the circuit with an expected intrinsic uncertainty reaching one part in 10^9, and a SET electrometer can detect $\approx 10^{-5}e$ in a 1-Hz bandwidth. As a result, these systems have opened the path toward the realization of quantum current standards based on highly accurate current sources or ultrasensitive electrometers for making sensors or electrons counters. It is noteworthy that in the shorter term the direct use of SET as a primary current standard will be relevant for a current range less than 1 nA (Figure 9.19). This concerns the calibration of sub-nanoammeters (commercial electronic devices or home-made integration bridges), which allow NMIs in the electrical domain to then calibrate their own secondary low-current standards and high-value resistance standards ($R > 1\ \mathrm{T\Omega}$). The improvement of the traceability of small currents should benefit instrument manufacturers (detectors or meters of small electrical signals) and the semiconductor industry (characterization of components and testing of wafers).

A SET device consisting of an electron pump servo-controlled by an electrometer enables one to realize a quantum capacitance standard as well. This provides an elegant alternative to the representation of the farad, which is presently reproduced by implementing either a calculable capacitor or the QHE (DC or AC) associated to a measurement chain linking resistance to capacitance. Undoubtedly, these SET devices will play an important role for the "mise en pratique" of the possible redefinition of the ampere and the farad within the framework of a new SI.

After a few theoretical considerations about the Coulomb blockade, using the basic SET transistor as an example, the electron pump intensively investigated in NMIs is briefly described in the following subsection. Connected to an external RF source generating an f frequency harmonic signal, this device transfers one electron charge per cycle. Consequently, the current amplitude is proportional to the elementary electron charge and the applied frequency: $I = ef$. The last subsection lists devices

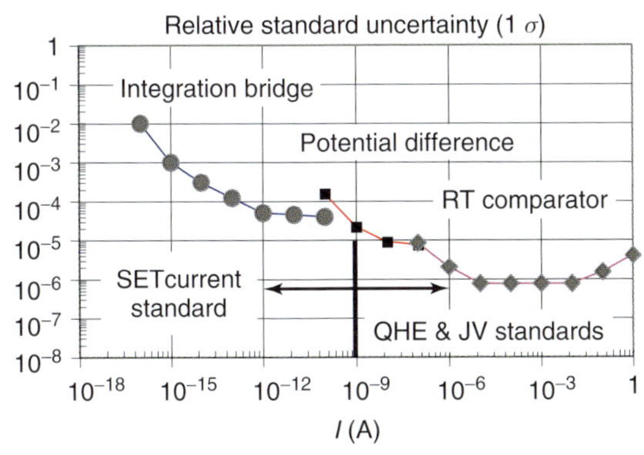

Fig. 9.19 Different methods (current comparator at room temperature, potential difference bridge, and integration bridge) used for the calibration of direct current less than 1 A and corresponding uncertainties. The integration method consists in the measurement of the rising time of the supplied current for a given voltage variation applied to a capacitance or inversely in the measurement of a varying voltage across a same capacitance along a known period.

insuring the transport or control of single charges, which are presently studied in NMIs for developing a quantum standard possibly at higher currents than the present electron pump.

9.4.2
Single Electron Devices

9.4.2.1 Double Tunnel Junction and Coulomb Blockade

The Coulomb blockade of electron tunneling, observed for the first time in disordered granular materials (Zeller and Giaver, 1969), takes place in a SET device when a metallic *island* is electrically insulated from the rest of the circuit. If the total capacitance of the island C_Σ is sufficiently small compared to e^2/kT, the energy change required for the addition or subtraction of one electron on the island becomes high enough to prohibit the tunneling transfer of other electrons. The first remarkable feature of such a device is that the island consists of billions of electrons but remains sensitive to the presence of a single additional electron. Second, the extra electron number on the island is necessarily an integer, whereas $Q_{1,2}$, charges of the first or second junctions, can be fractional charges. When thermally activated transport is suppressed (typically $T \ll 4$ K), tunneling of electrons through the circuit is possible only under certain bias conditions.

The SET transistor (see inset Figure 9.20) often used as an electrometer is the simplest SET device and consists of a single metallic island (usually aluminum) separated by thin insulating tunnel barriers (alumina) and coupled to a gate electrode through a capacitor (C_G). The energy needed for transferring an electron through the first or the second junction is the change of free energy $\Delta E^{\pm}_{1,2}$ of the complete circuit:

$$\Delta E^{\pm}_{1,2} = e^2/(2C_\Sigma) \pm eV_{1,2} \tag{9.32}$$

with $V_{1,2} = Q_{1,2}/C_{1,2}$ where $C_{1,2}$ are capacitances of the first and second junctions, respectively. From the above-mentioned equation, the Coulomb energy, $E_C = e^2/(2C_\Sigma)$, is recognized. Moreover, tunneling into an unoccupied state is possible if the electron gains kinetic energy coming from the decrease in the electrostatic energy of the system, and thus a threshold voltage occurs:

$$\Delta E^{\pm}_{1,2} \leq 0 \Rightarrow |V_b| \geq V_t = e/C_\Sigma \tag{9.33}$$

As shown in Figure 9.20 (inset), the transport properties of the transistor change periodically with the gate voltage V_G, or, in other words, with the charge induced on the island by the gate. The period corresponds to one electron induced on the island. V_G can be adjusted such that the electron transfer through the device is blocked and so the current is zero. Consequently, the tunneling can be stopped, thanks to two parameters: V_G and V_B.

Figure 9.20 also shows the measured I–V characteristic for a SET transistor with symmetrical junctions ($C_1 = C_2 = C_\Sigma/2$) and for two gate voltages where the Coulomb gap is maximum and minimum. In the case of blockade state, the conduction below the threshold voltage ($V_t = e/C_\Sigma$) is close to zero. By changing the gate voltage, the Coulomb gap can be completely suppressed and the curve appears almost linear.

9.4.2.2 Electron Pump

The SET pump, first investigated by Pothier *et al.* is a device allowing the transfer of electrons one by one at an adjustable clock

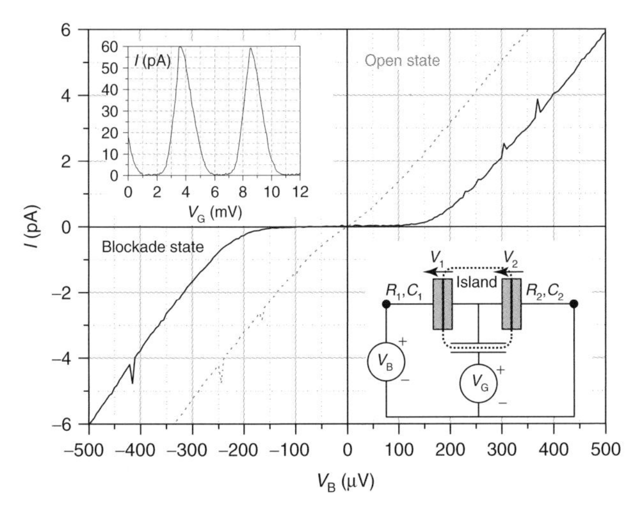

Fig. 9.20 $I - V_B$ characteristics of SET transistor (blue: $V_G = -4$ mV open state and red: $V_G = 1$ mV closed state). Left inset: $I - V_G$ characteristics obtained at bias voltage of 100 μV. Right inset: schematic view of a transistor: the triple-point regions (Feltin *et al.*, 2003). This diagram has been obtained at LNE by means of a CCC used as a current amplifier (Gay, Piquemal, and Genevès, 2000; Piquemal *et al.*, 2007).

frequency f and in a quasi-adiabatic way (Pothier *et al.*, 1991). The current through the electron pump can be expressed as follows: $I = e \times f$. The simplest electron pump consists of two metallic *islands* separated by three junctions (here we assume $C_1 = C_2 = C_3$). The gate voltages V_{G1} and V_{G2} through the gate capacitances C_{G1} and C_{G2} control the electric potential of each island.

The pump operation can be illustrated by means of the typical diagram given in Figure 9.21a, which displays the stability domains of the different states (n_1, n_2) in the $V_{G1} \otimes V_{G2}$ plane. The integer couple (n_1, n_2) denotes the number of extra charges present on the first and the second island. The points (Figure 9.21b), the so-called triple points, where conduction can take place are located at the intersection of three neighboring domains. Everywhere else, the pump is in a blockade state and the electron configuration (n_1, n_2) is stable. Dashed lines represent the boundaries between stability domains and form a typical honeycomb pattern. The pumping of electrons is based on these topological properties.

The controlled transfer of electrons is obtained in the following way: two periodic signals with the same frequency f but phase shifted by $\Phi \approx 90^\circ$ are superimposed on each applied DC gate voltage couple (V_{G10}, V_{G20}) as follows:

$$V_{G1} = V_{G10} + A \cdot \cos(2\pi ft) \tag{9.34}$$

$$V_{G2} = V_{G20} + A \cdot \cos(2\pi ft + \Phi) \tag{9.35}$$

When the DC voltages (V_{G10}, V_{G20}) correspond to coordinates of the point denoted by P in Figure 9.21b, the circuit follows a closed trajectory around P. The configuration changes from (0, 0) to (1, 0), then from (1, 0) to (0, 1), and returns to the initial state (0, 0). In the real space, the complete sequence involves the transfer of one electron through the pump.

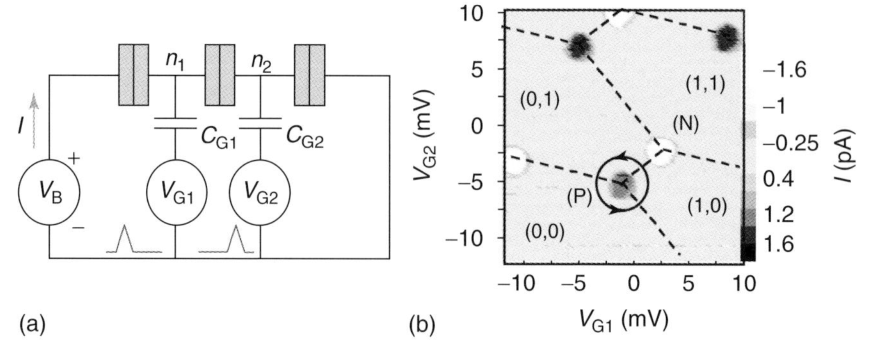

Fig. 9.21 (a) Schematic view of a R-pump device. (b) Stability diagram in $V_{G1} \otimes V_{G2}$ plane, which displays the stable configurations (n_1, n_2) of numbers of the extra electrons on each island. Boundaries between each domains (dashed lines) form a typical honey-comb pattern. The charges tunneling transfer as a function of gate voltages V_{G1} and V_{G2} takes place only in the triple-point regions (Feltin *et al.*, 2003). This diagram has been obtained at LNE by means of a CCC used as a current amplifier (Gay, Piquemal, and Genevès, 2000; Piquemal *et al.*, 2007).

The frequency is chosen low compared to the reciprocal of the tunnel rate ($f \ll R_j C$, where R_j is the junction resistance). This condition ensures that tunneling occurs near threshold (i.e., near the boundaries shown by dashed lines in Figure 9.21b), which means that the system remains close to its ground state at all times. This quasi-adiabatic condition minimizes heating due to hot electrons entering the islands well above the Fermi energy. By adding 180° to the phase shift Φ, the rotation sense is reversed in configuration space, and the electron-by-electron current takes place in the opposite direction (Feltin *et al.*, 2003).

The cross-capacitance effect, which is due to an unintentional capacitive coupling between gates and islands (not shown in Figure 9.21a), can be eliminated by means of an electronic device connected to both gate wiring inputs, which adds a fraction of the voltage applied to one gate to the other gate, with opposite polarity (Keller *et al.*, 1996). A detailed description on the adjustment of pumping parameters is given in Steck (2007) and Steck *et al.* (2008), and practical considerations are detailed in Keller (2009).

The accuracy of the charge transfer is partly limited by the cotunneling effect. This phenomenon involves simultaneous tunneling of electrons through multiple junctions. To avoid errors in the transport rate, increasing the number of the junctions is a first solution. By means of a seven-junction pump, the NIST has measured an error rate per cycle as low as one part in 10^8 (Keller *et al.*, 1996; Keller, Martinis and Kautz, 1998). However, PTB has proposed to keep three-junction pumps, the easiest to use, and to place on-chip resistive Cr-microstrips of typically 50 kΩ in series with the pump (Lotkhov *et al.*, 2001): this device is named an R-pump (Figure 9.22). As a result, dissipation of electron tunneling energy in the resistors suppresses undesirable effects of cotunneling and an increased accuracy can be achieved.

The $I - V_B$ curve, given in Figure 9.23, shows the typical current step and illustrates the current stability with bias conditions. This characteristic is crucial for the

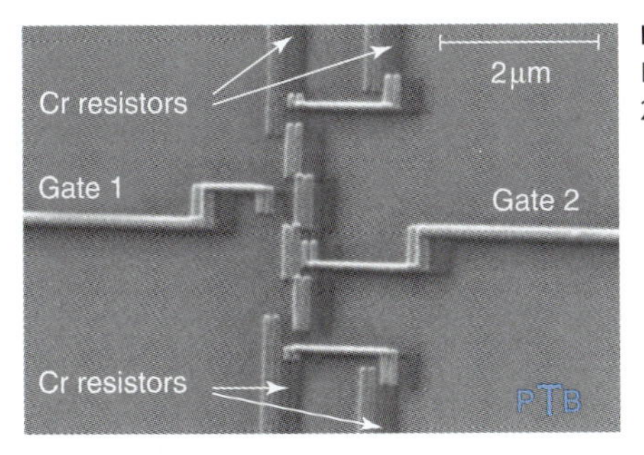

Fig. 9.22 SEM image of three-junctions R-pump fabricated by PTB (Lotkhov *et al.*, 2001).Illustration by courtesy of the PTB.

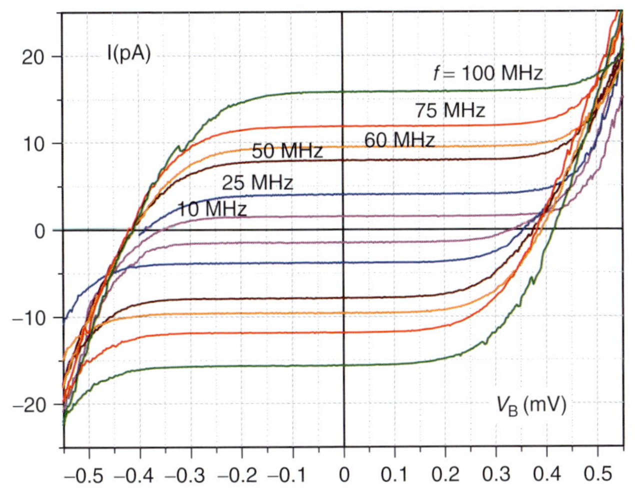

Fig. 9.23 Current steps obtained with a three-junctions R-pump operating at various pumping frequencies ($f = 10$, 25, 50, 60, 75, and 100 MHz) and connected to a CCC (Steck *et al.*, 2008).

development of current standards. Thus, stable current was obtained with a PTB R-pump connected to a CCC. An investigation of long-time measurements has shown that these pumps were able to generate a current $I = ef$ during more than 12 hours (Feltin *et al.*, 2003; Steck *et al.*, 2008).

9.4.2.3 **Other Single Charge Transport Devices**

The metallic tunnel junction pump previously described is the single charge transport device whose metrological properties are the best known so far and enables the on-going development of quantum current standards and the QMT (see the following section) to step across the uncertainty level of one part in 10^6. Coming close to the next uncertainty threshold, a few parts in 10^7 remain conceivable for the metallic pump except for the experiments based on a current measurement. The required current amplitude becomes higher than 100 pA, whereas the metallic pump supplies low current (16 pA for $f = 100$ MHz) and connecting large numbers of these devices in parallel is a very difficult task, without saying impossible.

Faced with such an issue, a growing number of devices are being investigated as promising candidates to realize a quantum current standard with amplitude larger than 100 pA. Some of them operate at frequencies up to the gigahertz range, whereas others work at lower frequencies but are scalable, that is, they can be put in parallel. These devices can be categorized into three types: semiconducting, superconducting, and hybrid.[2] The list below is far from exhaustive, but bibliographies can be found in the quoted references.

The first category includes surface-acoustic wave (SAW) driven pumps. A SAW propagating in a one-dimensional channel, either defined from a 2DEG in a heterostructure of GaAs/AlGaAs (Shilton *et al.*, 1996) or formed by a carbon nanotube (Shin *et al.*, 2006; Würstle *et al.*, 2007), can induce a quantized current.[3] Owing to the piezoelectric effect, a potential modulation is created, propagates through the device, and is superimposed on the energy barrier of the depleted one-dimensional channel. The minima of the energy modulation act like potential wells, which can propagate $N = 1, 2, 3, \ldots$ electrons per cycle through the barrier, generating a current $I = Nef$. Tunable barrier devices composed of a gated nanowire, from a silicon MOSFET (Fujiwara, Nishiguchi, and Ono, 2008) or GaAs/AlGaAs heterostructure (Blumenthal *et al.*, 2007), are also very promising semiconducting candidates to pump single charges accurately and with high speed. The transfer mechanism here relies less on the wave behavior of electrons tunneling through the barrier, but instead on their corpuscular behavior, the electrons being transported one by one by modulation of potential barriers (like "surfing on a wave"). Alternatively, fixed (nontunable) barrier devices based on silicon nanowire transistors with doping modulation have recently been proposed, taking advantage of the well-mastered CMOS technology (Hofheinz *et al.*, 2006).

The second category of high current sources concerns superconducting pumps. In principle, devices consisting of small-capacitance Josephson junctions separating superconducting islands coupled to gate electrodes are able to pump Cooper pairs one-by-one driven by a frequency higher than in the normal pump case. However, the tunneling of Cooper pairs is a much more complex phenomenon. Nevertheless, a current $I = \pm 2ef$ generated by a three-junction superconducting pump has been observed by several authors (Averin and Likharev, 1991; Geerligs *et al.*, 1991; Zorin *et al.*, 2000). However, the transfer of the Cooper pairs across the device is disturbed by various factors (Cooper pair cotunneling, quasi-particles poisoning, etc.), leading to an imperfect plateau in the $I-V$ curve. Solutions have been proposed to improve the accuracy of the superconducting pumps (Zorin *et al.*, 2000). A Cooper pair sluice, consisting of two mesoscopic SQUIDs forming between them a superconducting island fitted with a gate, is another approach to pump single Cooper pairs per cycle (Niskanen, Pekola, and Seppä, 2003; Vartiainen *et al.*, 2007).

2) One should add a metallic type if the RF SET device is considered. This is a SET transistor coupled to a radiofrequency resonant circuit. Although the pump actively generates a current, the RFSET device passively monitors a current source. Indeed, it can be used as a very sensitive single electron electrometer ($1.10^{-6}\,e\,H_z^{-1/2}$ obtained nowadays) and allows the very accurate counting of electrons crossing an array of junctions. However, the best measurements reported so far show a measured current less than 1 pA (Bylander, Duty and Delsing, 2005).

3) In the first case, current uncertainty of a few parts in 10^4 has been estimated, but no real flat plateau has been displayed (Fletcher, Janssen and Hartland, 2002).

The gate provides the possibility of coherent transfer of Cooper pair charges, one at a time, under the influence of an applied RF signal. Quantized currents of 10 up to 100 pA could be obtained with a calculated accuracy of one part in 10^7.

Different Josephson devices are currently investigated for the observation of the Bloch oscillations (e.g., Kuzmin and Haviland, 1991; Nguyen *et al.*, 2007). These are periodic oscillations in the voltage across a current-biased single Josephson junction at frequency $f = I/(2e)$ (Likharev and Zorin, 1985; Gallop, 2005). The Bloch voltage oscillations and the Josephson current oscillations are actually dual phenomena. Phase locking of Bloch oscillations by an external microwave signal could yield current steps in the $I-V$ characteristics. Very recently, it has been shown that the current range of 100 pA to 1 nA could be attainable (Nguyen *et al.*, 2007). Similarly, exploiting the effect of quantum phase slip, devices constituted of superconducting nanowires irradiated by microwave signal could deliver quantized steps in the $I-V$ curve, with $I = 2nef$, where f is the frequency (Mooij and Nazarov, 2006). Following the example of Josephson voltage standards, the two devices (Bloch oscillation and quantum phase slip) are thus expected to be promising candidates for realizing quantum current standards.

Last but not least, the third category deals with hybrid metal–superconductor devices. Recently invented by Pekola *et al.*, these devices are single-electron transistors with normal/insulating/superconducting tunnel junctions (Pekola *et al.*, 2008). There are two types: NISIN type where the island (fitted with a gate) is superconducting and the leads are normal metal and conversely SINIS type (normal island and superconducting leads). Well-developed current plateaux $I = Nef$ have been observed on the $I-V$ characteristics at different frequencies up to 20 MHz (Pekola *et al.*, 2008). Compared with superconducting pumps that would run faster but seem difficult to operate, the hybrid SINIS transistors are simpler in design and operation, thus making parallelization easier.

In conclusion to this short overview of high current sources, it must be noted that the best demonstration of electron pumping reported so far from these devices has shown an accuracy of one part in 10^4 at a pumping frequency of slightly higher than 500 MHz, that is, a current amplitude coming near 90 pA (Blumenthal *et al.*, 2007). However, this is only a starting point.

9.4.3 Quantum Metrological Triangle

SET, or more widely, single charge transport, provides the missing link for the QMT (Figure 9.24) by realizing a quantum current standard whose amplitude is only given by the product of the elementary charge and a frequency. The closure of the QMT experimentally consists here in applying Ohm's law $U = RI$ directly to the three phenomena SET, JE, and QHE. Another approach to closing the triangle proposes to apply $Q = CV$ by means of an ECCS.

In practice, the experiment determines the dimensionless product $R_K K_J Q_X$, expected to be equal to 2, where the constant Q_X is defined as an estimate of the elementary charge (Piquemal and Genevès, 2000), $Q_X = e|_{\mathrm{SET}}$, by analogy with the definitions of the Josephson and von Klitzing constants, $K_J = 2e/h|_{\mathrm{JE}}$ and $R_K = h/e^2|_{\mathrm{QHE}}$. Checking the equality $R_K K_J Q_X = 2$ with an uncertainty of one part in 10^8 will be a relevant experimental test of the validity of the three theories.

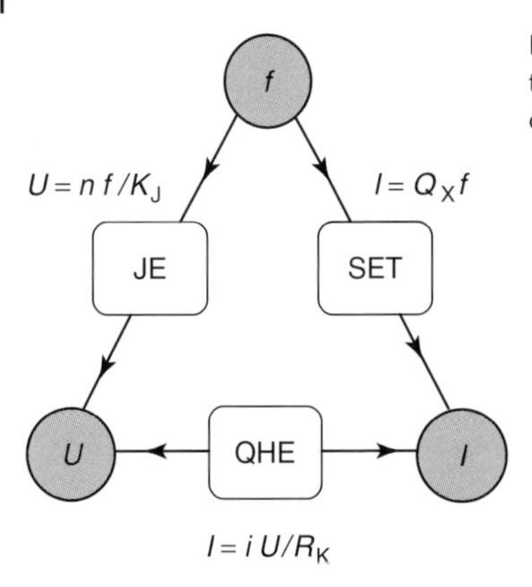

Fig. 9.24 Quantum-metrological triangle. Theory predicts that R_K, K_J, and Q_X correspond to the fundamental constants h/e^2, $2e/h$, and e.

Supported by strong theoretical arguments, the high level of agreement shown by numerous comparisons of quantum resistance and voltage standards involving different kinds of devices (see Sections 9.2 and 9.3.1) undoubtedly strengthens our confidence in the universal and fundamental aspects of K_J and R_K and hence in the equalities $K_J = 2e/h$ and $R_K = h/e^2$. However, one cannot obviously conclude that these relations are exact, that is, free of any correction factors.

Conclusions could already be drawn up if the results of QMT experiments were obtained with larger uncertainties (Mohr *et al.*, 2007; Piquemal *et al.*, 2007; Keller, 2008). As outlined by Keller, first, the closure of the QMT at a level of one part in 10^6 will constitute a first critical test of validity for SET (knowing that neither the JE nor the QHE is questionable at that uncertainty level). This has recently been completed by NIST with a relative standard uncertainty of 9.2 parts in 10^7 (Keller, Zimmerman, and Eichenberger, 2007).

The second threshold in uncertainty, which is of interest for closing the QMT, lies between about seven parts in 10^7 and three parts in 10^8. In that case, the resulting information will be mainly relevant for the JE and the SET (in Section 9.3.5.2, we have seen that from least squares adjustments of fundamental constant values the Josephson and von Klitzing relations have been checked with uncertainty of seven parts in 10^7 and two parts in 10^8, respectively).

9.4.3.1 QMT by Applying Ohm's Law and Using a CCC

The first way to close the QMT by applying Ohm's law on quantities provided by the QHE, JE, and SET consists in the direct comparison of the voltage U_J supplied by a Josephson junction array to the Hall voltage of a QHE sample crossed by a current I delivered by a SET current source. The current is amplified with a high accuracy by means of a CCC; see, for example, Piquemal *et al.* (2007). This comparison leads to the following relation:

$$U_J = R_H G I \tag{9.36}$$

where G is the gain or winding ratio of the CCC. Considering the JE, QHE, and SET relationships, the relation (12.28) becomes

$$n f_J / K_J = (R_K / i) G Q_X f_{\mathrm{SET}} \tag{9.37}$$

where n is the index of the voltage step delivered by the JAVS at the microwave frequency f_J, i is the index of the QHE plateau, and f_{SET} is the driving frequency of the SET current source. This leads to the dimensionless product:

$$R_K K_J Q_X = n(i/G) f_J / f_{\mathrm{SET}} \quad (9.38)$$

Another approach that leads to the same relation (12.30) consists in balancing the current delivered by the SET device against the current applied to a cryogenic resistor of high resistance value (100 MΩ) by a Josephson voltage standard. The current is detected either by a CCC operating as an ammeter (Elmquist, Zimmerman, and Huber, 2003) or by another system (Manninen *et al.*, 2008). Then a CCC is used for calibrating directly the 100 − MΩ resistance with the QHR standard (Elmquist *et al.*, 2005).

A QMT experimental set-up is sketched in Figure 9.25. The current amplifier consists of a CCC of high winding ratio, $G = N_1/N_2$, a DC SQUID with low white-noise level and low corner frequency f_c, and a secondary current source, servo-controlled by the SQUID in such a way that the latter works at null magnetic flux and the relation $I_2/I_1 = N_1/N_2$ is verified. To minimize the contribution from $1/f$ flicker noise, the polarity of the current to be amplified is periodically reversed. The Hall voltage is simultaneously compared to the voltage of a programmable JAVS, well suited here because of the low voltage level and the requirement of periodic reversal of polarity. The null detector is balanced by adjusting the operating frequency of the SET source f_{SET}. This frequency and the irradiation frequency of the Josephson array are both referred to a 10-MHz rubidium clock.

The great challenge of this experiment is to reduce the type-A uncertainties to the level of few parts in 10^7 and ultimately one part in 10^8, taking into account the low current delivered by the SET source. The largest type-B uncertainties are estimated to be on the order of one part in 10^8 or less and depend weakly on the current level (Piquemal *et al.*, 2007).

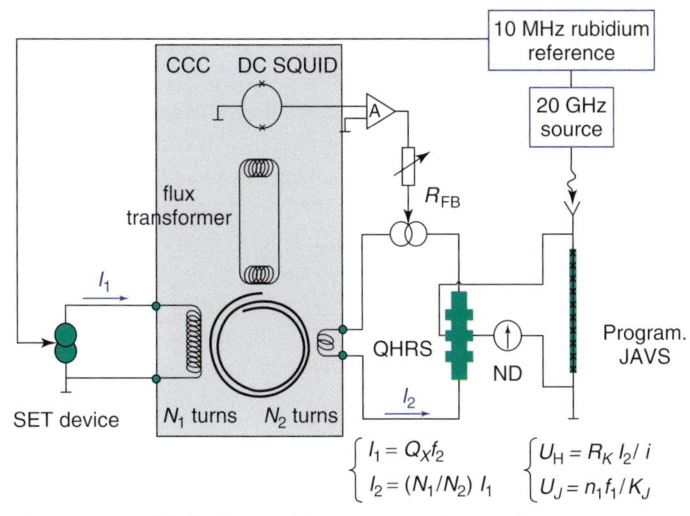

Fig. 9.25 Simplified scheme of the experimental set-up for closing the quantum metrological triangle.

The best results reported so far was a relative type A uncertainty of 3.9 parts in 10^6 on the measurement of a single electron current generated by a three-junction R-pump of close to 16 pA (Steck *et al.*, 2008). Nevertheless, some irreproducibility problems (on the order of one part in 10^4) have been observed from one series of measurements to another. Once this problem is solved, some other improvements have to be made to both the R-pump and the CCC used as current amplifier to reach an uncertainty below one part in 10^6. SET devices capable of supplying currents up to 100 pA and generating lower noise must be developed in order to attempt the ultimate uncertainty of one part in 10^8. A new amplifier with a better sensitivity is also needed. This can be obtained by increasing the CCC gain by a factor of 5 and by using a SQUID well suited to the experiment.

9.4.3.2 QMT and Electron Counting Capacitance Standard

A competitive approach to close the QMT implements a quantum capacitance standard from SET devices, the so-called ECCS. This is the method used at NIST to successfully close the QMT within an uncertainty slightly better than one part in 10^6. The principle simply relies on applying the definition of capacitance: the transfer of a well-known charge Q between the electrodes of a cryogenic capacitor with a capacitance C_{cryo} and the measurement of the potential difference ΔV between these electrodes: $C_{\text{cryo}} = Q/\Delta V$ (Williams, Ghosh, and Martinis, 1992; Keller *et al.*, 1999).

As shown in Figure 9.26, the NIST system consists of a seven-junction electron pump, a SET transistor/electrometer with a charge detection threshold of the order of $e/100$, and a cryogenic capacitor of 1.8-pF capacitance. Two mechanical cryogenic switches N1 and N2 allow two working phases:

- **N1 closed, N2 open:** In this phase, the cryogenic capacitance $C_{\text{cryo}}(\approx 1\ \text{pF})$ is charged with N electrons generated one by one through the pump. The process is stopped for a short time (20 seconds) to measure the voltage V_c^+. Then, the pump is forced to transfer N electrons in the opposite direction. Another stop occurs to measure a voltage V_c^- and so on. The successive voltages V_c^+ and V_c^- are compared to those of a JAVS and the differences $\Delta V = V_c^+ - V_c^-$ are calculated. The average of these differences $<\Delta V>$ gives the capacitance

$$C_{\text{cryo}} = Ne/ < \Delta V > = (N/nf_J)K_J Q_X \tag{9.39}$$

  from the relation $<\Delta V> = nf_J/K_J$, where n is the index of the voltage step provided by the binary Josephson array at a frequency f_J. The best relative standard deviation of C_{cryo} values obtained so far with electron counting is of the order of a few parts in 10^7 (Keller *et al.*, 1999).
- **N1 open, N2 closed:** In this second configuration, C_{cryo} is compared with the capacitance C_X of a capacitor at room temperature using a capacitance bridge. This capacitance comparison is carried out at a frequency in the kilohertz range, much higher than the effective frequency of electron counting (25 mHz) and at 15 V of rms voltage value (compared to 3.5 V in the first phase).

Two kinds of results could be obtained.

1. If C_X has been previously measured in terms of the second and R_K with a quadrature bridge, it can be written in

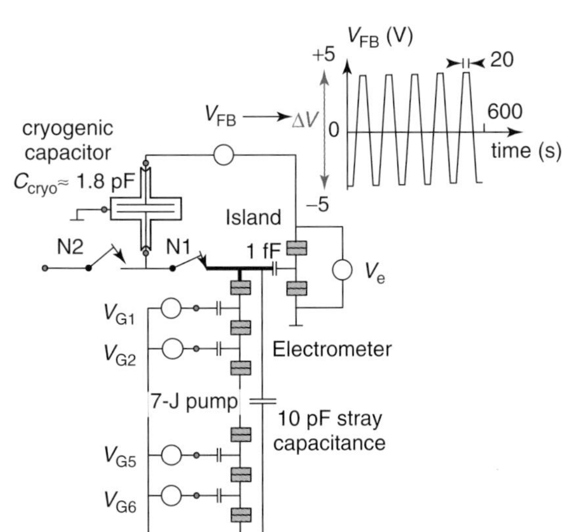

Fig. 9.26 Principle of electron counting capacitance standard (Keller *et al.*, 1999). Two operating modes: (1) N1 closed, N2 open: electrons are periodically pumped forward and backward and (2) N1 open, N2 closed: The cryogenic capacitance is compared to a reference capacitance placed at room temperature.

a simplified form as

$$C_X = A_1/(R_K f_q) \tag{9.40}$$

where A_1 is a dimensionless factor issued from the measurement and f_q is the balance frequency of the quadrature bridge. Combining the two last relations (12.31) and (12.32) leads to a new expression of the dimensionless product $R_K K_J Q_X$:

$$R_K K_J Q_X = A_1(n/N)(C_{\text{cryo}}/C_X)(f_J/f_q) \tag{9.41}$$

2. If the capacitance C_X has been directly compared to the capacitance variation ΔC of the Thompson–Lampard calculable capacitor (Section 9.2.6.1) and consequently known with a value expressed in SI units:

$$C_X = \{C_X\}_{\text{SI}}\ \text{F} = A_2\{\Delta C\}_{\text{SI}}\ \text{F} \tag{9.42}$$

where the quantity inside brackets $\{\}_{\text{SI}}$ is a dimensionless numerical value and, using the same notation as before, A_2 is a dimensionless factor. Then, from relation (12.31) an SI value of the product $K_J Q_X$ can be deduced:

$$\begin{aligned} K_J Q_X &= A_2(N/n)(C_{\text{cryo}}/C_X) \\ &\quad \times \{f_J \Delta C\}_{\text{SI}}\Omega^{-1} \\ &= A_3 f_J \Delta C_{\text{SI}}\Omega^{-1} \end{aligned} \tag{9.43}$$

where $A_3 = A_2(N/n)(C_{\text{cryo}}/C_X)$. It must be noted that the relation (12.35) gives rise to a measurement of a quantity whose value might be compared to the values of h/e^2 obtained elsewhere, and this is independent of the QHE.

Unlike the $U = RI$ approach, the experimental achievement of the relation (12.24) does not need a SET source supplying currents higher than a few picoamperes to close the triangle, at least with an uncertainty of one part in 10^7. This uncertainty level will be reached by this method if efforts are undertaken particularly on the capacitance measurement. A drastic improvement will be in the implementation of a coaxial capacitance bridge based on the two terminal-pair method (Kibble and Rayner, 1984). This will be absolutely required in order to reduce more the uncertainty and to reach the level of few parts in 10^8. Furthermore, a better knowledge of the frequency dependence of the cryogenic capacitor is needed. A model has been proposed to explain the observed logarithmic increase of the capacitance when the frequency decreases below a few 100 Hz (Zimmerman, Simonds, and Wang, 2006). Within this framework, developments of ECCS are also in progress at METAS and PTB (Scherer, Lotkhov, and Willenberg, 2008).

9.4.3.3 Determination of the Elementary Charge

The combination of all the three experiments, QMT, calculable capacitor, and watt balance (in the same laboratory or not) would lead to a first direct determination of the charge quantum involved in SET devices, the expected elementary charge, without assuming that $R_K = h/e^2$ and $K_J = 2e/h$ (Piquemal *et al.*, 2007; Keller *et al.*, 2008).

The determination of R_K from the complete experiment linking the Thompson–Lampard calculable capacitor to the QHR standard gives

$$R_K = A_4 \left\{ \left(\Delta C f_q \right)^{-1} \right\}_{\mathrm{SI}} \Omega \tag{9.44}$$

where A_4 is a dimensionless factor and f_q is again the frequency of the balanced quadrature bridge (Section 9.2.6.1).

The watt balance provides the SI value of the product $K_J^2 R_K$ (Section 9.3.5.1)

$$K_J^2 R_K = A_5 \{ f_J^2 / [Mgv] \}_{\mathrm{SI}} \Omega\, \mathrm{V}^{-2}\, \mathrm{s}^{-2} \tag{9.45}$$

where, as before, A_5 is a dimensionless factor and f_J is the Josephson frequency. M, g, and v correspond to the suspended mass, Earth's gravitational acceleration, and the constant velocity of the measurement coil.

From the two approaches of QMT ($U = RI$ or $Q = CV$), the measured SI value of Q_X is thus given by

$$Q_X = A_6 \{ [\Delta C f_q M g v]^{1/2} / f_{\mathrm{SET}} \}_{\mathrm{SI}}\, \mathrm{C} \tag{9.46}$$

or

$$Q_X = A_7 \{ [\Delta C M g v / f_q]^{1/2} \}_{\mathrm{SI}}\, \mathrm{C} \tag{9.47}$$

where A_6 and A_7 are dimensionless factors.

The value of Q_X, which results from the QMT experiment at NIST through relation (12.30), gives $Q_X = 1.6021763 \times 10^{-19}$ C with the relative standard uncertainty of 0.92 parts in 10^6 (Keller *et al.*, 2008). This value is in good agreement with the best-known value of e provided by the 2006 CODATA adjustment ($e = 1.602176487$ C, with an uncertainty of 2.5 parts in 10^8).

Acknowledgment

The author is very grateful to R. Behr, B. Jeckelmann, and M. Keller for their critical reading.

Glossary

Calculable Capacitance Standard: The calculable capacitance standard is an electro-mechanical system whose principle relies on a theorem linking direct capacitances per unit of length of two pairs of electrodes. The practical use of this theorem allows one at the same time to realize the farad in the SI and to maintain this unit. Moreover, it makes the SI realization of the ohm possible by means of a quadratic bridge used to compare impedances between capacitance and resistance.

Josephson Voltage Standard: The Josephson voltage standard is a quantum standard based on the AC Josephson effect, which occurs in a junction of two superconducting electrodes weakly coupled (Josephson junction) and links directly a voltage to a frequency through the Josephson constant K_J.

Quantum Hall Resistance Standard: This quantum standard implements the quantum Hall effect observed in a two-dimensional electron gas (2DEG) whose value of the Hall resistance is equal to the von Klitzing constant R_K divided by an integer number i for certain ranges of magnetic field.

Quantum Current Standard: The realization of a quantum current standard is expected from nanostructured devices whose working principle is based on single electron tunneling (or more generally single-charge transport) and quantization of charge enforced by Coulomb blockade. The amplitude of the delivered current is equal to the product of the transported charge quantum times the pumping frequency.

SI Realization of Unit: The meaning of SI realization of a unit, or direct determination of a unit in the SI, has to be distinguished from the simple relative measurement of a quantity expressed in this unit. The first case involves a whole combination of experimental operations, in general complex, which provides the standard representing this unit, or the associated fundamental constant, with a value expressed purely in terms of mechanical units (m, kg, and s), and as a function of the constants c (speed of light in vacuum) and μ_0 (magnetic constant of vacuum) whose values are exactly known by definition in the present SI. The second case consists in a comparison with a standard of the same quantity. The value of this standard, while very reproducible, may involve factors that are not known purely in terms of SI mechanical units.

SQUID: The SQUID is a superconducting ring interrupted by one or two Josephson junctions. The operation principle rests both on the quantization of the magnetic flux across a superconducting loop and on the Josephson effects. This is a very sensitive magnetic flux detector. The typical performance for the best commercially available device is a magnetic flux noise spectral density on the order of $3 \times 10^{-6}\Phi_0\ \mathrm{Hz}^{-1/2}$ and an energy resolution of about $1 \times 10^{-31}\ \mathrm{J\ Hz}^{-1}$ at 4.2 K in the white noise regime.

National Metrology Institutes and Other Organizations Cited

BIPM: Bureau International des Poids et Mesures

CIPM: Comité International des Poids et Mesures

CODATA: Committee on Data for Science and Technology[4)]

LNE: Laboratoire National de métrologie et d'Essais, France

METAS: Federal Office of Metrology, Switzerland

NIST: National Institute of Science and Technology, USA

NMIA: National Metrology Institute of Australia

4) CODATA is an international committee of the International Council of Scientific Unions (ISCUs). Established in 1969, its task is to improve the compilation, the critical evaluation, the stocking, and the research of data of primary importance for science and technology. The values of fundamental constants recommended by CODATA, for example, the 2006 CODATA values, are available at *http//www.physics.nist.gov/constants*

NMI/VSL: National Metrology Institute/ Van Swinden Laboratorium, The Netherlands

NPL: National Physical Laboratory, UK

PTB: Physikalisch Technische Bundesanstalt, Germany

References

Averin, D.V. and Likharev, K.K. (**1991**) in *Mesoscopic Phenomena in Solids* (eds B.L.Alsthuler, P.A. Lee and R.A. Webb), Elsevier, Amsterdam, Noth-Holland, pp. 173–271.

Benz, S.P. (**1995**) *Appl. Phys. Lett.*, **67**, 2714–2716.

Benz, S.P., Burroughs, C.J., Harvey, T.E., and Hamilton, C.A. (**1999**) *IEEE Trans. Appl. Supercond.*, **9**, 3306–3309.

Benz, S.P. and Hamilton, C.A. (**1996**) *Appl. Phys. Lett.*, **68**, 3171–3173.

Benz, S.P. and Hamilton, C.A. (**2004**) *Proc. IEEE*, **92**, 1617–1629.

Benz, S.P. et al., (**2008**) Proceedings of the conference in precision Electromagnetic measurements (A.H. Cookson and T. winter) Boolder, *IEEE*, 48–49, IEEE catalog number CFP08PEM-PRT.

Behr, R., Schulze, H., Müller, F., Kohlmann, J., and Niemeyer, J. (**1999**) *IEEE Trans. Instrum. Meas.*, **48**, 270–273.

Behr, R., Grimm, L., Funck, T., Kohlmann, J., Schulze, H., Müller, F., Schumacher, B., Warnecke, P., and Niemeyer, J. (**2001**) *IEEE Trans. Instrum. Meas.*, **50**, 185–187.

Behr, R., Funck, T., Schumacher, B., and Warnecke, P. (**2003a**) *IEEE Trans. Instrum. Meas.*, **52**, 521–523.

Behr, R., Kohlmann, J., Janssen, J.T., Kleinschmidt, P., Williams, J., Djordjevic, S., Lo-Hive, J.P., Piquemal, F., Hetland, P., Reymann, D., Eklund, G., Hof, C., Jeanneret, B., Chevtchenko, O., Houtzager, E., van den Brom, H.E., Sosso, A., Andreone, D., Nissilä, J., and Helistö, P. (**2003b**) *IEEE Trans. Instrum. Meas.*, **52**, 524–528.

Behr, R., Palafox, L., Funck, T., Williams, J.M., Patel, P., and Katkov, A. (**2006**) CPEM 2006 Conference Digest, Torino, pp. 440–441.

Behr, R., Palafox, L., Ramm, G., Moser, H., and Melcher, J. (**2007**) *IEEE Trans. Instrum. Meas.*, **56**, 235–238.

Bloch, F. (**1970**) *Phys. Rev. B*, **2**, 109–121.

Blumenthal, M.D., Kaestner, B., Li, L., Giblin, S., Janssen, T.J.B.M., Pepper, M., Anderson, D., Jones, G., and Ritchie, D.A. (**2007**) *Nat. Phys.*, **3**, 343–347.

Bounouh, A., Poirier, W., Piquemal, F., Genevès, G., and André, J.P. (**2003**) *IEEE Trans. Instrum. Meas.*, **52**, 555–558.

van den Brom, H.E., Houtzager, E., Brinkmeier, B.E.R., and Chevtchenko, O.A. (**2008**) *IEEE Trans. Instrum. Meas.*, **57**, 428–431.

Burroughs, C.J., Benz, S.P., Hamilton, C.A., Harvey, T.E., Kinard, J.R., Lipe, T.E., and Sasaki, H. (**1999**) *IEEE Trans. Instrum. Meas.*, **48**, 282–284.

Burroughs, C.J., Benz, S.P., Dresselhaus, P.D., Waltrip, B.C., Nelson, T.L., Chong, Y., Williams, J.M., Henderson, D., and Patel, P. (**2007a**) *IEEE Trans. Instrum. Meas.*, **56**, 289–294.

Burroughs, C.J., Rufenacht, A., Benz, S.P., Dresselhaus, P.D., Waltrip, B.C., and Nelson, T.L. (**2007b**) IEEE in Proceedings of the NCSLI, St Paul, Minnesota.

Büttiker, M. (**1988**) *Phys. Rev. B*, **38**, 9375–9389.

Bylander, J., Duty, T., and Delsing, P. (**2005**) *Nature*, **434**, 361–364.

CIPM (**1988a**) Représentation du volt au moyen de l'effet Josephson, Recommandation 1 (CI-1988), 77ème session.

CIPM (**1988b**) Représentation de l'ohm au moyen de l'effet Hall quantique, Recommandation 2 (CI-1988), 77ème session.

CIPM (**2000**) Rapport de la 22ème session du CCEM, 89ème session, 34.

Cladé, P., de Mirandes, E., Cadoret, M., Guellati-Kh'elifa, S., Schwob, C., Nez, F., Julien, L., and Biraben, F. (**2006**) *Phys. Rev. A*, **102**, 052109.

Clarke, J. and Braginski, I. (**2004**) *The SQUID Handbook*, Fundamentals and Technology of SQUIDs and SQUID Systems, vol. **1**, (2006) Applications, vol. 2, Wiley-VCH Verlag GmbH & Co, KgaA, Weinheim.

Clothier, W.K. (**1965**) *Metrologia*, **1** (2), 35–56.

Clothier, W.K., Sloggett, G.J., Bairnsfather, H., Currey, M.F., and Benjamin, D.J. (**1989**) *Metrologia*, **26**, 9.

Delahaye, F. (**1978**) *IEEE Trans. Instrum. Meas.*, **27**, 426–429.

Delahaye, F. (**1991**) *IEEE Trans. Instrum. Meas.*, **40**, 883–888.

Delahaye, F. (**1993**) *J. Appl. Phys.*, **73**, 7914–7920.

Delahaye, F. andJeckelmann, B. (**2003**) *Metrologia*, **40**, 217–223.

Delahaye, F., Kibble, B.P., and Zarka, A. (**2000**) *Metrologia*, **37**, 659–670.

Delahaye, F., Witt, T.J., Piquemal, F., and Genevès, G. (**1995**) *IEEE Trans. Instrum. Meas.*, **44**, 258–261.

Djordjevic, S., Séron, O., Solve, S., and Chayramy, R. (**2008**) *Metrologia*, **45**, 1–7.

Eichenberger, A., Jeckelmann, B., and Richard, P. (**2003**) *Metrologia*, **40**, 356–365.

Elmquist, R., Zimmerman, N.M., and Huber, W.H. (**2003**) *IEEE Trans. Instrum. Meas.*, **52**, 590–593.

Elmquist, R.E., Hourdakis, E., Jarret, D.G., and Zimmerman, N.M. (**2005**) *IEEE Trans. Instrum Meas.*, **54**, 525–528.

Feltin, N., Devoille, L., Piquemal, F., Lotkhov, S., and Zorin, A. (**2003**) *IEEE Trans. Instrum. Meas.*, **52**, 599–603.

Fletcher, N.E., Janssen, T.J.B.M., and Hartland, A. (**2002**) Proceedings of BEMC'2001, IEE Special Issue, Harrogate.

Fujiwara, A., Nishiguchi, K., and Ono, Y. (**2008**) *Appl. Phys. Lett.*, **92**, 42102.

Fulton, T.A. and Dolan, G.J. (**1987**) *Phys. Rev. Lett.*, **59**, 109–112.

Funck, T. and Sienknecht, V. (**1991**) *IEEE Trans. Instrum. Meas.*, **40**, 158–141.

Funck, T., Behr, R., and Klonz, M. (**2001**) *IEEE Trans. Instrum. Meas.*, **50**, 322–325.

Gallop, J.C. (**2005**) *Philos. Trans. R. Soc. Lond. A*, **363**, 2221.

Gallop, J.C. and Piquemal, F. (**2006**) in *SQUIDs Handbook*, Applications of SQUIDs and SQUID Systems, vol. **II**, Chapter 9 (A. Braginski and J. Clark), Wiley-VCH Verlag GmbH, Berlin.

Gay, F., Piquemal, F., and Genevès, G. (**2000**) *Rev. Sci. Instrum.*, **71**, 4592–4595.

Geerligs, L.J., Verbrugh, S.M., Hadley, P., Mooij, J.E., Pothier, H., Lafarge, P., Urbina, C., Estève, D., and Devoret, M.H. (**1991**) *Z. Phys. B*, **85**, 349.

Genevès, G., Gournay, P., Gosset, A., Lecollinet, M., Villar, F., Pinot, P., Juncar, P., Clairon, A., Landragin, A., Holleville, D., Pereira Dos Santos, F., and David, J. (**2005**) *IEEE Trans. Instrum. Meas.*, **54**, 850–853.

Gibbings, D.L.H. (**1963**) *Proc. Inst. Electron. Eng.*, **110** (2), 335–347.

Grabert, H. and Devoret, M.H. (eds) (**1991**) *Single Charge Tunneling Coulomb Blockade Phenomena in Nanostructures*, NATO ASI Series, Series B: Physics, vol. **294**, Plenum Press, New York.

Haddad, R.J. (**1969**) A resistor calculable from dc to $\omega = 105$ rad/s, MSc thesis, George Washington University, Washington, DC.

Hamilton, C.A., Burroughs, C.J., and Benz, S.P. (**1997**) *IEEE Trans. Appl. Supercond.*, **17**, 3756–3761.

Hamilton, C.A., Burroughs, C.J., and Kautz, R.L. (**1995**) *IEEE Trans. Instrum. Meas.*, **44**, 223–225.

Hamilton, C.A., Lloyd, F.L., Chieh, K., and Goeke, W.C. (**1989**) *IEEE Trans. Instrum. Meas.*, **38** (2), 314–316.

Hamilton, C.A. and Tang, Y.H. (**1999**) *Metrologia*, **36**, 53–58.

Hanneke, D., Fogwell, S., and Gabrielse, G. (**2008**) *Phys. Rev. Lett*, **100**, 120801.

Hartland, A., Jones, K., Williams, J.M., Gallagher, B.L., and Galloway, T. (**1991**) *Phys. Rev. Lett.*, **66**, 969–973.

Hartland, A. (**1992**) *Metrologia*, **29**, 175–190.

Harvey, I.K. (**1972**) *Rev. Sci. Instrum.*, **43**, 1626–1629.

Hassel, J., Seppä, H., Gronberg, L., and Suni, I. (**2001**) *IEEE Trans. Instrum. Meas.*, **50**, 195–198.

Hein, G., Schumacher B., and Ahlers, F.J. (**2004**) Conference Digest CPEM 2004, London, pp. 273–274.

Hofheinz, M., Jehl, X., Sanquer, M., Molas, G., Vinet, M., and Deleonibus, S. (**2006**) *Appl. Phys. Lett.*, **89**, 143504.

Houtzager, E., Benz, S.P., Van dem Brom, H.E., (**2008**) Proceedings of the conference in precision Electromagnetic measurements (A.H. Cookson and T. winter) Boolder, *IEEE*, 46–47.

Inglis, A.D. (**2004**) Conference Digest CPEM 2004, London, pp. 275–276.

Janssen, M., Viehweger, O., Fastenrath, U., and Hajdu, J. (**1994**) *Introduction to the Theory of the Integer Quantum Hall Effect*, VCH Verlagsgesellschaft, Weinheim.

Jeanneret, B., Rüfenacht, A., and Burroughs, C.J. (**2001**) *Trans. Instrum. Meas.*, **50**, 188–191.

Jeanneret, B., Overney, F., Callegaro, L., and Mortara, A. (**2008**) *Proceedings of the Conference on Precision Electromagnetic Measurements* (A.H. Cookson and T. Winter), IEEE, Boulder 74–75.

Jeckelmann, B., Inglis, A.D., and Jeanneret, B. (**1995**) *IEEE Trans. Instrum. Meas.*, **44**, 269–272.

Jeckelmann, B. and Jeanneret, B. (**2000**) *Rep. Prog. Phys.*, **64**, 1603–1655.

Jeffery, A.M., Elmquist, R.E., Lee, L.H., Shields, J.Q., and Dziuba, R.F. (**1997**) *IEEE Trans. Instrum. Meas.*, **46**, 264.

Josephson, B.D. (**1962**) *Phys. Lett.*, **1**, 251–253.

Kaneko, N., Oe, T., Domae, A., Urano, C., Itatani, T., Ishii, H., and Kiryu, S. (**2008**) in *Proceedings of the Conference on Precision Electromagnetic Measurements* (A.H. Cookson and T. Winter), IEEE, Boulder 692–693.

Katkov, A., Behr, R., and Lee, J. (**2008**) in *Proceedings of the Conference on Precision Electromagnetic Measurements* (A.H. Cookson and T. Winter), IEEE, Boulder 380–381.

Keller, M.W., Martinis, J.M., and Kautz, R.L. (**1998**) *Phys. Rev. Lett.*, **80**, 4530–4533.

Keller, M.W., Martinis, J.M., Zimmerman, N.M., and Steinbach, A.H. (**1996**) *Appl. Phys. Lett.*, **69**, 1804–1806.

Keller, M.W., Eichenberger, A.L., Martinis, J.M., and Zimmerman, N.M. (**1999**) *Science*, **285**, 1706–1709.

Keller, M.W., Zimmerman, N.M., and Eichenberger, A.L. (**2007**) *Metrologia*, **44**, 505–512.

Keller, M.W. (**2008**) *Metrologia*, **45**, 102–109.

Keller, M.W., Piquemal, F., Feltin, N., Steck, B., and Devoille, L. (**2008**) *Metrologia*, **45**, 330–334.

Keller, M.W. (**2009**) in *Quantum Metrology and Fundamental Constants*, European Physical Journal Special Topics (F. Piquemal and B. Jeckelmann), (**2009**) vol. **172**, pp. 297–309.

Kibble, B.P. (**1976**) in *Atomic Masses and Fundamental Constants*, vol. **5** (J.H. Sanders and A. Wapstra), Plenum Press, New York, pp. 545–551.

Kibble, B.P. and Rayner, G.H. (**1984**) *Coaxial AC Bridges*, Adam Hilger, Bristol.

von Klitzing, K., Dorda, G., and Pepper, M. (**1980**) *Phys. Rev. Lett.*, **45** (6), 494–497.

Kohlmann, J., Schulze, H., Behr, R., Krasnopolin, I.Y., Müller, F., and Niemeyer, J. (**2000**) *Appl. Supercond., Inst. Phys. Conf. Ser.*, **167**, 769–772.

Krasnopolin, I.Y., Behr, R., and Niemeyer, J. (**2002**) *Supercond. Sci. Technol.*, **15**, 1034.

Kuzmin, L.S. and Haviland, D.B. (**1991**) *Phys. Rev. Lett.*, **67**, 2890–2893.

Laughlin, R.B. (**1981**) *Phys. Rev. B*, **23** (10), 5632–5633.

Laughlin, R.B. and Pines, D. (**2000**) *Proc. Natl. Acad. Sci.*, **97**, 28–31.

Levinson, M.T., Chiao, R.Y., Feldman, M.J., and Tucker, B.A. (**1977**) *Appl. Phys. Lett.*, **31**, 776–778.

Likharev, K. and Zorin, A. (**1985**) *J. Low Temp. Phys.*, **59**, 347–382.

Likharev, K.K. (**1999**) *Proc. IEEE*, **87**, 606–632.

Lo-Hive, J., Djordjevic, S., Cancela, P., Piquemal, F., Behr, R., Burroughs, C., and Seppä, H. (**2003**) *IEEE Trans. Instrum. Meas.*, **52**, 516–520.

Lotkhov, S.V., Bogoslovsky, S.A., Zorin, A.B., and Niemeyer, J. (**2001**) *Appl. Phys. Lett.*, **78**, 946–948.

Manninen, A., Hahtela, O., Hokonen, P., Helistö, P., Kemppinen, A., Möttönen, M., Paalanen, M., Pekola, J., Satrapinski, A., and Seppä, H. (**2008**) in *Proceedings of the Conference on Precision Electromagnetic Measurements* (A.H. Cookson and T. Winter), IEEE, Boulder 630–631.

Mc Cumber, D.E. (**1968**) *J. Appl. Phys.*, **39** (7), 3113–3118.

Mills, I.M., Mohr, P.J., Quinn, T.J., Taylor, B.N., and Williams, E.R. (**2005**) *Metrologia*, **42**, 71–80.

Mills, I.M., Mohr, P.J., Quinn, T.J., Taylor, B.N., and Williams, E.R. (**2006**) *Metrologia*, **43**, 227–246.

Mohr, P.J., Taylor, B.N., Newell, D.B., Report, 18th meeting of the CCU (Comité Cousultratif sur les Unités), 11 – 13th June (**2007**).

Mohr, P.J., Taylor, B.N., and Newell, D. (**2008**) *Rev. Mod. Phys.*, **80**, 633–730.

Mooij, J.E. and Nazarov, Y.V. (**2006**) *Nat. Phys.*, **2**, 169–172.

Niemeyer, J., Hinken, J.H., and Kautz, R.L. (**1984**) *Appl. Phys. Lett.*, **45**, 478–480.

Niskanen, A.O., Pekola, J.P., and Seppä, H. (**2003**) *Phys. Rev. Lett.*, **91**, 177003.

Mueller, F., Behr, R., Palafox, L., Kohlmann, J., Wendisch, R., and Krasnopolin, I. (**2007**) *IEEE Trans. Appl. Supercond.*, **17**, 649–652.

Nguyen, F., Boulant, N., Ithier, G., Bertet, P., Pothier, H., Vion, D., and Esteve, D. (**2007**) *Phys. Rev. Lett.*, **99**, 1870051–1870054.

Novoselov, K.S., Geim, A.K., Morozov, S.V., Jiang, D., Katsnelson, M.I., Grigorieva, I.V., Dubonis, S.V., and Firsov, A.A. (**2005**) *Nature*, **438**, 197.

Oeh, T., Kaneko, N., Urano, C., Itatani, T., Ishii, H., and Kiryu S. (**2008**) in *Proceedings of the Conference on Precision Electromagnetic Measurements* (A.H. Cookson and T. Winter), IEEE, Boulder 692–693.

Overney, F., Jeanneret, B., and Jeckelmann, B. (**2003**) *IEEE Trans. Instrum. Meas.*, **52**, 574–578.

Palafox, L., Ramm, G., Behr, R., Ihlenfeld, W.G.K., and Moser, H. (**2007**) *IEEE Trans. Instrum. Meas.*, **56**, 534–537.

Pekola, J., Vartiainen, J.J., Möttönen, M., Saira, O.P., Meschke, M., and Averin, D.V. (**2008**) *Nat. Phys.*, **4**, 120.

Picard, A., Stock, M., Fang, H., Witt, T.J., and Reymann, D. (**2006**) *IEEE Trans. Instrum. Meas.*, **56**, 538–542.

Piquemal, F., Genevès, G., Delahaye, F., André, J.P., Patillon, J.N., and Frijlink, P. (**1993**) *IEEE Trans. Instrum. Meas.*, **42**, 264–268.

Piquemal, F., Blanchet, J., Genevès, G., and André, J.P. (**1999**) *IEEE Trans. Instrum. Meas.*, **48**, 296–300.

Piquemal, F. (**1999**) *Bull. BNM*, **116**, 5–57 (in French).

Piquemal, F. and Genevès, G. (**2000**) *Metrologia*, **37**, 207–211.

Piquemal, F., Devoille, L., Feltin, N., and Steck, B. (**2007**) in *Proceedings of the International School of Physics "Enrico Fermi", Course CLXVI, Recent Advances in Metrology and Fundamental Constants* (T.W. Hansch, S. Leschiutta, A.J. Wallard, and M.L. Rastello), Societa Italiana di Fisica-Bologna and IOS Press.

Poirier, W., Piquemal, F., Fhima, H., Bensaïd, N., and Genevès, G. (**1999**) Proceedings of 9ème Congrès International de Métrologie, Bordeaux.

Poirier, W., Bounouh, A., Hayashi, K., Fhima, H., Piquemal, F., Genevès, G., and André, J.P. (**2002**) *J. Appl. Phys.*, **92**, 2844–2854.

Poirier, W., Bounouh, A., Piquemal, F., and André, J.P. (**2004**) *Metrologia*, **41**, 285–294.

Poirier, W. and Schopfer, F. (**2009**) in *Quantum Metrology and Fundamental Constants,* European Physical Journal–Special Topics (eds F. Piquemal and B. Jeckelmann), IEEE, vol. **172**, pp. 207–245.

Pöpel, R., Niemeyer, J., Fromknecht, R., Meier, L., Grimm, L., and Dünschede, F.W. (**1991**) *IEEE Trans. Instrum. Meas.*, **40**, 298–300.

Pothier, H., Lafarge, P., Orfila, P.F., Urbina, C., Esteve, D., and Devoret, M.H. (**1991**) *Phys. B*, **169**, 573–574.

Prange, E. and Girvin, M. (**1987**) *The Quantum Hall Effect*, Springer-Verlag, New York.

Reymann, D. and Witt, T.J. (**1993**) *IEEE Trans. Instrum. Meas.*, **42**, 596–599.

Reymann, D., Witt, T., Eklund, G., Pajander, H., Nilsson, H., Behr, R., Funck, T., and Müller, F. (**1999**) *IEEE Trans. Instrum. Meas.*, **48**, 257–261.

Robinson, I.A. and Kibble, B.P. (**2007**) *Metrologia*, **44**, 427–440.

Rüfenacht, A., Burroughs, C.J., Benz, S.P., Dresselhaus, P.D., Waltrip, B., and Nelson, T. (**2008**) in *Proceedings of the Conference on Precision Electromagnetic Measurements* (eds A.H. Cookson and T. Winter), IEEE, Boulder 70–71.

Shapiro, S. (**1963**) *Phys. Rev. Lett.*, **11**, 80–82.

Shilton, J.M., Mace, D.R., Talyanskii, V.I., Galperin, Y., Simmons, M.Y., Pepper, M., and Ritchie, D.A. (**1996**) *J. Phys. Condens. Matter*, **8**, L531–L539.

Shin, Y., Song, W., Kim, J., Woo, B., Kim, N., Jung, M., Park, S., Kim, J., Ahn, K., and Hong, K. (**2006**) *Phys. Rev. B*, **74**, 1954151–1954154.

Scherer, H., Lotkhov, S.V., and Willenberg, G.D. (**2008**) in *Proceedings of the Conference on Precision Electromagnetic Measurements* (eds A.H. Cookson and T. Winter), IEEE, Boulder 278–279.

Schopfer, F. and Poirier, W. (**2007**) *J. Appl. Phys.*, **102**, 054903.

Schumacher, B. (**2007**) *Metrologia*, **44**, Tech. Suppl., 01004.

Schurr, J., Alhers, F.J., Hein, G., Melcher, J., Pierz, K., Overney, F., and Wood, B.M. (**2006**) *Metrologia*, **43**, 163.

Schurr, J., Bürkel, V., and Kibble, B. (**2008**) in *Proceedings of the Conference on Precision Electromagnetic Measurements* (eds A.H. Cookson and T. Winter), IEEE, Boulder 108–109.

Steck, B. Application en métrologie électrique de dispositifs monoélectroniques: vers une fermeture du triangle métrologique, (**2007**) PhD thesis. University of Caen.

Steck, B., Gonzalez-Cano, A., Feltin, N., Devoille, L., Piquemal, F., Lotkhov, S., and Zorin, A.B. (**2008**) *Metrologia*, **45**, 1–10.

Steiner, R.L., Williams, E.R., Liu, R., and Newel, D. (**2007**) *IEEE Trans. Instrum. Meas.*, **56**, 592–596.

Stewart, W.C. (**1968**) *Appl. Phys. Lett.*, **12** (8), 277–280.

Stone, M. (**1992**) *Quantum Hall Effect*, World Scientific, Singapore.

Sullivan, D.B. and Dziuba, R.F. (**1974**) *Rev. Sci. Instrum.*, **45**, 517–519.

Thompson, A.M. and Lampard, D.G. (**1956**) *Nature*, **177**, 888–890.

Trapon, G., Thévenot, O., Lacueille, J.C., and Poirier, W. (**2003**) *Metrologia*, **40**, 159–171.

Tsai, J.S., Jain, A.K., and Lukens, J.E. (**1983**) *Phys. Rev. Lett.*, **51** (4), 316–319.

Vartiainen, J.J., Möttönen, M., Pekola, J.P., and Kemppinen, A. (**2007**) *Appl. Phys. Lett.*, **90**, 082102.

Wicht, A., Hensley, J.M., Sarajilic, E., and Chu, S. (**2002**) *Phys. Scr.*, **T102**, 82.

Williams, E.R., Ghosh, R.N., and Martinis, J.M. (**1992**) *J. Res. Natl. Stand. Technol.*, **97**, 299–304.

Witt, T.J. (**2003**) *IEEE Trans. Instrum. Meas.*, **52**, 487–490.

Wood, B.M. and Solve, S. (**2008**) in *Proceedings of the Conference on Precision Electromagnetic Measurements* (eds A.H. Cookson and T. Winter), IEEE, Boulder, pp. 98–99. See also Reports of BIPM.EM-K10.a &. b comparison: “on-going direct 10~V Josephson on-site comparisons” available in *http://kcdb.bipm.org*.

Würstle, C., Ebbecke, J., Regler, M.E., and Wixforth, A. (**2007**) *N. J. Phys.*, **9**, 73.

Yamamori, H., Ishizaki, M., Shoji, A., Dresselhaus, P.D., and Benz, S.P. (**2006**) *Appl. Phys. Lett.*, **88**, 0425031–0425033.

Yoshioka, D. (**2002**) *The Quantum Hall Effect*, Springer-Verlag, Berlin.

Zeller, H.R. and Giaver, I. (**1969**) *Phys. Rev.*, **181**, 789.

Zhang, Y.B., Tan, Y.W., Stormer, H., and Kim, P. (**2005**) *Nature*, **438**, 201.

Zimmerman, N.M., Simonds, B.J., and Wang, Y. (**2006**) *Metrologia*, **43**, 383–388.

Zorin, A.B., Bogoslovsky, S.A., Lotkhov, S.V., and Niemeyer, J. (**2000**) Cooper pair tunneling in circuits with substantial dissipation: the three-junction R-pump for single cooper pairs, Cond-mat/0012177.

Further Reading

Gallop, J.C (**1990**) *SQUIDs, the Josephson Effects and Superconducting Electronics*, Adam Hilger, Bristol.

Hansch, T.W., Leschiutta, S., Wallard, A.J., and Rastello, M.L. (eds) (**2007**) *Proceedings of the International School of Physics “Enrico Fermi”, Course CLXVI, Recent Advances in Metrology and Fundamental Constants*, Societa Italiana di Fisica-Bologna and IOS Press.

Kadin, A.M. (**1999**) *Introduction to Superconducting Circuits*, John Wiley & Sons, Inc., New York.

Koch, H., Lübbig, H., and Luebbig, H. (eds) (**1992**) *Single-Electron Tunneling and Mesoscopic Devices*, Series in Electronics and Photonics, vol. **31**, Springer.

Petley, B.W. (**2001**) in *Proceedings of the International School of Physics “Enrico Fermi”, Course CXLVI, Recent Advances in Metrology and Fundamental Constants* (eds T.J. Quinn, S. Leschiutta, and R. Tavela), Societa Italiana di Fisica-Bologna and IOS Press, pp. 121–155.

Piquemal, F., and Jeckelmann, B. (eds) (**2009**) *Quantum Metrology and Fundamental Constants*, European Physical Journal – Special Topics, vol. **172**, pp 267–296.

Quinn, T.J. and Burnett, K. (eds) (**2005**) Royal Society Discussion Meeting: the fundamental constants of physics, precision measurements and the base units of the SI. *Philos. Trans. R. Soc. Lond.*, **A 363**, 2097–2327.

Rogalla, H. (**1998**) in *Handbook of Applied Superconductivity* (ed. B. Seeber), Institute of Physics Publishing, Bristol 1759–1775.

Weinstock, H. (**1996**) *SQUID Sensors: Fundamentals, Fabrication and Applications*, Kluwer Academic Publishers.

10
Metrology of Time and Frequency

Felicitas Arias and Andreas Bauch

Handbook of Metrology. Edited by Michael Gläser and Manfred Kochsiek

ISBN: 978-3-527-40666-1

10.1 Introduction

Everyday life can today scarcely be imagined without time measurement. This term designates not only the measurement of the length of time intervals (stopwatch) and the recording of the frequency of events during a particular interval of time (frequency measurement) but also the dating of events within a time scale (time of the day). Although it is sufficient for the former two tasks to define the unit of time, the latter task certainly is the most important one as regards everyday life. It presupposes conventions for the parameters of a time scale, that is, to say for the point of time at which the time scale begins, and a rule for how multiples of the basic scale unit are to be formed. For legal time or civil time in most countries, the *second* as the unit of time in the International System of Units (SI) Bureau International des Poids et Mesures (BIPM) (BIPM, 2006) forms the basic scale unit. Twenty-four hours consisting each of 60 minutes, which in turn consist each of 60 seconds form one day – this convention is part of our cultural heritage. The beginning of the day is fixed at 0 : 00 hours. For the counting of days, many calendars are in use in various cultures, with a predominance of the Christian Gregorian calendar in business life. Useful specifications as regards time, calendar, numbering of weeks, and notation of date and time can be found in the standard ISO 8601 and in the equivalent European standard EN28601 (EN, 1992). The *Further Reading* section contains references to books and articles that reflect the historical aspects of time keeping, the evolution of standard time, and the change from the astronomical determination of time to its determination in terms of atomic physics, which is only briefly presented in this chapter.

High-level metrology of time and frequency is also vital to the success of many fields of science and technology. Examples from atomic physics are atom–photon interactions, atomic collisions, and atomic interactions with static and dynamic electromagnetic fields. Geodesy, radioastronomy (very long baseline interferometry, VLBI), and millisecond-pulsar timing rely strongly on the local availability of stable frequency standards and access to global uniform timescales. The same is true for the operation of satellite-based navigation systems. Applications such as management of electric power networks and telecommunication networks also require synchronization of local timing sources or syntonization of locally maintained frequency sources with national or international standards. Strictly speaking,

Handbook of Metrology. Edited by Michael Gläser and Manfred Kochsiek

ISBN: 978-3-527-40666-1

synchronization (Greek, $\chi\rho$ $o\upsilon o\sigma$, time) of two clocks means adjusting the reading of one clock to that of the other, whereas syntonization (Latin, tonus) means equalizing the rate of the two clocks. It is common practice to use "synchronization" as a general term. In almost all these fields, atomic frequency standards (AFSs) have played an important role since decades. Today, ten thousands of rubidium AFS have been installed, several hundred commercial cesium atomic clocks are used in timing laboratories and in military and scientific institutes, and the number of hydrogen masers (*maser* from microwave amplification by stimulated emission of radiation) in operation surely exceeds 100. This reflects the fact that among all the standards used to implement the SI, AFSs have the best accuracy, precision, and repeatability, which is true because the frequency of the standards' output signal is related to an inherent property of free atoms.

For the realization of AFSs and in consequence, the current definition of the unit of time, advantage is taken of the fact that a transition between two atomic eigenstates with an energy difference ΔE is linked with the absorption or emission of electromagnetic radiation of the frequency $f = \Delta E/h$ (h is Planck's constant). Accordingly, the frequency f and the period $1/f$, respectively, of such a radiation are basically constant – in contrast to the period of the Earth's rotation and, still more so, the oscillation period of a pendulum. In 1955, the Englishman Louis Essen suggested using the transition between the hyperfine levels of the ground state of the element cesium as a potential reference transition. The first cesium atomic beam frequency standard (in short called *Cs clock*) had just started operation at the British National Physical Laboratory (NPL) in Teddington (Essen and Parry, 1955). From 1955 to 1958, in cooperation with the United States Naval Observatory (USNO), Washington, the duration of the unit of time then valid, that is, to say of the ephemeris second, was determined at 9 192 631 770 periods of the Cs transition frequency f_0 (Markowitz *et al.*, 1958). This measurement result formed the basis for the definition of the SI second, which was adopted in 1967 by the 13th General Conference for Weights and Measures (CGPM) (Terrien, 1968) and is still valid today:

> *The second is the duration of 9 192 631 770 periods of the radiation corresponding to the transition between the two hyperfine levels of the ground state of the ^{133}Cs atom.*

Following good metrological practice, the "atomic" second thus was chosen as long as that valid before. It is slightly shorter than the second of universal time observed during the last decades, but on the other hand, it is independent of all nonuniformities of the Earth's rotation and of the Earth's revolution around the Sun.

In the logic of the "atomic" approach, all standards using the same atomic species should in principle deliver the same frequency. This is, admittedly, not the case since the technical capabilities to transfer atomic properties to that of macroscopic electric circuits are imperfect. The estimated deviation of a standard's frequency from its nominal value is expressed as the standard's uncertainty. At the time of writing (spring 2008), the relative uncertainty has been reported lower than one part in 10^{16} in two cases (Fortier *et al.*, 2007; Rosenband *et al.*, 2008). Other optical standards have, however, been recognized to provide "secondary representations" of the second (Gill and Riehle, 2006). For about 10 standards values lower than one part in 10^{14} were reported in the course of time in the *BIPM Annual Reports on Time Activities*. Such remarkable values have been achieved just about five decades after the first atomic clock was operated.

10.2 Characterization of Clocks and Oscillators

The frequency of oscillators and clocks is subject to systematic and random variations with respect to the intended nominal output value. As the need for an adequate language with which to communicate the properties of the standards has been recognized since long, the subject is extensively covered in the literature, building to a large extent on work done under the auspice of the IEEE (Allan, 1988; IEEE, 1999). In this chapter, we briefly introduce the measures for accuracy and instability. The variety of such measures is apparently wider than in other fields of metrology since they have to describe non-Gaussian and nonstationary behavior of signals, which is quite common in the field.

We start with the expression for the (nearly) sinusoidal signal voltage generated by a frequency standard, given by

$$V(t) = [V_0 + e(t)] \sin\{2\pi \nu_0 t + \phi(t)\} \tag{10.1}$$

where $\nu_0, \phi(t)$, and $e(t)$ are the nominal frequency, the instantaneous phase fluctuations, and the amplitude variations, respectively. We introduce the instantaneous phase-time variations, $x(t) = \phi(t) / (2\pi\nu_0)$, and the instantaneous normalized frequency departure $y(t) = (d\phi/dt)/(2\pi\nu_0)$. In the Fourier frequency domain, frequency instability may be expressed by several spectral densities, such as $S_y(f)$ of $y(t)$, $S_\phi(f)$ of $\phi(t)$, or $S_x(f)$ of $x(t)$. It is, however, more popular to use time domain quantities, based on mean frequency values $\bar{y}(\tau)$, such as the Allan variance (Allan, 1966):

$$\sigma_y^2(\tau) = \left\langle \left(\bar{y}_{k+1}(\tau) - \bar{y}_k(\tau) \right) \right\rangle^2 \tag{10.2}$$

Here the mean values $\bar{y}(\tau)$ are contiguous (no dead time), and the brackets $\langle\rangle$ denote an infinite time average. In practice, a finite sum of terms is considered, and according to a rule of thumb, the number of samples should be 10 or larger. In the case of white frequency noise, the relative uncertainty for a measurement of $\sigma_y(\tau)$ based on m measurements is given by $(3 \times m - 4)^{1/2}/(m-1)$ (Lesage and Audoin, 1973).

In a double logarithmic plot of $\sigma_y(\tau)$ versus τ, one can discriminate among some causes of instability in the clock signal because they lead to different slopes. If shot-noise of the detected atoms is the dominating noise source, the frequency noise is white and $\sigma_y(\tau)$ decreases like $\tau^{-1/2}$. In this case, $\sigma_y(\tau)$ agrees with the classical standard deviation of the sample. Typically, however, one notes long-term effects due to colored noise processes, which indicate that parameters defining ν_0 are not sufficiently well controlled. The classical standard deviation would diverge with increasing τ in such a case, whereas $\sigma_y(\tau)$ remains bounded. The slope in the log–log plot then changes to zero or even becomes positive. Later, in Figure 10.7 we show examples of the frequency instability expected or observed for a variety of AFSs.

So far we dealt only with random perturbations of the output frequency, which admittedly may be difficult to distinguish from systematic effects for very long τ. The term *accuracy* is generally used to express the agreement between the clock's output frequency and its nominal value conforming to the SI second definition. The manufacturers of commercial devices state the accuracy as a range of clock frequencies to be expected for a given model, usually without giving details about the causes of potential frequency deviations. A detailed uncertainty estimate has to be provided for primary clocks and frequency standards for which a similar status shall be obtained. It deals with the quantitative knowledge of all effects, which may entail that the output frequency does not reflect the transition frequency of unperturbed atoms at rest. Given the lack of knowledge of the involved experimental parameters or the underlying mechanism, one estimates first the components of the uncertainty due to individual effects and finally states a combined uncertainty u (International Organization for Standardization (ISO), 1993). Table 10.1 contains examples for such uncertainty budgets.

10.3 Quartz Crystal Frequency Standards

Quartz crystal oscillators are the true work-horses in time and frequency metrology. The piezoelectric effect of quartz (SiO_2) is used to excite mechanical vibrations by applying electric oscillating fields to a piece of suitably cut single crystal of the material (Gerber and Ballato, 1984). Mechanical deformations in the material, such as flexure, extension, and shear, can be used. The quartz resonator design is aimed at minimizing the perturbation of the intrinsic properties by environmental factors. The "cut" – the orientation of the piece of quartz with respect to the crystallographic axes – is chosen such that the influence of temperature and stress upon the oscillating frequency is minimized. In an advanced electrodeless design – called *BVA* for *Boîtier Vieillissement Amé lioré*– the electric fields are not – as usual – applied through metallic electrodes deposited on the surface directly, but through auxiliary plates suspended at a few micrometers distance (Besson, 1977).

Tab. 10.1 Uncertainty budgets of two primary clocks with a thermal beam, JPO of LNE-SYRTE (Makdissi and De Clercq, 2001), CS2 (Bauch *et al.*, 2005), and two fountain clocks, FO2 of LNE-SYRTE and NIST-F1 (BIPM, 2007).[a]

Cause of frequency shift	JPO	CS2	FO2	NIST F1
Quadratic Zeeman effect	13	50	0.2	0.52
AC Stark effect caused by thermal radiation	5	10	0.6	2.8
AC Stark effect caused by fluorescence	24	0	–	$<10^{-4}$
Cavity phase difference	40	100	3	0.2
Quadratic Doppler effect	26	10	<0.1	0.2
Asymmetric population of hyperfine sublevels	23	1	<1.0	0.2
Collisions among cold atoms	0	0	2.9	1.3
Electronics, microwave leakage fields	20	30	0.5	1.2
Combined uncertainty	63	120	4.6	3.4

[a] All values are relative ones, given in parts in 10^{16}.

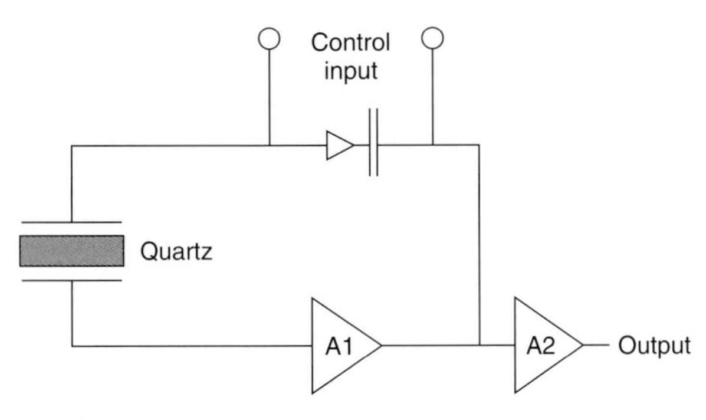

Fig. 10.1 Schematic representation of a quartz crystal oscillator including an external control input.

A quartz oscillator circuit scheme is depicted as Figure 10.1. The resonator is inserted in a feedback loop of amplifier A1, and the output is typically formed in a second amplifier A2. As part of the loop a voltage-controllable capacitor is shown, which serves for fine frequency tuning of the resonance frequency in voltage-controlled X-tal oscillators (VCXOs). VCXOs including a BVA quartz resonator are used as part of the microwave signal generation of many primary frequency standards and hydrogen masers. The very best of them exhibit a relative short-term frequency instability of below 1×10^{-13} for averaging times between 1 and 30 seconds. In general, an external control reference does not exist, but a variety of passive control means and compensation tricks have been designed to minimize frequency dependence on temperature, vibration, orientation of the mounted unit, radiation, and aging with time.

It has been estimated that more than 10^9 quartz crystal oscillators are produced annually for applications ranging from inexpensive watches and clocks to oscilloscope and counter time bases, mobile phones, and personal radio-navigation units. Given this range of applications, performances, and prices, we give no tables or graphs indicating performance, and so on, but rather refer to the *Further Reading* section. However, it should be noted that even cheap quartz oscillators built into radio-controlled wrist watches can keep time to a few parts in 10^7. This is why these clocks display "the correct time" even if the radio-control part does not work any longer as the battery voltage has decreased below a certain limit. Users only get alerted when the transit from standard time to daylight saving time, or the leap day are not displayed correctly.

10.4 Atomic Frequency Standards

10.4.1 Concept

It is commonly assumed that atomic properties such as energy differences between atomic eigenstates and thus atomic transition frequencies are natural constants and do not depend on space and time (apart from relativistic effects). They are determined by fundamental constants that describe the interaction of elementary particles. A transition between two eigenstates differing in energy by ΔE is accompanied by absorption or emission of electromagnetic radiation of frequency $f_r = \Delta E/h$. The

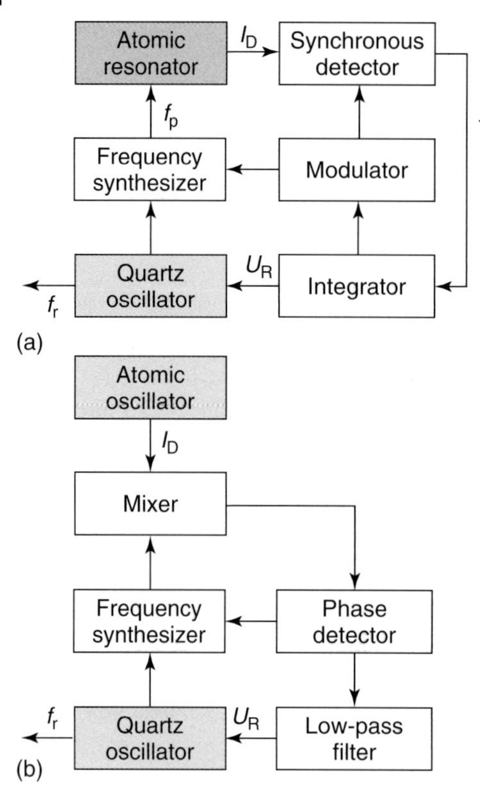

Fig. 10.2 Schematic representation of a passive (a) and an active (b) frequency standard; f_p characterizes the probing signal, f_r the output signal, I_D is the signal carrying the primary information from the atoms, and U_R is the control voltage to the quartz oscillator.

principle of operation of the so-called *passive* AFS is illustrated in Figure 10.2a. A signal at a probing frequency f_p is used to interrogate the atomic resonator, and the response of the latter is used to steer the quartz oscillator. In contrast, an *active AFS*, such as an active hydrogen maser, relies on emitted radiation, which is detected and which then steers a built-in electrical oscillator (Figure 10.2b).

The choice of the particular atomic transition as "clock transition" is directed not only by certain requirements on instability and accuracy but also by practical constraints regarding the ease of manufacture, operation and maintenance, and the reliability of the standard. In brief, it is required to have

- a small natural line width, that is, a low rate of spontaneous transitions among the energy levels involved in the clock transition;
- a long interaction time T_i of the atomic absorber with the probing radiation;
- sources of probing radiation, which deliver a spectrally narrow radiation so that no technical broadening of the observed resonance curve occurs; and
- atomic eigenstates that should be insensitive to electric and magnetic fields and whose separation in energy should be as large as technically accessible.

This list of requirements has dictated the development of virtually all standards discussed in this section.

10.4.2
Cesium Atomic Beam Frequency Standards

In the early 1950s, the element cesium was identified as a very suitable candidate to fulfill many of the above-mentioned requirements. The isotope ^{133}Cs that, favorably, is the only stable isotope of this element has a nuclear spin of 7/2 and its ground state consists of two hyperfine level manifolds, separated in frequency by about 9.2 GHz, now defined as $f_{Cs} = 9\,192\,631\,770$ Hz for atoms at zero field and at rest. It was very important, particularly in the earlier days when the laser was still unknown, that magnetic selection of cesium atoms in the different hyperfine states is possible and also that efficient detection of the atoms using surface ionization on a hot-wire detector can be performed. The two manifolds each comprise eight magnetic sublevels that have different deflection properties in strong magnetic fields. In Figure 10.3, the function is illustrated. A beam of atoms effuses from the oven and passes through the polarizer. Owing to the relatively high vapor pressure at moderate temperatures, intense thermal cesium atomic beams with relatively low mean velocity (about 200 m s^{-1}) can be produced, leading to an interaction time T_i of a few milliseconds even in small structures. In the so-called Ramsey cavity, made up of a U-shaped waveguide, the atoms are irradiated twice with a standing microwave probing field of frequency f_p (Ramsey, 1950). The analyzer discriminates between atoms that have made a transition and those that have remained in the initial state and directs atoms in one

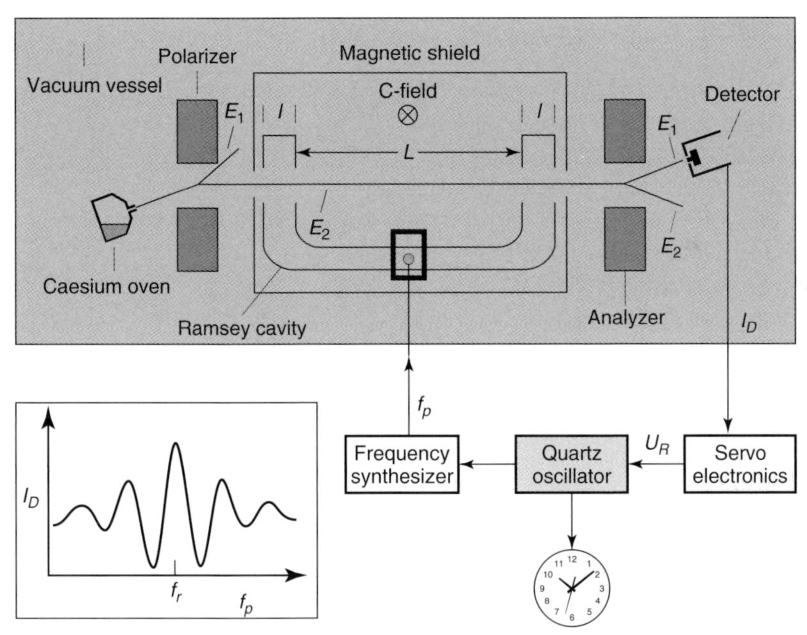

Fig. 10.3 (a) Schematic representation of a cesium atomic clock using magnetic state selection with dipole magnets; the grey-shaded rectangle is referred to as *atomic resonator* in Figure 10.2; E_1 and E_2 represent the two hyperfine energy states involved in the clock transition; and L is the separation of the two Ramsey interaction zones of extension l and (b) Included is a schematic representation of the detected resonance signal.

of the states to the hot-wire detector. A secondary electron multiplier behind some ion optics (for acceleration, deflection, and focusing) is typically used to amplify the initial electric current of some picoamperes to the range of microamperes for easier and faster processing of the beam signal I_D.

The transition between states with the magnetic quantum number (m_F) zero occurs at a frequency f_r close to f_{Cs} defined above, which is the most insensitive to the presence of a magnetic field and thus designated as *clock transition*. The transition frequencies between all other states are detuned by a sufficiently strong magnetic field ("C-field" in Figure 10.3). In Figure 10.3, the designation E_1 and E_2 was introduced for the atoms in the two $m_F = 0$ states. When f_p is tuned across f_r, I_D exhibits a resonance feature centered around f_r, which is shown schematically in Figure 10.3. In clock operation, the probing frequency is modulated around a central value. By phase-sensitive detection of I_D and subsequent integration, the control voltage U_R is generated, which tunes the quartz oscillator – as explained in Section 10.3 – so that f_p and f_r agree on average. The deviation between f_r and f_{Cs} is known within the accuracy of the standard and taken into account in the setting of the frequency synthesizer.

Following the above principles, cesium clocks have been produced commercially since the late 1950s, starting with the so-called Atomichron of the National Company (Forman, 1985; Cutler, 2005). The relative uncertainty of the currently best model is specified as 2×10^{-13} under good laboratory conditions, and its frequency instability is often below 1×10^{-14} for averaging times >5 days. The ensemble of such clocks operated in national timing institutes (Guinot and Arias, 2005) exhibits a scatter of monthly clock rates mostly between $\pm 2 \times 10^{-13}$, with the ensemble mean rate in very close agreement with that of International Atomic Time (TAI), which is determined with the help of primary clocks (see Section 10.7).

Several national metrology institutes have during the past decades competed in building the most accurate *primary* clock. Their basic principle is alike to that shown in Figure 10.3 – magnets have in some cases been replaced by laser interaction zones – but in details their construction allows the determination of all frequency-shifting effects with high accuracy at all times. In 2007 operation of three primary clocks with a thermal atomic beam has been reported. These are the French JPO (Jet de Pompage Optique) (Makdissi and De Clercq, 2001) using optical state selection and detection, and CS1 and CS2 of the Physikalisch-Technische Bundesanstalt (PTB) (Bauch *et al.*, 2005). Each of these standards realizes the SI second with a relative uncertainty of about 1×10^{-14} or slightly below.

10.4.3
Other Commercial Atomic Frequency Standards

The ground-state hyperfine transitions of the elements cesium, hydrogen, and rubidium (^{87}Rb) serve as references in today's commercially available AFSs. The energy level manifold in the ground state is created in all three atoms by the magnetic interaction of the single outer electron with the nucleus with nuclear spin I. A similar energy level configuration can be found in some singly ionized atoms, like $^{199}Hg^+$ and $^{171}Yb^+$, and these and some other ions have been used in laboratory standards. We defer the reader to the textbooks listed in the *Further Reading section* for details on these systems. Here we restrict ourselves to a brief description of the most popular devices.

10.4.3.1 The Active Hydrogen Maser

The ground-state hyperfine splitting of the hydrogen atom corresponds to a transition line at 1.4 GHz. In the active maser stimulated emission inside a high-Q cavity is used to detect the atomic transition (Ramsey, 1990). A schematic diagram of the hydrogen maser is shown in Figure 10.4. Molecular hydrogen is introduced in a dissociator consisting of an electrical discharge in a glass enclosure. The dissociation of the molecules takes place with a relatively high efficiency and a beam of atomic hydrogen is formed with the help of a collimator. The beam is directed along the axis of a six-pole magnet that focuses atoms in the upper hyperfine states into the entrance of the storage bulb. The bulb made of fused silica is placed inside a microwave cavity resonating at the hyperfine frequency and having the geometry that enables the highest possible Q (minimal energy dissipation in the walls). The inner surface of the bulb is coated with Teflon^{TM} that prevents recombination of the atoms into molecular hydrogen and relaxation of the atoms into the ground state so that a one-second lifetime of an atom in one particular level is achieved.

In a cavity at ambient temperature, a weak microwave field is always present. Atoms that entered the storage bulb emit their energy through the process of stimulated emission of radiation. The radiation emitted is added in phase to the existing radiation, a process that results in amplification, explaining the acronym maser. The energy provided by the atoms is dissipated in the walls of the cavity, but part of the generated field ($\approx 10^{-13}$ W) is extracted via the coupling loop. If the losses are small and the relaxation time sufficiently long, a continuous oscillation at the hydrogen transition frequency occurs. The signal coupled from the maser cavity is processed for phase locking a quartz oscillator to the maser signal as it was shown in Figure 10.2b.

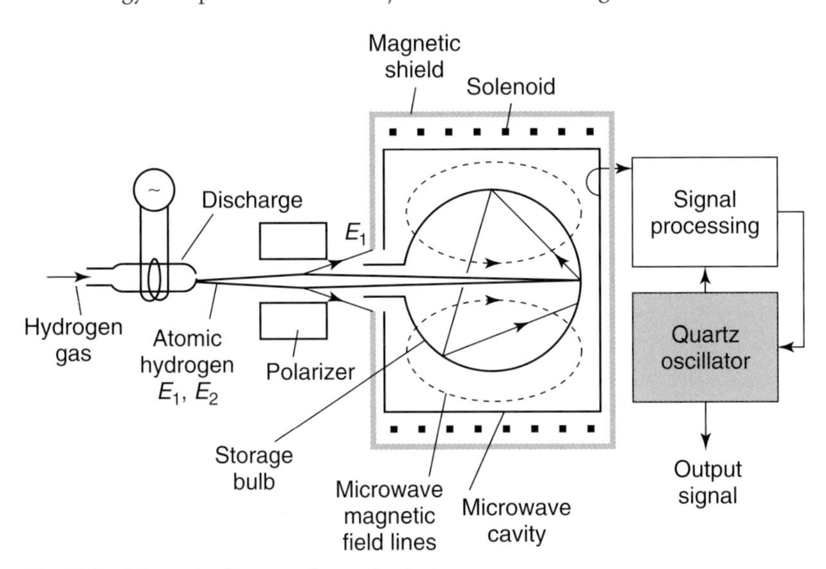

Fig. 10.4 Schematic diagram of an active hydrogen maser.

The collisions of the atoms with the walls of the storage bulb cause a large frequency shift, amounting to a few parts in 10^{11}. It is one of the arts of making masers to keep this shift stable. The same holds for the shift due to pulling of the realized frequency by the cavity resonance frequency. Both effects explain that the maser cannot be developed to become an intrinsically accurate frequency standard. However, its excellent frequency instability, as low as one part in 10^{15} at few hours averaging times, motivates the use of this rather complex and costly instrument. The very best results have been reported from masers kept in well-controlled environmental conditions (Parker, 1999).

10.4.3.2 The Rubidium Gas Cell Frequency Standard

Techniques involving optically pumped atomic vapor in closed cells have been successfully used in numerous applications where the requirements regarding frequency instability in the short and long term cannot be fulfilled with quartz oscillators, but where constraints on space, power consumption, or cost prevent the use of other AFS. The rubidium gas cell frequency standard illustrated in Figure 10.5 is an example. The reference transition used is

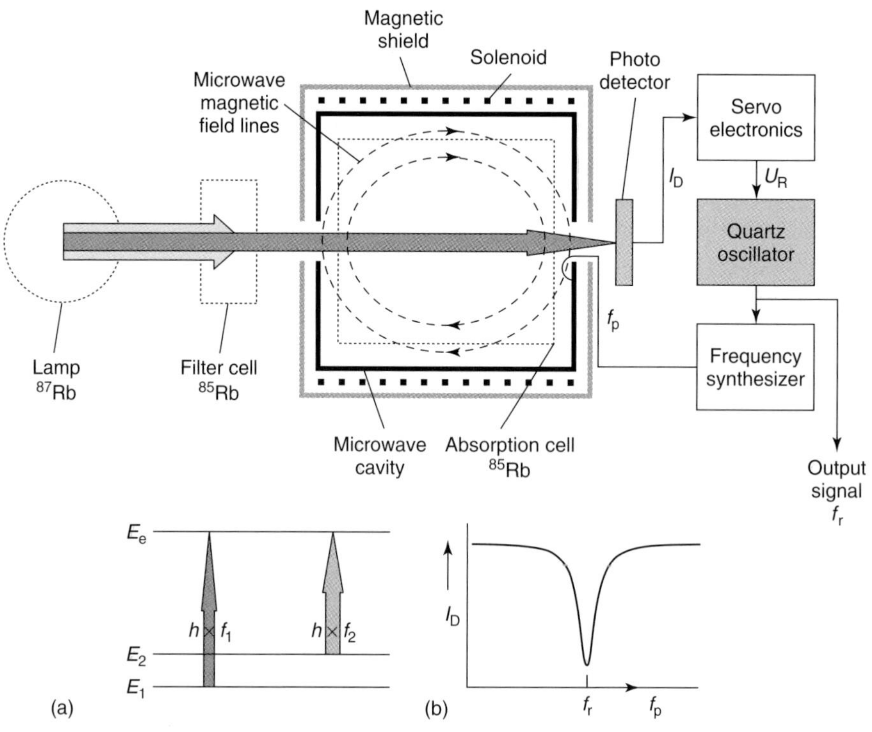

Fig. 10.5 Schematic diagram of a Rb gas cell AFS; inset (a) explains the two components of radiation for optical pumping, of which the component with frequency f_2 is blocked in the filter cell and inset (b) shows the Lorentz-shaped absorption line observed.

the ^{87}Rb hyperfine transition at 6.84 GHz. The simplified level scheme is given as inset (a) in Figure 10.5. Development of this kind of frequency standard started in the late 1950s in the United States, and the first commercial model became available in the 1960s (Packard and Swartz, 1962). At its heart is the absorption cell containing a vapor of ^{87}Rb and a buffer gas. Optical excitation on the D_2 line to the $^2P_{3/2}$ level (E_e in Figure 10.5) requires a radiation source at 780 nm wavelength. Atoms excited to the P states in the cell then relax to both ground-state levels either by spontaneous emission or by collisions with the buffer gas. Collisional relaxation is typical for gas cell standards. The light spectrum necessary to obtain population inversion is generated with an ^{87}Rb lamp and a filter containing ^{85}Rb vapor. The purpose of the filter is to clean the spectrum of the lamp from the lines at those frequencies corresponding to the transitions from the level E_2 to E_e in ^{87}Rb. Nitrogen was found to be a suitable buffer gas. Favorably, the buffer gas also reduces the diffusion velocity of the atoms to a few centimeters per second. The Doppler shift is thus reduced, the resonance line is narrowed, and the effect of wall collisions is also reduced thereby.

In a typical arrangement, as shown as Figure 10.5, the absorption cell is placed inside a low Q microwave cavity tuned to the resonance frequency of the ^{87}Rb atoms. The light transmitted through the cell is measured by means of a photodetector. Optical pumping of the atoms causes the cell to become transparent to the incident radiation since atoms are pumped out of the absorbing state. If microwave energy is fed to the cavity at the hyperfine frequency of the rubidium atoms, a field is created inside the cavity, which may stimulate transitions from the level E_2 to E_1. This results in a decrease in light intensity at the photodetector if the microwave field is resonant with the hyperfine transition. By sweeping the microwave frequency across the resonance, a signal is observed as sketched in the inset (b) of Figure 10.5. The signal can be used to control the frequency of the quartz oscillator according to the scheme of Figure 10.2a. The shape of the signal is a Lorentzian line broadened and shifted by several mechanisms such as buffer gas collisions, spin exchange interactions, optical pumping, and saturation caused by the microwave excitation.

The multitude of frequency-shifting effects prevents the Rb AFS to be an intrinsically accurate clock. The output signal at typically 10 MHz is calibrated with respect to some reference source. In the long term, the frequency typically drifts by a few parts in 10^{11} per month. The causes thereof are changes in the spectrum of the light passing through the gas cell and changes of pressure and composition of the buffer gas. The short-term frequency instability, however, is quite favorable, and the best available commercial Rb AFS are more stable than some commercial Cs AFS for averaging times up to 10^4 seconds. This combination of properties calls for disciplining the Rb AFS through an external reference with a long time constant. Signals of the global positioning system (GPS), but also of long-wave standard frequency transmitters, such as DCF77 in Europe (see Section 10.8), are well suited for the purpose.

In this contribution, we can only touch upon the three directions of development going on in the field. The use of laser diodes, having much narrower line spectra than provided by spectral lamps, could improve the efficiency of the optical pumping (Vanier and Mandache, 2007). The background light would be reduced and a substantial gain in signal-to-noise ratio feasible. Up to now, however, the issue of reliability has prevented the commercial usage of laser-pumped Rb AFS. The Rb AFS can be made rather compact, but still becoming part of a hand-held device is essentially excluded. Miniaturization of AFS has been discussed since about a decade, and several research groups work on the so-called *clock on the chip*, see (IEEE, 2005), for example. Very compact AFS could become useful in receivers for navigation systems signals and for secure communications. Another vivid subject is the use of the coherence in an atomic sample of Rb or Cs created by irradiating it with two laser fields separated in frequency exactly by the hyperfine splitting. The sample of atoms is then excited into a nonabsorbing state, explaining the term *coherent population trapping* (CPT) for this technique (Godone, Levi, and Micalizo, 2002; Vanier, 2005).

10.4.4
Fountain Clocks

In continuation of the endeavors to improve primary clocks further, a real breakthrough came when the first fountain clocks appeared in the late 1980s. Here only a short description of fountain operation and properties is given. Detailed account is given in the *Further Reading* section, which contains references to aid understanding of laser cooling, the key to making a fountain work. By laser cooling one can prepare state-selected samples of atoms (e.g. Cs and Rb) with thermal velocities of the order of a few millimeters per second – corresponding to 1 μK when expressed in temperature units – which can be manipulated as shown in Figure 10.6. The atom sample is launched upward with a velocity v_S. It comes to rest under the action of gravity at the height $H = v_s^2/(2\ g)$ above the starting

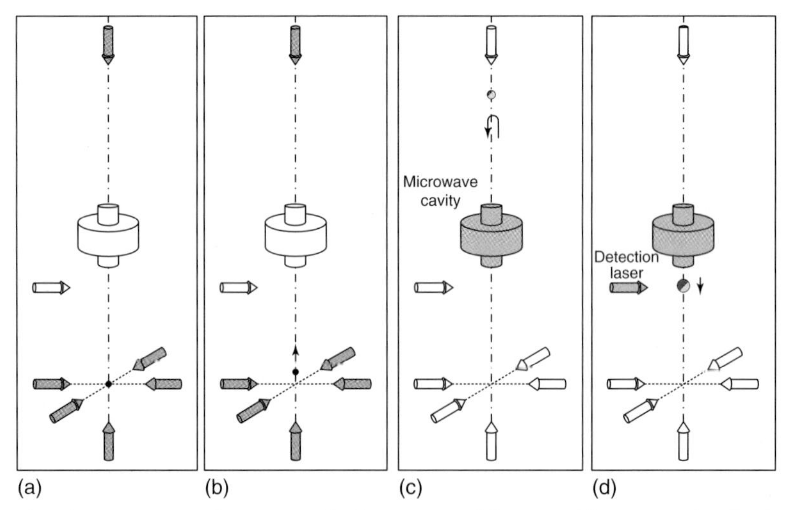

Fig. 10.6 Sequence of operation of a fountain frequency standard, illustrated in a time sequence from left to right; laser beams are indicated by arrows (white if they are blocked). (a) A cloud of cold atoms is loaded. (b) The cloud is launched by detuning of the frequency of the vertical lasers. (c) The cloud with an initially small volume and high density expands during ballistic flight. (d) After the second passage of the atoms through the microwave cavity, the state population is probed by laser irradiation and fluorescence detection.

point before it falls back. A starting velocity $v_s = 4.4\ \mathrm{m\ s^{-1}}$ leads to a height H of about 1 m and a total time of flight, back to the starting point, of about 0.9 seconds. On their way the atoms interact twice with the field sustained in the microwave cavity, on their way up, and then on their way down. The interaction time T_i becomes typically 0.5 seconds. The detection comprises the determination of the number of atoms in both hyperfine levels. Without cooling to kinetic energies equivalent to microkelvin temperature, the thermal expansion of the atomic cloud would be so large that the fraction of detected atoms would be too small to obtain a useful signal-to-noise ratio. During clock operation, the transition probability is determined changing the frequency f_p from cycle to cycle alternately on either side of the central fringe where the sensitivity to changes of f_p relative to f_n has its maximum. The difference of successive measurements is numerically integrated and a control signal is derived to steer the quartz or to adjust the output frequency of a synthesizer included in the generation of the microwave signal.

In Table 10.1, the contributions to the uncertainty resulting from the most significant causes of frequency shift for two primary clocks with a thermal beam and two fountains are combined. "0" indicates that the effect is nonexistent and <0.1 indicates that the component is by all means smaller than 10^{-17}. Because of the reduced line width, slightly below 1 Hz, and the reduced atomic velocity some systematic frequency-shifting effects are reduced by orders of magnitude for fountains. A new effect needs consideration. Ultracold atoms are used and the cross section for frequency-shifting collisions among cesium atoms becomes much larger than that of thermal cesium atoms. The collisional shift is proportional to the density of atoms in the cloud and depends on details of the state in which the atomic cloud is initially prepared and it currently leads to a significant uncertainty contribution, still a subject of detailed studies.

Rubidium (^{87}Rb) has been identified as another candidate species for use in a fountain since laser cooling and manipulation are technically feasible as well. The cross section for phase changing collisions among cold ^{87}Rb atoms is so much smaller than among cold cesium atoms that it can be hardly measured (Bize *et al.*, 2005). The corresponding uncertainty contribution for an Rb fountain is essentially zero. The rubidium hyperfine transition was thus recently recommended as a secondary representation of the second (Gill and Riehle, 2006).

In conclusion, Figure 10.7 shows the frequency instability of the AFS described in this section.

10.4.5 Optical Frequency Standards

Frequency standards in the infrared and in the visible range of the electromagnetic spectrum have been developed and used since decades. The most prominent use has been as wavelength standards in practical length metrology and for the realization of the meter. Since the new definition of the meter became effective in 1983, this SI unit should be realized according to guidelines given in the so-called "mise-en pratique" (Quinn, 2003). High-resolution laser spectroscopy both in the context of fundamental research and in the context of developing an all-optical clock is currently pursued by more than a dozen groups, in metrology institutes as well as in academic institutions. The atomic systems providing the optical reference transition are typically different from group to group. To reach the

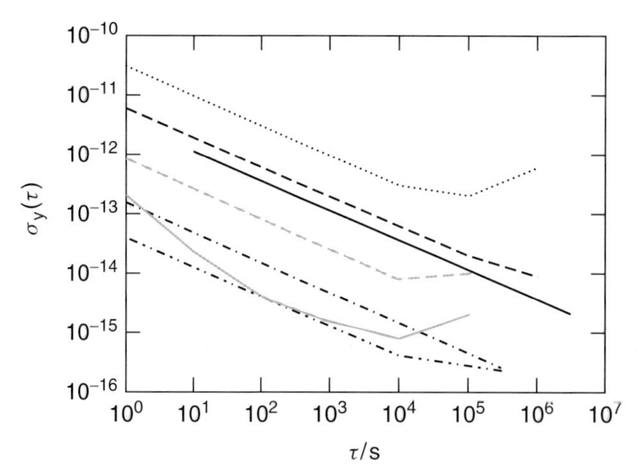

Fig. 10.7 Relative frequency instability $\sigma_y(\tau)$ of different AFS: (dashed black) specifications of the widely used commercial cesium AFS 5071A (high-performance option) of Symmetricom, (solid gray) average specification found for active hydrogen masers, (dashed gray) average specifications found for passive hydrogen maser (described in Vanier and Audoin, *Further Reading* section), (dotted black) typical data of Rb AFS, (solid black) measured instability in continuous comparison of CS1 and CS2 of PTB during eight years, (black dot-dashed) typical instability of a fountain frequency standard, and (black dot-dot-dashed) measured frequency instability of FO-2 (Bize *et al.*, 2005).

ultimate goal that one atomic transition in the optical frequency domain is accepted as basis for a new definition of the second, it would be necessary to identify one "best" candidate and – following good metrological practice – realize a couple of standards and verify that they agree at the expected level.

In the context of this Handbook, it seems adequate to give only a cursory overview of this currently highly dynamic field of research and point to the *Further Reading* section. The most accurate values for optical transition frequencies have up to now been obtained for forbidden transitions in trapped and laser-cooled single ions (Fortier *et al.*, 2007; Rosenband *et al.*, 2008). Ions can be localized in radiofrequency ion traps while only minimally perturbing the internal level structure. Combined with laser cooling, it is possible to reach the confinement to a spatial region smaller than the clock transition wavelength, so that the linear Doppler shift is eliminated and thus no line broadening occurs. Neutral-atom-based optical frequency standards have gained a particularly strong interest since it was shown that the trapping of atoms in standing light fields is possible without shifting the energy difference between the two states defining the reference transition (Katori, 2002). One expects frequency standards with even lower frequency instability than feasible with trapped ion standards.

10.5 Time and Frequency Measurement

10.5.1 Time or Frequency?

The frequency standards discussed so far, excluding for a moment the optical frequency standards, produce a standard frequency output, for example, at 10 MHz. Internally this frequency may be dived down to produce pulses at a rate of 1 Hz, the so-called one pulse per second (1 PPS) signal. For many applications it is interesting to know only the absolute frequency value and its instability in time. Measurement techniques adapted to that are described subsequently. If the epoch of the 1 PPS output matters, for example, when the signal reflects a local reference time scale, then some extra care is needed, as discussed in Section 10.5.3. If the standard's output is that of a stabilized laser, quite different techniques are used to compare it to other similar standards or to bridge the gap to the radiofrequency domain. This is subject of Section 10.5.4.

10.5.2 Frequency Counters and Phase Comparators

The standard instrument for the direct measurement of arbitrary frequencies f is the frequency counter. At its input, it has a discriminator circuit that defines an "event," such as a zero crossing with positive slope in the input signal. The number of events during a selected gate time t_g is measured, divided by t_g to give a mean frequency $\bar{y}(t_g)$, with a constant weighting factor for the frequency averaging over the time window t_g. Such measurement values can then be analyzed using Equation (10.2). The ± 1 count ambiguity often is the dominant measurement uncertainty, which amounts to $1/(f \times t_g)$ relatively. High-resolution counters utilize interpolation between the signal event and the preceding or following gate impulse in order to increase the resolution, see, for example (Kramer and Klische, 2001). Recently it was pointed out that the operation of modern "calculating" counters involves multiple averaging within a single gate time t_g in order to suppress spectral components in the signal out of the band defined by $1/t_g$. In consequence, the values $\bar{y}(t_g)$ determined that way are not suited to provide a correct estimate of $\sigma_y(t_g)$ as defined in Equation (10.2) rather than another measure for frequency instability (Dawkins, McFerran, and Luiten, 2007).

The limitation caused by the finite resolution can be overcome using some form of heterodyne system that works, however, only for fixed frequencies. A common technique includes the use of a transfer oscillator, offset in frequency by 10 Hz (Levine, 1999) or 1 Hz, which is mixed with the input signals of two standards to be compared. Time-interval measurements between the zero crossings of the two signals are performed, and the resolution of the measurement is enhanced by the ratio of the signal frequency to the beat frequency output by the mixer. Such dual-mixer time-difference measurements have been perfected to provide a relative measurement uncertainty below 1×10^{-14} for averaging times of 1 second (Šoidr, Čermak, and Barillet, 2004).

10.5.3 Time-Interval Counters

Time-interval measurements differ from frequency measurements with counters only to the extent that the "open gate" and "close gate" signals are produced by the two input signals, and that the measurement quantity is to first order the number or periods $1/t_b$ of the counter time base recorded between the two events. The trigger point is chosen near a point of maximum slope of the impulse, so as to minimize trigger point variations due to the finite rise time of the impulse. In many time-interval counters (TICs), the resolution is improved compared to $1/t_b$ by adding an analog interpolator such that sub-100 picosecond resolution is achieved. The measurement uncertainty is specified to below 1 nanosecond for several counter models. More sophisticated instrumentation allows time-interval measurement with picosecond resolution (Kalisz, 2004). Two time-difference measurements, Δt_1 and Δt_2 separated in time by t_{mess}, provide the mean frequency difference, $\bar{y}(t_{mess}) = (\Delta t_2 - \Delta t_1)/t_{mess}$, between the two sources providing the input to the counter for analysis as described in Equation (10.2). True time-interval measurements, however, have additional problems. The sharp 1 PPS signals are distorted by reflections from imperfectly terminated cables and circuits, and at the end of a long cable the shape of the impulse has changed due to the damping of high-frequency (HF) components in the cable. In general, the direct time-difference measurement system will very probably have a higher measurement noise than the time stability of typical atomic clocks for averaging times out to several hours. For long-term time comparisons, however, it is the method of choice due to its inherent simplicity. A word of caution should be added regarding the cables connecting the clocks. Cable delays are part of the measurement result. They have to be determined and monitored for true time comparisons. The properties of cables regarding delay changes and damping as a function of temperature differ substantially from type to type. Thus, the right choice has to be made, depending on the performance of the clock and the environment of the cable conducts.

10.5.4 Frequency Combs

To exploit the outstanding properties of optical frequency standards in terms of accuracy and stability, it is necessary to have a device at hand that can count optical frequencies. Only since a few years such a device, called an *optical frequency comb* and based on a mode-locked femtosecond-pulse laser, is widely available (Hollberg *et al.*, 2005, papers of Hall and Hänsch in the Further Reading section). Owing to the mode-locking, a single laser pulse propagates between the mirrors of the optical cavity in the laser. At the output, a pulse train with the repetition rate f_R is observed. In the frequency domain, this pulse train corresponds to a comb of discrete optical frequency modes. Measuring an (unknown) optical frequency relies primarily on the fact that the mode spacing is constant across the spectrum. The intensity of the modes, however, is not constant, the bandwidth being roughly given by the inverse pulse duration. The measurement principle as described subsequently requires that the comb spans a full octave. The first devices were based on mode-locked Ti : sapphire lasers in which the comb is centered around a wavelength of 800 nm and the pulse duration is not short enough to provide such an octave wide spectrum. The spectrum had to be broadened by passing the light through a photonic crystal fiber (Cundiff and Ye, 2003). Recently erbium-doped femtosecond fiber lasers with the wavelength centered around 1500 nm have become even more popular since they can be operated more easily and more reliably (Ma *et al.*, 2004).

The phase and group velocity of the traveling pulse in the laser are not equal in general, which causes a frequency offset f_{CEO} in the mode spectrum so that

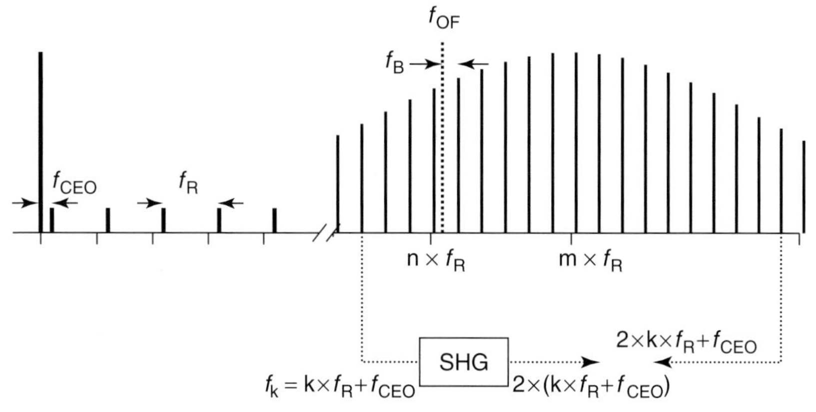

Fig. 10.8 Principle of the optical frequency measurement with a femtosecond comb. See the text for the various designations. SHG: second harmonic generation for frequency doubling. The beat notes between the signals at $2 \times (k \times f_R + f_{CEO})$ and $2 \times k \times f_R + f_{CEO}$, respectively, provide f_{CEO}. The other radiofrequency measurements are that of f_R and f_B.

the frequency of each mode is given by $f_M = f_{CEO} + p \times f_R$. The measurement of an optical frequency f_{OF} is illustrated in Figure 10.8. It comprises the measurement of the beat note f_B between the signal of unknown frequency and the nearest comb frequency (comb number p), the measurement of f_R and f_{CEO}, and the measurement of p. The measurement of f_{CE} is done by the so-called self-referencing technique (Jones *et al.*, 2000). Modes from the low-frequency (LF) tail of the comb are frequency doubled and a beat note with modes at roughly twice the frequency in the HF tail is produced. As illustrated in Figure 10.8, this gives access to f_{CEO}. Thus, the measurement of f_{OF} is reduced to the (straightforward) measurement of three frequencies in the radiofrequency part of the spectrum and of p. The mode number p can be determined using a coarse wavelength measurement with a standard wavemeter. The measurement principle has been used to determine optical frequencies in terms of the SI hertz provided by fountain clocks and ratios of optical frequencies (Lea, 2007; Fortier *et al.*, 2007; Gill and Riehle, 2006). In all cases, the measurement uncertainty has been determined by the properties of the standards involved and has not been limited by the measurement principle.

10.6 Time and Frequency Comparisons between Remote Sites

10.6.1 GNSS Time Comparison

The comparison of distant clocks has always been an important part of time metrology. On the one hand, it allows the assessment of the properties of (primary) frequency standards, for example, to judge whether they agree within their assigned uncertainty. On the other hand, such comparisons are needed for the calculation of a time scale on the basis of the readings of clocks located in different laboratories.

The most prominent example thereof is described in the following section. In a perfect world, the methods of time transfer do not compromise the frequency stability and accuracy of the clocks as seen at the remote site. This has not always been the case in the past, and again today the comparison of remote optical frequency standards is a challenge, to say the least.

Since signals from the satellites of the GPS – the first global navigation satellite system (GNSS) (see books on GPS in the *Further Reading* section) – started to be used since the late 1980s, the quality of time comparisons has been substantially improved. The signals broadcast by GPS satellites are derived from onboard atomic clocks and contain timing and positioning information. Specific GPS timing receivers have been developed for the purpose, which come in two distinct configurations:

- *Receiver type 1*: uses the received signal in space to discipline an inbuilt oscillator to GPS time, which then delivers a 1 PPS output or even a set of output signals (standard frequency signals, signals for telecommunication applications). This is further discussed in Section 10.8.
- *Receiver type 2*: determines the pseudorange of each satellite in view with respect to a reference signal provided externally and uses the correction data transmitted in the signal in space to provide output data in the form of local reference (local time scale) minus GPS time (the system time that is calculated from an ensemble of clocks in the satellites and on Earth). This type of receivers is needed for the purpose discussed in this section. Laboratories compare their local timescales with GPS time. Another option is the use of the International GNSS Service (IGS) time, a time scale calculated at the IGS (Dow, Neilan, and Gendt, 2005), which is used as the reference for the IGS products (Ray and Senior, 2003). To facilitate the data exchange for time transfer and dissemination, directives on a common format and standard formulae and parameters have been provided jointly by the BIPM and the Consultative Committee for Time and Frequency (CCTF) (Allan and Thomas, 1994).

The common-view (CV) method, proposed in the 1980s (Allan and Weiss, 1980), has been in use for the comparison of distant clocks in the last decades. This method of simultaneous reception of the same emitted signal at two Earth laboratories minimizes the impact of common errors in the GPS signals caused by errors in the satellite position, instabilities of the satellite clocks, and the effects of the intentional degradation (known as *selective availability*) that was applied to the GPS signal until May 2000.

Receivers of the first generation used for time comparison were single-channel, single-frequency C/A code (Coarse Acquisition) receivers. In this case, the information used is the code that is modulated on the GPS L1 frequency (1575.42 MHz). Receiver manufacturers developed later multichannel receivers, operating also at one frequency, but allowing simultaneous observations of several satellites at a time. The propagation of the signal is affected by atmospheric effects, producing a supplementary delay. The ionosphere provokes delays that introduce significant errors, particularly during periods of high solar activity. Dual-frequency reception eliminates the ionospheric delays. Multichannel, dual-frequency receivers have thus started to replace older equipment in a number of laboratories. Most of these are geodetic-type receivers that provide observations on two frequencies denominated L1 and L2. Such receivers and an evaluation method called *P3* (Defraigne and Petit, 2003) have helped to increase the accuracy of time transfer. GPS observations with single-frequency receivers are still made. They can be corrected for ionospheric delays by making use of IGS ionospheric maps (Dow, Neilan, and Gendt, 2005; Wolf and Petit, 1999) and for satellite positions using IGS postprocessed precise satellite ephemerides.

The method of common view can be favorably replaced by a new one, named *GPS all in view* (Petit and Jiang, 2009), which is actually simpler to implement. After exchange of the (standardized) data files among the laboratories, the individual observations may be corrected for the above-mentioned effects before averages over convenient intervals are formed. Subtraction of corresponding data allows the comparison of the local time scales or frequency standards.

The most advanced use of GPS signals is based on the carrier phase (Schildknecht *et al.*, 1990; Larson and Levine, 1999; Kouba and Héroux, 2001; Ray and Senior, 2005; Petit and Arias, 2009). This allows frequency comparisons among remote standards with the lowest transfer noise of all currently common methods (Bauch *et al.*, 2005). In Figure 10.9, time transfer data between PTB and USNO obtained in different methods are depicted.

The Russian global navigation satellite system GLONASS is not yet widely used, since the satellite constellation is still in the process of completion and suitable

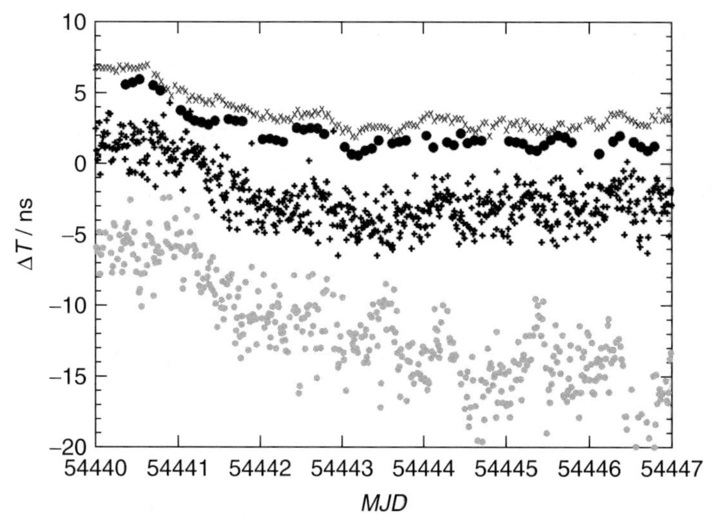

Fig. 10.9 Comparison of time scales UTC(PTB) with UTC(USNO) during one week (MJD 54440 = 2008-12-06), using four different technologies: GPS L1 C/A-code multichannel CV (gray dots) and postprocessed GPS CV P3 (symbol +), both based on 16-minutes averaging, TWSTFT in X-band, sessions of 15-minutes duration, nominally one per hour (symbol x), and TWSTFT in Ku-band, sessions of 2 minutes duration nominally every 2 hours (full dots). Offsets were introduced to separate the data, except for the latter that reflect the "true time comparison" with an uncertainty of 2 nanoseconds.

receivers have not been produced until recently. The system is potentially useful for accurate time transfer because of the unlimited access to the precision code by the civil community. Within a couple of years, the variety of new civil signals provided by GPS as well as by the future European satellite navigation system Galileo will enable even more accurate time transfer using this one way – receive only – method.

10.6.2
Two-Way Satellite Time and Frequency Transfer

Two-way satellite time and frequency transfer (TWSTFT) (Kirchner, 1991; Kirchner, 1999) is based on the exchange of signals through geostationary telecommunication satellites. Currently, TWSTFT is made using fixed satellite services in the Ku-band and the X-band. It is done by transmission of pseudorandom noise (PRN) binary phase-shift keying (BPSK) modulated carriers. The phase modulation is synchronized with the local clock's 1 PPS output. Each station uses a dedicated PRN for its BPSK sequence in the transmitted signal. The receiving equipment allows to generate the BPSK sequence of the remote stations and to reconstitute a 1 PPS tick from the received signal. This is measured by a TIC with respect to the local clock. Following a prearranged schedule, both stations of a pair lock on the code of the corresponding remote station for a specified period, measure the signal's time of arrival, and store the results. After exchanging the data records, the difference between the two clocks is computed. Details of the data reduction and the treatment of systematic effects involved have been elaborated in the Study Group 7 (Science Services) of the International Telecommunication Union (ITU) and published as an ITU-R recommendation (ITU, Radiocommunication Sector, 2003). TWSTFT has proven as the most appropriate means of comparing time scales and AFSs with an uncertainty in time of less than 1 nanosecond and with relative uncertainty for frequency of about 1 part in 10^{15} at averaging times of one day. This is why TWSTFT is used in the international network of time keeping institutions supporting the realization of TAI (the following section). It is performed operationally in at least 2 laboratories in the United States, 12 in Europe, and 7 in the Asia Pacific Rim region. For the very reasons, TWSTFT has been chosen as the primary means to synchronize the two precise timing facilities (PTFs), part of the Ground Mission Segment of the European satellite navigation system Galileo, as well as to support the measurement of the time difference between GPS time and the Galileo system time (GST) (Hlavác *et al.*, 2006).

10.6.3
Calibration of Time Transfer Equipment

Calibration of the signal delays in the laboratory's equipment is instrumental for performing accurate time transfer. Without that, assuming stability of the (unknown) delays, one can still perform frequency transfer. Campaigns of GPS time equipment differential calibration are organized by the BIPM. The typical uncertainty for a GPS link calibration is estimated at 5 nanoseconds. Regional metrology organizations have been invited to support the work of BIPM by performing similar campaigns in their member countries. TWSTFT equipment is calibrated by using a portable TWSTFT station and an uncertainty of ≈ 1 nanosecond has been repeatedly demonstrated (Piester *et al.*, 2008).

The experience on long-term stability of the installed equipment at that level is scarce, and variations at the nanosecond level have been observed in some cases.

10.7 International Atomic Time and Coordinated Universal Time

10.7.1 Work of BIPM

A mission conferred to the BIPM by the Conférence Générale des Poids et Mesures (CGPM) is that of providing the basis for a coherent system of measurements adopted worldwide, which should be traceable to the SI. One of the responsibilities mandated to the BIPM is that of maintaining and disseminating TAI and coordinated universal time (UTC) (see Glossary), and this is done by its Time, Frequency, and Gravimetry Section (in short "the BIPM" in the following).

TAI and UTC are postprocessed time scales; they are the result of worldwide cooperation of about 65 national metrology laboratories and astronomical observatories that operate commercial cesium standards and hydrogen masers. The data are regularly reported to the BIPM by timing centers that maintain a local realization of UTC. Results are calculated on the basis of data acquired on the previous month and are published in the monthly *BIPM Circular T* (*http://www.bipm.org/jsp/en/TimeFtp.jsp?TypePub=publication*).

All contributing laboratories operate industrial atomic clocks as described in Section 10.4 and are equipped with devices for their comparison at a distance. The BIPM keeps long-term track of the clocks behavior and assigns weights to clocks according to their individual stabilities. About 10 laboratories develop and maintain primary frequency cesium standards that realize the atomic second, with a nonnegligible number of cesium fountains that contribute more or less regularly to improve the accuracy of TAI.

The algorithm used for the calculation of TAI has been designed to guarantee the reliability, the long-term frequency stability, the frequency accuracy, and the accessibility of the scale. Nevertheless, the quality of TAI rests critically on the methods of clock comparison, which may bring significant instability mostly at short averaging time (5–10 days). The BIPM organizes the international network of time links to compare local realizations of UTC in contributing laboratories and uses them in the formation of TAI. The network of time links presently used by the BIPM is nonredundant and relies on the observation of GPS satellites (Section 10.6.1) and on TWSTFT (Section 10.6.2).

For nearly two decades, GPS C/A-code observations have provided the best tool for clock comparisons in TAI; thus, no test of its performance with respect to other methods was possible. Since about 2000, the situation is quite different with the introduction of the independent TWSTFT technique and resulted in a more reliable system. For the links where the two techniques are available, both GPS and TWSTFT links are computed; the best being used in the calculation of TAI and the other(s) kept as a backup. The GPS links using geodetic-type, dual-frequency code receivers (P3) have further increased the reliability of the system of time links, providing a method of assessing the performance of the TWSTFT technique. Comparisons of results obtained on the same baselines with the different techniques show equivalent performances for GPS geodetic-type dual-frequency receivers and TWSTFT equipment, at long averaging times, with root-mean-squared differences at or below 1 nanosecond (Petit and Jiang, 2004).

As in April 2008, most of the links in TAI are obtained by using GPS equipment (70% with GPS single-frequency receivers; 16% with GPS dual-frequency receivers), and about 14% of the links are provided by TWSTFT observations. It should be noted that the best techniques (GPS dual frequency and TWSTFT) link about 70% of the clock weight used for TAI.

The network of international time links follows a scheme that is well adapted for the "GPS all-in-view" (Petit and Jiang, 2009) solution that has replaced on a routine basis the classical common views of GPS satellites in October 2006 (refer to Section 10.6.1 for more information). Approximations to UTC, indicated by UTC(k) in national laboratories are compared to that of the laboratory that acts as a cross-over site. This practice has become possible, thanks to the availability of the IGS clock solutions and the IGS time scale IGST (Petit and Arias, 2009).

A two-step algorithm is used for the calculation of TAI. At the first step, the "échelle atomique libre" (EAL) (free atomic scale) is calculated from the weighted average of the frequencies of the participating atomic clocks. In a second step, measurements of primary frequency standards realizing the SI second made over the previous one-year period are introduced to calculate the relative departure of the second of TAI with respect to the SI second; if necessary, frequency corrections are applied in a process known as *frequency steering* to improve the accuracy of TAI and UTC.

Although TAI makes use of primary frequency standards measurements over the year preceding the month of calculation, another time scale called *TT(BIPMxx)* is calculated at the BIPM including all data from primary frequency standards (xx indicates the two last figures of the year it is produced) (Guinot, 1988; Petit, 2003). This improved time scale is necessary for studies such as timing of millisecond pulsars (Section 10.9) and may be extended backward since its origin was chosen to give continuity with the previous reference, dynamical time.

Results and information related to the formation of TAI and UTC are published in the monthly *BIPM Circular T*, in the key comparison in time available through the key comparison database on the BIPM web site, and in the *BIPM Annual Report on Time Activities*.

10.8 Time Dissemination

10.8.1 General Remark

During the last few decades, various procedures were introduced for time and frequency information to be disseminated both to the broad public and for scientific and technical applications. Therefore, the circulation of transfer standards in "key comparisons," as it is common in many fields of metrology is inappropriate here. The ITU and other scientific and technical organizations have recommended UTC as the appropriate international reference for time and frequency in most applications. Local realizations of UTC, UTC(k), kept in timing institutes k are often kept well within 1 microsecond of UTC. Therefore, for many applications the origin of a selected distribution service is less important than its reliable and technically simple availability. A distinction should be made between *time* and *frequency* distribution, even if the two quantities are closely related. Time distribution is directly affected by the delay in the transmission between the source and the user, whereas for frequency distribution the delay need not to be known but has to be stable. The various radio-based dissemination services have been described well in an ITU Handbook (ITU, Radiocommunication Sector, 1997a). Since its publication, the Internet has grown and today provides an additional resource of time-of-day information using specific protocols, as described in the last subsection.

10.8.2 GPS Time Dissemination

The use of Global Positioning System disciplined oscillators (GPSDOs) as frequency references in telecommunication networks and calibration laboratories has become very popular. A variety of products are available, which offer long-term frequency instability like that of a cesium clock. Access to UTC(USNO) – an excellent representation of UTC – through the information transmitted in the GPS signal in space can be obtained with low uncertainty, provided that the delays of antenna, cable, and signal processing have been determined. Time-of-day information in UTC can of course be extracted from the signal.

The use of GPSDOs in laboratories of national calibration services is popular, but rules on how traceability to national standards can be achieved may vary from country to country. As a basic requirement, national time and frequency laboratories k monitor the GPS constellation and provide time differences (*UTC*(k) – *GPS time*) on their web sites. The BIPM provides the differences (*UTC* – *GPS time*) in their monthly *Circular T*. GPSDO users should take notice of this information. It has been discussed that a kind of type approval would be necessary before a GPSDO could be used as a calibration reference. The rapid evolution of product lines due to the technical development has rendered this idea not very practical. It is by all means recommended that only such GPSDO be used whose specifications are complete regarding their short- and medium-term frequency instability, and whose signal delays of internal circuits and antenna are known if the GPSDO shall be used as a time reference. Additionally, the GPSDO should allow an assessment whether the disciplining process is active and successful before using the device. It is very much encouraged that inexperienced users seek advice from their national metrology institute before acquiring or using GPSDO in demanding applications.

10.8.3 Standard Frequency and Time Dissemination Services

Although GPS signals have found widespread use, there are still many users of the various national terrestrial services of dissemination of standard frequency and time signals in frequency bands allocated by the ITU. The current status of such services can be freely accessed on the ITU web portal (ITU, Radiocommunication Sector, 2008). Although the services in the HF band have mostly technical applications, the LF services including coded time information, which are operated in Germany, Japan, the United Kingdom, the United States, and Switzerland on a regular basis, have millions of users interested in time-of-day information with moderate accuracy requirements. In such applications, the signal propagation delay is simply ignored.

To give an example, the German DCF77 signal provides standards frequency with an uncertainty of 1 parts in $10^{12}(1\sigma)$ for one-day averaging. A pseudorandom noise-type phase modulation superimposed on the carrier allows identification of points in time to well below 0.1 milliseconds. Coded time-of-day information is provided as phase modulation and amplitude modulation. Table 5.2 provides access to the most relevant information on the existing services of that kind.

10.8.4 Internet and Telephone Time Services

To distribute time through the Internet, a hierarchical network of time servers has been deployed over the years, with the number of servers and users growing almost exponentially. Primary (stratum 1) servers are synchronized to national time standards either directly (operation of the servers in the metrology institutes) or via GPS. Secondary (stratum 2) servers and clients are synchronized to the primary servers via a hierarchical subnet. Redundancy of servers as well as a diversity of network paths ensures the reliability of the services. The most common network clock synchronization protocol that can read a server clock, transmit the reading to one or more clients, and adjust the lower stratum server clocks is the network time protocol (NTP), the current version is 4 (NTP, 2006; Levine, 2008).

The method of synchronization is briefly as follows: on request from a client, transmitted at time t_1, the server sends a message including its clock values at reception of the request, t_2, and at transmission of the message, t_3. The client records its own clock value, t_4, upon arrival of the message. This provides a measure of the total server–client round-trip propagation delay as $\delta t = (t_4 - t_1) - (t_3 - t_2)$, and the time difference between server and client clock as $\Delta t = [(t_2 - t_1) + (t_3 - t_4)]/2$.

Network paths and the associated delays can differ significantly for both directions, which limits the achievable uncertainty. The NTP-4 software includes algorithms for reducing jitter and discriminating between correctly and incorrectly received time information by selecting and averaging among multiple sources. The client clock is disciplined in time and frequency using an adaptive algorithm, which handles network time jitter and local oscillator frequency offset and drift.

Telephone time services allow computers and data acquisition instruments to be provided with the exact time via automatic accession. The accuracy lies between 1 second and a few milliseconds. Comparable services are available in various countries, such as Japan and the United States, but in particular different European institutes offer comparable services with an identical time code. Relevant information is provided at the BIPM web site at *http://www.bipm.org/jsp/en/TimeFtp.jsp?TypePub=scale.*

Several European time institutes agreed upon the "European telephone time code" which is described on the web, see, for example, *http://www.ptb.de/time.* The information generated and emitted consists in a series of ASCII characters, where the data valid for the second $n + 1$ are transmitted in second n. In addition, two ASCII characters – here especially <CR><LF> – are emitted immediately before the second change. Emission of this synchronizing signal allows compensation of the signal travel time from the transmitter to the receiver. For this purpose, the user must return the telephone time signals received to the telephone time code generator. This generator then measures the round-trip

Tab. 10.2 Low-frequency stations broadcasting continuously standard frequency and time signals, status spring 2008.[b]

Call sign	Country responsible institute	Location	Carrier frequency (kHz)	Relative uncertainty of the carrier frequency (1 σ, one-day average)	Emitted power (kW)	URL
DCF77	Germany PTB	Mainflingen 50°01′ north 09°00′ east	77.5	1×10^{-12}	30	*http://www.ptb.de/time*
HBG[b]	Switzerland METAS	Prangins 46°24′ north 6°15′ east	75	1×10^{-12}	20	*http://www.metas.ch/*
JJY	Japan NICT	Ohtakadoya-yama 37°22′ north 140°51′ east	40	1×10^{-12}	12.5	*http://jjy.nict.go.jp/index-e*
JJY	Japan NICT	Hagane-yama 33°28′ north 130°11′ east	60	1×10^{-12}	15	*http://jjy.nict.go.jp/index-e*
MSF	UK NPL	Anthorn 54°55′ north 03°15′ west	60	2×10^{-12}	17	*http://www.npl.co.uk/*
WWVB	USA NIST	Colorado Springs 40°40′ north 105°02′ west	60	1×10^{-11}	65	*http://tf.nist.gov*

[a] Detailed information is provided from the responsible institutes (see last column).
[b] It has been annonced that the operation of HBG will stop at the end of 2011.

time and then sends the signals earlier by half the amount determined so that they arrive at the use approximately at the right moment. Travel time correction allows uncertainties of time transmission by means of telephone modems of a few milliseconds to be achieved. The telephone services in some European countries are offered as services subject to charges and can be reached only from the respective country network. The facilities at the users consist of a switched-line modem and a computer on which the suitable software must be installed to establish the connection, evaluate the time code received, and synchronize the computer clock or other clocks that can be set via the computer.

10.9 Applications

10.9.1 General Metrology

The SI is based on seven basic quantities, each one having a defined unit. Linked to these base units are the numerous derived units. The unit of time – the second – can be reproduced with an uncertainty much lower than any other of the base units (BIPM, 2006). This is why other units are practically based on the time unit. Since 1983, the definition of the meter is based on the fixed value of the speed of light in vacuum. The measure of voltages and electrical resistances is transposed into a measure of frequency using the Josephson effect and the quantum Hall effect, respectively. The importance of the time unit in this respect may even become larger when the units of mass, electrical current, and temperature will be redefined as it is currently widely discussed (Mills *et al.*, 2006; Becker *et al.*, 2007).

10.9.2 Synchronization of Networks

Telecommunication networks are based on a hierarchy of clocks distributed in a number of levels, called *strata*. At the upper level accurate "master clocks," often commercial cesium clocks (see Section 10.4.2), are used. Descending the pyramidal hierarchy one finds clocks of lesser and lesser quality, such as rubidium clocks and quartz oscillators. Some network operating agencies completely rely on GPSDO (see Section 10.8.2) as master clocks, whereas others use them at Stratum-2 level. The operation of telecommunication networks is governed by recommendations of the ITU Telecommunication Sector. The relevant documents (ITU, Telecommunication Sector, 1997b) and (ITU, Telecommunication Sector, 2004) specify timing characteristics of primary reference clocks and timing requirements of slave clocks in network nodes, respectively.

Power distribution networks are another example of extended use of time and frequency sources. Tracing power outages in extended networks requires clocks synchronized to a common time reference, preferentially UTC, to better than 1 millisecond at network nodes. Disciplined quartz and rubidium AFSs have been deployed in large numbers at the nodes of the power grids. Some European network operators have even acquired cesium clocks providing autonomous frequency references for their network monitoring. In Europe, the members of the Union for the Coordination of Transmission of Electricity (UCTE) agree on standards for network and supply properties, including synchronization issues. As of 2008, UCTE comprises 34 operators in 24 European countries, serving 450 million customers with an average annual electric energy of 2500 TWh. On the UCTE web portal (*http://www.ucte.org*), immediately the current network frequency is displayed.

Another type of network is that of the ground reference stations of the satellite navigation systems, such as GPS, GLONASS, in the future complemented by the European Galileo, the Chinese Compass, and the Indian Regional Navigation Satellite System (IRNSS). In all cases, a network of ground stations at well-known positions is equipped with atomic clocks and serves as the reference for the reception and monitoring of the satellites signals. GST, just to give one example, will be realized in two PTFs, each comprising four cesium atomic clocks and two active hydrogen masers. The anticipated about 30 reference stations – called *Galileo Sensor Stations* – will be deployed worldwide and carry rubidium AFSs (Hlavác *et al.*, 2006).

10.9.3 Pulsar Research

Time metrology and astronomy have always been interdependent in various ways. Nowadays astronomical observatories are prominent users of the various time dissemination services for precise timing of their observations. Among the fields of astronomical research, the observation and timing of pulsars are rather special: during the last 30 years, the role of pulsars has caused some debate among the communities. Can pulsars be considered as "celestial clocks" for long-term monitoring of the performance of earth-bound atomic clocks or is pulsar timing with respect to terrestrial time scales primarily a method to study the properties of pulsars themselves, of the interstellar medium, astrometry, general relativity?

Pulsars are strongly magnetized, rapidly rotating neutron stars. The pulsars observed hitherto have masses of order 1.5 times the solar mass, diameters of about 20 km, and spin rotation periods between 1.34 milliseconds and 8 seconds. They have been designated as celestial clocks with highly invariable rotation periods since they have large moments of inertia and large stores of rotational energy. However, the exact mechanism by which a pulsar radiates the energy observed as radio pulses is still a subject of debate. The basic picture is that of a misaligned magnetic dipole with radiation from charged particles accelerated along the open field lines above the polar cap. The observed pulses fluctuate severely both in intensity and shape, thus typically a standard profile is generated by averaging over some thousand pulses, which is the characteristic for a given pulsar at a given observation frequency. Such time-averaged profiles have been used for high-precision "timing" (Stairs, 2003).

Timing means determination of the time of arrival with respect to the local time scale. To make a meaningful time comparison, one has to account for the propagation delay that requires the knowledge of several parameters, such as the momentary geometry, orbital motion of the pulsar if it is part of a binary system, orbital and rotational position of the earth, the interstellar medium, relativistic effects, and so on and, last but not the least, to relate the observations to a long-term stable time scale such as TT(BIPMxx) (Taylor, 1992; Lorimer, 2005). Some pulsars have quite low spin-period instabilities, which can be interpreted as "relative frequency instability" of 10^{-14} and below at averaging times of one year and longer. However, with the progress of atomic time keeping, the stability of TT(BIPM07), for example, is lower at all averaging times.

The probably most spectacular result obtained using pulsar timing is the indirect proof of the existence of gravitational radiation emitted by the binary system of two neutron stars, one of which is the millisecond pulsar denoted as PSR 1913 + 16 (for its celestial coordinates) (Hulse, 1994; Taylor, 1994). According to general relativity, the emission of gravitational waves should lead to orbital energy loss. This became indeed observable as an orbital period change.

10.9.4 Search for the Variation of Fundamental Constants

The question whether fundamental constants are really *constant* or whether they may show temporal variations within the evolution of the universe was raised as early as 1937 by Dirac. More generally, a violation of Einstein's equivalence principle (including variation of fundamental constants in space and time) is predicted by currently debated cosmological models such as inflation (Uzan, 2003; Karshenboim, 2000); see also the *Further Reading* section. It has been proposed to search for variations of dimensionless quantities such as Sommerfeld's fine structure constant $\alpha \approx 1/137$. Comparisons between AFSs can be used to place limits on the present-day magnitude of such a change. As the transition frequencies in different classes of transitions depend differently on α and on the Rydberg constant R_y, measuring the frequency of an electronic transition repeatedly with a cesium AFS may prove whether the product of some power of α and of the ^{133}Cs nuclear magnetic moment (μ_{Cs} expressed in units of the Bohr magneton μ_B) is constant or not. Clearly, commercial cesium AFSs are not accurate and stable enough for the purpose, which explains that such kind of study was in fact stimulated by the advent of cesium fountain clocks. Optical clock transition frequencies in mercury (Hg^+), hydrogen and ytterbium (Yb^+) were measured with reference to the SI hertz provided by cesium fountains (Lea, 2007). The measurement of optical frequency ratios has also supported such studies. Up to now, the data analysis has always provided "null" results – with relative uncertainties in the low 10^{-16} per annum recently (Rosenband *et al.*, 2008). Several research groups are engaged to continue this kind of measurement and study atomic systems, which promise to yield more stringent limits after a few years of data taking.

The last two sections were intended as a demonstration how the research into AFSs and time and frequency dissemination can help improving our understanding of the laws of physics in general. This proves the interdependence of metrology and fundamental science.

10.10 Disclaimer

The BIPM and the PTB as a matter of policy do not endorse any commercial product. The mentioning of brands and individual models seems justified here, because all information provided is based on publicly available material or data taken at PTB and it will help the reader to make comparison with own observations.

Glossary

This Glossary represents an excerpt of the one contained in ITU, Radiocommunication Sector (1997a) and concentrates on terms that are specific for time and frequency.

Clock Time Difference: The difference between the readings of two clocks at the same instant. Note: this term can be attributed also for time scale difference. Let a denote the reading of clock A, and b the reading of clock B at a time T of a reference time scale. The clock difference is expressed by $A - B = a - \mathrm{b}$. If $a - \mathrm{b}$ increases with time, then clock B is said to have a higher rate (or frequency) than clock A.

Coordinate Time: The concept of time in a specific coordinate frame, valid over a spatial region with varying gravitational potential. To give an example, international atomic time TAI is a coordinate time scale.

Coordinated Universal Time (UTC): Time scale, maintained by the BIPM and the International Earth Rotation and Reference Systems Service (IERS), which forms the basis of civil and legal times and of standard frequencies and time signals. It deviates from TAI only by an integer number of seconds. The acronym UTC was chosen in continuation of the historic designation of astronomical time scales based on universal time (UT).

Frequency Instability: The spontaneous and/or environmentally caused frequency change within a given observation period.

Frequency Standard: A generator, the output of which is used as a frequency reference.

International Atomic Time (TAI): Time scale established by the BIPM on the basis of data from atomic clocks operated worldwide in such a way that its scale unit conforms with the SI second on the rotating geoid.

Modified Julian Date(MJD): Convenient decimal day count. MJD is the number of days that have elapsed since midnight of an arbitrarily chosen date, 17 November 1858.

Primary Frequency Standard: A frequency standard whose frequency corresponds to the adopted definition of the second, with its specified accuracy achieved without external calibration of the device.

Proper Time: Local time realized with a clock, which is significant in a region of small extension in which any change of the gravitational potential can be neglected. The definition of the SI second defines a proper time unit.

Secondary Frequency Standard: A frequency standard that requires external calibration.

Time Scale: A system of unambiguous ordering of events, including the definition of its origin and its scale unit.

References

Allan, D.W. (**1966**) *Proc. IEEE*, **54**, 221–230.

Allan, D.W. (**1988**) Proceedings of the 42nd Annual Symposium on Frequency Control, Baltimore, MD, USA, pp. 419–425.

Allan, D.W. and Thomas, C. (**1994**) *Metrologia*, **31**, 69–79.

Allan, D.W. and Weiss, A.M. (**1980**) Proceedings of the 34th Annual Symposium on Frequency Control Ft. Monmouth, NJ, USA, pp. 334–346.

Bauch, A. *et al.* (**2005**) *Metrologia*, **42**, 43–54.

Becker, P. *et al.* (**2007**) *Metrologia*, **44**, 1–14.

Besson, R.J. (**1977**) Proceedings of the 31st Annual Symposium on Frequency Control Ft. Monmouth, NJ, USA, pp. 147–152.

BIPM (**2006**) *Le Systéme International d'unités (SI Sérvres, France)/The International System of Units (SI)*, 8th edn.

BIPM (**2007**) Report of LNE-SYRTE from September 2007 and NIST from October 2007, published at *http://www.bipm.org/jsp/en/TimeFtp.jsp?TypePub=data*; see also the BIPM Annual Report on Time Activities for 2007.

Bize, S. *et al.* (**2005**) *J. Phys. B: At. Mol. Opt. Phys.*, **35**, S449–S468.

Cundiff, S.T. and Ye, J. (**2003**) *Rev. Mod. Phys.*, **75**, 325–342.

Cutler, L. (**2005**) *Metrologia*, **42**, 90–99.

Dawkins, S.T., McFerran, J.J., and Luiten, A.N. (**2007**) *IEEE Trans. Ultrason. Ferroelectr. Freq. Control*, **54**, 918–925.

Defraigne, P. and Petit, G. (**2003**) *Metrologia*, **40**, 184–188.

Dow, J.M., Neilan, R.E., and Gendt, G. (**2005**) *Adv. Space Res.*, **36**, 320–326.

EN (**1992**) Specification for Representation of Dates and Times in Information Interchange, EN28601:1992, see *http://www.cen.eu/cenorm/aboutus/index.asp*.

Essen, L. and Parry, J.V.L. (**1955**) *Nature*, **176**, 280–282.

Forman, P. (**1985**) *Proc. IEEE*, **73**, 1181–1204.

Fortier, T.M. *et al.* (**2007**) *Phys. Rev. Lett.*, **98**, 070801.

Gerber, E.A. and Ballato, A. (**1984**) *Precision Frequency Control*, Academic Press, Orlando.

Gill, P. and Riehle, F. (**2006**) Proceedings of the 20th European Frequency and Time Forum, Braunschweig, pp. 282–288.

Godone, A., Levi, F., and Micalizo, S. (**2002**) *Coherent Population Trapping*, CLUT Editrice, Torino.

Guinot, B. (**1988**) *Astron. Astrophys.*, **192**, 370–373.

Guinot, B. and Arias, E.F. (**2005**) *Metrologia*, **42**, 20–30.

Hlavác, R., Lösch, M., Luongo, F., and Hahn, J. (**2006**) Proceedings of the 20th European Frequency and Time Forum, Braunschweig, pp. 391–398.

Hollberg, L. *et al.* (**2005**) *Metrologia*, **42**, S105–S124.

Hulse, R.A. (**1994**) *Rev. Mod. Phys.*, **66**, 699–710.

IEEE (**2005**) Proceedings of the 2005 Joint IEEE International Frequency and Control Symposium and the PTTI Systems and Applications Meeting, Vancouver, Canada.

International Organization for Standardization (ISO) (**1993**) Guide to the Expression of Uncertainty in Measurement, Geneva.

ITU, Radiocommunication Sector (**1997a**) Handbook on the Selection and Use of Precise Frequency and Time Systems, International Telecommunication Union, Geneva, Switzerland.

ITU, Telecommunication Sector (**1997b**) Timing characteristics of primary reference clocks, Recommendation ITU-T G.811.

ITU, Radiocommunication Sector (**2003**) The Operational Use of Two-Way Satellite Time and Frequency Transfer Employing PN Codes, Recommendation ITU-R TF.1153-2, Geneva, Switzerland.

ITU, Telecommunication Sector (**2004**) Digital Networks Digital Transmission Models, Recommendation ITU-T G.801.

ITU, Radiocommunication Sector (**2008**) Time Codes, Recommendation ITU-R TF-583, *http://www.itu.int/ITU-R/study-groups/docs/rsg7-583-en.doc* and Standard Frequency and Time Signals, Recommendation ITU-R TF-768, *http://www.itu.int/ITU-R/study-groups/docs/rsg7-768-en.doc*.

Jones, D.J. *et al.* (**2000**) *Science*, **288**, 635–639.

Kalisz, J. (**2004**) *Metrologia*, **41**, 17–32.

Karshenboim, S.G. (**2000**) *Can. J. Phys.*, **78**, 639–678.

Katori, H. (**2002**) *Proceedings of the 6th Symposium on Frequency Standards and Metrology* (P. Gill), World Scientific, pp. 323–330.

Kirchner, D. (**1991**) *Proc. IEEE*, **79**, 983–989.

Kirchner, D. (**1999**) *Review of Radio Science 1996–1999*, Oxford University Press, Oxford, pp. 27–44.

Kouba, J. and Héroux, P. (**2001**) *GPS Solut.*, **5**, 12–28.

Kramer, G. and Klische, W. (**2001**) Proceedings of the 2001 IEEE IFCS and PDA Exhibition, Seattle, WA, USA, pp. 144–151.

Larson, K.M. and Levine, J. (**1999**) *IEEE Trans. Ultrason. Ferroelectr. Freq. Control*, **46**, 1001–1012.

Lea, S. (**2007**) *Rep. Prog. Phys.*, **70**, 1473–1523.

Lesage, P. and Audoin, C. (**1973**) *IEEE Trans. Instrum. Meas.*, **IM-22**, 157–161.

Levine, J. (**1999**) *Rev. Sci. Instrum.*, **70**, 2567–2596.

Levine, J. (**2008**) Metrologia, **45**, S12–S22.

Lorimer, D.R. (**2005**) Binary and Millisecond Pulsars, published at *http://relativity.livingreviews.org/Articles/lrr-2005-7/*.

Ma, L.-S. *et al.* (**2004**) *Science*, **303**, 1843–1845.

Makdissi, A. and De Clercq, E. (**2001**) *Metrologia*, **38**, 409–425.

Markowitz, W., Hall, R.G., Essen, L., and Parry, J.V.L. (**1958**) *Phys. Rev. Lett.*, **1**, 105–107.

Mills, I.M. *et al.* (**2006**) *Metrologia*, **43**, 227–246.

NTP (**2006**) Network Time Protocol Version 4, Reference and implementation Guide, NTP Working Group Technical Report 06-6-1, University of Delaware, June 2006.

Packard, M.E. and Swartz, B.E. (**1962**) *IRE Trans. Instrum.*, **11**, 215–223.

Parker, T. (**1999**) *IEEE Trans. UFFC*, **46**, 745–751.

Petit, G. (**2003**) Proceedings of the 35th PTTI, San Diego, CA, USA, pp. 307–317.
Petit, G. and Jiang, Z. (**2004**) Proceedings of the 36th PTTI, Washington, DC, USA, pp. 31–39.
Petit, G. and Jiang, Z. (**2008**) *Metrologia*, **45**, 33–45.
Petit, G. and Arias, Z. (**2009**) *Geodesy*, **83**, 327–334.
Piester, D. *et al.* (**2008**) *Metrologia*, **45**, 185–198.
Quinn, T.J. (**2003**) *Metrologia*, **40**, 103–133.
Ramsey, N.F. (**1950**) *Phys. Rev.*, **78**, 695–698.
Ramsey, N.F. (**1990**) *Rev. Mod. Phys.*, **62**, 541–552.
Ray, J. and Senior, K. (**2003**) *Metrologia*, **40**, S270–S288.
Ray, J. and Senior, K. (**2005**) *Metrologia*, **42**, 215–232.
Rosenband, T. *et al.*, (**2008**) *Science*, **319**, 1808–1812.
Schildknecht, T., Beutler, G., Gurtner, W., and Rothacher, M. (**1990**) Proceedings of the 4th European Frequency and Time Forum, Neuchâtel, Switzerland, pp. 335–346.
Šoidr, L., Čermak, J., and Barillet R. (**2004**) Proceedings of the 18th EFTF, Guildford, UK on CD-ROM.
Stairs, I.H. (**2003**) Testing General Relativity with Pulsar Timing, published at *http://relativity.livingreviews.org/open?pubNo=lrr-2003-5page=node5.html.*
Taylor, J.H. (**1992**) *Philos. Trans. R. Soc. Lond. A*, **341**, 117–134.
Taylor, J.H. (**1994**) *Rev. Mod. Phys.*, **66**, 711–719.
Terrien, J. (**1968**) *Metrologia*, **4**, 43–48.
Uzan, J.-P (**2003**) *Rev. Mod. Phys*, **75**, 403–455.
Vanier, J. (**2005**) *Appl. Phys. B*, **81**, 421–442.
Vanier, J. and Mandache, C. (**2007**) *Appl. Phys. B*, **87**, 565–593.
Wolf, P. and Petit, G. (**1999**) Proceedings of the 31st PTTI, Dana Point, CA, USA, pp. 419–428.

Further Reading

Historical Time Keeping, Astronomical Time

Blaise, C. (**2001**) *Time Lord, Sir Sandford Fleming and the Creation of Standard Time*, Weidenfeld and Nicholson, London.
Ginzel, F.K. (**1914**) *Handbuch der Mathematischen und Technischen Chronologie*, J.C. Hinrich'sche Buchhandlung, Leipzig.
Howse, D. (**1997**) *Greenwich Time and the Longitude*, Philip Wilson, London.
Jespersen, J. and Fitz-Randolph, J. (**1999**) *From Sundials to Atomic Clocks: Understanding Time and Frequency*, second revised edition, Dover Publications, available also at URL *http://tf.nist.gov/timefreq/general/pdf/1796.pdf*.
Jones, T. (**2000**) *Splitting the Second: The Story of Atomic Time*, IOP Publishing, Bristol.
Nelson, R.A. *et al.* (**2001**) *Metrologia*, **38**, 509–529.
Seidelmann, P.K. (ed.) (**1992**) *Explanatory Supplement to the Astronomical Almanac*, University Science Books, Mill Valley.
Audoin, C. and Guinot, B. (**2001**) *The Measurement of Time*, Cambridge University Press, Cambridge.
Major, F.G. (**1998**) *The Quantum Beat*, Springer-Verlag, New York.
Riehle, F. (**2004**) *Frequency Standards, Basics and Applications*, Wiley-VCH Verlag GmbH, Weinheim.
Vanier, J. and Audoin, C. (**1989**) *The Quantum Physics of Atomic Frequency Standards*, Adam Hilger, Bristol.

In addition to that, we suggest to take notice of the educational resources provided by the IEEE UFFC Society on its web site: *http://www.ieee-uffc.org/edmain.asp*, which cover in particular detail the subject of quartz oscillators not treated in any detail in this chapter.

Time and Frequency Standards and Measurement Principles

Audoin, C. and Guinot, B. (**2001**) *The Measurement of Time*, Cambridge University Press, Cambridge.
Lombardi, M.A. (**2002**) NIST Time and Frequency Services, NIST Special Publication 432, 2002 Edition.
Major, F.G. (**1998**) *The Quantum Beat*, Springer-Verlag, New York.
Riehle, F. (**2004**) *Frequency Standards, Basics and Applications*, Wiley-VCH Verlag GmbH, Weinheim.
Riley, W.J. (**2008**) *Handbook of Frequency Stability Analysis*, NIST Special Publication 1065, NIST, July 2008.
Vanier, J. and Audoin, C. (**1989**) *The Quantum Physics of Atomic Frequency Standards*, Adam Hilger, Bristol.

In addition to that, we suggest to take notice of the educational resources provided by the IEEE UFFC Society on its web site: *http://www.ieee-uffc.org/edmain.asp*, which cover in particular detail the subject of quartz oscillators not treated in any detail in this chapter.

Frequency Instability

NIST (**1990**) Characterization of Clocks and Oscillators, NIST Technical Note 1337.
Sections 3 and 4 in the ITU Handbook (ITU, **1997**)
Desaintfuscien, M. (**2007**) *Data Processing in Precise Time and Frequency Applications*, Springer, Berlin , Heidelberg.

Laser Cooling, Fountains, and Optical Frequency Standards

Chu, S. (**1998**) *Rev. Mod. Phys.*, **70**, 685–706.
Cohen-Tannoudji, C. (**1998**) *Rev. Mod. Phys.*, **70**, 707–720.
Diddams, S.A., Bergquist, L.C., Jefferts, S.R., and Oates, C.W. (**2004**) *Science*, **306**, 1318–1324.
Haensch, T. (**2006**) *Rev. Mod. Phys.*, **78**, 1297–1309.
Hall, J. (**2006**) *Rev. Mod. Phys.*, **78**, 1279–1295.
Lea, S. (**2007**) cited above.
Luiten, A.N. (ed.) (**2000**) *Frequency Measurement and Control: Advanced Techniques and Future Trends*, Springer Topics in Applied Research, Springer, Berlin, Heidelberg.
Maleki, L. (ed.) (**2009**) *Proceedings of the 7th Symposium on Ferquency Standards and Metrology*, Pacific Grove, CA, USA, October 2008, World Scientific, New Jersey.
Metcalf, H.J. and van der Straten, P. (**1999**) *Laser-Cooling and Trapping*, Springer, New York.
Phillips, W.D. (**1998**) *Rev. Mod. Phys.*, **70**, 721–741.
Riehle, F. (**2004**) cited above.
Wynands, R. and Weyers, S. (**2005**) *Metrologia*, **42**, S64–S79.

Global Navigation Satellite Systems

Kaplan, E.D. and Hegarty, C.J. (eds) (**2006**) *Understanding GPS, Principles and Applications*, 2nd edn, Artech, Boston , London.
Parkinson, B.W. and Spiller, J.J.Jr (**1996**) *Global Positioning System: Theory and Applications*, Progress in Astronautics and Aeronautics, vol. **163**, AIAA, Washington, DC.
Xu, G. (**2003**) *GPS Theory, Algorithms and Applications*, Springer, Berlin , Heidelberg.

11
Temperature Measurement

Joachim Fischer

Handbook of Metrology. Edited by Michael Gläser and Manfred Kochsiek

ISBN: 978-3-527-40666-1

11.1 Introduction

The definition of temperature was first suggested by William Thomson, later named Lord Kelvin, and it is based on the amounts of heat entering and leaving an ideal heat engine:

$$\frac{Q_h}{T_h} = \frac{Q_c}{T_c} \tag{11.1}$$

where Q_h is the heat flowing into the engine from a hot reservoir at a temperature T_h, and Q_c is the heat flowing out of the engine to a cold reservoir at a temperature T_c. Thomson summarized that heat provides the basis of almost all power generation, in which the efficiency increases with increase in the temperature of the hot reservoir. Clausius observed that one consequence of Thomson's temperature definition is that the sum of all the heat flowing out of a heat engine multiplied by $1/T$ is identically zero for reversible processes and greater than zero for irreversible processes, that is,

$$\sum_i \frac{Q_i}{T_i} \geq 0 \tag{11.2}$$

Clausius called the new quantity, $S = Q/T$, entropy, being a sort of transformed energy (from the Greek word *trope*, which means transformation).

The temperature defined by Thomson was based on the conceptual device of the ideal heat engine; however, such a device is not necessary. The mathematician Carathéodory showed that in any system there exists a unique state variable that characterizes the reversibility of a process, and this variable is proportional to the integral of the heats associated with the process divided by a unique "integrating factor" (Chandrasekhar, 1967). Comparison of Carathéodory's result with Equation (11.2) shows that the state variable is the entropy defined by Clausius and the integrating factor is the temperature defined by Thomson. Carathéodory's result leads to the definition of temperature:

$$\frac{1}{T} = \frac{dS}{dU} \tag{11.3}$$

where U is the internal energy of the system and S is the conventional entropy.

Boltzmann subsequently showed that the entropy of a system is related to the number of ways the constituent atoms and molecules can be arranged into the observed macroscopic state:

$$S = k \ln P \tag{11.4}$$

Handbook of Metrology. Edited by Michael Gläser and Manfred Kochsiek

ISBN: 978-3-527-40666-1

where k is a constant and P is the probability of the system being in the observed state. In his original work, Boltzmann deduced the relation (Equation (11.4)) without the factor k. Naturally, this leads to a temperature measured in energy units (joule); hence, in principle, we do not need a separate base unit for temperature, the kelvin. However, such a temperature scale would have impracticably small ($\sim 10^{-20}$) and unfamiliar values. For this reason, Planck introduced a constant k, later named after Boltzmann, to provide the link to conventional definitions of entropy and temperature (Planck, 1921).

Equation (11.3) applied to various idealized systems yields thermodynamic relations that can be used to measure temperature. For example, for an ideal gas, we can derive the equation of state:

$$pV_m = N_A kT \tag{11.5}$$

where p is the pressure, V_m is the molar volume, and N_A is the Avogadro constant, the number of particles per mole.

The quantity kT, which occurs in the equations of state, is a characteristic energy determining the energy distribution among the particles of the system when it is in thermal equilibrium. Thus for unbound atoms, temperature is proportional to the mean translational kinetic energy. Temperature is linear and rational: equal intervals or ratios of temperature correspond to equal differences or ratios of mean kinetic energy, and a single definition is required to fix the magnitude of the temperature unit. All other temperature values must then be determined experimentally, using a suitable thermal system and equation of state.

Today, the kelvin is defined in terms of the temperature of the triple point of water (T_{TPW}) and the Boltzmann constant k is a measured quantity. However, there is a current proposal of the Comité International des Poids et Mesures (CIPM) to define a numerical value for k, from which it follows that all temperatures, including the TPW, must be measured. Of course, the adopted value for k will be such that the temperature values will as far as possible remain unchanged. The determination of the Boltzmann constant applying primary thermometers is discussed in Section 4.13.

The ideal gas as introduced with Equation (11.5) is as fictitious as the ideal heat engine. However, using a noble gas at a very low pressure, a thermometer can be constructed that really measures the so-called thermodynamic temperatures. In fact, the gas thermometer was the first thermometer to derive thermodynamic temperatures. Such kinds of thermometers, whose basic relation between the measurand and T can be written down explicitly without unknown, temperature-dependent constants being introduced, are generally called *primary thermometers* (see Sections 4.2–4.6 for further examples). Any accurate realization of the thermodynamic temperature is very time consuming and requires extreme metrological effort. A gas thermometer is appropriate only as a primary standard in fundamental laboratory measurements. On this ground, and to harmonize the differing national temperature scales for the day-to-day practical use, the so-called International Temperature Scales have been developed by the Consultative Committee of Thermometry (CCT) and adopted by the General Conference of Weights and Measures (CGPM). The International Temperature Scales reflect the most recent state of metrological accuracy and therefore they are replaced by new versions from time to time (see Sections 11.7 and 11.8). Instruments and methods to disseminate the scales to the users in science and industry are discussed in Sections 4.9–4.12.

With a few exceptions, thermodynamic temperatures are determined only to establish internationally agreed temperature scales. Only at very low temperatures, thermometers are calibrated with thermodynamic methods of measurement. The following primary thermometric methods are described in some detail: As an introduction, constant-volume gas thermometry is discussed in Section 11.2. More recent gas thermometric work is performed at Physikalisch-Technische Bundesanstalt (PTB), Germany, employing dielectric-constant gas thermometry (DCGT, Section 11.2.1) and at the French and Italian metrology institutes LNE-INM/CNAM, INRIM, the National Institute of Standards and Technology (NIST), USA, and the University College London (UCL), UK, using acoustic gas thermometry (AGT, Section 11.2.2). Noise thermometry work and Doppler-broadening techniques are discussed in Sections 11.3 and 11.4. The total and spectral radiation thermometry are described in detail in Sections 11.5 and 11.6.

11.2 Primary Gas Thermometry

The classical gas thermometer is based on the equation of state for an ideal gas (Equation (11.5)) and is reviewed here to introduce the principle of a primary thermometer. Gases have been used for thermodynamic thermometry since its inception, and we shall see that they are still preeminent for much of the range, although the techniques have changed. In classical gas thermometry, the virial equation of state for a gas is used to relate pressure p and molar density $\rho = (n/V)$, of the gas to T (Figure 11.1).

Although many gases exhibit a nearly ideal behavior at low temperatures, in view of the desired level of accuracy the small departures from the ideal behavior must be carefully considered. This is done by measuring the relevant property that is dependent on the density. Then, the ideal behavior is deduced by an extrapolation to zero density applying an appropriate virial expansion. At sufficiently low pressures and densities, the behavior of a real gas can be described by the virial expansion, where $B(T)$ and $C(T)$ are the second and

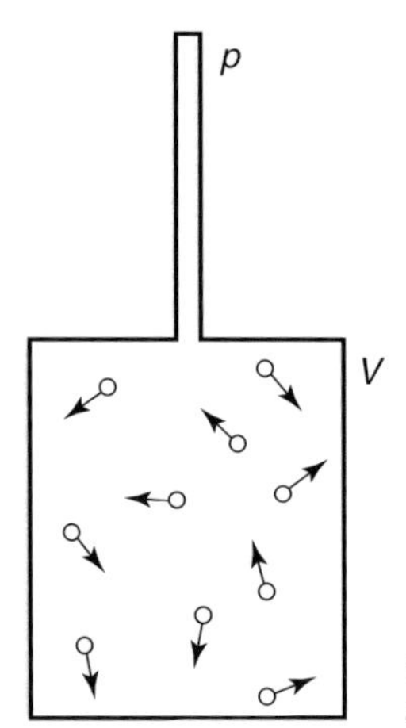

Fig. 11.1 Principle of constant-volume gas thermometry: pressure p and density n/V are the measurands.

third density virial coefficients, and R is the molar gas constant:

$$pV = nRT\left(1 + \frac{B(T)}{V_m} + \frac{C(T)}{V_m^2} + \cdots\right) \tag{11.6}$$

For practical reasons, in primary gas thermometry, the volume is kept constant and therefore the method is called constant-volume gas thermometry (CVGT). In the method of absolute pV-isotherm CVGT, the gas bulb at a constant but unknown temperature T is filled with a series of increasing amounts of gas to obtain a series of pressures p. pV/nR may be plotted as a function of $1/V_m$ according to Equation (11.6) The intercept of the resulting isotherm is the temperature T. For this method, it is not necessary to know the virial coefficients because the extrapolation to zero pressure is made by fitting a virial expansion to the experimental data. There are different ways of measuring the amount of gas. One method is to weigh a bulb of known volume with and without the gas sample and deduce the amount from the weight difference. This method has not been used in recent high-precision CVGT owing to the practical difficulties inherent in the weighing of low-density gases. It was used in historical measurements at the TPW to determine the molar gas constant R. In view of the careful analysis of these measurements in Colclough (1984), and also considering systematic effects related to gas sorption, to the measurement of the gas-bulb volume and to the pressure dilatation of the bulb, at present it does not seem possible to reach a level of relative uncertainty below about 10 parts in 10^6 (10 ppm).

Relative pV-isotherm CVGT is performed by measuring the pressures p_r and p of the same gas sample in the thermometer bulb at a known thermodynamic reference temperature T_r and the unknown temperature T. If the bulb volume V is assumed to be independent of temperature and pressure, in the case of an ideal gas, T is obtained from the following equation:

$$T = \frac{T_r p}{p_r} \tag{11.7}$$

However, for a real gas, the density virial coefficients have to be taken into account again. In addition, the volume of the gas bulb changes with the temperature due to the thermal expansion of the bulb material. The thermal expansion must therefore be known as accurately as possible, as it is one of the main sources of error in gas thermometry. A change in the gas pressure also causes the bulb to dilate, and this change in volume must either be calculated, or the gas bulb must be surrounded by a second bulb filled with gas at the same pressure as the gas in the first bulb. In this case, the dilation of the bulb due to pressure is negligible. CVGT has been employed to establish the thermodynamic basis of the International Temperature Scale of 1990 ITS-90 from very low temperatures up to 660°C and is used to interpolate the scale between 3 and 25 K; for details see Fischer and Fellmuth (2005a).

11.2.1 Acoustic Gas Thermometers

The utility of measuring the speed of sound in gases as a primary means of measuring temperature has long been appreciated. The parameters are all intensive; hence, in principle, quantities such as volumes and the amounts of gas contained in them need not be measured. Moreover, pressure need not be measured to the first order of

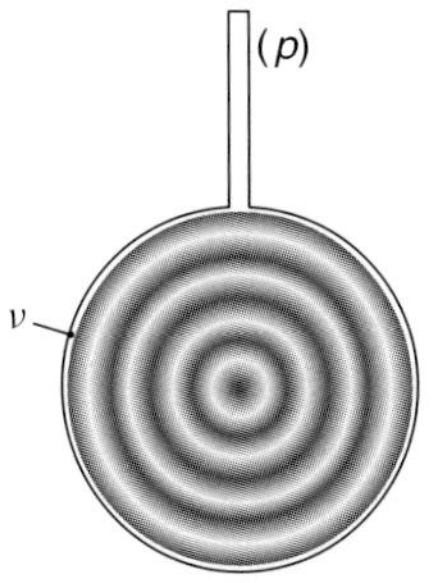

Fig. 11.2 Principle of acoustic gas thermometry: from the resonance frequency ν and the dimension of the resonator the speed of sound u is deduced.

accuracy, being only needed for evaluating the nonidealities.

For an ideal gas, the relation between the speed of sound u_0 and the thermodynamic temperature T is given by the following equation ($\gamma = C_p/C_V$ is the adiabatic exponent and M is the molar mass of the gas):

$$u_0 = \left(\frac{\gamma RT}{M}\right)^{1/2} \tag{11.8}$$

The pressure dependence of the speed of sound in a real gas can be expressed by a virial expansion:

$$u^2 = u_0^2\left(1 + \alpha p + \beta p^2 + \cdots\right) \tag{11.9}$$

that is, the influence of the pressure is of second order. The acoustic virial coefficients α and β can be expressed in terms of the density virial coefficients $B(T)$ and $C(T)$. As can be seen from Equation (11.8), the gas constant R and the molar mass M directly influence the result. Thus, the uncertainty of T depends on their uncertainties. The influence of R and M may be eliminated if the speed of sound is determined at the T_{TPW} with the same thermometer and the same gas, that is, if relative isotherms are measured. Applying Equation (11.9), u_0 is determined by extrapolation to zero density and T is calculated using Equation (11.8).

The main challenge of the method is the measurement of u with the necessary uncertainty, and in spite of significant advances this is still the limitation. In earlier works, a fixed-frequency, variable-path, cylindrical acoustic interferometer was used (Quinn, 1990; Pavese and Molinar, 1992). Such a system was used for measurements of T in the temperature range from 4.2 to 20 K with an uncertainty of 1–4 mK, and with a similar instrument the gas constant R was determined (Colclough *et al.*, 1979).

Nowadays, variable-frequency, fixed-path spherical resonators are preferred (Figure 11.2). Their figure of merit is about an order of magnitude higher than that of cylindrical resonators. Furthermore, boundary layer effects are smaller and the problems due to the excitation of different modes are fewer. Moldover *et al.* (1988) assembled a spherical resonator and used it in 1986 to redetermine the gas constant R with an estimated relative uncertainty of 1.7 ppm, a factor of 5 smaller than the uncertainty of the best previous measurement. They measured the acoustic resonance frequencies of the argon-filled resonator and the microwave resonance frequencies of the same cavity when evacuated. The microwave data were used to deduce the thermal expansion of the cavity. More recent measurements lead to new values of $T - T_{90}$ (T_{90} temperature according to the ITS-90) between 217 and

303 K (Moldover *et al.*, 1999). Ewing and Trusler (2000) performed thermodynamic measurements in the range from 90 to 300 K. The estimated standard uncertainty ranges from 0.9 to 1.3 mK.

Later, a new spherical resonator was developed at NIST including new acoustic transducers to cover the temperature range from 273 to 800 K. Results have been obtained for the melting point of gallium and the freezing points of indium and tin (Strouse *et al.*, 2003) and in the second experimental run between 271 and 552 K (Ripple, Strouse and Moldover, 2007). Independent acoustic measurements of the thermodynamic temperature have also been performed from 234 to 380 K (Benedetto *et al.*, 2004) and from 7 to 24.5 K and from 90 to 273 K by Pitre, Moldover and Tew (2006) using spherical resonators. All results agree within the remarkably small combined uncertainty.

11.2.2
Dielectric-constant Gas Thermometers

DCGT is based upon the variation with the temperature of an intensive property of the gas, the relative dielectric constant (permittivity) ε_r. The gas particles act in a capacitor as induced dipoles (Figure 11.3) with the static electric dipole polarizability α_0 according to the Clausius–Mosotti equation:

$$\frac{\varepsilon_r - 1}{\varepsilon_r + 2} = \frac{N}{V}\frac{\alpha_0}{3\varepsilon_0} \qquad (11.10)$$

where ε_0 is the exactly known electric constant and N/V is the number density. The basic idea of DCGT is to replace the density in the state equation of a gas by the dielectric constant. Thus, the state equation of an ideal gas (Equation (11.5)) can be expressed in the following form:

$$p = \frac{kT(\varepsilon - \varepsilon_0)}{\alpha_0} \qquad (11.11)$$

with $\varepsilon_0\varepsilon_r = \varepsilon$ and the approximation $\varepsilon_r + 2 \approx 3$ valid for ideal gases. Therefore, DCGT avoids the density determination of the conventional gas thermometry that is complicated by the gas contained in dead spaces not at the measuring temperature T and by gas adsorption in the system. Other advantages are that the pressure-sensing tubes can be of any convenient size and that the thermometric gas can be moved in or out of the thermometer cell without the need to measure the amount of the gas involved.

DCGT can be used for primary thermometry in two ways. Absolute measurements require that the polarizability α_0 be known with the necessary accuracy. Nowadays, this condition is fulfilled for helium, which became a model substance for

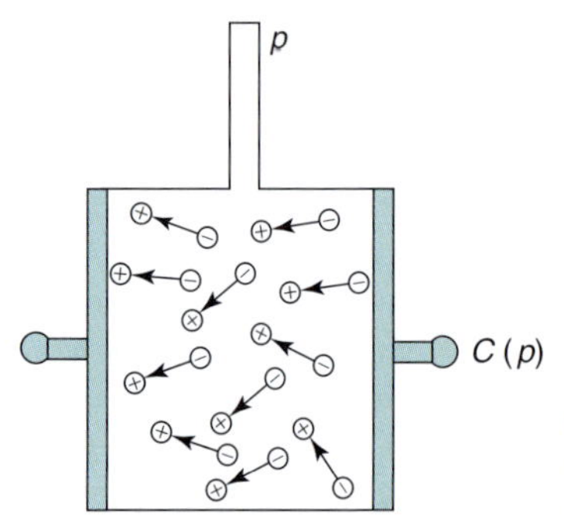

Fig. 11.3 Principle of dielectric-constant gas thermometry: the relative dielectric constant ε_r is determined by the change of the capacitance $C(p)$.

evaluating the accuracy of *ab initio* calculations of thermophysical properties. Recent progress has decreased the uncertainty of the *ab initio* value of α_0 well below one part in 10^6 (Łach, Jeziorski and Szalewicz *et al.*, 2004). However, one disadvantage of helium is its relatively small polarizability, which is, for instance, smaller than that of argon by a factor of 8. On the other hand, primary thermometry does not require a value of α_0 if measurements are made both at the T_{TPW} and at the measuring temperature T. The ratio of the two temperatures is given by

$$\frac{T}{T_{\mathrm{TPW}}} = \left(\frac{p}{p_{\mathrm{TPW}}}\right) \frac{(\varepsilon_{\mathrm{TPW}} - \varepsilon_0)}{(\varepsilon(T) - \varepsilon_0)} \qquad (11.12)$$

where the quantities measured at the TPW have the corresponding index.

By measuring the relative change in capacitance $(C(p) - C(0))/C(0) = \chi + (\varepsilon/\varepsilon_0)\kappa_{\mathrm{eff}} p$ of the capacitor as a function of gas pressure p, the slope $\varepsilon_0 kT/\alpha_0$ of the isotherms (Equation (11.11)) is determined. κ_{eff} takes into account the effective compressibility of a suitable capacitor used to measure the susceptibility $\chi = \varepsilon_r - 1$. A polynomial fit to the resulting p versus χ data points (Figure 11.4), together with the knowledge of the dependence of the dimensions of the capacitor on $p(\kappa_{\mathrm{eff}})$, yields the temperature. Because of very small susceptibility (for instance, 0.003 at 4 K and 0.05 MPa for helium), this technique is extremely demanding for measurement of capacitance changes. For a real gas, the interaction between the particles has to be considered by combining the virial expansions of the state equation (Equation (11.5)) and the Clausius–Mosotti equation (Equation (11.10)), see the dashed curve in Figure 11.4.

Recently, with this method, thermodynamic temperatures have been determined between 4 and 26 K (Gaiser, Fellmuth and Haft, 2008). Considering the progress in the measurement of pressure and capacitance changes, it is seen that DCGT has potential for both decreased uncertainty and increased application range. Measurements up to the TPW would be desirable because the density is measured *in situ* and hence the thermal expansion of the thermometer bulb does not cause problems as in CVGT. Furthermore, as an inverse application of DCGT, measurements at the TPW allow the determination of the Boltzmann constant k. The system described in Luther, Grohmann and Fellmuth (1996) would allow to achieve a standard uncertainty of k of about 15 parts in 10^6 and presently at PTB a system is under construction, which should yield an uncertainty that is an order of magnitude smaller (Gaiser, Fellmuth and Haft, 2008). Another inverse application of DCGT is

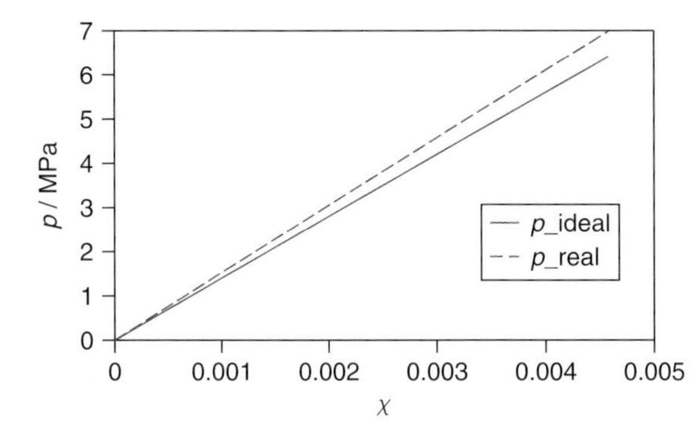

Fig. 11.4 Pressure versus dielectric susceptibility isotherms of helium at the TPW with (dashed curve p_real) and without (straight, full curve p_ideal) consideration of the interaction between the atoms.

the realization of a pressure standard near 1 MPa applying a toroidal cross capacitor filled with helium (Moldover, 1998). Since the interaction between the helium atoms at the realized pressure has to be considered, this project requires values calculated *ab initio* for both the polarizability α_0 and the second density virial coefficient.

11.3 Noise Thermometry

The noise thermometer is based on the temperature dependence of the mean square noise voltage, $\langle U^2 \rangle$, developed in a resistor. Nyquist derived Equation (11.13) from thermodynamic calculations (Nyquist, 1928)

$$\langle U^2 \rangle = 4kTR\Delta f \qquad (11.13)$$

valid for frequencies $f \ll kT/h$, where R is a frequency-independent resistance, Δf is the bandwidth, and h is Planck's constant. From the statistical nature of the measured quantity, long measuring times arise, which may be estimated from Equation (11.14):

$$\frac{\Delta T}{T} \approx \frac{2.5}{\sqrt{t\Delta f}} \qquad (11.14)$$

with t being the measuring time. One of the main problems is the accurate measurement of the very small voltages developed avoiding extraneous sources of noise and maintaining constant bandwidth and gain of the amplifiers. For details see Quinn (1990) and White, Mason and Saunders (2002). In the past, in the high-temperature range, the uncertainties of noise thermometry have not been comparable to those of gas-based techniques owing to limitations from the nonideal performance of electronic detection systems. The noise generated by the wires that connect the sensor to the amplifiers has to be eliminated from the measurement. This is conveniently done by making a four-wire connection to the sensor and using two amplifier chains. The most successful technique to date is the switched-input digital correlator pioneered by Brixy *et al.* (1992) and Edler, Kühne and Tegeler (2004), see Figure 11.5a. The correlator is implemented by digitizing the signals from the two channels and carrying out the multiplication and averaging function by software. This eliminates the amplifier and transmission line noise superimposed on the thermal noise signal. In practical applications, the thermometer switches between a reference noise source at a reference temperature T_0 and the noise source at

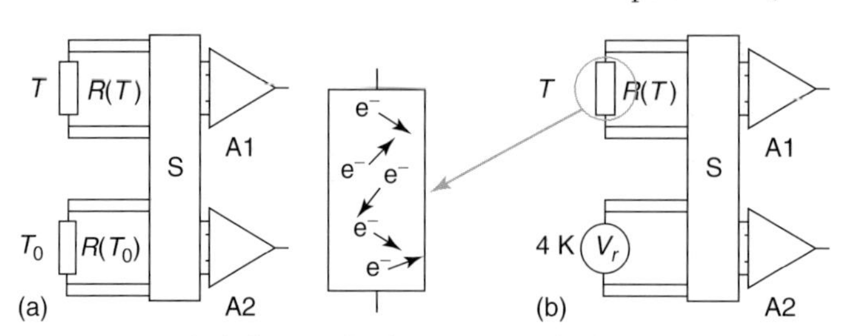

Fig. 11.5 (a) Block diagram for the conventional relative method with switched-input noise correlator; S, switching and A1, A2, amplification and digitization. (b) Block diagram for the new absolute method.

the unknown temperature T. The switching removes the effects of drift in the gain and bandwidth of the amplifiers and filters. A relative standard uncertainty of 2×10^{-5} at the zinc fixed point has been estimated.

The conventional switching approach does not contain sufficient free parameters to resolve the contradictory requirements for matching both the sensing resistances and the noise powers. Currently, a collaboration between NIST and MSL (Benz *et al.*, 2002) explores a new approach using the perfect quantization of voltages from the Josephson effect. This approach keeps the proven elements of the switched correlator, but separates the roles of the temperature reference and the voltage reference. The sensing resistor in the reference arm of the comparator is replaced by an AC Josephson voltage standard. A block diagram is shown in Figure 11.5b. The long-term goal of the project is to build a noise thermometer with an uncertainty of 1×10^{-5} over the temperature range of 83–430 K. First steps were a proof of concept at the fixed points of gallium and water (Nam *et al.*, 2003), and a relative measurement of the zinc freezing point (Labenski *et al.*, 2008). Noise thermometry has been successfully used in the low-temperature range and the interested reader is referred to the review in Fischer and Fellmuth (2005a).

11.4 Doppler-Broadening Thermometry

Doppler-broadening thermometry is a standard means of diagnostics for high-temperature plasmas (Griem, 1964) and is based on the Doppler shift of the frequency of an electromagnetic wave in a moving frame of reference as compared to a frame at rest. Consider the propagation of a laser beam through an absorption cell containing an ideal gas of atoms or molecules with uniform temperature T. The Gaussian–Maxwell probability density for the velocity of the molecules v is proportional to $\exp[-(v/v_0)^2]$ with $v_0^2 = 2kT/m$ for atomic mass m. Around the absorption frequency ν_0, this translates into the corresponding Doppler-broadened absorption line profile with the Doppler width

$$\Delta\nu_D = \left(\frac{2kT}{(mc_0^2)}\right)^{1/2} \cdot \nu_0 \tag{11.15}$$

This relation allows the determination of the temperature by spectroscopic measurement of a Doppler-broadened absorption line profile and determination of its width. In principle, the measurement can be done using standard laser-spectroscopic techniques. As a main advantage compared to other optical methods such as absolute radiation thermometry, the Doppler profile can be determined by relative radiation measurements since it is only its width that is of interest here. Moreover, laser frequencies can be controlled with extremely small uncertainties. However, at the 10^{-6} or 10^{-7} uncertainty level, various other sources of uncertainty will have to be investigated in detail. Apart from the quadratic Doppler effect, these include, among many others, the effects brought about by interatomic interactions, notably the additional line broadening (collisional, transit time, and saturation broadening) and the reduction in Doppler broadening caused by a finite mean free path length (Dicke narrowing). Because of these, measurements may have to be performed at a series of pressure values and extrapolated to pressure zero. It should

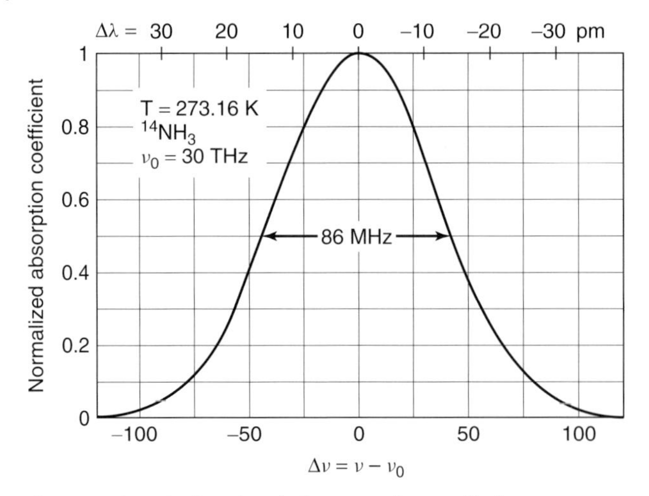

Fig. 11.6 Doppler-broadened absorption line profile for ammonia $^{14}NH_3$ at T_{TPW} assuming a central frequency $\nu_0 = 30\,\text{THz}$.

be noted that the extrapolation to zero pressure is problematic for reasons connected with the speed distribution of particles in monolayers on the cell's surface. Similarly, heating by the absorbed laser power, if it cannot be neglected at all, may require an extrapolation to vanishing laser power.

The possibility of obtaining accurate temperature values has been demonstrated (Daussy *et al.*, 2007) in an experiment using an ammonia line probed by a CO_2 laser spectrometer close to 30 THz (Figure 11.6). The absorption signal was recorded by splitting the laser beam in two and then propagating one of the two beams through the ammonia cell for spectroscopy while the other was used as a reference beam. The two beams were amplitude-modulated by two acousto-optic modulators at two different frequencies and then recombined and focused on a single photodetector. On the basis of these preliminary results the authors (Daussy *et al.*, 2007) expect that the Doppler-broadening measurements can contribute to the determination of the Boltzmann constant with an uncertainty close to 10^{-6}.

11.5
Total Radiation Thermometry

It is only with the development of the cryogenic radiometer that accurate measurements of the total radiation emitted from a black body have become possible. The total radiant exitance $M(T)$ of a black body at a temperature T is given by

$$M(T) = \frac{2\pi^5 k^4}{15h^3c^2}T^4 = \sigma T^4 \tag{11.16}$$

where σ is the Stefan–Boltzmann constant and c is the speed of light. To determine the total radiant exitance, it is necessary for practical reasons to make measurements over only a restricted solid angle rather than over a complete hemisphere. An aperture system must be interposed between the black body and the detector so that $M'(T) = gM(T)$,

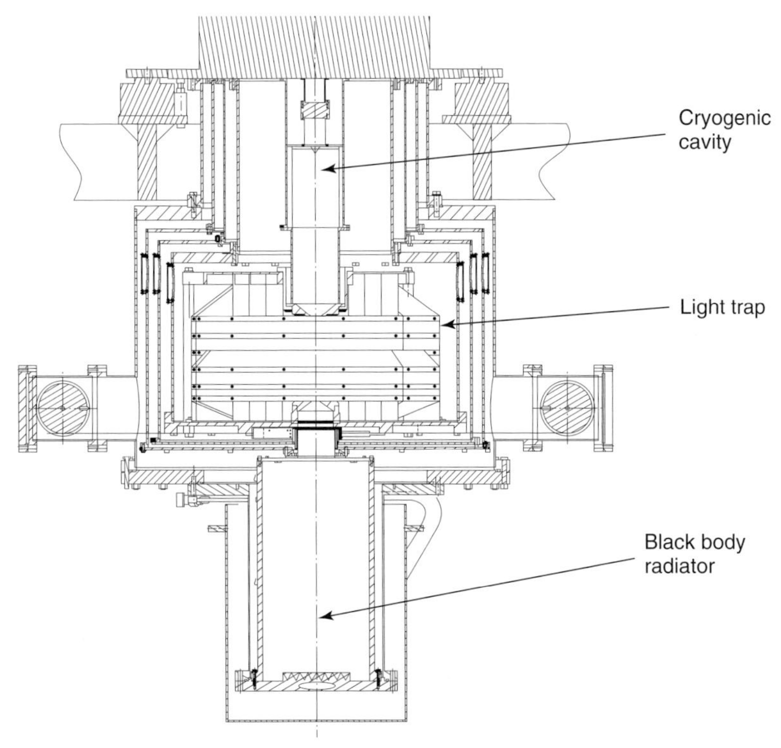

Fig. 11.7 Principle of cryogenic radiometer.

where g is the throughput of the optical system. Provided that g is independent of temperature we may express

$$\frac{M(T)}{M(T_{\mathrm{TPW}})} = \frac{M'(T)}{M'(T_{\mathrm{TPW}})} = \left(\frac{T}{T_{\mathrm{TPW}}}\right)^4 \tag{11.17}$$

where $M'(T)$ and $M'(T_{\mathrm{TPW}})$ are the quantities to be determined. Near room temperature, a measurement of the ratio $M'(T)/M'(T_{\mathrm{TPW}})$ to 1 part in 10^5 is sufficient to determine T to 1 mK or better. Quinn and Martin (1985) and Martin, Quinn and Chu (1988) have demonstrated that such measurements are possible and have obtained results between -130 and $100°$C. The principle is illustrated in Figure 11.7. The black body radiator at a temperature T irradiates an aperture system at liquid helium temperatures, which allows a beam of thermal radiation to enter the second black body held initially at a temperature of 2 K. The absorbing black body acts as a heat flow calorimeter or cryogenic radiometer. The radiant power absorbed in the calorimeter leads to a rise in its temperature until the radiant power absorbed is balanced to a close approximation by the heat flow along a poorly conducting heat link to a heat sink maintained at a very stable temperature near 2 K. The rise in temperature of the calorimeter is monitored. When equilibrium has been reached a shutter at liquid helium temperature is closed, cutting off radiation from the black body radiator. At the same time, sufficient electrical power is supplied to a heater on the calorimeter to keep it at the same temperature. Provided a number of conditions are met, this easily measurable electrical power is a very precise equivalent of the thermal radiative power.

For measuring the ratio $M'(T)/M'(T_{TPW})$, the following parameters of the system had to be evaluated: the emissivity of the radiator, its effective temperatures and the absorptivity of the calorimeter, the diffraction effects at the apertures, the effects of scattering of thermal radiation from surfaces between the apertures, the absorption of thermal radiation at the aperture edges, the departures from ideal geometry of the apertures, the equivalence of radiant and electrical heating of the calorimeter, the uncertainty in the measurements of the electrical power applied to the calorimeter, and the energy transfer from radiator to calorimeter by residual gas. Although some of these parameters had to be known absolutely, others needed to be known only to the extent that their dependence on the wavelength or temperature of the radiator was required. An absolute measurement of $M'(T_{TPW})$ for the determination of the Stefan–Boltzmann constant, however, would require an absolute knowledge of all of them (Quinn and Martin, 1985). The result of the determination of the Stefan–Boltzmann constant had an uncertainty of about 1.3 parts in 10^4.

Recently, NPL has built a new version of a total radiation thermometer (Figure 11.7), which is expected to measure σ with an uncertainty of 0.001% and thermodynamic temperatures between the Hg and Sn fixed points (Martin and Haycocks, 1998) with uncertainties of about 0.5 mK. It should be noted that the determination of σ with an uncertainty of 0.001% corresponds to the determination of the Boltzmann constant with an uncertainty of 2.5 parts in 10^6.

11.6 Spectral Radiation Thermometry

11.6.1 Relative Method Referenced to Known Temperature

As reviewed in Section 11.7, Equation (11.18) defines the ITS-90 in the temperature range above the silver point where $L_\lambda(\lambda, T)$ is the spectral radiance of a black body at a temperature T and wavelength λ, and c_2 is the second radiation constant. The measurement of radiance ratios can also be used to measure thermodynamic temperature if the temperature of one of the black bodies is known in terms of thermodynamic temperature as a reference. This method has been applied to determine the temperatures of the Al, Ag, and Au fixed points of the ITS-90 and is described in the following:

$$\frac{L_\lambda(\lambda, T_{90})}{L_\lambda(\lambda, T_{90,\mathrm{ref}})} = \frac{\exp\left(c_2/\left(\lambda T_{90,\mathrm{ref}}\right)\right) - 1}{\exp\left(c_2/(\lambda T_{90})\right) - 1} \tag{11.18}$$

The spectral radiation thermometer used by Jung (1984), Jung (1986), and Fischer and Jung (1989) in the range 410 – 962°C employed two simultaneously running black bodies B1 and B2. They are observed alternatively by means of a rotatable plane mirror P1 (Figure 11.8). One of the black bodies has the unknown temperature and the other has the reference temperature. The radiation thermometer alternatively focused the apertures of both black bodies onto a linear detector. An interference filter defined the wavelength. The unknown temperature is calculated from the photocurrent ratio and the reference temperature using Planck's radiation law and taking into account the spectral responsivity of the radiation thermometer. During the first run of measurements,

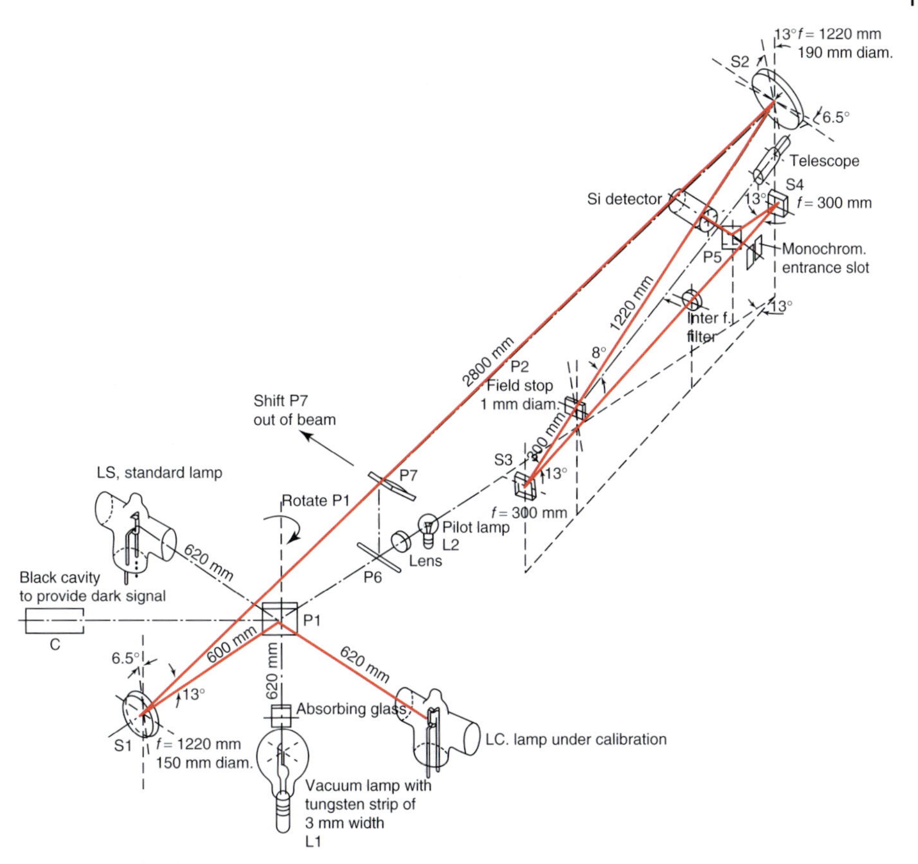

Fig. 11.8 Optical lay out of the spectral radiation thermometer of PTB. In place of the lamps LC, LS black bodies B1 and B2 can be mounted.

black body B1 had a reference temperature close to $T_{\mathrm{ref}} = 729$ K, a thermodynamic temperature as derived from gas thermometry (Guildner and Edsinger, 1976; Edsinger and Schooley, 1989). Black body B2, immersed in freezing aluminum, had the unknown temperature.

During the second run of measurements black body B2, constantly running aluminum freezes, served as reference radiator. The thermodynamic temperature T that corresponded to the temperature of black body B1 was unknown. The latter was set to values between 410 and 630°C using standard platinum resistance thermometers (SPRTs) while the corresponding thermodynamic temperatures were obtained from the measured photocurrent ratios and the redetermined aluminum point. The results gave rise to the SPRT reference function used for the ITS-90 in the high-temperature range. Finally, during the third run, the freezing points of silver and gold have been determined (Fischer and Jung, 1989). The reference black body contained aluminum and the other one contained freezes of silver and gold, respectively.

11.6.2
Absolute Method

The measurement of radiance ratios does not require one to know the spectral responsivity of the radiation thermometer absolutely. However, the determination of the spectral radiance without referencing to a source of known temperature, requires an instrument that has a known absolute spectral response with a well-defined geometric viewing system. This system is called, in its simplest version, a *filter radiometer* and has a set of two view-defining apertures with accurately known dimensions (Figure 11.9). Only recently has it become possible to measure the absolute spectral responsivity of a filter radiometer with sufficient accuracy to compete with the method used in the measurement of radiance ratios. This became possible through the use of cryogenic radiometers as primary reference standard defining a scale of spectral responsivity. The first cryogenic radiometer was constructed at NPL (Quinn and Martin, 1985) to measure the total radiation as described earlier. To establish scales of spectral responsivity, cryogenic radiometers use the spectral power of a monochromatic source such as a laser and subsequently calibrate the response of a transfer detector (Fox, 2002). The additional uncertainty of the calibration of the filter transmission has to be taken into account, which cannot be lower than the uncertainty of the cryogenic radiometer used for calibration. Altogether, this results in best relative uncertainties at the 10^{-4} level for this type of radiation thermometry and measurements at temperatures about 500°C and above (Fox, 2002; Taubert *et al.*, 2003; Yoon *et al.*, 2005).

At the time of establishment of the ITS-90, there were no direct measurements of the thermodynamic temperature T above 729 K of sufficiently low uncertainty that could be used to anchor the scale. Instead, the fixed points of Al, Ag, Au, and Cu were determined through ratios of radiance using radiation thermometers as described in Section 11.6.1. The temperature value assigned to the reference point at 729 K (456°C) was the mean of two separate NIST gas thermometry experiments (Rusby *et al.*, 1991). Since these two experiments differed by about 30 mK, this became the dominant uncertainty of ITS-90 realizations for all higher temperatures. This relatively large thermodynamic uncertainty of T_{90} propagates as T_{90}^2 and is thus around 50 mK at the gold point. To reduce these uncertainties, work is in progress applying acoustic thermometry (see Section 11.2.1) and spectral radiation thermometry.

Using absolute filter radiometry, measurements of $T - T_{90}$ in the temperature range from 660 to 962°C (Stock *et al.*, 1996) and subsequently down to zinc-point temperatures (Hartmann, Taubert and Fischer, 2002; Taubert *et al.*, 2003; Noulkhow *et al.*, 2009) have been performed at PTB. They

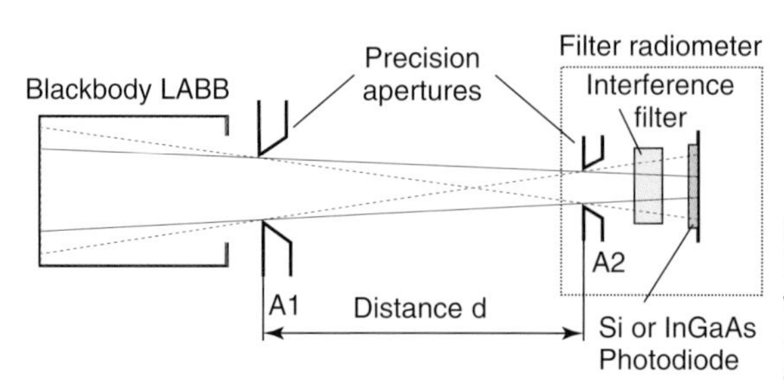

Fig. 11.9 Optical layout of the PTB primary radiation thermometer with the large-area black body LABB, two precision apertures A1 and A2, and the filter radiometer.

employed a large-area black body formed by two concentric sodium heat pipes so that the filter radiometer does not need imaging optics whose transmittance is to be calibrated (Figure 11.9).

A different approach is applied at NIST (Yoon *et al.*, 2005) and at NPL (Fox, 2002) where imaging filter radiometers were developed. With these systems, the thermodynamic temperatures of fixed-point black bodies with their small cavity apertures can be measured directly.

11.7 The International Temperature Scale of 1990

The International Temperature Scales reflect the most recent state of metrological accuracy and therefore they are replaced by new versions from time to time. The International Temperature Scale of 1927 (ITS-27) was the first to overcome the practical difficulties of the direct realization of thermodynamic temperatures by gas thermometry and the first universally acceptable replacement for the differing existing national temperature scales. Finally, on 1 January 1990, the ITS-90 (Preston-Thomas, 1990a) came into force. We note that the thermodynamic temperature T (K), may also be expressed in terms of Celsius temperature t according to

$$t(^\circ\mathrm{C}) = T(\mathrm{K}) - 273.15 \qquad (11.19)$$

The ITS-90 accordingly defines both international kelvin temperatures T_{90} and international Celsius temperatures t_{90} by the corresponding relation:

$$t_{90}(^\circ\mathrm{C}) = T_{90}(\mathrm{K}) - 273.15 \qquad (11.20)$$

Both the thermodynamic and the International Temperature Scale have the same units, the kelvin and the degree Celsius. Users sometimes prefer kelvin in the range below 273.15 K and degree Celsius above this point.

The thermodynamic basis of the ITS-90 is described in Rusby *et al.* (1991) and recommendations for its realization are given in Preston-Thomas, Bloembergen and Quinn (1990b). The ITS-90 extends upward from 0.65 K to the highest temperature practically measurable in terms of the Planck radiation law using monochromatic radiation. It is based on 17 well-reproducible thermodynamic states of equilibrium, the defining fixed points: boiling points (3–5 K with helium, 17 and 20.3 K with hydrogen), triple points (equilibrium hydrogen, neon, oxygen, argon, mercury, and water), melting point of gallium, and freezing points (indium, tin, zinc, aluminum, silver, gold, and copper). To these states numerical values of the temperature T_{90} are assigned. These are the values that have been determined by measurements of thermodynamic temperatures T in several national metrology institutes (Rusby *et al.*, 1991). They are considered to be the best estimates at the time the scale was adopted. The defining fixed points are listed in Table 11.1.

11.7.1 Interpolating the Scale

For the first time, the ITS-90 comprises a number of ranges and subranges, throughout each of which temperatures T_{90} are defined differently. Several of these ranges or subranges overlap. In the range from 0.65 to 5 K, relations of helium vapor pressure to temperature are used to interpolate. From 3 up to 25 K, an interpolating version of the CVGT is applied to define the scale. From the triple point of equilibrium hydrogen (13.8033 K)

Tab. 11.1 Defining fixed points of the ITS-90 with uncertainties $u(T_{90})$ of the best practical realization in terms of ITS-90 and uncertainties $u(T)$ of the thermodynamic temperature.

Fixed point	T_{90}(K)	$u(T_{90})$(mK)	$u(T)$(mK)
Cu	1357.77	15	60
Au	1337.33	10	50
Ag	1234.93	1	40
Al	933.473	0.3	25
Zn	692.677	0.1	13
Sn	505.078	0.1	5
In	429.7485	0.1	3
Ga	302.9146	0.05	1
H_2O	273.16	0.02	0
Hg	234.3156	0.05	1.5
Ar	83.8058	0.1	1.5
O_2	54.3584	0.1	1
Ne	24.5561	0.2	0.5
e-H_2	≈20.3	0.2	0.5
e-H_2	≈17.0	0.2	0.5
e-H_2	13.8033	0.1	0.5
^{4}He	4.2221	0.1	0.3

All values are quoted as standard uncertainties. Values have been taken from Table 1.2 of the Supplementary Information for the ITS-90 (Preston-Thomas, Bloembergen and Quinn, 1990b).

to the freezing point of silver (1234.93 K), SPRTs are used to interpolate between the fixed points. The temperature dependence of the resistivity of high-purity platinum is too complicated for a description based only on the calibration at the available fixed points of sufficient quality (Nicholas, 1999; Quinn, 1990), but it is highly reproducible from sample to sample. Therefore, the ITS-90 prescribes two general reference functions for the ranges below and above the TPW representing a "typical" SPRT (Preston-Thomas, 1990a). Values and derivatives of the two functions are continuous at the TPW. This allows to describe the characteristic of an SPRT by the sum of a reference function and an individual deviation function. The coefficients of the deviation function are deduced from the results of the calibration at the defining fixed points of the ITS-90. The calibration can be performed for 11 subranges, which overlap more or less, using different deviation functions. To avoid the influences of the dimensions of the platinum wire, both the reference and deviation functions are specified for the resistance ratio $W(T_{90}) = R(T_{90})/R(273.16\text{ K})$, where $R(T_{90})$ is the resistance at a temperature T_{90} and $R(273.16\text{ K})$ is the resistance at the TPW. A detailed analysis of the uncertainty in the realization of the SPRT subranges of the ITS-90 is given in White *et al.* (2007).

An acceptable SPRT must be made from pure, strain-free platinum and satisfy one of the relations to be checked at the triple point of mercury ($W(-38.8344\,^\circ\text{C}) \leq 0.844235$) or the melting point of gallium ($W(29.7646\,^\circ\text{C}) \geq 1.11807$). An acceptable

SPRT to be used up to the freezing point of silver (961.78°C) must also satisfy the relation: $W(961.78°C) \geq 4.2844$. The first two conditions guarantee a minimum purity of the platinum and the third aims at avoiding SPRTs with excessive leakage currents at high temperatures. Three types of different designs cover the SPRT range from 13.8 K to 962°C, where the first two types have similar sensor elements:

1. The capsule-type SPRT from 13.8 to 273 K (sometimes 430 K). Characteristic data are as follows: $R(0°C) = 25\ \Omega$, 5 mm diameter, 60 mm length, filled with 30 kPa helium at room temperature, the four platinum leads are taken out through a glass seal.
2. The conventional long-stem SPRT from −189 to 420°C (sometimes 630 or 660°C). Characteristic data are as follows: $R(0°C) = 25\ \Omega$, 7 mm diameter, 600 mm length, filled with dry air. The sensors of 0.07-mm-diameter platinum wire are supported by insulators of mica, alumina, or silica.
3. The high-temperature long-stem SPRT from 0 to 962°C (suited to −189°C). Characteristic data are as follows: $R(0°C)$ between 0.25 and 2.5 Ω, 7 mm diameter, 600–800 mm length, filled with 90% argon +10% oxygen at 20 kPa near room temperature. The sensors and leads are designed such as to minimize the strain on heating and cooling and the leakage through the insulation resistance. Platinum wire up to 0.4 mm diameter and insulators and sheaths of quartz are used to achieve this.

From 961.78°C, the freezing point of silver, up to the highest practicably measurable temperatures T_{90} is defined in terms of the ratio of the spectral radiances $L_\lambda(T_{90})$ and $L_\lambda(T_{90,\mathrm{ref}})$ of two black bodies according to Equation (11.18) One of them has the temperature T_{90} to be determined. The other has the reference temperature $T_{90,\mathrm{ref}}$ that stands for one of the freezing points of silver, gold, or copper. The ratio is measured by means of a spectral radiation thermometer. The temperature T_{90} is calculated from the measured ratio using Planck's law for monochromatic radiation. Appropriate designs of the apparatus and good current practice of their application are extensively described in "Supplementary Information for the ITS-90" (Preston-Thomas, Bloembergen and Quinn, 1990b).

11.8 The Provisional Low-Temperature Scale PLTS-2000

In October 2000, the CIPM adopted the provisional low-temperature scale PLTS-2000 (CIPM, 2001). It is based on noise and magnetic thermometry performed at three institutes (Rusby *et al.*, 2002; Fellmuth, Hechtfischer and Hoffmann, 2003). Considering the uncertainty estimates for the thermometers used and the spread of the results obtained, the relative standard uncertainty of the PLTS-2000 in thermodynamic terms has been estimated to range from 2% at 1 mK to 0.05% at 1 K.

The PLTS-2000 is defined from 0.9 mK to 1 K by a polynomial with 13 terms that describes the temperature dependence of the melting pressure of ^{3}He, which is shown in Figure 11.10. Furthermore, four natural features on the ^{3}He melting curve can be used as intrinsic fixed points of temperature and pressure because their temperature and pressure values are also

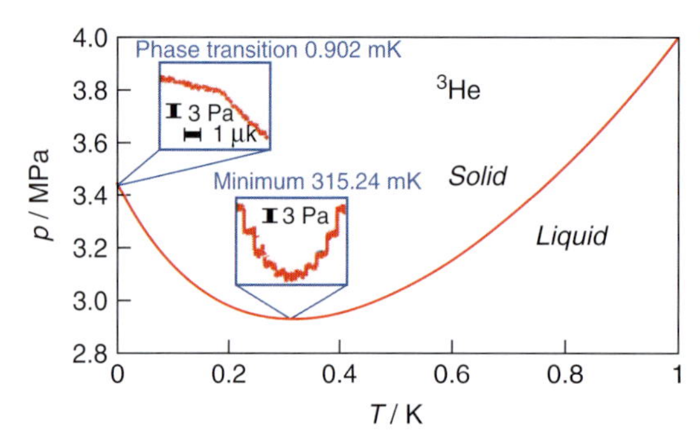

Fig. 11.10 Phase diagram of ^{3}He to define the provisional low-temperature scale PLTS-2000.

defined in the text of the PLTS-2000: the minimum pressure (315.24 mK, 2.93113 MPa), the transition to the superfluid "A" phase (2.444 mK, 3.43407 MPa), the "A-to-B" transition in the superfluid (1.896 mK, 3.43609 MPa), and the Néel transition in the solid (0.902 mK, 3.43934 MPa). The melting pressure of ^{3}He has been chosen as scale carrier for different reasons. First of all, it is a thermodynamic property of a pure substance, that is, no nonuniqueness due to a different interpolation behavior of thermometers occurs. The melting pressure can be reproduced much better than the readings of all other thermometers and a temperature range of about three decades is covered. Furthermore, apart from a narrow range near the minimum of the melting curve, a high resolution down to 0.1 μK can be achieved. The minimum is caused by the unusual fact that the entropy of the liquid phase becomes smaller than the entropy of the solid phase.

The occurrence of the intrinsic fixed points of temperature and pressure is of great importance for using the melting pressure of ^{3}He as scale carrier because melting-pressure thermometry at the highest level of accuracy requires the measurement of the pressure with an uncertainty, which is already close to that of the best national pressure standards. The use of the intrinsic fixed points decreases the uncertainty significantly since the dominating influence of the components caused by the calibration of the pressure balance applied as standard instrument of the highest quality and the hydrostatic-head correction is removed.

For the realization of the PLTS-2000, the melting pressure must be measured *in situ* because for temperatures below the minimum of the ^{3}He melting curve, the pressure-sensing capillary becomes blocked by solid ^{3}He. However, also above the minimum, the usual way of applying a ^{3}He melting-pressure sensor (MPS) is the "blocked capillary method," where the MPS is isolated during operation by a solid ^{3}He plug in the pressure-sensing capillary and thus the conditions are most stable. Up to now, exclusively homemade MPSs were used that are all capacitive diaphragm gauges, but differ in the detailed design (Rusby *et al.*, 2007). Recommendations concerning the facilities and procedures to be used for realizing the PLTS-2000 are given in Schuster, Hoffmann and Hechtfischer (2001). The state-of-the-art level of accuracy can be deduced from the uncertainty budgets presented in Engert, Fellmuth and Hoffmann (2004) and Peruzzi and de Groot (2003).

The dissemination of the PLTS-2000 is difficult because the parameters of the available secondary thermometers cause a large uncertainty of their reading. For instance, resistance and capacitance thermometers suffer from a high instability and relatively large thermal resistances especially at temperatures below about 0.6 K. Therefore, the smallest uncertainty of the scale dissemination, being comparable with that of the scale realization, is obtained by calibrating superconductive reference samples as transfer standards for the calibration of MPSs *in situ* in the users' cryostats. These superconductive transfer standards can also be used for checking the parameters of other secondary thermometers together with the measuring equipment. They are of great importance for approximating the PLTS-2000 in dilution refrigerators without a demagnetization stage that do not allow to realize the three intrinsic fixed points at the lower end of the definition range (below 3 mK). The additional uncertainty component caused by the approximation may be very small if an appropriate combination of reference samples is used. For instance, it is possible to decrease this component below 20 μK in the whole range by calibrating a MPS at the minimum of the melting curve as well as at the super-to-normal-conducting transitions of tungsten (15 mK) and molybdenum (9.2 K) samples (Schuster, Hoffmann and Hechtfischer, 2001).

11.9 Metal–Carbon Eutectic Phase Transitions

At extremely high temperatures, most materials become so reactive that the choice of a crucible material, which does not affect the fixed-point material, becomes severely limiting. There are many types of metals with melting temperature higher than copper, the highest fixed point of the ITS-90, but none had been used successfully as practical fixed point because contamination by the graphite crucible would cause depression of the melting and freezing temperatures.

Yamada *et al.* (1999) showed that the use of metal–carbon eutectics as the fixed-point material offers a solution to the combined problem of the graphite crucible and the fixed-point materials. Graphite, the crucible material, is already a component of the fixed-point alloy and cannot cause contamination. As shown in the binary phase diagram for Ni–C in Figure 11.11 (Massalski, 1990), the liquid phase takes its lowest temperature at the eutectic point, and there cannot be any further depression of the freezing point by carbon contamination. The molten metal, at slightly higher temperature than the eutectic point, would always be slightly richer in C content than the eutectic. However, solidification of graphite during the cooling would reduce the carbon content and the molten metal would reach eutectic composition when the freezing commences. Thus reproducible plateaus are observed. There is no way to adjust the composition ratio because carbon could endlessly be supplied from the crucible. Furthermore, graphite crucibles form black bodies of high emissivity suitable for radiation thermometer calibration, and also there is no concern about reaction with the surrounding furnace material, which is also graphite.

Possible high-temperature fixed points using metal–carbon eutectics are listed in Figure 11.12. The melting temperatures are conveniently spaced out of the temperature range from the copper point up to 2500°C. Metal–carbon eutectics, with no special precautions or procedures and with a wide variation in the cooling rates for the preceding freezes, have shown melting plateau repeatability of 20 mK (Woolliams *et al.*, 2006) in some cases. Evidently,

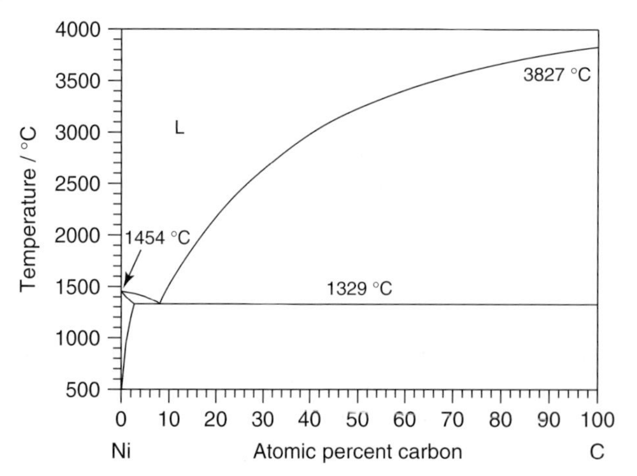

Fig. 11.11 Binary phase diagram of the eutectic system Ni–C, where L denotes the liquid phase.

material purity plays the major role as the metal–carbon eutectics appear to be more susceptible to impurities than pure-metal fixed points.

The eight metals in the box in Figure 11.12 are the ones that show the simplest binary phase diagrams with carbon and are under consideration to become supplementary fixed points of the ITS-90. For most of the others, carbides are formed, which make the phase diagram complex. For instance, Ti gives TiC, with a melting temperature of 3067°C and forms a eutectic with Ti and with graphite. The possible material combinations are Ti–TiC eutectic in a Ti (or TiC) crucible, and TiC–C eutectic in a C (or TiC) crucible. The former is not easy to realize as reaction between the crucible material and the furnace material is difficult to prevent. On the other hand, TiC–C eutectic in a graphite crucible requires no additional technical improvements and is of more interest because the eutectic point has a temperature above the 3000 K mark. Other candidates are ZrC–C eutectic at 2927°C and HfC–C eutectic at 3180°C (Woolliams *et al.*, 2006).

In the ITS-90, the defining fixed point with the highest temperature is the copper point. Above this, no fixed point is available that can be widely used as a practical calibration device. High-temperature fixed points using metal–carbon eutectics may change this situation. The eutectics can be held in graphite crucibles, which makes the cells practical devices for use at

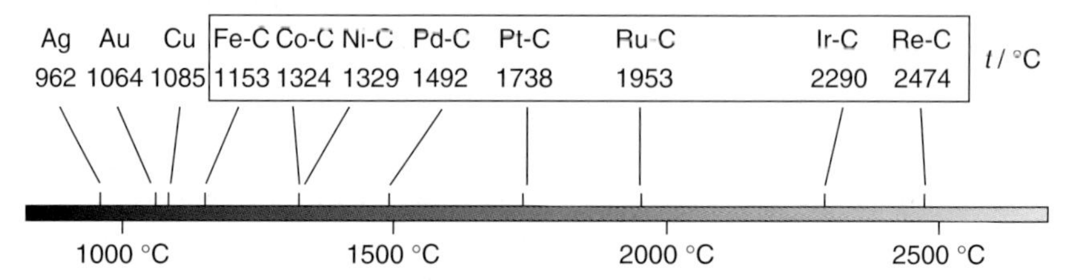

Fig. 11.12 Approximate melting temperatures of metal–carbon eutectic compositions. The three highest defining fixed points of the ITS-90 are shown on the left, outside the box.

very high temperatures. These fixed points can benefit the future high-temperature standards in various ways. They may play the role of transfer standards replacing the standard tungsten strip lamps in radiation thermometry or standard incandescent lamps in source-based radiometry. The temperature scale may be realized with much smaller uncertainty by interpolation at a selected number of these fixed points (Bloembergen *et al.*, 2003). Extension of this technique to above 3000 K is possible with metal carbide–carbon eutectics. However, the temperature values to be assigned to these new fixed points have to be determined beforehand with sufficiently low uncertainty by primary thermometry (Machin *et al.*, 2007).

11.10 Industrial Resistance Thermometers

Although it is not possible to develop precise formulae for the temperature dependence of the resistance of metals (Nicholas, 1999), the dependence exhibits an extremely stable and well-characterized form. Consequently, resistance thermometers are capable of quite reliable and economical industrial-quality temperature measurement. Besides nickel and copper, pure platinum proves to be the best material for most purposes; hence, the platinum resistance thermometer is the most common form of this instrument and most national standard specification codes, as well as the IEC-60751 (International Electrotechnical Commission, 1995a), are limited to platinum.

IEC-60751 defines the temperature–resistance relationship for industrial platinum resistance thermometers (IPRTs) in the temperature range of $0^\circ\text{C} \leq t \leq 850^\circ\text{C}$ by a simple quadratic equation:

$$R_t = R_0(1 + At + Bt^2) \tag{11.21}$$

For the range from -200 to 0°C an additional term is used:

$$R_t = R_0(1 + At + Bt^2 + C(t - 100)t^3) \tag{11.22}$$

R_0 is the resistance at the ice point, usually 100 Ω (Pt100 type). Values for the constants in these two equations are given for ITS-90, for example, in Bernhard (2004). Using these formulae, the IPRTs can be interchanged and will give temperature readings within the allowable tolerances without any calibration. For two classes of IPRTs, the temperature tolerances δt defined in IEC-60751 at a temperature t are as follows:

$$\text{class A} : \delta t = \pm(0.15 + 0.002|t(^\circ\text{C})|)^\circ\text{C} \tag{11.23}$$

$$\text{class B} : \delta t = \pm(0.30 + 0.005|t(^\circ\text{C})|)^\circ\text{C} \tag{11.24}$$

The tolerances of class A must not be exceeded in the temperature range $-200^\circ\text{C} \leq t \leq 650^\circ\text{C}$, and the class B values are valid in the whole usable temperature range specified by the manufacturer. In the new version of IEC-60751, which is currently in the final international adoption process, different classes with more temperature subranges, taking into account also the sensor construction, are proposed; for details see Bernhard (2004). In general, IPRTs should be considered as an alternative or replacement for liquid-in-glass thermometers when an uncertainty of better than 0.5°C is required. It is not possible to batch IPRTs with tolerances close enough to work at a level of uncertainty of about 0.1% or better; hence, they have to be calibrated individually. Then, they may have the capability to be used as laboratory standards with uncertainties approaching 10 mK at temperatures near room temperature.

Compared to thermocouples, IPRTs offer some advantages in sensitivity, although they are more limited in temperature range. An IPRT of the Pt100 type has an effective sensitivity of about 400 μV K^{-1} for the usually used measuring current of 1 mA. This is a factor of about 40 greater than the sensitivity of a type R or S thermocouple. The main disadvantage of an IPRT is the need of lead compensation, which usually requires three or four leads running to the sensor rather than two wires for thermocouples.

As described above, the platinum wire in the sensor of an SPRT is wound in a configuration that is as strain-free as possible. This makes it rather susceptible to vibration and shock and, therefore, IPRT sensors have to be designed differently. For industrial use, in addition, the sheathing must be more robust than for SPRTs. IPRT sensor constructions use often encapsulated Pt-wires wound around glass or ceramic materials whose thermal expansion coefficients match that of platinum. Cheaper sensors can be manufactured applying lithographic techniques. IPRTs of this type are available as thick-film or thin-film sensors, the latter also as an SMD. A comprehensive overview of the different constructions is given in Bernhard (2004). The direct contact between the platinum sensor element and the supporting material may cause strains or even an exceeding of the yield strength. The yield strength determines the maximum instability and corresponds to a temperature equivalent of the order of 0.5 K. Thus, the strains may cause a hysteresis of several tenths of a kelvin depending on the application range (Quinn, 1990).

So far, we have described only resistance thermometers based on platinum sensor elements. For low temperatures down to about 0.5 K, that is, below the operation range of PRTs, especially rhodium–iron resistance thermometers (RIRTs) have been developed (Bedford *et al.*, 1990). Since their resistance versus temperature characteristic is complicated and nonunique, they have to be calibrated individually at 10–30 temperatures in the range from 0.5 to 30 K depending on the desired uncertainty level. RIRTs are available as precision thermometers designed similar to capsule-type SPRTs, the long-term instability of which may be of the order of only a few tenths of a millikelvin, as well as industrial thermometers with wire sensors wound on a ceramic substrate or thin-film sensors. The room-temperature resistance is between 25 and 100 Ω.

In contrast to metal-based resistance thermometers, for instance, thermistors manufactured from semiconducting metal oxides and resistance thermometers made from germanium, carbon, or some other materials show high-negative temperature coefficients of resistance. Thermistors have relative temperature coefficients of about $-0.04\ K^{-1}$ near room temperature, compared with a 10-fold smaller and positive value of $+0.004\ K^{-1}$ for metals. They typically can be obtained commercially with room-temperature resistances ranging from 100 Ω to 1 M Ω and can be useful for temperature measurement from −200 to 300°C. Their high sensitivity permits use with inexpensive, direct reading instruments and, as the lead resistance is small compared to the sensor resistance, lead compensation is usually unnecessary. The characteristic of thermistors is very nonlinear and is described to a good approximation by the empirical relation based on the formula given in Steinhart and Hart (1968):

$$\frac{1}{T} = A + B\ln\left(\frac{R}{R_0}\right) + C\left[\ln\left(\frac{R}{R_0}\right)\right]^2 + D\left[\ln\left(\frac{R}{R_0}\right)\right]^3 \quad (11.25)$$

where R_0 is a convenient reference resistance. For limited temperature ranges, measurement uncertainties below 0.1°C can be easily obtained.

11.11 Thermocouples

In industry, thermocouples are the most widely used of all temperature sensors. Their basic simplicity and reliability have an obvious appeal for many applications. However, when accuracies greater than normal industrial requirements are called for, their simplicity in use is lost and their reliability cannot be assumed *a priori*.

The thermocouple is a device in which the difference between the electromotive forces (emfs) in its two arms gives an indication of the temperature difference between the hot and cold ends of the two wires (thermoelements). This difference is known as the *Seebeck effect*. The point of connection of the two dissimilar arms A and B, at a temperature T_1, is called the *measuring junction*, and the free ends are referred to as the *reference junction*. To permit measurements to be made of the thermoelectric potential difference between the free ends, an additional pair of identical conductors is attached, which leads to a detector. It is evident that the Seebeck effect is in no way a junction phenomenon but is, instead, a temperature-gradient phenomenon. For a proper understanding of the behavior of thermocouples, this cannot be overemphasized. The thermoelectric potential difference between the free ends, both at temperature T_0, is

$$E_{AB}(T_0 \to T_1) = \int_{T_0}^{T_1} (S_A(T) - S_B(T))dT \quad (11.26)$$

where $S_A(T)$ and $S_B(T)$ are the thermopowers of the respective conductors. $E_{AB}(T_0 \to T_1)$ is a unique function of T_0 and T_1 only if the thermoelements are homogeneous throughout the temperature gradient. The effect of the presence of an inhomogeneity is to add a small additional thermopower at the place in the thermoelement where the inhomogeneity occurs. The thermoelectric potential difference of an inhomogeneous thermocouple will be a function of its position as well as the temperature difference between the hot and cold ends. In using a thermocouple to explore the temperature gradient in a furnace, there is thus always the problem of deciding whether or not the change in potential difference, as the thermocouple is moved, indicates the presence of a temperature gradient in the furnace or an inhomogeneity in the thermocouple. Fortunately, it is usually possible to place an upper limit on the size of the effects of inhomogeneities.

Over 20 different types of thermocouples are extensively used and, of these, 8 have been standardized, with their emf–temperature relationships represented by internationally recognized standard reference functions (International Electrotechnical Commission, 1995b). The metals used in the standardized thermocouple types are given in Table 11.2 with the letters that identify them and the temperature ranges for which tolerances apply (International Electrotechnical Commission, 1989). In general, the tolerances are higher than those for IPRTs and are in the order of 1 K.

The thermocouples mentioned in Table 11.2 fall into three groups. The first is

Tab. 11.2 Thermocouple types in common use with their letter designations, typical compositions in percent weight, and temperature ranges for which tolerances apply.

Type	Thermoelements		Temperature range (°C)
	positive	Negative	
B	Pt 30Rh	Pt 6Rh	600 to 1700
R	Pt 13Rh	Pt	0 to 1600
S	Pt 10Rh	Pt	0 to 1600
K	Chromel (Ni 9.5Cr 0.5Si)	Alumel (Ni 5(Si, Mn, Al))	−40 to 1200
N	Nicrosil (Ni 14.2Cr 1.4Si)	Nisil (Ni 4.4Si 0.1Mg)	−40 to 1200
E	Chromel (Ni 9.5Cr 0.5Si)	Constantan (Cu 44Ni)	−200 to 900
J	Fe	Constantan (Cu 44Ni)	−40 to 750
T	Cu	Constantan (Cu 44Ni)	−200 to 350

the rare-metal group, comprising types B, R, and S, based on platinum and its alloys with rhodium. They are the most accurate of the tabulated thermocouples and may be used at higher temperatures, but they are more expensive and are particularly sensitive to contamination.

The second group consists of the two nickel-based thermocouples, types N and K. They are preferred for most applications not requiring the higher temperature limit or accuracy of the rare-metal thermocouples. The choice between types N and K depends especially on whether they are in the bare-wire or the mineral-insulated, metal-sheathed (MIMS) form. In the MIMS configuration, the thermoelements, their insulation, and a sealed sheath are integrated into a flexible cable. In contrast, bare-wire thermocouples, insulated in loose-fitting beads, are exposed to the local atmosphere and are vulnerable to the effects of oxygen, carbon, sulfur, and so on.

The third group, comprising the types E, J, and T, is based on the use of constantan as the negative arm. Constantan has the most negative value of thermopower; hence, the net thermopowers for these thermocouples are high.

Besides the standardized thermocouples, the elemental thermocouples – those whose thermoelements are pure elements – need to be mentioned. A detailed review of the recent developments of elemental thermocouples is given in Bentley (1998). A pure element is more homogeneous and thermoelectrically more stable than an alloy, because it is free of effects that arise from lattice ordering, selective volatilization, and oxidation. The most stable thermocouple is Au/Pt, although its use is limited to temperatures below 1000°C. The instabilities in Pt/Pd are greater, mainly due to inhomogeneity effects in the Pd wire. For example, the drift of a Pt/Pd thermocouple may be as low as 0.02°C for 200 hours at 963°C (Burns and Ripple, 1996). Even so, these changes are about 10 times smaller than that would occur in a type R or S thermocouple under the same conditions.

Above 1750°C the most useful thermocouples are those based on W and Re. There are several thermocouple combinations in this category, such as W and various alloys from W 3Re to W 26Re. Their main difficulty is that on heating above recrystallization temperatures they become brittle. The recrystallization temperature of W is 1200°C and it has the greatest embrittlement problem, whereas W 26Re contains enough Re so that there is no problem of this kind. The thermocouple combination W 5Re versus W 26Re is possibly the best choice. It should not be used in air, but in a vacuum, in an inert atmosphere or in dry hydrogen. Its upper limit for reliable operation is considered to be 2760°C.

11.12 Radiation Thermometers

The use of noncontact temperature measurement by radiation thermometers has increased considerably in the last few years. Accordingly, a wide variety of different designs of radiation thermometers for laboratory and industrial use is available. In contrast to the procedure prescribed in ITS-90, for most practical radiation thermometers, there is no direct comparison against a reference black body of known temperature. Rather, the thermometers make use of the stability of their built-in detectors and after a calibration of their sensitivity against a temperature standard of the ITS-90 they read directly in temperature.

Practical total radiation thermometers, operating at room temperature, are the simplest radiation thermometers consisting generally of a very basic imaging system restricting the field of view with an aperture. The detector is usually a thermopile bolometer, which measures the rise in temperature of a blackened disk and allows the detection of the whole radiated spectrum. Because of their comparatively low sensitivity, they are, on the other hand, the most sensitive of the radiation thermometers to emissivity variations of the target. For example, a 10% overestimation of the emissivity will result in a 2.5% underestimation of the absolute temperature.

Spectral-band radiation thermometers use a filter to selectively measure only thermal radiation in a narrow band as described in Section 11.4. A relative bandwidth ($\Delta\lambda/\lambda$) less than a few percent is typical. Usually a silicon-semiconductor detector is used. Spectral-band radiation thermometers are used to overcome surface emissivity problems. In this case, the center wavelength λ is chosen to be as short as possible while still giving enough signal. From Equation (11.18), it is derived that typically for this kind of radiation thermometers $c_2/\lambda T \cong (10 - 20)$, so that a 10% overestimation of the emissivity will lead to only a (0.5–1)% underestimation of the temperature.

The most common bands are around 0.65 µm for temperatures above 700°C, and 0.85–0.9 µm for temperatures above 500°C. The design of the high-precision radiation thermometer LP3, widely used in national metrology institutes, is shown in Figure 11.13 (Fischer *et al.*, 2002). For lower temperatures, longer wavelengths and different detectors are applied. For example, at 3.43 µm, many plastics have high absorptance. A radiation thermometer operating at this wavelength will thus see even a very thin plastic film as opaque and having high emissivity, allowing accurate temperature measurements during the manufacturing process at 200–300°C.

The thermal radiation emitted by objects around room temperature is comparatively low and exclusively in the infrared (IR). To obtain sufficient signal, most radiation thermometers will use a wider band of wavelengths. Atmospheric absorption of IR radiation limits the choice of bands to mainly two windows, the (3–5) µm band

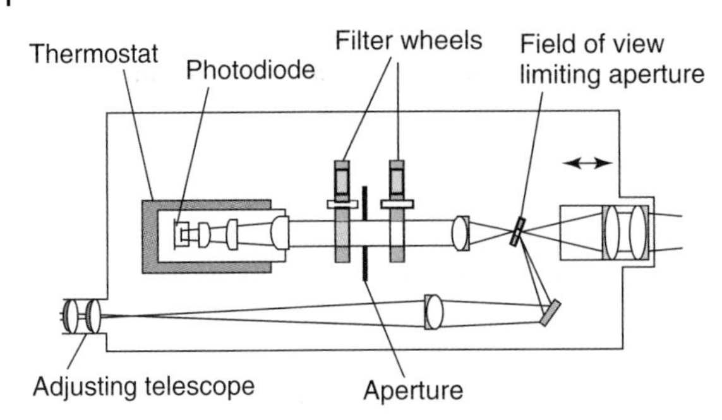

Fig. 11.13 Optical layout of the high-precision radiation thermometer LP3.

and the (8–14) μm band. Pyroelectric or thermoelectric detectors are used to detect the radiation and the wavelength range is often set by the transmission of the optics materials. Since the filter defining the spectral sensitivity, the optics, and the apertures will also radiate energy in this wavelength region, there must be some form of background temperature compensation within the instrument. Some instruments solve this problem by using a rotating sector disk providing a reference signal of known temperature. Recently, there has been considerable progress in the instrumentation for the calibration of low-temperature radiation thermometers (Hollandt *et al.*, 2003/2004).

11.13
New Definition of the Kelvin

Experimentally testing the fundamental laws of physics involves, in practice, the precise determination of the fundamental constants appearing in the laws. Since they are ultimately related to the physical units, the precise experimental realization of the latter is an unavoidable prerequisite for the progress in the development of our physical understanding of nature. Precision experiments relating the (space- and time-independent) fundamental constants to the system of units guarantee stability and reproducibility. The essence of current activities is that prototypes, which may vary uncontrollably with time and location, are replaced by abstract experimental prescriptions that relate the units to the constants. This ensures that the requirement of invariance with space and time is fulfilled. This approach is shown here for the definition of the kelvin and the Boltzmann constant. The unit of temperature T – the kelvin – is presently defined by the temperature of the TPW. Thus, the kelvin is linked to a material property. Instead, it would be advantageous to proceed in the same way as with other units: to relate the unit to a fundamental constant and fix its value. By this, no temperature value and no measurement method would be favored. For the kelvin, the corresponding constant is the Boltzmann constant k, because temperature always appears as thermal energy kT in the fundamental laws of physics.

As shown by the recently published CODATA values for fundamental physical constants (Mohr, Taylor and Newell, 2008), the 2006 recommended value of the Boltzmann constant k with $u_r(k) = 1.7 \times 10^{-6}$ is to a very large extent determined by the NIST result (Moldover *et al.*, 1988) and, therefore, is not yet regarded as sufficiently corroborated to replace the present definition of the kelvin. The important point to note here is that the measurement uncertainty of any value of k would be transferred to the value of T_{TPW}, if that k value were taken to be the exact value of the Boltzmann constant and used to define the kelvin. Hence, if the 2006 CODATA recommended value is fixed as the exact value of k in future, the best estimate of T_{TPW} would still be 273.16 K. However, this value would no longer be exact (as it is now as a result of the current definition of the kelvin) but would become uncertain by $u_r(T_{TPW}) = 1.7 \times 10^{-6}$, which corresponds to 0.46 mK.

Considering the uncertainty estimates in the sections mentioned earlier, the two most promising methods for the reduction in the uncertainty of k currently are DCGT (Section 11.2.2) and AGT (Section 11.2.1). Doppler-broadening thermometry (Section 11.4) utilizing radiation measurements has only recently been proposed for the purpose of determining the Boltzmann constant (Daussy *et al.*, 2007) and is presently under investigation at the University Paris North, France, with respect to the uncertainty that can possibly be achieved. In a joint project of the universities of Milan and Naples (Fellmuth and Fischer, 2007), Italy, a similar experiment is under development. They will determine the Doppler broadening of an absorption line of water vapor probed by a diode-laser-based spectrometer system in the near infrared. It seems to be possible that the DCGT method at PTB or the AGT work of different groups will have been advanced so far by the end of 2010, that they can contribute to an improved value of k or R with similar relative uncertainty as that obtained by Moldover *et al.* (1988) with the AGT in 1988. Thus, an improved value of the Boltzmann constant proposed for the definition of the kelvin would ideally have been determined by at least these two fundamentally different methods and be corroborated by other – preferably optical – measurements with larger uncertainty. We shall assume here that the experiments currently underway to measure R or k (Fellmuth, Gaiser and Fischer, 2006) will achieve a relative standard uncertainty by the end of 2010, which is a factor of about two smaller than the current u_r of approximately 2×10^{-6}, so that $u_r(T_{TPW})$ will be reduced to about 1×10^{-6}, corresponding to about 0.25 mK, and that this will be small enough for the redefinition of the kelvin to be adopted by the 24th CGPM in 2011.

In the discussion about the new definition of the kelvin, it should also be recognized that the "practical" ITS-90 is a defined temperature scale that assigns an exact temperature value T_{90} to each defining fixed point. Hence, the ITS-90 value of the TPW temperature will remain 273.16 K, that is, $T_{TPW-90} = 273.16$ K exactly. The value and uncertainty of T_{TPW} would only need to be taken into account if, for some critical reason, one has to know how well the thermodynamic temperature scale is represented by the ITS-90 at a particular temperature or in a particular temperature range. In fact, although the consistency of T_{TPW} as realized by different TPW reference cells can be as low as 50 μK and even less if the isotopic composition of the water used is taken into account (Strouse and Zhao, 2007), the uncertainties of the thermodynamic temperatures of all other defining fixed points, which are the basis for all practical thermometry, are significantly larger. In contrast to other units, the uncertainty of the realization of the kelvin varies greatly with temperature: at 1300 K, for instance, the uncertainty is roughly 100 times greater than at T_{TPW}. Hence, the fact that T_{TPW} will not be exactly known but have a standard uncertainty of 0.25 mK will have negligible practical consequences (Fischer *et al.*, 2007).

To put the new definition of the kelvin into practice, a *mise-en-pratique* has already been recommended to the CIPM by the CCT (2005). The *mise-en-pratique* will allow direct determination of thermodynamic temperatures particularly at temperatures far away from the TPW in parallel to the realization described in the International Temperature Scale. In the high-temperature range, this will considerably reduce the uncertainty of the realization of the kelvin for many purposes for which the need to refer back to the TPW is anomalous, such as radiation thermometry.

In Mills *et al.* (2006), new definitions are suggested for each of the units, kilogram, ampere, kelvin, and mole to be chosen by the CGPM in 2011. The following definition for the kelvin was proposed for the first time in Fischer *et al.* (2005b) and could be as simple as "The kelvin is the change of thermodynamic temperature that results in a change of the thermal energy kT by exactly $1.3806504 \times 10^{-23}$ joule."

Glossary

Black Body: Radiation source whose emission spectrum depends besides wavelength only on temperature and not on material properties.

Boltzmann Constant: Conversion factor between thermal and mechanical energy.

Calibration: Comparing the reading of an instrument with a standard.

Cryogenic Radiometer: Electrical substitution radiometer operated at cryogenic temperatures.

Emissivity: Ratio of radiance emitted by an object to the radiance emitted by a black body of the same temperature.

Gas Thermometer: Measures temperature by the variation in volume or pressure of a gas. The common apparatus is a constant-volume thermometer CVGT. Consists of a bulb connected by a capillary tube to a manometer.

Noise Thermometer: Measures temperature by the electronic noise generated by the thermal agitation of electrons inside an electrical conductor.

Primary Thermometer: Thermometer, whose basic relation between the measurand and temperature can be written down explicitly without having to introduce unknown, temperature-dependent constants.

Radiation Thermometer: Measures temperature by detecting heat radiation.

Resistance Thermometer: Measures temperature by using the dependency of its own resistance on temperature developed across the metal.

Secondary Thermometer: Thermometer that needs calibration against a primary thermometer.

Thermocouple: A device for measuring temperature based on the junction of two dissimilar metals that have a voltage output proportional to the difference between the hot junction and the cold junction (lead wires).

Thermodynamic Temperature: Temperature of a state of equilibrium measured by a primary thermometer.

References

Bedford, R.E., Bonnier, G., Maas, H., and Pavese, F. (**1990**) *Techniques for Approximating* *the International Temperature Scale of 1990*, BIPM, Sèvres, Pavillon de Breteuil.

Benedetto, G., Gavioso, R.M., Spagnolo, R., Marcarino, P., and Merlone, A. (**2004**) *Metrologia*, **41**, 74–98.

Bentley, R.E. (ed.) (**1998**) *Handbook of Temperature Measurement*, Springer, Singapore.

Benz, S.P., Martinis, J.M., Nam, S.W., Tew, W.L., and White, D.R. (**2002**) *International Symposium on Temperature and Thermal Measurements in Industry and Science (TEMPMEKO)*, vol. **8** (B. Fellmuth, J. Seidel, and G. Scholz), VDE Verlag GmbH, Berlin, pp. 37–44, ISBN 3-8007-2676-9.

Bernhard, F. (ed.) (**2004**) *Technische Temperaturmessung*, Springer, Berlin Heidelberg, New York.

Bloembergen, P., Yamada, Y., Yamamoto, N., and Hartmann, J. (**2003**) *Temperature Its Measurement and Control in Science and Industry*, vol. **7** (editor-in-chief D.C. Ripple), American Institute of Physics, Melville, pp. 291–296.

Brixy, H., Hecker, R., Oehmen, J., Rittinghaus, K.F., Setiawan, W., and Zimmermann, E. (**1992**) *Temperature Its Measurement and Control in Science and Industry*, vol. **6** (J.F. Schooley), American Institute of Physics, New York, pp. 993–996.

Burns, G.W. and Ripple, D.C. (**1996**) *International Symposium on Temperature and Thermal Measurements in Industry and Science (TEMPMEKO)*, vol. **6** (ed. P. Marcarino), Levrotto & Bella, Torino, pp. 171–176.

Chandrasekhar, S. (**1967**) *An Introduction to the Study of Stellar Structure*, Dover, New York, pp. 11–37.

CIPM (**2001**) The provisional low temperature scale from 0.9 mk to 1 k, PLTS-2000. *Comité international des poids et mesures*, vol. **68** BIPM, Sèvres, isbn 92-822-2182-2, pp. 129–130.

Colclough, A.R. (**1984**) Gas constant, X-ray interferometry, nuclidic masses, other constants, and uncertainty assignments, in *Precision Measurement and Fundamental Constants II*, NBS Special Publication, vol. **617** (eds B.N. Taylor and W.D. Phillips), National Bureau of Standards, Washington, DC, pp. 263–275.

Colclough, A.R., Quinn, T.J., and Chandler, T.R.D. (**1979**) *Proc. R. Soc. Lond.*, **A368**, 125–139.

Daussy, C., Guinet, M., Amy-Klein, A., Djerroud, K., Hermier, Y., Briaudeau, S., Bordé, Ch.J., and Chardonnet, C. (**2007**) *Phys. Rev. Lett.*, **98**, 250801.

Edler, F., Kühne, M., and Tegeler, E. (**2004**) *Metrologia*, **41**, 47–55.

Edsinger, R.E. and Schooley, J.F. (**1989**) *Metrologia*, **26**, 95–106.

Engert, J., Fellmuth, B.,and Hoffmann, A. (**2004**) *J. Low Temp. Phys.*, **134** (1-2), 425–430.

Ewing, M.R. and Trusler, J.P.M. (**2000**) *J. Chem. Thermodyn.*, **32**, 1229–1255.

Fellmuth, B. and Fischer, J. (eds) (**2007**) *Workshop on Progress in Determining the Boltzmann Constant*, Report PTB-Th-3, PTB, Braunschweig, ISBN 978-3-86509-684-5.

Fellmuth, B., Gaiser, Ch.,and Fischer, J. (**2006**) *Meas. Sci. Technol.*, **17**, R145–R159.

Fellmuth, B., Hechtfischer, D., and Hoffmann, A. (**2003**), *Temperature Its Measurement and Control in Science and Industry*, vol. **7** (editor-in-chief D.C. Ripple), American Institute of Physics, Melville, pp. 71–76.

Fischer, J. and Fellmuth, B. (**2005a**) *Rep. Prog. Phys.*, **68**, 1043–1094.

Fischer, J., Fellmuth, B., Seidel, J., and Buck, W. (**2005b**) International Symposium on Temperature and Thermal Measurements in Industry and Science (TEMPMEKO), vol. **9** (ed. D. Zvizdic), Laboratory for Process Measurement, Faculty of Mechanical Engineering and Naval Architecture, Zagreb, pp. 12–22.

Fischer, J. and Jung, H.J. (**1989**) *Metrologia*, **26**, 245–252.

Fischer, J., Neuer, G., Schreiber, E., and Thomas, R. (**2002**) *International Symposium on Temperature and Thermal Measurements in Industry and Science (TEMPMEKO)*, vol. **8** (eds. B. Fellmuth, J. Seidel,and G. Scholz), VDE Verlag GmbH, Berlin, isbn 3-8007-2676-9, pp. 801–806.

Fischer, J., Gerasimov, S., Hill, K.D., Machin, G., Moldover, M.R., Pitre, L., Steur, P., Stock, M., Tamura, O., Ugur, H., White, D.R., Yang, I. and Zhang, J. (**2007**) *Int. J. Thermophys.*, **28**, 1753–1765.

Fox, N.P. (**2002**) *International Symposium on Temperature and Thermal Measurements in Industry and Science (TEMPMEKO)*, vol. **8** (eds. B. Fellmuth, J. Seidel, and G. Scholz), VDE Verlag GmbH, Berlin, isbn 3-8007-2676-9, pp. 27–35.

Gaiser, C., Fellmuth, B. and Haft, N. (**2008**) *Int. J. Thermophys.*, **29**, 18–30.

Griem, H.R. (**1964**) *Plasma Spectroscopy*, McGraw-Hill, New York.

Guildner, L.A. and Edsinger, R.E. (**1976**) *J. Res. Natl. Bur. Stand. Sect.*, **A80**, 703–738.

Hartmann, J., Taubert, D.R., and Fischer, J. (**2002**) *International Symposium on Temperature and Thermal Measurements in Industry and Science (TEMPMEKO)*, vol. **8** (eds B. Fellmuth, J. Seidel, and G. Scholz), VDE Verlag GmbH, Berlin, isbn 3-8007-2676-9, pp. 377–382.

Hollandt, J., Friedrich, R., Gutschwager, B., Taubert, D.R. and Hartmann, J. (**2003/2004**) *High Temp. – High Press.*, **35/36**, 379–415.

International Electrotechnical Commission (**1989**) Thermocouples – Part 2 Tolerances IEC 60584-2, Genève.

International Electrotechnical Commission (**1995a**) Industrial platinum resistance thermometer sensors IEC 60751, Genève.

International Electrotechnical Commission (**1995b**) Thermocouples – Part 1 Reference tables IEC 60584-1, Genève.

Jung, H.J. (**1984**) *Metrologia*, **20**, 67–69.

Jung, H.J. (**1986**) *Metrologia*, **23**, 19–31.

Labenski, J.R., Tew, W.L., Benz, S.P., Nam, S.W., and Dresselhaus, P. (**2008**) *Int. J. Thermophys.*, **29**, 1–17.

Łach, G., Jeziorski, B., and Szalewicz, K. (**2004**) *Phys. Rev. Lett.*, **92**, 233001-1–233001-4.

Luther, H., Grohmann, K., and Fellmuth, B. (**1996**) *Metrologia*, **33**, 341–352.

Machin, G., Bloembergen, P., Hartmann, J., Sadli, M., and Yamada, Y. (**2007**) *Int. J. Thermophys.*, **28**, 1976–1982.

Martin, J.E. and Haycocks, P.R. (**1998**) *Metrologia*, **35**, 229–233.

Martin, J.E., Quinn, T.J., and Chu, B. (**1988**) *Metrologia*, **25**, 107–112.

Massalski, B.T. (ed.) (**1990**) *Binary Alloy Phase Diagrams*, vol. **1**, ASM International, Materials Park.

Mills, I.M., Mohr, P.J., Quinn, T.J., Taylor, B.N., and Williams, E.R. (**2006**) *Metrologia*, **43**, 227–246.

Mohr, P.J., Taylor, B.N., and Newell, D.B. (**2008**) *Rev. Mod. Phys.*, **80**, 633–730.

Moldover, M.R. (**1998**) *J. Res. Natl. Inst. Stand. Technol.*, **103**, 167–175.

Moldover, M.R., Boyes, S.J., Meyer, C.W., and Goodwin, A.R.H. (**1999**) *J. Res. Natl. Inst. Stand. Technol.*, **104**, 11–46.

Moldover, M.R., Trusler, J.P.M., Edwards, T.J., Mehl, T.J., and Davis, R.S. (**1988**) *J. Res. Natl. Bur. Stand.*, **93**, 85–114.

Nam, S.W., Benz, S.P., Martinis, J.M., Dresselhaus, P., Tew, W.L., and White, D.R. (**2003**) *Temperature Its Measurement and Control in Science and Industry*, vol. **7** (editor-in-chief D.C. Ripple), American Institute of Physics, Melville, pp. 37–42.

Nicholas, J.V. (**1999**) *International Symposium on Temperature and Thermal Measurements in Industry and Science (TEMPMEKO)*, vol. **7** (eds J.F. Dubbeldam and M.J. de Groot), IMEKO / NMi Van Swinden Laboratorium, Delft, pp. 100–105.

Noulkhow, N., Taubert, D., Meindl, P., and Hollandt, J. (**2009**) *Int. J. Thermophys.*, **30** 134–143.

Nyquist, H. (**1928**) *Phys. Rev.*, **32**, 110–113.

Pavese, F. and Molinar, G. (**1992**) *Modern Gas-Based Temperature and Pressure Measurements*, Plenum Press, New York and London.

Peruzzi, A. and de Groot, M. (**2003**) *Proceedings of the 2nd International Seminar and Workshop on Low Temperature Thermometry* (eds A. Szmyrka-Grzebyk and A. Kowal), Institute of Low Temperature and Structure Research Polish Academy of Sciences, Wrocław, isbn 83-906218-5-1, pp. 35–40.

Pitre, L., Moldover, M.R., and Tew, W.L. (**2006**) *Metrologia*, **43**, 142–162.

Planck, M. (**1921**) *Theorie der Wärmestrahlung*, 4th edn, Barth, Leipzig.

Preston-Thomas, H. (**1990a**) *Metrologia*, **27**, 3–10, 107.

Preston-Thomas, H., Bloembergen, P., and Quinn, T.J. (**1990b**) *Supplementary Information for the International Temperature Scale of 1990*, BIPM, Sèvres, Pavillon de Breteuil.

Quinn, T.J. (**1990**) *Temperature*, 2nd edn, Monographs in Physical Measurement, Academic Press, London, San Diego, New York, Boston, Toronto, Sydney, Tokyo.

Quinn, T.J. and Martin, J.E. (**1985**) *Philos. Trans. R. Soc. Lond. A*, **316**, 85–189.

CIPM (**2005**) Recommendation T 3 to the CIPM: Creation of a mise en pratique of the definition of the kelvin. *BIPM Com. Cons. Thermométrie*, **23** (Document CCT/05-32).

Ripple, D.C., Strouse, G.F., and Moldover, M.R. (**2007**) *Int. J. Thermophys.*, **28**, 1789–1799.

Rusby, R.L., Durieux, M., Reesink, A.L., Hudson, R.P., Schuster, G., Kühne, M., Fogle, W.E., Soulen, R.J., and Adams, E.D. (**2002**) *J. Low Temp. Phys.*, **126**, 633–642.
Rusby, R.L., Fellmuth, B., Engert, J., Fogle, W.E., Adams, E.D., Pitre, L., and Durieux, M. (**2007**) *J. Low Temp. Phys.*, **149**, 156–175.
Rusby, R.L., Hudson, R.P., Durieux, M., Schooley, J.F., Steur, P.P.M., and Swenson, C.A. (**1991**) *Metrologia*, **28**, 9 18.
Schuster, G., Hoffmann, A., and Hechtfischer, D. (**2001**) *Realisation of the Temperature Scale PLTS-2000 at PTB*, PTB, Braunschweig . PTB-ThEx-21, isbn 3-89701-742-3.
Steinhart, J.S. and Hart, S.R. (**1968**) *Deep Sea Res.*, **15**, 497–503.
Stock, M., Fischer, J., Friedrich, R., Jung, H.J., Werner, L., and Wende, B. (**1996**) *International Symposium on Temperature and Thermal Measurements in Industry and Science (TEMPMEKO)*, vol. **6** (ed. P. Marcarino), Levrotto & Bella, Torino, pp. 19–24.
Strouse, G.F., Defibaugh, D.R., Moldover, M.R., and Ripple, D.C. (**2003**) *Temperature Its Measurement and Control in Science and Industry*, vol. **7** (editor-in-chief D.C. Ripple), American Institute of Physics, Melville, pp. 31–36.
Strouse, G.F. and Zhao, M. (**2007**) *Int. J. Thermophys.*, **28**, 1913–1922.
Taubert, D.R., Hartmann, J., Hollandt, J., and Fischer, J. (**2003**) *Temperature Its Measurement and Control in Science and Industry*, vol. **7** (editor-in-chief D.C. Ripple), American Institute of Physics, Melville, pp. 7–12.
White, D.R., Ballico, M., del Campo, D., Duris, S., Filipe, E., Ivanova, A., Kartal Dogan, A., Mendez-Lango, E., Meyer, C.W., Pavese, F., Peruzzi, A., Renaot, E., Rudtsch, S., and Yamazawa, K. (**2007**) *Int. J. Thermophys.*, **28**, 1868–1881.
White, D.R., Mason, R.S., and Saunders, P. (**2002**) *International Symposium on Temperature and Thermal Measurements in Industry and Science (TEMPMEKO)*, vol. **8** (eds B. Fellmuth, J. Seidel and G. Scholz), VDE Verlag GmbH, Berlin, isbn 3-8007-2676-9, pp. 129–134.
Woolliams, E., Machin, G., Lowe, D., and Winkler, R. (**2006**) *Metrologia*, **43**, R11–R25.
Yamada, Y., Sakate, H., Sakuma, F., and Ono, A. (**1999**) *Metrologia*, **36**, 207–209.
Yoon, H.W., Gibson, C.E., Allen, D.W., Saunders, R.D., Litorja, M., Brown, S.W., Eppeldauer, G.P., and Lykke, K.R. (**2005**) International Symposium on Temperature and Thermal Measurements in Industry and Science (TEMPMEKO), vol. 9 (ed. D. Zvizdic), Laboratory for Process Measurement, Faculty of Mechanical Engineering and Naval Architecture, Zagreb, pp. 59–70.

Further Reading

Bentley, R.E. (ed.) (**1998**) *Handbook of Temperature Measurement*, Springer, Singapore.
Bernhard, F. (ed.) (**2004**) *Technische Temperaturmessung*, Springer, Berlin Heidelberg, New York.
Michalski, L., Eckersdorf, K., Kucharski, J., and McGhee, J. (**2001**) *Temperature Measurement*, 2nd edn, John Wiley & Sons, Ltd, Chichester, New York, Weinheim, Brisbane, Singapore, Toronto.
Nicholas, J.V. and White, D.R. (**2001**) *Traceable Temperatures. An Introduction to Temperature Measurement and Calibration*, 2nd edn, John Wiley & Sons, Ltd, Chichester, New York, Weinheim, Brisbane, Singapore, Toronto.
Preston-Thomas, H., Bloembergen, P., and Quinn, T.J. (**1990b**) *Supplementary Information for the International Temperature Scale of 1990*, BIPM, Sèvres, Pavillon de Breteuil.
Quinn, T.J. (**1990**) *Temperature*, 2nd edn, Monographs in Physical Measurement, Academic Press, London, San Diego, New York, Boston, Toronto, Sydney, Tokyo.

12
Metrology in Medicine

Rainer Macdonald and Stephan Mieke

Handbook of Metrology. Edited by Michael Gläser and Manfred Kochsiek

ISBN: 978-3-527-40666-1

12.1 Introduction

Medical diagnosis is, in general, a complex process of identifying a medical condition or disease by its signs, symptoms, and from the results of diagnostic procedures (Pignone and McPhee, 2006). In practice, scientifically obtained knowledge is combined with personal experience and intuition of the medical doctor for this purpose. Consequently, medical diagnosis has been traditionally relying strongly on nonquantitative investigations on one hand: results of a detailed interview of the patient and his or her medical history, visual inspection, a qualitative comparison of observed with known disease patterns or the results of tissue histology are examples for this kind of input. On the other hand, quantitative results of measurement are more important for medical diagnosis (Muir Gray, 1997). Examples are measurement of blood pressure or body temperature, the determination of substances like glucose or hormones in the blood, as well as results of quantitative imaging modalities like X-ray radiography or magnetic resonance tomography, to mention a few. Measurements that are relevant to medical diagnosis are discussed as "medical measurements" in the following. It must be noted that measurements related to medical therapy like, for example, in dosimetry for radiotherapy or measurement for the quantization of drugs to be applied, and so on, are not treated in the following.

Although medical measurements are only pieces within the complex process of medical decision making in general, their contribution toward diagnosis is increasing. Therefore, the accuracy and reliability of medical measurements are of direct consequence for the health of each individual patient.

However, this is only one part of the story. Medical knowledge as well as guidelines for medical decisions are often based on statistical analysis and conclusions of clinical studies (Brownson *et al.*, 2002). In general, medical measurements are incorporated within these studies and correlated with other medical findings. Consequently, each medical decision for an individual may be influenced by the results of previous studies, including the data from medical measurements. Moreover, statistics incorporating data from medical measurement are also important for socioeconomic decisions within the health care system, and are therefore of tremendous impact for the health care expenses.

As a consequence, the results of medical measurements that are relevant for each individual as well as for clinical statistics must be comparable and reliable in time

Handbook of Metrology. Edited by Michael Gläser and Manfred Kochsiek

ISBN: 978-3-527-40666-1

and space despite all the uncertainties and problems with their clinical and medical interpretation. It is not necessary to note that this is an international problem at all, although there is more than just the need to improve the metrological base of those studies to make them comparable and valuable internationally. Hence, measurement quality assurance tools like calibrations, reference measurement methods, or certified reference materials must be provided, implemented and established if not available, or must be improved if not sufficient. These concepts shall include, of course, metrological traceability as well as commutability (ISO, 2003) for *in vitro* diagnostics. An integral part of establishing traceability is the estimation of measurement uncertainty. In the field of medical measurement, there is no common practice, indicating that traceability is presently not well established.

In this article, different aspects of metrology in medicine relevant to medical diagnosis are discussed. After this introduction, a few examples of standards for metrology in medicine are given in Section 2, focusing on physical measurands on physiological-based quantities. In Section 3, the challenges and requirements for metrology in medical laboratories are discussed. Section 4 concludes this article with a brief summary together with an outlook on some future trends.

It should be mentioned that the choice of topics discussed in the following is not exhaustive. It is rather a selection of fields in metrology for medical diagnosis.

12.2
Physical vs. Physiological Quantities: Standards for Metrology in Medicine

If medical measurement is directly concerned with quantities like length, volume, mass, time, electric current or electric potential difference, temperature, pressure, and so on, it seems that the main concepts of other areas in metrology, for example, industrial metrology, can be adapted and used more or less straightforwardly. In principle, traceability up to primary standards may be possible for these quantities at first glance. However, although the traceability chain is well established in classical metrology for these quantities, this route cannot always be followed for metrology in medicine. In many cases, primary standards that are "using a primary reference measurement procedure" (VIM, 2007) have to be substituted by well-accepted physiological test signals such as standard flow-time waveforms in spirometry or clinically tested devices, because the measurand is not directly accessible, or even no physical quantity represents the measurand at all. Automated sphygmomanometers, for example, estimate the systolic and diastolic blood pressures from cuff pressure oscillations originating from compressed arteries. To test the accuracy of the cuff pressure measurement alone is not sufficient, because interpretation of the physiological signals, that is, cuff pressure oscillations, plays an important role in the whole measurement procedure requiring, for example, sophisticated software beside the physical pressure measurement.

Table 12.2 summarizes three different approaches toward traceability in medical metrology. The term "*medi I*" refers to procedures identical to the classical ones in metrology. In "medi II" the primary standard in its role within the traceability chain is replaced by physiological signals of a database, such as, for example, standard flow-time waveforms in spirometry (cf. Section 6.2.2). The physiological signals of a database are reference data in this case,

Tab. 12.1 Traceability in classical vs. medical metrology.

Traceability in classical metrology	Traceability in medical metrology		
	medi I	**medi II**	**medi III**
Primary standard	Primary standard	Database of physiological test signals	One clinically tested device
Reference standard	Reference standard	Simulator/test generator (technically traceable to primary standards)	Transfer standard (technically traceable to primary standards)
Working standard	Working standard		
Device (e.g., balances)	Device (e.g., ergometers)	Device (e.g., spirometers)	Device (e.g., air puff eye-tonometer)

which cannot replace the primary standard physically, since it is not the realization of a given quantity. On the other hand, it is, by convention, the root of the traceability chain. The signals are generated by a simulator, and used to calibrate the device. Since reference databases and adequate simulators are not available always, another procedure (medi III) was developed to establish traceability in these cases. In this procedure, one sample device of a certain manufacturer has to be clinically tested to become the root of the traceability chain for this type of device. By convention, a procedure must be established on how to perform the clinical test. Often, the test results are compared with those of the widely accepted "gold standard" for this measurand. Transfer standards help to calibrate each device by comparing the relevant measurement parameters with the clinically tested device. It should be noted that "medi III" can establish traceability only for a certain type of device of a certain manufacturer.

The Medical Device Directive (MDD, 1993) valid in all member states in the European Union has stated particular requirements for medical devices with a measuring function, such as the indication of the limits of accuracy in the user's guide and the use of legal units. Additionally, the CE (Comformité Europeénne) marking expressing the conformity with all relevant European Directives, should always to be placed in collaboration with independent test houses, so-called notified bodies. This indicates the importance of these devices for the society as seen by the European Commission. Moreover, German legislations requires the metrological check in addition to the European regulations for medical devices with a measuring function in use. In a special ordinance (MPBetreibV, 1998), several medical devices in use, for example, crank ergometers, medical thermometers, noninvasive sphygmomanometers, and eye-tonometers have to undergo periodic metrological checks to ensure their performance. In Germany, detailed guidelines (LMKM, 2002) were issued on how the metrological checks have to be performed. In some other countries, for example, in the Czech Republic, Slovakia, Romania, Brazil, and Japan, similar requirements exist.

In the following sections, selected examples for traceability in medical metrology are discussed to highlight the peculiarities of the three approaches in Table 12.1.

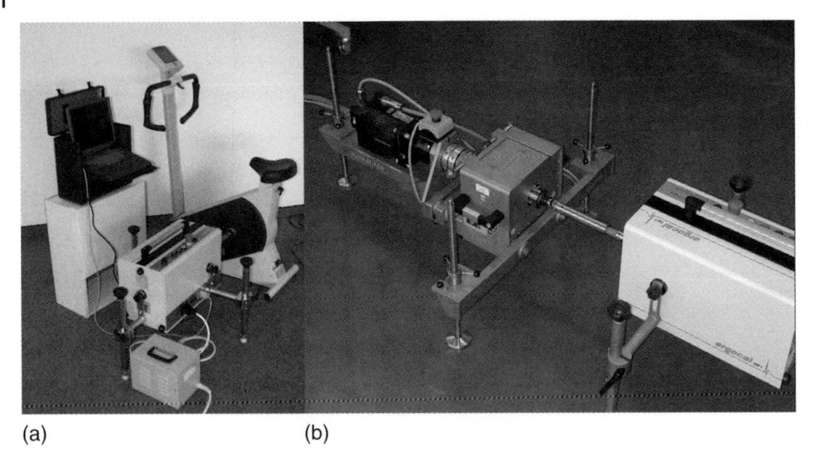

(a) (b)

Fig. 12.1 Setup for a metrological check. (a) Testing the accuracy of the dynamic rotatory power of a bicycle ergometer by a working standard (in front of the ergometer). (b) Calibration of the working standard (lower right corner) by the primary PTB-standard for dynamic rotatory power (upper left).

12.2.1 Crank Ergometers

Crank ergometers for medical use are examples where traceability to primary standards is performed in the classical way (cf. medi I in Table 12.1). A crank ergometer is a bicycle-like exercise machine that is used to measure the work performed by exercising. According to German regulations (MPBetreibV, 1998), crank ergometers in use must be checked by independent test houses every two years. The crank ergometers (Figure 12.1) should maintain the (relative) error limits of 5% for the dynamic rotatory power. For the check, so-called calibrators, that is, working standards, have to be used with a measurement uncertainty of less than or equal to 2%. Often these calibrators have measurement uncertainties of less than 1%. The primary standard for dynamic rotatory power to test working standards at Physikalisch-Technische Bundesanstalt has a relative uncertainty of less than or equal to 0.3%. The primary standard for dynamic rotatory power itself is based on the primary standards for torque, length, and time.

12.2.2 Spirometers

A spirometer is an apparatus for measuring the volume of air inspired and expired by the lungs, that is, a precision differential pressure transducer for the measurements of respiration flow rates. Spirometers are among the few medical devices with measuring function that are tested with generators or simulators. These simulators generate physiological signals, evaluated by the device when performing the measurement (cf. Figure 12.2). The American Thoracic Society has issued about 50 different flow-time waveforms (ATS, 1995), representing the whole spectrum of patients on which spirometric measurements are usually performed. For each flow-time waveform, the results to be determined by the

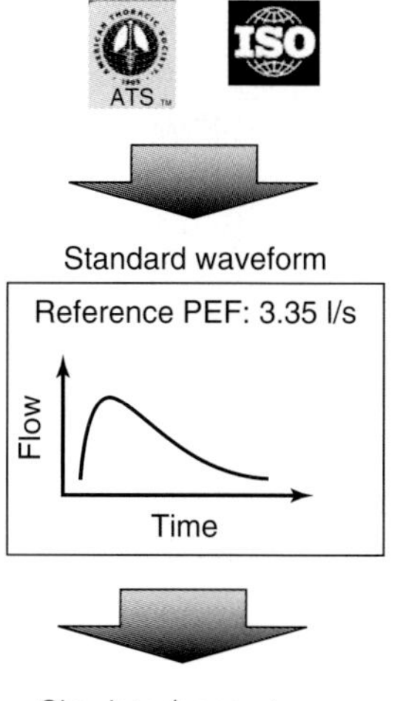

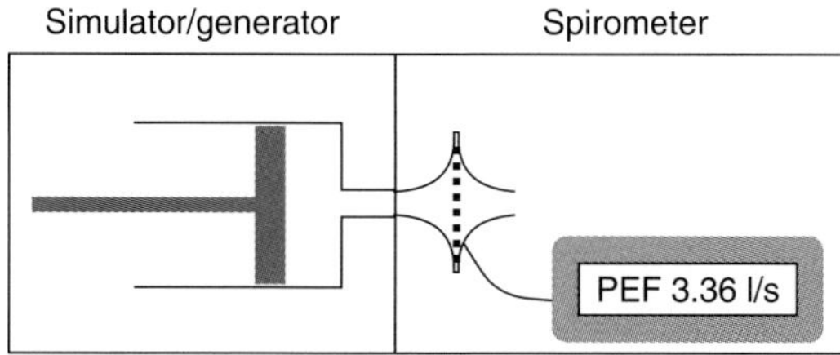

Fig. 12.2 Drawing to illustrate the way traceability is established for spirometer measurements: By convention of the American Thoracic Society (ATS) and the International Organization for Standardization (ISO) about 50 standard flow-time waveforms are specified, which are utilized to test the accuracy of spirometers, for example, peak expiratory flow (PEF).

spirometer are known. Thus, to prove the accuracy of the device under testing, it is connected to a specially designed pump system (simulator), which generates the required waveforms. The results displayed by the spirometer are compared with the well-known reference values. The simulator itself has to be calibrated with regard to its technical specifications, that is, it is traceable to national standards. It is important to note that traceability for the determination of, for example, the peak expiratory flow (PEF) is not to a physical quantity but to dynamic physiological test signals (i.e., flow-time waveforms) taken from a database. To become equivalent and acceptable as a primary standard (medi II, cf. Table 12.1) within the traceability chain, the database has to be representative for the application of the medical device and must match with the method used by the device. Furthermore, the procedure including the physiological test signals must be in general agreement and accepted by medical societies or standard organizations to proceed in this way.

12.2.3 Air Puff Eye-Tonometers

Eye-tonometers are devices to indirectly measure the intraocular pressure inside the human eye. In air puff eye-tonometers, this is realized in a noncontact fashion by measuring the flattening of the cornea caused by an air puff. Following the principles of "medi III," one air puff eye-tonometer, representative for the production of a certain device (type) of a certain manufacturer, has to be selected to perform a clinical test by comparing its measurement results with those of a "gold standard." The "gold standard" for the indirect measurement of the intraocular pressure is the measurement

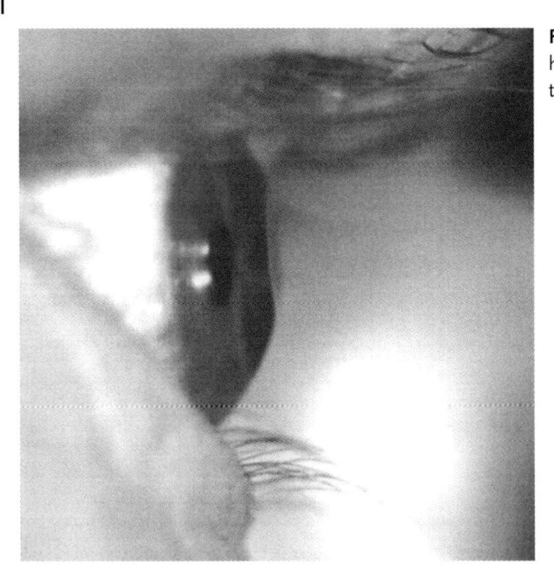

Fig. 12.3 Flattened cornea of a human eye due to the air puff of a tonometer.

with a well-defined applanation tonometer, that is, with the help of a special calibrated disinfected probe attached to a slit lamp biomicroscope, which is used to flatten the central cornea in a contact mode to a fixed amount. The tonometer and the procedure to be followed are described in an international standard (ISO, 2001). If the clinically tested air puff eye-tonometer stays within certain error limits, it becomes equivalent to a primary standard within the traceability chain. A transfer standard (Figure 12.3) is subsequently used to compare the relevant parameters for the measurement with each produced, maintained, or repaired device of the same type. The transfer standard simulates the main measurement. The air puff on the eye results in a flattening of the cornea (cf. Figure 12.3), which is detected by the tonometer with the help of an optical system. Looking for reflection at the flattened cornea, the transfer standard simulates this effect with the help of a swinging mirror (Figure 12.4). When adjusting the torque for the mirror movement to defined values, the measurement at the transfer standard allows a comparison of different air puff eye-tonometers of the same type and manufacturer. In practice, the transfer standard is attached to the clinically tested device, certain torques for the mirror movement are chosen and the resulting simulated intraocular eye pressures are recorded. When repeating these measurements several times, the mean value becomes reference value for the chosen transfer setting, that is, torque of the mirror movement. Each air puff eye-tonometer of the same manufacturer and the same type has to display the same or very similar values as the reference values for the same setup. It is important to note that the stability of the transfer standard is a key element of this procedure to ensure comparability. It should be also noted that the "gold standard" in this case is based on very few measurements performed on dead eyes by comparing results of invasive measurements of intraocular eye pressure with those obtained with noninvasive applanation tonometry. By a convention of the medical societies and opinion leaders this procedure was accepted.

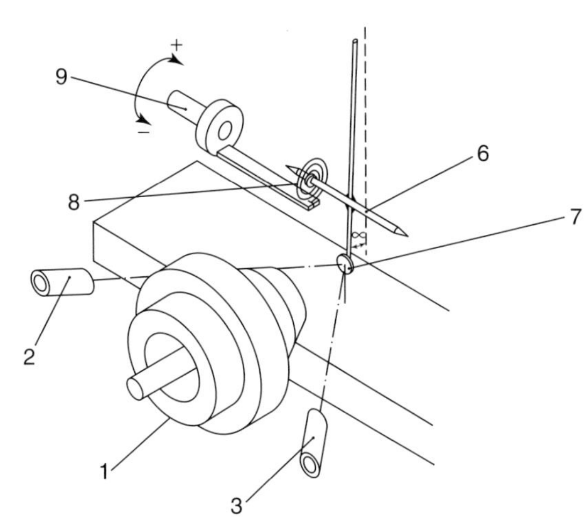

Fig. 12.4 Main components of a transfer standard (no. 6–9) when attached to an air puff eye-tonometer (no. 1–3). Through the nozzle (1) of the tonometer an air puff is released, which moves the mirror (7) to the back. Depending on the adjustable torque of a spring (8), the time required by the air puff to move the mirror (7) to a reflecting position for the optical system (2,3) of the tonometer is detected and analyzed by the tonometer. From this analysis the displayed value is derived.

12.3 Chemical and Biological Measurands

Compared to physical quantities, the situation is quite different for chemical as well as biological measurements, which are concerned with the identification and quantization of substances like the concentration of glucose or cholesterol in blood or other body fluids, the pH-value, the concentrations of hormones, the activity of enzymes, and so on, pp. One key problem in this field is that a complete definition of the measurand is often not as straightforward as with the physical quantities. Furthermore, the number of entities is huge, matrix effects have to be considered, and stability of reference materials is not sufficient in many cases, if available at all. A simplified scheme for traceability in clinical chemistry (Dube, 2001) is shown in Figure 12.5.

Presently, primary standards are not available for the majority of clinically relevant analytes. It has been estimated (Greenberg, 2001) that many *in vitro* diagnostic laboratories perform routinely up to 600 different measurements on the amount of substance. Full calibration systems and traceability are, however, only available for about 5% of these analytes. In general, traceability is often clearly below primary standards or primary methods at the level of reference materials and reference methods or even below. The need to expand the range of reference measurement procedures and reference materials of higher order has been stated by the International Joint Committee for Traceability in Laboratory Medicine (JCTLM), a joint body of the Comité International des Poids et Mesures (CIPM), the International Federation of Clinical Chemistry and Laboratory Medicine (IFCC), and the International Laboratory Accreditation Co-operation (ILAC).

12.3.1 Metrological Controls in Laboratory Medicine

For fulfilling the comparability needs in analytical measurements in laboratory medicine, a sophisticated system of quality assurance is required as part of a more general quality management system (Pyzdek and Keller, 2003). For this purpose, different approaches have been introduced to define the quality required and to

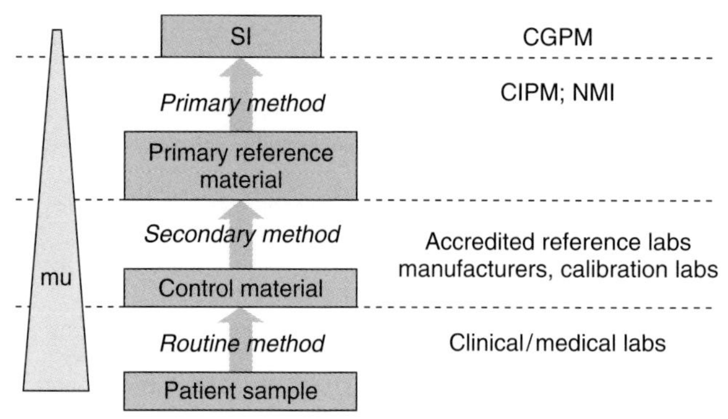

Fig. 12.5 Simplified scheme of traceability in clinical chemistry and laboratory medicine. CGPM: Conférence Générale des Poids et Mesures, CIPM: Comité International des Poids et Mesures, NMI: National metrology institute, mu: uncertainty of measurement.
Adapted from Dube (2001).

derive specifications for the quality control of analytical results (Levey and Jennings, 1950; Westgard *et al.*, 1977; Westgard *et al.*, 1981; Westgard, 1992). The important tools for this purpose include intralaboratory metrological controls by frequently analyzing control samples with known target values ("internal quality assessment") as well as interlaboratory surveys, including reference laboratories ("external quality assessment"), if available. In Germany, minimum requirements for internal and external quality assessment of quantitative measurements in medical laboratories have been established within the (RiLiBÄK, 2001; RiLiBÄK, 2008) Guidelines of the German Medical Association ("Richtlinien der Bundesärztekammer," called *The Guidelines* in the following) for more than 30 years now. Since 1998, legislation of these Guidelines is based on the European Council Directive on In Vitro Medical Devices (IVDD, 1998), which has been transferred into German law by the Medical Devices Act and the Ordinance on the Installation, Operation, and Use of Medical Devices (MPBetreibV, 1998). In particular, the minimum requirements for the accuracy of measurements and validation of measurement results (Macdonald, 2006) by giving maximum permissible deviations from target values are defined in The Guidelines. The measurands and analytes concerned have been selected according to their frequency of investigation and based on their medical importance. The maximum permissible deviations of measurement have been committed taking into account and balancing the actually acknowledged medical requirements as well as the available state-of-the-art analytical techniques and control materials.

It must be noted that metrological controls carried out with the help of control samples are not without problems, in particular because the properties of the control samples often do not match the relevant properties of patient samples sufficiently. Reasons for this are as follows: the need to supply stabilized or even artificial samples rather than native patient samples, on one hand, and that it is impossible to provide control samples that properly reflect the different matrix effects usually present in biological samples, on the other hand. The latter problem is hardly to be resolved with control samples anyhow, since native samples may exhibit large intra and interindividual variations. The only solution to this problem would be to compare measurements performed with a routine method to that performed with a reference method on exactly the same patient sample – close in time and space for stability reasons – which is practically not possible in most cases and hence is not suitable for frequent metrological controls.

12.3.1.1 External Quality Assurance in Hematology

Knowledge about the concentration of different cells in blood is important for medical diagnosis and therapy control in many cases. Reference procedures to determine reference values for the measurands of the complete blood count (CBC) for interlaboratory surveys (round robin tests) organized by German medical associations have been developed, and frequent participation of medical laboratories in these round robin tests is mandatory according to the above-mentioned Guidelines for external quality control since the year 2002. For the CBC hemoglobin concentration, hematocrit, the concentration of red blood cells (erythrocytes), white blood cells (leukocytes), and platelets (thrombocytes) is determined (DIN 58932-3, 1994; DIN 58931, 1995; 1995; DIN 58932-2, 1998; DIN 58932-4, 2003; DIN 58932-5, 2007).

Reference procedures for cell counting based on serial dilutions have been developed (Ruhenstroth-Bauer and Zhang, 1960; Wales and Wilson, 1961; Helleman, 1970) to correct for coincident losses as well as cell losses, for example, due to adhesion. In resistance pulse measurements (Coulter, 1956), the signal amplitude is proportional to the cell volume (Thom, 1972; Kachel, 1990) and typical count rates of 100 Hz to 500 Hz can be achieved, which are much higher than with microscopic techniques. The determination of cell concentrations with highest accuracy by measuring serial dilutions and extrapolation to zero volume fraction of the sample in the measurement suspension was also adopted (Lewis, England, and Kubota, 1989; Helleman, 1990) by the International Council for Standardisation in Haematology (ICSH). Besides resistance pulse detection, optical measurement techniques have been developed for cell counting in so-called flow cytometers since the late 1960s (Valet, 2003) allowing cell analyses at even higher throughput with count rates in the kilohertz frequency range (Shapiro, 2003; Laerum and Bjeknes, 1992). Hence, a large number of cells can be counted in a short time, leading to excellent statistical precision. In these devices, cell-induced light scattering (Ost, Neukammer, and Rinneberg, 1998) or laser-induced fluorescence signals caused by dye-labeled cells passing through a capillary are analyzed. Combined with gravimetric volume measurement (Rinneberg, Neukammer, and Ost, 1995) and application of sophisticated data analysis methods, the accuracy for state-of-the-art reference methods is at least three times better than the maximum permissible measurement deviation for routine laboratories.

The reference procedures for CBC were also applied to investigate the suitability of reagents to stabilize blood samples for internal quality assurance of hematological laboratories (Springer *et al.*, 1999).

Figure 12.6 shows a scheme of a laser flow cytometer as an example. A diluted blood sample is fed into the flow cell

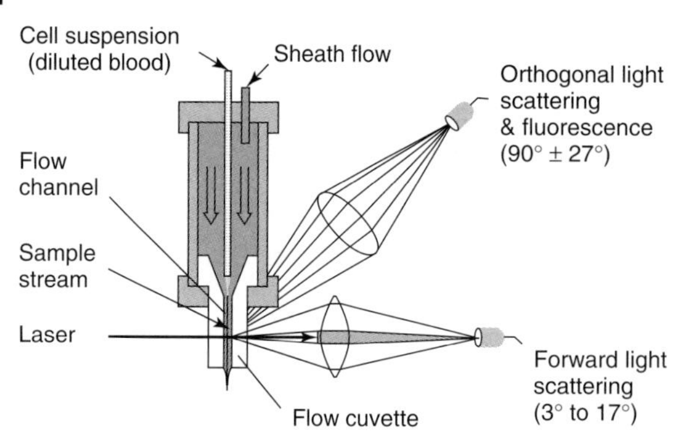

Fig. 12.6 Schematic drawing of a laser flow cytometer for differentiation and counting of blood cells by light scattering and fluorescence detection (courtesy of J. Neukammer, PTB). Details are explained in the text.

through a capillary and hydrodynamically focused by a sheath flow. As a result of hydrodynamic focusing, the blood cells pass one or more focused laser beams in single file and are detected, identified, and counted by analyzing the light scattering in orthogonal as well as in forward direction.

Differentiation of cells is possible if their scattering cross sections, which depend on volume, shape, granularity, and (average) refractive index, are different. Improved cell differentiation is achieved by detecting angular distributions of scattered light (Neukammer *et al.*, 2003). If differentiation is not possible by these physical signatures alone, fluorescent staining of the cells and detection of laser-induced fluorescence is required, which is discussed below.

Figure 12.7 shows a typical result – a so-called scattergram – of flow-cytometric cell differentiation, obtained with a diluted blood sample by employing a laser wavelength of 413.1 nm. Each point in the diagram corresponds to at least one event, that is, one blood cell. The coordinates are given by the light scattering intensity detected in orthogonal ($90^\circ \pm 27^\circ$) vs. the forward (between 3.3° and 17°) direction, respectively. Integrated differential scattering cross sections were obtained by calibration with monodisperse polystyrol spheres and comparison with the generalized Mie theory as described by Gouesbet, Lock, and Gréhan (1995). Clearly, clusters of scattering events resulting from erythrocytes (red blood count, RBC), and leukocytes with subpopulations of lymphocytes (Ly), monocytes (M), and granulocytes (G) are discernable.

The laser wavelength of 413.1 nm almost coincides with the absorption maximum of hemoglobin. Owing to this strong absorption, the light scattering intensity caused by red blood cells is reduced compared to other wavelengths, whereas the scattering cross section for white blood cells only slightly changes with wavelength. Because of the significantly reduced light scattering intensity of red blood cells, much less frequent (approximately 1 : 1000) white blood cells can be detected and differentiated simultaneously to red blood cells without lysis, that is, chemical destruction of red blood cells. The application of this method is of particular interest when analyzing blood samples containing lysis resistant erythrocytes, for example, blood from newborn infants or from anemic patients.

The reference procedure for the determination of blood cell concentrations as described in DIN 58 932-(2003) is based on measurements of a dilution series. The conventional true value for the concentration of blood cells is determined

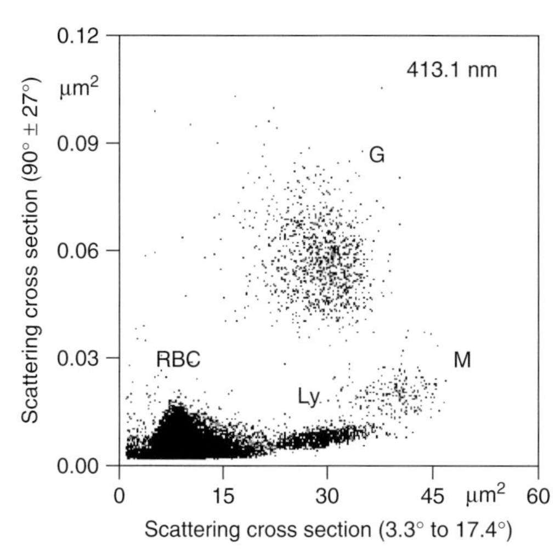

Fig. 12.7 Scattergram obtained with a diluted blood sample in a laser flow cytometer (Ost, Neukammer, and Rinneberg, 1998). Each dot corresponds to (at least) a single scattering event, that is, a single cell, the coordinates of which are given by the measured scattering cross-section in orthogonal vs. forward direction. RBC: Red blood cells, Ly: Lymphocytes, M: Monocytes, G: Granulocytes.

accordingly by

$$C = \frac{N}{V} \qquad (12.1)$$

where N is the conventional true value of the number of cells, for example, erythrocytes, in the sample volume V obtained by extrapolation to vanishing volume fraction $\phi_i = N_i/N \to 0$ of the sample within the measurement suspension of index i. Assuming particles with zero volume as well as Poissonian probability distribution for the temporal separation of particles, the correction is performed according to

$$\ln \frac{\overline{N}_{ri}}{\phi_i} = \ln N - \phi_i \cdot N \cdot p \qquad (12.2)$$

where $\overline{N}_{ri}$ is the average of repeated measurements of the recorded number of cells of suspensions i and $p = \tau/t$ is a coincidence parameter depending on the pulse resolution time τ and the total measuring time t. The pulse resolution time is determined by the interaction time of the cells in the sensor region of the particle counter as well as additional electronic dead times. Since τ, and hence p, is not only determined by the size of the sensor region but also depends on the rheological properties of the cells or particles, calibration of the instrument to correct for dead time losses is not possible. As a consequence, serial dilutions have to be prepared and analyzed for each blood sample individually.

By plotting $\ln\left(\overline{N}_{ri}/\phi_i\right)$ versus ϕ_i (cf. Equation 12.2) the value of N is obtained as the sectioning point of the y-axis with a linear regression line of the experimental data. According to Equation 12.2, the coincidence parameter can be derived from the slope of the regression line. To reduce the uncertainties of measurement, volumes as well as volume fractions are determined gravimetrically with high precision. The mass densities ρ of the blood sample and all solvents are measured with the mechanical oscillator method.

The uncertainties of measurement of the PTB reference methods for the CBC are shown in Table 12.2. Although these uncertainties of measurement seem to be fairly high compared to those achievable in other areas of metrology, it should be noted that this is the state of the art for these measurands and the corresponding reference methods define national

Tab. 12.2 Expanded uncertainty of measurement for the PTB reference procedure for the determination of concentration of blood cells in blood, hemoglobin concentration and hematocrit.

Measurand	Range of measurement	Expanded relative or absolute measurement uncertainty ($k = 2$)
Concentration of erythrocytes in human blood	$0,5 \cdot 10^6\ \mu L^{-1}$ to $7 \cdot 10^6\ \mu L^{-1}$	1.5%
Concentration of leukocytes in human blood	$<5 \cdot 10^3\ \mu L^{-1}$	$0,2 \cdot 10^3\ \mu L^{-1}$
	$5 \cdot 10^3\ \mu L^{-1}$ to $50 \cdot 10^3\ \mu L^{-1}$	4%
Concentration of thrombocytes in human blood	$< 30 \cdot 10^3\ \mu L^{-1}$	$3 \cdot 10^3\ \mu L^{-1}$
	$30 \cdot 10^3\ \mu L^{-1}$ to $1000 \cdot 10^3\ \mu L^{-1}$	10%
Concentration of hemoglobin in human blood	$25\ gL^{-1}$ to $250\ gL^{-1}$	1.5%
Hematocrit (packed cell volume)	0.1 to 0.7	1.5%

(German) standards of highest order at present. To improve the accuracy, various quantities causing systematic deviations shall be considered. The concentration of blood cells is modified, for example, by adhesion and agglomeration, the age of the sample, the anticoagulants used, and mechanical stress or lysis. Since these influences depend on the properties of cells, the uncertainties of measurement given in Table 12.2 refer to typical values, that is, they are valid for cells with properties that do not deviate significantly from those of normal blood samples. If, however, pathological blood is analyzed, for example, from leukaemia patients, white blood cells might be considerably more sensitive against lysis and this effect must be quantified and considered properly. In anemic patients, results for the white blood count are compromised owing to interference of this cell population by lysis-resistant erythrocytes.

The reference methods for flow-cytometric enumeration of blood cells discussed so far are based on identification and differentiation of cells by their physical properties like volume, shape, or granularity. To extend the applicability of reference methods to cell populations where these quantities are not sufficient, identification has to be complemented by immunological staining methods. In particular, platelets in blood samples – even of healthy subjects – may cover a volume range from 0.5 fl to 500 fl. Microplatelets cannot be detected in impedance counters and their light-scattering amplitude might be below the detection threshold. Signals of large platelets, on the other hand, interfere with white blood cells when measuring physical quantities. Consequently, reference methods for the determination of the concentration of platelets (DIN 58 932-5, 2007) relying on immunological staining using CD61/CD41 antibodies (Dickerhoff and von Ruecker, 1995) are under investigation.

Apart from platelet identification, reference methods based on immunological staining procedures offer the potential to develop reference procedures for the concentration of cells relevant for the immune status, that is, monocytes, T-helper and T-suppressor cells, B-cells, and natural killer (NK) cells, the concentrations of which are relevant to HIV-diagnosis.

12.3.1.2 Primary Methods in Clinical Chemistry

Primary methods are mandatory for the traceability of results of measurements on the amount of substance in clinical chemistry for a number of analytes according to the requirements of the above-mentioned *Guidelines*. These analytes comprise not only ions of inorganic elements like Na, K, Ca, Mg, and Cl but also organic compounds like cholesterol, glucose, and creatinin. Primary methods rather than primary standards are used in clinical chemistry since reference materials, taking into account the huge variability of different matrices – like blood, urine, or liquor – are hard, if not impossible, to realize and not available in most cases.

The main core of the primary methods developed for the analytes given above is (Siekmann, Hoppen, and Breuer, 1970; Siekmann, 1991) isotope dilution-mass spectroscopy (IDMS), which is employed for organic compounds in blood in on-line combination with methods for chromatographic separation like gas chromatography, high pressure liquid chromatography (HPLC), and capillary electrophoresis. For measurements with multi-isotopic inorganic elements, IDMS is applied together with thermal surface ionization and ion-chromatographic separation. In addition, gravimetric and titration measurement procedures are developed as reference methods for mono-isotopic sodium and chloride, which cannot be measured easily by IDMS.

According to the definition (Kaarls and Quinn, 1997) of the Commité Consultatif pour la Quantité de Matière (CCQM) of the Bureau International des Poids et Mesures (BIPM), IDMS is a so-called primary ratio method, that is, it measures the value of a ratio of an unknown to a standard of the same quantity, and its operation must be completely described by a measurement equation, which is discussed below. The scheme of IDMS procedure is drawn in Figure 12.8.

Practically, a sample of measured volume containing the analyte in a matrix (e.g., human serum) is spiked with a defined amount of the same analyte labeled with a suitable isotope as an internal standard

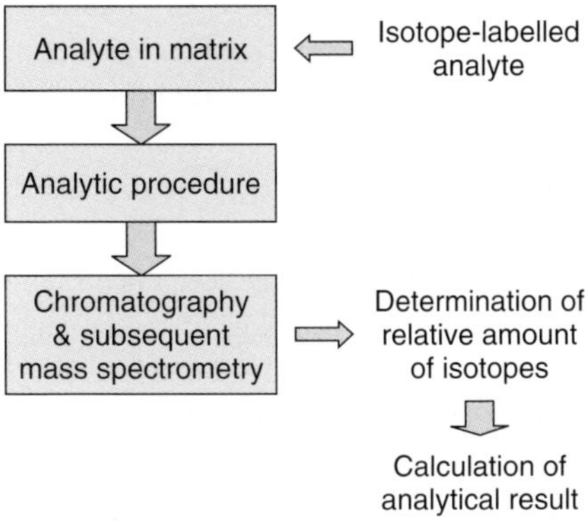

Fig. 12.8 Scheme of the procedure of isotope diluted mass spectroscopy (IDMS) combined with on-line chromatography.

in IDMS. Stable, nonradioactive isotopes like ^{2}H, ^{13}C, ^{15}N, or ^{18}O are preferred for labeling. Afterward, the sample is cleaned using chemical methods and the analyte is separated using chromatographic procedures as mentioned above. Since all errors arising during the analytic procedure are compensated perfectly because of using an internal standard, IDMS measurement can be traced to a primary standard of mass.

During mass-spectroscopic chromatography, two characteristic signatures are recorded continuously, one corresponding to the analyte and the other corresponding to the isotope-labeled internal standard which is shifted by 2, 3, or 4 mass numbers in the spectrum (cf. Figure 12.9).

From these mass chromatograms, the relative amount of isotopes is derived by dividing the relative areas under the curves of the spectral signatures, that is,

$$R = \frac{A_M}{A_{M*}} \tag{12.3}$$

where A_M is the area under the curve obtained with substance M, and A_{M*} is that obtained with the isotope-labeled substance. The absolute amount of substance x is then derived with the help of a second-order polynomial calibration curve (Pickup and McPherson, 1977)

$$R = ax^2 + bx + c \tag{12.4}$$

as

$$x = -\frac{b}{2a} \pm \sqrt{\left(\frac{b}{2a}\right)^2 + \frac{R-c}{a}} \tag{12.5}$$

where the sign in front of the square root has to be chosen properly according to the sign of the constant a. The coefficients a, b, and c of the calibration curve must be determined with at least three different calibrators containing well-defined concentrations of the native and the isotope-labeled analyte.

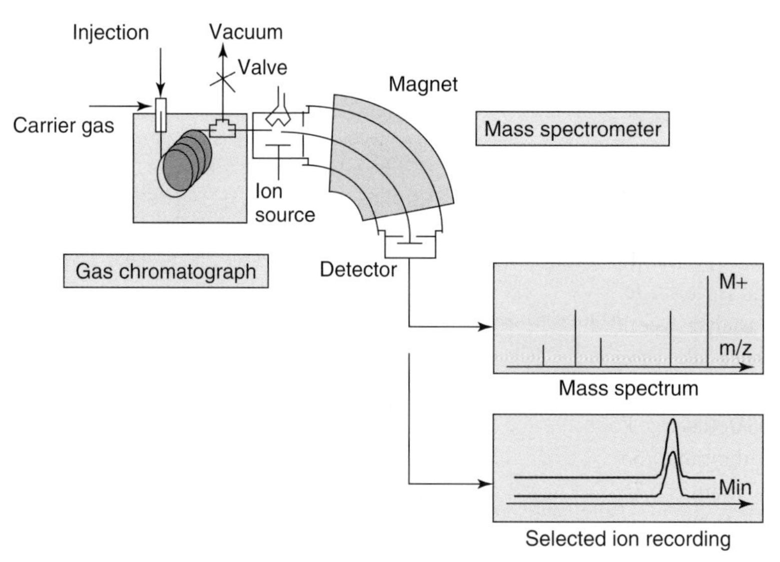

Fig. 12.9 Drawing of a combined gas chromatography and isotope dilution-mass spectroscopy (IDMS) setup (courtesy of L. Siekmann, Inst. Clin. Chemistry and Pharmacology, Univ. Bonn).

Tab. 12.3 Measurands and uncertainty of measurements of the PTB primary methods for clinical chemistry (part 1, inorganic).

Measurand	Range of measurement (μg/g)	Expanded relative measurement uncertainty ($k = 2$)($\times 10^{-3}$)
Mass fraction of Na in human serum	2700 to 3200	6 to 4
Mass fraction of Ka in human serum	130 to 240	9 to 7
Mass fraction of Li in human serum	4 to 18	11 to 9
Mass fraction of Ca in human serum	87 to 136	9 to 7
Mass fraction of Mg in human serum	18 to 45	10 to 8
Mass fraction of Cl (chloride) in human serum	3100 to 4100	6 to 4

IDMS as a primary ratio method in clinical chemistry has been reported for uric acid (Siekmann, 1985a), urea (Keßler and Siekmann, 1999), creatinine (Siekmann, 1985b), cholesterol (Siekmann, Hüskes and Breuer, 1976), and total glycerol (Siekmann, Schönfelder, and Siekmann, 1986), all in human serum, and furthermore, for the hormones cortisol (Siekmann, 1982), progesterone, aldosterone, testosterone (Siekmann, 1979), estriol, and estradiol-17β (Siekmann, 1978).

Tables 12.3 and 12.4 show the measurands and the uncertainty of measurement of the PTB primary methods for clinical chemistry that are used to provide the SI units for the analytes mentioned above to clinical reference laboratories which have been accredited by the German Calibration Service ("Deutscher Kalibrierdienst," DKD) as a calibration laboratory. In addition, primary methods for pH measurement (Buck *et al.*, 2002; Spitzer, 2005) and for electrolytic conductivity (Brinkmann *et al.*, 2003) have been developed which are also relevant for traceability in clinical chemistry.

12.3.2 Biomeasurements for Molecular Medicine

During the recent years, tremendous progress has been made in understanding the causes of disease on the cellular and even molecular level. This opens up new perspectives and options for medical diagnosis as well as therapy, which are often summarized under the key words "molecular medicine" (Kulozik *et al.*, 2000; Ganten, 2003). The advantages of molecular approaches for medical diagnosis are that they have superior detection sensitivity and specificity compared to conventional methods. It is clearly visible already today that key innovations in medicine during the next 10 years will be achieved in this field.

Employing molecular and cellular approaches to medical diagnosis is another step toward a more scientifically based medicine, in particular requiring sophisticated experimental techniques, including measurement. This has been stated very clearly by Elias Zerhouni, Director of the National Institutes of Health, USA (Zerhouni, 2003): "Future progress in medicine will require quantitative knowledge about

Tab. 12.4 Measurands and uncertainty of measurements of the PTB primary methods for clinical chemistry (part 2, organic).

Measurand	Range of measurement	Expanded relative measurement uncertainty ($k = 2$)(%)
Amount-of-substance cholesterol in human serum	3 mmol/l to 9 mmol/l	1.5
Amount-of-substance glucose in human serum	5 mmol/l to 16 mmol/l	1.5
Amount-of-substance creatinin in human serum	0.045 mmol/l to 0.55 mmol/l	1.5
Amount-of-substance creatinin in urine	2.5 mmol/l to 7 mmol/l	1.5
Amount-of-substance urea in human serum	5 mmol/l to 30 mmol/l	2.0
Amount-of-substance uric acid in human serum	0.2 mmol/l to 0.7 mmol/l	2.5
Amount-of-substance progesterone in human serum	3 nmol/l to 100 nmol/l	3.0
Amount-of-substance cortisol in human serum	100 nmol/l to 800 nmol/l	1.8
Amount-of-substance digitoxin in human serum	10 nmol/l to 50 nmol/l	3.0
Amount-of-substance digoxin in human serum	0.5 nmol/l to 5 nmol/l	3.0

the many interconnected networks of molecules that comprise cells and tissues, along with improved insights into how these networks are regulated and interact with each other." In other words, measurements are needed on the level of (bio) molecules, for example, proteins, enzymes, deoxyribonucleic acid (DNA), and ribonucleic acid (RNA) as well as on the level of cells to unravel these interconnected networks, regulative mechanisms, and so on, to gain the quantitative knowledge mentioned above and to harness it for medical applications. This includes measurements *in vitro* as well as *in vivo*. It must be noted, however, that the definition of measurement in biology and biomedicine is often quite different compared to the classical definition used in metrology. Although measurement according to the classical definition is the estimation of attributes of objects such as its length or weight, and so on, relative to a unit of measurement, there are more general definitions used in discussions of "biomeasurement," which are based on mathematical measurement theory and statistics (Stevens, 1946; Finkelstein, 1993; Hand, 1996). According to these general definitions, measurement is the process of assigning numbers or other symbols to attributes of things in such a way that relationships of the numbers or symbols reflect relationships of the attributes of the things being measured. This definition covers determination and description of quantitative as well as qualitative attributes, and assignment of numbers or symbols to scales of measurement – such as metric or topological scales – is only a particular way of

Tab. 12.5 Measurements in bioscience (BIA, 2001).

Target of measurement	To be measured
Nucleic acid	Sequence of bases Length of base sequence Amount (quantification)
Protein	Identity *amino acid/peptide fragment sequence* Amount (quantification) Size *peptide fragment size, mass* Function *receptor, signal transduction, binding* Activity *enzyme catalysis, antibody affinity* Structure *primary through quaternary*
Cell/tissue	Identity *cell typing, profiling, growth characteristics* Quantity *cell counting* Size Viability *growth, response* Functionality *gene expression, metabolism* Interaction *adhesion, recognition, toxicity*

following it. To illustrate the different kinds of measurements in biology, Table 12.5 shows a selection of targets of biomeasurement – relevant to molecular diagnosis as well – which have been discussed more generally in BIA (2001). Obviously, several of the measurands follow the classical definition as, for example, the length of a nucleic acid base sequence, the size or the mass of a protein, the amount-of-substance of a nucleic acid or a protein, as well as the quantity (count) of cells. It should be noted, however, that even these "classical" biomeasurands often need to be measured under challenging conditions or require high precision because the dimensions or the volumes may be fairly small. Furthermore, the target of measurement is present in a matrix together with many other biomolecules or substances which may be important for the function or the activity of the target, but introduce severe bias at the same time.

Secondly, more complex attributes are "measured," like the identity or the functionality of proteins and cells as well as the viability and interaction of cells. These attributes seem to be qualitative rather than quantitative at first glance. Consequently, the general definition of measurement may be more appropriate to these kinds of attributes since it covers not only the metric scales of the classical definition but also nominal or ordinal levels of measurement. However, it is not clear at present, if and how the concept of traceability including uncertainty of measurement can be generalized or adapted to such nonquantitative levels of measurement.

On the other hand, some of the apparent qualitative attributes given in Table 12.5 may be measured quantitatively in fact. The identity of a protein may be expressed, for example, by its sequence of amino acids or peptide fragments. The function of a protein may be measured by its capability to bind to a receptor, that is, by measuring the binding force or the binding kinetics. The viability of cells may be expressed in terms of cell division rate or the response time following a sudden change in its environment. The cellular functionality may be measured by the gene expression or the metabolic rate. The challenging questions for metrology in this field are how to find out which of the attributes on the molecular and cellular levels can be measured quantitatively, how to define appropriate measurands, and how the concept of traceability can be established to improve comparability and reliability.

To help answer these questions, flow-cytometric methods for cell differentiation, counting, and sorting as discussed above are developed and applied, including sophisticated biomolecular labeling and staining techniques as well as quantitative fluorescence measurements (Resch-Genger *et al.*, 2005; Neukammer *et al.*, 2005). Furthermore, microscopic methods are necessary for validation of cells or their localization within tissue. Advanced microscopic techniques (Eigen and Rigler, 1994; Basché, Nie, and Fernandez, 2001; Rigler and Elson, 2001; Zander, Enderlein, and Keller, 2002) are also needed for the detection, localization, and quantification of extra or intracellular biomolecules like monoclonal antibodies, DNA, or mRNA. For this purpose, single molecule fluorescence microscopic measuring techniques (Rüttinger, 2006a) like Förster resonant energy transfer (FRET) are explored to measure nanometric distances between molecules under native conditions (Rüttinger *et al.*, 2006b; Krämer, Rüttinger, and Koberling, 2007), which can be applied, for example, to distinct between specific and nonspecific binding of biomolecules or to monitor conformational changes of proteins. Furthermore, fluorescence correlation spectroscopy (FCS) is studied to develop quantitative methods for ultrasensitive analytics and local measurements of molecule concentrations in the picomolar range or to study dynamic processes of biomolecules.

The concept of molecular diagnosis is also discussed for *in vivo* applications like molecular or targeted imaging (Weissleder *et al.*, 1999; Weissleder and Mahmood, 2001; Weissleder, 2006). The rationale for this kind of imaging is that many pathogenetic paradigms of disease like changes in perfusion, angiogenesis, inflammation, or apoptosis are expressed rather early and specifically by molecular or cellular targets, for example, cell surface receptors, endothelian receptors, matrix proteins, proteolytic enzymes or macrophages, and so on. Since these targets can be probed with appropriate ligand molecules like antibodies, peptides, or DNA/RNA-fragments, which can be conjugated with radioactive, magnetic, or fluorescent labels, their detection is feasible. An example (Becker *et al.*, 2001) for a targeted fluorescence imaging of a tumor is shown in Figure 12.10. The tumor shows up as a bright fluorescence spot in Figure 12.10c. Note that in the control imaging (Figure 12.10d), only two amino acids in the peptide demarcating the tumor have been exchanged in position within the sequence, but the tumor is no longer visible, which demonstrates the high specificity of this molecular diagnostic approach. Figures 12.10a,b show the fluorescence contrast before injection of the contrast agent as compressor.

The metrological tasks that are of interest in molecular imaging include the development of not only measuring methods sensitive enough to detect the labels with sufficient signal-to-noise ratio but also those that help answer the fundamental question as to whether, in addition to the presence of a disease, its degree also can be correlated or even quantified by the number of target molecules expressed.

The development of quantitative molecular imaging methods is a prerequisite to study and answer that question, and hence to support progress toward the vision of a more quantitative medical diagnosis.

12.4
Summary and Outlook

Metrology in medicine is a prerequisite to assure that measurement results relevant

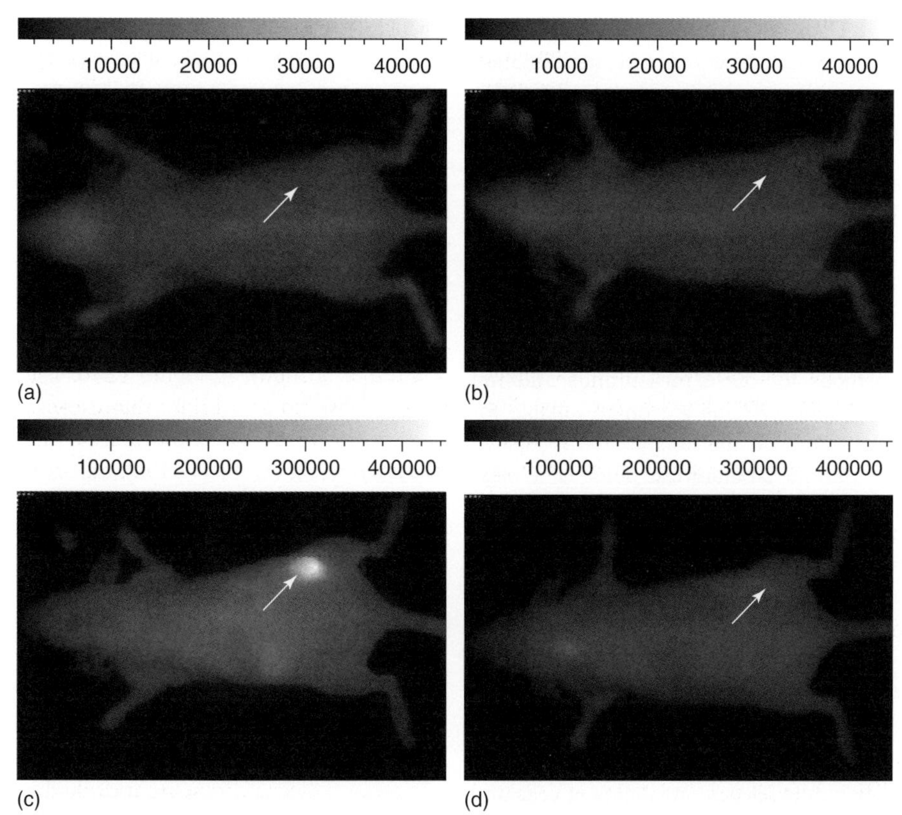

Fig. 12.10 Fluorescence mediated molecular imaging of tumors in nude mice before (a,b) and 6 hours after (c,d) intravenous injection of indotricarbocyanine (ITCC) octreotate (a,c) or ITCC-(M^2M^7) octreotate (b,d) at a dose of 0.02 μmol/kg body weight. In ITCC-(M^2M^7) octreotate two amino acids at positions 2 and 7 in the peptide have been exchanged. Arrows indicate position of the tumors, gray scales are given in photon counts. For details cf. Becker *et al.* (2001).

to medical diagnosis are accurate, reliable, and comparable between different places and over time to achieve optimum care for each individual patient as well as for the most efficient use of available funds in the health care system. For this purpose, measurement quality assurance tools based on calibrations, reference measurement methods, or certified reference materials are needed.

It has been shown in this article that the classical principle of traceability by an unbroken chain to internationally agreed standards or reference values is not well established in medical measurements because things are often too complex for this approach to be followed. Even if medical measurement is concerned with physical quantities well known in classical metrology, the conventional route of traceability can often be not realized and new concepts are needed. Examples have been given in which a database of physiological test signals or a specific clinically tested and validated device provides the highest available standards for traceability in medicine at present.

As far as chemical and biological measurands are concerned, the situation is even more complicated since definition of the measurand is not easy in many cases: the number of entities is huge, matrix effects have to be taken into account, and stability and availability of reference materials are practical problems. On the other hand, only a few primary methods are available. As a result, traceability is clearly below the primary standards or methods for these measurands and full calibration systems are only available for about 5% of the analytes investigated by routine laboratories. This is a clear need to expand the range of reference measurement procedures and reference materials of higher order.

In the future, complex measurements on biological entities will become increasingly important for molecular medicine and molecular diagnosis. The metrological questions that are of interest in this field are how to define measurands to describe biologically relevant attributes of cells and molecules quantitatively and how to establish measurement quality assurance tools as calibration systems including traceability. On the other hand, it is not clear if quantitative measurements based on the classical concepts are always best suited. Perhaps, the definition of measurement should be broadened to cover not only quantitative but also qualitative attributes, although, it is not clear at present if and how the concept of traceability including uncertainty of measurement can be generalized and adapted to nonquantitative levels of measurement.

It should be mentioned, finally, that metrology for health has been identified as one of the four grand challenges in metrology within the European Metrology Research Programme (EMRP), a partnership between 16 European member states and 3 nonmember states, in 2007. Metrology in medicine is clearly a part of that challenge toward a more reliable, efficient, and scientific exploitation of medical diagnostic methods.

Acknowledgment

Helpful discussions and critical reading of the manuscript or parts of it by J. Neukammer and B. Ebert are gratefully acknowledged.

Glossary

Angiogenesis: Growth of new blood vessels from preexisting vessels.

Antibody: Gamma globulin proteins that are found in blood or other body fluids of vertebrates, and are used by the immune system to identify and neutralize foreign objects, such as bacteria and viruses. Also known as *immunoglobulins*.

Apoptosis: A form of programmed cell death in multicellular organisms.

Complete Blood Count (CBC): A test that gives information about the cells in the blood of a patient. Also known as full blood count (FBC) or full blood exam (FBE).

Deoxyribonucleic Acid (DNA): Nucleic acid that contains the genetic instructions used in the development and functioning of all known living organisms and some viruses.

Ergometer: Exercise machine, an apparatus for measuring the work performed by exercising.

Flow Cytometer: A device for counting, examining, and sorting microscopic particles (cells) suspended in a stream of fluid. It allows simultaneous multiparametric analysis of the physical and/or chemical characteristics of single cells flowing through an optical and/or electronic detection sensor.

Fluorescence Correlation Spectroscopy (FCS): A technique used by physicists, chemists, and biologists to experimentally characterize fluorescent species (proteins, biomolecules, pharmaceuticals, etc.) and their dynamics.

Förster Resonance Energy Transfer (FRET): Energy transfer mechanism between two chromophores. The transfer efficiency depends very strongly on the distance between the chromophores, which can be utilized for colocalization or distance measurements.

Isotopic Dilution: A technique to increase the precision and accuracy of chemical analysis by adding a known amount of an isotope to the sample.

Mass Spectrometry: An analytical technique that measures the mass-to-charge ratio of charged particles.

Molecular Medicine: A new scientific discipline to investigate molecular and genetic mechanisms of cellular function and interaction for purposes of diagnosis and therapy of diseases, based on molecular anatomy, physiology, and biochemistry.

Ribonucleic Acid (RNA): Nucleic acid made from a long chain of nucleotide units, consisting of a nitrogenous base, a ribose sugar, and a phosphate. RNA is very similar to DNA, but differs in a few important structural details.

Spirometer: An apparatus for measuring the volume of air inspired and expired by the lungs. It is a precision differential pressure transducer for the measurements of respiration flow rates.

Sphygmomanometer: A device to measure blood pressure noninvasively; comprises an inflatable cuff to restrict blood flow. Manual techniques require a stethoscope used by an observer to detect Korotkoff sounds; automated techniques mostly detect and analyze cuff pressure oscillations to estimate the blood pressure values.

Tonometer: A device to measure the intraocular pressure in the eye indirectly.

References

ATS (**1995**) American thoracic society – standardization of spirometry. *Am. J. Respir. Crit. Care Med.*, **152**, 1107–1136.

Basché, T., Nie, S., and Fernandez, J.M. (**2001**) Single molecules. *Proc. Natl. Acad. Sci. U.S.A.*, **98**, 10527–10528.

Becker, A., Hessenius, C., Licha, K., Ebert, B., Sukowski, U., Semmler, W., Wiedenmann, B., and Grötzinger, C. (**2001**) Receptor-targeted optical imaging of tumors with near-infrared fluorescent ligands. *Nat. Biotechnol.*, **19**, 327–331.

BIA (**2001**) *Better Measurement for Biotechnology*, BioIndustry Association, London, *www.bioindustry.org*.

Brinkmann, F., Dam, N.E., Deák, E., Durbiano, F., Ferrara, E., Fükö, J., Jensen, H.D., Máriássy, M., Shreiner, R.H., Spitzer, P., Sudmeier, U., Surdu, M., and VyskoŠil, L. (**2003**) Primary methods for the measurement of electrolytic conductivity. *Accredit. Qual. Assur.*, **8**, 346–353.

Brownson, R.C., Baker, E.A., Leet, T.L., and Gillespie, N. (**2002**) *Evidence-Based Public Health*, Oxford University Press, New York.

Buck, R.P., Rondinini, S., Baucke, F.G.K., Camoes, M.F., Covington, A.K., Milton, M.J.T., Mussini, T., Naumann, R., Pratt, K.W., Spitzer, P., and Wilson, G.S. (**2002**) The measurement of pH-definition, standards and procedures. *Pure Appl. Chem.*, **74**, 2169–2200.

Coulter, W.H. (**1956**) High speed automatic blood cell counter and cell size analyzer. *Proc. Natl. Electron. Conf.*, **12**, 1034–1040.

Dickerhoff, R. and von Ruecker, A. (**1995**) Enumeration of platelets by multiparameter flow cytometry usind platelet-specific antibodies and fluorescent reference particles. *Clin. Lab. Haematol.*, **18**, 163–172.

DIN 58 931 (**1995**) Bestimmung der Hämoglobinkonzentration im im Blut – Referenzmethode (in German).

DIN 58 932-3 (**1994**) Bestimmung der Blutkörperchenkonzentration Blut: Bestimmung der Konzentration der Erythrozyten, Referenzmethode (in German).

DIN 58 932-2 (**1998**) Bestimmung der Blutkörperchenkonzentration im Blut, Teil 2: Kennzeichnende Größen für Erythrozyten (Erythrozytenindizes) (in German)

DIN 58 932-4 (**2003**) Determination of the Concentration of Blood Corpuscles in Blood – Part 4: Reference Procedure for the Determination of the Concentration of Leucocytes (in German and English).

DIN 58 932-5 (**2007**) Determination of the Concentration of Blood Corpuscles in Blood – Part 5: Reference Procedure for the Determination of the Concentration of Thrombocytes (in German and English).

DIN 58 932-5 (**2007**) Determination of the Concentration of Blood Corpuscles in Blood – Part 5: Reference Procedure for the Determination of the Concentration of Thrombocytes (in German and English).

DIN 58 933-1 (**1995**) Bestimmung des Volumenanteils der Erythrozyten im Blut, Teil 1: Zentrifugations methode als Referenzmethode (in German).

Dube, G. (**2001**) Metrology in chemistry – a public task. *Accredit. Qual. Assur.*, **6**, 3–7.

Eigen, M. and Rigler, R. (**1994**) Sorting single molecules: application to diagnostics and evolutionary biotechnology. *Proc. Natl. Acad. Sci. U.S.A.*, **21**, 5740–5747.

Finkelstein, L. (**1993**) Theory and philosophy of measurement, in *Handbook of Measurement Science* (P.Sydenham), John Wiley & Sons, Ltd, Chichester, New York.

Ganten, D. (**2003**) *Grundlagen der molekularen Medizin*, Springer, Berlin, Heidelberg, New York (in German).

Gouesbet, G., Lock, J.A., and Gréhan, G. (**1995**) Partial-wave representations of laser beams for use in light-scattering calculations. *Appl. Opt.*, **34**, 2133–2143.

Greenberg, N. (**2001**) Calibrator traceability: the industry impact of the IVD directive's new requirements. *IVD Technol.*, **7**, 18–27.

Hand, D.J. (**1996**) Statistics and the theory of measurement. *J. R. Stat. Soc. Ser. A*, **159**, 445–492.

Helleman, P.W. (**1970**) A study of the phenomenon of count loss in the Coulter counter, in *Standardisation in Haematology* (eds G. Astaldi, C. Sirtori, and G. Vanzetti), Franco Angeli Editore, pp. 161–168.

Helleman, P.W. (**1990**) More about coincidence loss and reference methods. *Phys. Med. Biol.*, **35**, 1159–1162.

ISO 8612 (**2001**) Ophthalmic Instruments – Tonometers.

ISO 17511 (**2003**) In Vitro Diagnostic Medical Devices – Measurement of Quantities in Biological Samples – Metrological Traceability of Values Assigned to Calibrators and Controle Materials.

IVDD (**1998**) Directive 98/79/EC of the European Parliament and of the Council of 27 October 1998 on in vitro diagnostic medical devices, *Off. J.*, **L 331**, 0001–0037.

Kaarls, R. and Quinn, R.J. (**1997**) The Comité Consultatif pour la Quantité de Matière: a brief review of its origin and present activities. *Metrologia*, **34**, 1–5.

Kachel, V. (**1990**) Electrical resistance pulse sizing: Coulter sizing, in *Flow Cytometry and Sorting*, 2nd edn (eds M.R. Melamed, T. Lindmo, and M.L. Medelsohn), Wiley-Liss, New York, pp. 45–80.

Keßler, A. and Siekmann, L. (**1999**) Measurement of urea in human serum by isotope dilution mass spectrometry: a reference procedure. *Clin. Chem.*, **45**, 1523–1529.

Krämer, B., Rüttinger, S., and Koberling, F. (**2007**) Single pair Förster resonance energy transfer with multiparametric excitation and detection. *IEEE J. Sel. Top. Quantum Electron.*, **13**, 984–989.

Kulozik, A.E., Hentze, M.W., Hagemeier, C., and Bartram, C.R. (**2000**) *Molekulare Medizin*, Walter de Gruyter, Berlin, New York (in German).

Laerum, O.D. and Bjeknes, R. (**1992**) *Flow Cytometry in Hematology*, Academic Press, London.

Levey, S. and Jennings, E.R. (**1950**) The use of control charts in the clinical laboratory. *Am. J. Clin. Pathol.*, **20**, 1059–1066.

Lewis, S., England, J.M., and Kubota, F. (**1989**) Coincidence correction in red blood cell counting. *Phys. Med. Biol.*, **34**, 1239–1246.

LMKM (**2002**) Leitfaden zu messtechnischen Kontrollen von Medizinprodukten mit Messfunktion, *www.ptb.de* (in German).

Macdonald, R. (**2006**) Quality assessment of quantitative analytical results in laboratory medicine by root mean square of measurement deviation. *J. Lab. Med.*, **30**, 111–117.

MDD (**1993**) Council Directive of 14 June 1993 concerning medical devices, *Off. J.*, **L 169**, 1.

MPBetreibV (**1998**) Medizinprodukte-Betreiberverodnung, *BGBI*, **I**, S.1762 (in German).

Muir Gray, J.A. (**1997**) *Evidence-Based Healthcare*, W.B. Saunders Company, New York.

Neukammer, J., Gohlke, C., Höpe, A., Wessel, T., and Rinneberg, H. (**2003**) Angular distribution of light scattered by single biological cells and oriented particle agglomerates. *Appl. Opt.*, **42**, 6388–6397.

Neukammer, J., Gohlke, C., Krämer, B., and Roos, M. (**2005**) Concept for the traceability of fluorescence (beads) in flow cytometry: exploiting saturation and microscopic single molecule bleaching. *J. Fluoresc.*, **15**, 433–441.

Ost, V., Neukammer, J., and Rinneberg, H. (**1998**) Flow cytometric differentiation of erythrocytes and leukocytes in dilute whole blood by light scattering. *Cytometry*, **32**, 191–197.

Pickup, J.F. and McPherson, C.K. (**1977**) A theory of stable-isotope dilution mass spectrometry, in *Quantitative Mass Spectrometry in Life Sciences* (eds A.P. de Leenheer and R.R. Roncucci), Elsevier Scientific Publishing Company, Amsterdam.

Pignone, M. and McPhee, S.J. (**2006**) Approach to the patient & health maintenance, in *Current Medical Diagnosis & Treatment* (eds L.M.Tiery, S.J. McPhee, and M.A. Papadakis), Mc Graw-Hill, New York.

Pyzdek, T. and Keller, P.A. (**2003**) *Quality Engineering Handbook. Quality and Reliability*, vol. **60**, Marcel Dekker, New York.

Resch-Genger, U., Hoffmann, K., Nietfeld, W., Engel, A., Neukammer, J., Nitschke, R., Ebert, B., and Macdonald, R. (**2005**) How to improve quality assurance in fluorometry: fluorescence-inherent sources of error and suited fluorescence standards. *J. Fluoresc.*, **15**, 337–362.

Rigler, R. and Elson, E.S. (eds) (**2001**) *Fluorescence Correlation Spectroscopy – Theory and Applications*, Springer, Berlin, Heidelberg.

RiLiBÄK (**2001**) Richtlinie der Bundesärztekammer zur zur Qualitätssicherung quantitativer laboratoriumsmedizinischer Untersuchungen, *Dtsch. Ärztebl.*, **98**, B2356–B2367.

RiLiBÄK (**2008**) Richtlinie der Bundesärztekammer zur Qualitätssicherung laboratoriumsmedizinischer Untersuchungen, *Dtsch. Ärztebl.*, **105**, A341–A355.

Rinneberg, H., Neukammer, J., and Ost, V. (**1995**) Laser- Durchflußzytometer als Normalmesseinrichtung zur Bestimmung der Konzentration von Erythrozyten, Leukozyten und Thrombozyten. *Lab. Med.*, **19**, 238–242 (in German).

Rüttinger, S. (**2006a**) Confocal microscopy and quantitative single molecule techniques for metrology in molecular medicine, Dissertation, Technischen University, Berlin, pp. 1–141.

Rüttinger, S., Macdonald, R., Krämer, B., Koberling, F., Roos, M., and Hildt, E. (**2006b**) Accurate single-pair Förster resonant energy transfer through combination of pulsed interleaved excitation, time correlated single-photon counting, and fluorescence correlation spectroscopy. *J. Biomed. Opt.*, **11**, 024012.

Ruhenstroth-Bauer, G. and Zhang, D. (**1960**) Automatische Zählmethoden: Das Coulter'sche Partikelzählgerät. *Blut*, **6**, 446–462 (in German).

Shapiro, H.M. (**2003**) *Practical Flow Cytometry*, 4th edn, Whiley-Liss, New York.

Siekmann, L. (**1978**) Isotope dilution-mass spectrometry of oestriol and oestradiol-17β: an approach to definitive methods. *Z. Anal. Chem.*, **280**, 122–123.

Siekmann, L. (**1979**) Determination of steroid hormones by the use of isotope dilution-mass spectometry: a definitive method in clinical chemistry. *J. Steroid Biochem.*, **11**, 117–123.

Siekmann, L. (**1982**) Determination of cortisol in human plasma by isotope dilution-mass spectrometry: a definitive method in clinical chemistry. *J. Clin. Chem. Clin. Biochem.*, **20**, 883–892.

Siekmann, L. (**1985a**) Determination of uric acid in human serum by isotope dilution-mass spectrometry: a definitive method in clinical chemistry. *J. Clin. Chem. Clin. Biochem.*, **23**, 129–135.

Siekmann, L. (**1985b**) Determination of creatinine in human serum by isotope dilution-mass spectrometry: a definitive method in clinical chemistry. *J. Clin. Chem. Clin. Biochem.*, **23**, 137–144.

Siekmann, L. (**1991**) Gaschromatographie-Massenspektrometrie als Referenz- und definitive Methode in der Klinischen Chemie, in *Analytiker-Taschenbuch*, Bd. 10 (ed. H. Günzler (Hrsg.)), Springer Verlag, Berlin, Heidelberg (in German).

Siekmann, L., Hoppen, H.O., and Breuer, H. (**1970**) Zur gaschromatographisch-massenspektrometrischen Bestimmung von Steroidhormonen in Körperflüssigkeiten unter Verwendung eines Multiple Ion Detectors (Fragmentographie). *Z. Anal. Chem.*, **252**, 294–298 (in German).

Siekmann, L., Hüskes, K.P., and Breuer, H. (**1976**) Determination of cholesterol in serum using mass fragmentography – a reference method in clinical chemistry. *Z. Anal. Chem.*, **279**, 145–146.

Siekmann, L., Schönfelder, A., and Siekmann, A. (**1986**) Isotope dilution-mass spectrometry of total glycerol in human serum - a reference method in clinical chemistry. *Z. Anal. Chem.*, **324**, 280–281.

Spitzer, P. (**2005**) Traceable measurements of pH, in *Traceability in Chemical Measurement*, Springer, Berlin, Heidelberg, pp. 206–211.

Springer, W., Prohaska, W., Neukammer, J., Höpe, A., and von Rücker, A. (**1999**) Evaluation of a new reagent for preserving fresh blood samples and its potential usefulness for internal quality controls of multichannel hematology analyzers. *Am. J. Clin. Pathol.*, **111**, 387–396.

Stevens, S.S. (**1946**) On the theory of scales of measurement. *Science*, **103**, 677–680.

Thom, R. (**1972**) Method and results by improved electronic cell sizing, in *Modern Concepts in Haematology* (eds G.Izak and S.M. Lewis), Academic Press, New York, pp. 191–200.

Valet, G. (**2003**) Past and present concepts in flow cytometry: a European perspective. *J. Biol. Regul. Homeost. Agents*, **17**, 213–222.

VIM (**2007**) ISO/IEC Guide 99, International Vocabulary of Metrology. Basic and General Concepts and Associated Terms (VIM).

Wales, M. and Wilson, J.N. (**1961**) Theory of coincidence in Coulter particle counters. *Rev. Sci. Instrum.*, **32**, 1132–1136.

Weissleder, R. (**2006**) Molecular imaging in cancer. *Science*, **312**, 1168–1171.

Weissleder, R. and Mahmood, U. (**2001**) Molecular imaging. *Radiology*, **219**, 316–333; references herein.

Weissleder, R., Tung, C.H., Mamood, U., and Bogdanov, A. Jr (**1999**) In vivo imaging of tumors with protease-activated near-infrared fluorescent probes. *Nat. Biotechnol.*, **17**, 375–378.

Westgard, J.O. (**1992**) Charts of operational process spefications (OPSpecs charts) for assessing the precision, accuracy, and quality control needed to satisfy proficiency testing performance criteria. *Clin. Chem.*, **38**, 1226–1233.

Westgard, J.O., Barry, P.L., Hunt, M.R., and Groth, T. (**1981**) A multi-rule shewart chart for quality control in clinical chemistry. *Clin. Chem.*, **27**, 493–501.

Westgard, J.O., Groth, T., Aronsson, T., and de Verdier, C.-H. (**1977**) Combined Sheward-Cusum control chart in clinical chemistry. *Clin. Chem.*, **23**, 1881–1887.

Zander, Ch., Enderlein, J., and Keller, R.A. (eds) (**2002**) *Single Molecule Detection in Solution – Methods and Applications*, Wiley-VCH Verlag GmbH, Berlin.

Zerhouni, E. (**2003**) The NIH roadmap. *Science*, **302**, 63–72.

Further Reading

Hawkins, D. (**2005**) *Biomeasurement: Understanding, Analysing and Communicating Data in the Biosciences*, Oxford University Press, New York.

Landay, A.L., Ault, K.A., Bauer, K.D., and Rabinovitch, P. (**1993**) *Clinical Flow Cytometry*, New York Academy of Sciences, New York.

Richter, W. and Güttler, B. (**2005**) A national traceability system for chemical measurements, *Traceability in Chemical Measurement*, Springer, Berlin, Heidelberg, pp. 141–146.

Sargent, M. and Harrington, C. (**2002**) *Guidelines for Achieving High Accuracy in Isotope Dilution Mass Spectroscopy (IDMS)*, Royal Society of Chemistry, Cambridge.

Trent, R.J. (**2005**) *Molecular Medicine: An Introductory Text*, Elsevier, Amsterdam.

Watson, J.W. (**1991**) *Introduction to Flow Cytometry*, Cambridge University Press, Cambridge.

Handbook of Metrology

Edited by
Michael Gläser and
Manfred Kochsiek

Related Titles

Stock, R. (ed.)

Encyclopedia of Applied Nuclear Physics

ISBN: 978-3-527-40742-2

Stock, R. (ed.)

Encyclopedia of Applied High Energy and Particle Physics

2009

ISBN: 978-3-527-40691-3

Wilkening, G., Koenders, L.

Nanoscale Calibration Standards and Methods

Dimensional and Related Measurements in the Micro- and Nanometer Range

2005

ISBN: 978-3-527-40502-2

Riehle, F.

Frequency Standards

Basics and Applications

2004

ISBN: 978-3-527-40230-4

Trigg, G. L. (ed.)

Encyclopedia of Applied Physics

The Classic Softcover Edition

2004

ISBN: 978-3-527-40478-0

Brown, T. G., Creath, K., Kogelnik, H., Kriss, M. A., Schmit, J., Weber, M. J. (eds.)

The Optics Encyclopedia

Basic Foundations and Practical Applications. 5 Volumes

2004

ISBN: 978-3-527-40320-2

Gåsvik, K. J.

Optical Metrology

2002

ISBN: 978-0-470-84300-0

Keithley, J. F.

The Story of Electrical and Magnetic Measurements

From 500 BC to the 1940s

2001

ISBN: 978-0-7803-1193-0

Handbook of Metrology

Edited by
Michael Gläser and Manfred Kochsiek

Volume II

WILEY-VCH Verlag GmbH & Co. KGaA

The Editors

Dr. Michael Gläser
Physikalisch-Technische
Bundesanstalt
Bundesallee 100
38116 Braunschweig
Germany

Prof. Dr.-Ing. Manfred Kochsiek
Physikalisch-Technische
Bundesanstalt
Bundesallee 100
38116 Braunschweig
Germany

Library of Congress Card No.:
applied for

British Library Cataloguing-in-Publication Data
A catalogue record for this book is available from the British Library.

Bibliographic information published by the Deutsche Nationalbibliothek
The Deutsche Nationalbibliothek lists this publication in the Deutsche Nationalbibliografie; detailed bibliographic data are available on the Internet at ⟨http://dnb.d-nb.de⟩.

Composition Laserwords Private Ltd., Chennai, India
Printing and Binding T.J. International Ltd., Padstow, Cornwall
Cover Design Schulz Grafik-Design, Fußgönheim

Printed in the Federal Republic of Germany
Printed on acid-free paper

ISBN: 978-3-527-40666-1

Contents

Handbook of Metrology. Edited by Michael Gläser and Manfred Kochsiek

ISBN: 978-3-527-40666-1

List of Contributors

Felicitas Arias
Bureau International des
Poids et Mesures
Pavillon de Bretevil
92312 Sèvres cedex
France

Andreas Bauch
Physikalisch-Technische Bundesanstalt
Bundesallee 100
38116 Braunschweig
Germany

John H. Cantrell
University of Tennessee
NASA Langley Research Center
3 East Taylor Street
Hampton, VA 23681-2199
USA

Simona M. Cristescu
Radboud University
Heijendaalseweg 135
6525AJ Nijmegen
The Netherlands

Horst Czichos
University of Applied Sciences
BHT, Berlin
Luxemburger Straße 10
13353 Berlin
Germany

Richard S. Davis
Bureau International des
Poids et Mesures
Pavillon de Bretevil
92312 Sèvres cedex
France

Christian Fabjan
Institute of High Energy Physics of the
Austrian Academy of Science and
University of Technology
HEPHY
Nikolsdorfer Gasse 18
1050 Vienna
Austria

Joachim Fischer
Physikalisch-Technische Bundesanstalt
Bundesallee 100
38116 Braunschweig
Germany

J. L. Flowers
National Physical Laboratory
Teddington Middlesex
TW11 OLW
UK

Handbook of Metrology. Edited by Michael Gläser and Manfred Kochsiek

ISBN: 978-3-527-40666-1

Eckhart Förster
X-ray Optics Group
Institute of Optics and Quantum Electronics
Friedrich Schiller University Jena
Max-Wien-Platz 1
07743 Jena
Germany

Michael Gläser
Physikalisch-Technische Bundesanstalt
Bundesallee 100
38116 Braunschweig
Germany

Claus Grupen
Department of Physics
Siegen University
Walter-Flex-Str. 3
57072 Siegen
Germany

James E. Hardy
Engineering Science and Technology Division
Managed by UT-Battelle
LLC for the U.S. Department of Energy
under contract DE-AC05-00OR22725
P.O. Box 2008
Oak Ridge, TN 37831-6003
USA

Frans J. M. Harren
Radboud University
Heijendaalseweg 135
6525AJ Nijmegen
The Netherlands

Massimo Inguscio
LENS-European Laboratory for Non-Linear Spectroscopy and
Dipartimento di Fisica
Universita di Firenze
via N. Carrara 1
I-50019 Sesto Fiorentino-Firenze
Italy

Bryan P. Kibble
National Physical Laboratory
10 Warwick Close
Middlesex
Hampton TW12 2TY
UK

Manfred Kochsiek
Physikalisch-Technische Bundesanstalt
Bundesallee 100
38116 Braunschweig
Germany

Rainer Macdonald
Physikalisch-Technische Bundesanstalt
Bundesallee 100
38116 Braunschweig
Germany

Thomas H. Markert
Center for Space Research
Massachusetts Institute of Technology
Cambridge, MA 02139
USA

Stephan Mieke
Physikalisch-Technische Bundesanstalt
Bundesallee 100
38116 Braunschweig
Germany

Stefan Persijn
National Metrology Institute
Van Swinden Lab
Thijsseweg 11
2629 JA Delft
The Netherlands

B. W. Petley
National Physical Laboratory
Teddington Middlesex
TW11 OLW
UK

François Piquemal
Laboratoire National de Métrologie
et d'Essais (LNE)
29, rue Roger Hennequin
Trappes
France

Kenneth A. Rubinson
The Five Oaks Research Institute
Bethesda, MD 20817
USA

and

Department of Biochemistry and
Molecular Biology
Wright State University
Dayton, OH 45435
USA

Francis R. Ruppel
Alstom Power Inc.
1409 Centerpoint Boulevard
Knoxville, TN 37932-1962
USA

Giulia Rusciano
Dipartimento di Scienze Fisiche
Università di Napoli "Federico II"
Complesso Universitario
Monte Sant'Angelo
Via Cintia-80126
Napoli
Italy

Antonio Sasso
Dipartimento di Scienze Fisiche
Università di Napoli "Federico II"
Complesso Universitario
Monte Sant'Angelo
Via Cintia-80126
Napoli
Italy

Bernd R. L. Siebert
Physikalisch-Technische
Bundesanstalt
Bundesallee 100
38116 Braunschweig
Germany

Klaus-Dieter Sommer
Physikalisch-Technische
Bundesanstalt
Braunschweig und Berlin
Germany

Jörn Stenger
Physikalisch-Technische
Bundesanstalt
Bundesallee 100
38116 Braunschweig
Germany

Introduction

Michael Gläser and Manfred Kochsiek

Handbook of Metrology. Edited by Michael Gläser and Manfred Kochsiek

ISBN: 978-3-527-40666-1

Metrology is the science and technology of measurement. It is as old as human culture, because measurements were necessary even in ancient times for manufacturing tools, even if they were simple like axes, lances, and plows, or for construction of houses. Weighing of goods was common – at least since 5000 years – and this has been corroborated by the discovery of an old Egyptian balance beam. Uniform standards were first necessary only for particular manufacturing projects. A stone axe, for example, was probably made first by processing the stone and drilling the hole in the shaft. The shape of the hole was the standard for carving the shaft. The measurement was made by comparing the size of the shaft with that of the hole of the stone. Even though this is a very simple procedure, it shows the principle of measurement: comparing a measure of an object with that of a standard. An advanced measuring procedure is the comparison of multiples or submultiples of the standard with the object to be measured, as it is necessary for the construction of houses, temples, pyramids, or other buildings. In a larger community, uniform standards were necessary in commerce or technology, when goods were sold by weight or parts of a construction were premanufactured at a distant place. Ancient cultures with central rulers often kept national standards of weight, length, and volume in a temple of the capital. Even before the creation of the meter convention in 1875, cities kept their binding standards in the town hall or somewhere near the market place. Measurement of time has been important for agriculture since early times. The success and survival of a community sometimes depended on the time of seeding and harvesting. Ancient calendars of the Babylonian or the Maya cultures are well known. In Europe, the Julian calendar since Caesar's time and the Gregorian calendar since the sixteenth century successively optimized the measure of the length of the year by algorithms based on the day as a counting digit. The Babylonian divided the day into 12 hours at daylight and 12 hours at night, a division that is still common today. The only change now is that all the 24 hours of the day have the same length. Measurement and units of other quantities like density, force, velocity, acceleration temperature, electric quantities, frequency, amount of substance, energy, or power came in use at the end of the middle ages, when scientific research was progressively modernized. During the French revolution, efforts were made to unify the units of measurement not only in France but also on an international basis. However, it was only in 1875 that an internationally accepted system of units, the International System of Units

Handbook of Metrology. Edited by Michael Gläser and Manfred Kochsiek

ISBN: 978-3-527-40666-1

(SI) was accepted by 17 countries in what was called the *Meter Convention*. There are 52 members and 26 associated countries today in this convention. The SI comprises seven base units: meter, kilogram, second, ampere, kelvin, mole, and candela. Besides the kilogram, which is still defined by the mass of a material object, the base units are defined such that they can be realized at any place and at any time. The other units of measurement are derived from the base units as products and quotients. Two "dimensionless" units, the radian and steradian, are also part of the SI.

Metrology comprises (i) the calibration of a measurement standard or a measuring instrument from the realization of the definition of a unit, usually through a chain of several intermediate standards; (ii) the development and specification of reliable measuring methods and instruments for particular measurement quantities and measurement ranges, according to modern requirements in science, technology, commerce, environment, or health; (iii) the evaluation of measurement data, including reasonable corrections, to obtain the required measurement result; (iv) the evaluation of the measurement uncertainty by taking into account statistical rules and all influencing parameters; and (v) finally, from the knowledge of recent physical and technical research, projecting a new definition of a unit, based on a better constant of nature and developing a corresponding design for realizing such a new unit. Legal metrology is a particular branch of metrology. It deals with national regulations for the use of units, standards, and measurement procedures as well as institutions like National Metrology Institutes and verification offices. Calibration laboratories are working under the auspices of national accreditation bodies and mostly according to international rules, which, among others, require the traceability of their standards to national standards. Measurement means comparison of a quantity between an object and a standard. The standard can be a ruler, a weight, a zener diode, or an atomic clock, for example. For the comparison, an instrument, often named *comparator*, is required. The measuring method of such an instrument comprises the conversion of the quantity to be measured, preferably in a linear function, to a visible or recordable, quantitative signal. Today, the majority of such signals make use of electrical quantities like voltages, currents, or resistances that are digitized and indicated as a decimal number and a unit on a display. There is always a sophisticated sensor that makes the conversion between the physical quantity and the corresponding electrical signal and provides such values. The quality of the sensor is crucial for the reliability and accuracy of the indicated value. Normally, a manufacturer of such instruments allows only as many digits to be indicated as the accuracy of measurement allows. Some scientific instruments can be adjusted for the zero point and the slope of the indication for allowing the user to calibrate the instrument – at least for its linear slope. Electronic balances of high accuracy, for example, are calibrated by the user for the zero point and the slope by using an internal reference weight. The result of a measurement is given by a number, a unit, and the associated uncertainty. The uncertainty may be given as a number with a unit or as a relative value. For the evaluation of the uncertainty, the instrument's specification or class, the ambient conditions like temperature, humidity, or pressure, the number of repeated measurements and other parameters that influence the measured data, for example, the position of the sensor and its adaption to the measured object, are taken into account. A report or certificate of the measurement result proves the traceability of the reference standards to the national or international standards.

This Handbook contains articles that are also published in the Encyclopedia of Applied Physics (Wiley-VCH) and it is divided into two parts. It contains articles dealing with general topics of measurement and articles on particular subjects in mechanics and acoustics, electricity, optics, temperature, time and frequency, chemistry, medicine and particles. The contributions of the second part are summarized as follows.

"Uncertainty" starts with basic concepts like the probability density function (PDF), Bayes' theorem, modeling and the use of Taylor series, then describes the propagation of PDFs or uncertainties, the standard procedure of the *Guide to the Expression of Uncertainty in Measurement* (GUM), the Monte Carlo Method, and finally selected advanced topics in uncertainty evaluation.

"Laser Spectroscopy" deals with the analysis of atomic or molecular structure by using laser radiation. After a historical introduction, the width and broadening mechanism of spectral lines are discussed, aspects that are crucial for spectral resolution. Then, various methods of high-resolution spectroscopy and high-sensitivity laser spectroscopy are presented, followed by methods of time-resolved spectroscopy, ultrahigh-resolution spectroscopy using laser cooling and atom traps, and, finally, single molecule spectroscopy.

"Chemical Analysis Metrology" first explains what metrology means in chemical analysis and the key role of validation, and lists related terms used in chemical analysis. Subsection "Approaching the Truth" introduces the concepts of standard deviation, precision, bias, and error. The following sections deal with standards, sampling, sampling preparation, the signal and noise, detection limits, the working concentration limits, and propagation of uncertainty

"Chemical Analysis" presents a comprehensive review on methods for specifying the elemental and molecular composition of materials. A tabulated overview on the methods and general rules for assessing measurement results is followed by descriptions of the various chemical and physical assay methods, such as analyses of known chemical reactions, spectrometrical methods from gamma to radio-frequency spectrometry, mass spectrometry, electrochemical methods, chromatography, and surface-sensitive and thermal methods.

"Photoacoustic Spectroscopy" first explains the method as an indirect technique measuring an effect of light absorption rather than light absorption itself and its advantages compared with regular spectroscopy. Then, the various devices and equipment are presented, followed by applications for characterizing solids and gases in chemical, biological, material and environmental analyses, and in human health research.

"Radiation Detectors" starts with a view on the historical development of radiation detectors and mentions the application fields of such instruments. Detailed descriptions of historical radiation detectors, modern gas detectors, scintillators, solid-state detectors, Cherenkov detectors, and calorimeters are presented next.

"X-Ray Spectrometry" are introduced as instruments that measure wavelengths or equivalent energies and fluxes of radiations with wavelengths in the range between ultraviolet and gamma radiation with equivalent photon energies between about 100 eV and 100 keV. Various instruments of the two general categories of spectrometers, the dispersive and nondispersive (energy-dispersive) spectrometers, distinguished by techniques based on the wave or particle property of the radiation, are described in detail. Finally, applications, for example, in plasma physics, astrophysics, or material analysis, and the appropriate spectrometers are discussed

"Calorimetric Particle Detectors" first points out their properties and advantages as energy detectors for charged and neutral particles compared with momentum and velocity detectors, which are appropriate only for charged particles. The following two sections present calorimetric methods in nuclear physics and particle physics for measuring photons, leptons, and hadrons. The performance of calorimeters with respect to energy resolution and provision of further information on time, space, and angular resolution is discussed. The most important construction principles, instrumentation techniques, and calorimeter facilities at the LHC are finally described.

We hope the readers of the Handbook of Metrology will find the information contained herein new and useful.

13 Uncertainty

Bernd R. L. Siebert and Klaus-Dieter Sommer

Handbook of Metrology. Edited by Michael Gläser and Manfred Kochsiek

ISBN: 978-3-527-40666-1

13.1 Introduction

The value of a quantity Q, as obtained by measurement or computation, can be written as a product of its numerical value $\{Q\}$ and a unit $[Q]$ (see Chapter 2). The unit is a reference quantity and the numerical value, denoted here by q, is a number and given by $q = Q/[Q]$. Consider for instance the length L_{edge} of a Teflon plate. The numerical value of the quantity L_{edge}, that is, $l_{\text{edge}} = \{L_{\text{edge}}\}$, depends on the unit chosen. Throughout this chapter it is tacitly assumed that coherent SI units are used.

Experience shows that repeated measurements of the length of the *same* Teflon plate in the *same* laboratory by the *same* operator yield slightly different values, and if the plate is transferred to another laboratory again other slightly different values are observed. This situation raises the following questions:

1. Is any of these many numerical values the "true" value?
2. Are there reliable estimates of the "true" numerical value?

There are two fundamentally different approaches to answer these questions. The VIM (BIPM, IEC, IFCC, ILAC, ISO, IUPAC, IUPAP, and OIML, 2008) calls them *error approach* and *uncertainty approach*.

In the *error approach* the true value of a quantity is an unknown constant and the result of measurement is a random variable. Classical statistics is applied to random errors, whereas systematic errors are treated separately (Cox and Harris, 2006a).

The *uncertainty approach* is laid down in the "Guide to the Expression of Uncertainty in Measurement" (GUM) (ISO, 1993), and it has been clarified recently in "Supplement 1 to the GUM" GUM-SUP (JCGM, 2008). Here, as in Bayesian inference, the quantity is considered as a random variable whose values are represented by a probability density function (PDF) derived from available knowledge, whereas the measured values are considered as facts and they are part of the available knowledge (Cox and Harris, 2006a). The users of this approach are called *Bayesians.*

To point out the different underlying concepts of probability, users of the error approach are called *"frequentists"* and the users of the uncertainty approach "Bayesians." The Bayesian answer to the first of the above stated questions is: A completely defined scalar quantity does have a "true" value, but it is generally never exactly known due to incomplete knowledge. The frequentist answer is: A

Handbook of Metrology. Edited by Michael Gläser and Manfred Kochsiek

ISBN: 978-3-527-40666-1

completely defined scalar quantity does have a "true" value, but it is in general masked by insufficient statistics and unknown systematic errors. The answers of both approaches to the second of the above-stated questions are in principle the same: Reliable estimates are in general possible; however, the "reliability" is defined differently.

It would be futile to join the sometimes fierce debate between frequentists and Bayesians. Instead, two papers that provide a fair summary on either view are recommended. The author of the first of these is a statistician (Gleser, 1998), and the authors of the other paper are physicists (Lira and Wöger, 2006).

Since the GUM is endorsed by the world's leading metrological institutions, this chapter focuses predominately on Bayesian approach. However, the *error approach* is briefly summarized and compared with the *uncertainty approach* in Section 13.1 and the Appendix.

Uncertainty is a measure of the reliability and trustworthiness of the result of a measurement or computation. Computation is explicitly mentioned, since virtual experiments play an increasing role and approximate algorithms are often needed in evaluation procedures. The uncertainty associated with the stated value of a quantity is the standard deviation of the (Bayesian) PDF for that quantity. In applications subject to mutual recognition agreements, it is required to state a coverage interval in addition to uncertainty.

Cox's theorem (Cox, 1946) states that any measure of belief is isomorphic to a probability measure. This theorem justifies the concept of Bayesian probability. On the basis of the given facts and prior knowledge about reasonably possible values, the principle of maximum information entropy (PME) and Bayes' theorem can be used for constructing or updating a PDF for a quantity. Most measurands, termed *output quantities,* depend on several input quantities. Their inter-relation is described by a model of the measurement. Supplement 1 to the GUM (JCGM, 2008) and related publications, see Bich and Cox (2006a) and Bich, Cox, and Harris (2006b), endorse these statements and allow to identify two basic concepts:

1. The knowledge about the possible values of a quantity can be expressed by a Bayesian PDF for that quantity.
2. A model of measurement determines the propagation of the PDFs for the input quantities to the output quantities (measurands).

Section 13.1 introduces a running example that is used in Sections 13.3.1 and 13.3.2 for evolving the basic concepts of the GUM in detail. Assistance on formulating the model for the evaluation of uncertainty is given in Section 13.3.3. Section 13.3.4 introduces the concept of using a Taylor series for obtaining linear approximation of nonlinear models. Section 13.4 introduces the Markov formula as a general tool for propagating PDFs and shows that the central limit theorem (CLT) provides answers to the second of the above-stated questions. Section 13.5 treats uncertainty propagation using standard GUM procedure and discusses its limitations. How to use a Monte Carlo method (MCM) for propagating PDFs is explicated in Section 13.6, and some advanced applications are presented in Section 13.7. The Appendix provides some details on the error approach and the Glossary lists the often-used symbols, acronyms, and scientific terms.

13.2 Running Example

The running example considers the use of a calibrated aluminum ruler to measure the area A_{plate} of a "perfectly" rectangular thin Teflon plate at temperature $T_{\text{plate}} = 20^\circ$C. The calibration certificate for the ruler states that the calibration has been performed at a temperature $T_{\text{ruler}} = 21^\circ$C, that the value of the measurement deviation Δl_{CAL} is 0 mm, and that the associated uncertainty $u(\Delta l_{\text{CAL}}) = 10^{-2} \times l_{\text{IND}}$, where l_{IND} is the length indicated by (read of) the ruler. The expanded uncertainty $U_{0.95}$, assuming a Gaussian PDF, has the value $2 \times 10^{-2} \times l_{\text{IND}}$. Figure 13.1 illustrates the running example and two major sources of uncertainty. For better comprehensibility, the running example is somewhat oversimplified. The data given are selected to discuss important features of uncertainty analysis and are not in all cases realistic. This is also true for the selected means of measurement. Despite these simplifications, the model allows to discuss a large fraction of practical problems faced in determining uncertainty.

In general, the first step of any measurement is to carefully define the measurand, to select a principle and a method of measurement, and, based on these, to set up a measurement procedure. In this running example, the measurand is well defined and the procedure to be followed is simple. The ruler is placed parallel to the edges of the Teflon plate and readings are taken at the adjacent lower and upper edges. The length of the first edge $L_{\text{edge},1}$ is then inferred from the difference

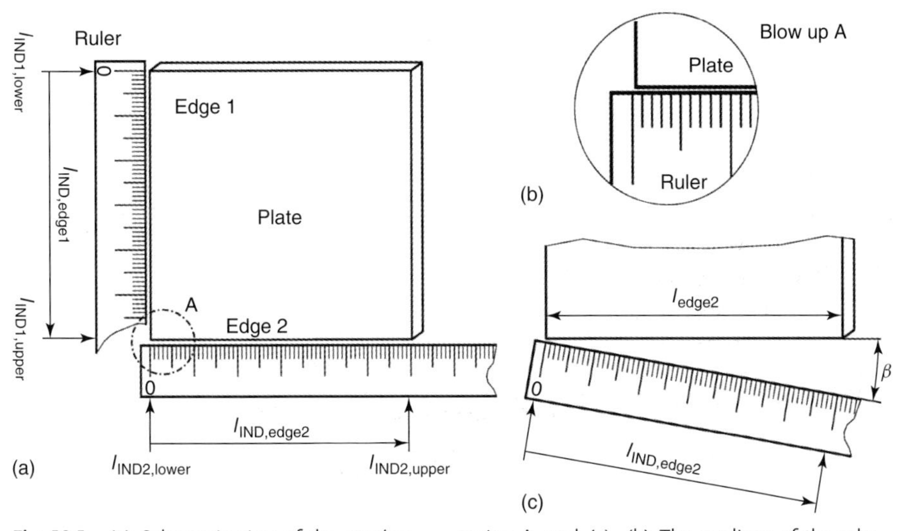

Fig. 13.1 (a) Schematic view of the *running example*: The area A_{plate} of a "perfectly" rectangular thin Teflon plate is inferred from the product of the edge lengths $l_{\text{IND,edge1}}$ and $l_{\text{IND,edge2}}$. These lengths are determined as differences of two readings, that is, $l_{\text{IND,edge}} = l_{\text{IND,upper}} - l_{\text{IND,lower}}$. Two major contributions to the overall uncertainty for the measured value of the area are illustrated in (b), a blow up of the encircled region A, and (c). (b) The readings of the ruler have a finite resolution. In the case shown, it is only possible to state that the value is closer to zero than to 0.5 mm. Therefore, the reading is zero. (c) The reading overestimates the edge length by the factor $1/\cos\beta$ if the ruler and edge are not parallel. This is called *cosine error*.

of the readings: $l_{\mathrm{IND,\,edge,1}} = l_{\mathrm{IND,upper}} - l_{\mathrm{IND,lower}}$, and the length of an edge perpendicular to the first edge, $l_{\mathrm{IND,edge,2}}$, is obtained analogously. Finally, the area of the plate is inferred from the product of $l_{\mathrm{IND,edge,1}}$ and $l_{\mathrm{IND,edge,2}}$.

The second task is to identify possible sources of uncertainty. The measurement deviation Δl_{CAL} of the ruler and the resolution of reading the ruler δl_{RES} are sources of uncertainty. Figure 13.1b illustrates the finite resolution of ruler readings. As in any length measurement, temperature is an important input quantity, whence the incomplete knowledge of the temperatures of the plate and ruler is a source of uncertainty. Finally, the chosen measurement procedure requires placing the ruler parallel to the edge of the measurand. As illustrated in Figure 13.1c, the ruler and edge are in practice not perfectly parallel, and the quantity "angle between the ruler and the edge of the plate" B is a further source of uncertainty; B is understood as upper case β. Other sources of uncertainty are not considered for keeping the running example simple. However, it should be noted that in real applications all possible sources of uncertainty should be listed in this step, even those that are expected to contribute little to the combined uncertainty.

Next, a model of the measurement for evaluating the uncertainty is formulated. Modeling is often the most difficult part in determining uncertainty. A theory on modeling does not exist; however, guidance on how to proceed systematically is available (Sommer and Siebert, 2006), and useful examples can be found in an EA-publication (EA, 1999). For more details on modeling see Section 13.3.3.

For greater ease in writing it is useful to introduce a *submodel* for the length as inferred from the readings:

$$\begin{aligned} L_{\mathrm{edge},i,\mathrm{reading}} = {} & (l_{\mathrm{IND,upper},i} + \delta l_{\mathrm{RES,upper},i} \\ & + \Delta l_{\mathrm{CAL}}(l_{\mathrm{IND,upper},i})) \\ & - (l_{\mathrm{IND,lower},i} + \delta l_{\mathrm{RES,lower},i} \\ & + \Delta l_{\mathrm{CAL}}(l_{\mathrm{IND,lower},i})) \end{aligned} \quad (13.1)$$

where $i = 1$ pertains to the length and $i = 2$ to the width of the plate. After taking the readings $l_{\mathrm{IND,lower},i}$ and $l_{\mathrm{IND,upper},i}$, they are *known constants*. The measurand $L_{\mathrm{edge},i,\mathrm{reading}}$ is the output *quantity*, and δl_{RES} and Δl_{CAL} are input *quantities*.

The models for the length of the first and second edges of the Teflon plate are identical and, using the submodel for $L_{\mathrm{edge},i,\mathrm{reading}}$, given by

$$\begin{aligned} & L_{\mathrm{edge},i} \cdot (1 + \alpha_{\mathrm{Teflon}} \cdot (T_{\mathrm{plate},i} - 20^{\circ}\mathrm{C})) \\ & \quad = L_{\mathrm{edge},i,\mathrm{reading}} \cdot (1 + \alpha_{\mathrm{aluminum}} \cdot (T_{\mathrm{ruler},i} \\ & \quad -21^{\circ}\mathrm{C})) \cdot \cos B_i \end{aligned} \quad (13.2)$$

The thermal expansion coefficients $\alpha_{\mathrm{aluminum}}$ and α_{Teflon} do depend on temperature and uncertainties are associated with their values; however, to keep the running example simple, they are considered here to be known constants. The measurand $L_{\mathrm{edge},i}$ is the output quantity and $L_{\mathrm{edge},i,\mathrm{reading}}$, $T_{\mathrm{plate},i}$, $T_{\mathrm{ruler},i}$, and B_i are input quantities. Equation 13.2 is intuitively clear, but for further use it is helpful to rearrange it such that $L_{\mathrm{edge},i}$ is directly given as a function of its input quantities:

$$L_{\mathrm{edge},i} = L_{\mathrm{edge},i,\mathrm{reading}} \times \frac{(1 + \alpha_{\mathrm{aluminum}} \cdot (T_{\mathrm{ruler},i} - 21^{\circ}\mathrm{C})) \cdot \cos B_i}{(1 + \alpha_{\mathrm{Teflon}} \cdot (T_{\mathrm{plate},i} - 20^{\circ}\mathrm{C}))} \quad (13.3)$$

For further simplification, the temperatures in the ruler and plate are taken as equal. The model for the area of the Teflon plate is then given by

$$A_{\text{plate}} = L_{\text{edge},1} \cdot L_{\text{edge},2}$$
$$= L_{\text{edge},1,\text{reading}} \cdot L_{\text{edge},2,\text{reading}}$$
$$\times \frac{(1 + \alpha_{\text{aluminum}}(T - 21^\circ\text{C}))^2}{(1 + \alpha_{\text{Teflon}}(T - 20^\circ\text{C}))^2}$$
$$\times \cos B_1 \cdot \cos B_2 \qquad (13.4)$$

The area of the Teflon plate A_{plate} is the output quantity, and $L_{\text{edge},1}$ and $L_{\text{edge},2}$ are here input quantities obtained as output quantities of the submodels defined in Equations 13.2 and 13.3.

The model functions for the lengths as inferred from the readings, that is, $L_{\text{edge},i,\text{reading}}$, depend *linearly* on the input quantities $\delta l_{\text{RES,lower}}$, $\delta l_{\text{RES,upper}}$, and Δl_{CAL}. However, the model functions for the length of the first and second edges of the Teflon plate and even more so the model function for the area of the Teflon plate are not linear functions of the input quantities. The standard GUM procedure then uses a first-order Taylor expansion to obtain a linear model function; see Section 13.3.4.

13.3 Basic Concepts

13.3.1 Knowledge about a Quantity

How can knowledge be gained about the length L_{edge} and the area A_{plate} of a Teflon plate? A first answer is "by measurements." The information gained by measurement is necessary. However, as shown in Section 13.2, this information alone is not enough, because the measurement is influenced by other quantities, for instance by the temperature at the time of measurement, and it depends on the measurement procedure, which in turn depends on the measurement principle and the method of measurement.

The GUM distinguishes between information obtained by repeatedly taking values in the course of a given measurement procedure and information obtained by "scientific judgment." To determine the standard uncertainty, the GUM uses a Type A evaluation for the first and a Type B evaluation for the latter information. The formal procedure of Type A evaluation is given in Section 13.4.2. Examples of information assessed by Type B evaluation are previous measurement data, experience with or general knowledge of the behavior and properties of materials and instruments, and manufacturer's specifications or data given in calibration or other certificates, for instance the measurement deviation Δl_{CAL} in the running example.

The resolution of the ruler reading for a length measurement δl_{RES} might be inferred via a Type A evaluation. To that purpose, one would use the ruler repeatedly to obtain values of $l_{\text{IND,lower}}$ or $l_{\text{IND,upper}}$. Under the – most unrealistic – assumption that all other influencing quantities remain unchanged during these measurement series, one could infer the resolution from the variance s^2 of the readings. However, even if this assumption were justified, possible influences of the measurement deviation Δl_{CAL} could not be detected by such a procedure. Therefore a Type B evaluation of the resolution is suggested instead. The temperature T, too, could be repeatedly measured and a Type A evaluation used. In practice, however, one tries to avoid the expenditure of these measurements by relying on an air conditioner that allows assuring temperatures in a small interval $[t_{\min}, t_{\max}]$ such that the expectation of the temperature is given by $t = (t_{\min} + t_{\max})/2$. A Type A procedure could be used to obtain information about the angle B between the ruler and the plate. However, since such repeated measurements would be influenced by both δl_{RES} and Δl_{CAL}, a Type B evaluation is selected here, too.

For a Type B evaluation of δl_{RES}, it may be given that the distance between two marks on the ruler is 1 mm. An observer, depending on his eyes and experience, is able to say that the end of the edge coincides with a point somewhere in the first or the second half of this interval. He would then describe the resolution by stating l_{IND} = reading $\pm$ 0.25 mm, and the reading would have the values x.0 or x.5 mm, where x is an integer. In the case of instruments with a digital display, the resolution of the reading would correspond to half of the smallest possible difference between the two readings. Similarly, one can use a Type B evaluation for the angle B; for the running example, it is assumed that the possible values are contained in the interval $[\beta_{min}, \beta_{max}]$. Again, the values of β_{min} and β_{max} depend on the skill and experience of the observer, but one may assume that $|\beta_{min}| = \beta_{max}$.

The information gathered above, irrespective of the type of assessment, provides knowledge that allows to localize the possible values of the examined quantities but not to state one and only one value. Assessing this information required already available knowledge or beliefs. For instance, when using Type A evaluations one "believes" that repeatedly measured values are *representative* "samples" of the possible values, and when using Type B evaluations one "believes" the calibration certificate or, with respect to the thermal expansion one uses and "believes", the appropriateness of linear function of temperature.

The following section introduces the concept of Bayesian probability and shows that the knowledge about the possible values of a quantity can be represented by a PDF.

13.3.2 Probability Density Function (PDF) for a Quantity

Bayesian probability can be thought of as betting or believing based on given facts and incomplete knowledge. In that sense one would have differently strong beliefs on different possible values. This means one has a measure of belief for the possible values and, based on Cox's theorem (Cox, 1946), can treat this measure as probability. If one considers a continuous range of possible values, $\xi \in (\xi_{min}, \xi_{max})$, one uses a PDF for the quantity X for expressing the belief.

It is necessary to consider *all given facts*. One should be aware that one's knowledge is in general incomplete. This means that any procedure used to combine given facts and incomplete prior knowledge must allow that *new information* leads to a *less incomplete* posterior knowledge.

The PDF for the quantity X is written as $g_X(\xi)$. If more than one quantity is considered, vector notation is used, $\boldsymbol{X} = (X_1, \ldots, X_N)^T$, and the joint PDF for $\boldsymbol{X}$ is written as $g_{\boldsymbol{X}}(\boldsymbol{\xi}), \boldsymbol{\xi} = (\xi_1, \ldots, \xi_N)^T$. However, $\boldsymbol{X}$ is not considered as a vector, but rather as a set of quantities.

The first subsection describes the general properties of a PDF and Sections 13.3.2.2 and 13.3.2.3 explain how to *assign* or *derive* a PDF based on available incomplete knowledge and measured values.

13.3.2.1 General Properties of a PDF

For convenience the same symbol X is used for a quantity *and* the random variable characterized by the PDF for the quantity. Confusion is unlikely, since the symbol for the random variable, if not explicitly expressed as $g_X(\xi)$, is used in conjunction with the "operators" E (expectation), Var (variance), or Cov (covariance), as indicated in the expressions below that summarize the general properties of PDFs:

$$\int_{-\infty}^{\infty} g_X(\xi)d\xi = 1 \quad \text{Normalization} \quad (13.5)$$

$$\int_{-\infty}^{\infty} g_X(\xi)\xi d\xi = \mathrm{E}X \quad \text{Expectation} \quad (13.6)$$

$$\int_{-\infty}^{\infty} g_X(\xi)\xi^2 d\xi = \mathrm{E}X^2 = \mathrm{Var}X + (\mathrm{E}X)^2 \quad \text{Variance} \quad (13.7)$$

$$\int_{-\infty}^{\infty} g_{X_1,X_2}(\xi_1,\xi_2)(\xi_1 - x_1)(\xi_2 - x_2)d\xi_1 d\xi_2 = \mathrm{Cov}(X_1X_2) \quad \text{Covariance.} \quad (13.8)$$

The relation between PDFs and GUM conventions can be stated as

$$x = \mathrm{E}X = \int_{-\infty}^{\infty} g_X(\xi)\xi d\xi \text{ and } u^2(x) = \mathrm{Var}X = \int_{-\infty}^{\infty} g_X(\xi)(\xi - x)^2 d\xi \quad (13.9)$$

$$r(x_1,x_2) = \frac{\mathrm{Cov}(X_2X_2)}{u(x_1)u(x_2)} \quad \text{Correlation coefficient.} \quad (13.10)$$

The "operators" E, Var, and Cov are simply abbreviations for the underlying integrals and the rules of the calculus of these operators follow the rules for these integrals. Thus, for instance, if c is a constant, then $\mathrm{E}c\mathrm{X} = c\mathrm{E}X$ and

$$\begin{aligned}
\mathrm{Var}X &= \mathrm{E}((X - \mathrm{E}X)^2) \\
&= \mathrm{E}(X^2 - 2Xx + x^2) \\
&= \mathrm{E}(X^2) - 2x\mathrm{E}X + x^2 \\
&= \mathrm{E}(X^2) - x^2 \\
\mathrm{Cov}(X_1,X_2) &= \mathrm{E}((X_1 - \mathrm{E}X_1)(X_2 - \mathrm{E}X_2)) \\
&= \mathrm{E}(X_1X_2) - x_1\mathrm{E}X_2 \\
&\quad - x_2\mathrm{E}X_1 + x_1x_2 \\
&= \mathrm{E}(X_1X_2) - x_1x_2 \qquad (13.11)
\end{aligned}$$

For ease in calculations, it is often possible and then helpful to represent random variables in the standard form:

$$X = x + u(x)\ X_{\mathrm{std}} \text{ with } \mathrm{E}X_{\mathrm{std}} = 0 \text{ and } \mathrm{E}X_{\mathrm{std}}^2 = 1 \qquad (13.12)$$

where the subscript "std" is a placeholder, for example, X_G and X_R would indicate a Gaussian or a rectangular PDF in the standard form. This representation is especially useful for PDFs for which all odd central moments vanish. Besides Gaussian and rectangular PDFs, the often-used PDFs are the triangular (folding two rectangular PDFs having the same variance), the trapezoidal (folding two rectangular PDFs having different variances, useful to account for two resolutions, e.g., upper and lower reading on the ruler), and the U-shaped PDFs (sinus of frequency if the phase is uniformly distributed in the interval $[0,\pi]$) and, especially in radiation physics, the Poisson distribution. Other PDFs are found in textbooks (e.g., Burry, 1999).

The GUM and quite a number of standards request not only to state expectation ("best" estimate of the value) and uncertainty (square root of the variance) associated with it, but also the so-called *expanded uncertainty*. To explain this concept, there is a need to introduce the cumulative distribution function (CDF):

$$G_X(\xi) = \int_{-\infty}^{\xi} g_X(\xi')d\xi' \text{ or, implicitly,}$$
$$g_X(\xi) = \frac{d}{d\xi}G_X(\xi) \qquad (13.13)$$

The CDF $G_X(\xi)$ is the probability that the value of X is less than or equal to ξ. Therefore, the probability that $\xi \in [\xi_{min}, \xi_{max}]$ is given by $G_X(\xi_{max}) - G_X(\xi_{min})$. If a PDF is symmetric about its expectation, that is, if $g_X(x-\xi) = g_X(x+\xi)$, then $G_X(\xi_{max}) - 1/2 = 1/2 - G_X(\xi_{min})$. For instance, Gaussian or rectangular PDFs are symmetric. The *expanded uncertainty* U_p is then defined as the half-width of the interval $[\xi_{min}, \xi_{max}]$ that covers (contains) the value of X with a stated probability p. The GUM recommends $p = 0.95$ for most of the applications. If the PDF is not symmetric, then one may distinguish betweeen $U_{p,-}$ and $U_{p,+}$, and search for the shortest coverage interval,

$$U_{p,-} = x - G_X^{-1}(\xi_{min}) \text{ and } U_{p,+} = G_X^{-1}(\xi_{max}) - x \text{ with } G_X(\xi_{max}) - G_X(\xi_{min}) = p \qquad (13.14)$$

under the constraint that $[\xi_{min}, \xi_{max}]$ is the shortest coverage interval. In the case of symmetric PDFs, the so-called *coverage factor* k_p is the ratio $U_p/u(x)$; for a Gaussian PDF, $k_{0.95} = 1.96$, but this is generally rounded off to 2. For a rectangular PDF, $k_{0.95} = 1.64$.

Classical probability is defined as the ratio of the number of "successes" to the number of trials. Considering the event "*value is contained in the confidence interval*" as success, one expects that the fraction of "successful" measurements repeated under strict repeatability conditions converges asymptotically to 95% (i.e., $p \times 100\%$). This concept of a confidence interval, used in error analysis, is fundamentally different from the concept of a coverage interval which is constructed to contain the value of the quantity with a stated coverage probability.

13.3.2.2 **Principle of Maximum Entropy (PME)**

The principle of maximum (information) entropy (PME) derived by Shannon (1948) can be used to assign a PDF to a quantity (Jaynes, 1957). Let us assume we know only that the values of a quantity X are bounded, that is, $\xi \in [\xi_{min}, \xi_{max}]$. Then the PME requires a rectangular (also called *uniform*), PDF, characterized by $EX = (\xi_{min} + \xi_{max})/2$ and $\text{Var}X = (\xi_{min} + \xi_{max})^2/12$. This is quite plausible since any other shape would require more information. The general case is that one knows the expectation of functions of the possible values of X, left-hand-side relation in Equation 13.15. Then the PME defines the PDF for X, that is, $g_X(\xi)$, by maximizing the information entropy H, right-hand-side relation in Equation 13.15, under the constraints $\{\varphi\}$:

$$Ef_\nu(\xi) = \int_{-\infty}^{\infty} g_X(\xi)f_\nu(\xi)d\xi = \varphi_\nu, \nu \in [1, n],$$
$$H = -\int_{-\infty}^{\infty} g_X(\xi)\ln g_X(\xi)d\xi \qquad (13.15)$$

This is a variation problem that may be solved using Lagrange multipliers. The two most important cases in practice are the following:

- If one *only* knows that values of a quantity are within an interval bounded by the values ξ_{min} and ξ_{max}, then the PME yields a rectangular-shaped PDF, with properties described above.

- If one *only* knows EX and Var$X$, for example, from a table in a handbook or from somebody else's measurements, then the PME yields for $\varphi_1 = \mathrm{E}X = x$ and $\varphi_2 = \mathrm{Var}X = u^2(x)$, a Gaussian PDF, characterized by EX and Var$X$.

For discrete values, the PME is expressed as

$$\mathrm{E}f_\nu(x) = \sum_{i=1}^{N} p_i f_\nu(x) = \varphi_\nu, \nu \in [1, n],$$
$$H = -\sum_{i=1}^{N} p_i \ln p_i \tag{13.16}$$

Consider, for example, a coin. The probability that after tossing the coin the head or tail faces upward is denoted by p or $1 - p$, respectively. The other information is not available, that is, constraints are not given. One obtains

$$H = -p \ln p - (1-p)\ln(1-p) \text{ and } \frac{dH}{dp} = -\ln p + \ln(1-p) = 0 \tag{13.17}$$

Thus, $p = 1/2$ yields an extreme value of H. Since the second derivative of $H(p)$ at $p = 1/2$ is negative, it follows that H assumes its maximum at $p = 1/2$:

$$\frac{d^2H}{dp^2} = -\frac{1}{p} + \frac{1 \times (-1)}{1-p} = \frac{-1}{p(1-p)} = -4 < 0 \Rightarrow H(p{=}1/2) = H_{\max} = -\ln 1/2 \tag{13.18}$$

The values of the minima of H are 0, and they are obtained for $p = 1$ or $p = 0$. These results are plausible. Without any additional information one has to assume that tosses show heads and tails with equal probability. The maximum of entropy coincides here with absolute indifference.

13.3.2.3 Bayes' Theorem

Bayes' theorem allows *"updating"* of a PDF $g_X(\xi|\mathrm{I})$ that reflects already given *prior* information I if new data D are available:

$$g_X(\xi|D, I)\ d\xi = Cl(\xi|D, I) g_X(\xi|I) d\xi \tag{13.19}$$

where $C^{-1} = \int_{-\infty}^{\infty} l(\xi|D, I) g_X(\xi|I) d\xi$, $g_X(\xi|D, I)$ is called the *posterior*, and $l(\xi|D, I)$ is the likelihood function, or often simply called *likelihood.* The likelihood reflects the belief about obtaining the data given that the hypothesis $X \in (\xi, \xi + d\xi)$ is true and C is a normalization constant.

To demonstrate Bayes' theorem again a coin is considered. As in Section 13.3.2.2, one "believes" that a tossed coin can only show head or tail, but now uses the information that the outcome of any single tossing can be modeled by a binomial PDF and considers the *probability for seeing a head* as quantity X and acquires useful facts, that is, new data D, by tossing the coin, say M times. Prior to the first tossing one knows only, or rather believes, that the possible values of X are bounded, that is, $\xi \in [0, 1]$, and assigns, according to the PME, a rectangular PDF to X, that is, $g_X(\xi|I) = 1.0$ for $\xi \in [0, 1]$ and 0 for $\xi \notin [0, 1]$. Obviously, the statistical model to be used in stating the likelihood is here a binomial PDF. The number of heads seen in M tosses is k; using this in Equation 13.15 leads then to

$$g_X(\xi|D, I) d\xi = C \binom{M}{k} \times \xi^k (1-\xi)^{M-k} g_X(\xi|I) d\xi \tag{13.20}$$

where $g_X(\xi|D, I)$ is the *posterior* PDF for X that results from combining the new information D, k heads in M tosses, with the prior rectangular PDF for X with

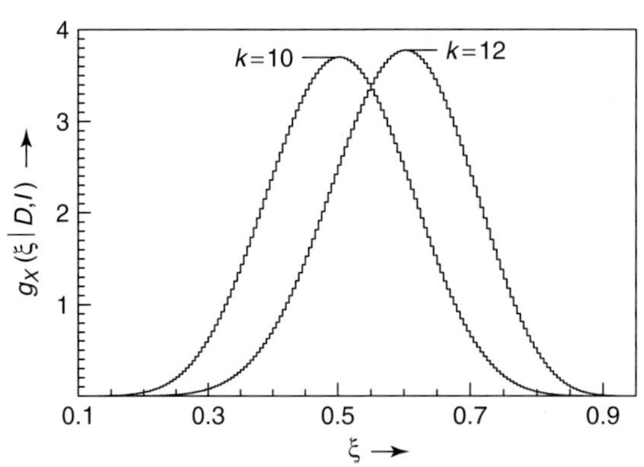

Fig. 13.2 Probability density function (PDF) for the probability of "tossing head" after obtaining the data *D*: 10 heads in 20 tosses or *D*: 12 heads in 20 tosses, using the noninformative prior *I*: *p* is bound by 0 and 1 that via PME leads to a uniform prior PDF.

$EX = 0.5$ and $\mathrm{Var}X = 1/12$. The constant C can be obtained by normalization. The likelihood, M over k multiplied with $\xi^k(1-\xi)^{M-k}$, expresses here the belief that the value of X is ξ if k heads are seen in M tosses.

The binomial coefficients M over k are constants and $g_X(\xi|\mathrm{I})$ has the value 1 for $\xi \in [0,1]$. Therefore Equation 13.20 can be reduced to $g_X(\xi|D,I) = C'\xi^k(1-\xi)^{M-k}$. Figure 13.2 shows $g_X(\xi|D,I)$ for two cases where $M = 20$ and $k = 10$ or 12 as approximated at $\xi_i = (i-1) \times 0.0025$ with $i \in [1,401]$. The expectations, that is, EX approximated by the sum over the 401 values $g_X(\xi_i|D,I)\xi_i$, are 0.5000 and 0.5909, and the associated uncertainties, that is, the square root of $\mathrm{Var}X$ approximated by the sum over the 401 values $g_X(\xi_i|D,I)(\xi_i - EX)$, are then 0.1043 and 0.1025 for $k = 10$ and 12, respectively. The expectation for the former is 0.5 and the associated uncertainty is 0.2887. Figure 13.3 shows the resulting CDFs and, using blowups, the shortest coverage intervals $U_{0.95}$. For $k = 10$, $U_{0.95}$ is symmetric about the expectation and

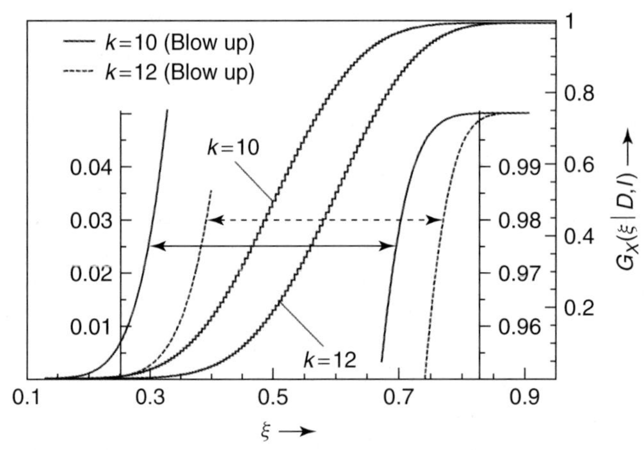

Fig. 13.3 Same as Figure 13.2, but cumulative distribution function (CDF). The abscissa pertaining to all curves are shown. The right ordinate pertains to the fully shown CDF. The inserted ordinates pertain to the blow ups. The arrows indicate the coverage intervals.

$U_{0.95} = 0.2023$. For $k = 12$, $U_{0.95}$ is slightly asymmetric about the expectation, the coverage interval is [0.3924, 0.7895], such that $U_{0.95,-} = 0.1985$ and $U_{0.95,+} = 1.986$.

With the Bayesian approach one answers the question: "What does one know about the probability of "tossing head" if one *has obtained* k heads in M tosses *and believes* that only tail or head can occur in any tossing?" This is, mutatis mutandis, the fundamental question in determining the uncertainties: "Which values of a quantity are reasonably possible in view of both, the data given *and* the accepted ('believed') prior information?"

13.3.3 Model for the Evaluation of Uncertainty

13.3.3.1 Overview of Tools

Stating a model for a measurement is needed not only for uncertainty evaluation but also for understanding the measurement, including all influences, relationships, and effects. Without such a model it is hardly possible to interpret the results of a measurement.

Unfortunately, models representing a measurement can rarely be established completely and perfectly. In general, models should pragmatically reflect the "purpose-relevant" behavior of the measurement to be evaluated and, furthermore, be as simple and understandable as possible.

A formalized procedure for modeling does not exist. However, there are various means available to support modeling and representing technical processes, states, and facts. For that purpose, one can use verbal models, graphical models, and mathematical models (Sommer, 2009).

Verbal models are not suitable for structuring and quantifying technical systems and processes, but they are useful, or even essential, for identifying input and influence quantities and thereby for establishing graphical models, which comprise signal-flow charts, block diagrams, structure diagrams or state graphs, and other tools (Sommer, 2009). Verbal models also contain the information available on the input and influence quantities, usually the PDF assigned for a quantity and its expectation and the uncertainty associated with it.

Graphical models are capable of comprehensibly depicting interconnections of the relevant quantities as "flow charts." However, graphical models are hardly suitable for representing quantitative relationships between the quantities of interest. But they support the formulation of mathematical models for that purpose.

Mathematical models might be structured into data models, such as data sequences and frequency distributions, and analytical models or connective models, for example, neuronal networks. Evaluating measured data and uncertainty requires a mathematical model, which represents both, the measurement process and all quantities and parameters that may influence the measurement result.

When using the standard GUM procedure, cf. Section 13.5, it is helpful if the mathematical model provides the output quantity explicitly as a function of input and influence quantities. When using Monte Carlo procedures, cf. Section 13.6, any form of the model is acceptable. However, in both cases it is mandatory that the model assigns *one and only one* possible value to the output quantity for a given set of possible values of the input and influence quantities. Conversely, a given possible value of the output quantity in general may result from different sets of possible values of the input and influence quantities!

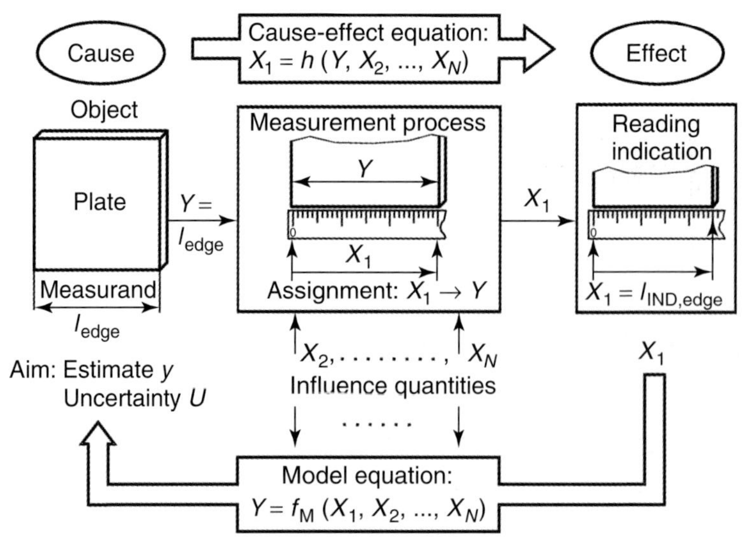

Fig. 13.4 Visualization of the measurement process in the running example. The output quantity Y, here the measurand "length of an edge of the plate at a specified temperature" denoted by L_{edge}, *causes* via the *measurement process* the *effect* that the instrument produces a *reading indication*. The reading is here considered as the value of the input quantity X_1 that is assigned to Y. However, the value indicated by x_1, here $l_{IND,edge}$, is not necessarily the same as the value of the output quantity y, here l_{edge}. The *cause–effect equation* $X_1 = h(Y, X_2, \ldots, X_N)$ takes account of the *influence quantities* $X_i, i \in [2, N]$, that perturb the measurement process inevitably. To infer l_{edge} from $l_{IND,edge}$, a *model equation* $Y = f_M(X_1, X_2, X_N)$ is to be derived that relates the input quantity and the influence quantities to the output quantity. The model equation is then also used to propagate the uncertainties associated with the input and influence quantities to the uncertainty associated with y, here l_{edge}, and to compute the expanded uncertainty U.

13.3.3.2 **Tools Applied to Running Example**

Using the running example, Figure 13.4 visualizes the measurement process and describes the basic tasks in modeling. It shows that a measurement can be conceived as a *cause–effect-chain*. The analysis of this chain leads to the *cause–effect equation* that is then transformed to the *model equation*.

In the running example, a *verbal model* is used for defining the measurand and for identifying input and influence quantities and the information given about these quantities. The first four paragraphs of Section 13.2 contain the verbal model for the running model.

Figures 13.1a–c and 13.4 are examples of using *graphical models* for depicting the "action" of input quantities and visualizing influencing quantities.

Figure 13.4 also shows the "flow" of the measurement signals starting from the measurand and ending with the reading indication. These figures are supplemented by an explicit *graphical flow chart*, see Figure 13.5. The graphical flow chart consists of only three different modules: source (SRC), transposition (TRANS), and indication (IND). In general, any of these modules can appear several times and contain branches. Depending on the method of measurement, one obtains characteristic generic

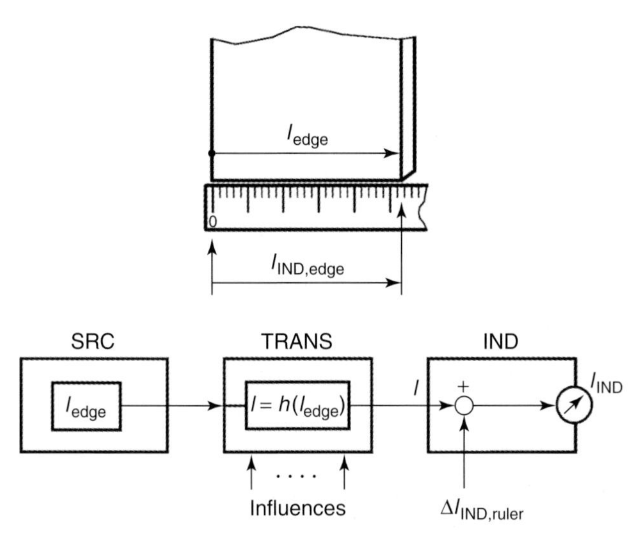

Fig. 13.5 Sketch of the *cause–effect-chain* consisting of source (SRC), transposition (TRANS), and indication (IND). SRC is essentially the measurand, TRANS "handles" the perturbations due to the influence quantities and IND is generally considered as first input quantity.

flow charts. Sommer and Siebert (2006) provides examples of generic flow charts for the cause-and-effect relationship, for instance, for

- a calibration by means of substitution (comparison with a standard, conjoining chains);
- a calibration by direct comparison of two indicating measuring instruments (forking chain);
- a compensation method (closed loop);
- the application of "patched standards".

Figure 13.6 presents examples of graphically supporting the mathematical formulation models or parts of models. Ideally, graphical representations allow us to directly read off the mathematical formulation; cf. bottom part of Figure 13.6.

The step from the *cause–effect equation* $X_1 = h(Y, X_2, \ldots, X_N)$ to the *model equation* $Y = f_M(X_1, \ldots, X_N)$ is, in the running example, achieved by a simple rearrangement; see Equations 13.1–13.4. If such a simple rearrangement is not possible, standard tools of numerical analysis can

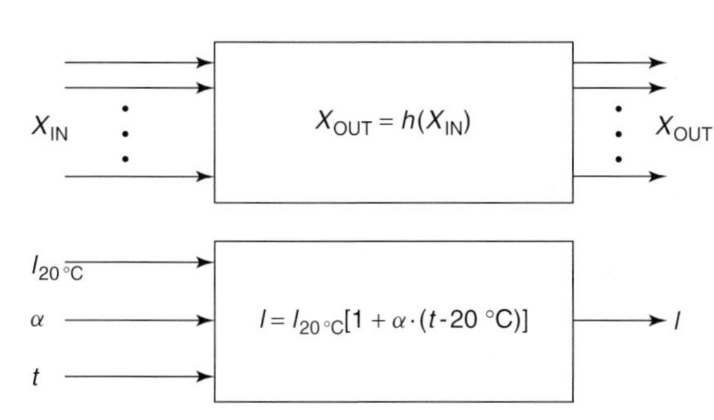

Fig. 13.6 The upper graph visualizes solving the cause–effect equation and the bottom graph shows the action of the TRANS module in the running example that describes the influence of the plate temperature on the measurand.

be applied. For examples, one can refer to the calibration of paired temperature sensors for heat meters that requires mathematical operations such as a matrix inversion (Tegeler, Heyer, and Siebert, 2008) or an evaluation of uncertainty values of building acoustic single-number quantities (Goydke, Siebert, and Scholl, 2003), and also Section 13.5.3.

In general, using *submodels* helps to reduce the complexity of modeling. In the running example, submodels for the quantities $L_{\text{edge},i,\text{reading}}$, $i = 1$ or 2, have been used. However, if, as in this case, the quantities represented by submodels are correlated, one needs to compute and use the correlation coefficient in the standard GUM procedure, see Equation 13.43, or circumvent the problem as shown in Section 13.5.

13.3.4 Use of Taylor Series

13.3.4.1 Formalism

The model for the evaluation of uncertainty can often be expressed by a function $Y = f(X_1, \ldots, X_N)$. The possible values η of Y are given by $\eta = f(\xi_1, \ldots, \xi_N)$, where ξ_i are the possible values of X_i. If the model function is not linear, one can use a first-order Taylor series to obtain a linear approximation. The full Taylor series is given by

$$\eta = f(\xi) = \sum_{k=0}^{\infty} \frac{1}{k!} \left(\sum_{i=1}^{N} \Delta\xi_i \frac{\partial}{\partial \xi_i} \right)^k \times f(\xi)|_{\forall_i \xi_i = x_i} \tag{13.21}$$

where f is the model function, $\boldsymbol{\xi} = (\xi_1, \ldots, \xi_N)^T$, ξ_i is a possible value of X_i, $\Delta\xi_i = \xi_i - x_i$, $x_i = \mathrm{E}X_i$, and k is the order of expansion. Linearization means to keep only the terms for $k = 0$ and 1. These are given by

$$\eta_0 = f(x) \text{ and } \eta_1 = \sum_{i=1}^{N} \frac{\partial f(\xi)|_{\forall_i \xi_i = x_i}}{\partial \xi_i} (\xi_i - x_i) \tag{13.22}$$

Upon using the symbol c_i for the first partial derivatives evaluated at $\xi_i = x_i$, as is done in the GUM, one obtains

$$\eta_{\text{lin}} = \sum_{k=0}^{1} \eta_k = f(x) + \sum_{i=1}^{N} c_i(\xi_i - x_i) \Rightarrow y_{\text{lin}} = f(x), \text{ since } \forall_i \int g_X(\xi)(\xi_i - x_i) d\xi = 0 \tag{13.23}$$

The above relations hold also for the quantities

$$Y_{\text{lin}} = \sum_{k=0}^{1} Y_k = f(x) + \sum_{i=1}^{N} c_i(X_i - x_i) \Rightarrow \mathrm{E}Y_{\text{lin}} = y_{\text{lin}} = f(x), \text{ since } \forall_i \mathrm{E}(X_i - x_i) = 0 \tag{13.24}$$

where $\mathrm{E}Y_0 = \mathrm{E}f(x) = f(x)$, since $f(x)$ is a constant, and $\mathrm{E}X_i = x_i$.

The first partial derivatives c_i are called *sensitivity coefficients*. The combined uncertainty using this linearization for the possible values of the measurand is then given by

$$u^2(y_{\text{lin}}) = \int g_X(\boldsymbol{\xi}) \left(\sum_{i=1}^{N} c_i(\xi_i - x_i) \right)^2 d\boldsymbol{\xi} = \sum_{i=1}^{N} c_i^2 u^2(x_i) + 2 \sum_{i_1=1}^{N} \sum_{i_2=i_1+1}^{N} \times c_{i_1} c_{i_2} u(x_{i_1}) r(x_{i_1}, x_{i_2}) u(x_{i_2}) \tag{13.25}$$

Equation 13.25 expresses the law of uncertainty propagation as stated in the

GUM. An alternative compact presentation is

$$u^2(y_{lin}) = \boldsymbol{c}^{\mathrm{T}} \boldsymbol{U}_x \boldsymbol{c} \quad (13.26)$$

where $\boldsymbol{c}^{\mathrm{T}} = (c_1, \ldots, c_N)$ and the elements of the uncertainty matrix $\boldsymbol{U}_x$ are

$$\begin{aligned} u_{i_1 i_2} &= \int g_X(\boldsymbol{\xi})(\xi_{i_1} - x_{i_1})(\xi_{i_2} - x_{i_2}) d\boldsymbol{\xi} \\ &= u(x_{i_1}, x_{i_2}) = u(x_{i_1}) r(x_{i_1}, x_{i_2}) u(x_{i_2}) \end{aligned} \quad (13.27)$$

The submodel for $L_{\mathrm{edge},i,\mathrm{reading}}$ in the running example, see Equation 13.1, is already a linear model. The first partial derivatives (sensitivity coefficients) are given by

$$\begin{aligned} c_{\delta l_{\mathrm{RES,upper},i}} &= 1, c_{\delta l_{\mathrm{RES,lower},i}} \\ &= -1, c_{\Delta l_{\mathrm{CAL}}(l_{\mathrm{IND,upper},i})} \\ &= 1 \text{ and } c_{\Delta l_{\mathrm{CAL}}(l_{\mathrm{IND,lower},i})} = -1 \end{aligned} \quad (13.28)$$

13.3.4.2 Check on the Validity of a Linear Model

If the model depends on one variable only, the corresponding remainder term allows for judging on the quality of the approximation. However, most model functions depend on more than just one variable. Therefore, using the model $Y = X_1/X_2$ as an example, a general approach is shown that can be used for any number of variables in the model function. Second-order sensitivity coefficients are denoted by $c_{i,i'}$. The model used for demonstration is similar to but less complicated than the model for the length of an edge, cf. Equation 13.3. The sensitivity coefficients are given by

$$c_1 = \frac{1}{x_2} 1, c_2 = \frac{-x_1}{x_2^2}, c_{1,1} = 0, c_{2,2} = \frac{x_1}{x_2^3} \text{ and } c_{1,2} = c_{2,1} = \frac{-1}{x_2^2} \quad (13.29)$$

Using the standard form of the PDFs for X_1 and X_2, cf. Equation 13.12, one obtains the possible values of Y from

$$\begin{aligned} \eta_{\mathrm{qdr}} &= \sum_{k=0}^{2} \eta_k, \eta_0 = \frac{x_1}{x_2}, \eta_1 = \sum_{i=1}^{2} c_i(\xi_i - x_i), \\ &\text{and } \eta_2 = c_{2,2}(\xi_2 - x_2)^2 \\ &+ c_{1,2}(\xi_1 - x_2)(\xi_2 - x_2) \end{aligned} \quad (13.30)$$

where the subscript "qdr" denotes the approximation of the model function using up to second-order terms. For simplicity it is assumed that $\mathrm{Cov}(X_{1,\mathrm{std}}, X_{2,\mathrm{std}}) = 0$. The second-order term then adds only $c_{2,2}u^2(x_2)$ to the expectation. Therefore,

$$\begin{aligned} y_{\mathrm{qdr}} &= \sum_{k=0}^{2} y_k = \frac{x_1}{x_2} + \sum_{i=1}^{2} c_i \mathrm{E}(\xi_i - x_i) \\ &+ c_{2,2} \mathrm{E}(\xi_2 - x_2)^2 \\ &= \frac{x_1}{x_2} + 0 + \frac{x_1}{x_2^3} u^2(x_2) \\ &= \frac{x_1}{x_2}(1 + w^2(x_2)) \end{aligned} \quad (13.31)$$

where $w(x_2) = u(x_2)/x_2$.

The contributions of first and second order terms to the uncertainty associated with y_{qdr} are in this simple example given by

$$\begin{aligned} u^2(y_1) &= \frac{x_1^2}{x_2^2}(w^2(x_1) + w^2(x_2)) \text{ and} \\ u^2(y_2) &= \frac{x_1^2}{x_2^2}[w^2(x_2)(w^2(x_1) + 2w^2(x_2))] \end{aligned} \quad (13.32)$$

In the computation of $u(y_2)$, a Gaussian PDF is assigned to X_2 such that $\mathrm{E}(\xi_2 - x_2)^4 = 3u^4(x_2)$.

The uncertainty needs to be stated with two significant digits only. Therefore, the second-order term is to be taken into account, if $w^2(x_2)$ is larger than $0.05 \times u(y)$, cf. Equation 13.32, or if $u(y_1)$ is larger than $0.05 \times u(y_1)$. Thus, the second-order contributions can be neglected if

$$\frac{x_1}{20 \times x_2}\sqrt{w^2(x_1) + w^2(x_2)} > w^2(x_2)$$
$$\text{and } \frac{w^2(x_1) + w^2(x_2)}{400} > w^2(x_2)(w^2(x_1) + 2w^2(x_2)) \quad (13.33)$$

For greater ease in discussion, $w(x_1)$ is set to be equal to $\alpha w(x_2)$. This substitution leads to

$$\frac{x_1}{20 \times x_2}\sqrt{\alpha^2 + 1} > w(x_2)$$
$$\text{and } \frac{\sqrt{\alpha^2+1}}{20\sqrt{\alpha^2+2}} > w(x_2) \quad (13.34)$$

Equation 13.34 yields, for the limiting cases $\alpha \ll 1$ and $\alpha \gg 1$,

$$\alpha \ll 1 : \frac{x_1}{20 \times x_2} > w(x_2) \text{ and } \frac{1}{20\sqrt{2}} > w(x_2),$$
$$\alpha \gg 1 : \frac{x_1 \alpha}{20 \times x_2} > w(x_2)$$
$$\text{and } \frac{1}{20\sqrt{2}} > w(x_2) \quad (13.35)$$

where the expressions following the colon reflect the conditions for neglecting changes in the expectation of Y and the expressions following the "and" those for neglecting changes in the uncertainty for y. Equation 13.35 shows that for the model considered here, second-order terms ought to be taken into account if $w(x_2)$ is greater than $(20 \times \sqrt{2})^{-1}$, irrespective of the value of α.

The above discussion becomes very tedious for more complicated model functions. A generally applicable approach is to employ Monte Carlo techniques, cf. Sections 13.6 and 13.6.4 in particular.

Using the same samples $\xi_{1,r}$ and $\xi_{2,r}$, where r is the sample number, one computes

$$\eta_{\text{lin},r} = \frac{x_1}{x_2} + \sum_{i=1}^{2} c_i(\xi_{i,r} - x_i),\ \eta_{2,r}$$
$$= \sum_{i=1}^{2}\sum_{i'=1}^{2} c_{i,i'}(\xi_{i,r}-x_i)(\xi_{i',r}-x_{i'}) \text{ and } \eta_r$$
$$= \frac{x_1 + (\xi_{1,r} - x_2)}{x_2 + (\xi_{2,r} - x_2)} \quad (13.36)$$

where $\xi_{i,r}$ is a random sample of the corresponding PDF in standard form multiplied by $u(x_i)$; for details on computing expectation and variance of Y_{lin}, Y_2, and Y see Section 13.6.3. The results are not masked by statistics and therefore allow to decide whether the linear model is sufficiently accurate.

13.4 Propagation of PDFs and Uncertainties

13.4.1 Markov Formula

The model $Y = f(X_1, \ldots, X_N)$ that relates the input quantities to the output quantity is given. For greater ease in writing, we use the vector notation: $\boldsymbol{X} = (X_1, \ldots, X_N)^{\text{T}}$. This allows construction of the CDF for the output quantity Y by summing up the probabilities for all possible combinations of possible values of $\boldsymbol{X}$, that is, $\boldsymbol{\xi} = (\xi_1, \ldots, \xi_N)^T$, under the constraint that $\eta \le f(\boldsymbol{\xi})$. To express this formally, we use the Heaviside step function:

$$G_Y(\eta) = \int_{-\infty}^{\infty} g_X(\xi) H(\eta - f(\boldsymbol{\xi})) d\boldsymbol{\xi},\ H(z) = \begin{cases} 1, z \ge 0, \\ 0, \text{otherwise.} \end{cases} \quad (13.37)$$

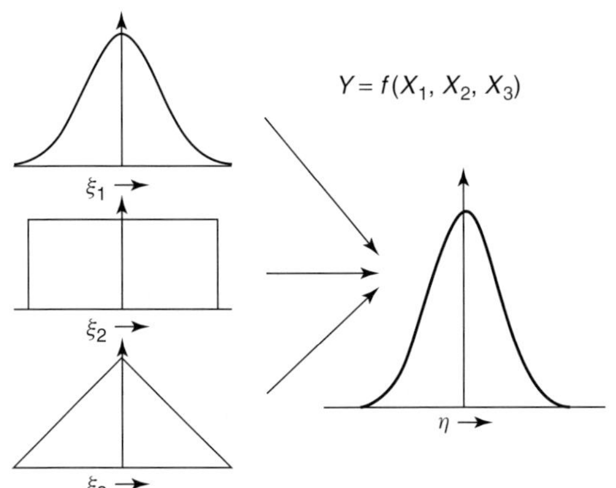

Fig. 13.7 The knowledge of possible values ξ_i of each input quantity X_i is expressed by a PDF and the knowledge of their inter-relation with the values η of the output quantity (measurand) Y is expressed by the model function f. The linear combination of many PDFs tends asymptotically to a Gaussian. In this figure it is assumed that the PDF for Y is a Gaussian.

The derivative of $H(z)$ is the Dirac delta function $\boldsymbol{\delta}(z)$; therefore,

$$\begin{aligned}\frac{\mathrm{d}}{\mathrm{d}\eta}G_Y(\eta) &= g_Y(\eta)\\ &= \int_{-\infty}^{\infty} g_X(\xi)\frac{d}{d\eta}H(\eta - f(\xi))d\boldsymbol{\xi}\\ &= \int_{-\infty}^{\infty} g_X(\boldsymbol{\xi})\delta(\eta - f(\boldsymbol{\xi}))\mathrm{d}\boldsymbol{\xi}\end{aligned} \tag{13.38}$$

The highlighted parts represent the so-called *Markov formula*. Thus, the PDF for the output quantity is determined by the PDFs for the input quantities via the Markov formula. The formula is valid for more than one output quantity, too.

Figure 13.7 illustrates the Markov formula. The integral in the formula encompasses all possible combinations (ξ_1, ξ_2, ξ_3) and produces a contribution with the statistical weight $g_X(\xi_1, \xi_2, \xi_3)$ to $\eta = f(\xi_1, \xi_2, \xi_3)$. Figure 13.7 shows this integration schematically as a propagation of the PDFs of the input quantities to the PDF for the output quantity, which results under the constraint formulated by the δ-function. An algorithm to carry out this propagation is described in Section 13.6.

Figure 13.8, using a Gaussian PDF, summarizes the properties of a PDF for use in uncertainty evaluation. The definitions given in relations (Equations 13.5–13.10) apply to $g_Y(\eta)$. The PDF for Y is not needed

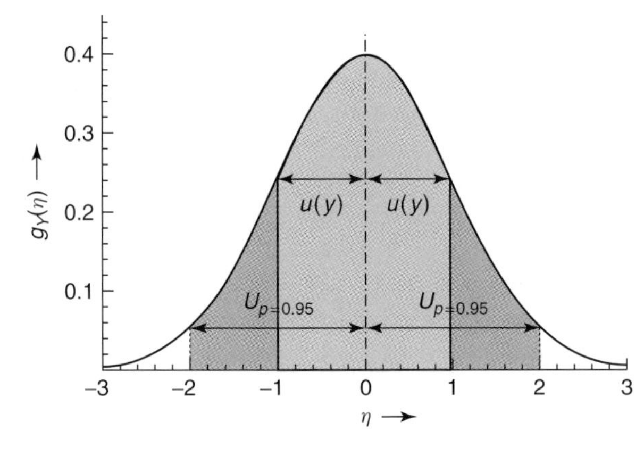

Fig. 13.8 The PDF for Y as obtained in Figure 13.7 encodes the knowledge of possible values of Y. The expectation of Y is taken as the best estimate y of Y and the square root of the variance as the standard uncertainty $u(y)$ associated with y. The expanded uncertainty U_p is here the half-width of an interval that covers the value of Y with a stated probability (here 95%).

explicitly for the determination of y and $u(y)$, since they can be obtained directly by integrating formula over η:

$$\begin{aligned} y &= \int_{-\infty}^{\infty} g_Y(\eta)\eta d\eta \\ &= \int_{-\infty}^{\infty} g_X(\boldsymbol{\xi}) f(\boldsymbol{\xi}) d\boldsymbol{\xi} \quad \text{and} \\ u^2(y) &= \int_{-\infty}^{\infty} g_Y(\eta)(\eta - y)^2 d\eta \\ &= \int_{-\infty}^{\infty} g_X(\boldsymbol{\xi})(f(\boldsymbol{\xi}) - y)^2 d\boldsymbol{\xi} \end{aligned} \tag{13.39}$$

If the model function $Y = f(X)$ is linear and if all x_i and $u(x_i)$ are given, then Equation 13.39 yields the results obtained in Equations 13.23 and 13.24 for y and Equation 13.25 for $u(y)$. If in addition the PDFs for all X_i are Gaussians, then the PDF for Y is also a Gaussian for which the coverage intervals for any coverage probability are known. In all other cases coverage intervals can be determined if $g_Y(\eta)$ has been computed.

In Section 13.6, methods for computing $g_Y(\eta)$ for any model and any physically meaningful PDFs for the input quantities will be given. However, if the number of input quantities becomes very large, statements on $g_Y(\eta)$ can be made with the help of CLTs.

13.4.2
Sampling and Central Limit Theorem

The GUM uses the term Type A evaluation of uncertainty if the knowledge about a quantity X is obtained from a usually small number M of measurements. This is a relict of the error approach. The PDF derived for that quantity depends on the model of the distribution from which these measurements are "*sampled*" and the GUM assumes in general a Gaussian. The measurement is first characterized by

$$\begin{aligned} \tilde{x} &= \frac{1}{M}\sum_{r=1}^{M} \xi_r \quad \text{and} \\ s_x^2 &= \frac{1}{M-1}\sum_{r=1}^{M} (\xi_r - \tilde{x})^2 \end{aligned} \tag{13.40}$$

where the ξ_r are the measured values; throughout this paper, and the tilde is used to indicate results obtained by a finite number of random samples of values of quantities. A Bayesian analysis (Lira and Wöger, 2006) leads to a relation between the PDF for X and the Student or t-distribution T with a degree of freedom $\nu = M - 1$:

$$\begin{aligned} T &= \frac{\sqrt{M}}{s_x}(X - \tilde{x}), \\ g_{T,\nu}(\tau) &= \frac{\Gamma(M/2)}{\sqrt{\pi\nu}\Gamma(\nu/2)}\left[1 + \frac{\tau^2}{\nu}\right]^{-M/2} \quad \text{with} \\ \tau &= \frac{\sqrt{M}}{s_x}(\xi - \tilde{x}) \end{aligned} \tag{13.41}$$

Since $ET = 0$ and $\mathrm{Var}T = \nu/(\nu - 2)$, it follows that $EX = \tilde{x}$ and $u^2(\tilde{x}) = (s_x^2/M)\cdot\mathrm{Var}T$.

In the Bayesian approach, s_x^2/M is merely a practical abbreviation, whereas in classical statistics it is taken as an estimator for the variance of X. The GUM still follows classical statistics in this case and lists values of the coverage factor k_p for a coverage probability $p = 0.95$ that are to be multiplied with $s_x/\sqrt{M}$ to obtain the expanded uncertainty. The values of $k_{0.95}$ for ν ranging from 1 through 13 are

ν	1	2	3	4	5	6	
$k_{0.95}$	12.71	4.30	3.18	2.78	2.57	2.45	
	7	8	9	10	11	12	13
	2.36	2.31	2.26	2.23	2.20	2.18	2.16

For more values see ISO (1993).

Bayesian analysis for repeated measurements assuming a rectangular or a Poissonian PDF is available(Lira and Wöger, 2006). For any PDF X "effective" coverage factors can be obtained by using Monte Carlo techniques. One takes a sufficiently large number of $M = \nu + 1$ samples and computes their mean values $\tilde{x}_r$, $r \in [1, M]$. These M values are then sorted to form a frequency distribution that converges asymptotically to the CDF for $X_{M \in X}$ that allows computing coverage intervals for any desired coverage probability.

Apart from few exceptions, the GUM assumes implicitly that the PDF for the output quantity Y as given by a weighted sum of input quantities is a Gaussian. This is only correct if the PDFs for all input quantities are Gaussians. However, if many input quantities are summed, one can infer from the CLT that this is asymptotically true.

On the basis of different assumptions, there exist several formulations of the CLTs. In metrology, one can assume that the PDFs for measured or computed quantities have finite expectations x_i, finite variances $u^2(x_i)$ and finite third moments. If these assumptions hold, then, following a textbook (Burry, 1999) but transcribing to GUM terminology, one can state the following CLT:

$$Y = \sum_{i=1}^{N} c_i X_i \text{is } \textit{asymptotically} \text{ normally}$$

$$\text{distributed with } y = \sum_{i=1}^{N} c_i x_i \text{ and}$$

$$u^2(y) = \sum_{i=1}^{N} c_i^2 u^2(x_i) \tag{13.42}$$

Thus, it is reasonable to assume a Gaussian for the sum of many input quantities, irrespective of their PDFs. This is the rationale behind the standard GUM procedure.

If correlations between input quantities are relevant, the last relation in Equation 13.42 is to be replaced by

$$u^2(y) = \sum_{i=1}^{N} c_i^2 u^2(x_i) + 2 \sum_{i=1}^{N} \sum_{i'=i+1}^{N} \times c_i u(x_i) r(x_i, x_{i'}) c_{i'} u(x_{i'}) \tag{13.43}$$

where $r(x_i, x_{i'})$ is the correlation coefficient as defined in Equation 13.10. Equation 13.43 governs the standard GUM procedure.

13.5 Standard GUM Procedure

13.5.1 Overview

The *standard GUM Procedure* is laid down in the "GUM", (ISO, 1993). A major goal of the authors of the GUM was to replace different existing approaches to uncertainty evaluation by *one* internationally accepted procedure. To achieve broad acceptance, the GUM is focused on simple and clear procedures for practical applications. It treats mainly linear model functions or model functions for which their first-order Taylor expansion provides an acceptable approximation within the range spanned by $x_i \pm u(x_i)$; for more details see Section 13.6.4. The standard GUM procedure, therefore, propagates *not the PDFs* for the input quantities, *but the uncertainties* associated with their expectations. This is justified for linear models as shown in Equation 13.42.

Given the sensitivity coefficients c_i, the expectations x_i and the uncertainties $u(x_i)$ associated with them, one can compute the expectation y and the combined uncertainty $u(y)$ associated with it, see Equations 13.23 and 13.24 for y and Equation 13.25 for $u(y)$. If the PDFs for all input quantities are Gaussians, then the PDF for the measurand Y as given by a linear model is Gaussian, too, and the expanded uncertainty can be computed using the rounded coverage factor $k_{p=0.95} = 2$; the exact value is 1.96. If some or even all of the PDFs for the input quantities are not Gaussians, then the CLT, see Equation 13.42, allows approximation of the PDF for the measurand Y by a Gaussian, if the number of input quantities N is *sufficiently* large and if for any non-Gaussian $c_i u(x_i)$ is *considerably* smaller than $u(y)$. The vague terms "*sufficiently*" and "*considerably*" are discussed below.

If one or more quantities were subject to Type A evaluation, then one computes the effective degree of freedom ν_{eff} using the Welch–Satterthwaite formula as the approximation:

$$\nu_{\text{eff}} = \frac{u_c^4(y)}{\sum_{i=1}^{N} \nu_i^{-1} u_i^4(y)} \quad \text{where}$$
$$u_c^2(y) = \sum_{i=1}^{N} c_i^2 u^2(x_i) \text{ and } u_i(y) = c_i u(x_i) \tag{13.44}$$

The resulting expanded uncertainty is then given by $U = k_{p=0.95}(\nu_{\text{eff}}) \cdot u(y)$. In general, uncertainties and the expanded uncertainty are given only up to two significant digits, that is, with a "resolution" of 5%. See Section 13.7.3 for a discussion on correlated repeatedly measured quantities.

The standard GUM procedure can also be summarized as a step-by-step procedure as shown in Figure 13.9. The first three steps, especially the second one, are in practice quite demanding, whereas the steps 4, 5, and the first part of step 6

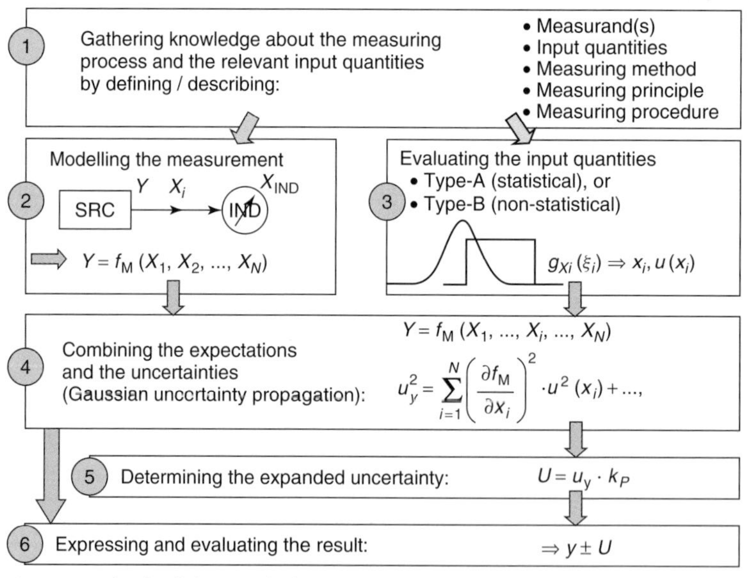

Fig. 13.9 Sketch of the standard GUM procedure as a step-by-step procedure. The abbreviations used in step 2 stand for source (SRC) and indication (IND); the value read from an analog instrument or displayed by a digital instrument is here called an *indication*.

(expressing the result) can easily be performed using an appropriate software. Commercial software is available; however, anyone acquainted with FORTRAN, or any other higher programming language, will find it easy to encode the standard GUM procedure. EXCEL is often used, too. It is versatile, especially so in conjunction with embedded routines written in a higher programming language. The second part of step 6 is especially important in the process of designing or optimizing an experiment or a measurement procedure.

13.5.2 Description of Input and Output Quantities and Their Inter-relation

This subsection summarizes the first three of the six steps depicted in Figure 13.9. The input and output quantities for the *running example* and their inter-relation, that is, the model for evaluating uncertainty, have been described in Section 13.1. It is a good practice to explicitly document the sources of information for each quantity in a table. As far as the readings of instruments or the values of standards are concerned, it is practical to refer to corresponding chapters in the quality management document. For each calibrated instrument used, this document should include the name of the calibration laboratory, the date of calibration, and, if given, information on possible drifts. If the drift is significant then it should be contained in the model. Analogously, one would note the source, that is, a standards laboratory, a handbook, or other literature, for standards or constants.

In any case, one should compile a table that contains the so-called uncertainty budget. That table lists all input and output quantities. It is helpful to express these in GUM notation. Table 13.1 pertains to the submodel $L_{edge,1}$ in the running example,

Tab. 13.1 Uncertainty budget for the quantity $L_{edge,1}$ – Part I

Quantity	Symbol		Best estimate	SI unit	Uncertainty	PDF[a]
	Physics	GUM				
Reading	$l_{lower,1}$	$l_{IND,1}$	10	mm	constant	–
Resolution of reading	$\delta l_{RES,lower,1}$	X_1	0^b	mm	0.25/3	R
Reading	$l$upper,1	$l_{IND,2}$	20	mm	constant	–
Resolution of reading	$\delta l_{RES,upper,1}$	X_2	0^b	mm	$0.25/\sqrt{3}$	R
Measurement deviation	Δl_{CAL}	X_3	0	mm	$10^{-2} \times l_{IND}$	G
Temperature (ruler and plate)	T	X_4	21^c	°C	$2/\sqrt{3}^c$	R
Thermal expansion coefficient	α_{Teflon}	${\alpha_T}^d$	6.8×10^{-5}	1/°C	constant	–
Thermal expansion coefficient	$\alpha_{aluminum}$	${\alpha_A}^d$	2.3×10^{-5}	1/°C	constant	–
Angular deviation	B_1	X_5	[e]	[e]	[e]	R
Length of edge 1	$L^{edge,1}$	Y	Y	mm	$u(y)$	?

Note: The values entered are not all realistic; they have been chosen here for ease in computation or for discussion of possible simplifications. In the last column, headed "PDF," R denotes a rectangular and G a Gaussian PDF.

[a] For Type A evaluations one enters t (Student distribution) and the degree of freedomν.
[b] In general, the expectation of the resolution of an instrument has the numerical value 0.
[c] Only introduced for use in Part II of the budget table.
[d] Assumed mean temperature T in the laboratory; T is kept within ±2 °C.
[e] These values will be discussed in Section 13.5.3.

cf. Equation 13.4. It lists the quantities, and their symbols used in the normal context and for ease in GUM convention. Furthermore, it lists the numerical values for the best estimates (expectations), the SI unit used, the numerical values of the uncertainties associated with the best estimates, and the PDF assigned for the quantity. The column "source of information" is omitted for lack of space.

The model function for $Y = L_{\text{edge},1}$ as given in Equation 13.2 can be expressed using the symbols in GUM notation:

$$Y = \frac{(l_{\text{IND},2} - l_{\text{IND},1} + X_2 - X_1 + X_3(^*l_{\text{IND},2}) - X_3(^*l_{\text{IND},1})) \cdot (1 + (2.3 \times 10^{-5}/^\circ\text{C}) \cdot (X_4 - 21^\circ\text{C})) \cdot \cos X_5}{1 + (6.8 \times 10^{-5}/^\circ\text{C}) \cdot (X_4 - 20^\circ\text{C})} \tag{13.45}$$

where $^*l_{\text{LIND},i} = l_{\text{LIND},i} + X_i$, $i = 1, 2$. Since $x_i = 0\,\text{mm}$, $^*l_{\text{LIND},i} = l_{\text{LIND},i}$ as far as the standard GUM procedure is concerned; however, when using the MCM one would take $^*l_{\text{LIND},i} = l_{\text{LIND},i} + \xi_{i,r}$, where $\xi_{i,r}$ is a sample of X_i, cf. Section 13.6.

13.5.3 Computation, Reporting, and Analysis

This subsection summarizes the last three steps shown in Figure 13.9. To *compute the uncertainty* $u(y)$ one needs to compute the sensitivity coefficients for all input quantities, cf. Section 13.3.3:

$$c_1 = -1\frac{(1+(2.3 \times 10^{-5}/^\circ\text{C}) \cdot (x_4-21^\circ\text{C})) \cdot \cos x_5}{(1+6.8 \times 10^{-5}/^\circ\text{C}) \cdot (x_4 - 20^\circ\text{C})} = \frac{-\cos x_5}{1 + 6.8 \times 10^{-5}}$$
$$\cong -(1 - 6.8 \times 10^{-5}) \cdot \cos x_5 \tag{13.46}$$

the same procedure yields $c_2 = (1 - 6.8 \times 10^{-5}) \cdot \cos x_5$.

As stated above, $^*l_{\text{LIND},i}$ is simplified to $l_{\text{LIND},i} + X_i$. This is justified here, since X_3 depends linearly on l_{IND} and the PDFs for X_1 and X_2 are symmetric about their expectations and $x_1 = x_2$. Simply taking the first partial derivative with respect to X_3 would yield $c_3 = 0$, however this would be false. Since the PDF for X_3 is rectangular, it can be expressed as $X_3 = x_3 + 10^{-3} \times l_{\text{IND},i} X_{R,\text{STD}}$. Therefore, one can replace the terms $X_3(l_{\text{IND},2}) - X_3(l_{\text{IND},1})$ by $X_{3,*}$ and set $u(x_{3,*}) = 10^{-2} \times |20 - 10\,\text{mm}|$. Within the approximations made for c_1 and c_2 one obtains then $c_{3,*} = (1 - 6.8 \times 10^{-5}) \cdot \cos x_5$.

To assess c_4 one can, using implicitly a first-order Taylor series, remove the nominator in Equation 13.45:

$$\frac{1 + 2.3 \times 10^{-5}/^\circ\text{C}) \cdot (X_4 - 21^\circ\text{C})}{1 + 6.8 \times 10^{-5}/^\circ\text{C}) \cdot (X_4 - 20^\circ\text{C})}$$
$$\cong 1 - (4.5/^\circ\text{C}) \cdot (X_4 - 21^\circ\text{C}) - (6.8/^\circ\text{C})) \times 10^{-5} \tag{13.47}$$

Using this linearization with respect to the input quantity X_4 (temperature) one obtains $c_4 = -\Delta l_{\text{IND}} \times 4.5 \times 10^{-5}(^\circ\text{C})^{-1} \cos x_5$, that is, the difference of the thermal expansion coefficients multiplied with the product of Δl_{IND} and $\cos x_5$;

where for ease in writing $l_{IND,2} - l_{IND,1}$ is expressed as Δl_{IND}.

The assessment of c_5 is somewhat complicated within the standard GUM procedure. The first derivative of $\cos X_5 = \sin X_5$ and $\sin x_5 = 0$. Therefore, the GUM requests to use the second-order term of the Taylor expansion, that is, $-1/2\cos X_5$ and $\cos x_5 = 1$. However, the computation of the uncertainty requires then to evaluate the integral over $(\cos\xi - 1)^4$. There are two alternatives to circumvent this problem. One alternative is to consider $\cos B$ as uniformly distributed between $\cos\beta_{max}$ and $\cos 0 = 1$, noting that $\cos\beta$ is symmetric about $\beta = 0$. Assume $\beta_{max} = 0.1$ rad, then $\cos\beta_{max} = 0.995$ and the expectation is 0.9975. Therefore, one may replace $\cos X_5$ by $X_{5,*}$ and use $x_{5,*} = 0.9975$ and $u(x_{5,*}) = 0.0025/\sqrt{3}$, and $c_{5,*} = \Delta l_{IND}\cdot(1 - 6.8 \times 10^{-5})$. The other alternative is to approximate $\cos X_5$ by the first two terms of the corresponding Taylor series. This yields $\cos X_5 = 1 - 1/2X^2$, and hence, for a rectangular PDF

$$\mathrm{E}\cos X \cong \mathrm{E}\left[1-\frac{1}{2}X^2\right]=1-\frac{1}{2}u^2(x) \text{ and}$$
$$\mathrm{Var}X^2 = 0.8 \times u^4(x) \quad (13.48)$$

$x_{5,*} = 1 - 1/2 \times (0.1/\sqrt{3})^2 = 0.9983$ and $u(x_{5,*}) = 0.0016$. As the first alternative is more widely used, it is selected here too.

The expectation of Y, that is, y, can now be computed by inserting the expectations of the X_i, that is, x_i in Equation 13.45: Using all the above approximations in Equation 13.45 yields

$$\mathrm{E}Y = y = \Delta l_{IND}\frac{1}{(1 - 6.8 \times 10^{-5})}0.9975$$
$$= 9.97568\,\text{mm} \quad (13.49)$$

These results are now added to the budget table and used for computing $u(y)$. Table 13.2 shows this part of the budget table omitting the columns "Quantity" and "Symbol Physics." The added columns are "sensitivity coefficients c_i," "$u(y_i) = c_i u(x_i)$," and the so-called uncertainty index, "$u^2(y_i)/u^2(y)$." The resulting combined uncertainty is rounded off to two significant digits and the resulting best estimate is rounded off accordingly.

Finally, the uncertainty $u(y)$ can now be computed using Equation 13.27. Since all correlation coefficients have the value zero, one obtains

$$u^2(l_{edge,1}) \equiv u^2(y) = c_1^2u^2(x_1) + c_2^2u^2(x_2)$$
$$+ c_{3,*}^2u^2(x_{3,*}) + c_4^2u^2(x_4)$$
$$+ c_{5,*}^2u^2(x_{5,*}) \quad (13.50)$$

The next task is to determine the *expanded uncertainty*, that is, to perform step 5. Without further calculation one does not know the PDF for Y. Since repeated measurements have not been used, there is no need to compute the effective degree of freedom ν_{eff}, cf. Equation 13.44, and a Gaussian PDF for Y is assumed. Therefore, the coverage factor has the value 2.0. Next, the first part of step 6 is to state the result for the measurement using the ruler:

The best estimate of the "length of edge #1" is 9.98 mm, assuming a Gaussian PDF; the value of the "length of edge #1" is contained in the interval [9.52, 10.44 mm] with a (coverage) probability of $p = 0.95$.

The *analysis of the results*, that is, the second part of step 6, is supported by inspecting the uncertainty index, that is, the last column in Table 13.2. The contribution of the resolution of the ruler readings accounts for 80% of the combined uncertainty. Thus in practice, if the resulting uncertainty cannot be accepted, one would use for

Tab. 13.2 Uncertainty budget for the quantity $L_{edge,1}$ – Part II.

GUM symbol for quantity	Best estimate	SI unit	Uncertainty	PDF*	Sensitivity coefficients	$u(y_i) = c_i u(x_i)$	$u^2(y_i)$	$u^2(y_i)/u^2(y)$
$l_{IND,1}$	10	mm	Constant	–	–	–	–	
X_1	0	mm	$0.25/\sqrt{3}$	R	−0.99743	−0.143987	0.020726	0.402
$l_{IND,2}$	20	mm	Constant	–	–	–	–	
X_2	0	mm	$0.25/\sqrt{3}$	R	0.99743	0.143987	0.020726	0.402
$X_{3,*}$	0	mm	$10 - 2 \times \Delta l_{IND}$	G	0.99743	0.099757	0.009895	0.192
$X_{4,*}$	21	°C	$2/\sqrt{3}$	R	-4.488×10^{-4}	0.000518	$>10^{-6}$	$>10^{-5}$
α_T	6.8×10^{-5}	1/°C	Constant	–	–	–	–	–
α_A	2.3×10^{-6}	1/°C	Constant	–	–	–	–	–
$X_{5,*}$	0.9975	–	$0.0025/\sqrt{3}$	R	9.9932	0.014435	0.000208	0.004
Y	9.98	mm	0.23	?	–	$0.0025/\sqrt{3}$	0.051624	–

instance a micrometer instead of a ruler. The influence of the different thermal expansion coefficients and of the angle between the ruler and the edge are negligible. This seems to justify the used approximations.

The ruler is now exchanged by micrometer. Assume that the resolution is exactly twice as good, that is, $u(x_1) = u(x_2) = 0.125/\sqrt{3}$. This results in $y = 9.976$ and $u(y) = 0.143$. Then, the computation is repeated and the new result, now deemed acceptable, is stated:

The best estimate of the "length of edge #1" is 9.976 mm, assuming a Gaussian PDF; the value of the "length of edge #1" is contained in the interval [9.690, 10.262 mm] with a (coverage) probability of $p = 0.95$.

13.5.4 Limitations

Although the running example is simple, some approximations were needed to achieve linearity, and since two input quantities with rectangular PDFs dominate the combined uncertainty, one could expect a significantly non-Gaussian PDF for Y, the shape of which cannot be computed within the standard procedure of the GUM.

Furthermore, the linearization of the model for area of the Teflon plate, see Equation 13.4, would require even more approximations, and the number of sensitivity coefficient to be calculated would increase and the correlation because of the common measurement deviation might even require using higher-order terms of the Taylor series. This would lead to fairly complicated and error-prone computations.

Finally, the standard GUM procedure does not treat explicitly models for more than one output quantity. Assistance can be found in a German standard (DIN, 1999) and the book by Weise and Wöger, see further reading, which is also written in German. However, in the most simple case one can determine functions such that $Y_j = f_j(\boldsymbol{X}), j \in [1, m]$. If all $f_i(\boldsymbol{X})$ are linear functions, one can use the standard procedure of the GUM to obtain the $u(y_j)$, and the correlation coefficients $r(y_j, y_{j'})$ are given by

$$r(y_j, y_{j'}) = \frac{\mathrm{Cov}(f_j(\boldsymbol{X}) f_{j'}(\boldsymbol{X}))}{u(y_j)u(y_{j'})} \qquad (13.51)$$

If the standard GUM procedure is not applicable or if its use is very complicated, one has to resort to numerical or analytical procedures (Cox and Siebert, 2006b). The most versatile numerical method is MCM as it can be used generally and not only in special cases, as is the case with virtually all other methods such as for instance convolution using fast Fourier transforms, see Korczynski M.J., Cox M. G and Harris P. M. (2006).

When using MCM, a linearization of the model is not necessary and the correct computation of the expanded uncertainty, irrespective of the resulting PDF for Y, is performed easily, cf. Equation 13.14. MCM is applicable even if it is not possible to provide a differentiable model function, for example, Goydke, Siebert, and Scholl (2003), and in general the most expeditious approach if the modeling requires mathematical operations such as a matrix inversion, for example, Tegeler, Heyer, and Siebert (2008).

Furthermore, using MCM allows solving multivariate problems elegantly, and it provides useful tools to examine approximations such as those discussed above for treating the temperatures of the ruler and the plate, and the angle between the ruler and the edge of the plate.

13.6
Monte Carlo Method

13.6.1
Overview

The MCM is an effective and versatile tool for determining the PDF for the measurands. This method provides a consistent Bayesian approach to the evaluation of uncertainty. It allows using prior information about the output quantities (Elster, 2007). In essence, MCM simulates repeated measurements under strict repeatability conditions, by combining possible values of the input quantities and obtaining possible values of the output quantities via the given model. Although this method is in principle straightforward, care is required in representing and validating the results obtained using the method. However, as the number of samples is usually large, the CLT can be used to infer the convergence of this procedure.

Knowledge about the input quantities is acquired in the same manner as with the standard GUM procedure. However, since sensitivity coefficients are not needed, models for describing the inter-relation between the input quantities need not be given in the closed form. The only requirement is a procedure to infer *one and only one* set of values of the output quantities from any set of possible values of the input quantities. However, a given set of output quantities may result from various sets of input quantities.

Using a *pseudorandom number generator*, the possible values of the input quantities $\xi_{i,r}$ are sampled from the PDFs assigned to these quantities and combined to yield possible values of the output quantities η_r. These values can be sorted to form a frequency distribution that, upon normalization, is expected to converge to the PDF for the output quantities. The usual statistical techniques are then used to infer the expectations of the output quantities and the uncertainty matrix for these estimates. In the case of one or two output quantities, MCM can compute the shortest coverage intervals or smallest coverage regions.

The remainder of this section provides a brief introduction only; details are found in Cox and Siebert (2006b) and JCGM (2008).

13.6.2
Representative Draws (Sampling)

The basis for any Monte Carlo code is a generator that produces pseudorandom numbers ρ_r that are statistically uniformly distributed in the interval (0, 1). The period $M_{\max}$ (the total number of draws that can be made before the starting point of the sequence is reached again) should be as large as possible, and the sequence should not possess any autocorrelation. If this requirement is fulfilled, then statistically independent sub-sequences can be generated.

Figure 13.10 shows two basic sampling techniques for achieving representative draws. *Weighted* sampling covers the selected range of arguments uniformly such that the uncertainty of the resulting frequency distribution is the same for each bin. However, unless the range of arguments is finite, as is the case for instance for a rectangular PDF, great care is needed in selecting range boundaries. *Analog* sampling asymptotically covers the full range and is easier to handle, as the statistical weight of every sample is equal to M^{-1}; where M is the number of samples drawn. For many PDFs routines exist that generate analog samples (Cox and Siebert, 2006b). Weighted sampling and analog sampling are identical in the case

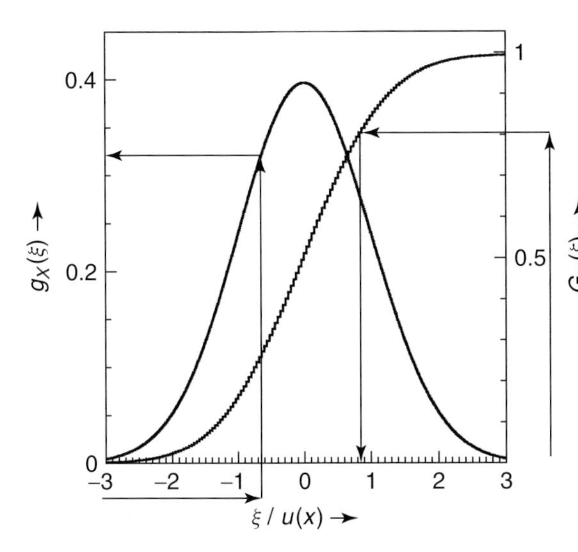

Fig. 13.10 Illustration of sampling methods. Left part of the figure, *weighted* sampling: (1) Select random number $\rho \in [0, 1]\xi = x_{\min} + \rho(x_{\max} - x_{\min})$, (2) $w(\xi) = g_X(\xi)$. Right part of the figure, *analog* sampling: (1) Select random number $\rho \in [0, 1]$, $G_X(\xi) \equiv \rho$, (2) $\xi = G_X^{-1}(\xi)$.

of a rectangular PDF. Both techniques are representative as the frequency distributions achieved asymptotically approach the corresponding PDFs. A frequency distribution is represented by the normalized content of K adjacent intervals, termed *bins*. The boundaries of these bins are usually equidistant: $x_k = x_{\min} + (k-1) \cdot \Delta x$; where $\Delta x = (x_{\max} - x_{\min})/K$. Any ξ_r obtained as weighted sample of the PDF for X can be sorted into one and only one kth bin by requesting $x_k \leq \xi_r < x_{k+1}$ for $k < K - 1$ and $x_k \leq \xi_r \leq x_{\max}$ for $k = K$.

Denoting the content of the kth bin by $\tilde{g}_k$, one obtains with weighted sampling M_k contributions to the kth bin and

$$\begin{aligned}\lim_{M_k \to \infty} \tilde{g}_k &= M_k^{-1} \sum_{r=1}^{M_k} g_X(\xi_r) \\ &\to \int_{x_k}^{x_{k+1}} g_X(\xi) d\xi \\ &= G_X(x_{k+1}) - G_X(x_k); \\ \lim_{M \to \infty} M_k &= \frac{K}{M} \end{aligned} \tag{13.52}$$

With *analog sampling* one can use the same bin structure with the additional convention that x_1 is set to $-\infty$ and x_{K+1} is set to ∞. The number of contributions to the kth bin is given by

$$\begin{aligned}\lim_{M \to \infty} M_k &= M(G_X(x_{k+1}) - G_X(x_k)) \\ &\Rightarrow \lim_{M_k \to \infty} \tilde{g}_k = M^{-1} M_k \\ &= G_X(x_{k+1}) - G_X(x_k) \\ &= \int_{x_k}^{x_{k+1}} g_X(\xi) d\xi \end{aligned} \tag{13.53}$$

Figure 13.11 summarizes the MCM. The simple model $Y \equiv f(X) = X_1 + X_2 + X_3$ is used for demonstration. A Gaussian, a rectangular, and a triangular PDF are assigned to X_1, X_2, and X_3, respectively. Furthermore, $x_i = \mathrm{E}X_i = 0$ and $u(x_i) = (\mathrm{Var}X_i)^{1/2} = 1$. A representative sample of the possible values of the output quantity Y, that is, η_r, is then given by

$$\begin{aligned}\eta_r &= f(\xi_r) = \xi_{1,r} + \xi_{2,r} + \xi_{3,r}, \\ w_r &= \prod_{i=1}^{3} w_{i,r} \text{ for the weighted and} w_r \\ &= M^{-1} \text{ for the analog sampling.}\end{aligned} \tag{13.54}$$

The representative samples η_r can be used to generate a frequency distribution.

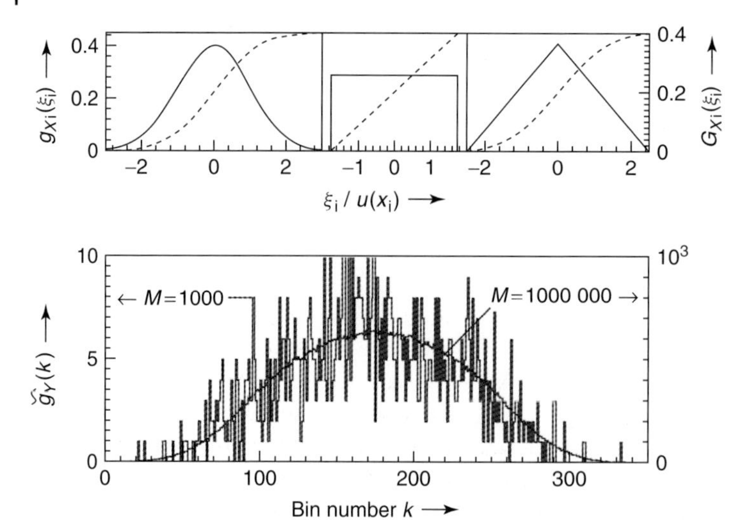

Fig. 13.11 Illustration of computing a frequency distribution. The upper graphs show a Gaussian, a rectangular and a triangular PDF and (right ordinate) the corresponding CDFs. Representative draws $\xi_{G,r}$, $\xi_{R,r}$, and $\xi_{D,r}$ are made and the corresponding model function formed. Doing so yields a representative draw, $\eta_r = f(\xi_{G,r}, \xi_{R,r}, \xi_{D,r})$, from the PDF for Y. Sorting the complete set of such draws yields a histogram representation of the frequency distribution via the relation $k = K(\eta_r - \eta_{\min})(\eta_{\max} - \eta_{\min})^{-1} + 1$. The bottom graph shows the resulting frequency distribution obtained for $M = 10^3$ and (right ordinate) $M = 10^6$ draws. Taken from Cox and Siebert (2006b).

In the same manner as shown for the input quantities, a bin structure is used and the content of the bin to which η_r belongs is enhanced by w_r. Normalization is obtained by dividing the content of each bin by the sum over all weights. This frequency distribution converges with increasing sample size to the PDF for the output quantity and contains all given information.

Analog sampling is preferable as it requires less effort in programming than weighted sampling. Therefore, only analog sampling will be considered below. More details on weighted sampling can be found in Siebert (2008).

In the case of *more than one output quantity* one simply takes samples of $\eta_r = (\eta_1, ..., \eta_m)$. While using MCM one does not need linear functions, and the only requirement is, as mentioned above, that for any set of samples of the input quantities one and only one set of output quantities is produced. In general, one can generate a frequency distribution for each output quantity that converges with increasing sample size to the marginal PDF for the output quantity considered.

If all or some of the input quantities are *correlated*, then one can use a Cholesky decomposition of the uncertainty matrix $\boldsymbol{U}_x$ if the PDFs for all correlated input quantities are Gaussians. If all $u(x_i)$ and the correlation coefficients $r(x_i, x_j)$ are given, then the uncertainty matrix can be written as

$$\boldsymbol{U}_x = \begin{pmatrix} u^2(x_1) & \mathrm{Cov}(X_1X_2) \ldots \mathrm{Cov}(X_1X_N) \\ \mathrm{Cov}(X_2X_1) & u^2(x_1) \ldots \mathrm{Cov}(X_2X_N) \\ \ldots\ldots\ldots\ldots\ldots & \\ \mathrm{Cov}(X_NX_1) & \mathrm{Cov}(X_NX_2) \ldots u^2(x_N) \end{pmatrix},$$

$$\mathrm{Cov}(X_iX_j) = u_i(x_i)r(x_i, x_j)u(x_j) \qquad (13.55)$$

The Cholesky decomposition is given by $U_x = R^T R$, where R is an upper triangular matrix. As an example consider two correlated input quantities:

$$R^T = \begin{pmatrix} u(x_1) & 0 \\ u(x_2) r(x_1, x_2) & u(x_2)\sqrt{1 - r^2(x_1, x_2)} \end{pmatrix}$$

$$= \begin{pmatrix} u(x_1) & 0 \\ u(x_2) \sin\alpha & u(x_2)\cos\alpha \end{pmatrix} \quad (13.56)$$

where $r(x_1, x_2) = \sin\alpha$ is used for greater ease in computation. For details on the decomposition algorithm see Appendix B in Cox and Siebert (2006b) and JCGM (2008). The samples of the two input quantities are then obtained by

$$\boldsymbol{\xi}_r = x + \boldsymbol{R}^T \boldsymbol{\xi}_{G,r} \Rightarrow \begin{pmatrix} \xi_{1,r} \\ \xi_{2,r} \end{pmatrix}$$

$$= \begin{pmatrix} x_1 + u(x_1)\xi_{1,G,r} \\ x_2 + u(x_2)\xi_{1,G,r}\sin\alpha \\ +u(x_2)\xi_{2,G,r}\cos\alpha \end{pmatrix} \quad (13.57)$$

where $\boldsymbol{\xi}_{G,r} = (\xi_{1,G,r}, \ldots, \xi_{N,G,r})^T$ and $\xi_{i,G,r}$ is a sample from a Gaussian with expectation zero and standard deviation one. The Use of correlation coefficients is inevitably paid for by the loss of information. "Good physics" should always try to identify the source of the correlation and include the corresponding quantities in the set of input quantities. Appendix B in Cox and Siebert (2006b) describes a method that supports this goal.

13.6.3 Computation, Reporting, and Analysis

When using analog sampling, one can compute $\tilde{y}$ and $\tilde{u}(y)$ as unbiased estimators for y and $u(y)$, respectively:

$$\tilde{y} = \sum_{r=1}^{M} \frac{1}{M}\eta_r \quad \text{and}$$

$$\tilde{u}^2(y) = \frac{1}{M-1}\left(\sum_{r=1}^{M} \eta_r^2 - M\,\tilde{y}^2\right) \quad (13.58)$$

Furthermore, from a set $\{\eta_r, r \in [1, M]\}$one can infer a coverage interval $[\tilde{y}_{M,\text{low}}, \tilde{y}_{M,\text{high}}]$ that contains the $p{\cdot}M$ values, where p is the coverage probability; the subscript M is necessary, as the procedures described next provide biased estimators if M is finite. One procedure is to examine the resulting frequency distribution, cf. Equation 13.14. Another possibility is to obtain the CDF by sorting the set $\{\eta_r, r \in [1, M]\}$ monotonously increasing such that the value of CDF($\eta_{r\text{-sort}}$) is given by r-sort/M. The latter method is recommended in JCGM (2008). Using a coverage probability of 0.95 one takes $\eta_{r\text{-sort}=M/40}$ as an estimate for $\tilde{y}_{M,\text{low}}$ and $\eta_{r\text{-sort}=M+1-M/40}$ as an estimate for $\tilde{y}_{M,\text{high}}$.

In the case of *more than one output quantity* one computes $\tilde{y}_j$ and $\tilde{u}(y_j)$ as in Equation 13.58 and

$$r(\tilde{y}_j, \tilde{y}_{j'}) = \frac{1}{M-1}\frac{\sum_{r=1}^{M}(\eta_{j,r} - \tilde{y}_j)(\eta_{j',r} - \tilde{y}_{j'})}{u(y_j)u(y_{j'})} \quad (13.59)$$

Furthermore, using the procedures described above, one can infer a coverage interval $[\tilde{y}_{j,M,\text{low}}, \tilde{y}_{j,M,\text{high}}]$ that contains $p{\cdot}M$ values of the corresponding marginal PDF. These intervals provide some information; however, the hypercube formed by the corresponding lengths of these intervals would not be very informative, since most of its volume would be void. Therefore, it is sometimes suggested to compute hyper ellipses instead since these would in most cases have a considerable smaller volume than the hypercube. In applied physics, unless safety aspects are an issue, the design aim for coverage region is to provide a region with the smallest possible volume. A well-suited procedure for the coverage region for the two output quantities is available (Cox and Siebert, 2006b).

It is important to clearly distinguish the uncertainty associated with y, estimated by $\tilde{u}(y)$, from the uncertainty associated with the computed value $\tilde{y}$ that according to the CLT vanishes asymptotically with increasing sample size M.

To determine the convergence of $\tilde{y} \rightarrow y$, $\tilde{u}(y) \rightarrow u(y)$, $\tilde{y}_{M,\mathrm{low}} \rightarrow y_{M,\mathrm{low}}$ and $\tilde{y}_{M,\mathrm{high}} \rightarrow y_{M,\mathrm{high}}$, the GUM Supplement 1 (JCGM, 2008) recommends computing H samples of these statistical functions. Each of the H samples is based on M possible values of Y. We consider only $\tilde{y} \rightarrow y$ explicitly, although the procedure shown can be applied to $\tilde{u}(y)$, $\tilde{y}_{M,\mathrm{low}}$ and $\tilde{y}_{M,\mathrm{high}}$, too.

The CLT, cf. Equation 13.42, guarantees that

$$y = \lim_{H\rightarrow\infty}\tilde{y}_H = \sum_{h=1}^{H}\frac{1}{H}\tilde{y}_h \text{ where}$$
$$\tilde{y}_h = \frac{1}{M}\sum_{r=1}^{M}\eta_{r,h} \text{ and } u^2(\tilde{y}_H) = \frac{1}{H^2}\sum_{h=1}^{H}u^2(\tilde{y}_h) \tag{13.60}$$

Analogous relations hold for $\tilde{u}(y)$, $\tilde{y}_{M,\mathrm{low}}$ and $\tilde{y}_{M,\mathrm{high}}$.

These convergence properties can be used to set up convergence criteria. A numerical result is deemed to have stabilized if twice the standard deviation associated with it is less than the numerical tolerance δ associated with the standard uncertainty $u(y)$ (JCGM, 2008). By convention, the uncertainty is generally stated with two digits (three digits are only used if the leading digit has the value 1). The value can be written as $z \times 10^{-l}$, for instance, cf. Table 13.2 – Part II, $u(y) = 0.23 \Rightarrow u(y) = 23 \times 10^{-2}$. The numerical tolerance δ is then given by $1/2 \times 10^{-l}$; that is, in our example, $\delta = 1/2 \times 10^{-2}$.

Sufficient convergence for $\tilde{y} \rightarrow y$ is reached if $2 \times u(\tilde{y}_H) < \delta$. In GUM Supplement 1 (JCGM, 2008), it is recommended to estimate $u(y)$ not by the average of the H values of the $u(\tilde{y}_h)$ but by

$$\tilde{y} = \frac{1}{HM}\sum_{h=1}^{H}\sum_{r=1}^{M}\eta_{r,h} \text{ and}$$
$$\tilde{u}^2(y) = \frac{1}{HM-1}\sum_{h=1}^{H}\sum_{r=1}^{M}(\eta_{r,h}^2 - HM\tilde{y}^2) \tag{13.61}$$

If the relative uncertainty, $\tilde{w}_H = \tilde{u}_H(y)/\tilde{y}$, is not considerably larger than the smallest value of the mantissa on the used computer, then the formula for $\tilde{u}(y)$ becomes numerically instable. For a numerically safe algorithm see Cox and Siebert (2006b).

The results of the Monte Carlo calculation are stated in the same manner as in the first part of step 6 of the GUM standard procedure; this step is therefore omitted here. However, the second part of step 6, that is, the evaluation and analysis, is somewhat different. First of all, MCMs produce the PDF for the output quantity. This PDF together with the corresponding Gaussian PDF, which would be assumed when using the GUM standard procedure, should be plotted and interpreted. Figure 13.12 shows such a plot for the quantity length of an edge. Clearly, the PDFs obtained with MCM are almost Gaussians, and as expected low resolution (ruler) leads to broader peaks than those of high resolution (micrometer).

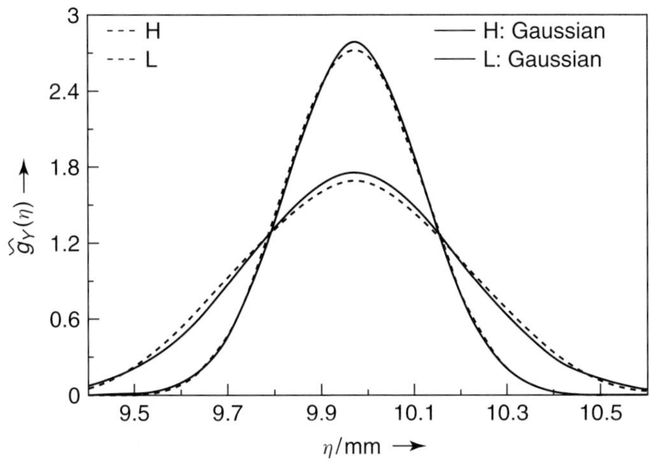

Fig. 13.12 Frequency distributions for possible values of the quantity "length of an edge" and (solid lines) corresponding Gaussian PDFs. H pertains to high and L to low resolutions of the length measurements. The broader curves with lower peak values pertain to L.

MCMs are also perfect tools for studying different assumptions and models. An example for studying the influence of correlation is shown in Figure 13.13. In one calculation, denoted by C, the correlation of the length measurements using either the *same* ruler, denoted by L (low resolution), or the *same* micrometer, denoted by H (high resolution) for either edge length measurement was accounted for, and in another calculation, denoted by U, this correlation was not considered. Inspection of Figure 13.13 shows that the correlation enhances the uncertainty if positively correlated quantities, that is, $r(x_i, x_j) > 0$, appear as products in the model; enhancement would also result from a sum, but a difference or a ratio would lead to a decreased uncertainty. The opposite holds if two quantities are negatively correlated.

Another example is modeling of the nonparalleled ruler and the edge. The PDF for quantity "angle between the ruler and the edge of the plate" B in the running example was approximated in the GUM-approach by a rectangular PDF for cos B. The expectation of this PDF has the value 0.9975, and the associated uncertainty has the value 0.00144. Using Monte Carlo one obtains $\mathrm{E}(\cos B) = 0.9983$ and $u(\cos\beta) = 0.00145$. In view of the combined uncertainty, these differences are negligible.

In general, for an examination, one computes $\eta_{r,\text{Variant 1}}$ and $\eta_{r,\text{variant 2}}$ as possible values as obtained from two models, variant 1 and variant 2, using the same sample values for all input quantities that contribute to either model. This procedure allows "seeing" even small differences between the models that would be blurred by statistics, if independent Monte Carlo runs were made.

13.6.4
Standard GUM versus Monte Carlo

If a linear Model is given, one should of course use the standard GUM procedure.

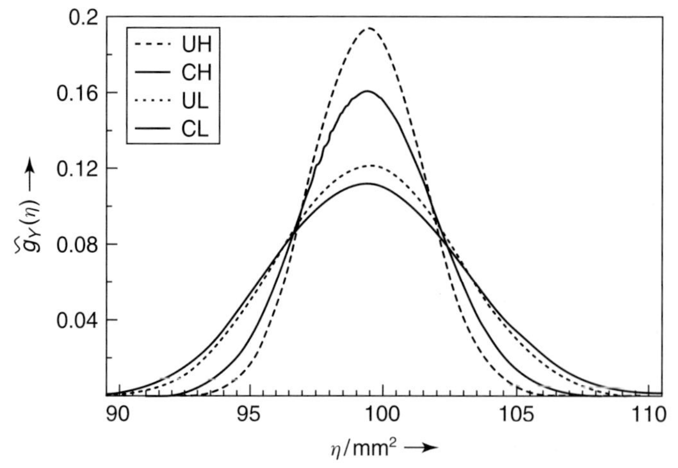

Fig. 13.13 Frequency distributions for possible values of the quantity "area of a Teflon plate." U indicates that the correlation of edge lengths because of using the same ruler is not considered, C denotes that it is considered, and H and L indicate the same as in Figure 13.12.

In all other cases one should first check the degree of nonlinearity. To that purpose consider for all $i \in [1, N]$

$$\delta_i = \max(|f(x) + c_i(x_i \pm u(x_i)) - f(x_1, \ldots, x_i \pm u(x_i), \ldots, x_N)|) \tag{13.62}$$

where max selects either + or −. The sum over the N δ_i is suitable as a measure of nonlinearity; if it is larger than $0.05{\cdot}u(y)$ one should consider either including second-order terms of the Taylor expansion or turning to MCMs.

When using MCM there is not much work if one also computes

$$\eta_{r,\text{lin}} = \sum_{i=1}^{N} c_i(\xi_{i,r} - x_i) \Rightarrow y_{\text{lin}} \text{ and } u(y_{\text{lin}}) \text{ and } \delta_r = \eta_{r,\text{lin}} - \eta_r \tag{13.63}$$

The plot of the frequency distribution of the δ_r provides information, on the basis of which one can decide whether the linear model is sufficient. If the absolute value of the mean of δ_r is larger than $0.05{\cdot}u(y)$ or if $|1 - u(y)/u(y_{\text{lin}})| > 0.05$, one should not use the linear model.

Finally, using MCM and assuming a linear model one can compute

$$\text{Cov}(Y, X_i) = \sum_{j=1}^{N} c_j \text{Cov}(X_j, X_i) = \sum_{j=1}^{N} c_j u(x_j) r(x_j, x_i) u(x_j) \tag{13.64}$$

If none of the input quantities are correlated, the expression reduces to $\text{Cov}(Y, X_i) = c_i u^2(x_i)$. The GUM standard procedure should not be applied if the covariance computed using MCMs deviates significantly from the right-hand-side expression in Equation 13.64. Agreement does not allow concluding that the GUM standard procedure is appropriate.

13.7 Selected Advanced Topics

13.7.1 Product Model

In some cases, the model of the measurement can be written as the product of a constant x_0 and correction factors X_i:

$$Y = f(X) = y_0 \prod_{i=1}^{N} X_i \text{ with } \mathrm{E}X_i = 1. \tag{13.65}$$

The Taylor series, see Equation 13.21, for the possible values of η of the measurand has terms from 0th to Nth order:

$$\eta_0 = x_0,\ \eta_1 = x_0 \sum_{i=1}^{N} \frac{(\xi_i - x_i)}{x_i},$$
$$\eta_2 = x_0 \sum_{i_1=1}^{N} \sum_{i_2=i_1+1}^{N} \frac{(\xi_{i_1} - x_{i_1})}{x_{i_1}} \frac{(\xi_{i_2} - x_{i_2})}{x_{i_2}}, \ldots,$$
$$\eta_N = x_0 \prod_{i=1}^{N} \frac{(\xi_i - x_i)}{x_i} \tag{13.66}$$

Linearization means to keep only the terms for $k = 0$ and $k = 1$, that is,

$$\sum_{k=0}^{1} \eta_k = x_0 \left(1 + \sum_{i=1}^{N} \frac{(\xi_i - x_i)}{x_i}\right)$$
$$\Rightarrow y_{k\leq 1} = x_0,$$
$$\text{since } \forall_i \int g_X(\xi) \frac{(\xi_i - x_i)}{x_i} d\xi_i$$
$$= \mathrm{E}\left[\frac{X_i - x_i}{x_i}\right] = 0 \tag{13.67}$$

and the uncertainty associated with $y_{k\leq 1}$ is given by

$$\mathrm{Var}Y_{k\leq 1} = u^2(y_{k\leq 1}) = y_0^2 \left(\sum_{i=1}^{N} w^2(x_i) + 2\sum_{i_1=1}^{N} \sum_{i_2=i_1+1}^{N} w(x_{i_1}) \times r(x_{i_1}, x_{i_2}) w(x_{i_2})\right) \tag{13.68}$$

where $w(x_i)$ is the relative uncertainty $u(x_i)/x_i$ and $r(x_{i_1}, x_{i_2})$ is the correlation coefficient.

The worst case happens if all correlation coefficients are set to $+1$. In this case one finds that

$$w^2(y_{\mathrm{lin}}) = \left(\sum_{i=1}^{N} w^2(x_i) + 2\sum_{i_1=1}^{N} \sum_{i_2=i_1+1}^{N} w(x_{i_1}) \times 1 \times w(x_{i_2})\right) = \left(\sum_{i=1}^{N} w(x_i)\right)^2 \tag{13.69}$$

The most simple product model is $Y = X_1 X_2$. For discussion, the representation introduced in Equation 13.12 is helpful. One obtains

$$y = \mathrm{E}Y = \mathrm{E}(X_1 X_2)$$
$$= \mathrm{E}((x_1 + u(x_1)X_{\mathrm{STD},1})(x_2 + u(x_2)X_{\mathrm{STD},2}))$$
$$= x_1 x_2 + \mathrm{Cov}(X_{\mathrm{STD},1} X_{\mathrm{STD},2}) \tag{13.70}$$

which reduces to $x_1 x_2$, if $X_{\mathrm{STD},1}$ and $X_{\mathrm{STD},2}$ are not correlated. For this uncorrelated case the uncertainty can be computed exactly:

$$u^2(y) = \mathrm{Var}Y = \mathrm{E}((Y - y)^2)$$
$$= \mathrm{E}((x_1 + u(x_1)X_{\mathrm{STD},1}) \times (x_2 + u(x_2)X_{\mathrm{STD},2}) - x_1 x_2)^2)$$
$$= x_2^2 u^2(x_1) + x_1^2 u^2(x_2) + u^2(x_1) u^2(x_2) \tag{13.71}$$

For comparison, the standard procedure of the GUM, using only the first partial derivatives of Y with respect to the X_i, would produce only the first two terms in the second line of Equation 13.71. Therefore, if both, $x_1$and x_2, are zero, the resulting uncertainty would vanish. However, using second-order terms, the correct result would be obtained.

Finally, we consider the model $Y = X^2$ and assume $x = 0$. This example is often used to criticize GUM, since the linear model would lead to $y = x^2 = 0$ and $u^2(y) = 4x^4u^4(x) = 0$. Again, when also using the second-order term, the correct results are obtained. For $x = 0$ one obtains $Y = X_{\mathrm{STD}}^2$ and

$$\begin{aligned} y &= \mathrm{E}Y = \mathrm{E}(u^2(x)X_{\mathrm{STD}}^2) = u^2(x) \\ \text{and } u^2(y) &= \mathrm{E}(u^2(x)X_{\mathrm{STD}}^2 - u^2(x)^2) \\ &= u^4(x)(\mathrm{E}X_{\mathrm{STD}}^4 - 1) \end{aligned} \tag{13.72}$$

The value of the fourth moment of X_{STD} depends on its shape. For instance, for a Gaussian PDF its value is 3 and for a rectangular PDF 9/5.

13.7.2
Least-Squares Problems

The model for the determination of uncertainties associated with fitted parameters in the calibration curves is often given by an overdetermined set of linear equations.

For instance, the calibration of a ruler or a micrometer could lead to the following set of equations:

$$\left.\begin{aligned} \Delta l_{\mathrm{IND},1} &= l_{\mathrm{IND},1} - l_1 = y_0 + l_1 y_1 + l_1^2 y_2 \\ &\ldots \\ \Delta l_{\mathrm{IND},N} &= l_{\mathrm{IND},1} - l_N = y_0 + l_N y_1 + l_N^2 y_2 \end{aligned}\right\} \Rightarrow x = \begin{pmatrix} 1 & l_1 & l_1^2 \\ \ldots & & \\ 1 & l_N & l_N^2 \end{pmatrix} \cdot y \Rightarrow x = Ay \tag{13.73}$$

where the $l_{\mathrm{IND},i}$ are measured values, the l_i are known values of standards, for instance of gauge blocks, and y_0 through y_2 are the fitting parameters that are here the *output* quantities. In practice, it is often assumed that $u(l_i) \ll u(l_{\mathrm{IND},i})$. In that case one obtains the well known least-squares solution within the standard GUM procedure:

$$y = \boldsymbol{U}_y \boldsymbol{U}_x^{-1} \boldsymbol{A}^{\mathrm{T}} x \text{ where } \boldsymbol{U}_y = (\boldsymbol{A}^{\mathrm{T}} \boldsymbol{U}_x^{-1} \boldsymbol{A})^{-1} \tag{13.74}$$

If the assumption $u(l_i) \ll u(l_{\mathrm{IND},i})$ cannot be made, it is a practice to combine all uncertainties in each line of Equation 13.76 to an effective uncertainty $u_{\mathrm{eff}}(l_{\mathrm{IND},i})$. This, however, is straightforward only if one considers only the parameters y_0 and y_1. In this case the effective uncertainty is given by

$$u_{\mathrm{eff}}^2(l_{\mathrm{IND},i}) = u^2(l_{\mathrm{IND},i}) + y_1^2 u^2(l_i) \tag{13.75}$$

Since y_1 is a fitting parameter, which depends on the uncertainty matrix, see Equation 13.73, at least one iteration is necessary. Clearly, if higher order fitting parameters are used, this procedure becomes difficult. It becomes even more difficult if the model is not a polynomial of l, but a nonlinear function of l. Consider for instance $x_i = y_0 \exp(y_1 l_i)$, which would lead to $\ln x_i = \ln y_0 + y_1 l_i$.

In such cases, the use of Monte Carlo offers a straightforward solution. The model given in Equation 13.73 is used for demonstration. As in Section 13.5 we transcribe it for ease in writing to GUM nomenclature. The symbols ξ, ζ, and η_i are used to denote representative samples of l_{IND}, l, and y_0 through y_2. One obtains

$$\begin{pmatrix} \xi_{1,r} \\ \dots \\ \xi_{2,r} \end{pmatrix} = \begin{pmatrix} 1 & \zeta_{1,r} & \zeta_{1,r}^2 \\ \dots & & \\ 1 & \zeta_{N,r} & \zeta_{N,r}^2 \end{pmatrix} \cdot \begin{pmatrix} \eta_{1,r} \\ \eta_{2,r} \\ \eta_{3,r} \end{pmatrix} \Rightarrow \begin{pmatrix} \eta_{1,r} \\ \eta_{2,r} \\ \eta_{3,r} \end{pmatrix}$$

$$= \left(\begin{pmatrix} 1 & \zeta_{1,r} & \zeta_{1,r}^2 \\ \dots & & \\ 1 & \zeta_{N,r} & \zeta_{N,r}^2 \end{pmatrix} \begin{pmatrix} 1 & \zeta_{1,r} & \zeta_1^2 \\ \dots & & \\ 1 & \zeta_{N,r} & \zeta_{N,r}^2 \end{pmatrix}^{\mathrm{T}} \right)^{-1} \begin{pmatrix} \xi_{1,r} \\ \dots \\ \xi_{2,r} \end{pmatrix} \quad (13.76)$$

As described in Section 13.6, these *simulated repeated measurements* are then evaluated using standard statistical procedures that yield the expectations of Y_i and the corresponding uncertainty matrix.

13.7.3 Repeated Measurements and Correlation

If repeated simultaneous measurements of more than one quantity are taken, special attention should be given to possible correlations. In some cases, the measured quantities might not be independent of each other, for example, two lengths would both depend on temperature. If a very large number of repetitions of simultaneous length measurements under varying temperatures were made, then the two lengths would show correlation. However, this physical correlation will be masked by statistics if only a small number of repetitions are made, and vice versa. For instance, even if the temperature was kept absolutely stable, the two lengths would appear as correlated in a small series of repeated measurements.

Consider that X_1 and X_2 are simultaneously measured M times and $Y = c_1X_1 + c_1X_2$. The evaluation of the M measured pairs $(\xi_{1,r}, \xi_{2,r})$ is very simple if one evaluates the M values of $\eta_r = c_1\xi_{1,r} + c_2\xi_{2,r}$:

$$y = \frac{1}{M}\sum_{r=1}^{M} \eta_r$$
$$\text{and } u^2(y) = \frac{1}{M-1}\sum_{r=1}^{M} (\eta_r - y)^2 \quad (13.77)$$

and assuming that the η_r pertain to a normal PDF, one could use the coverage factor for the effective degree of freedom $\nu = M - 1$. Alternatively, one could first determine $x_1, x_2, u(x_1), u(x_2)$ and $r(x_1, x_2)$:

$$x_i = \frac{1}{M}\sum_{r=1}^{M} \xi_{i,r}, u^2(x_i)$$
$$= \frac{1}{M-1}\sum_{r=1}^{M} (\xi_{i,r} - x_i)^2$$
$$\text{and } r(x_1, x_2) = \frac{1}{M-1} \times \sum_{r=1}^{M} \frac{(\xi_{1,r} - x_1)(\xi_{2,r} - x_2)}{u(x_1)u(x_2)} \quad (13.78)$$

and, using the standard GUM procedure arrive at

$$y = c_1x_1 + c_2x_2$$
$$\text{and } u^2(y) = c_1^2u(x_1) + c_2^2u(x_2) + 2c_1u(x_1)r(x_1, x_2)c_2u(x_2) \quad (13.79)$$

To determine the coverage factor one would need to compute the effective degree of freedom; however, the Welch–Satterthwaite formula, cf. Equation 13.44, is not applicable if correlation is to be considered. This means that the procedure used in Equation 13.77 should be preferred. This is justified since the values for y and $u(y)$ are the same for either procedure since

$$\begin{aligned}
y &= \frac{1}{M}\sum_{r=1}^{M}\eta_r \\
&= \frac{1}{M}\sum_{r=1}^{M}(c_1\xi_{1,r} + c_2\xi_{2,r}) \\
&= c_1x_1 + c_2x_2 \\
u^2(y) &= \frac{1}{M-1}\sum_{r=1}^{M}(\eta_r - y)^2 \\
&= \frac{1}{M-1}\sum_{r=1}^{M}(c_1(\xi_{1,r} - x_1) \\
&\quad +c_2(\xi_{2,r} - x_2))^2 \\
&= \frac{1}{M-1}\sum_{r=1}^{M}(c_1^2(\xi_{1,r} - x_1)^2 \\
&\quad +c_2^2(\xi_{2,r} - x_2)^2 \\
&\quad +2c_1(\xi_{1,r} - x_1)c_2(\xi_{2,r} - x_2) \\
&= c_1^2u^2(x_1) + c_1^2u^2(x_2) \\
&\quad +2c_1u(x_1)r(x_1, x_2)c_2u(x_1).
\end{aligned} \tag{13.80}$$

Repeated measurements are important in practice, for instance to check the stability of an apparatus. However, great care is needed to assure unchanged repeatability conditions. Furthermore, one should clearly distinguish variations stemming from the spectral nature of an examined quantity from varying conditions. The latter may average out in small time scale, for example, noise, or in a large time scale, for example, changes of local gravity over one day.

In summary, an experiment or measurement is only fully understood if all significant influences have been identified. This is not the case if the uncertainties obtained from repeated measurements contribute significantly to the overall uncertainty.

Acknowledgment

We wish to thank JCGM for their continued effort for the further development and the harmonization of uncertainty determination and the BIPM for its generous open access policy on its webpage.

Appendix: Error Approach

The VIM (BIPM, IEC, IFCC, ILAC, ISO, IUPAC, IUPAP, and OIML, 2008) explains the *error approach* (sometimes also called *traditional approach or true value approach*) in its introduction:

The objective of measurement in the error approach is to determine an estimate of the true value that is as close as possible to that single true value. The deviation from the true value is composed of random and systematic errors. The two kinds of errors, assumed to be always distinguishable, have to be treated differently. No rule can be derived regarding how they combine to form the total error of any given measurement result, usually taken as the estimate. Usually, only an upper limit of the absolute value of the total error is estimated, sometimes loosely named "uncertainty."

The term *"bottom up"* is sometimes used for the GUM-approach when comparing it with the error approach. The "bottom" of any uncertainty evaluation in accordance with the GUM is the model; from that point one works "upwards" by determining the uncertainties associated with the input quantities and combines them weighted appropriately with their sensitivity coefficients to the combined standard uncertainty. This requires a detailed understanding of the measurement and assures its traceability to SI-units. However, in industrial practice and especially in environmental surveillance, many measurements are to be made as a basis for decisions such as "this product conforms with a specification" or "the content of lead in this probe of liquid waste is below the allowed limit." In such cases, it might be quite uneconomical to perform an uncertainty evaluation for each single measurement since it is only required that the measurement procedure is appropriate in the context of a given application and set limits and tolerances.

In principle, mass spectrometry could traceably determine the content of lead in a given sample. However, the costs for routine tests in environmental surveillance would be unacceptable. Therefore, one looks for cheaper methods. Such methods exist, but they depend on the matrix in which the lead is embedded. Therefore, such a method may have a bias, that is, the laboratory component of bias under repeatability conditions, which is not known exactly. Using the standard error model: $y = m + B + e$; where y is the value the measurand, m is the general mean (expectation), B is the bias, and e is the random error occurring in every measurement under repeatability conditions. One tries to determine the "closeness" of values measured with such a method to the "true value."

Therefore, the term *"top down"* is used for the error approach. The "top" is accuracy that describes the closeness of the agreement between the result of a measurement and a "true value." Then, one distinguishes between the *reproducibility*, that is, the closeness of the agreement between the result of successive measurements of the same measurand carried out under the changed conditions of measurement, and *repeatability*, that is, the closeness of the agreement between the result of a measurements of the same measurand carried out under the same conditions of measurement. The "changed conditions" for the reproducibility may include principle and method of measurements, observer, measuring instrument, reference standard, location, condition of use, and time. These changes are often realized in an interlaboratory experiment. The "same conditions" for repeatability are believed to be given in each laboratory.

Clearly, the reliability of the reproducibility is the Achilles' heel of this approach unless several well-experienced laboratories take part in the corresponding interlaboratory experiment. For that reason, the ISO standard 5725 provides explicit guidance for this approach. It encompasses six parts:

- Part 1: *General Principles and Definitions;* Corrigendum 1-1998
- Part 2: *Basic Method for the Determination of Repeatability and Reproducibility of a Standard Measurement Method;* Technical Corrigendum 1: 5/15/2002
- Part 3: *Intermediate Measures of the Precision of a Standard Measurement Method;* Corrigendum 1 10/15/2001
- Part 4: *Basic Methods for the Determination of the Trueness of a Standard Measurement Method*

- Part 5: *Alternative Methods for the Determination of the Precision of a Standard Measurement Method;* Corrigendum 1: 8/15/2005
- Part 6: *Use in Practice of Accuracy Values,* Corrigendum 1:10/15/2001.

Helpful guidance for using ISO 5725 approaches is provided in a book by Perruchet and Priel (2000).

Glossary

The glossary lists the most often used symbols, acronyms, and scientific terms. The main sources are the "Guide to the Expression of Uncertainty in Measurement" (GUM) (ISO, 1993), the VIM (BIPM, IEC, IFCC, ILAC, ISO, IUPAC, IUPAP, and OIML, 2008), and "Supplement 1 to the GUM" (JCGM, 2008).

Principal Symbols

$\boldsymbol{X}$ (input) Quantity or its statistics

$\boldsymbol{x}$ Expectation of the PDF for X

$\mathbf{X}$ Set of quantities: $X = (X_1, \ldots, X_N)^T$

$\mathbf{x}$ Set of expectations: $x = (x_1, \ldots, x_N)^T$

ξ Possible value of X

$\boldsymbol{\xi}$ Set of possible values of $X, \xi = (\xi_1, \ldots, \xi_N)^T$

Y, $\mathbf{Y}$ Same as X and $\boldsymbol{X}$, but for output quantities

y, $\mathbf{y}$ Same as x amd $\boldsymbol{x}$, but for output quantities

η, $\boldsymbol{\eta}$ Same as ξ and $\boldsymbol{\xi}$, but for output quantities

f_M, f Model function; the index M is only used if the context requires it.

$g_X(\boldsymbol{\xi})$ Joint PDF for quantities

$\boldsymbol{h(Y,..)}$ Cause–effect equation

$\boldsymbol{u(x)}$ Uncertainty associated with x

U_x Uncertainty matrix, see Equation (13.27)

$\boldsymbol{U(y)}, U_Y$ Same as $u(x)$ and U_x, but for output quantities

$\boldsymbol{W(x)}$ Relative uncertainty associated with x

$\boldsymbol{W(y)}$ Relative uncertainty associated with y

k_p Coverage factor for a coverage probability

$\boldsymbol{p}$ Coverage probability usually set to 0.95

U_p Expanded uncertainty

$\boldsymbol{i}$ Index for input quantities $i \in [1, N]$

c_i Sensitivity coefficient, see Equation (13.28)

$c_{i,i'}$ Second-order sensitivity coefficients

$\boldsymbol{j}$ Index for output quantities $j \in [1, m]$

$\boldsymbol{r}$ Running index for subsamples, $r \in [1, M]$

$\boldsymbol{H}$ Running index for samples of size M, $h \in [1, H]$ in "adaptive" Monte Carlo

Acronyms

BIPM: The International Bureau of Weights and Measures (*Bureau International des Poids et Mesures*) was set up by the Metre Convention in 1875 and is located in Sèvres near Paris, France. The task of the BIPM is to ensure worldwide uniformity of measurements and their traceability to the International System of Units (SI). The BIPM also forms joint committees with other international organizations, for instance the JCGM.

CDF: Cumulative Distribution Function.

CLT: Central Limit Theorem.

GUM: Guide to the Expression of Uncertainty in Measurement, see ISO (1993)

GUM-SUP: Supplement I to the GUM, see JCGM (2008)

JCGM: The Joint Committee for Guides in Metrology has two working groups (WG). WG 1 promotes the use of the GUM, prepares supplemental guides for its broad application. WG 2 revises and promotes the use of the VIM.

MCM: Monte Carlo Method.

PDF: (Probability) Density Function.

PME: The Principle of Maximum (Information) Entropy states that the *use* of all available information and *exclusion* of any nonavailable information maximize information entropy(. For its formal definition see Section 13.3.2.2.

SI (International System of Units): System of units, based on the International System of Quantities, their names and symbols, including a series of prefixes and their names and symbols, together with rules for their use, adopted by the General Conference on Weights and Measures (CGPM) and available at BIPM.

VIM: (International Vocabulary of Metrology): Basic and general concepts and associated terms; see BIPM, IEC, IFCC, ILAC, ISO, IUPAC, IUPAP, and OIML (2008)

Scientific Terms

Bayesian Probability: Degree of belief in a statement based on the information given.

Bayes' Theorem: The probability that two propositions A and B are true, that is, $p(A \wedge B)$ is given by the product of the probability $p(A)$ and the probability $p(B|A)$; where $|A$ means that A is true. It follows that $p(B)p(A|B) = p(A)p(B|A)$. Since A and B can be exchanged, this relation can be rearranged to $p(A|B) = p(B|A)p(A)/p(B)$ which is Bayes' theorem:

The probability for A given B is equal to the probability of B given A multiplied by the ratio of the probabilities for A and B.
Note: "given" is here equivalent to "true." The theorem can be used to *link* prior knowledge and new facts to obtain posterior knowledge that is consistent with both, the new facts and the prior knowledge, cf. Section 13.3.2.3.

Bias: Estimate of a systematic error.

Bin: A small interval of a variable of interest. For numerical representations of a function of that variable its range is covered by adjacent bins. The content of a bin is equal to the mean value of the function in that interval, which in turn is equal to the integral of that function over the interval divided by the length of the interval.

Calibration: Operation that, under specified conditions, in a first step, establishes a relation between the quantity values and measurement uncertainties provided by measurement standards and corresponding indications with associated measurement uncertainties and, in a second step, uses this information to establish a relation for obtaining a measurement result from an indication.

Cause–Effect-chain: A *chain* of *causes* that leads from the primary cause, the measurand Y, to the final *effect*, which is the quantity that is indicated by the instrument and is termed *input quantity* X_1. Both measurand and input quantity can be each a set of quantities. In general, there are additional chain links that as influence quantities $X_2, X_3, \ldots, X_N$ transform or perturb the primary cause.

Cause–Effect Equation: For many measurement problems it is possible to express the final *effect* as a function of the output quantity and the influence quantities: $X_{\text{input}} = h(Y, X_2, X_3, \ldots, X_N)$ transform or perturb the primary cause in the cause–effect-chain.

Central Limit Theorem(s): Based on different assumptions central limit theorems describe the asymptotic behavior of sums of random samples. For its formal definition see Section 13.4.2.

(Central) Moments: Property of a PDF. The kth (absolute) moment m_k is defined by $\mathrm{E}X^k$ and the corresponding central moment μ_k is defined by $\mathrm{E}[(X - m_1)^k$. For any PDF, the condition $m_0 = 1$ holds. Generally used names are location parameter for m_1, variance for μ_2, and standard deviation for $\sqrt{\mu_2}$.

Coherent Unit: *Unit* that is defined as product of the number 1 and powers of the base units.

Confidence Interval: Interval within which the value of the parameter is expected to lie with a stated (frequentist) probability p. This term is used with the error approach, see also coverage interval. Interpretation: If $M \gg 1$ measurements are taken, then Mp values are expected to fall in the confidence interval.

Coverage Interval: Interval containing the value of a quantity, based on the information available, with a stated coverage probability p. Interpretation: The PDF for a quantity represents the given incomplete knowledge. The integral over the PDF for the quantity from the lower to the upper boundary yields the stated coverage probability. See also Equation (13.14).

Cumulative (Probability) Distribution Function: Function giving, for every value ξ, the probability that the random variable X is less than or equal to ξ. Note: The distribution function is often called *cumulative distribution function* and the abbreviation CDF is used.

Error: Measured quantity value minus a reference quantity value. See also systematic and random errors.

Error Approach: The true value of a quantity is perceived as an unknown constant and the result of a measurement is a random variable. Classical statistics is applied to evaluate random errors, whereas systematic errors are treated separately. See also uncertainty approach.

Expanded Uncertainty: In the case of a symmetric PDF for Y, the half-width of the coverage interval. The symbol U is used to represent it and is defined as the product of coverage factor and combined uncertainty: $U = k_p u(y)$. See also Equation (13.14).

Expectation: The probability weighted mean value. Of note is that the expectation is given by $x = \mathrm{E}X = \sum_{i=1}^{n} p_i \xi_i$ or $x = \mathrm{E}X = \int g_X(\xi)\xi \mathrm{d}\xi$ for discrete probabilities or a probability density function, respectively.

Frequency Distribution: In this chapter, it is used as an approximation to a probability density function using Monte Carlo. The range of possible values is subdivided into adjacent bins such that any sample value belongs to one and only one bin. The content of that bin is enhanced by the statistical weight of the sample. After taking the samples, normalization is obtained by dividing the content of every bin by the sum of the statistical weights of all samples.

Frequentist (Classical) Probability: Limit of the ratio of the number of times an event occurs to the number of experiments in which it could occur.

Influence Quantity: A quantity that is assigned to an intermediate link in the cause–effect-chain.

Information Entropy: A measure of given Information. For its formal definition see Section 13.3.2.2.

Input Quantity: A quantity that is assigned to the last link in the cause–effect-chain.

International System of Units (SI): System of units, based on the International System of Quantities, their names and symbols, including a series of prefixes and their names and symbols, together with rules for their use, adopted by the General Conference on Weights and Measures (see Chapter 2).

Likelihood Function: A factor used in Bayes' theorem to reflect the belief about obtaining the data given that the hypothesis is true. In *classical statistical inference* likelihood is a parameter, associated with an *assumed* model, proportional to the probability of the data given the truth of the model. The likelihood is not necessarily a probability.

Linear Model (of measurement): A (measurement) model (for the evaluation of uncertainty) given by $Y = c_1X_1 +, \ldots, +c_NX_N$. If the original physical model is not linear, a Taylor series expansion, see Section 13.3.3, is used to obtain a linear approximation.

Measurand: A particular physical quantity subject to measurement. In general it is assigned to the first link in the cause–effect-chain, see also output quantity.

Measurement Function: If the (measurement) model can be stated in closed form, it is also called *measurement function* or *model function*; see also model equation.

Measurement Principle: Phenomenon serving as a basis of a measurement, for example, use of the thermoelectric effect to measure the temperature.

Measurement Method: Generic description of a logical organization of operations used in a measurement. Often used methods are substitution, differential, or null measurement methods. One also distinguishes between direct and indirect measurement methods.

(Measurement) Model (for the evaluation of uncertainty): An inter-relation between input and output quantities for the propagation of PDFs or uncertainties of input quantities to obtain the PDFs for the output quantities or the uncertainties associated with the values of those quantities. The model is also used with the numerical values of those quantities, and it is used to provide one and only one set of possible values of the output quantities $\eta_1, \ldots, \eta_N$ for any set of possible values of the input quantities $\xi_1, \ldots, \xi_N$. Note: A given set of values of the output quantities can be due to a multitude of possible values of the input quantities. In general, modeling is a Bayesian learning process.

Measurement Procedure: Detailed description of a measurement according to one or more measurement principles and to a given measurement method, based on a measurement model and including any calculation to obtain a measurement result. Of note is that a measurement procedure can include a statement concerning a target measurement uncertainty.

Measurement Unit (also called unit of measurement or unit): Real scalar quantity, defined and adopted by convention, with which any other quantity of the same kind can be compared to express the ratio of the two quantities as a number. See also International System of Units.

Model: (Measurement) model (for the evaluation of uncertainty).

Model Equation: For many measurement problems, it is possible to obtain Y explicitly from the cause–effect equation: $Y = f_M(X_1, X_2, X_3, \ldots, X_N)$; where $X_1 \equiv X_{\text{input}}$. The equation is called *model equation* and the function f_M is called *model function.*

Model Function: See model equation; the term measurement function is used synonymously.

Monte Carlo Method: Method for propagating PDFs for input quantities to yield the joint PDF for one or more than one output quantities. Knowledge of the (joint) PDF allows determining expectations, variances, and covariances of or associated with the output quantities.

Mutual Recognition Agreements: International agreements on the mutual acceptance of the results of measurements and tests by accredited laboratories.

Numerical Value of a Quantity: Ratio of that quantity to the unit used.

Systematic (Measurement) Error: Component of measurement error that in replicate measurements remains constant or varies in a predictable manner. Note: A systematic error cannot be determined by repeated measurements.

Output Quantity: A quantity that is assigned to the first link in the cause–effect-chain.

(Probability) Distribution: Random variable function giving the probability that a random variable takes any given value or belongs to a given set of values.

Probability Density Function (PDF): Derivative, when it exists, of the distribution function, see Bayesian and frequentist probability.

Possible Values: Values of a quantity that are consistent with the (incomplete) knowledge about the quantity.

Probability: See Bayesian and frequentist (classical) probability.

Quantity: Property of a phenomenon, body, or substance, where the property has a magnitude that can be expressed as a number and a reference. See also the glossary entries influence, input and output quantity.

Reading Indication: The effect of an instrument resulting from the cause–effect-chain is in the uncertainty analysis considered as input quantity. The reading, that is, the value indicated is the expectation of the PDF for that quantity.

Type A Evaluation (of measurement uncertainty): Evaluation of a component of measurement uncertainty by a statistical analysis of measured quantity values obtained under defined measurement conditions.

Type B Evaluation (of measurement uncertainty): Evaluation of a component of measurement uncertainty by means other than a Type A evaluation, for example, evaluations based on information.

Uncertainty (of measurement): Parameter associated with the result of a measurement that characterizes the dispersion of the values that could reasonably be attributed to the measurand (ISO, 1993).

Uncertainty Approach: A Bayesian distribution density (PDF) derived from available knowledge quantity represents this knowledge on the reasonably possible values of that quantity. The measured values are considered as facts and they are part of the available knowledge and a distinction between random and systematic errors is not necessary. See also error approach.

Unit: See measurement unit

Value of a Quantity: Number {Q} and reference [Q] together expressing magnitude of a quantity, see also numerical value of a quantity.

Yield: Often used synonymously with reading indication for the indication of counters, for example, in radiation physics.

References

NOTE: All links to the World Wide Web have been last checked in July 2009.

Bich, W. and Cox, M.G. (**2006a**) Foreword to the special issue on statistical and probabilistic methods for metrology. *Metrologia*, **43** (4).

Bich, W., Cox, M.G., and Harris, P.M. (**2006b**) Evolution of the "Guide to the expression of uncertainty in measurement". *Metrologia*, **43**, 161–166.

BIPM, IEC, IFCC, ILAC, ISO, IUPAC, IUPAP, and OIML (**2008**) International Vocabulary of Metrology – Basic and General Concepts and Associated Terms (VIM), *http://www.bipm.org*.

Burry, K. (**1999**) *Statistical Distributions in Engineering*, Cambridge University Press, Cambridge.

Cox, R.T. (**1946**) Probability, frequency, and reasonable expectation. *Am. J. Phys.*, **14**, 1–13.

Cox, M.G. and Harris, P.M. (**2006a**) SSfM Best Practice Guide No. 6, Uncertainty Evaluation.Technical Report DEM-ES-011, National Physical Laboratory, Teddington, ISSN 1744–0475, *http://www.npl.co.uk/*.

Cox, M.G. and Siebert, B.R.L. (**2006b**) The use of a Monte Carlo method for evaluating uncertainty and expanded uncertainty. *Metrologia*, **43**, 178–188.

DIN (**1999**) DIN 1319-4. *Grundlagen der Meßtechnik Teil 4: Auswertung von Messungen, Messunsicherheit*, Beuth Verlag, Berlin.

EA (**1999**) Expression of the Uncertainty of Measurement in Calibration, European co-operation for Accreditation: EA-4/02, *http://www.europeanaccreditation.org*.

Elster, C. (**2007**) Calculation of uncertainty in the presence of prior knowledge. *Metrologia*, **44**, 111–116.

Gleser, L.J. (**1998**) Assessing uncertainty in measurement. *Stat. Sci.*, **13**, 277–290.

Goydke, H., Siebert, B.R.L., and Scholl, W. (**2003**) Considerations on the evaluation of uncertainty values of building acoustic single-number quantities. *Proceedings of the 5th European Conference on Noise Control [CD-ROM], ISBN 88-88942-00-9.*

ISO (**1993**) *Guide to the Expression of Uncertainty in Measurement*, International Standards Organisation, Geneva. With minor correction available as JCGM 100 from *http://www.bipm.org*.

Jaynes, E.T. (**1957**) Information theory and statistical mechanics. *Phys. Rev*, **106**, 620–630, paper available: *http://bayes.wustl.edu/etj/etj.html*.

JCGM (**2008**) Evaluation of Measurement Data – Supplement 1 to the "Guide to the Expression of Uncertainty in Measurement" – Propagation of Distributions using a Monte Carlo Method, *http://www.bipm.org*.

Korczynski, M.J., Cox, M.G., and Harris, P.M. (**2006**) Convolution and uncertainty evaluation, in *Advanced Mathematical and Computational Tools in Metrology VII*, Series on Advances in Mathematics for Applied Sciences, Vol. 72 (eds P. Ciarlini, E. Filipe, A.B. Forbes, F. Pavese, C. Perruchet, and B.R.L. Siebert), World Scientific, Singapore, pp. 188–195.

Lira, I. and Wöger, W. (**2006**) Comparison between the conventional and Bayesian approaches to evaluate measurement uncertainty. *Metrologia*, **43**, 249–259.

Perruchet C. and Priel M. (**2000**) *Estimer L'incertitude: Mesures, Essais*, Afnor, La Plaine Saint-Denis.

Press, W.H., Flannery, B.P., Teukolsky, S.A., and Vetterling, W.T. (**1989**) *Numerical Recipes – The Art of Scientific Computing* Cambridge University Press, Cambridge, newer editions providing FORTRANoCare available, *http://www.nrbook.com/nr3/*.

Shannon, C.E. (**1948**) A mathematical theory of information. *Bell Syst. Technol. J.*, **27**, 623–656.

Siebert, B.R.L. (**2008**) GUM-based uncertainty in radiation protection – survey on methods and practices, in *Proceedings of the Workshop "Uncertainty Assessment in Computational Dosimetry: a Comparison of Approaches"*, EURADOS Publication, available from *office@eurados.org*.

Siebert, B.R.L., Ciarlini, P., and Sibold, D. (**2006**) Monte Carlo study on logical and statistical correlation, in *Advanced Mathematical and Computational Tools in Metrology VII*, Series on Advances in Mathematics for Applied

Sciences, Vol. **72** (eds P. Ciarlini E. Filipe A.B. Forbes F. Pavese C. Perruchet B.R.L Siebert) World Scientific, Singapore, pp. 237–244.

Sommer, K.-D. (**2009**) Modelling of measurements, system theory and uncertainty evaluation, in*Data Modelling for Metrology and Testing in Measurement Science*, Series: Modelling and Simulation in Science, Engineering and Technology (eds F. Pavese A. Forbes), A Birkhäuser Book.

Sommer, K.-D. and Siebert, B.R.L. (**2006**) A systematic approach to the modelling of measurements for uncertainty evaluation. *Metrologia*, **43**, 200–210.

Tegeler, E., Heyer, D., and Siebert, B.R.L. (**2008**) Uncertainty of the calibration of paired temperature sensors for heat meters. *Int. J. Thermophys.*, **29**, 1174–1183.

Further Reading

Preface: Bich, W. and Cox, M.G. (eds.) (**2006**) Special issue on statistical and probabilistic methods for metrology. *Metrologia*, **43**(4).

This special issue of Metrologia is devoted to the generic topic of statistical and probabilistic methods for metrology. It features paper on the evolvement of the GUM in general and Supplement I, in particular. A wide range of topics are covered, such as Monte Carlo methods for propagating distributions, use of a generalization of the sensitivity coefficients in the GUM to correlated quantities, and considerations on obtaining best estimates when the model is nonlinear. In addition special problems are treated, such as systematic modeling, key comparisons, and decision making.

BIPM (**2006**) *Le Système International d'Unités (SI)*, 8th édn, Bureau International des Poids et Mesures, Sèvres, *http://www.bipm.org*.

D'Agostini, G. (**2003**) *Bayesian Reasoning in Data Analysis – A Critical Introduction*, World Scientific Publishing.

The author introduces to Bayesian reasoning and its applications to data analysis. The basic ideas of this approach to the quantification of uncertainty are presented using examples from research and everyday life. Applications covered include parametric inference; combination of results; treatment of uncertainty due to systematic errors and background; comparison of hypotheses; unfolding of experimental distributions; upper/lower bounds in frontier-type measurements. Approximate methods for routine use are derived and are shown often to coincide – under well-defined assumptions! – with "standard" methods, which can therefore be seen as special cases of the more general Bayesian methods.

EA (**2003**) Guidelines on the Expression of Uncertainty in Quantitative Testing, European co-operation for Accreditation: EA-4/16, *http://www.european-accreditation.org*.

Grabe, M. (**2005**) *Measurement Uncertainties in Science and Technology*, Springer, Berlin.

The author claims to revise Gauß' error calculus ab initio. He treats random and unknown systematic errors on an equal footing and proposes "well-defined measuring conditions" as a prerequisite for defining consistent confidence intervals with basic statistical concepts. The resulting measurement uncertainties are as robust and reliable as required by modern-day science, engineering, and technology. The approaches proposed in this book do not conform with GUM (ISO, 1993); however, it may be of interest to those aiming at the worst case error analysis.

ISO (**1994**) 5725. *Accuracy (Trueness and Precision) of Measurement Methods and Results*, 3rd edn, International Standards Organisation, Geneva, for Details see Appendix 2.

Sivia, D.S. (**2005**) *Data Analysis – A Bayesian Tutorial*, revised 2nd edn, Clarendon Press, Oxford.

This book is intended as tutorial, and it provides a logical and unified approach to the whole subject of data analysis. It explains the basic principles of Bayesian probability theory and illustrates their use with a variety of examples ranging from elementary parameter estimation to image processing. It covers topics such as reliability analysis, multivariate optimization, least-squares and maximum likelihood, error propagation, hypothesis testing, maximum entropy, and experimental design.

Weise, K. and Wöger, W. (**1999**) *Meßunsicherheit und Messdatenauswertung*, Wiley-VCH Verlag GmbH, Weinheim.

The authors present the foundations of data analysis and metrology. In general, the analysis of measurement data requires statistical tools. The authors base these tools on Bayesian probability theory. Elucidated in this work are in-depth considerations of measurement deviations, the principle of maximum (information) entropy, Bayes' theorem and coverage intervals. In summary, the authors provide a consequent Bayesian approach for the determination of uncertainty for one or many output quantities. Many ideas presented in this book are used in the GUM supplement (JCGM, 2008)

14
Laser Spectroscopy

Massimo Inguscio, Giulia Rusciano, and Antonio Sasso

Handbook of Metrology. Edited by Michael Gläser and Manfred Kochsiek

ISBN: 978-3-527-40666-1

14.1
Brief Historical Introduction

Most of what we know of atomic and molecular structure has been derived from spectroscopic analysis of the radiation emitted when atoms or molecules are excited. From measurements of the wavelengths and intensities of observed spectral components, pioneering spectroscopists, with diligent and patient work, laid the basis upon which the exploration of the physics of microsystems would begin.

The foundation of spectral analysis dates back to about 1860, with Kirchhoff and Bunsen. The first instruments for wavelength measurement were spectrographs, which, in their modern versions, continue to occupy an important position in the spectroscopy laboratory. Basically, investigation can be performed by spectrally analyzing the light emitted by the sample after proper excitation (excitation spectroscopy, Figure 14.1a) or by monitoring the absorption of spectrally selected wavelengths (absorption spectroscopy, Figure 14.1b).

The first approach to treating the large amount of information accumulated during the early decades of these investigations consisted of finding some regularity in the sequences of spectral lines. Liveing and Dewar, around 1880, emphasized physical similarities between the spectral lines of alkali elements and overtones in acoustics. In 1883, Hartley found an important numerical relationship that allowed, from the large number of lines in any given spectrum, the isolation of those groups of lines (multiplets) that were undoubtedly related. It soon became evident that the frequency ν was more representative than the wavelength λ. Nevertheless, because spectroscopists measure directly not frequency but wavelength, the wavenumber $\tilde{\nu} = 1/\lambda$ is preferred to frequency.

Spectral series formulas date from a discovery by Balmer (1885). He reproduced very closely the wavelengths of the nine then-known lines in the visible spectrum of hydrogen (Balmer series) using a simple formula:

$$\lambda_n = 3645.6n^2/(n^2 - 4)\text{Å} \tag{14.1}$$

n being a variable integer number assuming the values 3, 4, 5, ... for, respectively, the first, second, third, and so on lines in the spectrum. Other series of hydrogen were investigated in the ultraviolet by Lyman in 1906 and, two years later, in the infrared by Paschen.

A more general formula applicable to other series and elements was found by Rydberg. Using the comparatively large mass of wavelength data then available for alkali elements, Rydberg found that

Handbook of Metrology. Edited by Michael Gläser and Manfred Kochsiek

ISBN: 978-3-527-40666-1

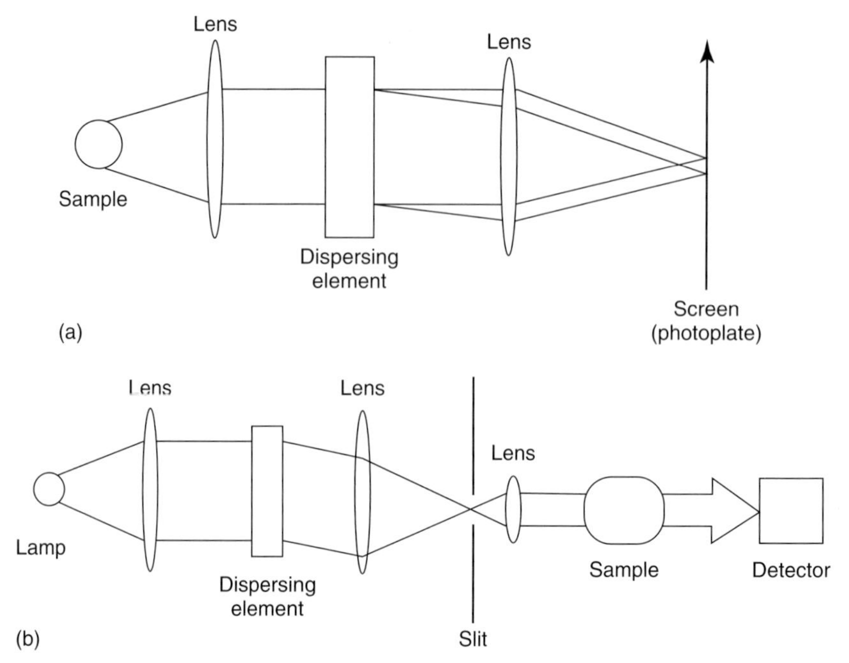

Fig. 14.1 Principle of (a) excitation and (b) absorption spectroscopy.

the wavenumbers of the observed spectral sequence were described by the formula

$$\tilde{\nu}_n = \tilde{\nu} - R_\infty/(n+\mu)^2 \qquad (14.2)$$

where μ and $\tilde{\nu}$ are constants depending upon the series, $\tilde{\nu}$ representing the high-frequency limit to which the lines in the series converge. The constant R_∞ in Equation (14.2), now called the *Rydberg constant*, was found to be independent of the series considered and showed slight variations from one atom to another. The level scheme of the hydrogen atom is shown in Figure 14.2, and a photograph of the Balmer spectral series is shown in Figure 14.3.

These experimental outcomes, not explicable in terms of classical physics, extensively contributed at the beginning of this century to the formulation of quantum mechanics. In particular, Niels Bohr was to apply the idea of quantum mechanics to the hydrogen atom, to arrive at a theoretical formula for its spectrum and explain the experimental observations. According to Bohr's model, electrons move around the nucleus with definite energies. When an electron makes a transition from one allowed stationary state with energy E_u to another of lower energy E_l, the frequency of the emitted radiation is determined by the condition that

$$h\nu = E_u - E_l \qquad (14.3)$$

where h is the Planck constant ($h = 6.63 \times 10^{-34}$ J s). Conversely, it is necessary to absorb a quantum of electromagnetic energy – *a photon* – with energy $h\nu$ to induce a transition from a lower to a higher energy level. Measurements of the wavelengths of spectral lines are therefore related to the separations between energy levels, while the measured intensities are proportional to the transition probabilities.

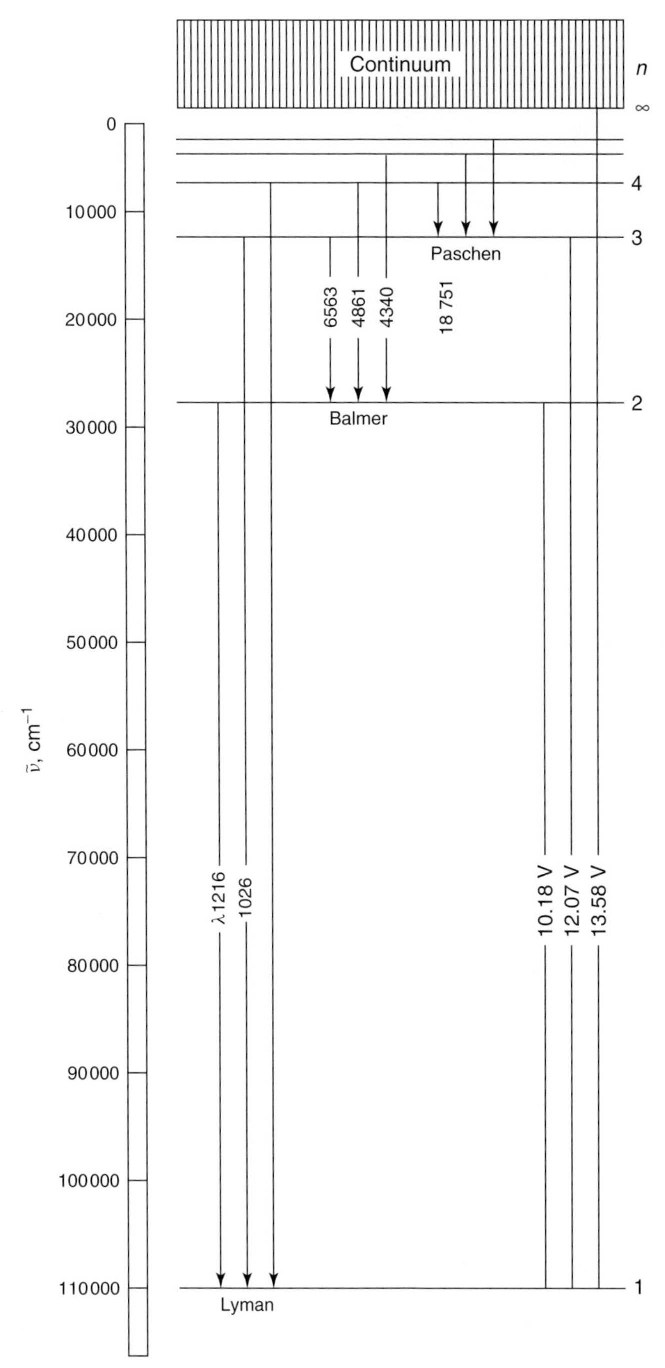

Fig. 14.2 Energy-level diagram of the hydrogen atom, showing some of the lines of the principal series.

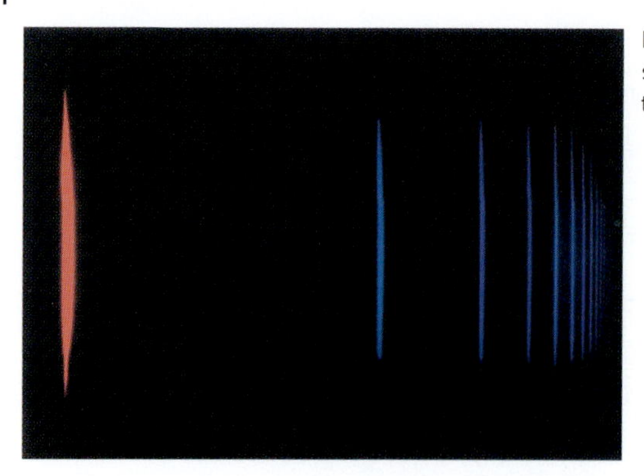

Fig. 14.3 Photograph of the Balmer series of lines in the hydrogen spectrum. (Courtesy of T. Hänsch.)

From this basic idea of Bohr, experimental and theoretical investigations of atomic and molecular structure have evolved considerably over the intervening decades. Newer and more accurate experimental observations have stimulated the development of deeper and more sophisticated theories, and theoretical predictions have, in turn, challenged experimentalists to seek ever finer structures in the spectra of atoms and molecules. This process gained a stimulating impetus from the discovery of the laser, which opened a new era in spectroscopy. The peculiarities of laser radiation (high intensity, spectral purity, short light pulses, etc.) have made it possible to explore the interaction between radiation and matter in a completely new light reaching, in the spectra of atoms and molecules, the ultimate limits of resolution.

Besides fundamental physics, spectroscopic investigations are of direct importance in an increasing variety of applications in fields such as astrophysics, combustion, and plasma physics. Optical absorption and emission spectroscopy, for example, are used for the remote sensing of atmospheric pollutants and for medical tests.

In the following, the main techniques of modern laser spectroscopy are examined. For discussions in detail, the reader is referred to more specific and exhaustive textbooks as, for instance, Corney (1977), Demtroeder (1981), Svanberg (1992), and Letokhov and Chebotayev (1977).

14.2 Widths of Spectral Lines

The more deeply we understand the structure of an atom or molecule, the higher is the resolution with which we need to examine its absorption and emission spectra. The limit of resolution is determined by the width of the spectral lines. According to the Rayleigh criterion, two close lines with equal intensities are resolved if the dip produced in the overlapping profile drops to 0.8 of the maximum intensity. In general, the spectral resolving power R of any dispersing instrument is defined by

$$R = \lambda/\Delta\lambda = \nu/\Delta\nu \tag{14.4}$$

where $\Delta\lambda$ is the minimum separation of two closely spaced lines that are just resolved.

As matter of fact, the primary limit to the spectral resolution is usually the optically dispersing element used in the spectroscopic apparatus. If the experimental resolution is extremely high, however, spectral lines will still exhibit an intrinsic width. In other words, the lines in discrete spectra are never strictly monochromatic but show a spectral distribution $I(\nu)$ around the central frequency $\nu_0 = (E_u - E_l)/h$, where E_u and E_l are the energies of the upper and lower levels of the transition under investigation. The width of such a line profile is usually expressed in terms of the so-called full width at half maximum (FWHM), which corresponds to the frequency interval $\Delta\nu$ where

$$I(\nu_0 - \Delta\nu/2) = I(\nu_0 + \Delta\nu/2) = I(\nu_0)/2 \quad (14.5)$$

14.2.1 Mechanisms of Line Broadening

There are several mechanisms that cause a broadening of the emission or absorption lines of atoms or molecules. These are usually divided into two categories, homogeneous and inhomogeneous broadening, depending upon whether the probabilities of absorption or emission of radiation with frequency ν are equal (homogeneous) or are not equal (inhomogeneous) for all the atoms of the sample.

14.2.1.1 Homogeneous Broadening

The natural (also named *radiative* or *spontaneous*) width is an example of homogeneous broadening and represents the intrinsic width of a spectral line. The finite lifetime τ of an energy level gives rise, according to the Heisenberg uncertainty principle $\Delta E\tau \sim h$, to an uncertainty in the energy of the level. As a consequence, the radiation emitted when an electron jumps from a level of lifetime τ_l to a level of lifetime τ_u is composed of different frequencies, and the relative frequency distribution gives rise to a lineshape described by a Lorentzian curve,

$$\begin{aligned} I(\nu) &= I(\nu_0)L(\nu - \nu_0) \\ &= I_0\gamma/[4\pi^2(\nu - \nu_0) + (\gamma/2)^2] \end{aligned} \quad (14.6)$$

where I_0 is the total intensity of the line, $\gamma = 2\pi\,\Delta\nu_{\text{nat}}$, where $\Delta\nu_{\text{nat}}$ is given by

$$\Delta\nu_{\text{nat}} = 1/2\pi\tau_{\text{u}} + 1/2\pi\tau_{\text{l}} \quad (14.7)$$

Typically, the lifetime of radiative levels is of the order of tens of nanoseconds, and hence typical natural widths are of a few megahertz.

Nonradiative de-excitation processes, such as collisions with other atoms or with the walls of the gas container, also play an important role in the shape of spectral profiles. Collisional (or pressure) broadening is proportional to the pressure p of the collision partner species, and the total homogeneous width is given by the sum of two terms:

$$\gamma_{\text{horn}} = \gamma_{\text{rad}} + \gamma_{\text{coll}} = \gamma_{\text{rad}} + ap \quad (14.8)$$

In some cases, the interaction time of atoms with the radiation field can be shorter than the spontaneous lifetime, giving rise to time-of-flight broadening. In such circumstances, the linewidth is determined by the interaction time T, and in the limiting case the width becomes

$$\gamma = 1/2\pi T \quad (14.9)$$

14.2.1.2 Inhomogeneous Broadening

The main cause of inhomogeneous broadening for atoms and molecules in the gas phase is the Doppler effect. The central frequency ν_0 emitted by an excited atom moving with velocity $\boldsymbol{v}$ relative to the rest

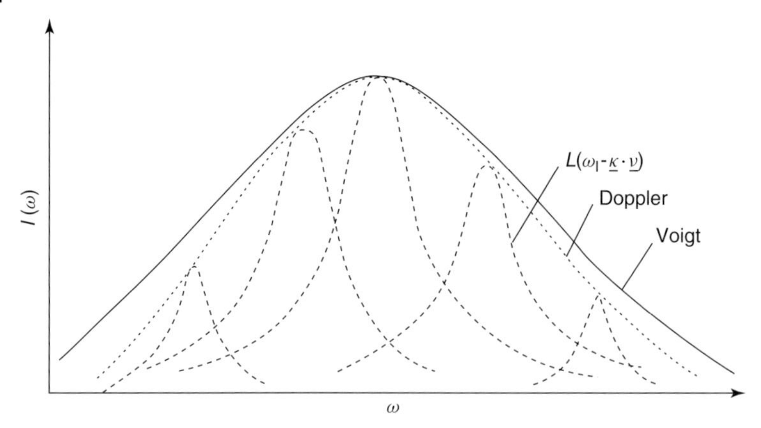

Fig. 14.4 Voigt profile obtained from the convolution of Gaussian and Lorentzian profiles. (From Demtroeder,1981.)

frame of the observer will be shifted by a quantity $\Delta\nu$ given by

$$\Delta\nu = \nu - \nu_0 = k \cdot v \quad (14.10)$$

where the wave vector $\mathbf{k}(|\mathbf{k}| = 2\pi/\lambda)$ indicates the direction of the emitted photon. At thermal equilibrium, the molecules of a gas follow a Maxwellian velocity distribution. The whole lineshape is hence given by the superposition of differently Doppler-shifted emitters, and the observed lineshape is Gaussian:

$$I(\nu) = I_0(0.94/\Delta\nu_{\mathrm{D}})\exp -[(\nu - \nu_0)^2/(0.36\Delta\nu_{\mathrm{D}}^2)] \quad (14.11)$$

where the FWHM is given by

$$\Delta\nu_{\mathrm{D}} = 7.16 \times 10^{-7}\nu_0(T/M)^{1/2} \quad (14.12)$$

with T the gas temperature (expressed in kelvin) and M the molar mass (expressed in grams). For example, for the sodium D_1 line, the Doppler width, at $T = 500$ K, $\Delta\nu_D = 1.7$ GHz, is about 2 orders of magnitude broader than the radiative width.

In our discussion of Doppler broadening, we have neglected the homogeneous broadening of each velocity class of atoms. This means that we should properly write the whole lineshape as a superposition of different Lorentzian curves, each centered at the Doppler-shifted frequency and weighted by a Gaussian curve. The result of such a convolution of a Lorentzian and a Gaussian is called a *Voigt profile* (see Figure 14.4). In practice, when $\Delta\nu_D \gg \Delta\nu_{\mathrm{hom}}$, the lineshape is well described by a Gaussian curve.

14.3 High-Resolution Spectroscopy

The intrinsic width of a spectral line is determined by spontaneous emission. Especially when the gaseous sample under investigation is rarefied to render collisions negligible, the Doppler effect will be the first obstacle to overcome.

At first sight, the easiest approach to reducing Doppler broadening might appear to be by cooling the sample. However, the slow dependence of $\Delta\nu_D$ upon temperature (see Equation (14.12)) means that, for instance, to reduce $\Delta\nu_D$ by 2 orders of magnitude, the gas temperature

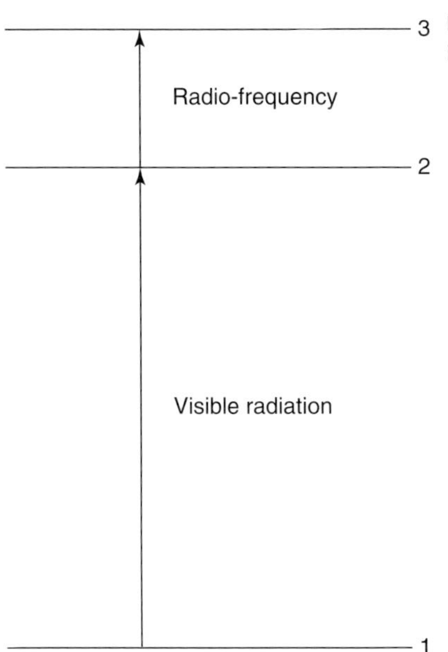

Fig. 14.5 Level scheme for double resonance involving optical and radio-frequency radiation.

must be decreased by 4 orders of magnitude. This is practically feasible only for atomic and molecular species with high vapor pressures.

Another approach that has been used to reduce Doppler broadening is the use of atomic beams. If a well-collimated beam of particles is irradiated perpendicularly by resonant radiation, then the first-order Doppler effect is removed ($\mathbf{k} \cdot \mathbf{v} = 0$). Because of the finite collimation of the beam, of course, the Doppler effect will still be present due to the small velocity component perpendicular to the atomic beam. If θ is the beam divergence, the residual Doppler effect is given by $\Delta\nu_{D-\text{res}} = \Delta\nu_D \tan\theta$.

It is worth noting that the relativistic second-order Doppler effect, depending upon $(v/c)^2$, is not eliminated even with a perfectly collimated beam. This quadratic Doppler effect can play an important role in ultrahigh-resolution spectroscopy, and we shall see later some novel techniques based upon laser cooling of the atoms that aim to overcome this limit. It should also be noted that, because the atoms interact with the laser radiation downstream of the nozzle, the method of atomic beam spectroscopy is not appropriate for the study of species in excited states with short lifetimes.

Doppler-free resolution is also possible using classical light sources. This has been achieved, for instance, using double-resonance schemes in which two successive transitions, one in the visible and another in the radio-frequency region, are excited simultaneously (see Figure 14.5). The radio-frequency transition will be between two closely spaced energy levels, such as fine-structure or hyperfine levels; the Doppler width of these transitions, as explained previously, will be negligible, and the spacing of the levels, if not their absolute positions, may thus be measured with high resolution.

Level splitting may also be investigated with the methods of level-crossing (Hanle effect) or quantum-beat spectroscopy. These techniques are based on the interference between different atomic radiative transition channels, which lead to a spatial distribution of the intensity and polarization of the emitted radiation. These techniques are particularly valuable for accurate measurements of fine and hyperfine structures and of the Zeeman and Stark splittings of atomic and molecular levels.

The advent of tunable laser sources has allowed the development of nonlinear techniques that eliminate Doppler broadening without suffering from these restrictions. These techniques are applicable to atoms and molecules in cells and hence have an immediate practical advantage with respect to beam methods, which require complex and expensive apparatus.

Doppler-free laser spectroscopic techniques are mainly based on two different schemes: saturation spectroscopy and Doppler-free two-photon spectroscopy. In the former, narrow-band cw laser radiation "marks" a small group of atoms within a narrow range of axial velocities. In two-photon spectroscopy an atom undergoes a transition between two quantum states of the same parity through the absorption of two photons, whose energies add up to the required energy. If the photons are absorbed from counter-propagating beams, the Doppler effect is cancelled. The basic principles of these sub-Doppler techniques are outlined using a rather simple approach in the following paragraphs.

14.3.1 Saturation Spectroscopy

We start with a single optical transition between two energy levels 1 and 2. Initially, we suppose that such a transition is homogeneously broadened. We recall that in linear (low-intensity) absorption spectroscopy, the absorption of the intensity I as a function of the optical thickness z of the sample is given by the well-known Beer–Lambert law

$$I(z) = I(0)\exp(-\alpha z) \tag{14.13}$$

where α is the absorption coefficient, a quantity depending upon microscopic properties of the medium under investigation. For low light intensities, the population densities of the upper (N_2) and lower (N_1) levels of the investigated transition are only slightly changed, and the coefficient α is thus independent of intensity:

$$\alpha = \sigma_{12}[N_1 - (g_1/g_2)N_2] \tag{14.14}$$

where σ_{12} represents the absorption cross section and g_1 and g_2 the degeneracies of the upper and lower levels, respectively. At higher intensities, N_1 may decrease significantly while the upper-state density N_2 increases, ultimately rendering the atomic sample completely transparent ($\Delta N = N_2 - N_1 = 0$). Since N_2 and N_1 are both functions of intensity, the absorption coefficient itself becomes nonlinearly intensity dependent. In this case, it can be shown that the population difference ΔN can be written as

$$\Delta N = \Delta N_0/[1 + S(\nu)] \tag{14.15}$$

where ΔN_0 is the population difference in the absence of laser radiation and $S(\nu)$ is the so-called saturation parameter,

$$S(\nu) = S_0 \frac{(\gamma/2)^2}{(\omega - \omega_0)^2 + (\gamma/2)^2} \tag{14.16}$$

where γ is the homogeneous breadth of the transition and $\omega = 2\pi\nu$ the optical angular

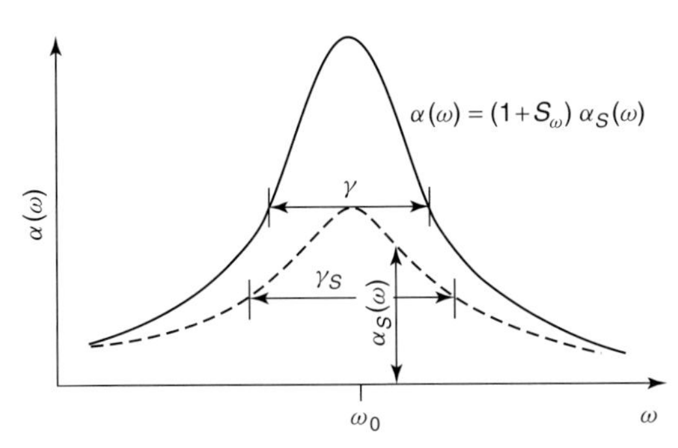

Fig. 14.6 Homogeneous line-shape affected by saturation. (From Demtroeder,1981.)

frequency. The saturated absorption coefficient $\alpha_s(\nu)$ thus becomes

$$\alpha_s(\nu) = \frac{C}{1 + I/I_s} \frac{(\gamma_s/2)^2}{(\omega - \omega_0)^2 + (\gamma_s/2)^2} \tag{14.17}$$

where C is a constant. Comparing Equation (14.17) with the unsaturated absorption profile of Equation (14.14), we see that saturation decreases the absorption by a factor $1 + S(\nu)$ and broadens the line profile by a factor $(1 + S_0)^{1/2}$, as shown in Figure 14.6.

Let us now consider the effect of the motion of the atoms. If the laser beam is a monochromatic wave of angular frequency ω and wave vector k, only atoms having a velocity component toward the laser in the interval $v \pm \Delta v$ given by the relation

$$(v \pm \Delta v)k = \omega - \omega_0 \pm \gamma \tag{14.18}$$

will absorb it. This causes a hole in the velocity distribution of the lower level and a corresponding bump in the upper-level distribution. As we have seen, this group of atoms will undergo saturation, and hence the hole (or dip) will exhibit a saturated profile. If the laser beam, after crossing the sample, is reflected back into the cell, then two symmetric holes (or dips) will be produced in the velocity distribution. As the laser frequency approaches the resonant frequency ν_0 of the transition, the two holes will collapse into one, and the two counterpropagating beams will simultaneously interact with the same group of atoms (see Figure 14.7a). Tuning the laser frequency across the Doppler width, it can be shown (in the weak-field approximation) that the absorption coefficient will be given by

$$\alpha_s(\omega) = \alpha D(\omega) \times \left[1 - \frac{S_0}{2}\left(1 + \frac{(\gamma_s/2)^2}{(\omega - \omega_0)^2 + (\gamma_s/2)^2}\right)\right] \tag{14.19}$$

which represents a Doppler profile modified by the presence of a minimum at the center of the line (the Lorentzian term in Equation (14.19)) as shown in Figure 14.7b.

In the previous analysis we have considered an atomic system at thermal equilibrium where $N_1 < N_2$. The same discussion can be extended to systems where a population inversion has been produced. This is, for instance, the case of a laser medium with an inhomogeneously broadened gain profile. If the length of the single-mode laser cavity is tuned so that the laser frequency $\nu_1 = \nu_0$, the output power will show a minimum around the central frequency ν_0.

Historically, saturation spectroscopy at optical frequencies dates from 1963 (three years after the discovery of the laser) when McFarlane, Bennet, and Lamb (1963) and,

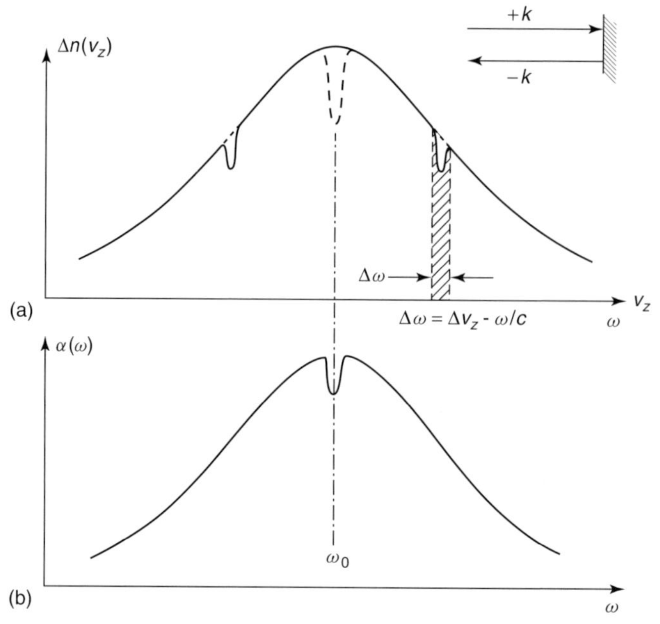

Fig. 14.7 (a) Holes in the velocity distribution of the lower level produced by two counterpropagating laser beams when the laser frequency is off resonance and in resonance. (b) The absorption profile as a function of the laser tuning.

independently, Szöke and Javan (1963) demonstrated the saturation minimum at the center of the emission curve of a single-mode He–Ne laser. Nevertheless, saturation spectroscopy became extensively used as a spectroscopic technique only with the advent of tunable dye lasers. Hänsch, Levenson, and Schawlow (1971) developed an experimental arrangement that provided a Doppler-free profile with a good signal-to-noise ratio. This scheme is shown in Figure 14.8. The primary beam is divided in two using a partially reflecting mirror (beam splitter). One of the two beams has a higher intensity than the other, by about a factor of 10, and is usually named the *pump* or *saturating beam*, while the weaker is called the *probe* or *analysis beam*. The two beams are sent in opposite directions through the cell, and the intensity of the transmitted probe is detected using, for example, a photodiode. When the laser is tuned across the absorption line, the probe beam sees the hole burned by the saturating beam, and the recorded lineshape consists of a Gaussian with a dip at the center. To isolate the saturating contribution, a phase-sensitive detection scheme can be used. For this purpose, the intensity of the pump beam is mechanically chopped at a given frequency – a few hundred hertz – and the signal processed using a lock-in amplifier.

A typical saturation spectrum is shown in Figure 14.9 for the neon transition 1s4 $(J = 1) - 2p4(J = 0)$ at 607.4 nm, obtained with a single-mode dye laser. The lower level of this transition is metastable and is populated by an electrical glow discharge. Figure 14.9a shows the Doppler-broadened absorption profile obtained with a single

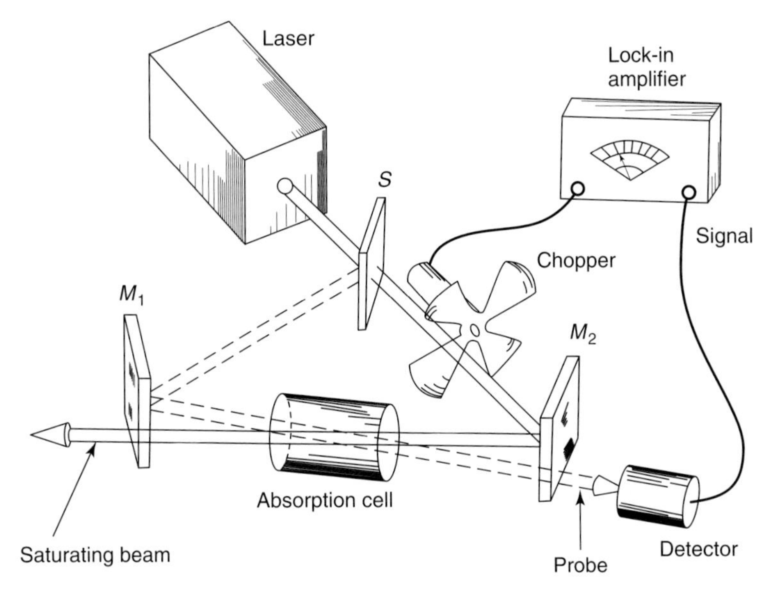

Fig. 14.8 Typical experimental arrangement for saturation spectroscopy. (From Hänsch and Schawlow, 1975.)

laser beam, from which a 2-GHz Doppler width may be estimated, corresponding to a gas temperature of about 500 K. A slight asymmetry is apparent, due to the presence of two closely spaced transitions corresponding to the two isotopes ^{20}Ne and ^{22}Ne present in the discharge with natural abundances of 91 and 9%, respectively. When counterpropagating beams are used, Lamb dips are produced at the resonances of the two isotopes (Figure 14.9b). Finally, a Doppler-free recording may be obtained using a phase-sensitive detection scheme (Figure 14.9c); in this case, the lines from two isotopes are completely resolved, and their separation is about 1.7 GHz.

14.3.2 Intermodulated Spectroscopy

As we have seen in the preceding section, saturation spectroscopy monitors changes in the absorption of the probe beam caused by a pump beam. This can significantly limit the sensitivity when weak transitions are to be studied. In such cases it is better to use "transverse" detection schemes where different observables are detected, such as fluorescent, optogalvanic (OG), and optoacoustic signals, treated in the following sections. For all these techniques, the detected signal is over a zero ground, and the sensitivity offered by it is orders of magnitude more than that offered by absorption spectroscopy.

Intermodulated spectroscopy is a technique derived from saturation spectroscopy (Sorem and Schawlow, 1972) in which one of these transverse detection schemes is used. The basic principle is illustrated in Figure 14.10. Here, the intensities of the pump (I_{pump}) and probe (I_{probe}) beams are of the same order of magnitude, and the beams are modulated at frequencies f_1 and f_2, respectively, that is,

$$\begin{aligned} I_{\text{pump}} &= I_0[1+\cos(2\pi f_1 t)], \\ I_{\text{probe}} &= I_0[1+\cos(2\pi f_2 t)] \end{aligned} \tag{14.20}$$

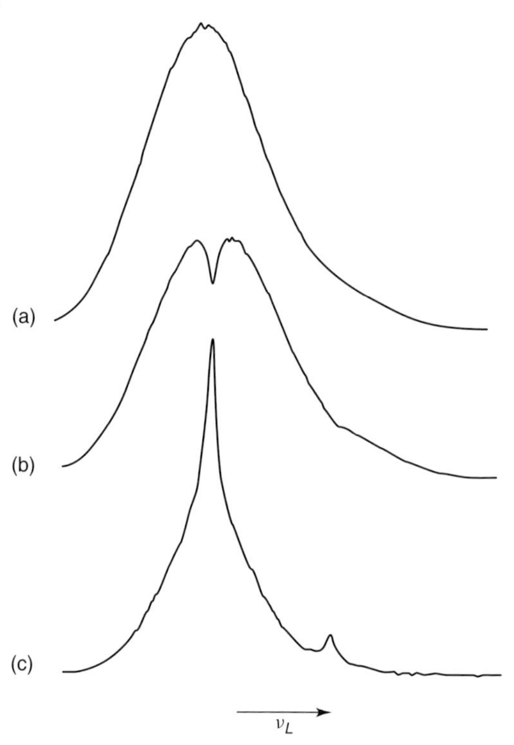

Fig. 14.9 Recording of the neon transition at 607.4 nm using a tunable single-mode dye laser. In (a) the line is recorded in presence of a single beam (Doppler broadened), while in (b) the effect of the saturation produced by the pump beam is visible within the Doppler profile as a homogeneous dip. (c) Doppler-free recording obtained by means of a phase-sensitive detection scheme.

The intensity of the fluorescent (or OG) signal I_{fl} induced simultaneously by the two beams is

$$I_{\mathrm{fl}} = CN_{\mathrm{sat}}(I_{\mathrm{pump}} + I_{\mathrm{probe}}) \tag{14.21}$$

where C is a constant that takes into account the transition probability and the detection efficiency, while N_{sat} is the saturated population density of the lower level and can be written as

$$\begin{aligned} N_{\mathrm{sat}} &= N_0(1 - S_0) \\ &= N_0[1 - a(I_{\mathrm{pump}} + I_{\mathrm{probe}})] \end{aligned} \tag{14.22}$$

N_0 being the unsaturated population density. By use of Equations (14.20) and (14.22), and assuming $I_{\mathrm{pump}} = I_{\mathrm{probe}} = I_0$, it may readily be shown that Equation (14.21) becomes

$$\begin{aligned} I_{\mathrm{fl}} = {} & CI_0 a_0(\omega - \omega_0) \\ & \times[1 + 1/2\cos(2\pi f_1 t) \\ & + 1/2\cos(2\pi f_2 t)] \\ & + C' I_0^2 a_0(\omega - \omega_0) \\ & \times[1 + 1/4\cos^2(2\pi f_1 t) \\ & + 1/4\cos^2(2\pi f_2 t) + \cos(2\pi f_1 t) \\ & + \cos(2\pi f_2 t) + \cos(2\pi f_1 t)\cos(2\pi f_2 t)] \\ & \times L(\omega - \omega_0) \end{aligned} \tag{14.23}$$

This shows that the fluorescence signal is composed of several terms modulated at different frequencies. The first two terms depend linearly upon intensity: they are modulated at frequencies f_1 and f_2, respectively, and originate from the pure interaction of each laser beam. Their dependence upon the laser frequency is, therefore, given by the Doppler-broadened absorption $a_0(\omega - \omega_0)$. The term $\cos(2\pi f_1 t)\cos(2\pi f_2 t)$, quadratically dependent upon intensity, is the most interesting because it gives rise to

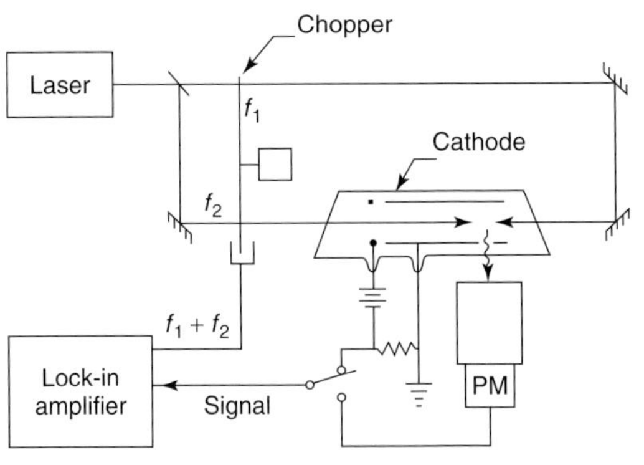

Fig. 14.10 Scheme for intermodulated spectroscopy. The sample consists of an electrical discharge, and the measurement is made by recording the fluorescence or optogalvanic signal at the sum or difference frequency produced by the interference of the modulation frequencies of the two laser beams.

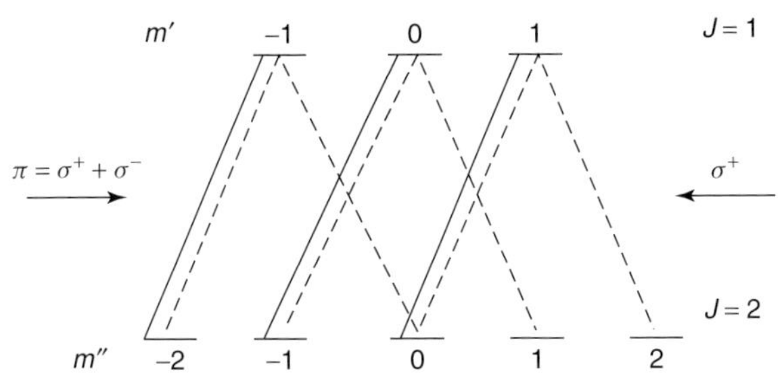

Fig. 14.11 The $\sigma+$ pump beam induces transitions with $\Delta m_j = +1$ creating a birefringence in the gaseous sample. The birefringence is monitored with a second counterpropagating beam (probe beam) normally extinguished between two crossed polarizers.

two terms modulated at the beat frequencies $(f_1 + f_2)$ and $|f_1 - f_2|$. These quadratic terms correspond to the simultaneous interaction of the two counterpropagating laser beams with the same velocity class of atoms that occurs when the laser frequency is tuned near the center of the line. Therefore, if the detected fluorescence signal is processed by a lock-in amplifier set to one of the two beat frequencies, a Doppler-free signal will be obtained.

14.3.3 Polarization Spectroscopy

Polarization spectroscopy is based on the light-induced birefringence and dichroism of an absorbing gas. These effects were first observed by Fornaca, Gozzini, and Strumia (1963) using classical radiation sources; polarization spectroscopy using single-mode tunable lasers was demonstrated by Wieman and Hänsch (1976).

In a polarization spectroscopy experiment, the laser beam is split into two beams of different intensities, the pump beam being circularly polarized by means of a quarter-wave plate while the probe beam is polarized linearly. As with the previous Doppler-free techniques, the two beams pass in opposite directions through the sample cell. The transmitted probe beam passes through a high-extinction-ratio polarizer that is set to maximum extinction, before falling upon the detector.

The basic idea of polarization spectroscopy can be understood with the help of Figure 14.11, which shows a transition between two levels having angular momenta of $J_1 = 2$ and $J_2 = 1$. Both levels are degenerate with $2J + 1$ Zeeman sublevels labeled with quantum number $m = J \ldots + J$, describing the projection of J along a quantization axis, which, in our case, is the laser beam direction. Transitions $m'' \rightarrow m'$ follow the selection rule $\Delta m = \pm 1$ for right ($\sigma+$) and left ($\sigma-$) circularly polarized light. Thus, if the pump beam is $\sigma+$ polarized only, some of the lower sublevels are partially depleted (the sublevels with $m = 2, 1$, and 0). The angular momentum of the light is hence transferred to the atomic sample, producing a nonuniform population of the m sublevels, which corresponds to a macroscopic anisotropy of the sample. Such induced birefringence can be "read" by the linearly polarized probe beam, which may be considered to be a superposition of $\sigma+$ and $\sigma-$ circular polarizations. In the absence of the pump beam, and when the laser frequency is resonant with the transition under investigation, the intensity transmitted by the crossed polarizer is

$$I_{(\text{pumpoff})} = I_0(\chi + \theta^2 + b^2)\exp(-\alpha l) \tag{14.24}$$

where I_0 is the probe beam intensity, χ is the finite extinction coefficient of the polarizer, θ is the angular deviation from orthogonality, and l is the thickness of the sample. Finally, b is a term that takes into account the residual birefringence introduced by stress induced in the windows of the cell.

When the laser frequency is tuned to within the homogeneous width of the line, the $\sigma+$ and $\sigma-$ components experience different absorption coefficients $\alpha+$ and $\alpha-$ and refractive indices $\eta+$ and $\eta-$ because of velocity-selective pumping by the $\sigma+$ pump beam. As a consequence, after passing through the sample, the resulting probe polarization will generally be elliptical, with the minor axis rotated with respect to the transmission axis of the analyzer (x-axis in Figure 14.12). In the limit of low saturation and a thin sample, it can be shown that the transmitted intensity I after the crossed polarizer is

$$I = I_0\left[\chi + \theta^2 + b^2 - \left(\frac{\theta\Delta\alpha Lx}{2(1+x^2)}\right) + \left(\frac{Lb\Delta\alpha}{2(1+x^2)}\right) + \left(\frac{(\Delta\alpha L)^2}{16(1+x^2)}\right)\right] \tag{14.25}$$

where $x = (\omega - \omega_0)/\gamma$ is the laser detuning relative to the homogeneous width and $\Delta\alpha$ is the difference between the absorption coefficients of the $\sigma+$ and $\sigma-$ probe components. In Equation (14.25), only $\Delta\alpha$ appears, because $\Delta\eta$ is related to $\Delta\alpha$ through the Kramers–Kronig dispersion relations.

Equation 14.25 shows the spectral profile of the polarization signal; it is given by a constant term $I_0(\chi + \theta^2 + b^2)$, a dispersion-shaped term, and a Lorentzian. By an appropriate choice of the experimental conditions, a pure Lorentzian or dispersive shape can be obtained.

14.3.4 Velocity-Selective Optical-Pumping Spectroscopy

Polarization spectroscopy is a particular case of a more general class of Doppler-free techniques based on velocity-selective optical pumping (Pinard *et al.*, 1979). The population distribution among the Zeeman sublevels can be expressed in terms of a power series in $m (m = -|J| \ldots + |J|)$:

$$D = \sum_k \sum_m N_m m^k \tag{14.26}$$

where N_m is the probability of occupation of the sublevel m. The zero-order term ($k = 0$) is the population P and is given by

$$P = \sum_m N_m \tag{14.27}$$

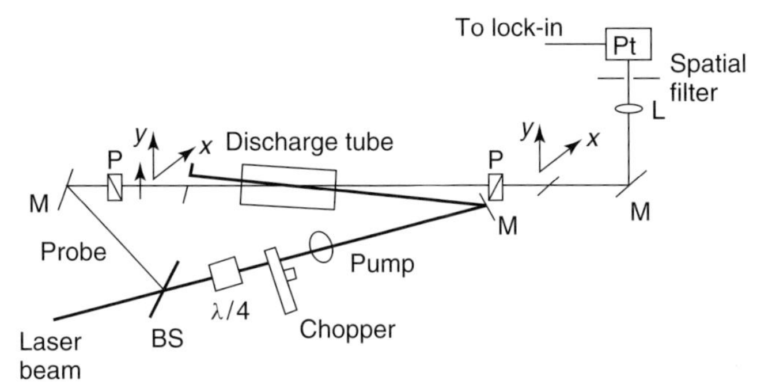

Fig. 14.12 Experimental arrangement for polarization spectroscopy.

while the first ($k = 1$) and second ($k = 2$) orders are the orientation O and alignment A expressed by

$$O = \sum_m N_m m \tag{14.28}$$

$$A = \sum_m N_m m^2 - \frac{1}{3}J(J+1) \tag{14.29}$$

Generally, the population distribution will be a superposition of orientation and alignment. Nevertheless, for particular light polarizations, a pure orientation or alignment can be achieved. For instance, pure orientation is created with circularly polarized light (atoms accumulated in the sublevels with $m = J(\sigma^-)$), while pure alignment may be induced using linearly polarized light.

14.3.5 Polarization-Intermodulated Excitation Spectroscopy (POLINEX)

The intermodulated technique was improved by Hänsch and coworkers by modulating the polarizations, rather than the amplitudes, of the counterpropagating beams. This method is referred to as the *polarization-intermodulated excitation*, or *POLINEX* (Hänsch and Toschek, 1970; Teets *et al.*, 1977). When the laser is tuned to within the homogeneous width, the total excitation rate is still modulated at the sum or difference frequency, but the combined absorption of the two beams in general depends upon their relative polarization. When the two laser beams share the same polarization, they will be preferentially absorbed by atoms of the same orientation, with a resultant increase in saturation; when the polarizations are different, the two beams will tend to interact with atoms of different orientation and, as a consequence, the total saturation decreases.

One advantage of POLINEX over intermodulation spectroscopy is that the cross-over signals are frequently of opposite signs and can therefore be easily distinguished even in complicated spectra. It should be noted, however, that if amplitude modulation is unintentionally present together with the polarization modulation, spurious signals can be produced.

14.3.6 Multiphoton Spectroscopy

In many spectroscopic techniques, two or more photons are involved. As a simple example, we consider the case where a level b is selectively excited by a first laser of frequency ω_1. This level represents a platform from which a second photon can be absorbed to reach a higher excited level c. Such a two-step excitation approach has been extensively used for many double-resonance schemes where the two photons lie in a wide spectral region (radio frequency, microwave, infrared, and visible). We now, however, focus our attention on a class of multiphoton spectroscopic techniques (Bruzzese, Sasso, and Solimeno, 1989) in which the jump from the lower level a to the excited level c occurs through the absorption of two or more photons (see Figure 14.13) without intermediate levels.

The first theoretical study considering the possibility of inducing transitions through two-photon absorption was performed in 1931 by Goeppert-Mayer (1931). Her idea was, at the time, experimentally unfeasible because of the low intensity of the available radiation sources: The higher-order terms of the perturbation theory, which describe multiphoton processes, become relevant only when the intensity of the electromagnetic field is sufficiently large with respect to the Coulombian atomic field. Since then, however, the development of maser and, successively, laser sources has offered the possibility of carrying out two-photon and more generally multiphoton processes, thereby providing powerful tools to investigate a new class of phenomena.

The first experimental evidence of multiphoton processes was produced by Brossel, Cagnac, and Kastler (1954), who used an intense radio-frequency field, and by Battaglia, Gozzini and Polacco (1959) with microwaves. In 1962, just two years after the operation of the first laser, Abella (1962) observed two-photon absorption between the $6^2S_{1/2}$ ground state and the $9^2D_{3/2}$ state of cesium by tuning a Q-switched ruby laser at $\lambda = 693.5$ nm by temperature control of the active medium. However, most spectroscopic applications based on multiphoton absorption followed the later development of pulsed tunable dye lasers.

The main advantages offered by two- (or more) photon processes are the following:

1. The selection rules for transitions induced by multiphoton absorption are different from those for single-photon processes. For instance, in the electric dipole approximation, two-photon transitions can occur only between levels having the same parity. As a consequence, it is possible to induce transitions from the ground state not otherwise feasible with single-photon absorption.
2. Two-photon excitation can be achieved by using visible or ultraviolet radiation instead of the higher-frequency radiation required for an equivalent single-photon transition.
3. If the two photons are absorbed from opposite directions, the first-order Doppler effect can be removed, and the resolution thus improved.

It is possible to demonstrate that the lineshape of a two-photon transition is given by the superposition of two curves: a Lorentzian curve, representing the Doppler-free term, and a Gaussian curve, taking into account the absorption of two photons traveling in the same direction (see Figure 14.14).

Doppler-free two-photon spectroscopy was first proposed by Vasilenko, Chebotaev, and Shishaev (1970) at Novisibirsk and demonstrated experimentally by Cagnac, Grynberg, and Biraben (1973) at Paris and by Bloembergen, Levenson, and Salour (1974) at Harvard. The investigated atom was chosen to be sodium because of its favorable energy-level scheme,

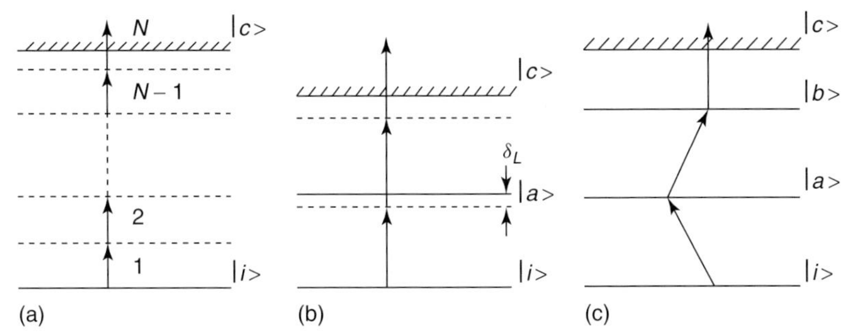

Fig. 14.13 Schematic representation of multiphoton processes: (a) nonresonant, (b) quasi-resonant, and (c) multistep excitation. Dashed lines represent virtual states.

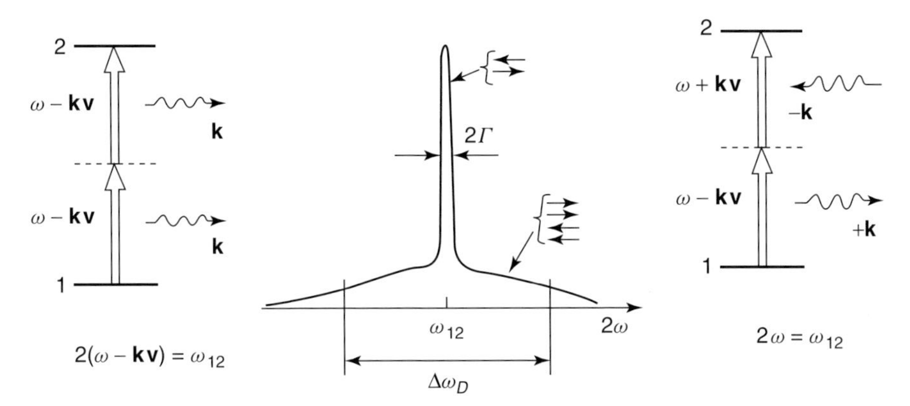

Fig. 14.14 Typical Doppler-free two-photon spectrum. The residual Doppler background is due to the absorption of two photons traveling in the same direction.

the two-photon transitions $3S - 5S$ at 602.23 nm and $3S - 4D$ at 578.73 nm both being in a spectral range where tunable dye lasers are easily available. Moreover, for both transitions, the intermediate $3P$ level makes the transition probabilities relatively large. Figure 14.15 shows a typical recording for the transition $3S_{1/2} - 5S_{1/2}$, where the two-photon absorption was monitored by observing cascading fluorescence from the $4P$ to the $3S$ state at 330.3 nm. This demonstrates another advantage of this technique, whereby an induced transition can be studied by detecting fluorescence at a wavelength quite different from that of the excitation radiation. Both the lower $3S_{1/2}$ and upper $5S_{1/2}$ levels consist of two hyperfine levels ($F = 3$ and 4). Because of the two-photon selection rules, only hyperfine transitions with $\Delta F = 0$ can be induced, giving rise to two spectral components separated by the combined hyperfine splittings of the lower and upper levels.

14.3.6.1 Doppler-Free Multiphoton Transitions

The Doppler effect can also be canceled when three or more photons are absorbed if a suitable geometry of laser beams is used. Consider the case of an atom irradiated by different beams having wave vectors $\mathbf{k}_i$. The Doppler shift for each interaction is $k_i \cdot v$,

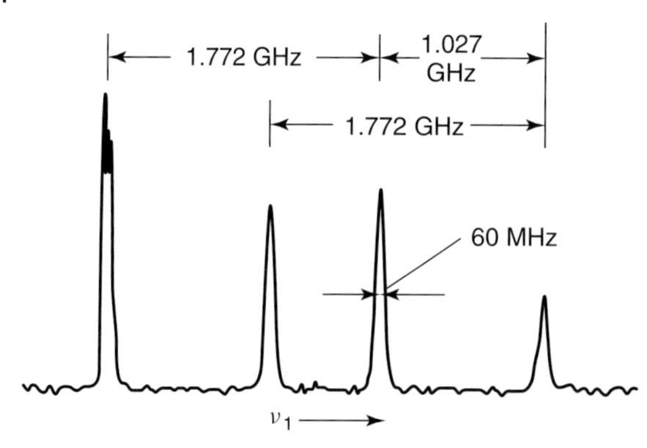

Fig. 14.15 The excitation spectrum for two-photon excitation of the 4*D* level in sodium. The peaks correspond to the hyperfine transitions between the ground state $3S_{1/2}(F = 1, 2)$ and the final state $4D_{3/2,5/2}$. The 1.772-GHz splittings are due to the hyperfine splitting of the ground state, while the 1.027-GHz splitting reflects the 4*D* fine structure. (From Bjorkholm and Liao, 1974.)

and hence, if the atom absorbs n photons, the total Doppler shift is given by $\Sigma_i k_i \cdot v$, where the sum extends to all n absorbed photons. So, if

$$\sum_i k_i = 0 \tag{14.30}$$

the n-photon transition is unaffected by the Doppler effect. In the particular case of photons of the same frequency, if $n = 2$, the two laser beams are counterpropagating ($k_1 = -k_2$), while for $n = 3$, the laser beams are directed along the bisectors of an equilateral triangle.

The first Doppler-free three-photon experiment, in which two photons were absorbed and one emitted, was made by Grynberg, Biraben, Bassini, and Cagnac.

14.3.6.2 Applications of High-Resolution Spectroscopic Techniques to the Hydrogen Atom: Lamb-Shift and Rydberg-Constant Measurements

Spectroscopic investigations of atomic hydrogen allow the testing of fundamental physical theory and the measurement of physical constants, with high accuracy. Doppler-free two-photon spectroscopy has been used by several groups to measure the Rydberg constant and to determine the Lamb shifts of the hydrogen energy levels accounted for by quantum electrodynamics.

The Rydberg constant is of particular importance because it is composed of only fundamental constants [$R_\infty = me^4/(8\varepsilon_0^2 h^3 c)$], so that a very precise measurement of R_∞ allows the consistency of other constants to be checked. The Rydberg constant can be determined by measuring the energy separation between two levels having different principal quantum numbers. R_∞ is related to these separations through the Bohr formula, with corrections for the isotope shift, Dirac fine-structure effects, and quantum-electrodynamical radiative effects.

Figure 14.16 shows a saturation spectrum of the Balmer-α line of atomic hydrogen produced in a gas discharge. Thanks to the high resolution offered by saturation spectroscopy, the separation between the $2S_{1/2}$ and $2P_{1/2}$ levels (Lamb shift) was thus determined. Furthermore, absolute wavelength measurements of the strong $2P_{3/2} - 3D_{5/2}$ component provided a new and more accurate value of the Rydberg constant. Nevertheless, the accuracy achievable in such experiments is generally dependent upon the characteristics of the atomic sample; indeed, a gas discharge introduces further broadenings and shifts of atomic lines due to collisions with other atoms and electrons.

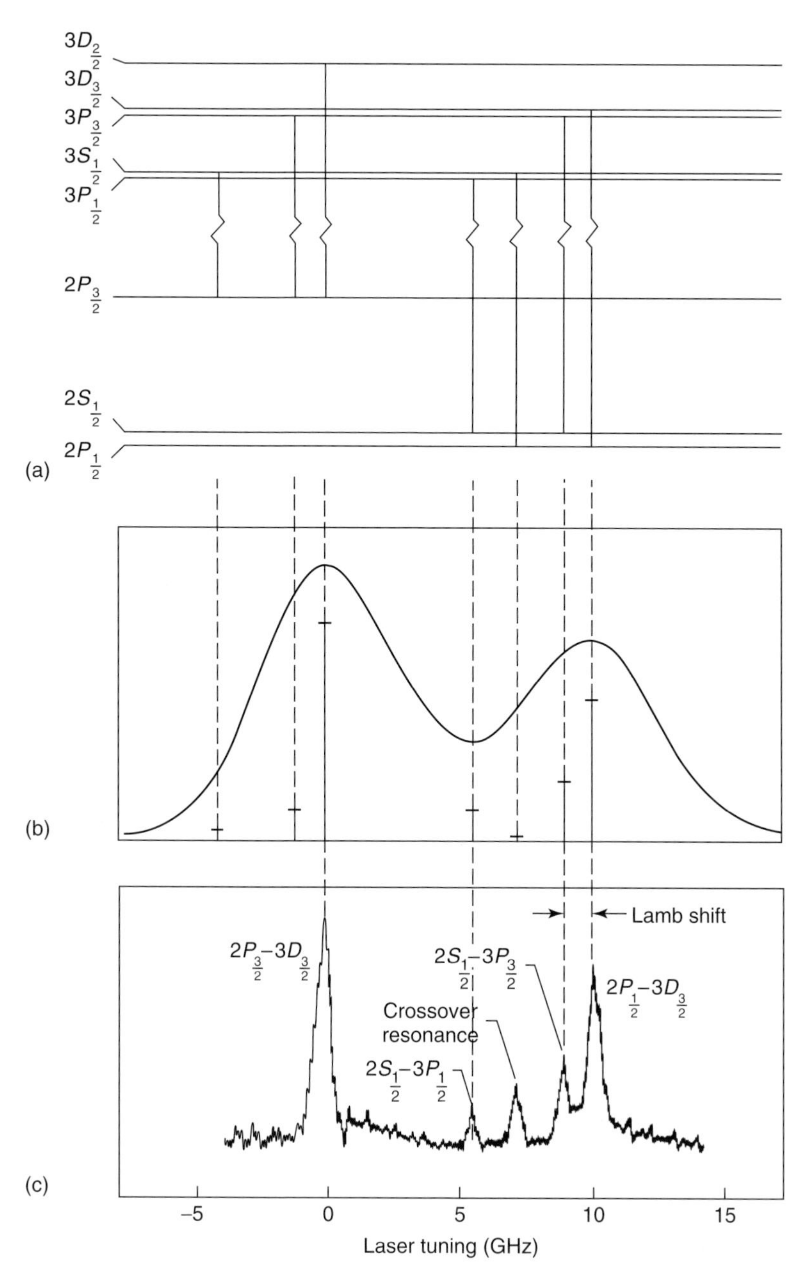

Fig. 14.16 Balmer-α line of atomic hydrogen: (a) energy levels with fine-structure transitions, (b) Doppler-limited emission line profile, and (c) saturation spectrum with optically resolved Lamb shift. (From Hänsch et al., 1975b.)

Atomic beams would, on the other hand, present an ideal atomic sample if collisions and other perturbations could be kept to negligible levels and if first-order Doppler broadening could be eliminated. To this end, the group of Lichten at Yale performed an optical study of hydrogen very similar to the historic experiment by Lamb at radio frequencies (Amin, Caldwell, and Lichten, 1981). Molecular hydrogen was dissociated by means of a hot wire, and the metastable $2S$ level was populated through collisions with a transverse electron beam. The atomic beam then interacted with a laser beam of low intensity. Absorption of the laser beam was monitored by detecting a variation in the flux of metastable atoms after the interaction with the laser beam. From an absolute wavelength measurement, Lichten derived one of the most accurate values of the Rydberg constant, $R_\infty = 109\,737.315\,73(3)\,\text{cm}^{-1}$.

Atoms with a single optical electron have also been used to determine the Rydberg constant. The idea is to investigate highly excited levels (Rydberg states) in which electrons experience a hydrogen-type Coulomb field. The advantage offered by this approach is that lasers with high spectral purity are unnecessary because the energy difference between closely spaced Rydberg levels corresponds to radiation wavelengths in the sub-millimeter range.

The first excitation of the $2S$ level by two-photon absorption was performed by Hänsch *et al.* (1979) and Hänsch (1977) at Stanford. The two-photon $1S-2S$ transition is particularly difficult because, in contrast to the $3S-5S$ transition in sodium, no intermediate levels are

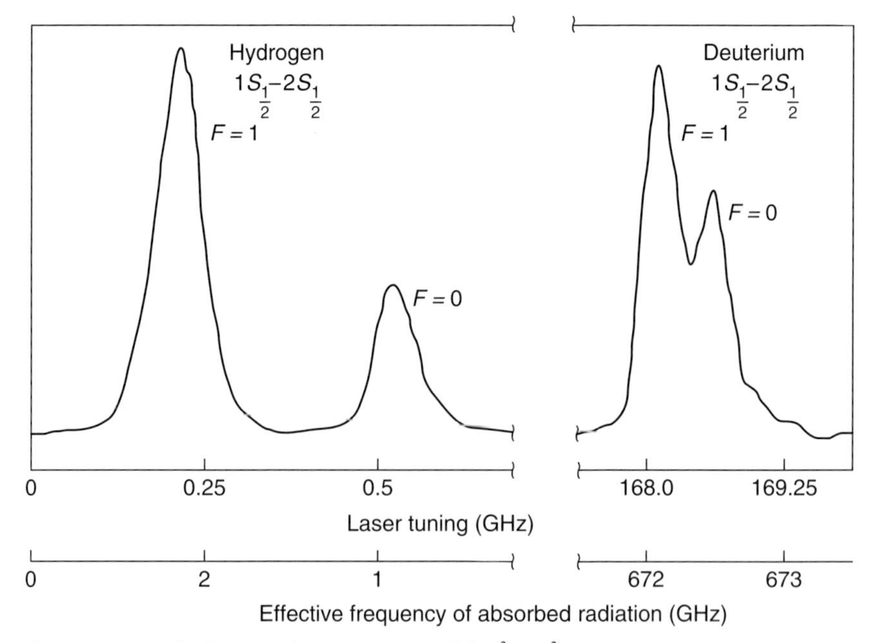

Fig. 14.17 Doppler-free two-photon spectrum of the $^3P_0 - ^3P_1$ transition of (a) atomic hydrogen and (b) deuterium. (From Hänsch et al.,1979.)

present. Furthermore, the 243-nm laser radiation required for this experiment is in a difficult spectral region and could be produced only by frequency doubling a 486-nm pulsed dye laser by means of a lithium fluoride nonlinear optical crystal. The two-photon transition was observed through collision-induced fluorescence on the Lyman-$\alpha(2P - 1S)$ transition at 121 nm. Frequency calibration was performed by sending part of the laser at 486 nm into a second discharge cell and observing the saturated spectrum of the Balmer-β line. In the absence of the Lamb shift, the frequency of the $1S - 2S$ transition should be four times that of the $2S - 4P$ line. The frequency difference between the two resonances (see Figure 14.17) is thus determined by the Lamb shifts of the $1S$ and $2S$ states. Using a previously measured value of the $2S$ Lamb shift, the Lamb shift of the ground state was determined to be 8151(30) MHz for hydrogen and 8177(30) MHz for deuterium.

The first version of the $1S - 2S$ two-photon measurement used pulsed lasers, and the linewidth of the observed transition was thus limited by the broad spectral width of these lasers. More recently, Hänsch *et al.* have investigated the same transition with much improved resolution by combining cw lasers with Fabry–Pérot built-up cavities.

14.4 High-Sensitivity Laser Spectroscopy

Laser radiation has been used extensively in spectroscopy to achieve not only high resolution but also high sensitivity, in order to investigate very weak transitions or to monitor species present at trace level.

The absorption coefficient α is related to microscopic parameters of the investigated atomic or molecular line by the relation

$$\alpha = [2\pi^2|\mu|^2\nu_{12}(N_1 - N_2)/3\eta\varepsilon_0 ch]g(\Delta\nu) \tag{14.31}$$

where ν_{12} is the frequency of the transition, μ is the dipole moment of the transition, η is the refractive index, and $g(\Delta\nu)$ is the lineshape. To determine α, we use the well-known Beer law, which describes the radiation absorbed by a length L of the sample:

$$I(L) = I(0)\exp[-\alpha L] \tag{14.32}$$

In the case of weak absorption, ($\alpha L \ll 1$), Equation (14.32) can be approximated by the first-order power series

$$I(L) = I(0)(1 - \alpha L) \tag{14.33}$$

Hence, the absorption coefficient can be determined from the fractional loss of the incident intensity:

$$\alpha L = [I(0) - I(L)]/I(0) \tag{14.34}$$

For weak absorption, however, this method cannot be very sensitive, because it measures small differences above a large background, and the minimum detectable absorption is thus limited by the amplitude fluctuations of the radiation source.

The advent of easily modulable radiation sources, such as lasers, has unleashed a wide variety of spectroscopic techniques that have increased the sensitivity enormously, allowing the detection of even a single atom. In the following section, we discuss some of these techniques.

The first approach is based upon the modulation, at a frequency f, of the laser amplitude (for instance, by using a mechanical chopper) and successive phase-sensitive signal detection by using a lock-in amplifier tuned to the same modulation frequency f. The advantage of this phase-sensitive method of detection is to restrict the noise to a narrow band centered around the modulation frequency f. In terms of the signal-to-noise ratio, frequency modulation is more sensitive than amplitude modulation because the frequency-independent background due to light scattered, for instance, from cell windows, is blocked. We note that in some applications it may be more practical to modulate the frequency of the absorption curve rather than that of the laser amplitude, as is the case, for example, with intracavity magnetic resonance spectroscopy.

Other more sensitive methods make use of observations related to the energy absorbed rather than the difference in that transmitted $(I(0) - I(L))$. Induced fluorescence, for instance, is proportional to the number of photons absorbed and provides a very sensitive detection scheme (excitation spectroscopy). In some cases, the absorbed energy can be transformed into thermal energy, giving rise to acoustic waves that are readily detected by a sensitive microphone (photoacoustic spectroscopy). Alternatively, if the atomic or molecular sample is prepared in a discharge or flame, absorption can be monitored by detecting the variation in the discharge current or ionization in the weak plasma represented by the flame (OG spectroscopy). This method can be generalized to the case of neutral samples, in which an atom or molecule is ionized through the resonant absorption of two or more photons. The electron–ion pairs thus produced are detected with unitary efficiency, and this method (resonance ionization spectroscopy) has been demonstrated to detect a single atom. A further technique achieves "amplification" of the absorption by including the sample within a laser cavity (intracavity spectroscopy).

14.4.1 Excitation Spectroscopy

Consider a spectroscopic arrangement in which the laser radiation is resonant with an atomic transition $1 \to 2$. The number of photons absorbed per unit time and unit path length, n_{abs}, is proportional to the incident photon flux, the absorption cross section σ_{12}, and the atomic density N_1 in the lower level. The number of photons emitted per unit time from the excited level, n_{fl}, is proportional to the number of photons absorbed, n_{abs}:

$$n_{fl} = \eta n_{abs} = N_2 A_2 \quad (14.35)$$

where $A_2 = \Sigma_m A_{2m}$ is the total spontaneous transition probability from the level 2 to all the levels with $E_m < E_2$. The quantum efficiency η is

$$\eta = A_2/(A_2 + R) \quad (14.36)$$

R being the collisional de-excitation rate. In the absence of collisions, all the atoms in level 2 decay emitting photons ($\eta = 1$) that can be collected with a given geometrical efficiency δ by a photomultiplier. Photons collected on the photocathode surface give rise to the emission of electrons at a rate

$$N_e = n_{abs}\eta\eta_e\delta = N_1\sigma_{12}N_{phot}\eta\eta_e\delta \quad (14.37)$$

where η_e is the quantum efficiency of the photocathode – that is, the ratio between the numbers of emitted electrons and incident photons. Modern photocathodes reach quantum efficiencies in excess of 40%, and, by use of photon-counting techniques, it is possible to measure

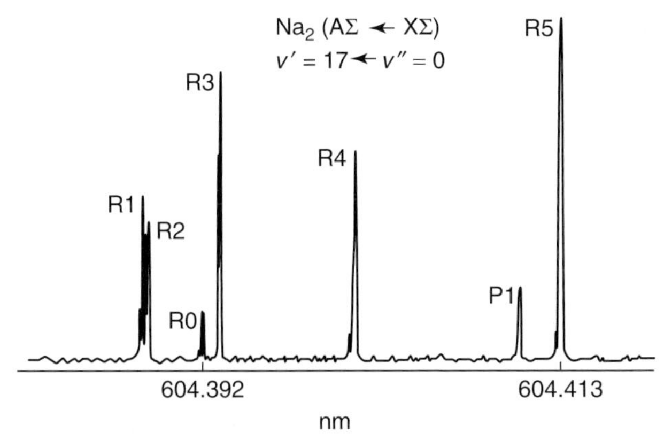

Fig. 14.18 Excitation spectrum of molecular sodium obtained in a molecular beam with radiation at 604 nm. (From Demtroeder, 1981.)

absorption rates photons per second. With a flux of photons $N_{phot} = 10^{18}$ photons s^{-1} (corresponding to a laser beam of 1 W at 500 nm), this implies that it is possible to detect a relative absorption of $\Delta I/I = 10^{-14}$. This impressive sensitivity can be further enhanced by placing the sample inside a laser cavity where the light flux is 1 or 2 orders of magnitude larger.

When the laser radiation is tuned across an absorption line, the total fluorescence intensity $I_{fl} = N_{phot}\sigma_{ik}N_i$ reproduces the absorption spectrum. Although the positions of the lines in the excitation spectrum are coincident with those of the absorption spectrum, the relative intensities can be different. Indeed, the quantum efficiency of both the transitions and the photocathode are dependent upon the laser wavelength. The geometrical efficiency may also vary: indeed, excited atoms with relatively long lifetimes may diffuse out of the interaction volume before emitting any fluorescence photons.

Excitation spectroscopy has been used in atomic- and molecular-beam spectroscopy where the density of particles is rather low. As an example, in Figure 14.18 we show a typical spectrum of molecular sodium, excited with radiation at 604 nm. The molecular density in the beam was $N = 10^8$ cm^{-3} while the absorption length was 10 mm. In the case of atomic sodium, absolute atomic densities down to 102 atoms cm^{-3} have been measured. The technique has also been successfully used to investigate radicals and short-lived intermediate products in chemical reactions.

Excitation spectroscopy becomes less sensitive in the infrared region, where the photocathode efficiency decreases. In addition, vibrational–rotational levels excited by infrared photons have lifetimes several orders of magnitude longer than do electronic states. As a consequence, excited molecules escape from the detection volume before they radiate. For such investigations, optoacoustic spectroscopy can be more advantageous.

14.4.2
Optoacoustic Spectroscopy

Although the optoacoustic effect was observed for the first time more than a century ago (Bell, 1881), spectroscopic applications began only with the development of laser sources (Tam, 1986). The technique is based on the fact that energy absorbed by molecules can be converted into rotational, vibrational, and translational energy. At thermal equilibrium, this energy is distributed over all the degrees of freedom, leading to sample heating. If the laser intensity is modulated at audio frequencies, sound waves of the same frequencies are induced inside the irradiated cell where they can be detected with sensitive microphones.

Optoacoustic spectroscopy is an unconventional detection scheme (Ernst and Inguscio, 1988) in the sense that microscopic processes, in the form of atom–photon or molecule–photon interactions, are detected through macroscopic changes of the system as a whole (in this case, an acoustic wave). Several advantages are offered with respect to conventional spectroscopic techniques where usually light is monitored. In this case, the detector is only sensitive to the radiation absorbed, and the problem of stray light is completely removed. Applications of optoacoustic spectroscopy are numerous and have concerned the monitoring of species at low concentrations in gases, liquids, and solids.

The sensitivity of optoacoustic spectroscopy can be enhanced if the cell is shaped to form a resonant acoustic cavity at the modulation frequency, allowing the detection of concentrations as low as one part in 10^9. This has also allowed the study of molecules at low pressures, where collisional broadening is small in comparison with Doppler broadening.

14.4.3
Optogalvanic Spectroscopy

The absorption of photons can alter the electrical equilibrium of an irradiated sample. In such cases, charge variations can be monitored, instead of the absorption or emission of light. This is the basic idea of OG spectroscopy, which is useful when the atoms or molecules under investigation are in a discharge (a weak plasma). In this case, resonant laser radiation can affect the electrical equilibrium of the discharge itself, producing an increase or decrease in the conductivity of the discharge known as the *optogalvanic effect* (*OGE*). The latter was first described by Penning (1928), who noted a variation in the impedance of a neon discharge when it was irradiated with emission from an adjacent neon discharge. Extensive and practical applications of the OGE had to await the introduction of tunable lasers (Barbieri, Beverini, and Sasso, 1990).

The mechanisms that translate the OGE into current perturbation are rather complex because of the large number of processes occurring within the discharge, but as a first approximation, the effect can be explained by considering the difference between the ionization cross sections σ_i for the two states involved in the optical transition. Such a simple model explains the fact that optical excitation leads to a change in the number of charges (electrons and ions), which in turn causes a change in the discharge impedance. In a typical OG detection scheme, the laser beam is amplitude modulated by a chopper, and variations in current are detected by means of a lock-in amplifier. The experimentally observed perturbation of the discharge characteristics induced by laser radiation is usually sufficiently small that the OGE can be considered linearly proportional to the number of photons absorbed.

Many features make OG spectroscopy attractive when compared with other conventional spectroscopic techniques in the ultraviolet, visible, and near-infrared regions of the spectrum. First, it, is inexpensive since it does not require the use of such devices as a monochromator or photomultiplier tube. A glow discharge is an inexpensive way of obtaining fairly large densities of excited states in volatile elements, especially in metastable states, and gaseous states of refractory elements are easily produced by sputtering in hollow-cathode discharges. A remarkable population of atoms in excited levels is present in the discharge as a result of electron–neutral-atom collisions, allowing transitions between excited levels to be investigated. OG spectroscopy permits one to record atomic and molecular lines that would otherwise be measurable only in an atomic or molecular beam. In molecular spectroscopy, OG techniques are the most natural way to study atomic and molecular species and radicals produced from parent molecular compounds that are present in the discharge.

OG spectroscopy is a relatively cheap technique and it is intrinsically more sensitive than that based upon absorption, yielding a signal against a zero background. In comparison with fluorescence techniques, OG spectroscopy offers the advantage of being unaffected by either the luminosity of the discharge or the stray scatter of excitation radiation.

14.4.4 Intracavity Spectroscopy

Spectroscopic sensitivity can be enhanced if the sample is placed within the laser cavity. Indeed, at a laser output power P_{out}, the power inside the laser cavity is

$$P_{\text{inside}} = (1/T)P_{\text{out}} = qP_{\text{out}} \tag{14.38}$$

T being the transmissivity of the output coupling mirror. By suitably shaping the laser cavity, a region can be created where the beam is focused (the waist) and the intensity thus further increased. For $\alpha L \ll 1$ the absorbed intensity is

$$\Delta I = (1/T)\alpha L I_{\text{out}} \tag{14.39}$$

that is, $1/T$ times the absorption rate observed outside the cavity. For a typical transmissivity of 0.02 ($R = 98\%$), the amplification factor is 50.

If the cell cannot be placed inside the active resonator, an external passive resonator may be used. By matching the laser output to a fundamental mode of the passive resonator and making as low as possible the losses g, the power circulating inside the resonator can be $1/q$ times the laser output power. Passive resonators represent an improvement over multipass cells and, since the atoms or molecules experience two counterpropagating beams, a Doppler-free scheme may be adopted.

Intracavity absorption can be monitored through the laser-induced fluorescence or using other detection schemes already discussed (optoacoustic, optogalvanic, etc.). Further sensitivity enhancement can be obtained if the output power is monitored and the laser is running close to threshold. Indeed, because of the nonlinear response of the laser near threshold, small changes in the losses Δg caused by the absorbing intracavity sample lead to a dramatic variation of the laser output. If G_0 is the unsaturated gain and g the total loss, it can be shown that the amplification factor q is

$$q = G_0/(G_0 - g)(g + \Delta g) \simeq G_0/g(G_0 - g) \tag{14.40}$$

where it has been supposed that $\Delta g \ll g$. Equation (14.40) shows that the sensitivity can be greatly enhanced when the

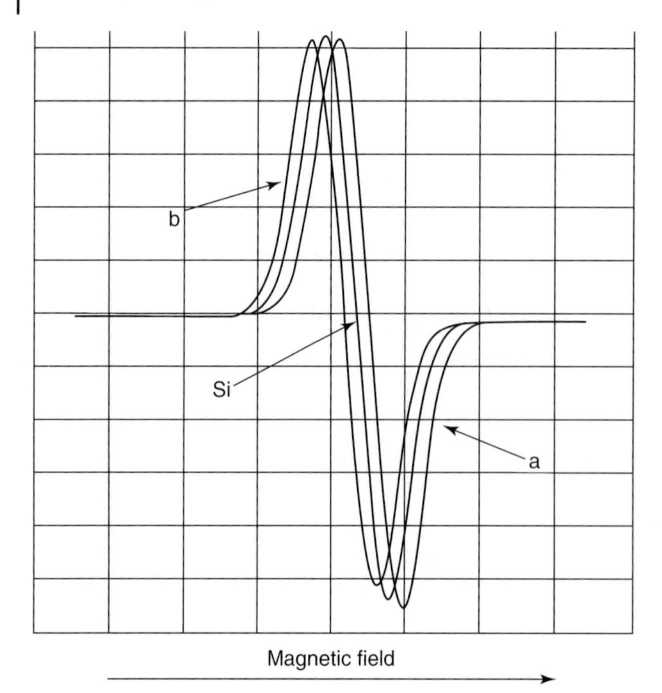

Fig. 14.19 Laser magnetic resonance measurement of the silicon ground-state fine-structure component ${}^3P_0{}^3P_1$ of the 129.5-mm line CH_3OH. In (a) and (b), the far-infrared cavity was slightly detuned to frequencies respectively higher and lower than the transition to the ground state. (From Evenson, 1981.)

threshold condition is approached ($G_0 \to g$). Nevertheless, an upper limit is imposed by the huge instabilities that a laser exhibits when such a condition is reached. Amplification factors of 10^5 have been obtained, allowing the study of very weak transitions having oscillator strengths below 10^{-12} or with very low concentrations below 10^8 atoms cm^{-3}.

An interesting demonstration of the high sensitivity reached by this technique is shown in Figure 14.19, which shows the fine-structure ${}^3P_0 -{}^3 P_1$ component of the ground state of silicon. Atoms of silicon were produced by dissociation of silane (SiH_4) in a microwave discharge, and the resonance condition was achieved by applying a magnetic field of 1121.2 g. It should be noted that there is an excellent signal-to-noise ratio in spite of the extremely low transition probability $A = 8 \times 10^{-6}\ s^{-1}$.

14.4.5 Fast-Modulation Spectroscopy

We have already seen that a significant improvement in sensitivity over direct absorption techniques may be achieved by modulating the laser amplitude at a frequency f and using a phase-sensitive detection scheme. This principle can be extended to modulating the laser frequency (wavelength modulation) at a relatively high frequency (kilohertz) and by an amount typically several times smaller than the width of the absorption line of the species under investigation (wavelength-modulation spectroscopy, WMS). Scanning the laser wavelength and using AC detection at the modulation frequency, or twice the modulation frequency, provides a detected signal that is the first- or second-derivative of the absorption lineshape. When the modulation frequency is comparable with or larger than the linewidth (megahertz or gigahertz range), the two modulation sidebands may be resolved, and the technique is known as *frequency-modulation spectroscopy* (*FMS*). Such techniques are particularly suited to semiconductor diode lasers, whose wavelengths may be modulated at very high frequencies simply by modulating the injection current. In this fashion, absorptions down to 10^{-9} cm^{-1} can be measured. The advantages of these techniques derive essentially from the high modulation frequencies, which allow the predominantly low-frequency $1/f$ noise to be rejected. It is worth noticing that, since the signal is detected at the modulation frequency f (or at $2f$), the bandwidth of the used detector must be higher than f. Detectors with such large bandwidth are expensive and mechanically difficult to handle. These problems can be overcome by using a two-tone frequency-modulation approach (TTFMS, two-tone frequency-modulation spectroscopy). In this case, the laser is modulated at two frequencies ($f_{1,2} = f_0 \pm \Omega/2$), which are of the order of the linewidth, and

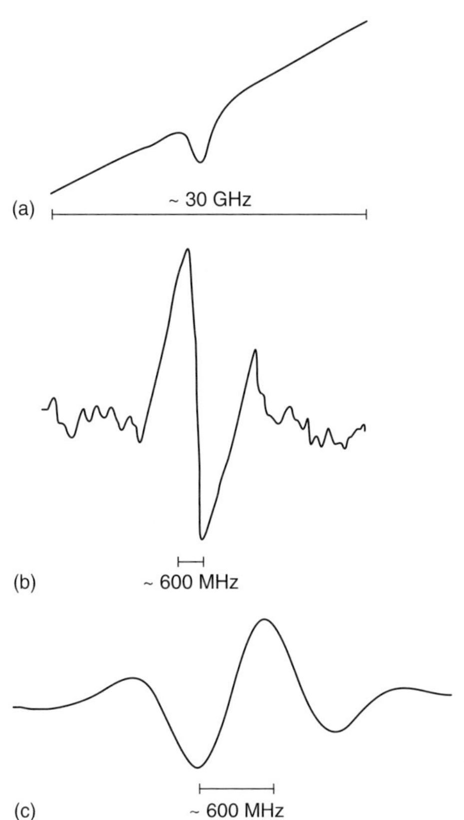

Fig. 14.20 (a) Pure absorption signal on the third overtone of methane (CH4) at 133 hPa. Derivative lineshapes recorded at 66.6 Pa with (b) WMS and (c) two-tone FMS.

absorption is detected at the beat frequency Ω, which is typically of the order of few megahertz. In this scheme, the $1/f$ noise is still quite small (although larger than in FMS), and the use of large bandwidth detectors can be avoided.

In Figure 14.20, the same transition in methane has been observed using absorption, wavelength-modulation, and frequency-modulation techniques. The signal-to-noise ratio in the frequency modulation (FM) case clearly represents an improvement of several orders of magnitude. Moreover, the sloping baseline is removed, and high discrimination against signals that do not show wavelength dependence is introduced.

14.5
Time-Resolved Spectroscopy

In the preceding sections, we have discussed a variety of spectroscopic techniques in which information about atomic or molecular structures is obtained by analyzing the response of the system as a function of the optical frequency. In this section, by contrast, we consider the temporal response of a system when it is irradiated by short laser pulses. We will see that spectroscopic techniques based on the temporal and frequency domains can provide complementary information.

The availability of short and intense laser pulses allows the transfer of a large fraction of an irradiated sample of ground-state atoms or molecules into excited states. By use of stepwise excitations with synchronized lasers, highly excited levels can be reached. By monitoring of the temporal decay of fluorescence from a given level, the lifetime of that level can be measured. If two closely lying states are excited coherently, a new class of phenomena can be studied, such as quantum beats, whereby atomic interference is observed as a temporal modulation of the fluorescent intensity. The recent development of ultrashort laser pulses, of picosecond or sub-picosecond duration (Shapiro, 1977), makes possible the investigation with extremely high time resolution of ultrafast relaxation processes occurring during the excitation and deactivation of molecular states.

The interaction between atoms and very short, intense laser pulses also gives rise to strong nonlinear processes. Atoms in very intense laser fields can be ionized by absorbing tens of photons. Furthermore, within the same laser pulse, the ions produced can be further ionized several times (multicharge ionization), while quasi-free photoelectrons may absorb further photons (above-ionization spectroscopy). In such strongly nonlinear regimes, high-order harmonics reaching into the X-ray region can be generated.

14.5.1
Lifetime Measurements

Lifetime measurements are of particular interest in spectroscopy, for the lifetime is directly related to the transition probability, and hence, such measurements make possible the testing of quantum mechanical calculations. Anomalies among the lifetimes within a molecular band can also provide information about perturbations due to coupling with other series. Finally, lifetimes are of more applied interest – for instance, in laser physics.

In Section 14.2 we showed that the radiative width of a line is related to the lifetimes of the lower and upper levels of the transition. We also mentioned that the homogeneous width is usually affected by other broadening effects as saturation, time of flight, collisions, and so on. Even if these further broadening mechanisms are removed, to determine from a transition linewidth the lifetime of a given level requires knowledge of the lifetime of the other level (unless it is the ground state). Unlike frequency-domain spectroscopy, time-resolved measurement allows the lifetime to be evaluated directly. Several experimental approaches have been developed.

14.5.1.1 **Phase-Shift Method**

This method is based on the use of a cw laser, sinusoidally modulated in amplitude at a frequency f and resonant with a transition $i \to k$ of frequency ω_{ik}. The rate equation for the final state k is

$$dN_k/dt = \sigma F(N_i - N_k) - N_k/\tau \qquad (14.41)$$

where σ is the absorption cross section, F the photon flux, and τ the lifetime of the level k. The difference $N_i - N_k$ in Equation (14.41) takes into account absorption and stimulated emission, while N_k/τ represents the spontaneous emission decay. The photon flux is

$$F = (I_0/\hbar\omega_{ik})(1 + \alpha \sin 2\pi ft) \cos \omega_{ik}t \qquad (14.42)$$

The induced resonant fluorescence is modulated at the same frequency but is shifted in phase with respect to the forcing excitation field by an amount ϕ, related to the level lifetime τ by

$$\tan \phi = 2\pi f \tau \qquad (14.43)$$

By measuring the phase shift ϕ with a lock-in amplifier, the lifetime τ can thus be estimated.

This method fails when two or more levels with different lifetimes are simultaneously populated. In this case, the induced fluorescence for each level presents different phase shifts, which are not easily determinable. The method also suffers in the event of stimulated emission (see Equation (14.41)), which is especially likely to be present when laser sources are used.

14.5.1.2 **Pulse Excitation**

Another approach to lifetime measurement makes use of pulsed lasers able to excite the atoms or molecules in a time short compared with the lifetime of the investigated level. The fluorescence decay can be observed in the absence of stimulated emission in two different ways.

With the first method the fluorescence is monitored directly by means of a transient digitizer or a boxcar. To reconstruct the whole exponential curve using either technique requires many fluorescence photons per excitation pulse, and nonlinearities in the photomultiplier response present a further limitation to this technique. These difficulties can be overcome by a delayed-coincidence method, which operates in the regime of extremely low intensity. In this technique, single-photon counts are recorded while the repetition rate of the excitation laser pulse is chosen to be as high as possible. At the heart of the experimental scheme is a time-to-amplitude converter. A trigger signal synchronized to the laser pulse starts a voltage ramp at time t_0, which is stopped by the first fluorescence photon after the excitation pulse, detected at time t_1. The ramp height is thus proportional to the time interval $t_1 - t_0$ and is stored by a multichannel analyzer. After each excitation pulse, a given channel will be increased by one unit, and the voltage distribution recorded by the multichannel analyzer thus yields the decay curve directly (see Figure 14.21).

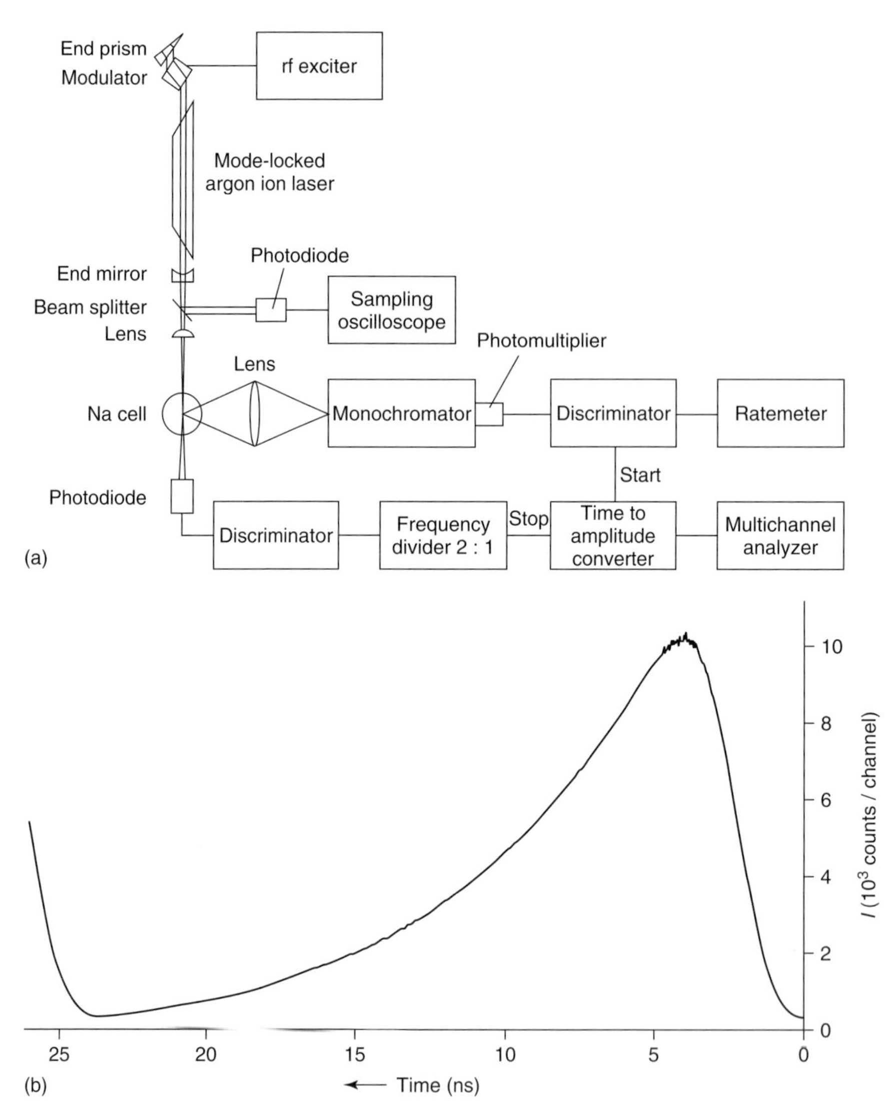

Fig. 14.21 (a) Schematic diagram of the delayed-coincidence technique. (b) Fluorescence decay, reconstructed by a multichannel analyzer. (from Demtroeder, 1981.)

It is important that the detection probability be kept below one fluorescence photon per excitation pulse to avoid so-called pileup. If two or more photons were generated, the photons with short delays would be overrepresented, and the decay curve would be altered.

Although a high repetition rate is required to provide a reasonable measurement time, the pulse frequency also has an upper limit, for the time interval between two pulses must be longer than the lifetime of the investigated level. Moreover, the pulse frequencies must not exceed the reciprocal of the dead time of the electronic chain, which is typically 100 ns.

The decay of an excited state can also be monitored with a second laser (probe) tuned to a transition sharing a common level with the investigated transition (pump-and-probe technique). When the pump pulse alters the population of this level, the probe pulse can monitor how quickly the population is restored through relaxation processes if a controlled delay is introduced between pump and probe pulses. The pump-and-probe technique is particularly useful for investigating ultrafast processes (picosecond and femtosecond range) where conventional techniques fail.

14.5.2 Quantum-Beat Spectroscopy

If two or more closely spaced atomic or molecular levels are simultaneously excited by a short laser pulse at $t = t_0$, the total wave function will be given by the superposition of the wave functions of the excited states:

$$\psi(0) = \sum_k a_k \Phi(0) \tag{14.44}$$

where a_k is the probability amplitude that the light pulse has prepared the atom in level $|k\rangle$. The time-dependent intensity of the fluorescence emitted when the excited levels decay into the final state $|f\rangle$ is given in terms of the matrix element by

$$I(t) = CI|\langle \Phi_f | \varepsilon r | \psi(t) \rangle|^2 \tag{14.45}$$

where ε is the polarization vector and $\psi(t)$ is the temporal evolution of the total wave function, given by

$$\psi(t) = \sum_k a_k |\Phi_k(0)\rangle \exp[-(iE_{kf}/h + \gamma_k/2)t] \tag{14.46}$$

Inserting Equation (14.46) into Equation (14.45) and assuming equal the decay rates of levels 1 and 2 ($\gamma_1 = \gamma_2 = \gamma$), we have

$$I(t) = C\exp(-\gamma t)(A + B\cos\omega_{21}t) \tag{14.47}$$

where A is the sum of the transition probabilities $1 \to f$ and $2 \to f$, B is the interference term, and $\omega_{21} = (E_2 - E_1)/h$. The time-dependent fluorescence is hence given by the superposition of an exponential curve and an oscillation with frequency ω_{21}. The measurement of such an oscillation allows the determination of very closely spaced energy levels. Quantum-beat spectroscopy therefore allows Doppler-free resolution.

14.6 Ultrahigh-Resolution Spectroscopy

Many of the processes by which spectral lines are broadened are due to the random motion of the atoms. Let us consider an atom of mass M and velocity v. On account of its motion, the atom absorbs photons at a frequency ω' given by the relation

$$\omega' = \omega_0 + kv - \omega_0 v^2/2c^2 \tag{14.48}$$

where $\omega_0 = (E_b - E_a)/h$ is the resonance frequency and the second and third terms represent, respectively, the first- and second-order Doppler shifts.

In the preceding sections we have seen that the linear (first-order) Doppler shift can be canceled by use of atomic beams or by means of one of the numerous Doppler-free spectroscopic techniques, but such methods fail to eliminate the second-order Doppler shift, which, as shown in Equation (14.48), depends quadratically upon the atomic velocity. Another effect strictly connected to the atomic motion is the previously discussed transit-time (or time-of-flight) broadening. In many cases, these two broadenings determine the ultimate limit of spectral resolution. This is the case, for instance, for transitions involving long-lived levels for which the radiative width is very narrow. Full resolution of such narrow transitions is of importance in testing the fundamental laws of physics and for many metrological applications such as atomic clocks. To achieve maximum resolution, it is necessary to reduce the atomic velocities, since this reduces the second-order Doppler shift and increases the interaction time between the atoms and the laser field.

When an atom interacts with a light beam, the emitted and absorbed photons carry a lot of information about the atomic structure, this being the essence of spectroscopy. But the interaction of photons with atoms can also be used to manipulate the kinetic status of the atoms, that is, their velocity. This phenomenon, usually manifested as so-called laser cooling, was first suggested independently by Hänsch and Schawlow (1975) for neutral species and by Wineland and Dehmelt (1975) for trapped ions.

In the last few years, there has been increasing interest in the use of near-resonant photon scattering to cool and trap atoms and ions. This interest is motivated in part by the physics governing such phenomena and in part by the wide variety of atomic and molecular physics experiments to which it may be applied. The production of ultracold atomic samples alone, having temperatures of a few tens of microkelvins, has opened up a rich area for experiments in atomic and molecular physics. One of the main applications is in ultrahigh-resolution spectroscopy, since for very slow atoms the first- and second-order Doppler shifts and the pressure shift are eliminated, and Doppler-free techniques are no longer required.

The production of cold atoms moreover offers the possibility of revisiting the physics of collisions in a new light. There are two principal reasons for investigating collisions in atomic traps: The first is to understand, and perhaps to minimize, the role of such collisions because they represent a significant loss channel; the other is to explore the fundamental physics of ultracold collisions between atoms with large de Broglie wavelengths, for which hitherto unseen effects can be expected.

14.6.1 Laser Cooling

The principle of laser cooling is best illustrated by a two-level atomic system (Meystre and Stenholms, 1985; Chu and Wieman, 1989). We consider an atom of mass M moving along the z axis with a velocity v. A photon of frequency ω traveling in the opposite direction can be absorbed if ω is appropriately Doppler shifted with respect to the resonance frequency ω_0.

The photon momentum $h\nu/c$ changes the velocity of the atom by

$$\Delta v = h\nu/Mc \qquad (14.49)$$

We suppose that the excited atom can decay radiatively only back to the initial state. During this process, the atom velocity is modified when it emits a photon by spontaneous or stimulated emission. In the presence of a single laser beam, stimulated photons will be emitted in the same direction as the incoming photons, and as a consequence, the momentum transferred by the absorbed photon will be canceled. In contrast, the spontaneous photons will be emitted in random directions; the average momentum transfer is in this case zero, and there is thus overall a net momentum transfer from the radiation to the atoms. Because this process depends upon spontaneous emission, the radiation pressure force that results is often known as the *spontaneous force.*

The recoil depends on the energy of the photon ($h\nu$) and the mass M of the atom ($\Delta v = h\nu/cM$). For instance, for sodium ($M = 23$ u) and $\lambda = 589$ nm, the velocity change per photon absorbed is about 3 cm s^{-1}. If the initial atomic velocity is typical of room-temperature distributions (about 103 m s^{-1}), the number of scattered photons required to stop the atoms will be $n_{\text{phot}} = v/\Delta v = 3 \times 10^4$. The cooling transition a $\rightarrow$ b must therefore be closed, in the sense that atoms in upper level b may decay radiatively only to the initial state a. For sodium, the $^3S_{1/2} -^3 P_{3/2}$ transition is suitable for such a cooling scheme, even though the presence of hyperfine structure renders the transition incompletely closed (optical pumping).

During laser cooling, atoms undergo a maximum deceleration given by

$$a = h\nu/2Mc\tau \qquad (14.50)$$

where τ is the lifetime of the upper level of the cycling transition and the factor 2 takes into account the fact that atoms spend only half of their time in the upper level. Again for sodium, $a = 10^6$ m s^{-2}. This strong acceleration is sufficient to bring to rest a thermal sodium atom in 1 ms and over a distance of 0.5 m, which is quite reasonable on the laboratory scale.

14.6.2 Atom Traps

In the preceding section, we have discussed a method of stopping the atoms in an atomic beam. Another aspect of atomic manipulation is confinement to a region of space where the atoms can then be investigated.

In 1985, Chu *et al.* demonstrated the possibility of cooling neutral sodium atoms in so-called optical molasses. Magneto-optical traps (MOTs) represent a refinement of this idea whereby atoms are localized at higher densities and cooled to temperatures limited by the atomic recoil energy.

14.6.2.1 Magneto-Optical Traps

The MOT (Raab *et al.*, 1986) depends upon two features: the Zeeman interaction with an inhomogeneous magnetic field, and the radiative selection rules that govern transitions between magnetic levels in an atom.

The basic principle of the MOT can be illustrated by considering a hypothetical atom with a spin $S = 0$ ($m_s = 0$) ground state and a spin $S = 1$ ($m_s = -1, 0, +1$) excited state. In a weak, inhomogeneous magnetic field $B(z) = \text{bz}$, the energy levels will be split through the Zeeman effect by an amount $\Delta E = \mu m_s bz$. In the presence of two counterpropagating waves with

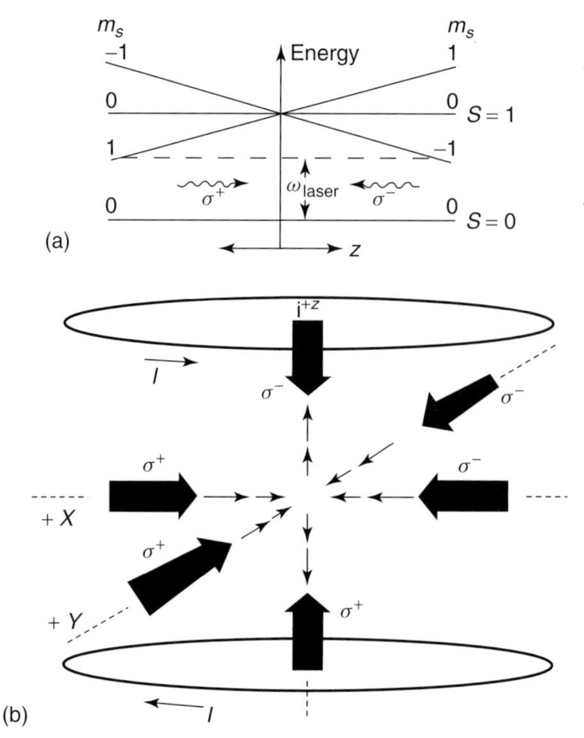

Fig. 14.22 Arrangement for a magneto-optical trap. (a) The energy-level scheme for an atom having spin $S = 0$ in the ground state and $S = 1$ in the excited state. (b) Experimental arrangement of a three-dimensional trap. (From Raab *et al.*, 1986.)

$\sigma+$ and $\sigma-$ circular polarizations (see Figure 14.22), with a frequency tuned below the atomic resonance, an atom at $z\rangle 0$ will absorb more $\sigma-$ photons than $\sigma+$ photons since the laser frequency is closer to the $\Delta m_s = -1$ transition frequency. Similarly, for an atom at $z < 0$, the Zeeman shift is reversed, and the atom will absorb more $\sigma+$ photons. Because of the momentum carried by these photons, the position-dependent differential absorption leads to the presence of a force, which pushes atoms into the region $z = 0$. The scheme can be readily extended to three dimensions by using three pairs of counterpropagating beams along the x-, y-, and z-directions and a spherical quadrupole magnetic field, as shown in Figure 14.22.

14.7 Single-Molecule Spectroscopy

All the experiments described above involve a huge number N of molecules, ranging from millions to billions At the same time, the theoretical description of the experimental results start from the behavior of a single molecule, after which an averaging over the number of molecules N is performed to compute the observable physical quantity.

Recently, the availability of new tools, such as optical tweezers (OTs) (Ashkin and Dziedzic, 1987) or atomic force microscopes (AFMs) (Sarid, 1991), have allowed the visualization and the manipulation of single molecules and, hence, have opened a new area of spectroscopy, which is aimed at studying the dynamics of single molecules and their interactions with the molecules and complexes surrounding them. This approach complements information obtained in traditional bulk assays, opening a real new view of the nanoworld.

There are several important advantages of looking at the dynamics and spectroscopy of single molecules, most of them become particularly evident when applied to biosystems. Single-molecule (SM) experiments allow exploring biochemical processes at an unprecedented level, offering a quantitative description of biological processes, previously only described in a qualitative fashion.

The main difference between SM and traditional bulk methods lies in the kind of average performed when investigating the properties of the analyzed system. Traditional spectroscopic techniques investigate the average behavior of a statistically significant number of molecules. Each member of this population undergoes a series of events, which are generally temporally uncorrelated with the behavior of the other members. Within this approach, the sequential dynamics and the correlation between successive steps are inevitably lost. Single-molecule spectroscopy (SMS) experiments, instead, allow experimentalists to access processes by following one molecule at a time throughout the course of events, removing any averaging effect and extolling the heterogeneity in the molecular dynamics.

Time-resolved measurements on single molecules also provide powerful insights into the rates and mechanisms of kinetic processes that are fundamentally stochastic. In bulk applications, even if all members of the ensemble, in principle, could be synchronized to begin at the same time, they would rapidly run out of phase with one another, obscuring the kinetic steps. Instead, in SMS measurements, there is no need for synchronization of molecules undergoing time-dependent transitions. In this way, it is possible to characterize the kinetics of molecular reactions and observe possible intermediates, which are hidden in bulk experiments. This point is particularly important in the investigation of biomolecules, where heterogeneity easily arises, for instance, for proteins or oligonucleotides in different folding states, different configurations, or different states of an enzymatic cycle. In this framework, D. Grunwald *et al.* have demonstrated that it is possible to directly observe single protein or mRNA molecules in a living cell's nucleus (Grunwald *et al.*, 2006). These studies suggest that a deeper insight of single-protein activity in live cells during transcription, replication, or translation will soon emerge. Pioneering experiments on single biomolecules have been proposed by the group of S. Xie, at the Harvard University. Just as representative example, this group has analyzed, in a living cell, the molecular processes that convert gene sequence into functional proteins, with a SM sensitivity (Elf, Li, and Xie, 2007).

A particular branch of biology that has taken advantage of SM experiments is the study of molecular motors. The first molecular motor to be studied was the myosin–actin system. Myosin uses the energy of ATP hydrolysis to move in one direction along an actin filament. During muscle contraction, fibers of myosin molecules pull on the anchored actin filaments, bringing them closer together. Until recently, most of our understanding of muscle activity (ATP consumption under load, contraction rates, and so on) had been obtained from studies of muscle fibers, that is, assemblies of myosin and actin filaments. The SM experiments have gone further, demonstrating that motor proteins act in a discrete, stepwise way with very high efficiency. In particular, T. Yanagida and coworkers have used a fluorescence microscope to observe simultaneously the displacement of the actin fiber and the hydrolysis of a fluorescently tagged ATP molecule by the myosin. They demonstrated that a single ATP molecule was consumed for three myosin steps, instead of only one step, as previously believed (Ishii and Yanagida, 2000). It is probably unnecessary to emphasize that bulk measurements of overall ATP consumption do not provide that kind of detailed molecular information.

Although biology constitutes certainly the most intriguing field of application of SMS, numerous other applications can be found. SM optical studies are being applied to an increasing number of problems in chemistry, physics, and medicine.

A single molecule can be used as a reporter of its local nanoenvironment, that is, a probe for the distribution of functional groups, ions, and even electrostatic charge in its immediate surrounding (Moerner, 1994).

Other applications are related to the possibility to regard single molecules as source for single photons, which are very interesting for quantum cryptography (Benjamin, 2000). Not all of these effects are as commonly observed as others: spectral diffusion, photon-bunching and antibunching, (dynamic) Stark shifts, Markovian dynamics.

In SMS experiments, it is also possible to measure small energies and detect large Brownian deviations in molecular reactions, offering new opportunities to scrutinize the basic foundations of statistical mechanics (Ritort 2007).

SM detection was first reported by Moerner and Orrit (1999). The emission of a single terrylene molecule in pentacene matrices at very low temperatures was detected by taking advantage of the inhomogeneous broadening of the emission of organic molecules at 1.5 K. The detection of single dye molecules in solution at room temperature was instead firstly demonstrated by Shera *et al.* (1990). These pioneering experiments experiment yielded a variety of new observations: direct observation of spectral diffusion (spectral shifting) of a single molecule owing to spontaneous changes of the surrounding environment, light-induced spectral shift, and Markovian dynamics.

Simultaneous spatial localization on nanometric scale and interrogation of single molecules found at "fixed" positions in solids at room temperature was first demonstrated by Betzig and Chichester (1993), by using a near-field scanning optical microscope (NSOM) (Betzig and Chichester, 1993). It was realized shortly thereafter that more conventional and user-friendly far-field optical microscopic methods, such as confocal and wide-field microscopy, could also be used to detect single molecules.

The impressive growth in the number of papers published per year that began immediately after these initial demonstrations is surely indicative of the general importance of SM methods and the potentiality of the "road ahead" opened by SM experiments (Rigler, Orrit, and Basché, 2001b).

At the basis of SM experiments, there is the possibility to guarantee that it is possible to distinguish the signal coming from an assigned molecule from both that coming from other molecules and from the background. Different strategies are usually employed to fulfill these requirements. Addressing of the first point is generally accomplished by dilution, in order to reach the condition in which only one molecule is in resonance in the probed volume of the laser. For instance, at room temperature, one needs to work with roughly 10^{-10} mol l^{-1} concentration with a probed volume of about 10 μm^3, which roughly corresponds to the diffraction-limited focal region of a visible laser. Other approaches rely on the limitation of the "active" volume in the sample, that is, the part of the sample that is illuminated by the excitation light and from which light is collected and sent to the detector. The latter condition can be achieved by combining the spectroscopical investigation with confocal detection schemes, or even with near-field optical excitation, allowing the illumination of sample volumes much smaller than the diffraction limit.

Up to now, SM sensitivity has been achieved mainly by fluorescence-based optical techniques. Effective fluorescence cross sections can reach the 10^{-16} cm^2 per molecule for high quantum yield fluorophores, which guarantees an ultrahigh sensitivity, even at room temperature. Just as in bulk spectroscopy, fluorescence measurements provide information about the size of the molecule, its rotation dynamic and, more importantly, changes in the size or conformation of the molecule as the investigated process proceeds. This last topic has been largely used for studying the rearrangement of biomolecules upon interaction with other molecular partners or as a response to an environmental change. In general, one can gain insight on the investigated phenomenon by looking to fluorescence amplitude changes, the wavelength of the emitted photons, their polarization state or the arrival time with respect to the excitation. All these observables clearly depend on the intrinsic properties of the emitting molecule.

The time distribution of photons emitted by single molecules presents interesting properties. For instance, as observed by Bernard *et al.*, when high excitation rates are applied, photons are emitted in bunches; in this case, emission events, observed at microsecond scale, are separated by dark periods, whenever the molecules are shelved through intersystem crossing in a triplet state (Bernard *et al.*, 1993). Photon antibunching has been instead reported by Basché *et al.* (1992) at the nanosecond scale: after the emission of a photon, a single molecule occupies definitely its ground electronic state, and the probability of an immediate further emission event becomes zero.

Many fluorescence-based techniques have been proven to be useful to observe single molecules, their interactions, and their structural changes. In particular, SM fluorescence resonance energy transfer (FRET) and fluorescence-correlation spectroscopy (FCS) have been largely employed. More recently, an enhanced version of the Raman spectroscopy (SERS, surface-enhanced Raman scattering), has demonstrated to be suited for SM studies. Owing to their importance in SM investigations, the basic principles of these techniques are briefly summarized in the following.

14.7.1 Fluoresce Resonance Energy Transfer

In FRET analysis, a donor chromophore in its excited state transfers energy by a nonradiative, long-range dipole–dipole coupling mechanism to an acceptor chromophore when their relative distance becomes comparable or shorter than a characteristic length, normally referred to as the *Förster radius* R_F (Figure 14.23), named after the German scientist, Theodor Förster, who

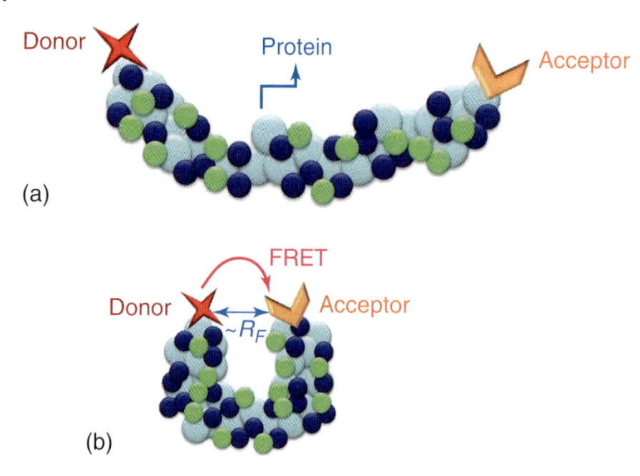

Fig. 14.23 Basic scheme for protein structural rearrangement by FRET spectroscopy (a). FRET occurs when the donor and acceptor are at a distance comparable to R_F (b).

discovered this effect (Förster, 1948). Typically R_F is less than 10 nm. Resonance energy can yield a significant amount of structural information concerning the donor–acceptor pair. In SM experiments, FRET measurements are employed to estimate the distance between two sites on a macromolecule, previously marked with the donor–acceptor chromophores, and to monitor the effects of conformational changes on this distance. In this type of experiment, the "observable" is the degree of energy transfer, which is used to calculate the distance between the donor and acceptor sites. SM FRET experiments have been widely used to investigate protein folding–unfolding processes. It is worth noticing that in "traditional" FRET analysis, conducted on an ensemble of proteins, it is impossible to synchronize the folding reactions, which consequently leads to a host of dynamic information. SM experiments, instead, can provide the distance between a pair of points on a polypeptide chain as folding progresses, allowing the observation of reaction intermediates as the protein rearranges itself to reach the bottom of the free energy ladder. A typical SM FRET application was presented by Deniz *et al.* (2000); they reported SM protein-folding studies on the guanidinium chloride denaturation of freely diffusing chymotrypsin inhibitor 2 (CI2) , a protein which folds by a two-state mechanism. Folded and denatured subpopulations were directly observed and identified in energy transfer efficiency histograms.

14.7.2
Fluorescence-Correlation Spectroscopy

FCS is a fluorescence-based technique used to experimentally characterize fluorescent species, such as biomolecules, and so on, and their dynamics. The working principle is quite simple: light is focused on a sample and the measured fluorescence-intensity fluctuations (due to diffusion, chemical reactions, aggregation, etc.) are analyzed using the temporal autocorrelation (Figure 14.24). FCS is a valuable method to measure molecular diffusion constants. It provides two important parameters: the number of molecules in the investigation volume and the translational diffusion coefficient of molecules. FCS under SM conditions has attracted considerable attention in during the past few years. Since its conceptual introduction by Magde, Elson, and Webb (1972), and the theoretical background for the analysis of translational

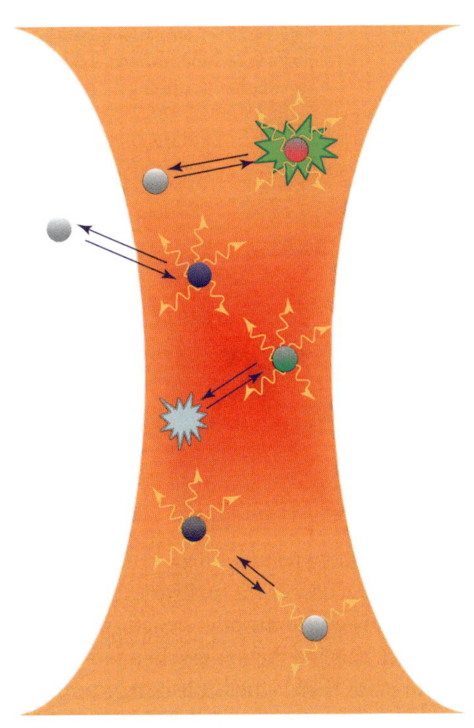

Fig. 14.24 Molecular mechanisms that give rise to fluorescence fluctuations in FCS analysis, including movements, conformational changes and photochemical reactions.

motion and chemical kinetics provided by Elson and Magde, many applications have been found. The group of Rigler were the first to report the use of FCS to study the translational diffusion of rhodamine 6G at the SM level (Rigler and Elson, 2001a). FCS may reveal new structural and dynamical features of biomolecules that cannot be detected by conventional measurements. For instance, rapid DNA sequencing schema based on FCS detection of a single nucleic acid has been reported by several groups. Various modifications from the basic scheme have been developed. E. Gratton presented raster image correlation spectroscopy (RICS) as a method to quantify both mobile and immobile molecule populations, distinguish between binding and diffusion, and measure translational diffusion on timescales from microseconds to minutes (Chen *et al.*, 2002). By combining FCS with the analysis of the fluorescence-intensity distribution, the group of E. Gratton has studied the mobility and clustering of EGFP (enhanced green fluorescent protein) labeled paxillin, an adhesion adapter protein, in the cell cytoplasm and during the formation of cell adhesions (Digman *et al.*, 2005).

14.7.3 Surface-Enhanced Raman Spectroscopy

SM detection and its simultaneous chemical identification represent the ultimate goal of SM studies. In "bulk" investigations, chemical structure can be achieved by infrared spectroscopy, that is, absorption and Raman spectroscopy, the second technique being generally preferred for samples in aqueous solution. In Raman analysis, indeed, the light incident on a sample is inelastically scattered and gets shifted in frequency by the energy of

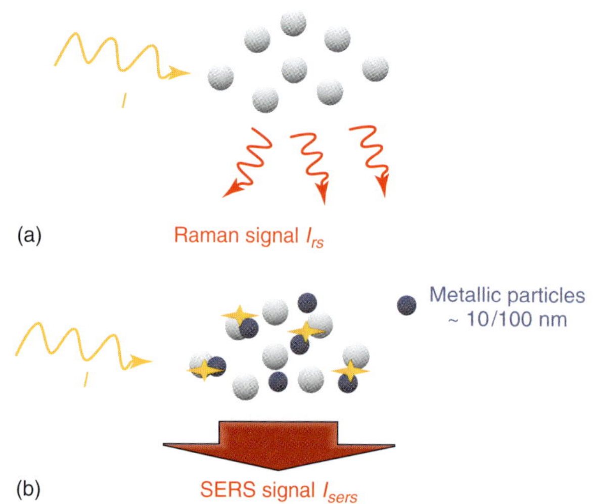

Fig. 14.25 Comparison between normal (a) and surface-enhanced (b) Raman scattering.

the sample characteristic vibrational transitions. The unfavorable aspect of this optical technique lies in the very small process cross section, which ranges between 10^{-30} and 10^{-25} cm^2 per molecule, the larger values occurring under some favorable resonance conditions (resonance Raman spectroscopy, coherent anti-Stokes Raman spectroscopy, etc.). This situation is dramatically changed when the analyte is placed close to metallic nanostructures. In these conditions, an enhancement of the Raman cross section up to 14 orders of magnitude can be observed, referred to as *surface-enhanced Raman scattering* (SERS), allowing, therefore, even the detection of a single molecular scatter (Figure 14.25). It is generally accepted that more than one effect contributes to this high enhancement. For this purpose, the enhancement mechanisms are roughly divided into "electromagnetic" and "chemical" effects. The first effect arises from enhanced local optical field at the site of the molecule near the metallic nanostructure, because of the excitation of plasmonic resonances in the metallic nanostructure and its coupling with the Raman probe. The chemical enhancement results, on the other hand, by the interaction of the analyte with the free electrons of the metallic structure. In particular, this interaction is schematized as a four-steps metal–molecule charge-transfer process: (i) photon annihilation and promotion of an electron into a hot excited state (hot electron); (ii) transfer of the hot electron to the lowest unoccupied molecular orbital (LUMO); (iii) electron transfer from the LUMO back to the metal; (iv) relaxation of the electron to its fundamental state and consequent emission of the Stock photon. Generally, SERS-active substrate consists of silver or gold nanoparticles, colloidal cluster, or island films. However, the SERS effect tends to be spatially localized in so-called "hot" areas, where extremely high electromagnetic fields are found. The first, unexpected, experimental evidence of the SERS effect was originally found by Fleischmann, Hendra, and McQuillan (1974) from pyridine adsorbed on electrochemically roughened silver in 1974 but its origin was not recognized. The SERS explanation was instead proposed by Van Duyne three years later (Jeanmaire and van Duyne, 1977). The enhanced Raman cross section can reach 10^{-16} cm^2 per molecule, which is comparable to effective fluorescence cross section of common dyes and sufficient for SM detection. As a matter of facts, in SM experiments the analyte, provided at picomolar concentration, is added to a silver/gold colloidal solution and a microscope objective is used for both laser excitation and collection of the Raman-scattered light. At this level of concentration, the average number of molecule in the focus volume is of the order of the unit. Of course, Brownian motion of molecules in and out of the probed volume is translated into strong fluctuations of the height of the Raman pecks, when a sequence of spectra is acquired. This effect was observed by the group of Kneipp *et al.*, 1997 by acquiring a sequence of 100 spectra, collected from a 30-pl volume of a $\sim 10^{-14}$ M crystal violet solution; at such a level of concentration, the average number of molecules in the probed volume was 0.6. By analyzing the statistical occurrence of the height of an assigned Raman peak, it turns out that the distribution is a superposition of four Gaussian curves. The gradation of the areas of the four statistical peaks is roughly consistent with a Poisson distribution for an average number of 0.5 molecules. This reflects the probability to find 0, 1, 2, or 3 molecules in the scattering volume during the actual measurement. SERS has found application for spectroscopic detection and identification of single nucleotides. As demonstrated in 1998, the nucleotide bases show well-distinguished surface-enhanced Raman spectra. Thus, after cleaving single native nucleotides from a DNA or RNA strand into a medium containing an SERS-active substrate, direct detection and identification of single native nucleotides should be possible owing to the real "fingerprint" effect associated with the Raman spectrum of an assigned sample (Kneipp *et al.*, 1998). SERS studies have also been performed on living cells. In these experiments, colloidal silver particles were incorporated inside the cells and the intracellular distribution of drugs in the whole cell was monitored, in order to study the antitumor drugs/nucleic acid complexes (Nabiev, Morjani, and Manfait, 1991). From this study it turns out that cells incubated with colloidal gold continue to grow normally and there is no evidence of any apoptotic activity or cell detachment from the growth surface when compared with control monolayers. These outcomes demonstrate the feasibility of measuring SERS of native constituents within a single viable cell using colloidal gold particles as SERS-active nanostructures inside a living cell. This paves the way for monitoring the small chemical changes in the cell, which could be the precursors of larger morphological changes, which later become clearly visible by a simple observation of the cell by an optical microscope. In this way, exciting opportunities might be opened up for the early diagnosis of cellular diseases, such as cancer. In addition, performing spectroscopy in the local optical fields of the nanostructures provides the most exciting capabilities of the effect in biophysics. Indeed, in SERS, lateral resolution is determined not by the diffraction limit, but by the spatial confinement of the local fields. For special nanostructures, this field confinement can be of the order of 10 nm, suggesting the possibility for spatially selective Raman probing of parts of large biomolecules.

Glossary

Absorption Spectroscopy: The study of radiant energy that is characteristically absorbed by a particular atom or molecule.

Doppler Linewidth: The width of a spectral line caused by the Doppler effect.

Doppler-Free Spectroscopy: Spectroscopic techniques that allow elimination of the Doppler broadening. They are essentially based on two main approaches: velocity-selective saturation or counter-propagating two-photon absorption.

Emission Spectroscopy: The study of radiant energy that is characteristically emitted by atoms or molecules after a suitable excitation.

Fluorescence-Correlation Spectroscopy (FCS): A fluorescence-based technique used to experimentally characterize fluorescent species. The working principle is quite simple: light is focused on a sample and the measured fluorescence-intensity fluctuations (due to diffusion, chemical reactions, aggregation, etc.) are analyzed using the temporal autocorrelation.

Fluoresce Resonance Energy Transfer (FRET): A spectroscopic technique, based on the energy transfer between two proper chromophores, when their relative distance becomes comparable or shorter than a characteristic length, normally referred as *Förster radius* R_F.

Intracavity Spectroscopy: High-sensitivity spectroscopic techniques where the atomic or molecular sample under investigation is inserted into a laser cavity or a passive cavity having a high quality factor.

Laser: An acronym for 'light amplification by stimulated emission of radiation". A device that emits a high-intensity, narrow-spectral-width, and highly directional beam of light by stimulated emission in an atomic or molecular system where a population inversion between two levels has been produced.

Laser Beam: The bright stream of light emitted by a laser. The near-zero divergence makes this beam extremely interesting for a wide variety of applications (surgery, communications, printing, physics, chemistry, etc.).

Laser Cooling: A method based on the scattering of near-resonant photons to slow atoms in an atomic beam to very low velocity.

Laser Trapping: A method based prevalently on laser radiation to trap neutral and ionized atoms in a small volume at extremely low kinetic energy and, hence, low temperatures.

Linewidth: The spread in wavelength of a spectral line. Usually it is measured as the full width at half maximum (FWHM).

Multiphoton Spectroscopy: Spectroscopic techniques based on the simultaneous absorption of two or more photons.

Natural Linewidth: The width of a spectral line caused by the finite lifetime of the levels.

Optogalvanic Spectroscopy: Spectroscopic techniques where atomic or molecular transitions are monitored through the variation of current of an electrical discharge when this is irradiated by resonant radiation.

Optoacoustic Spectroscopy: Spectroscopic techniques where atomic or molecular transitions are monitored through the detection of an acoustic wave produced by absorption of radiation.

Raman Spectroscopy: A technique based on inelastic scattering of radiation. It provides information on the molecular electronic structure of the analyzed sample.

Saturation: The condition reached in a two-level system in presence of resonant radiation when the populations of the two levels become equal.

Surface-Enhanced Raman spectroscopy (SERS): An enhanced version of Raman spectroscopy, which occurs when the analyte is in the proximity of nanosized metallic structures.

Single-Molecule Spectroscopy (SMS): Spectroscopic applications aimed at studying the dynamics of single molecules and their interactions with the molecules and complexes surrounding them.

Spectral Line: The response of a gas-phase sample obtained from the emission or absorption of radiation when the frequency is tuned around an atomic or molecular transition.

Spectroscopy: The branch of physics concerning the measurement of the emission and absorption spectra of atoms and molecules. According to the frequency range of the electromagnetic radiation, we speak of vacuum ultraviolet, ultraviolet, visible, infrared, or microwave spectroscopy.

Spectrum: The set of lines recorded after emission or absorption of radiation.

Time-Resolved Spectroscopy: Spectroscopic techniques that allow one to investigate fast relaxation processes involving the levels excited with a short-pulse laser.

Two-Photon spectroscopy: Spectroscopic technique where a transition forbidden in the dipole approximation can be induced by the absorption of two counterpropagating photons.

References

Abella, I.D. (**1962**) *Phys. Rev. Lett.*, **9**, 453–455.

Amin, S.R., Caldwell, C.D., and Lichten, W. (**1981**) *Phys. Rev. Lett.*, **47**, 1234–1238.

Ashkin, A. and Dziedzic, J.M. (**1987**) *Science*, **235**, 1517–1520.

Balmer, J. (**1885**) *Annalen der physik und Chemie*, **25**, 80–85.

Barbieri, B., Beverini, N., and Sasso, A. (**1990**) *Rev. Mod. Phys.*, **62**, 603–644.

Basché, Th., Moerner, W.E., Orrit, M., and Talon, H. (**1992**) *Phys. Rev. Lett.*, **69**, 1516–1519.

Battaglia, A., Gozzini, A., and Polacco, E. (**1959**) *Nuovo Cimento*, **14**, 1076–1081.

Bell, A.G. (**1881**) *Philos. Mag.*, **11**, 308.

Benjamin, S. (**2000**) *Science*, **290**, 2273–2274.

Bernard, J., Fleury, L., Talon, H., and Orrit, M. (**1993**) *J. Chem. Phys.*, **98**, 850–859.

Betzig, E. and Chichester, R.J. (**1993**) *Science*, **262**, 1422–1425.

Bjorkholm, J.E. and Liao, P.F. (**1974**) *Phys. Rev. Lett.*, **33**, 128–131.

Bloembergen, N., Levenson, M.D., and Salour, M.M. (**1974**) *Phys. Rev. Lett.*, **32**, 867–869.

Brossel, J., Cagnac, B., and Kastler, A. (**1954**) *J. Phys. Radium*, **15**, 6–8.

Bruzzese, R., Sasso, A., and Solimeno, S. (**1989**) *Riv. Nuovo Cimento*, **12** (7), 1–105.

Cagnac, B., Grynberg, G., and Biraben, F. (**1973**) *J. Phys. (Paris)*, **34**, 845–858.

Chen, Y., Müller, J.D., Ruan, Q.Q., and Gratton, E. (**2002**) *Biophys. J.*, **82**, 133–144.

Chu, S. and Wieman, C. (eds) (**1989**) *J. Opt. Soc. Am. B*, **6**, 2018–2278.

Corney, A. (**1977**) *Atomic and Laser Spectroscopy*, Clarendon Press, Oxford.

Demtroeder, W. (**1981**) *Laser Spectroscopy, Springer Series in Chemical Physics*, Springer-Verlag, Berlin Heidelberg.

Deniz, A.A., Laurence, T., Beligere, G.S., Dahan, M., Martin, A.B., Chemla, D.S., Dawson, P.E., Schultz, P.G., and Weiss, S. (**2000**) *Proc. Natl. Acad. Sci. U.S.A.*, **97**, 5179–5182.

Digman, M.A., Brown, C.M., Sengupta, P., Wiseman, P.W., Horwitz, A.R., and Gratton, E. (**2005**) *Biophys. J.*, **89**, 1317–1327.

Elf, J., Li, G.W., and Xie, X.S. (**2007**) *Science*, **316**, 1191–1194.

Ernst, K. and Inguscio, M. (**1988**) *Riv. Nuovo Cimento*, **11** (2), 1–66.

Evenson, K.M. (**1981**) *Faraday Discuss. R. Soc. Chem.*, **71**, 7–14.

Fleischmann, M., Hendra, P.J., and McQuillan, A.J. (**1974**) *Chem. Phys. Lett.*, **26**, 163–166.

Fornaca, G., Gozzini, A., and Strumia, F. (**1963**) in *Electronic Magnetic Resonance and Solid Dielectric* (eds R. Servant and A. Charru), North Holland Publishing Company, Amsterdam, p. 554.

Förster, T. (**1948**) *Ann. Phys.*, **437**, 55–59.

Goeppert-Mayer, M. (**1931**) *Ann. Phys. (Leipzig)*, **9**, 273–294.

Grunwald, D., Spottke, B., Buschmann, V., and Kubitscheck, U. (**2006**) *Mol. Biol. Cell.*, **17**, 5017–5027.

Hänsch T.W. (**1977**) in *Proceedings of the International School of Physics "Enrico Fermi," Course LXIV, Nonlinear Spectroscopy* (ed. N. Bloembergen), North-Holland, Amsterdam, p. 17.

Hänsch, T.W., Lee, S.A., Wallenstein, R., and Wieman, C. (**1975b**) *Phys. Rev. Lett.*, **34**, 307–309.

Hänsch, T.W., Schawlow, A., and Series, G.W. (**1979**) *Sci. Am.*, **240** (3), 72–86.

Hänsch, T.W., Levenson, M.D., and Schawlow, A.L. (**1971**) *Phys. Rev. Lett.*, **26**, 946–949.

Hänsch, T.W. and Schawlow, A. (**1975**) *Opt. Commun.*, **13**, 68–69.

Hänsch, T.W. and Toschek, P. (**1970**) *Z. Phys.*, **236**, 213–244.

Ishii, Y. and Yanagida, T. (**2000**) *Single Mol.*, **1**, 5–16.

Jeanmaire, D.L. and van Duyne, D.L.R.P. (**1977**) *J. Electroanal. Chem.*, **84**, 1–20.

Kneipp, K., Kneipp, H., Kartha, V.B., Manoharan, R., Deinum, G., Itzkan, I., Dasari, R.R., and Feld, M.S. (**1998**) *Phys. Rev. E*, **57**, R6284.

Kneipp, K., Wang, I., Kneipp, H., Perelman, L.T., Itzkan, I., Dasari, R.R., and Feld, M.S. (**1997**) *Phys. Rev. Lett.*, **78**, 1667–1670.

Letokhov, V.S. and Chebotayev, V.P. (**1977**) *Nonlinear Laser Spectroscopy*, Springer Series in Optical Science, Springer-Verlag, Berlin Heidelberg New York.

Magde, D., Elson, E.L., and Webb, W.W. (**1972**) *Phys. Rev. Lett.*, **29**, 705–708.

McFarlane, R.A. Jr., Bennet, W.R. Jr., and Lamb, W.E. (**1963**) *Appl. Phys. Lett.*, **2**, 189–190.

Meystre, P. and Stenholms, S. (eds) (**1985**) *J. Opt. Soc. Am. B*, **2**, 1706–1860.

Moerner, W.E. (**1994**) *Science*, **265**, 46–53.

Moerner, W.E. and Orrit, M. (**1999**) *Science*, **283**, 1670–1676.

Nabiev, I.R., Morjani, H., and Manfait, M. (**1991**) *Eur. Biophys. J.*, **19**, 311–316.

Penning, F. (**1928**) *Physica (The Hague)*, **8**, 13–23.

Pinard, M. and Aminoff, C.G. (**1979**) *Phys. Rev. A*, **19**, 2366–2370.

Raab, E., Prentiss, M., Cable, A., Chu, S., and Pritchard, D. (**1986**) *Phys. Rev. Lett.*, **59**, 2631–2634.

Rigler, R. and Elson, E.L. (**2001a**) *Fluorescence Correlation Spectroscopy: Theory and Applications*, Springer, Berlin.

Rigler, R., Orrit, M., and Basché, T. (eds) (**2001b**) *Single Molecule Spectroscopy–Nobel Conference Lectures*, Springer, Berlin.

Ritort, F. (**2007**) *Comptes Rendus Physique*, **8**, 528–539.

Sarid, D. (**1991**) *Scanning Force Microscopy*, Oxford Series in Optical and Imaging Sciences, Oxford University Press, New York.

Shapiro, S.L. (**1977**) *Ultrafast Light Pulse, Topics in Applied Physics*, vol. 18 , Springer, Berlin Heidelberg New York.

Shera, E.B., Seitzinger, N.K., Davis, L.M., Kellera, R.A., and Soper, S.A. (**1990**) *Chem. Phys. Lett.*, **174**, 553–557.

Sorem, M.S. and Schawlow, A.L. (**1972**) *Opt. Commun.*, **5**, 148–151.

Svanberg, S. (**1992**) *Atomic and Molecular Spectroscopy*, Springer Series on Atoms and Plasma, Springer-Verlag, Berlin Heidelberg.

Szöke, A. and Javan, A. (**1963**) *Phys. Rev. Lett.*, **10**, 521–524.

Tam, A.C. (**1986**) *Rev. Mod. Phys.*, **58**, 381–431.

Teets, R.E., Kowalski, F.V., Hill, W.T., Carlson, N.W., and Hansch, T.W. (**1977**) in *Advances in Laser Spectroscopy I*, SPIE Proceedings No. 113 (ed. A.H. Zewail), SPIE, Bellingham, p. 409.

Vasilenko, L.S., Chebotayev, V.P., and Shishaev, A.V. (**1970**), *Zh. Eksp. Teor. Fiz. Pis'ma Red.* **12**, 161–165 (*JETP Lett.* **12**, 113–116).

Wieman, C. and Hänsch, T.W. (**1976**) *Phys. Rev. Lett.*, **36**, 1170–1173.

Wineland, D. and Dehmelt, H. (**1975**) *Bull. Am. Phys. Soc.*, **20**, 637.

15
Chemical Analysis Metrology

Kenneth A. Rubinson

Handbook of Metrology. Edited by Michael Gläser and Manfred Kochsiek

ISBN: 978-3-527-40666-1

15.1 Introduction

Chemical analysis is the study and practice of investigating the atomic and molecular composition of materials. The materials analyzed include all the natural and synthesized materials of the contemporary world as well as perhaps less obvious materials such as the components of flames or the materials at a surface layer that lie at an interface between liquids, solids, and gases. Chemical analysis can also involve determining the spatial distribution of chemical species on or within a material. The great majority of materials are chemically complicated; this means that they are generally composed of a large number (tens to millions) of different elements and molecular species.

Chemical analysis, as a member of the measurement sciences or metrologies, can illustrate the structure of metrological investigations particularly well since it links the abstract structure of the logic of measurement with its practical application to materials in all forms. The general steps to analyze a material are illustrated in Figure 15.1. The final results are numbers, for, say, the contents of specific chemical species in a bulk sample, or a set of numbers of content and location that might be presented as a picture showing the range of concentrations of chosen chemical species over a surface or within a volume.

15.2 Chemical Analysis and Metrology

The logic of metrology is, in fact, the scientific method, and for chemical analysis it appears as a process called *validation*. Validation means demonstrating that chemical analyses measure what the analyst says they measure on a specific type of sample. For instance, an analyst might develop a new procedure to measure the amount of a particular substance, the analyte, in samples of bovine livers. But then, he or she must validate the new method by showing that its results agree with those obtained on a well-characterized sample using other, already accepted procedures.

To validate an analysis for some element(s) or molecules in a material, it is necessary to take samples from it. This sampling must be done so that the samples can be shown to be representative of the part of the material being analyzed. For example, the analysis may be of a whole truckload of plastic resin or of corrosion products on the surface of a steel beam or of the genome of a microscopic worm or of the distribution of elements over various nanometer-sized structures on a

Handbook of Metrology. Edited by Michael Gläser and Manfred Kochsiek

ISBN: 978-3-527-40666-1

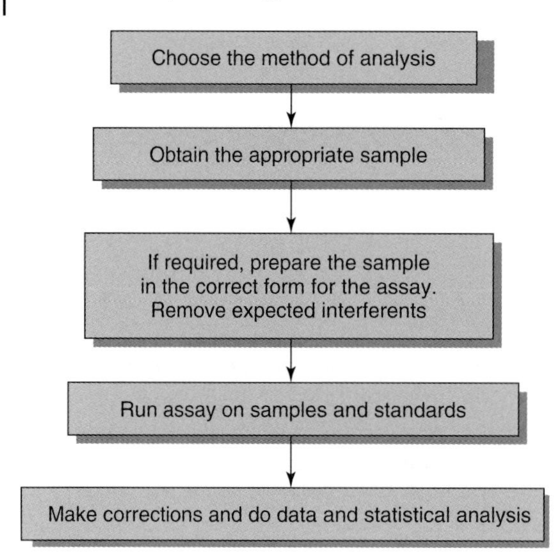

Fig. 15.1 Diagram showing how the methodology of chemical analysis is used to measure analytes, which are the chemical species that have to be quantitated. Each of the steps has potential errors that may either be random or may be regular and a part of the various techniques. The metrological methods allow the errors to be identified and eliminated or corrected mathematically to provide a measurement closest to the "true" value and with known uncertainty.

microchip. Sampling can be of two types: one where the material remains intact, and the other where samples are removed from the material. The first is called *nondestructive sampling*. The latter is often called *destructive sampling*, although the amount of material removed may be so small that the "damage" can only be seen with an electron microscope.

The physical form of the samples must match the ability of the assay to make a valid measurement on that type of specimen. For example, some methods require the analyte to be in a solution, some require solids, and some require the sample to be brought into the gas phase. This means a solid sample that might be usable directly for the method requiring a solid would need to be dissolved for an assay that requires a solution. If the assay requires the analyte to be in the gas phase, one of a number of techniques can be used to transform the atoms or molecules of the solid into a gas.

As can be seen from Figure 15.1, the assay is an integral part of an analysis, but it is only one part. In addition, with contemporary instrumentation to run the assays, the other experimental steps of the analysis provide the greatest challenge to obtaining analytical results that report the "true" content of the material. A separate article in this Encyclopedia (Chapter 16 and references therein) describes some of the wide variety of chemical assay methods. Here we consider the other factors that are involved in the design of effective chemical analyses. The essence of such analyses is that the true values of the contents of materials are not known unless they are artificially made from pure components. For regular materials that are to be analyzed, a set of experiments must be designed to determine the elements and/or molecular species present and to show that the answers lie close to the content existing in the material. We can assume that the true value of the contents can be found if we could perform a perfect analysis.

Before continuing to describe the paths to approaching a reliable measurement of analyte content that approaches the imaginary perfect analysis, some caution is in order about the nomenclature used around the world and at different times in the past. Different terms have been used for ideas that are approximately the same;

"approximately" means that they may differ in meaning in ways that are subtle compared to the level of understanding this article seeks to convey. These are collected in Table 15.1. Similar terms are collected together in the entries on the left of the table, and general definitions are given on the right. The first entry of each class is the one used here.

This article is being written in a period of transition in the language used to describe the results of a series of measurements of chemical content when compared to the known results for a standard material or from an imaginary perfect analysis. The terms *error* and *bias* used by analysts have meanings that are without the emotional content interpreted by the general public. Error is not a mistake, but simply an expected result of the measurement process. Bias is not an unreasonable dislike, but only the amount a result that is measured by a given technique differs from an accepted value from a standard material with known content. As a result, various groups are suggesting that *random error* be replaced with the term *imprecision*, and *bias* be replaced by *systematic effect*. Nevertheless, the language of error and bias will be with us for many more years and are used here because of their simplicity.

15.3 Approaching the Truth

How sure are you that the experimentally obtained value is close to the true value? This query states the fundamental question of chemical analysis (and all metrology). A corollary to this is, How sure are you that the value you have obtained is the same (or different) from the value obtained on the same sample at a different time or by another person? This latter question inquires about the reliability or robustness of the measurement. What is desired is the interchangeability of results among individuals at different times and in different laboratories or field sites.

How can you find out the answer to the fundamental question? One way is to repeat the whole analytical procedure for several replicate samples or replicates. All the replicates are obtained from the same original source. If the answers are similar for many replicate analyses, a reasonable expectation is that the next measured result is *going to be near the same values* as the previous ones.

However, for any analysis, the measurement of each replicate (each run) provides a somewhat different value. What, then, is the numerical result? One meaningful value is the arithmetic mean of the individual results: the sum of all of the results divided by the number of determinations in the series. This is expressed mathematically by

$$\overline{X} = \frac{\sum_i X_i}{N} \tag{15.1}$$

where

$\overline{X}$ represents the arithmetic mean,
X_i represents the numerical result of the *i*th run included in the sum, and
N is the total number of runs.

Often the arithmetic mean is simply called the *mean, mean value*, or *average*.

For each individual replicate sample, the *deviation from the mean* d_i is defined as the difference between each measured value x_i and the mean value for all the measurements,

$$d_i = X_i - \overline{X} \tag{15.2}$$

Tab. 15.1 Groups of closely related terms used in chemical analysis.

Differences in meaning are often subtle	General description
True value Conventional true value Assigned value Target value	The content of the analytes that would be measured if the analysis were perfect. However, inaccuracy and imprecision always exist in a laboratory quantitative measurement.
Precision (can be qualitative or quantitative) Imprecision (quantitative) Random error	The closeness of agreement between multiple measurements of the same material. Usually stated as some multiple of the standard deviation.
Accuracy (now qualitative) Mean error Error of the measurement (quantitative) Inaccuracy (quantitative)	Closeness of the measured result to the "true value," which would be the result found with a perfect measurement.
Standard deviation Standard uncertainty Uncertainty of the measurement	The quantitative measure of dispersion between multiple measurements of the same material. A characteristic parameter of a Gaussian distribution.
Material Specimen	The material that is being subjected to chemical analysis. A sample is a representative part of the material analyzed. Analytes are the chemical entities being determined in the analysis.
xx% confidence limit Uncertainty of the measurement Expanded uncertainty	The range within which one can be xx% confident that the next measurement of the same material will occur. Originally based on the integrated area under a Gaussian distribution.
Bias Determinate error Mean error Systematic error Systematic effect Difference from the "truth" Difference from an imaginary perfect analysis	A quantitative measure of the inaccuracy assuming no imprecision. These errors should be eliminated in developing an analysis or the results mathematically adjusted to eliminate them as much as possible.

The distribution of the individual replicate measurements usually turns out to be randomly distributed about the mean, and a statistically meaningful measure of the variability of the set of measurements is the standard deviation. The value does not depend on the sign of each of the deviations from the mean. In the equation below, N is again the number of replicates.

$$\text{Standard deviation} = s = \frac{\sqrt{d_1^2 + d_2^2 + \cdots + d_N^2}}{N-1} \quad (15.3)$$

Occasionally, the relative standard deviation is referred to as the *coefficient of variation*, and the value of s^2 is called the *variance*. If the standard deviation is expressed as a fraction or percentage of the mean value, it is classified as a relative standard deviation, a value by which the random errors for different levels of analyte can be compared.

Each part of a chemical analysis as outlined in Figure 15.1 has a variability associated with it that contributes to the precision of the overall process; this imprecision of previous steps is propagated through to the end. This overall precision is related to those of steps 1, 2, 3, ... by

$$\sigma_{\text{overall}} = \sqrt{\sigma_1^2 + \sigma_2^2 + \sigma_3^2 + \cdots} \quad (15.4)$$

To use this simple equation, the random variations of one step cannot influence the random variations of any of the others: they are uncorrelated. Such correlations are not considered here. This equation is based on a Gaussian error curve described immediately below and applies to a large number of experiments (order of 10^2). For the few replicates usually run, the equation for overall precision uses the applicable standard deviation in the same form:

$$s_{\text{overall}} = \sqrt{s_1^2 + s_2^2 + s_3^2 + \cdots} \quad (15.5)$$

The distribution of random errors about the mean value often can be described well by a Gaussian error curve, as illustrated in Figure 15.2. The curve is also known as the normal error curve or normal distribution curve. The x-axis of the plot is a scale of experimental values such as mass. The height of the curve in y is a plot of the probability that an experimental result will have the value noted on the x-axis. The mean value lies at the peak of the curve, is labeled μ, and equals the value of $\overline{X}$ in the limit of a very large number of trials. It may be noted that the variation can be both positive and negative from the mean value.

The Gaussian probability curve is *established* from a large number of experiments and the value of σ is found. This value of σ is related to s where s applies for a small number of runs and σ applies in the limit for a very large number of replicate samples from the material that is being analyzed. Both s and σ are called the *standard deviation*.

When this *random* scatter of the values of analytical results is small (s small), we say the precision is high, or, alternatively, the imprecision is small. However, denoting the precision of our data does *not* answer the question, "How close to the true value is the mean value of the analyses?", that is, "How accurate is the measurement?"

The difference between precision and bias can be shown visually as in Figure 15.3. Four different combinations of precision and bias are illustrated. The values from individual experiments are shown as dots along the horizontal axis. The top two plots show results of determinations that are precise, with the values clustered around

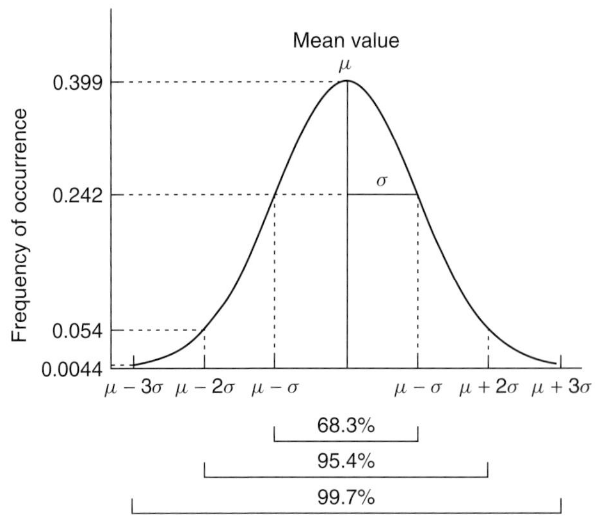

Fig. 15.2 The properties of a Gaussian distribution, which represents the random distribution of results about a mean value μ. The standard deviation σ is a quantitative measure of the precision and with minor adjustments approximates the results obtained from only a few measurements. (The standard deviation is symbolized by s then.) Once the curve is established for a specific analysis by analyzing replicates from a specific representative sample, we expect future measurements of replicates to have 68% probability of lying within one standard deviation, 95% of lying within two standard deviations, and more than 99% of lying within three standard deviations. These percentages are called the 68, 95, and 99% confidence limits. The equation for the curve is $f(x) = \frac{1}{\sigma\sqrt{2\pi}} \exp\left[-\frac{(\mu-x)^2}{2\sigma^2}\right]$ where the area under the curve is unity. Here x is the measured result and any $(\mu - x)$ is the deviation from the mean for that result.

the mean value. However, only for the top line can we say the measurements provide an accurate analysis. The mean value of the second set of data is clearly the result of a clustered set of measurements, but the deviation from the "true value" is much larger than the scatter of the measurements. So even though the results shown on the second line have high precision (or low imprecision), the overall result has a large bias and is not accurate. However, because of the relatively small, random scatter, the cause of this offset could be found by a suitable series of experiments.

The difference between the "true value" of the analyte and the mean value of a series of analytical results is called the *mean error*, E_m. E_m is not to be confused with the standard deviation of the set of measurements. In practice, the mean error cannot be perfectly known for certain because the "true value" is not known. However, we assume that we can approach the "truth" by running a suitable set of validation experiments that involve standards and blank samples as discussed in the next section.

The mean error falls in the category called *determinate errors* (note that the word is not determinant but determin*ate*) or systematic errors or systematic effects. Determinate means to originate from a fixed cause. A determinate error may result in an analysis where the values are higher than the true value every time or lower

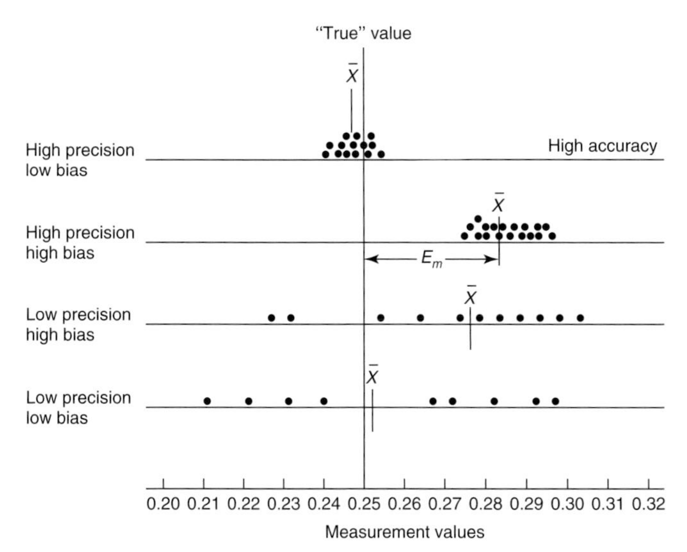

Fig. 15.3 Precision and bias have rigorous meanings. Precision is a measure of the clustering of multiple, repeated measurements about an average value, and bias is a measure of the closeness to the "true" content. The four limiting cases are illustrated with dots representing individual results for a number of replicate samples. The situations on the third and bottom line may be acceptable for ultratrace measurements – when the content of analytes are below parts per billion (parts per 10^9) because of inherent uncertainties in measuring such low levels, but such results are not useful otherwise.

than the true value every time for all the replicate analyses. This is in contrast to the standard deviation as a measure of random or indeterminate variations. Both random and determinate errors can occur in every stage of an analysis.

Results such as those shown in the lower two lines of the figure are less clear. From the distribution on line three, one might guess that the measured value is above the "true" one, but the statistical analysis would not provide much confidence in that conclusion, and the numerical value is highly uncertain. However, whether such data are useful or not depends on the type of analysis being done. If the analyte is a major component of a large sample, these results suggest that more work might be done to decrease the randomness of the measurements. On the other hand, if the units of the measurement values are parts per billion (parts in 10^9), this might be as narrow a distribution of results as could be expected, and the true value lies within the range of probable values. Similarly, the set of values on the fourth line may be considered of low accuracy or acceptable accuracy depending on the circumstances.

Errors that cause inaccuracy must be either eliminated in the experimental technique or at least accounted for by a correction in an analysis as far as possible. This idea may seem elementary; however, in practice, it can be difficult to account for all the major causes of inaccuracies. Nevertheless, this is one of the most important parts of a validation. In chemistry, unlike some physics measurements, the errors are not open to *a priori* calculations due to the complexities of the material and of the sample preparation chemistry (see Section 15.5). For example, in preparing the sample for the assay, some of the analyte may be lost (called a *low recovery of the analyte*), which, of course, changes the amount that will be measured subsequently.

15.4 Standards

To develop any new analytical methodology and to be sure its answer is correct when applied, calibration is required. The method is calibrated with standards, that is, samples that contain *known* amounts of the species to be assayed. If the analytical results from the standards agree with their known content with low uncertainty, the accuracy is confirmed. One special standard sample that must be used is a blank. An ideal blank contains all the components *except* the assayed species. The blank sample is passed through all the steps of the procedure, just like a regular sample. In biochemical analysis, the word *control* is often used instead of blank since the substance or solution may be so complicated that the word *blank* seems to be inapplicable.

Standards of chemical content are used to calibrate assays and also to discover whether all of the analyte originally present remains in the sample during any necessary conversion in form such as from a solid to a solution. The utilization of standards can be separated into two general types: (i) external standards and (ii) those in which the standard is added in some form into each sample to be analyzed.

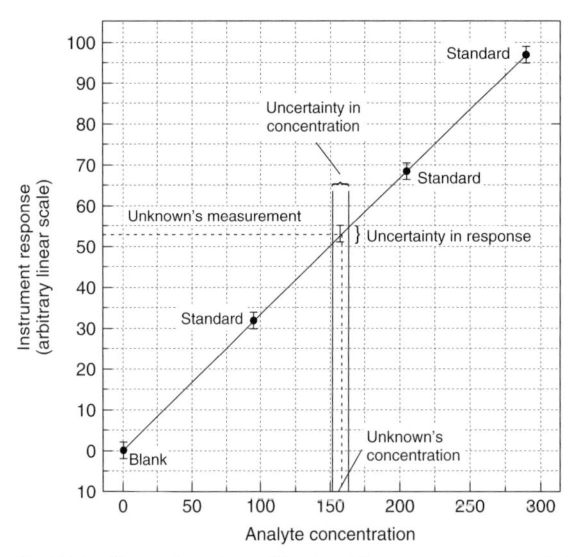

Fig. 15.4 Illustration of a calibration. The instrument response is calibrated versus the chemical content with standards, and then the content of the unknowns are found from their measurement. It is best to have the unknown within the range of the standards. The slope of the line is called the *sensitivity*, and the error bars show the uncertainty in the measurement. Usually such bars represent either one or three standard deviations around the mean value. As a result of the uncertainties of individual measurements, the slope and intercept also have uncertainties that can be stated quantitatively. The uncertainty in content equivalent to the indicated uncertainty in the response is indicated by the thin, solid lines. The concentration range is determined by the projections onto the calibration line of the top and bottom of the error bar.

An external standard is one that is analyzed separately from the replicate unknowns being tested. A series of such standards contain varying amounts of the analyte, together with a matrix that is as similar as possible to that of the samples. (The matrix is everything else in a material that is not an analyte.) A calibration plot such as illustrated in Figure 15.4 can be obtained from a series of external standards, and then the response of the instrument to the unknown samples can be found from its place on the plot. The slope of such a plot – the change in instrument response as a function of concentration – is called the *sensitivity of the assay*.

External standards can be used to calibrate an assay method when the components of the matrix, including any reagents required for the preparation, do not cause any interference. External standards can also be used if the analyst has enough control of the conditions so that the interferents' contribution to the measurement can be kept constant. A correction for any interferents' determinate error can then be made.

For those standards that are added to every sample, a known amount of standard is added each time. The addition is called a *spike*. These spikes are used in three different ways.

The first category is called a *standard addition*. Standard additions are fixed amounts of analyte that are added to each sample after an initial measurement is made on the sample. The assay measurement is then repeated after each addition. Sequential additions and remeasurements are usually done one or two times. By extrapolation as described in Figure 15.5, the amount of analyte originally in the sample can be found. In essence, this is similar to the external standard except that there is no blank; the lowest concentration is that of the sample. The content of the sample is then found by extrapolating back to the equivalent of a blank. For the

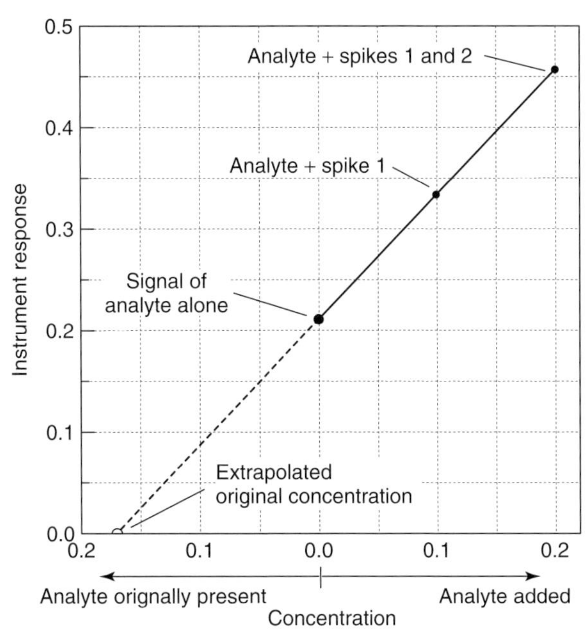

Fig. 15.5 Illustration of using an internal standard spike. This technique is necessary when control of all the conditions of the analysis is difficult. However, to obtain the minimum error in the results, the method should allow a precise subtraction of a blank from each sample that, in essence, leads to a zero-valued baseline. The solid line and data points appear similar to an external standard calibration from zero such as that in Figure 15.4.

method to work, it must be possible to correct the instrument response to a true zero in the absence of the analytes, which usually means that the extrapolation is linear as shown in the figure.

The second type, an internal standard, does not contain the analyte itself. The standard is sufficiently different chemically from the analyte that, while it is detected in the same experiment, it does not interfere with the assay. Nevertheless, the standard also is sufficiently similar that its recovery reflects the recovery of the analyte so that any losses can be compensated. The responses to the analytes are compared with those of the internal standard, and from the ratio of their responses, the analyte content can be calculated.

The third use of a spike is to determine recoveries of analytes. This is done by adding it to the sample matrix before the sample preparation step. A comparison of the instrument's response for the spiked sample and for an external standard at the same, known analyte level determines the recovery of the analyte (what percentage remains of the original amount) and whether or not interferences are present.

Adding one or more spikes to every sample is more difficult than using a few external standards, but there are three situations where it is preferred:

1. When the solid or liquid matrix of a sample is either unknown or so complex that an external standard cannot be used with confidence.
2. When the chemistry of the preparation or of the assay method is complex or highly variable.
3. When the assay depends on highly precise instrumental conditions that are difficult to control.

As an example, these problems (1, 2, and 3) can all occur when an assay involves injecting the sample into a flame or plasma. The chemistry is quite complicated, and the flame or plasma conditions are sometimes difficult to regulate with the desired precision. Internal standards are also used when especially high precision is desired.

15.5 Sampling: Good Samples are Representative and Homogeneous

The two aims of the sampling methods are to ensure that the sample is representative of the material being analyzed and that the samples that are analyzed in the laboratory are homogeneous. The term *representative sample* means the sample is the same overall as the material from which it is taken. The chemical mix in the sample should be a miniature replica of the contents of the whole from which it came. *Homogeneous* means that the sample is the same throughout. An analysis cannot be more precise than the least precise operation, and sampling is usually the major source of error in real-world analyses. At the same time, its uncertainty is often the most difficult to characterize and quantitate.

Almost all materials to be sampled, with the exception of small volumes of liquids or gases, are heterogeneous in composition at some scale of length. For example, granite is a heterogeneous solid composed of agglomerations of crystals of varying compositions and sizes. Another heterogeneous material to be sampled is a field of grass to be tested for natural nutrients such as phosphorus or for pollutants such as polychlorinated biphenyls (PCBs).

Figure 15.6 is an illustration that could belong to the latter sampling problem. Let us assume that the average concentration over the whole field is desired, and the

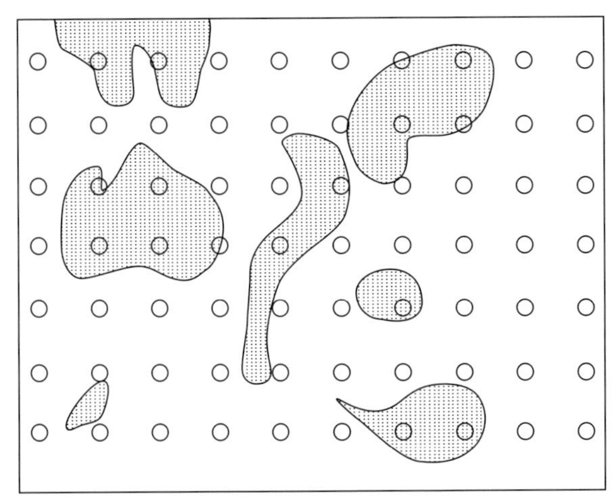

Fig. 15.6 A representation of a heterogeneous material to sample. Regular sampling of heterogeneously distributed analytes provides the best representation of their average values. A subset of these points chosen randomly should not inject bias into the sampling.

distribution of the material is irregular, such as the shaded areas of the figure. The general rule for all such heterogeneous arrangements is to make the sampling regular in geometry, such as taking aliquots at the sites represented by the circles. This is true for any length scale: the figure might just as well illustrate a sampling for a field a 100 m on a side, a slice of brain tissue 1 cm on a side, or an area of a nanocomposite 1 μm on a side. For the field, it may be too costly to obtain and analyze the indicated number of samples. The number can be reduced to a subset of the indicated sites but they must be selected randomly in order not to introduce bias in the sampling.

If a map of the material shaded is desired, only a rough distribution could be drawn from sampling with those intervals shown. With more effort, a finer grid could be used, and a more precise distribution could be found that more closely represents the actual one.

A generalization can be made about sampling to find the average of a material. To obtain a final, representative sample, the numerous samplings of the whole need to be obtained as just described. Following the primary sampling, mixing these together must be done thoroughly and randomly so that any fraction of the mix is as similar as possible. Again, the randomness prevents generating bias in the final representative sample, which then must be prepared in the form needed for the specific assay chosen. This sample preparation is discussed in the next section.

15.6 Sample Preparation

Five general principles apply to all sample preparations no matter how simple or complex. The sample preparation should

1. be done without losing any of the analyte(s), that is, have 100% recovery of the analyte;
2. include bringing the analyte(s) into the best chemical form for the assay method to be used;
3. include, perhaps, removing some interferents in the matrix;
4. be done without adding any interferents; and
5. include, if necessary, diluting or concentrating the analytes to bring their concentrations into the best range for the assay method used.

In item 1 of the list, ideally, a recovery of 100% would be desired. However, 100% recovery is not necessary to obtain an excellent analysis if the recovery can be reproduced sample to sample. For example, worthwhile work can be done with recoveries as low as 30%, although there is some tendency toward lower precision as the recovery level decreases.

Items 3 and 4 also require some further explanation. Suppose, when developing a chemical analysis, problems arise from chemical species that cause the analytical result to have a higher or lower value than if they were absent. These species are called *interferents*. For example, the method to measure nickel in a sample of stainless steel could be used to determine nickel in a plating bath. However, high levels of boric acid might interfere with the measurement. If so, the boric acid would be an interferent and, as a result, the procedure validated for stainless steel is not validated for the plating bath. The method could be modified for the bath either by removing the boric acid during the preparation step or by compensating the results for the interferent's effect. Then the modified method must be validated.

Sample preparation – also called *sample treatment* or *sample pretreatment* – is now the most time-consuming and most labor-intensive step of an analysis. In addition, preparing the sample for an assay requires, by far, the most expertise in chemistry of any part of an analysis. Usually, extensive practice or automation is needed to maximize the precision. These five requirements often involve trade-offs. For example, a treatment that might tend to lose less of the analyte in the preparation process might suffer from more interferences.

15.7 The Signal and Noise

All analytical methods rely on a signal for the analysis such as a change in electrical voltage. Noise is an unwanted random or nearly random time-dependent change in that signal. Noise should not, however, be confused with a constant signal that is present even for the blank. Such a constant value is called a *background* or *baseline*. The level of the background may change relatively slowly with time, an effect that is called *drift*, which also may be classified as a type of noise. For drift, the relatively long time over which it occurs usually means that correcting for a drifting baseline is usually quite easy.

It should be noted that not all signals are electrical ones. A signal can also be, for example, a change in color that can be seen by eye such as the endpoint of a titration where a small added volume causes the change. However, the color change does not have interference from randomly fluctuating noise. The limit to seeing the color change arises from the limitations of our eyes, which will cease to recognize the color change when an instrument may still easily do so. It is how noise limits an instrument's ability to detect a change in signal that is of concern here.

Incidentally, the term *signal* is often used in two different ways. The first, that of electronics, defines the signal as the voltage or current measured. In the second, that of chemical analysis, the signal indicates the difference between a voltage or current measured in the presence of a sample compared to that of a blank. The ideas presented in the rest of this section apply to both definitions of signal.

To illustrate the effect of noise, a simple example is shown in Figure 15.7. The signal, which is the difference in the voltage

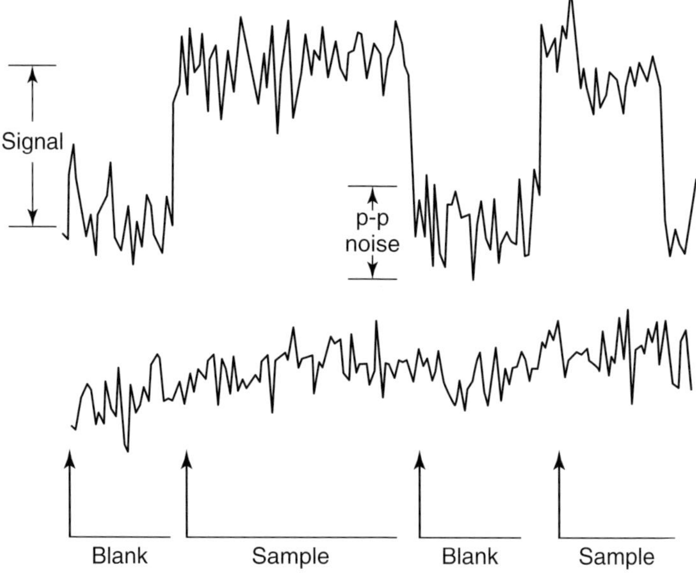

Fig. 15.7 Illustration of signals and noise. The sample and blank are alternately measured over time. The sample content produces a change in instrument output that is nearly the same level as the inevitable noise. The top trace results from an analyte content five times that of the bottom. Clearly the bottom content cannot be detected with the noise at the level here. Peak-to-peak is abbreviated p-p.

output in the presence of the blank and the sample, is easy to identify. However, if the signal level were reduced by a factor of 5 while the noise remains the same, finding the difference between the background and the sample is essentially impossible. If, on the other hand, the noise were also smaller, we could again then register the change between blank and sample. So the limit to measuring the analyte in the sample is not the size of the signal alone or the size of the noise alone, but the ratio of the two: the signal-to-noise (S/N) ratio.

Quantitatively, the noise part of S/N can be calculated by taking an approximate average of the differences between the upper and lower peaks of the noise (the peak-to-peak voltage) as a measure of the noise, or, alternatively, the average voltage can be used. The average is found by assuming the noise to be similar to a sine wave, and the average to be the root-mean-square (rms) voltage. In the equation, $V_{\text{p-p}}$ is the voltage difference between the upper and lower peaks of the sine wave.

$$\text{r m s voltage} = (0.707/2)\,V\text{p-p} = 0.35\,V\text{p-p} \tag{15.6}$$

It turns out that S/N ratios using both rms and peak-to-peak measurements are given in the literature. Fortunately, most authors state which one they are using.

The signal part of the S/N is the voltage difference between the average baseline and the average plateau. The averaging, in essence, has removed the noise.

15.8 Detection Limits

From the short discussion above, it follows that the precision and bias of an instrumental assay must depend on the magnitude of the analyte signal above the background and on the noise level. The analyte signal is related to the chemical content through the sensitivity of the instrument response. The illustration in Figure 15.7 showed this; the signal from the analyte in the dilute sample is not detectable above the noise level of the instrument. The analyte content in the sample is below the detection limit. A numerical value quantifying the detection limit can be calculated for each analyte determined by nearly any analytical method that might be used. From a compilation of such detection limits, the best analytical method(s) for a given analyte or set of analytes may be suggested for a given problem. Even if not working near the detection limit, a lower detection limit for a specific analyte suggests which is likely to be a better method to use. More importantly, by comparing the limits of detection for a desired analyte and for a potential interferent, a larger ratio suggests that the interferent will cause less difficulty.

The value of a detection limit is a number that is either a measure of weight such as nanograms of mercury or of a concentration such as 2×10^{-7} M cadmium. Both kinds of detection limits show the levels of analytes producing a *signal* that is *two to three times the rms noise level* (about equal to the peak-to-peak noise level). In general, the detection limit is an extrapolated value and is calculated for the best possible set of conditions with minimal interferences. The lowest limit that can *actually be measured* typically is three to five times the detection limit, and quantitation even for simple samples, such as a single substance dissolved in a solvent, is usually possible only when the content is greater than about 10 times the detection limit. This is called the *quantitation limit*. For complex samples, that is, ones with a lot of other species in them such as tomato sauce, the quantitation limit becomes 20–100 times the detection limit of the analytes alone with minimal interferences present.

The detection limits for the instrument alone – which are found by using pure analytes without any interferences – are called, reasonably, the *instrumental detection limits*. Instrumental detection limits can be extraordinary. For example, an inductively coupled plasma coupled to a mass spectrometer (ICP-MS) has instrumental detection limits for some elements in the parts per quadrillion range – parts per 10^{15}. However, for samples that require preparation by, say, digestion with ultrapure acid, a more correct detection limit is usually just below a part per billion – a part per 10^9. This is still outstanding, but far from that of the instrument alone under the best conditions.

When comparing published detection limits, it is often necessary to see whether the detection limit cited refers only to the assay or to the entire analysis. It is not uncommon for workers to include the effects of preconcentration steps in the detection limits. For instance, if an analyte is concentrated by a factor of 150 during the sample preparation step, the detection limit that is found experimentally might have been divided by 150 for that type of original sample. Without a clear statement of what part(s) of the analysis are accounted for in the calculated detection limit, direct comparison between methods and equipment is made difficult. If no details of the detection limit calculation are included in a report, comparisons are ineffectual.

15.9 The Working Concentration Range

Every assay method has a range of concentrations within which it will give results with the greatest precision and accuracy. For an intuitive example, consider a test tube filled with a solution of black dye. When looking through it, the dye's color in the solution is apparent. However, if the dye solution is diluted sufficiently, at some point, we will not be able to see the color. If, then, that colorless solution was diluted by a factor of 2, no change would be apparent. For this last dilution, the assay (by eye) of dye content lies outside of the usable range of our eyes. It is below the limit of detection.

Now let us think about the other end of the concentration range. Some of the same black dye is put into a solution so that no light comes through it at all. By comparing that with a tube with dye that has double that concentration, we can see that both appear the same. Such a concentrated solution is also of no use to determine changes in the amount of dye present. An upper limit to the useful concentration exists as well. The range of concentrations between the upper, useless range and the limit of detection is called the *working concentration range*: the range in which an analytical method can be used. Different assay methods have different working concentration ranges, and, in most cases, the range is specific for the analyte. Somewhere within the working concentration range, there is an optimum range in which the precision and accuracy will be best for the given assay. It lies neither at the limit of detection nor at the upper part of the working range, but somewhat removed from both extremes.

15.10 Propagation of Uncertainty

Uncertainty and bias inevitably accompany every step of a chemical analysis. Through the process of validation, the causes of the bias can be understood and either eliminated or the results adjusted to compensate quantitatively for their presence. What, then, remains is the uncertainty for each step, which can be characterized by a standard deviation. The uncertainties of the final results can be found by combining the uncertainties of the parts through the following equation:

$$\sigma_{\text{total}} = \left(\sigma^2_{\text{sampling}} + \sigma^2_{\text{sample preparation}} + \sigma^2_{\text{assay}}\right)^{1/2} \tag{15.7}$$

which has the same form as the equation for overall precision as in Section 15.2 and with the same limitation that the uncertainties of one step do not affect the uncertainties of another. Depending on the number of replicates, either σ or s is used. The enduring aim of a validated chemical analysis is to have only the uncertainty of random variations in the result and to make such uncertainty as small as possible.

Further Reading

Ellison, S.L.R., Rosslein, M., and Williams, A. (**2000**) *EURACHEM/CITAC Guide: Quantifying Uncertainty in Analytical Measurement*, EURACHEM/CITAC.

Eurolab (**2002**) Measurement Uncertainty in Testing. Technical Report No. 1/2002, European Federation of National

Associations of Measurement, Testing, and Analytical Laboratories, *www.eurolab.org/docs/e1_11-01_02_7871.pdf*.

Hibbert, D.B. (**2007**) *Quality Assurance for the Analytical Chemistry Laboratory*, Oxford University Press, Oxford.

Hibbert, D.B. (**2007**) Systematic errors in analytical measurement results. *J. Chromatogr. A*, **1158**, 25–32.

Horwitz, W. (**1982**) Evaluation of analytical methods used for regulation of foods and drugs. *Anal. Chem.*, **54**, 67A–76A.

Inczédy, J., Lengyel, T., and Ure, A.M. (**1998**) *Compendium of Analytical Nomenclature*, 3rd edn, Blackwell Scientific for the IUPAC, Oxford.

Magnusson, B. and Ellison, S.L.R. (**2008**) Treatment of uncorrected measurement bias in uncertainty estimation for chemical measurement. *Anal. Bioanal. Chem.*, **390**, 201–213.

Maroto, A., Boqué, R., Riu, J., and Rius, F.X. (**1999**) Evaluating uncertainty in routine analysis. *Trends Anal. Chem.*, **18** (9-10), 577–584.

Ramsey, M.H. (**2002**) Appropriate rather than representative sampling, based on acceptable levels of uncertainty. *Accred. Qual. Assur.*, **7**, 274–280.

Ramsey, M.H. and Ellison, S.L.R. (**2007**) *Measurement Uncertainty Arising from Sampling: A Guide to Methods and Approaches*, Eurachem, *www.eurachem.org/guides/ufs_2007.pdf: Eurachem/EUROLAB/CITAC/Nordtest/AMC*.

Ramsey, M.H. and Thompson, M. (**2007**) Uncertainty from sampling, in the context of fitness for purpose. *Accred. Qual. Assur.*, **12**, 503–513.

Rubinson, K.A. and Rubinson, J.F. (**2000**) *Contemporary Instrumental Analysis*, Prentice-Hall, Upper Saddle River.

16
Chemical Analysis

Kenneth A. Rubinson

Handbook of Metrology. Edited by Michael Gläser and Manfred Kochsiek

ISBN: 978-3-527-40666-1

16.1 Introduction

Chemical analysis is the study and practice of learning about the elemental and molecular compositions of materials. The materials analyzed include all the natural and synthesized materials of the contemporary world as well as perhaps less obvious materials such as the components of flames or the materials at the atomic layer that indicates the interface among liquids, solids, and gases. The practice of chemical analysis involves all of the following: deciding on the various techniques that should be used; proving that the techniques do, indeed, provide the expected information under the conditions present; and demonstrating that the chemical content that is found does, in fact, answer the germane questions. In other words, contemporary chemical analysis invokes the use of the principles of chemistry that describe chemical behavior and structures, the applications of physics to design the instruments used for measurements, and the tools of statistics and logic to weave the technology to societal desires.

16.2 Choosing the Analytical Methods

The great majority of materials are chemically complicated; this means that they generally are composed of a large number (up to millions) of different elements and molecular species. Materials may be solids, liquids, gases, solutions, glasses, flames, and other forms of matter such as rubbers and gels. When one does an analysis to determine the amounts of species and their identities, the work implicitly includes the necessity of proof. Chemical analysis is, then, the study and practice of learning about materials – their elemental and/or molecular composition – and the tools of statistics and logic to determine the level of certainty in the results. Chemical analysis can also involve determining the spatial distribution of chemical species on the surface of or within a material.

A specific element or compound that is being assayed is called an *analyte*. To determine the presence of one or more analytes and the levels at which they are present requires a series of methods of

Handbook of Metrology. Edited by Michael Gläser and Manfred Kochsiek

ISBN: 978-3-527-40666-1

Tab. 16.1 Primary general applicability of analytical assay methods.

Method	Element	Multielement	Molecule	Surface	Polymer
Absorption spectrometry (UV–visible)	x	x	x		x
Anodic stripping voltammetry	x	x			
Atomic absorption spectrometry	x				
Auger electron spectrometry	x	x		x	
Electron microprobe	x	x		x	
Electron paramagnetic resonance	x	x	x		x
Electrochemical methods	x	x	x		
Emission spectrometry (UV–visible)	x	x			
Fluorescence (UV–visible)			x	x	x
Gas chromatography			x		x
Infrared, Raman spectrometry			x	x	x
Liquid chromatography	x	x	x		x
Mass spectrometry (MS)	x	x	x	x	x
Gas chromatography/MS			x		
Plasma MS	x	x			
Mössbauer spectrometry	x			x	
Neutron activation analysis	x	x			
Nuclear magnetic resonance			x		x
Thermal methods			x		x
Titrations	x		x		x
X-ray diffraction			x	x	x
X-ray fluorescence spectrometry	x	x			
X-ray photoelectron spectrometry	x	x		x	

analysis; this series is called the *methodology* or *protocol of the analysis.*

In Table 16.1, some of the assay techniques that are used for various types of analyses are indicated. This classification is quite rough, and should be understood to be neither exclusive nor complete.

16.2.1 Standards and Validation

Chemical analysis is usually carried out by indirect measurements such as mass, time, volume, and voltage. Such values are converted into measurement of content through calibration. In this sense, to calibrate means to ascertain the relationship between the content of the sample and the response found with the assay method.

As a result, an extremely important requirement is to demonstrate that any analytical procedure measures what the analyst says it measures on a specific type of sample. This is called *validation of the analytical method.* Validation generally means that the methodology must first produce correct results on accepted standards. Chemical standards are pure substances, mixtures, solutions, gases, or materials such as alloys or biological substances that are used to calibrate and determine the validity of each step of a methodology. These steps are sampling, sample preparation for the assay, choice of assay method(s), and statistical treatment of the data to determine the level of certainty in the result. This set of techniques making up the chemical analysis constitute the metrology of chemical analysis, which is covered in greater detail in a separate article in the Encyclopedia of Applied Physics.

16.2.2 Precision, Bias, and Accuracy

We can describe the difference between the precision and the accuracy of results in chemical analysis with the following analogy. Consider a target in which the bull's-eye at the center is analogous to the true result of an analysis. In shooting at the target, we might find that we have a number of hits clustered far off of the bull's-eye. The shooting is, then, precise, but has a bias. The precision of the results from a series of samples of the same material indicates how close they are to each other, that is, the reproducibility. The value of the precision does not describe the bias of the methodology, which is only determined with the aid of standards. Only if the results are closely clustered (labeled as *low imprecision*), and when the bias can be eliminated, does the average of the results lie in or near to the bull's-eye (the "truth"). The analysis is then said to be *accurate*. Presenting a statistical uncertainty in the measured result is an essential part of the process, for example, 2.53 ± 0.04 ng.

16.2.3 Interferences and Representative Samples

The necessity for proof of a measurement of some analyte in, for example, a semiconductor or the air over a city results from the fact that no method of analysis, whether based on specific chemical reactions or those of applied physics, is free from interferences. Interferences (or interferents) are elements and compounds that are not only being measured but are also present in the material (in this context called the *matrix*) that cause, to some extent, a change in the measurement of the analyte. In other words, the measured quantity of analyte alone (say, in pure water) may not be equal to the same quantity measured in the presence of an interferent (say, in seawater).

The second problem is that the sample that is measured may not truly represent the material that is being analyzed. If one is analyzing for certain organic compounds in, say, a lake, then a decision must be made whether to sample from the mud at the bottom, the sand at the shore, and/or water at different locations and depths. It is highly likely that the organic compounds are not evenly distributed, and some care must be taken on how to report the results.

16.2.4 Destructive and Nondestructive Analyses

Another point to consider in any kind of analysis is whether or not the sample is to be destroyed. For instance, it is doubtful whether we would want to dissolve a priceless Egyptian statuette in acid in order to determine the brass formulation. To keep a sample intact, or cause no unacceptable change in its properties, it is necessary to use nondestructive analytical methods. The material may be modified on the atomic or molecular level to some extent, but otherwise the modification is undetectable. To cause as small a change as possible in the original sample, nondestructive tests are carried out with assay methods that are also used for microanalyses or trace analyses. One method that is useful for both micro- and trace analyses is neutron activation analysis (NAA). Two other microanalytical techniques are the electron microprobe and X-ray photoelectron spectroscopy (XPES, also called *electron spectroscopy for chemical analysis* or *ESCA*). In addition, a number of spectroscopic methods that only require collecting light reflected from the surface can be used to test nondestructively, for example, the composition of paints.

A larger number of techniques require destruction of the sample. Any methodology that requires the digestion with acid or other homogenizing methods must be classified as a destructive analytical methodology. A middle ground in classification arises when a small enough sample is needed for analysis so that sufficient material can be collected, say, from drilling a small pit in an obscure place on a valuable object. The method used for analysis might be destructive, but the sample is so small that the object is essentially intact.

16.2.5 Sample Classification by Size and Analyte Level

The size of a laboratory sample should not be confused with the concentration level of analyte in a sample. A descriptive classification scheme for these two properties is presented in Figure 16.1. The sample masses are divided into macro and micro with a vague boundary. The concentrations range from major through minor to trace

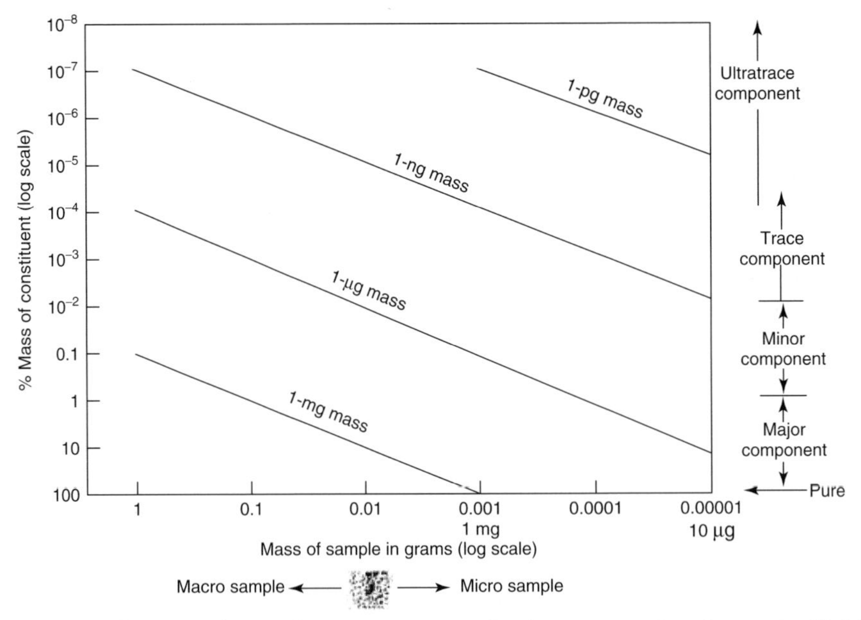

Fig. 16.1 Illustration showing the names given to the ranges of sample size and ranges of analyte content. The horizontal axis has a logarithmic scale of sample size. On the vertical axis is a logarithmic scale of the percentage by weight of a constituent in a sample. The diagonal lines correspond to the mass of the analyzed component. Samples are classified by the relative amount of an element or compound in it. The component can range from major (1–100%), through minor (0.01–1%) and trace (less than 0.01% = 100 ppm), to ultratrace (in the range of parts per 10^9 or less). (Modified from Sandell, 1959.)

and ultratrace. The figure caption contains further details.

When reading the analytical literature of trace and ultratrace analysis, everyone should be aware of a dichotomy in the nomenclature parts per billion (ppb) and parts per trillion (pptr). One group (primarily in the United States) ppb is parts in 10^9, and pptr is parts in 10^{12}, while for another group (primarily Europe) ppb is parts in 10^{12}, and pptr is parts per 10^{18}.

When the content of a component decreases into the parts per million (ppm) range and lower, eliminating interferences from the sample matrix becomes more difficult. As a result, disagreements among laboratories may increase until meaningful exchanges of information on content are not possible. An illustration of the trend is shown in Figure 16.2.

During analyses of trace components, every reagent, such as an acid used to dissolve a solid, must contribute as little interference as possible. Even the containers made of glass or plastic can contaminate solution samples. Such contributions of interfering species is one of the reasons for the trends shown in Figure 16.2. Analyses at the lowest concentration levels require that the laboratory spaces in which samples are handled be clean and filled with filtered air, and that reagents are highly pure and are stored in special containers. At these levels, even the containers that carry the samples to the laboratory must be clean and tested for suitability for the specific sample and for the assay to be run.

16.2.6
Sensitivity and Detection Limit

Two terms that are closely associated with measurements in trace and microanalyses are the sensitivity and the detection limit. The definitions of these terms in a chemical context are similar to our everyday language. If an instrument or other assay methods is more sensitive, it is able to measure smaller changes in content or concentration. *Sensitivity* is defined as the change in response for a given change in concentration or amount; a larger number means the method is more sensitive.

The detection limit is a numerical indication of the level (either concentration or mass) below which we cannot ascertain whether some analyte is present or not. Detection limits of methodologies are extrapolated from experiments done under optimized conditions. (See Table 16.2 for

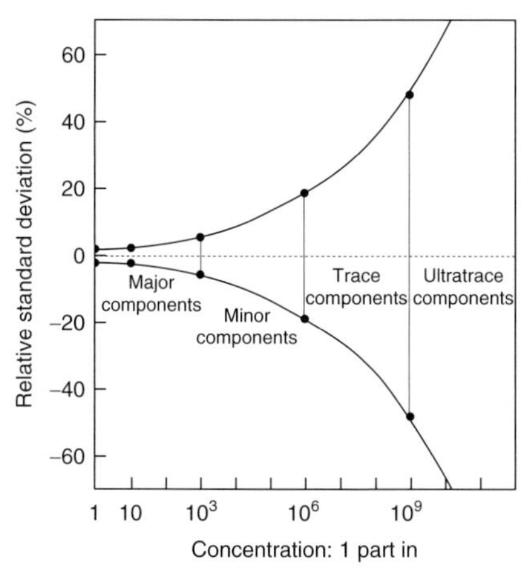

Fig. 16.2 When comparing results among analysts and among different laboratories, a tendency exists to exhibit an increase in the uncertainty of the results with decreasing concentrations of analyte. The horizontal axis is a logarithmic plot of relative concentration (ppb is parts in 10^9, pptr is parts in 10^{12}); the vertical axis is the percent relative standard deviation of the results, a common measure of uncertainty. This plot shows a general trend that has been demonstrated, not a result for a specific analyte or method. (Derived from Horwitz, 1982.)

Tab. 16.2 Characteristic concentrations.[a] and detection limits[a] of analytical techniques for representative elements in parts per 10^9. (Abbreviations and conditions are explained in the key.)

Element	Characteristic concentrations MAS	Detection limits DPP	Detection limits ASV	Characteristic concentrations AAS	Characteristic concentrations ETA–AAS	Detection limits ICP–ES	Detection limits ICP/MS	Detection limits NAA	Detection limits XRF
Ag	15	60	0.005	20	0.2	7	0.04	0.1	1300
Al	3	0.03	0.5	30	2	3	0.1	1	1600
As	15	1	0.5	100	2	60	0.4	1	300
Au	10	2	0.5	100	0.5	30	0.08	2	600
Ba	–	–	–	200	10	0.2	0.02	3	100
Br	20	0.03	–	–	–	–	–	0.2	600
Ca	20	15	–	20	1	0.02	5	200	100
Cd	15	0.5	0.005	15	0.05	2	0.07	10	200
Cl	–	0.3	–	–	–	–	–	2	200
Co	15	1	0.05	70	1	3	0.01	10	30
Cr	15	7	–	60	0.5	2	0.02	–	200
Cs	–	–	–	200	–	–	0.02	200	200
Cu	10	0.5	0.01	60	1	2	0.03	2	300
F	–	–	–	–	–	–	–	20	1000
Fe	20	1	0.1	60	1	7	0.2	–	300
Ge	3	–	–	1000	–	15	0.08	3	–
Hg	25	–	0.2	2200	100	60	0.08	3	300

I	–	0.03	–	–	–	–	0.01	0.2	100
In	15	0.5	0.01	300	10	60	0.01	0.1	–
K	–	–	–	20	0.5	60	–	10	30
La	60	–	–	–	–	3	0.01	1	200
Mg	15	15	–	3	0.03	0.2	0.1	10	600
Mn	1	3	–	30	0.2	1	0.04	0.05	100
Mo	30	6	–	300	5	3	0.08	60	600
Na	–	–	–	6	2	0.5	0.06	1	–
P	1	–	–	–	–	60	–	60	600
Pb	25	1	0.01	200	0.5	7	0.02		600
Pt	200	–	–	600	20	30	0.08	10	600
Rb	–	–	–	100	–	–	0.02	100	–
S	3	–	–	–	–	–	–	–	300
Si	10	–	–	2000	7	10	10	60	100
Sn	10	1	0.02	1000	0.4	30	0.03	200	20
Sr	–	–	–	100	1	0.1	0.02	60	300
Ti	10	3	–	100	–	2	0.06	60	100
Tl	20	3	0.01	200	3	200	0.05	–	–
U	30	3	–	–	1	30	0.02	1	100

(continued overleaf)

Tab. 16.2 *(continued)*

Element	Characteristic concentrations MAS	Detection limits DPP	Detection limits ASV	Characteristic concentrations AAS	Characteristic concentrations ETA–AAS	Detection limits ICP–ES	Detection limits ICP/MS	Detection limits NAA	Detection limits XRF
V	15	–	1	700	1	7	0.03	0.1	100
W	60	–	–	–	–	7	0.06	1	–

[a] As a result of historical practices, the detection limits and characteristic concentrations are calculated in a variety of ways depending on the assay method involved. The detection limits and characteristic concentrations apply to ideal samples. This means that the samples have no untoward matrix effects and no interferences. The values reported here are for samples directly assayed by the various methods without any preconcentration, which means without making the analyte more concentrated by, for example, evaporating solvent prior to the measurement. For realistic samples, detection limits are about 100 times higher than those noted here.

Key

A blank value (–) means either not a method for trace (less than ppm) analysis of the element or not reported.

MAS: Molecular absorption spectrometry in conjunction with colorimetric reagents. The values correspond to an absorbance of 0.025 for a 1-cm path length with a sample consumption of 1 ml.

DPP: Differential pulse polarography, assuming a sample consumption of 10 ml.

ASV: Anodic stripping voltammetry, with a sample size of 10 ml and a plating time of 10 minutes. Various media are used for electrodes. Ni and Co are determined by cathodic stripping of their oxides on Pt.

AAS: Atomic absorption spectrophotometry, with flame atomization and pneumatic nebulizer. The values correspond to an absorbance of 0.0025 with a 10-cm path and a 1-ml sample.

ETA–AAS: Electrothermal atomization–atomic absorption spectrometry. The values correspond to an absorbance of 0.025 with a 50-μl sample.

ICP–ES: Inductively coupled plasma–emission spectrophotometry. Values are for a pneumatic nebulizer.

ICP/MS: Inductively coupled plasma/mass spectrometry. This is with 10-seconds integration. All ions determined in the positive-ion mode.

NAA: Neutron activation analysis, with a sample volume of 100μl, thermal neutron flux of 10^{12} n cm^{-2} s^{-1}, and irradiation time of 10 hours followed by immediate counting.

XRF: X-ray fluorescence spectrometry, with a dispersive instrument for a sample of 1 ml over a surface area of 7–9 cm^2.

Sources: Commission on Microchemical Techniques and Trace Analysis (1982) and Morrison (1979). Manufacturer's literature for ICP/MS.

examples of analyses of the elements.) As a result, detection limits for more typical samples and conditions are factors of 10 to hundreds greater than those in the table. However, the values in the table apply more broadly. Reading across the table, the smaller the value of the detection limit generally means that the associated method will be more effective for that element at any concentration. Another term that is similar to detection limit is the characteristic concentration. More details are in the table notes.

16.2.7 Imaging, Hyphenated, and Multidimensional Techniques

In a simple color photograph, the picture can be considered to be a two-dimensional distribution (or map) of colors and their levels of brightness at each point or pixel (short for picture element). A similar map can be made for the response of a chemical measurement on the surface of a material. For example, the different responses of separate zinc and tin crystallites in an alloy could be mapped in two dimensions and a picture of the microscopic structural distributions recorded and analyzed. Another example would be to determine a full color analysis (a spectrum) for each pixel of the image of a material surface. Both these examples are called *spectral imaging*, and the latter, with the full spectra recorded, is also called *hyperspectral imaging*. Additionally, by recording a series of adjacent slices parallel to the surface, a three-dimensional map of the content can be constructed.

Hyphenated methods – also called *integrated* or *combination methods* – provide powerful tools by combining different analytical techniques. Examples are combining a separation of sample components (Section 16.7) by, for example, gas chromatography (GC) followed by identification and quantitation of each separated component using, for example, mass spectrometry (MS) (Section 16.5). The total process is called *gas chromatography–mass spectrometry* and (*GC–MS*) or *GCMS*.

A technique is multidimensional when a similar technique is used two or more times in sequence. In an example of separations, for instance, where different chemical components of a sample are separated in time, "slices" of this output are taken and each further separated with different conditions. Multidimensional methods differ from hyphenated ones in that the second level of analysis is similar in kind to the first. One example of two-dimensional chromatographies has samples that can physically be run on a square plate, first in one dimension and then perpendicular to that. Another example is two-dimensional GC, which is often abbreviated GC–GC or GC×GC, where each time slice of the first separation is relaunched into a second separation. A final example is MS, where molecules from samples are separated by their masses, and each of the masses produced can be chemically broken down by colliding it with inert gas atoms, and the fragments then analyzed by MS. This set of operations is called *two-dimensional MS*, or *two-dimensional* MS^2. In addition, the process can be repeated yet again on the second set of fragments, which results in the analysis of the fragments of the fragments: the overall measurement is labeled MS^3.

Finally, the multidimensional analysis can be part of a hyphenated or of an imaging method. A hyphenated method that has a multidimensional part includes GC–MS^2. Also, an image of organic molecules on a surface might be obtained by scanning the surface with a laser to desorb the molecules and observe the location-dependent material by MS or MS^2.

One common facet of imaging, hyphenated, and multidimensional methods is that they produce huge amounts of data. Fast computers and efficient algorithms are needed to process and make sense of this flood of data.

16.3 Assay Methods Based on Known Specific Chemical Reactions

The three types of assay methods described in this section – gravimetric, titration, and kinetic – are based on various specific chemical reactions that occur between a specific chemical form of the analyte and a chemical reagent that is added. In most cases, the reactions are done in a homogeneous solution, although some kinetic methods may be based on gas-phase processes. On the whole, these techniques are not inherently highly specific. An exception to the relative lack of specificity occurs when assaying for the total acid content of a solution by determining the quantity of hydrogen ion, H^+. This assay is relatively straightforward by titration.

16.3.1 Gravimetric Methods

Gravimetric analysis, as its name suggests, depends on a measurement of weight to determine the quantity of analyte in a sample. Because of the capabilities of weighing technology, the working precision of gravimetric analyses might be in the range of 1 part in 10 000 and, in special cases, 1 part in 100 000. However, the chemical purity and inhomogeneity of weighed materials limit the precisions of gravimetric analyses to 0.1–0.3%.

There are a number of different gravimetric assay methods. One of these is electrodeposition or electrogravimetry. This technique involves the electrochemical conversion of metal ions in solution into the metallic form which plates out onto an electrode. (Electrodes are electrical conductors such as platinum, gold, graphite, or electrically conducting polymers.) The amount of metal present is then found by the difference in weight of the electrode before and after the plating operation. Electrogravimetry is generally used to determine the quantity of a specific, known metal – for example, cadmium or zinc – in a macrosample. It can be highly precise (better than ±0.02%) and accurate. However, for routine determinations, when somewhat lower precision is needed (in the range ±0.2 to ±1%), the method has been supplanted significantly by spectroscopic methods such as atomic absorption (AA) (see Section 16.4.3.4).

A second gravimetric method involves measuring a sample's changes in mass upon heating at a controlled rate in a controlled atmosphere. This is called *thermogravimetry* (*TG*). Uses for TG include determining the water content of numerous types of samples and determining the carbon and ash content of coal. This technique is described further in Section 16.9.

A third technology for gravimetric determinations is to "weigh" changes in material deposited on a vibrating surface. A piezoelectric material (one that deforms with the application of a voltage across it) is vibrated at an ultrasonic frequency which is a resonant vibration frequency of the system. (The resonant vibration frequency is the frequency at which the solid would vibrate freely if hit sharply.) Since the resonant frequency changes when mass is added or removed from the surface of the device, the measurement of the frequency change provides a measure of mass. A number of different methods are used to produce the vibrating surface, a surface acoustic wave (SAW). Among the materials used is a quartz crystal that distorts under the influence of a voltage across it, which is used in a quartz crystal microbalance (QCM) that can be used both in air and submerged in a liquid.

16.3.2 Titrations

For a titration assay, a volume of a standardized solution (one of known concentration of the reagent) is added to a solution containing an unknown amount of analyte in order to ascertain the quantity of analyte in the solution. When the standardized reagent solution is added to the analyte solution, the correct term to describe the process is a volumetric titrimetric assay. The chemistry of the reaction between the analyte and the titrant must be relatively specific and well understood.

All titrations require a technique to indicate the equivalence point – the point when the standardized titrant has just finished reacting with all of the analyte that was present. At the equivalence point, the volume of the titrant (the solution added) can be related to the quantity of the analyte. However, to discern the equivalence point some physical change must be observed. The center of that change is the end point of the titration. Ideally, the end point and the equivalence point coincide. The measurement of the end point may be done by any technique that is sensitive to the disappearance of the analyte and the changes in the solution that occur because of that fact.

A number of different types of reactions fulfill the requirements for titrations. These are as follows:

1. reactions of acids with bases;
2. reactions in which one species is oxidized (one or more electrons removed) and one reduced (one or more electrons added) – these are called *oxidation–reduction reactions*;
3. reactions in which a precipitate is formed from two or more soluble species;
4. reactions in which a coordination complex is formed (a coordination complex is a soluble species formed when two separate, soluble ions or molecules bind tightly together in solution); and
5. reactions in which tightly bound stable polymer pairs (the analyte polymer and the reactant polymer) or colloids are formed. Colloids are stable, small agglomerations of molecules that do not settle out of solution.

Titration is not a method used for trace analysis. In general, the lower limit for the use of titration for precise determinations is for solutions containing approximately one-tenth of a milligram of the analyte. The amount and precision of titrant addition are not a problem. Contemporary automatic titrators can add tens of microliters with volumes known to better than a microliter. Overall, imprecision in the range of $\pm 0.1\%$ is possible with moderate care using well-calibrated titrants.

16.3.3 Analyses Based on Chemical Kinetics

Kinetics is the area of chemistry in which the rates of reactions between reagents are studied. Kinetic methods of analysis are those in which analyte content is found by measuring the rate of a chemical reaction. The analyte can have one of the following two different chemical actions:

1. The analyte reacts with the reagent and is transformed in the reaction. This is called the *direct kinetic method.*
2. The analyte acts as a catalyst for a reaction. This is called the *catalytic method.* (A catalyst is a material that speeds the rate of chemical reaction between other reactants but is not itself used up.)

Experimentally, there is little difference between the direct and catalytic methods. The rates of reaction must be determined under a set of standard, reproducible conditions. The rates are determined by monitoring the change in the reagent or product concentrations over time. Often, spectroscopic methods are chosen to follow the reaction, for example the increase or decrease in concentration of a colored material. The time dependence is then related to the initial concentration of the analyte. The great majority of kinetic assays are run in solution. Usually good temperature control is required.

There are four possible reasons for developing kinetic assays.

1. Kinetic assays can be extraordinarily specific. Assays based on enzyme reactions are examples. Often, only a few species that are similar in chemical structure to the analyte will interfere.
2. Kinetic assays also can be quite sensitive and, thus, useful for trace analysis. For example, in a catalytic kinetic assay, the analyte as catalyst can produce many times its concentration in some product.
3. Kinetic assays can be used when the rate of a reaction is so slow that waiting for equilibrium to be reached would be impractical. The rate of reaction itself can be observed relatively quickly.
4. A useful assay reaction might be irreversible: equilibrium cannot be attained. Nevertheless, the rate of reaction can be observed to quantify the analyte.

Since all chemical and biochemical clinical and diagnostic methods that use enzymes as reagents are kinetic methods, the majority of all chemical and instrumental analyses are, in fact, carried out using kinetic methods since millions of clinical analyses are run every day.

16.4 Spectrometric Assay Methods

16.4.1 Fundamentals

The measurement of the absorption and emission of light by materials is called *spectrophotometry*. Spectrophotometry is usually shortened to spectroscopy or spectrometry. The latter term is used here. In spectrometry, the words absorption and emission have the same meaning that they have in everyday use. Absorption means to take up, and emission means to give off. In spectrometry, the word "light" is used as a general term for electromagnetic radiation (simply because it is less cumbersome than "electromagnetic radiation"), and we use light as a generic term here.

Spectrometry can be used to assay both the elemental and molecular content of materials. The instruments that are used in spectrometry are referred to as *spectrophotometers* or *spectrometers,* the former associated more with light that is visible to our eyes. The term spectrometer is more common and is used here. The two fundamental variables that are measured in all spectrometric assays are the wavelength or frequency of the radiation and the amount of power at the wavelength. This is called the *spectrum* of the material in some range of wavelength.

Light energy will be absorbed (and/or emitted) by a specific chemical species in a sample only over certain wavelength ranges; at other wavelengths, the species will be transparent (or not emitting). The wavelength at which the energy is absorbed or emitted is a characteristic property of the specific chemical species and not the amount of material. This wavelength property gives spectrometry a measure of chemical specificity. In uncomplicated cases, changes in measured power – or, alternatively, "brightness" or light intensity – are proportional to changes in the amount of the species that interacts with (or produces) the light at the characteristic wavelength(s). As a result, measurement of the power at these characteristic wavelengths allows quantitative analysis of that analyte. This is the basis for all analytical spectrometry.

Because of the different spectra for different species, instead of separating analytes physically, the wavelengths of light associated with each may be measured separately, if the wavelengths do not overlap. Atomic emission spectrometry (AES) utilizes this property. In other cases, the spectra of components in a mixture do overlap. However, the total spectrum can often be separated into contributions from each species present, and the mixture components quantitated without separating them physically.

16.4.1.1 Energy in Spectrometry

An atom or molecule can take up energy from and give off energy to its surroundings in the following three forms: (i) as light; (ii) as heat; and (iii) as kinetic energy (the velocity) of any emitted particles such as electrons.

Different names are given to the spectrometries depending on the form of energy that impinges on the atoms or molecules and the form(s) that leave. These general names are listed in Table 16.3. The first two entries are emission and absorption. In the third, the luminescence can be divided into two types: fluorescence and phosphorescence. On an experimental basis, these can be separated by the differences in time over which emission takes place. When a phosphorescent material is illuminated, the emission may persist for an appreciable time after the illumination is removed. The fluorescence emission ceases virtually immediately. Phosphorescence is usually observed in the ultraviolet–visible range. Fluorescence is used analytically in the infrared, visible, ultraviolet, and X-ray regions. The last but one entry in Table 16.3 relates to electron emission. This is called the *photoelectric effect*, and is the basis for a number of electron spectroscopies, which are most important in the analysis of surfaces.

Tab. 16.3 The transformations of light energy interacting with atoms or molecules.

Energy in as	Energy out as	Spectrometry
Heat	Light	Emission
Light	Heat	Absorption
Light	Light	Phosphorescence, fluorescence
Light	Moving electrons	Photoelectron spectroscopies
Bonding energy	Light	Chemiluminescence

16.4.2 General Types of Spectrometry

Experimental spectrometry can be classified into three general types depending on the components of the spectrometers and their relative geometries. These three classifications are emission spectrometry (ES), absorption spectrometry, and fluorescence spectrometry.

16.4.2.1 Emission Spectrometry

A generalized representation of the equipment used in ES is shown in Figure 16.3. Emission from the sample itself provides the light, and the amount of light emitted at an analyte's characteristic wavelengths is a function of the analyte concentration. The illustration shows the light originating from a luminous flame, but gas discharges, such as are seen in a neon sign, can be used as well. In a few special cases, the energy is introduced from chemical-bond rearrangements; this is chemiluminescence, see the last entry in Table 16.3.

The transducer (a less general term is the detector) converts the light into a signal that we can relate to the concentration of the analyte. A common property of such transducers is that they respond to a range of wavelengths. For example, a transducer for the visible region responds to light of a wide range of colors – much wider than the emissions from specific analytes. Each analyte's emissions are produced at a set of relatively narrow ranges of wavelengths. To measure this set of emissions, each of these narrow ranges of wavelengths must be separated so they can be measured individually. For convenience, each narrow range of wavelength may be referred to as a *single wavelength*,that is, monochromatic, and the device used to isolate these single wavelengths of the spectrum is a monochromator. In ES, the monochromator is placed between the emitting sample and the transducer. Alternatively, after the wavelengths are spread out (dispersed), they can be measured simultaneously (see Figure 16.4).

16.4.2.2 Absorption Spectrometry

The second general spectral method involves measuring the amount of light at various wavelengths that pass through a sample. The geometry of the equipment is illustrated in Figure 16.4, where a separate radiant source generates light. Overall, the instrument measures the amount of light absorbed by the analyte at each wavelength. Both scanning and multichannel types are used.

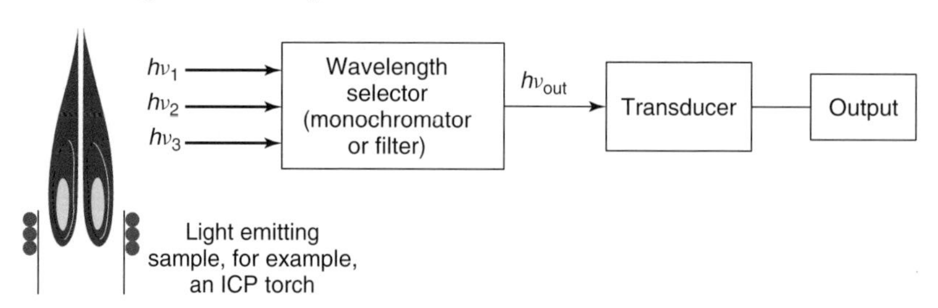

Fig. 16.3 Generalized pictures of equipment designed to measure light emission. Shown is a scanning spectrometer. The light from the source is collected and passed through a monochromator, which only allows a specific wavelength to pass through. The transducer converts the light energy at that wavelength to an electrical signal, which is amplified and displayed as the instrument output. The wavelengths are scanned and associated intensities collected over time.

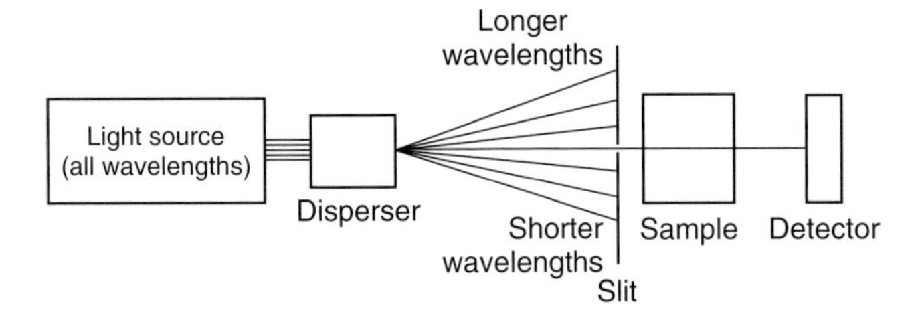

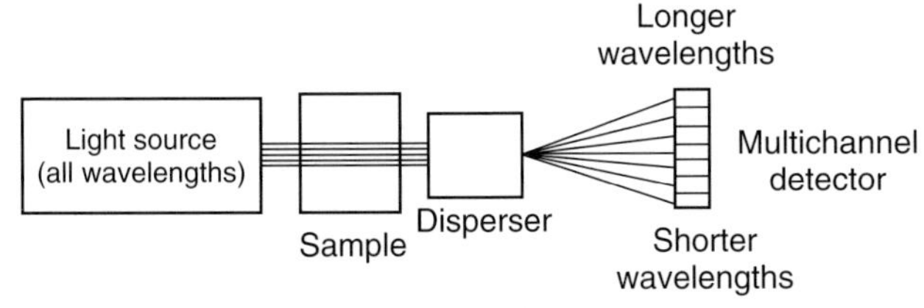

Fig. 16.4 Generalized picture of scanning instrument designed to measure sample absorption or transmittance. (a) A scanning spectrometer where the source light passes into a monochromator, which allows only a specific wavelength to pass through. The light passes through a sample that absorbs a part of the energy, and that which remains is converted to an electrical signal in the transducer. The monochromator is not needed if the light source is generated at a single wavelength such as a specific radiofrequency or a laser light emission. (b) A polychromator (multichannel) system where all the wavelengths are collected at once by a multichannel detector.

Most sources produce light with unwanted wavelengths in addition to the ones that are desired. (The exceptions to this general characteristic are radiofrequency sources and lasers.) Passing the light through either a monochromator or a filter selects the wavelength that is desired for the assay. The picture shows the monochromator between the source and the sample. However, in some instruments, the monochromator is placed between the sample and the transducer.

Absorbance and Concentration The amount of light absorbed by a sample can be reported as the absorbance of a sample. Experimentally, it can be shown to be directly proportional to, a constant that is a property of the material itself as well as the wavelength at which the measurement is being made; b, the length of the path through which the light travels in the sample; and c, the concentration of the material that absorbs the light. Algebraically, the relation is written as

$$A = \epsilon bc \tag{16.1a}$$

This equation describes Beer's law. When the concentration c is in (moles per liter), then ϵ will have the units (liters per mole centimeter), and is called the *molar extinction coefficient* or *molar absorptivity*. The value of b, the path length, must be expressed in units consistent with the units of ϵ. Labeling the units in Equation 16.1a, we get

$$A(\text{unitless}) = \epsilon(l(\text{mol}^{-1}\text{cm}^{-1}))b(\text{cm})c(\text{mol})^{-1} \tag{16.1b}$$

where l stands for liter, and a mole equals 6.022×10^{23} particles. A rough measure of the values of in the visible range with light passing through the sample's 1-cm

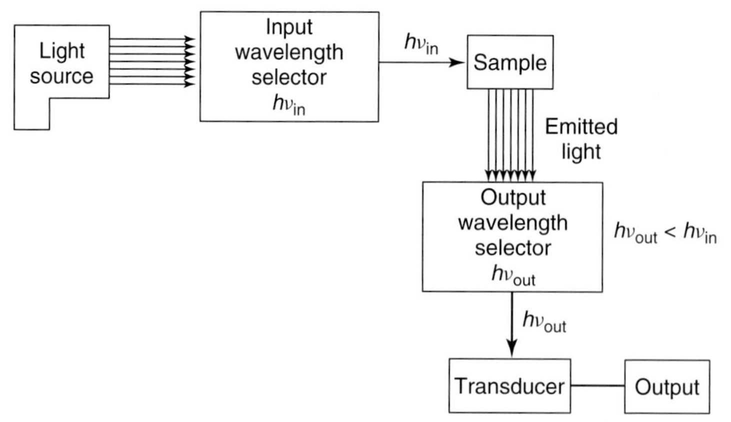

Fig. 16.5 Generalized picture of the equipment designed to measure sample luminescence in the wavelength regions from infrared through X-ray. The source light is collected and passed through a monochromator where a specific wavelength passes and impinges on the sample. The light emitted from the sample is collected and passed through a second monochromator set at a different wavelength. The light energy at that wavelength is converted to an electrical signal in the transducer. This is amplified and displayed as the instrument output. Various types of filters may be used in place of the monochromators.

path length are weakly colored, 10–100; moderately colored 100–10 000; intensely colored 10 000–200 000 in units (liters per mole centimeter).

16.4.2.3 Fluorescence/Phosphorescence Spectrometry

When light impinges on a sample and is absorbed, the energy can be reemitted in all directions as light of longer wavelengths. This emission of radiation is called *luminescence*. Measurement of the luminescence from analytes is done with the general instrument geometry shown in Figure 16.5. A light source for the appropriate energy region is required. The incident light is monochromatic after passing through the first monochromator. The light that comes from the sample is measured at an angle which is not in line with the source–sample axis. This angle is often 90°, but that is not the only angle possible. Figure 16.6 shows an example of both the spectrum of absorption and the spectrum of an analyte's reemission.

For chemical analysis, the division of luminescence into its two types – fluorescence and phosphorescence – is not too important. If the luminescence is present and is proportional to the analyte content, it can be used to quantify the substance. However, if a time-dependent spectrometric method is used, the light due to phosphorescence and that due to fluorescence may be separated. On this practical basis, phosphorescent reemission generally can last a much longer time after the light source no longer provides radiation to the phosphorescing sample. For both, measuring the quantity of analyte involves measuring the intensity of the emitted light at one or more wavelengths.

The organizing principle of the remainder of this section is simply the decreasing wavelength of the region of the spectrum that is used: γ-ray, X-ray, ultraviolet, visible, infrared, microwave, and radiofrequencies.

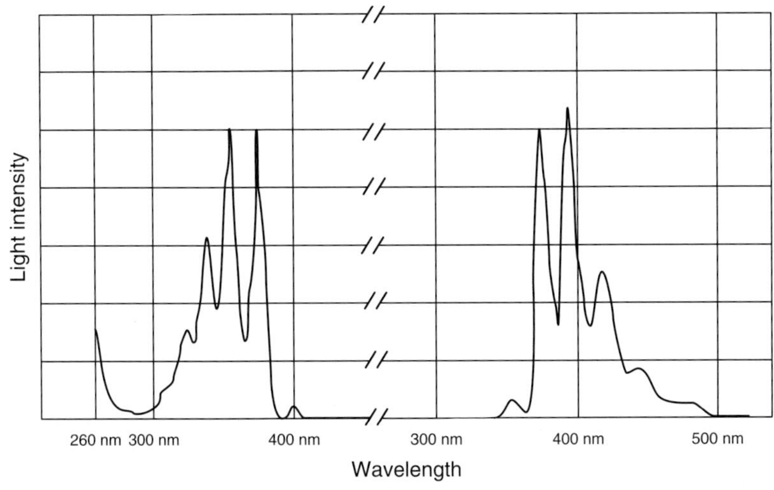

Fig. 16.6 Illustration of the absorption and reemission of radiation by molecules. The energy is absorbed with a wavelength dependence illustrated at the left. Absorbed energy is reemitted with a characteristic spectrum as shown on the right. The emission spectrum is at lower energy than the absorption spectrum; some energy is lost as heat. Sometimes light may be reemitted without changes in the spectrum, but merely with a decrease in intensity.

16.4.3 Specific Spectrometric Methods

16.4.3.1 Gamma Spectrometry: Neutron Activation Analysis

Unstable atomic nuclei that undergo spontaneous nuclear transformations with a concomitant emission of energy are called *radioactive*. Owing to the high energies of the electromagnetic radiation and the high kinetic energies (energy of motion) of the particles emitted upon nuclear decay, individual decay events are measured and counted. The emitted radiation, both electrons (e^-) and photons (the quantum of light), is dangerous physiologically and relatively difficult to block. As a result, analytical methods using the measurement of the radioactive decay have, for the most part, been developed for trace- or microanalyses.

Most radioactive isotopes used in chemical analyses are synthesized. This means that by an appropriate nuclear transformation, a nucleus is converted from a stable form to an unstable, radioactive form. Such transformations can be accomplished for a wide variety of elements by means of a reaction between a neutron and some other elementary particle with a stable atomic nucleus.

Sample irradiation with neutrons followed by measurement of the radioactivity characteristic of the atomic species present is called *neutron activation analysis* (*NAA*). The neutron source is a nuclear reactor in which the sample is placed. The radioactivity produced is characteristic in energy for each type of radioactive nucleus. A typical spectrum can be seen in Figure 16.7. The number of counts of each wavelength is used to quantitate the elemental contents of the material.

Virtually all NAAs are carried out by comparing the radioactivity induced in a sample with that of a standard. To make precise comparisons, the samples must have the same geometry. When these conditions are satisfied, the relative intensities

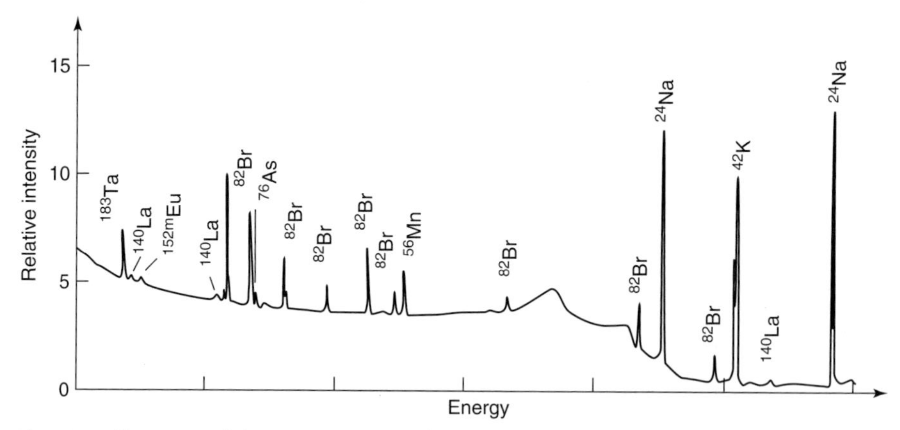

Fig. 16.7 Illustration of the γ-ray spectrum of a neutron-activated sample of tobacco leaf. A large number of different elements can be determined simultaneously in this manner. In this spectrum the lowest peak is 312.7 keV and the highest is 1732.9 keV. (Data from Ahmad *et al.*, 1979.)

of the characteristic radiation of the sample and standard are simply proportional to the total masses of the elements in each. In that way, an analysis utilizing γ-emissions is similar to any other ES.

An optimally sensitive and accurate NAA requires that isotopes be formed which have half-lives in a middle range of a few minutes to a few years. (The half-life is the time for half of the nuclei present to decay. For example, after two half-lives, a quarter of the original is left.) A list of the detection limits of NAA of some representative elements is presented in Table 16.2.

16.4.3.2 Gamma Spectrometry: Mössbauer Spectrometry

Mössbauer spectrometry is a γ-radiation spectrometry in which different, narrow ranges of wavelengths are characteristic of specific nuclear isotopes. The absorption and emission of radiation arise from spectroscopic levels of atomic nuclei and lie in the "hard X-ray/soft γ-ray" region of the spectrum. The precise wavelength depends on the chemical form of the atom.

A majority of elements of the periodic table have at least one isotope which is expected to be measurable by Mössbauer spectrometry. However, relatively few elements have been systematically studied and are, thus, useful in analysis. The isotopes that are used are listed in Table 16.4. One of the most commonly investigated elements is iron. A Mössbauer spectrum is illustrated in Figure 16.8.

Tab. 16.4 Most commonly studied Mossbauer isotopes and γ-emission energies).

Isotope	*E* (keV)
^{57}Fe	14.39
^{99}Ru	90
^{119}Sn	23.88
^{121}Sb	37.15
^{125}Te	35.48
^{129}I	27.75
^{129}Xe	39.58
^{193}Ir	73.08
^{197}Au	77

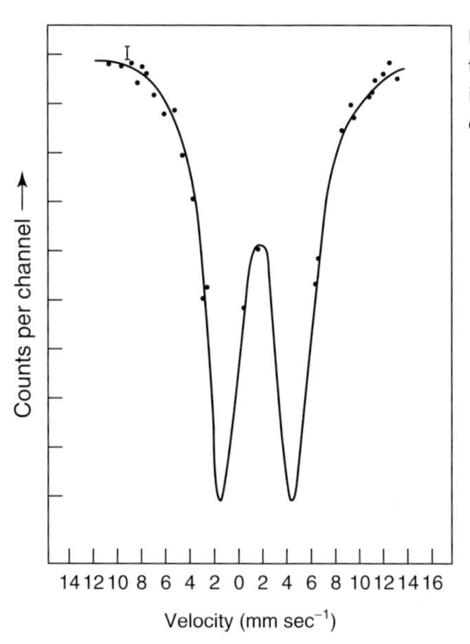

Fig. 16.8 An ^{57}Fe Mössbauer spectrum showing the two lines typical for iron. (This doublet – that is, two lines – results from a single line split by the charge distribution on the nucleus of the atom.)

Mössbauer spectra may be obtained in either absorption or emission, but usually are run in absorption. Emission spectra are usually used to investigate surfaces. The samples must be solids. The spectral energy levels are highly sensitive to the oxidation state of the atom and the chemical environment around the Mössbauer nucleus.

The absorption, which results from the Mössbauer effect, is quite small. Under normal experimental conditions, usually, only a few percent of the energy irradiating a concentrated solid sample is absorbed in the narrow-wavelength range. As a result, spectra are collected and averaged over times from minutes to days in order to see the absorption bands clearly among the noise.

Mössbauer spectrometry is used primarily for qualitative analysis of materials containing a known element in an unknown chemical form. Quantitation can be done, but its precision is subject to a number of assumptions. Only one isotope/element can be investigated at a time.

16.4.3.3 X-ray Spectrometry: X-ray Fluorescence, X-ray Photoelectron Spectrometry, and Particle-Induced X-ray Emission

Radiation in the X-ray region of the electromagnetic spectrum arises from transitions within the quantized energy levels of the most strongly held electrons of atoms. After an excitation, each element emits at characteristic wavelengths that can be used to identify the elements present in a sample (see Figure 16.9). The wavelengths of X-ray emission lines follow a regular pattern with the nuclear charge. The higher the nuclear charge is, the higher are the energies of the respective transitions.

After correcting for substantial background radiation, the intensities of the X-ray fluorescence are used to quantitate the amounts of the analytes that are present. In common with other luminescence methods, a number of elements may be determined during one experiment. X-ray fluorescence is used to determine the elemental composition of materials

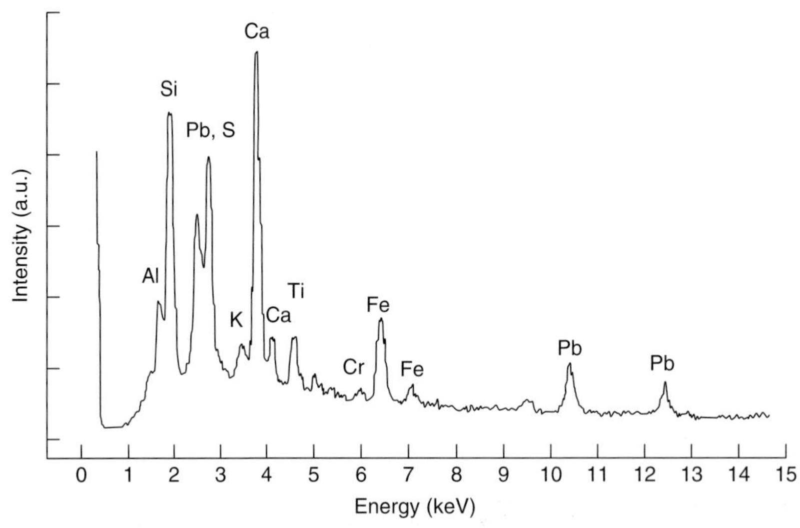

Fig. 16.9 Energy-dispersive X-ray spectrum of air particulates excited by X-rays generated by bombarding aluminum with electrons. Each point is the output summed over time (in counts of X-ray photons) in each of a number of channels of a computer memory. The peaks characteristic of the elements present and are labeled. The concentration of each element can be calibrated. (Redrawn from Wawros, Talik, and Pastuszka, 2003.)

since the X-ray spectra change little with changing chemical forms of elements. One of the method's main strengths is that it can be used for analysis of solids without sample dissolution.

The chemical elements which may be quantified by X-ray fluorescence are those heavier than neon, that is, sodium upward. Also, since almost all elements produce X-rays under the experimental conditions, the "contrast" is not especially high. As a result, X-ray fluorescence generally is limited to higher concentration levels – greater than about 0.1%, as can be seen from the detection limits listed in Table 16.2.

X-rays are relatively strongly absorbed by materials; on average, about 90% of the energy is absorbed within about 0.2 mm (see Table 16.7 in Section 16.8 below). Because of this strong absorption (and also because of possible surface contamination), samples must be prepared with reproducible physical properties.

This complicated spectral interference is one of the principal reasons that carefully prepared standards are important for high-precision (±1%) quantitative spectrometry. However, mathematical methods have been worked out to enable matrix corrections to be determined with fewer standards or to use some of the nonfluorescent, scattered radiation to calibrate the absorption coefficient. Contemporary instruments have both the hardware and software built in to do the corrections automatically after measuring the intensities of the peaks of all the elements present.

Another possible process that can occur when a material is irradiated with X-rays is the ejection of electrons from the atoms with energies characteristic for each element. These are called *photoelectrons*, and the ejection is called the *photoelectric effect*.

The depth from which these electrons can escape is only a few atomic diameters. As a result, these electrons can be used for surface analysis. The analytical technique is called *XPS* or *XPES*) and is described further in Section 16.8 on surface-sensitive techniques.

In contrast to this shallow probe, high-energy particles – with kinetic energies in the million electron volt range – penetrate into the bulk of a sample and cause the emission of both particles and X-ray fluorescence. The X-rays, since they can escape from the bulk, can be used to analyze the atomic species in the first half-millimeter or so of the material. This is particle-induced X-ray emission (PIXE). If the accelerated particles are protons, the technique is proton-induced X-ray emission, also called *PIXE*.

16.4.3.4 **Ultraviolet, Visible, and Near-Infrared Spectrometry**

Analytical methods using luminescence and absorption spectra in the ultraviolet through near-infrared regions (100–2000-nm wavelength), including the visible-light region, are among the most popular and broadly used in chemical analysis. Among the reasons for this popularity is that a large number of analyses of this type were developed before instrumental techniques were available, and the instruments allowed greater sensitivity, lower detection limits, and better precision. Experimentally atomic and molecular assays are often quite different. Atomic spectrometry is used for elemental analysis of analytes. An important condition of such an analysis is to break down the sample into its atomic components. On the other hand, an important condition of analytical molecular spectrometry is to leave the molecules intact.

Measurements are made of the relative amount of light absorbed (absorption) or emitted (emission) or the amount of light reemitted (fluorescence) by components of a sample. Essentially all elements can be assayed by AES or atomic absorption spectrometry (AAS). Also, essentially all molecules can be assayed by molecular absorption and/or fluorescence in the UV–visible or near-infrared spectral region. In this spectral range, atomic analysis is most commonly done on the atoms in the gas phase (usually in a flame or plasma) while molecular analytes tend to be dissolved in a liquid solvent.

Atomic Analysis Atomic spectra are strongly influenced by the atomic environment, and reproducibility in the spectra can best be obtained by producing isolated atoms (atomizing) in the gas phase. AA and atomic emission can be used for all levels of analyte from major to ultratrace by using a variety of methods to prepare and atomize the samples. Table 16.5 lists some sources such as flames, furnaces, plasma torches, and electric arcs that are used to heat and atomize nonvolatile samples.

AE or AES is inherently a multielement technique. The emission source must possess an extremely stable temperature to obtain the best precision. Also, hotter sources generally produce more complete atomization. As a result, in ES, a plasma source is now preferred to a flame source since plasma sources are both hotter than flames and more homogeneous in relative temperature. The major drawback to ES is difficulties with interferences. Nevertheless, quantitation of $\pm 1\%$ is possible with good control.

Unlike AES, AAS is inherently a method to assay a single element at a time since a light source producing a wavelength specific for one element is used. However,

Tab. 16.5 Sources of free atoms.

Method used to produce isolated atoms	Used in
Direct-current arc between two carbon rods in an inert atmosphere such as in argon	Emission
Radiofrequency-induced plasma in argon or helium (inductively coupled plasma)	Emission, fluorescence
Sample heated rapidly inside a hollow graphite tube (graphite furnace)	Absorption
Acetylene/air flame	Absorption, emission
Acetylene/oxygen flame	Emission

the offsetting benefit is that there are nearly no problems with interferences from other elements, which is a common problem in emission. Precision in AA assays better than $\pm 1\%$ is possible.

Molecular Analysis It is difficult to make generalizations about molecular analysis in this wavelength region because the spectroscopic properties of molecules vary so greatly. The basis for the analyses is Beer's law stated in Equation 16.1a. The value of A, the absorbance, can be measured below the 0.001 range with some contemporary instruments. Then, the lowest concentrations detectable depend on the molar absorptivity ε and the length of the light path (limited to about 10 cm for routine liquid analysis). With mirrors, the light's path for gases can be many meters by using multiple reflections through a relatively short cell. Precision in such absorption analyses depends more on the methodology than on the instruments. For instance, measurable variations in the amount of light passed can be seen with slight changes in angular position (order of a few degrees) of a sample. However, with rigid geometries or reproducible placement of samples, variations in results can be as good as $\pm 0.1\%$ for each sample.

Few generalizations can be made about molecular analyses with fluorescence except that the analytes must fluoresce. The practical limit of detection for highly fluorescent molecular species is in the picogram (10^{-12} g) range. That lower limit is practically routine since fluorescence from individual molecules can be measured, although such measurements are generally not done in the context of quantitating the fluorescent molecules.

For both absorption and fluorescence analyses, more than one molecular species can be analyzed in a mixture if standards for the analytes are available. In this case, spectral measurements are taken at a number of different wavelengths, and the concentrations of each of the analytes can be calculated from those data and the spectral properties of the standards. Fluorescence and absorption are two extremely popular methods for monitoring chromatographic separations (see Section 16.7) and quantifying the analytes separated chromatographically, and a majority of such detectors are multichannel systems that measure the entire spectrum continuously. Fluorescence is also used to monitor capillary electrophoresis (Section 16.7.4) as a part of DNA sequencing. Four different colors are used for the four different nucleotide bases.

16.4.3.5 Vibrational Spectrometry: Infrared Spectrometry and Raman Spectrometry

Infrared (IR) spectrometry and Raman spectrometry are used to probe the characteristic vibrations of molecular species in solids, liquids, and gases, as well as the vibrations of atoms and ions in crystals, glasses, and other condensed phases. The frequencies of vibrations of molecules correspond to the frequencies of infrared radiation. However, Raman spectrometry shows these vibrational frequencies as a set of lines separated from the illuminating laser line by those individual vibrational frequencies. (Experimentally, Raman spectrometry might more correctly be classified with visible spectrometries.) Infrared spectrometry for analysis is usually run as an absorption spectrometry (as Figure 16.4) while Raman is experimentally a fluorescence method (as Figure 16.5).

Vibrational spectra appear with numerous characteristic bands over the infrared wavelength range from 2.5 to 100 μm. (In chemistry, more commonly, the line energies are given in wave numbers, abbreviated as cm^{-1}, the inverse of the wavelength in centimeters. The infrared is from $\sim$ 100 to 4000 cm^{-1}.) Individual vibrational absorption bands can often be associated with specific chemical groups such as carbonyls, C=O, organic chlorides, –C–Cl, or ethers, –C–O–C–. As for all quantitative spectrometries, the wavelengths of the bands depend on properties of the molecules while the intensities can be used to determine concentrations. Quantitation to ±1% can be achieved.

Vibrational spectra are often complicated. Seldom can every band be assigned as originating from vibrations of specific groups in the molecule. However, an incomplete understanding of the spectra does not detract from their usefulness in both qualitative and quantitative analysis.

Since no two molecular species have the same vibrational spectrum – with every band matching in peak position (wave number), intensity, and widths of the absorption bands – infrared spectra can be used to identify molecular components of samples by matching the spectrum of an unknown with a library spectrum. (A library spectrum is one obtained from a pure species of known identity – from a standard.) An example of such spectrum matching is shown in Figure 16.10. The matching can be done by lining up spectra as shown in the figure or with computerized pattern-matching techniques.

Infrared spectrometry together with MS (and gas chromatography with mass spectrometry (GC/MS)) and nuclear magnetic resonance (NMR) form the basis for contemporary organic chemical qualitative analysis: identifying the molecular structure of unknown compounds and the compositions of mixtures.

Infrared spectra can be obtained on samples as small as a few micrograms although it is far easier to work with milligram samples. In this manner, infrared spectra are used in qualitative analysis of solids, liquids, and gases. In general, quantitation can be done relatively easily when individual bands are completely separated. With the best contemporary instruments, infrared spectra with precisions as good as for UV–visible spectra are obtained: easily ±1% and below. Raman spectral quantitation depends strongly on the properties of the sample and the stabilities of the laser

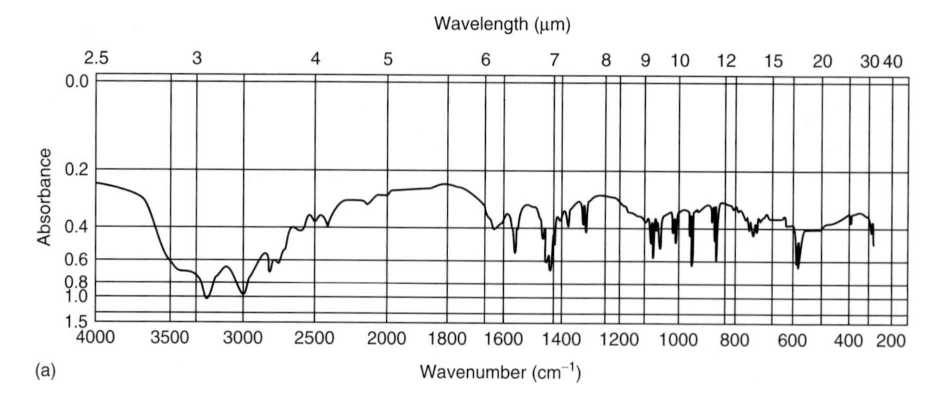

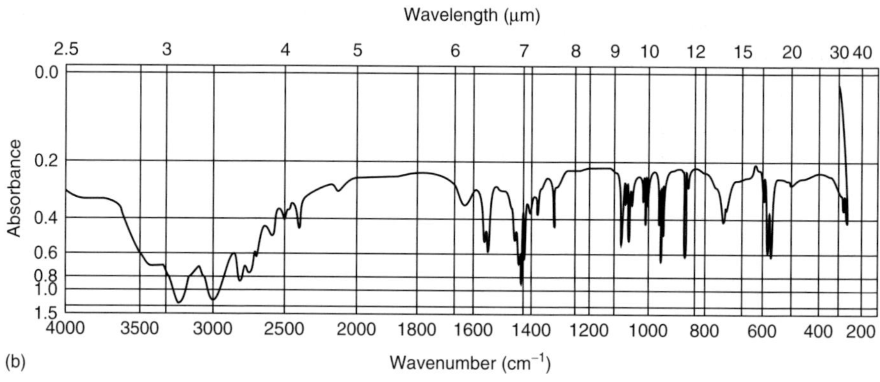

Fig. 16.10 Infrared spectra can be used to identify materials. The (a) is of some unidentified solids collected from a boiler. The (b) is of the tars formed on heating HCl salts of a corrosion inhibitor. There is one-to-one correspondence of all the features in the spectra. Since IR spectra are highly sensitive to structure, the identity of the two samples is quite certain.

light sources. As a result, precision can also be in the ±1% range.

Raman spectrometry tends, in general, to be limited to solids and relatively concentrated solutions. However, the intensity of the spectrum can be enhanced for the gas phase using a far more complicated experimental setup. This is the technique called *coherent anti-Stokes Raman spectrometry* (*CARS*), which is used to investigate, for example, the compositions of flames. Also, Raman spectral intensities can be enhanced by adsorbing to certain specially prepared metal surfaces, a method called *surface-enhanced Raman spectrometry* (*SERS*). Neither of these methods is as commonly used as simple Raman scattering.

16.4.3.6 Microwave Methods: Microwave Spectrometry, Electron Paramagnetic Resonance

Microwave Spectrometry In the microwave region of the spectrum, gas-phase molecules absorb and emit radiation due to changes in the energies of molecular rotations. The frequencies of the narrow spectral absorption lines are able to be measured with high accuracy, and can be used to identify the presence of numerous small molecules and the individual isotopic species such as $^{12}C{=}^{16}O$ and $^{13}C{=}^{16}O$. Rotational spectrometry is not a widespread analytical technique, but it has been used to identify a number of chemical species in astronomy.

Electron Paramagnetic Resonance (EPR) Spectrometry Electron paramagnetic resonance (EPR) spectrometry (or electron spin resonance (ESR) spectrometry) is an absorption spectrometry in the microwave region – 2–200 GHz – for which the sample is placed in a magnetic field. Experimentally, the frequency is fixed and the magnetic field is scanned. The magnetic field required is directly proportional to the microwave frequency. It is in the range around 0.3 T (3000 G) when the frequency is 10 GHz and 1 T when the frequency is 30 GHz. In EPR spectrometry, magnetic field positions are listed as "g values." The g values of a large number of compounds are near 2.0.

The primary requirement for a chemical entity to be measurable by EPR is that it contains a net electron spin. Such compounds include organic radicals (also called *free radicals*) such as $CH_3^{\bullet}$ and metal ions in specific oxidation states. As an example of the metals, cobalt as Co^{3+} has no EPR spectrum, but as Co^{2+} it does. As a result, EPR may be used to determine some oxidation states and quantities of a number of metal ions. However, not all paramagnetic metal ions exhibit EPR spectra. Some do so only in special chemical environments and the spectra of others may only appear at cryogenic temperatures. For analytical purposes, EPR spectra are run on solids and on liquid and frozen solutions.

A number of atomic constituents can be identified from the number of lines that appear in a characteristic spectrum. These lines are due to the interaction of the electron spin with the spin of nuclei in the molecule. Figure 16.11 illustrates the EPR spectrum of vanadium in its + 4 oxidation state.

As a result of the techniques used to increase instrument sensitivity, EPR spectra are presented as derivative spectra such as in Figure 16.11. The spectra are plots of the derivative of the energy absorbed

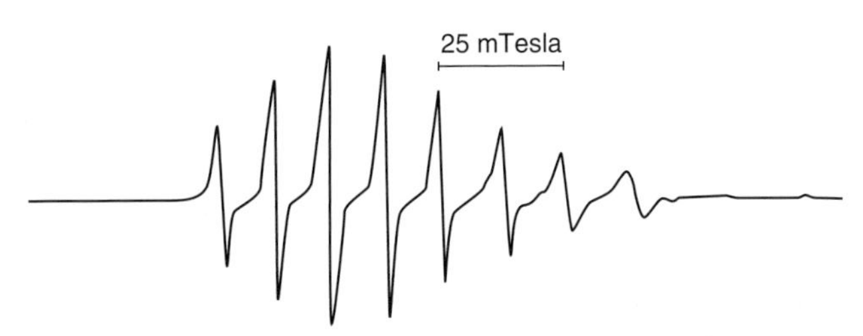

Fig. 16.11 Electron paramagnetic resonance spectrum of vanadium in its + 4 oxidation state in water solution. The eight individual lines result from the interaction of the electron with the nuclear spin. This is called *nuclear hyperfine interaction*. The spectrum is a mathematical derivative (the slope) of the absorption spectrum, and the different heights of the lines are due to the tumbling of the molecule.

versus a linear scale of magnetic field. The peaks (both positive and negative) of EPR spectra are at the points where a regular absorption spectrum has a maximum slope (either positive or negative, respectively). Precision in quantitation by EPR tends to be in the range ±5%. The detection limits are reached around 10^{12} spins for organic radicals (around 10^{-12} moles) and a factor of 10–100 higher for metal ions.

16.4.3.7 Radiofrequency Spectrometries: Dielectric Absorption, Nuclear Magnetic Resonance

Dielectric Absorption Absorption of radiofrequency electromagnetic energy by materials can occur through the interaction with the electric field or the magnetic field of the electromagnetic radiation. The former is the basis for dielectric absorption, and the latter is a property of NMR. The frequencies and magnitudes of dielectric absorption are used to characterize polymers and are the basis for some instrumental methods of moisture determination. However, in general, the absorption bands are not specific enough to be used alone to identify and quantify different molecular species.

Nuclear Magnetic Resonance (NMR) NMR spectrometry is one of the most powerful methods for identifying the chemical structures of organic – including biological and polymer – compounds. The reasons for this utility are primarily threefold. First, NMR spectra for nuclei – such as carbon (^{13}C) and proton (^{1}H) – are sensitive to an atom's chemical type, for example, $-CH_3$, $-CH_2-$, $=CH_2$, $-COH$ differ significantly for both the carbons and hydrogens. Second, the frequencies at which each of the chemically different nuclei appear in the spectra are characteristic for each. For example, hydrogens in methyl groups, $-CH_3$, appear consistently in one range of the spectra. And third, the chemical structures and magnetic resonance properties of atoms immediately adjacent to the resonant groups also influence the spectra. For example, the spectrum of a methyl group attached to a $-CH_2-$ group, that is, CH_3-CH_2-, differs in an obvious way from a methyl group attached to a $-CHCl-$ group although they both show up in the same "methyl region." As a result, it is possible to ascertain molecular structures from NMR spectra alone. The spectrum of a compound does change its appearance somewhat with different magnitudes of the magnetic field. However, these changes are well understood, and usually present no problems in interpretation.

NMR spectra are most easily obtained on species in solution, but more advanced techniques allow useful spectra of solids to be measured.

Because of the sensitivity to small chemical differences, the spectra can be complex, with large numbers of lines overlapping. An example is shown in Figure 16.12. Simple NMR spectra can be used in quantitation, and internal standards are recommended. Multidimensional NMR is still qualitative in the chemical sense; it is used more for identification. The atoms most commonly investigated in chemical systems by NMR are hydrogen, carbon, phosphorus, and fluorine. The NMR phenomenon is limited to specific isotopes of the heavier elements: ^{13}C, ^{31}P, and ^{19}F.

Obtaining NMR spectra involves measuring the interaction of radiofrequency radiation with a sample material that resides in a strong magnetic field (see, *inter alia*, Nuclear Magnetic Resonance in the Encyclopedia of Applied Physics). The radiation used is in the 100-MHz to 1

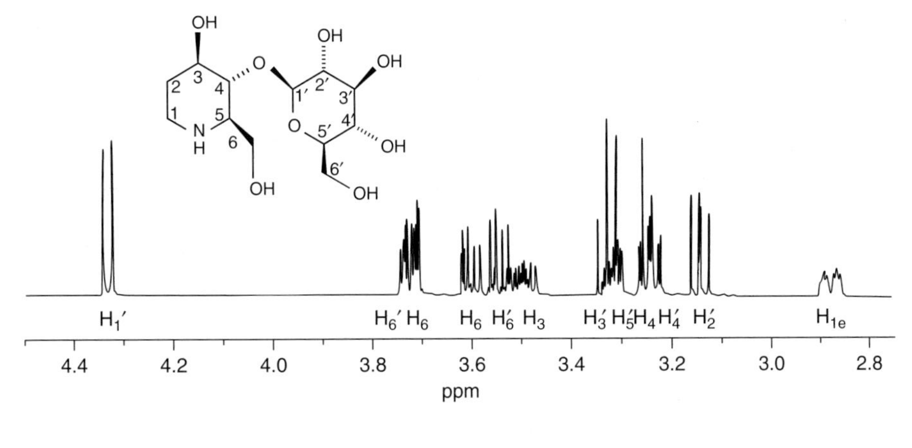

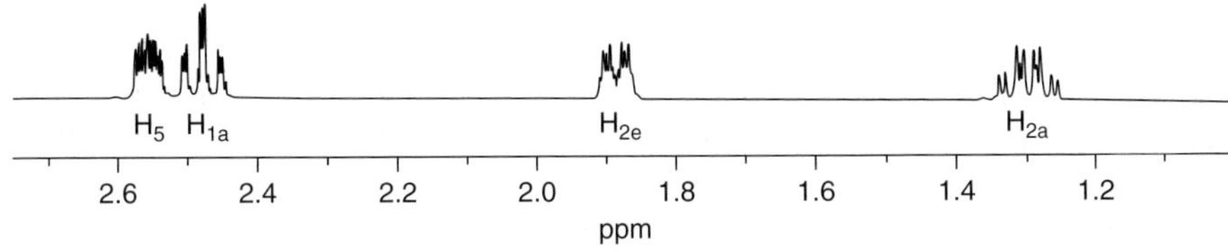

Fig. 16.12 A ^1H nuclear magnetic resonance spectrum of the molecule pictured. The assignments of the features in the spectrum to the various protons are noted below the spectra. The numbers refer to the carbon atoms to which the protons are bound. The numbers of the ring on the right are primed. By convention, the protons are not drawn but are understood to be attached. The bonds drawn as broadening lines indicate the bonds directed out of the plane of the paper. The dotted lines indicated bonds directed below the plane of the paper. The horizontal axis is numbered in ppm, which refers to parts per million of the radiofrequency used. (From Derome, 1987. Reproduced by permission.)

GHz range: the lower range is the frequency of FM radio in the United States and the higher end is near a common frequency for cell phones. The magnetic fields are large; the most sophisticated instruments use some of the largest constant, homogeneous magnetic fields that can be generated. Essentially all contemporary instruments collect the data as the sample's response to a train of high-power radiofrequency pulses that are a few microseconds long. The data obtained are mathematically transformed (a Fourier transform) to produce a spectrum such as shown in Figure 16.12. This method of data collection has the effect that all lines of the NMR spectrum are measured simultaneously. This manner of operation – so-called Fourier transform nuclear magnetic resonance (FT-NMR) – increases the sensitivity and improves the detection limit of the experiment. However, even with the best contemporary instruments, NMR is not really a trace method. To obtain proton spectra at least 10 μg is required, and for carbon more than 1 mg of a species is needed. These limits can be lowered somewhat by using miniature coils surrounding a cell on the order of 1 μl in volume. The analyte must be relatively concentrated.

NMR of Solids To obtain optimum NMR spectra, the analyte molecules should be

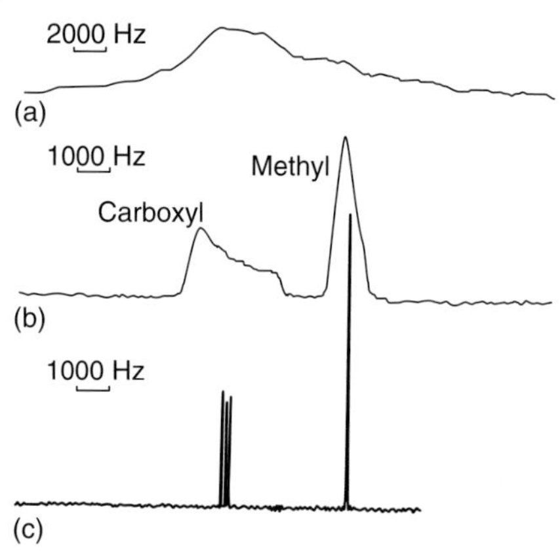

Fig. 16.13 Solid-state spectra of calcium acetate hemihydrate, $Ca(CH_3CO_2)_2 \cdot \frac{1}{2}H_2O$. (a) The solid-state spectrum; (b) the result when the various nuclear spin-nuclear spin interactions are removed by specific radiofrequency irradiation; and (c) when the sample is spun at the magic angle as well. (From Harris, 1983. Reproduced by permission.)

tumbling rapidly relative to the magnet. Part of the reason for this is to average out the interactions with neighboring molecules. In solids, neither the tumbling nor environmental averaging occurs within a sample, and the spectra tend to be broad and nearly featureless, with little chemical information (see Figure 16.13). However, moderate spectra can be obtained through two fundamental tricks. One is to spin the sample at a specific angle – 54°44′, called the *magic angle* – at a rate in excess of 200 KHz. This is called *magic-angle spinning* (*MAS*). The second is to average the magnetic interactions of the surroundings using strong radiofrequency fields at specific frequencies related to the resonance frequencies of the H nuclei present in the sample. This is the technique of proton decoupling. A spectrum is shown in Figure 16.13. A relatively large sample is needed. More commonly used instruments require tens of milligrams, while instruments with microcoils need a thousandfold less.

Two-Dimensional NMR A simple, one-dimensional NMR spectrum (such as shown in Figure 16.12) shows the magnitude of resonance (vertical) as it changes with a scale related to frequency (horizontal). The horizontal measure, ppm, is parts per million of the radiofrequency: 1 ppm equals 500 Hz for the 500-MHz instrument used to obtain the spectrum. A 2D-NMR spectrum, such as that shown in Figure 16.14, shows the resonance changes related to two frequency scales. In this case, both the horizontal scale (labeled on the bottom in ppm) and the vertical scale (labeled v_1) are identical to the ppm scale of the spectrum in Figure 16.12. The 2D spectrum shown is drawn as a contour plot. The usual 1D spectrum runs along the diagonal from lower left to upper right. The peaks that occur off the diagonal (which appear at the intersections of the vertical and horizontal positions of peaks on the diagonal) are called *cross peaks*. Where cross peaks occur, they show which of the peaks of the 1D spectrum are magnetically connected: their nuclei interact magnetically and are, thus, spatially close together. Two-dimensional NMR allows dissection of complex spectra into their component contributions, and because of the localized nature of the magnetic interactions, provides structural information. The results are qualitative in a chemical analysis sense.

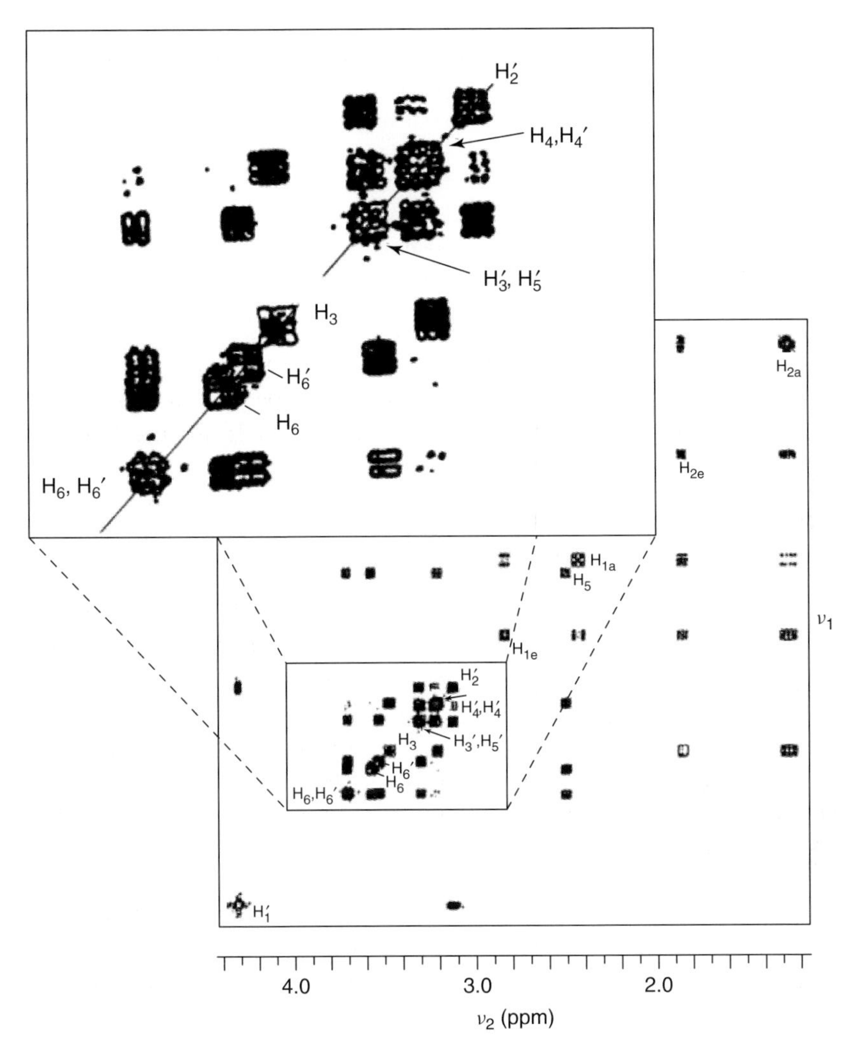

Fig. 16.14 The output from one type of 2D NMR as a contour plot. This is a spectrum of the same chemical as used for the 1D NMR in Figure 16.12. The 1D-NMR spectrum lies on the diagonal from lower left to upper right. The off-diagonal peaks lie on the vertical and horizontal locations of the two peaks that are magnetically connected. (From Derome, 1987. Reproduced by permission.)

Multidimensional NMR Other advanced methods – called *collectively multidimensional NMR* – involve three- and four-dimensional NMR spectroscopy. As for two-dimensional NMR, each added dimension involves a different range of frequencies and indicates how peaks of different nuclei are connected in space. Three- and four-dimensional NMR can be used to unravel far more complicated spectra to provide information that can be used to determine reliably the chemical structures of proteins up to molecular masses of about 30 000 Da. The methods require that the protein be partly enriched in ^{13}C and ^{15}N and require days of data collection on the most sophisticated instruments and extensive, intricate data analysis.

16.5 Mass Spectrometry

MS is the name for a collection of different techniques that are used to measure the relative numbers and the masses of ions that are in the gas phase. The methods of MS are among the most versatile and powerful tools of chemical analysis. This broad usefulness is due in part to having a wide choice of capabilities to transform atoms and molecules for analysis into gas-phase ions, and the sensitivity of ion detectors. Ions from hydrogen to as large as proteins can now be brought into the gas phase with varying amounts of excess energies, that is, energies beyond that needed to vaporize them. This energy causes the molecules to break up and/or ionize. Small amounts of excess energy – soft ionization – leaves the molecules relatively intact. Larger amounts cause fragmentation in amounts proportional to the excess. There are benefits to both having the molecules intact and causing fragmentation.

Once the ions are formed in the gas phase, their separation is determined by the ratio of the mass of the ion to its electric charge; this is written as m/z, the mass-to-charge ratio. A mass spectrum of the fragments of a small molecule is shown in Figure 16.15.

Analysis of organics for both qualitative (identification) and quantitative (the amount present), the method of choice is now the combination of GC (see Section 16.7) with a mass spectrometer as detector of separated species (GC/MS). The chromatography section of this instrument is used to separate the individual species of a sample, and the mass spectrometer allows identification from the pattern of the mass spectrum. Computerized methods of pattern identification are used. Similar mass spectral techniques are employed with liquid chromatography (LC) when separating, identifying, and quantifying less-volatile compounds. MS can be used to quantitate nanogram amounts of

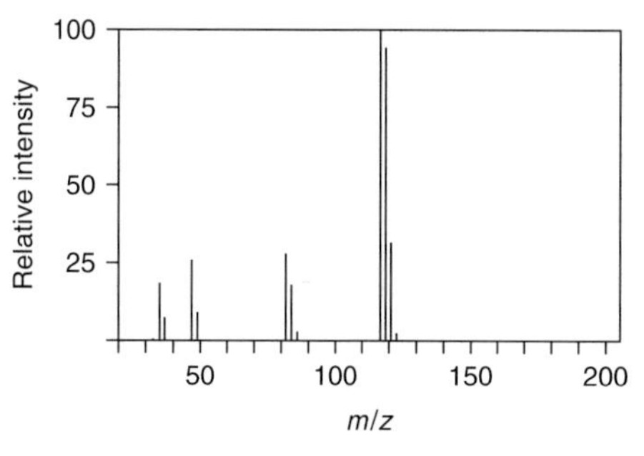

Fig. 16.15 A mass spectrum of carbon tetrachloride, CCl_4. The four peaks at the highest m/z are all the isotopic forms of CCl_3^+ : $^{12}C^{35}Cl_3$, $^{12}C^{37}Cl_3$, $^{12}C^{37}Cl_3$, $^{12}C^{37}Cl^{35}Cl_2$, $^{12}C^{37}Cl_2^{35}Cl$, $^{12}C^{37}Cl^{35}Cl_2$, with a small fraction contributed by the ^{13}C set as well.

molecules and atoms. With internal standards, precisions of ±0.5% are achievable.

16.5.1 Mass Spectrometers

Mass spectrometers, regardless of the design of the various parts, are constructed to separate ions of gas-phase molecules and atoms according to their masses. All mass spectrometers are constructed to carry out four operations as the following:

1. introduction of the sample as a gas;
2. ionization of the molecules in the gas;
3. separation by mass; and
4. detection of the separated species.

All the operations are done under high vacuum – generally around 10^{-9} Pa (1 Pa = 0.0075 torr = 0.0075 mm Hg.) The first two steps occur in the ion source (see Figure 16.16). (Some techniques of sample introduction inherently produce ionized molecules, which combines steps 1 and 2.) The third step is done by mass analyzers. The fourth step involves a transducer, usually an electron multiplier.

Bringing the atoms and molecules into the gas phase can be done in numerous ways; heating is the simplest. Others involve impact on the material with other particles, such as argon atoms, which dislodges molecules from the surrounding matrix of a sample; this is fast atom bombardment (FAB). If ions are used to dislodge the analyte, the ions expelled are called *secondary ions*, and the technique is called *secondary-ion mass spectrometry* (*SIMS*) (see Section 16.8.2). Ions of an analyte may also be formed by using a fast pulse of an ultraviolet laser to ablate a light-absorbing, low molecular mass matrix holding the analyte molecules. The gas-phase ions then pass into mass spectrometer. This is matrix desorption/ionization mass spectrometry (MALDI), which can bring polymers and proteins into the gas phase for analysis.

The methods for vaporization/ionization noted above require that the sample be placed into a high-vacuum chamber. A number of sources start with the samples at atmospheric pressure where they are brought into the gas phase and then pass

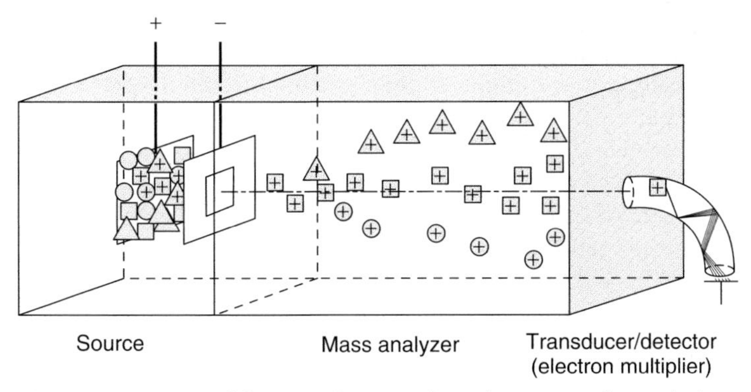

Fig. 16.16 Diagram of the major features of a mass spectrometer. The sample molecules travel from the left, where they are introduced, to the right. Shown in the source are three types of ions with different masses indicated by circles, triangles, and squares. They are separated in the mass analyzer, which may be a magnetic sector, a quadrupole mass filter, time-of-flight delay path, one of many forms of ion traps, or a small, highly precise cyclotron (FT-MS). The arrival of the ions is detected after the separation; here the detector is shown at the right.

into the vacuum for separation through a series of orifices with progressively lower pressures. The general method is called *atmospheric pressure ionization* (*API*). The most commonly used member of these atmospheric sources has a solution containing the sample pass through a hot (temperatures up to 10 000 K) plasma torch where the sample is atomized (broken into atoms) and the atoms ionized. Coupling the inductively coupled plasma with mass spectrometry (ICP/MS) is an extremely powerful method used for trace and ultratrace elemental analyses where nanograms per milliliter of most elements are easily quantitated (see Table 16.2)

In another external sample ionizer, a solution containing analytes can be ejected as an aerosol through a small, highly charged needle into warm air. The electric charges on the droplets cause the aerosol to break up into smaller and smaller droplets eventually yielding gas-phase ions such as proteins, which are drawn into the spectrometer. This source is called an *electrospray* and the general method is called *electrospray ionization* (*ESI*). The electrospray itself can be used as a solvent source to sample surfaces from which the solvent together with the analytes it picks up from the surface are drawn into the mass spectrometer. In that form it is called *desorption electrospray ionization* (*DESI*). A "dry" method of atmospheric ionization for small, relatively stable organics causes the molecules to be vaporized and ionized directly from a surface by placing the sample into a beam of excited helium atoms in front of the pickup orifice. An example of such an analysis is cocaine from paper money. The method is called *direct analysis in real time* (*DART*). These methods are only semiquantitative.

If ions are not produced by the above processes themselves, the molecules can be ionized in a number of different ways. One of the two most common methods is hitting the gas-phase molecules with accelerated electrons: this is electron-impact (EI) ionization. In the second method, chemical ionization (CI), a reagent gas such as methane (CH_4) is ionized by the accelerated electrons and, through subsequent reactions, forms a molecule ion, CH_5^+. The CH_5^+ then donates a proton, H^+, to the analyte to form a molecular ion with a mass one unit higher than before the protonation. (The extra mass mass unit is the mass of the added proton.) A molecule tends to break down into molecular fragments less with CI than with EI.

16.5.2
Multidimensional Mass Spectrometry: MS^n

It is somewhat like a picture puzzle to ascertain the identity of a molecule from the mass spectrum alone, especially if some material is not well purified. For instance, the fragments CO and C_2H_4 both have nominal mass of 28. However, if these fragments are further broken down, their identities become more certain; CO breaks into C and O while C_2H_4 produces CH_3 and CH fragments, among others. This analytical process can be carried out with a tandem mass spectrometer, such as is illustrated diagrammatically in Figure 16.17. In effect, a second mass spectrum is run on the breakdown products of a single peak of the first mass spectrum. MS/MS is an immensely powerful technique for identifying molecular species.

Such processes can be continued: measurement/breakup/measurement/breakup and so forth. This repetitive process is abbreviated MS^n as in MS–MS–MS, and so on, and can be observed with mass spectrometers based on various ion trap

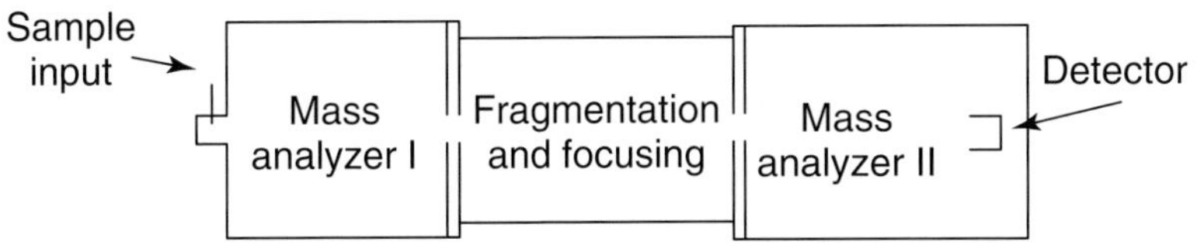

Fig. 16.17 A block diagram of one type of tandem mass spectrometer. The ions travel from the source through the first mass analyzer. Those that pass through this first mass spectrometer enter a region filled with a gas and surrounded by ion-focusing structures. The fast moving ions collide with the gas molecules and fragment. The ion fragments pass into the second mass analyzer, where they are separated by mass and detected.

designs. The ions are held in a vacuum in a small space inside the ion trap where by radiofrequency fields of specific geometries and frequencies, they are both held and manipulated to break up and then ejected to a detector.

16.6 Electrochemical Methods

In this section, some of the analytical instrumental methods used to measure electrical currents, charge, and/or electric potentials that originate from chemical reactions are presented. The analytes determined with electrochemical techniques must be contained in solutions containing unreacting ions called *spectator ions*. Many atomic and molecular species can be assayed from major to ultratrace levels. The subsections below deal with electrochemical methods in the following order: potentiometry, conductimetry, coulometry, amperometry, and voltammetry.

Briefly, electric potential differences (voltages, potentials) can arise in chemical systems due to differences in concentrations of ions and due to differences in the abilities of different species to give off or take up electrons (oxidation and reduction, respectively). The techniques based on measurements of these potentials when minimal current is allowed to flow are those of potentiometry.

When the content is quantified by measuring the changes in the electrical resistance of solutions of ions (electrolyte solutions), the techniques are those of conductimetry.

A potential can be applied to an electrode in a solution causing electrons to be exchanged with electroactive analytes (able to be oxidized or reduced) in the solution. When the analyte loses an electron it is said to be oxidized; when the analyte gains an electron, it is said to be reduced. We can relate the magnitude of the current and the time it flows to the amount of analyte present. The product current × time is the total charge transferred. The amount of the charge transferred for oxidation or reduction is the basis of a number of different techniques called *coulometric methods*.

One group of techniques combines the simultaneous measurement of a potential and current as they change with time. These are called the *voltammetric methods*. They are, perhaps, experimentally the most elegant and yet simple methods for trace and ultratrace analyses.

A straightforward electrochemical technique that does not fit well into the section's organization is electrogravimetry. It was discussed with other gravimetric methods in Section 16.3.1.

16.6.1 Potentiometry

The concentrations of analytes that are ions or change ion concentrations can be measured in solution with potentiometric methods, that is, those where the electric potentials of chemical reactions (their voltages) are obtained. The concentrations are related to the voltage logarithmically; for example, the voltage change for a 100-fold increase in concentration is twice the voltage change for a 10-fold increase in concentration.

Such a measurement requires a potential to which the measured one can be referred – a reference potential. Common references are the saturated calomel electrode (SCE) and the silver–silver chloride electrode. It is difficult to measure a cell potential to better than about $\pm 1\,\mathrm{mV} = \pm 0.001\,\mathrm{V}$, which represents an uncertainty in concentration greater than 4%. This is the limit of precision for potentiometry. If more precision is needed, potentiometry can be used to detect end points of titrations (Section 16.3.2). The electrodes used for such measurements are designed to respond to one or a few ions specifically. The generic name for these devices is ion-selective electrodes. A well-known example is the pH electrode, which responds to the amount of H^+ in the solution. The operation of this electrode depends on the response of a surface that binds specifically to the analyte ion; this is an ion-selective surface. The oxidation state of the H^+ remains unchanged by the measurement. For potentiometric measurements, the current flow is kept as small as possible. For example, a common pH meter will measure the potential when the current flowing is on the order of 100 pA.

To find the analyte concentration, the voltage in the presence of the analyte must be calibrated with a solution containing a standard, known concentration of the same analyte. There are two ways to do this. First, the concentration is found by comparing the voltages from the standard and the analyte and recognizing the logarithmic relationship to find the analyte concentration. Second, we can extrapolate from a set of voltages measured after adding a series of known amounts of the analyte into the original solution after its potential has been measured. This latter technique is known as *standard addition* or, alternatively, *spiking*. Both these techniques can also be used for all electrochemical and spectroscopic analyses.

16.6.2 Conductimetry

Experimental values of conductivities (the amount of current an ionic solution can carry when a voltage is applied) almost always are found using rigidly fixed electrodes placed in the analyte solution. One type of conductivity cell is illustrated in Figure 16.18. Quantitation using conductance measurements can be done simply by calibrating the instrument with known concentrations of the ion-forming analyte and then measuring the conductivities of the solutions to be analyzed. The use of conductimetry is limited to analytes that are ions in solution. Conductimetry is used both to measure multiple species together, such as for monitoring contamination in streams and rivers, and for a single species such as continuous monitoring of an acid concentration.

Quantitative conductivities can be determined to about $\pm 1-2\%$ if the resistance between the electrodes is in the range of 1000Ω and the temperature is thermostated to within $\pm 0.1^\circ$. With proper standards and temperature control, special

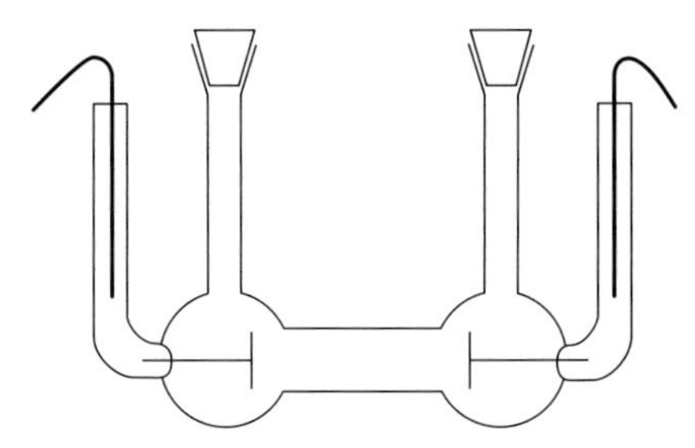

Fig. 16.18 Illustration of a conductivity cell. Usually the electrodes are platinum foil covered with a finely divided platinum on the surface. From its color, it is called *platinum black*, and it provides a large surface area – much larger than the dimension of the electrode. Samples fill the cells so that the geometry is the same for all runs.

high-precision conductimetry can be done with errors in the range of 0.1%. However, such precision is not routine.

16.6.3 Methods that Use Oxidation and Reduction

The remainder of the electrochemical methods discussed here depends on oxidation and/or reduction of the analytes with measurement of the amount of electric current that flows during the process. Either the current that flows or the total charge required for the oxidation/reduction process is measured. They are directly proportional to the analyte content. The types of materials that can be assayed include both elements and molecular species. When molecules contain organic groups that can be reduced electrochemically, they can be assayed by these methods. A number of electroactive groups are listed in Table 16.6.

16.6.3.1 Coulometry

A coulometric assay relies on measuring the total charge that is needed to complete an electrochemical oxidation or reduction. This total charge is relatable directly to the total amount of material electrolyzed. It is relatively easy to measure a total charge of millicoulombs over periods of minutes. A general calculation shows this to be representing about 1μ g of analyte.

The apparatus that is used is designed to keep a constant potential at the electrode where the analytical reaction (oxidation or reduction) occurs. The apparatus is a general piece of electrochemical equipment called a *potentiostat*, and it can also be used for electrogravimetry, coulometry, and the various forms of amperometry.

Coulometric assays are carried out in stirred solutions; at one time or another, all the analyte must come in contact with the electrode, and stirring hastens the process. An alternative to stirring is to use only a

Tab. 16.6 Bond types that can be reduced for electrochemical measurements.

C–C	C–N	C–O	C–S	C–X[a]
C=C	C=N	C=O	C=S	
C≡C	C≡N			
	N–N	N–O	N–S	
	N=N	N=O		
		O–O	O–S	O–X
			S–S	S–X
Condensed benzenoid rings				
Some heterocyclic rings				

[a] X = halogen

Source: *Chem. Eng. News* (1968), March 18, p. 96.

thin layer of solution so that the inherent diffusion of analyte is sufficient to result in a uniform concentration throughout the solution volume within a few minutes without stirring.

16.6.3.2 Amperometry and Voltammetry

Amperometric analyses are achieved by measuring the electrical current that passes into a solution and causes oxidation or reduction of the analytes at a fixed potential. The assay depends on quantitative measurements of the limiting current, i_L, which is proportional to the concentration of the species being oxidized or reduced at the electrode (see Figure 16.19). A relatively powerful but specialized use for fixed-potential amperometry is in LC. An electrode is set at a potential that will cause oxidation or reduction of the analytes, and the electrode is placed so that the analytes pass over it after they are separated on an LC column. The resulting increase and decrease in current as a band of material passes by can be used to quantify the amounts of each separated analyte. This is called *liquid chromatography with electrochemical detection* (*LC*/*EC*). Nanogram amounts of electroactive compounds can be determined inexpensively compared to other methods with comparable limits of detection.

A second, general set of amperometric methods involves measuring the changes in the current (as amperes or amps) as the voltage is changed; hence the name *voltammetry*. These voltammetric methods include polarography (a voltammetric method using mercury as its electrode), differential pulse voltammetry, and anodic stripping voltammetry (ASV). Figure 16.19 shows representative data of the current for an electroactive species as the applied potential varies from zero to more negative potentials. At first, little current flows since the analyte does not react in this voltage range. Then, the current increases relatively rapidly with voltage, and finally a plateau is reached where there is little further increase in current over the projected baseline as the potential is changed further. This is the limiting current, i_L. The curve shown in Figure 16.19 results from

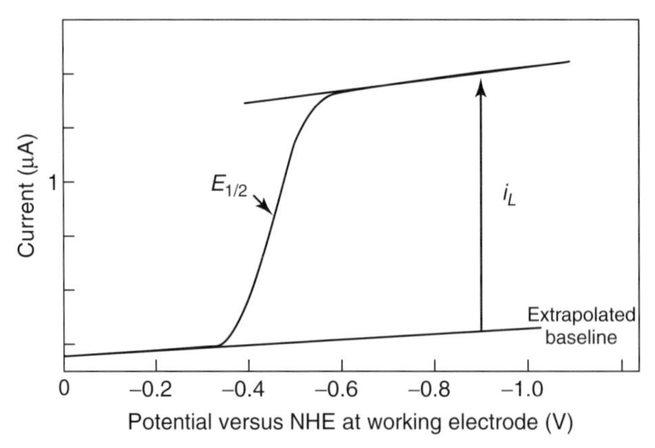

Fig. 16.19 Illustration of the output of one type of voltammetric experiment. The current flow is plotted versus electrode potential. At the left, only a background current is seen. Then, the electroactive species begins to react, finally reaching the limiting current. The midpoint voltage is characteristic of the analyte. The limiting current after the subtraction of the projected background is proportional to its concentration.

the simplest case of voltammetry, where the potential at the electrode changes with a constant increase or decrease with time. This is a voltage scan.

An increase in the flow of current can be measured more precisely by a careful combination of pulsing the potential together with selective data acquisition. The background current and the current produced when electrons are transferred after a voltage jumps are collected at specific times before and after the voltage jumps, superimposed on a more slowly changing voltage. One method that works this way is differential pulse voltammetry. (With a small mercury drop as the electrode, it is called *differential pulse polarography* (*DPP*)). Results of such determinations appear as seen in Figure 16.20. By comparing the area under the curve or just its height with the results from standards, one obtains the analyte content. DPP is a method suitable for trace analysis of both atomic and molecular species. The detection limits for various elements appear in Table 16.2.

16.6.3.3 Anodic (and Cathodic) Stripping Voltammetry

ASV is a technique that can be used to assay a number of metals and nonmetals at the trace and ultratrace levels. The technique contains a step that first concentrates the analyte, which allows far lower detection limits for many metals than without such a step (compare DPP and ASV in Table 16.2). For example, for metals, the ions are first reduced to their metallic forms, which dissolve in the small mercury drop electrode (the mercury/metal solution is called an *amalgam*). The concentrated, dissolved metals are subsequently reoxidized, and the resulting current provides a measure of their concentrations. Differential pulse techniques give the best detection limits. An example of the type of data obtained is shown in Figure 16.21. When the analyte is first oxidized to concentrate it and then subsequently reduced for the assay, the technique is called *cathodic stripping*; both are used primarily for elemental analysis.

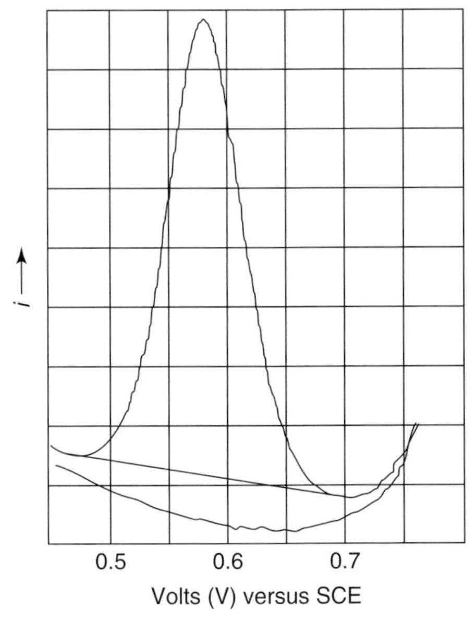

Fig. 16.20 Illustration of a differential pulse voltammogram and, underneath, the background current. The scale is about 5 nA per division. The shape of the curve is the slope of a curve such as seen in Figure 16.19. (From Rubinson and Rubinson, 2000.)

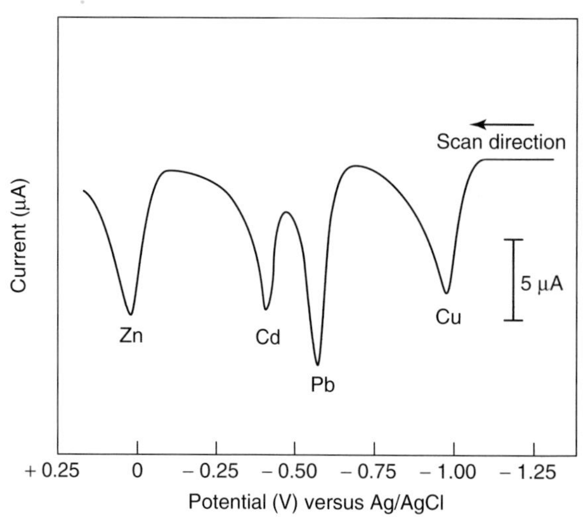

Fig. 16.21 An anodic stripping voltammogram of a solution 2.5 ppm in copper and zinc and 5 ppm in lead and cadmium. The sample was reduced and concentrated as an amalgam and then scanned as shown. The peak areas are proportional to the concentrations. (From Rubinson and Rubinson, 2000.)

16.7 Separations and Chromatography

To isolate quantities of pure substances involves the science and art of separations. Separation methods are based on the differing chemical properties of the components of a sample. In contemporary chemical analysis, separation is synonymous with chromatography. Figure 16.22 diagrammatically shows one type of chromatographic instrument in which the separation occurs within a hollow tube – the chromatographic column – containing the adsorbent or stationary phase that has a flowing fluid in contact with it. The stationary phase in column chromatography consists of either the column surface itself or particles of a solid material to which the analyte molecules bind. The analyte is placed at the beginning of the column and washed through with the liquid eluent. The individual components elute at the other end at earlier or later times depending on their differences in binding to the stationary phase. The analytes that spend more time on the stationary phase elute later.

The reasons for the wide applicability of chromatographic methods involve the variety of conditions that can be used to separate components of materials. For instance, we can use different mobile phases: gases, liquids, and supercritical fluids for LC, GC, and supercritical phase chromatography, respectively. Further flexibility arises from having many types of stationary phases (minerals and numerous kinds of polymers alone or with surfaces modified with molecules having varying properties). In addition, the geometries of the stationary phases can differ (inside columns of various sizes, on plates) or the scale of the separation can vary (large for preparative, smaller for analytical).

In contemporary analysis, the components that have been separated are detected as they elute from the stationary phase by measuring changes in any of a number of different chemical or physical properties. Virtually every instrumental and wet analytical chemical method has been used to detect and identify the components separated chromatographically. Four examples among these are electrical conduction, light absorption, ion production in flames, and atomic or molecular mass. The response of the detector with time as an eluted band passes – that is, a rise and fall (or

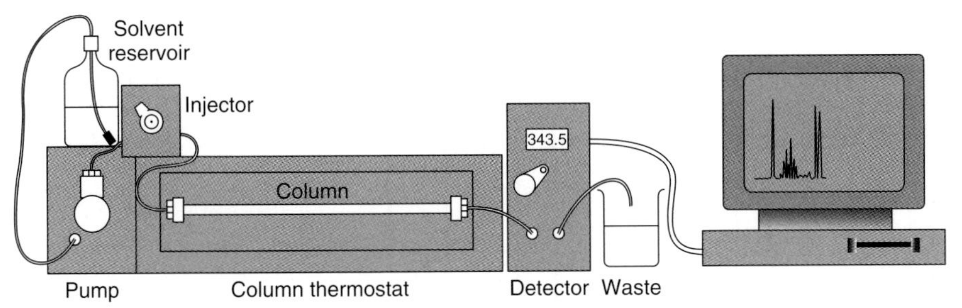

Fig. 16.22 Diagram of the components of a column liquid chromatograph. The sample of known volume is injected onto the column where the components are separated as they flow at a rate fixed by a pump. They elute and are detected, where a plot versus time is the chromatogram, as seen on the screen. Another chromatogram is shown in Figure 16.23.

fall and rise) in the property from the baseline level – produces a series of peaks as shown in Figure 16.23. This chart of detector response versus time is called a *chromatogram*.

16.7.1 Liquid Chromatography

16.7.1.1 Column Chromatography

Within that broad definition of LC are a myriad of techniques, and many of them are associated with more than one name. The names focus on many different aspects of LC methods. For instance, one of the primary classification schemes of LC is by the physical shape of the solid support such as column chromatography, thin-layer chromatography (TLC) (with the solid support forming a layer on a plate), capillary LC (with the inner surface of the capillary as the solid support), and flat-bed chromatography. Classification is also based on the efficiency of the separations such as high-performance liquid chromatography (HPLC) or high-performance thin-layer chromatography (HPTLC). Sometimes names of LC methods identify solutes which are separated and detected such as ion chromatography (done with column methods). One of the reasons that LC has become ubiquitous in analytical laboratories is the development of HPLC. The solid phase is composed of carefully packed particles that are between 10 and 2 μm in diameter. The size of the particles makes it necessary to pump the mobile phase through the column at pressures up to thousands of pounds per square inch (the commonly used, nonmetric pressure measurement, $1\,\text{lbf in.}^{-2} \approx 6895\,\text{Pa}$). HPLC pumps keep a highly precise eluent flow rate regardless of the conditions in the column. As a result, after calibration with standards, the positions of the peaks in time can be used for identification, and the peak heights and peak areas can be used to quantify analyte species without needing frequent recalibration.

Usually one of four basic types of chromatographic supports is used in LC. The general type of chemical interaction that occurs between the solid support and the solutes gives each separation a different name. The classifications are denoted as follows:

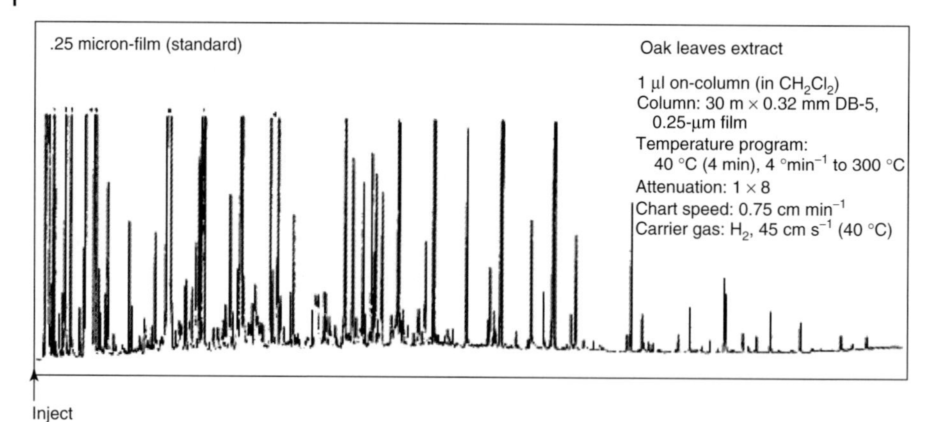

Fig. 16.23 Gas chromatogram of the volatile components of oak leaves. This was done on a capillary column. As complex as this appears, statistical arguments suggest that more than one component is under many if not most of the peaks. (Courtesy of J&W Scientific, Inc., Rancho Cordova, California.)

1. Normal phase or adsorption. The stationary phase is highly polar (i.e., large electric dipoles at the surface) and the eluent is somewhat less polar.
2. Reversed phase. The stationary phase is less polar than the eluent.
3. Ion exchange. The stationary phase has ionic charges on it and the separation is based on differences in charge.
4. Gel filtration, also called *gel permeation* or *exclusion*. The stationary phase has pores into which the analytes can migrate. Separation is by size.

Less commonly used are some solid supports that are designed to interact with specific chemical functional groups, such as thiols (–SH) or diols (–COH–COH–), that are on the solutes being separated. With a similar strategy of specific interactions, solid phases can be made with molecules that bind to one specific protein and can pull it out of a mixture. Chromatography with such protein-specific groups on the stationary phase is called *affinity chromatography*.

Running a chromatographic separation can take up to hours. In many cases, the process can be speeded up at least 10-fold if the analyte can withstand the conditions needed, which involve higher pressure to provide faster flow, higher temperatures, and shorter columns. However, the speed also requires detection methods that respond faster and possibly to smaller amounts of material. Nevertheless, this is an important trend.

Quantitation in column LC with automatic sample injection and with the use of internal standards can approach ±0.1%. Precision of ±1% is relatively straightforward with current instruments.

16.7.1.2 Separations of Chiral Molecules

There exist numerous molecules, particularly most biological ones, which, while having exactly the same compositions and same atoms attached to each other, nevertheless have different structures. Their differences are similar to the differences between right and left hands. As is true for hands, the molecular structures are mirror images of each other. These are chiral molecules. Significant progress has been made in recent years in both GC and HPLC separations of chiral molecules, that is, separating the right-handed isomers from the left-handed isomers.

The approaches are threefold. In one, a reaction is carried out binding a chiral reagent to the analytes, and the now-different reaction products separated with an ordinary chromatographic method. In the second, the analytes are chromatographed on a chiral stationary phase, and, since the two different structures are retained differently, they are separated. The third approach is a variation on the second; the stationary phase is not chiral, but the mobile phase has a chiral species added to it. This chiral species interacts reversibly with the chiral analytes, which allows a separation of the various products to occur on a nonchiral column. All three methods have been developed for HPLC. However, only the first two are used for GC chiral separations.

16.7.1.3 **Thin-Layer Chromatography**

In TLC, the adsorbent particles comprising the stationary phase are laid down as a layer a fraction of a millimeter thick attached to a solid backing plate of aluminum, glass, or polymer. Separations carried out on these stationary phases are called *TLC*.(Unfortunately, sometimes thick-layer chromatography, with layers 0.5–2 mm thick and primarily used for preparations, is also abbreviated TLC.) The primary advantage of TLC is that a number of samples and standards can be analyzed simultaneously side by side on the same plate. Using column chromatography, individual samples and standards must be analyzed sequentially. One other advantage of TLC is that samples that are difficult to resolve may be developed in two different solvents run in perpendicular directions: a two-dimensional separation. In this way, two or more species that run together in one solvent system may be separated from each other in the perpendicular direction in a second solvent system. One disadvantage of TLC is that quantitation tends to be less precise than in column chromatography. The greatest precision on intact plates is obtained with calibrated imagers.

16.7.1.4 **Field-Flow Fractionation**

When liquids flow without turbulence through a tube, the liquid at the walls is immobile, and that in the center flows the fastest. Since any suspended particles flow along with the fluid, the particles that reside in the center flow along the fastest and those nearer the wall more slowly. This is the basis for field-flow fractionation (FFF), which is carried out with the fluid flowing through a closed thin rectangular flow channel. FFF is the name of a set of techniques that use different kinds of forcing fields applied perpendicular to the flow direction across the thin direction to separate solutes of high molecular weight. The small separation set up by the field is, in effect, amplified by the flow of the fluid. The forcing fields can be set up by centrifuging (sedimentation), or by applying a voltage (electrical), or by applying a temperature difference (thermal). They find applications in the separation of macromolecules such as proteins, synthetic polymers, biological cells, and microscopic beads of various materials. Detection of the fractions is done with the same methods as for other LC methods: spectroscopic, electrochemical, and chemical.

16.7.2
Gas Chromatography

GC is an analytical method which has been studied and used intensively for more than 60 years in solving a wide variety of problems. Most classes of compounds that can be carried in the gas phase have been studied, and good separation conditions found for them.

GC equipment is quite different from that for HPLC. GC requires precise control of the flow of gases instead of liquids. The column, almost always, is significantly longer and/or narrower. Packing materials and stationary phases differ. The detectors differ both in construction and operation.

GC may be arbitrarily divided and named in terms related to the types of columns and column packings used. If the internal diameter of the column is in the range of 0.2–0.5 mm, this is called *capillary-column gas chromatography*, or *capillary GC*. With only a few exceptions, GC now means capillary GC when used in the context of analyses.

The analytes dissolve into and evaporate from the stationary phase multiple times while passing through the column. For analysis of organics on capillary columns, the material of the stationary phase generally is one of a number of different high-boiling-point liquids or stable solids that is coated on the inside surface of the capillary. With both these types of packings, the technique is called *gas chromatography (GC)*. However, the latter is often called *gas–liquid chromatography (GLC)*.

GC is routinely used for analyses of gases and for organic compounds with boiling points below about 450 °C. With boiling points in that range, the compounds have vapor pressures that are high enough so they will be carried in the gaseous mobile phase while the temperature is still low enough not to decompose either the analytes or the column materials. As a result, high-boiling materials are, perhaps, better separated by HPLC.

On the other hand, the GC column may be cooled during a separation of low-boiling materials such as methane or hydrogen. For such materials, GC can be run conveniently under conditions down to dry ice temperature, −80 °C.

The most broadly applicable detector for GC is a mass spectrometer (Section 16.5). With GC/MS, it is possible to identify most peaks that are of adequate size with good probability through the use of pattern-recognition methods. GC/MS is the method of choice for general organic chemical analysis, that is, when the identities of many different species are desired.

Other detectors, such as the electron-capture detector (ECD), are more sensitive and more specific than MS, and the specificity is often a benefit rather than a drawback. The ECD, which is relatively specific for halogens (F, Cl, Br, and I), is one of the most sensitive known; its detection limit is approximately 10^{-13} g s^{-1} of material passing into it. Even with capillary GC, in which the total sample may be in the microgram level, it is possible to obtain ppm determinations.

Numerous other types of detectors exist. Their use depends on the types of materials being analyzed and the sizes of the samples that are available. Two of the most common are the thermal conductivity detector, where the eluting gas flows over a hot wire and the temperature changes when eluents pass through, and the flame ionization detector. The latter works by burning the output material and causes gaseous ions of the separated components to form as they arrive, and the ions' quantities are measured. However, GC/MS is so powerful and flexible that MS has become the general detector of choice.

16.7.3 Supercritical Fluid Chromatography

In the chemistry of molecules that can be liquids at some temperature, there exists a temperature above which the substance will not condense to form a liquid-gas boundary at any pressure. This temperature is called the *critical temperature*. Above this temperature, the material is neither a liquid nor a gas; it is conveniently called a *supercritical fluid*. The mobile phase of supercritical fluid chromatography (SFC) is of this type. The most commonly used fluids are carbon dioxide and various lower molecular weight fluorocarbons. The equipment used for SFC is similar to GC, but the pressure throughout the column and the detector must be kept high enough to keep the conditions supercritical.

SFC is useful for certain classes of less-volatile analytes. Among these are molecules with hydrocarbon chains 12–50 carbons long such as the simple hydrocarbons themselves and detergents and fatty acids with such longer chains. Methods of separations and detection are well developed for these classes of compounds.

16.7.4 Electromigration

Electromigration is the movement of particles suspended in solution (most often water) in an electric field. The main electromigration technique that appears in chemical analysis is electrophoresis. Separations of charged particles from the size of individual ions of atoms to particles tens of micrometers in diameter are carried out by electrophoresis.

Electrophoresis separates various analytes based on their differences in rates of migration across the applied voltage. More highly charged particles migrate faster than less charged ones of the same size, and smaller particles migrate faster than larger ones with the same charge. The various analytes that are separated appear as zones as in other separation methods, and the name zone electrophoresis is used synonymously with electrophoresis alone. Electrophoresis is carried out either in a tube filled with solvent, or on a porous support – such as a plate with a gelled solution on it. The gel is, in fact, porous to the molecules which separate as they migrate through it. Gel electrophoresis is one of the primary techniques of biomacromolecule analysis. A popular technology involves carrying out electrophoretic separations in capillary tubes tens of micrometers in diameter – the diameter of hair. This produces high-resolution separations of small ionic species in micro samples. Perhaps the best known use of electrophoresis is the separation of sets of short DNA strands, which provides the experimental basis of most of contemporary genetic analysis.

16.8 Surface-Sensitive Methods

The difference between "regular" spectrometry and surface spectrometry is due not to the thinness of the sample, but to the thinness of the layer from which the analytical information is obtained. For example, in order to measure properties of a surface layer only, we can excite the spectral emission with particles, such as high-velocity electrons and ions. These can only penetrate through short distances (less than a few nanometers) of a solid. The name for the effective distance varies; two of these names are the penetration depth and the mean free path (see Table 16.7). The same limitation to penetration holds for the particles – such as electrons – expelled from the atoms residing in the interior.

With this idea in mind and from the information in Table 16.7, the following general statement can be made. To limit the spectrometric measurements to surface layers requires only that either energy is added using a bombarding charged particle or atom or a charged particle or atom is given off from the sample. Surface analyses are usually carried out under high vacuum, 10^{-10} Pa and less. (1 Pa = 9.87×10^{-6} atm = 7.50×10^{-3} torr.)

Other surface investigatory methods have been discussed in earlier sections although the surface analytical capabilities were not specifically emphasized. These methods are SIMS, and ultraviolet, visible, infrared and Raman, and X-ray spectrometry, done either at low angles of incidence or with total internal reflection methods. These methods can all be quantitative with precisions about ±20%.

Surface analyses heave two significant points of difficulty. Both result from the thinness of the sampled region – the "surface" of the assayed material. First, the quantity of sample is always small, and, as a result, the methods all involve microanalytical techniques. Second, the sample preparation methods are crucial to obtain a meaningful analysis. For instance, if a thin layer of oil about 5–10 *atoms* thick covers a metal sample, surface analysis may show only organic components in the sample.

16.8.1 Electron Spectrometry

16.8.1.1 Photoelectron Spectrometry

Electrons can be caused to be emitted from atoms and molecules upon irradiation with light. This is photoemission, and the electrons emitted are called *photoelectrons*. Each electron emitted has a kinetic energy (E_k) which is related to the wavelength of the exciting light and to the bonding energy of the electron that is emitted. Weakly held electrons have higher kinetic energies when emitted; strongly held ones have lower kinetic energies. An increase in irradiation intensity only increases the number of emitted electrons and not the individual kinetic energies.

When ultraviolet radiation is used, the spectrometry is called *ultraviolet photoelectron spectrometry* (*UPS* or *UV-PES*). If X-ray radiation is the source of ionizing energy, then the spectrometry is called *XPES* or *XPS*. Another name for XPS is ESCA. The two originally were separate classifications but are now joined together because a bright enough source of light tunable through both regions is available as synchrotron radiation.

Tab. 16.7 Approximate sampling depths of radiation (photons) and particles.

Name	Energy (eV)	Penetration depth (nm)
Infrared	0.5	100
Visible–UV	2–3	20
X-ray	1000	1000
Electron	2	0.5
Electron	1000	2
Ion	1000	2
Ion (proton)	10^7	10 000

For analysis of chemical content, however, it is XPS that is more often used, and its sensitivity tends to increase with atomic number of the elements. The strongly held electrons that are excited by the radiation are only slightly affected by chemical bonding. As a result, the values of the electron energies can be used to characterize the composition of the sample surface layer, but only major differences in chemical type can be seen. For example, sulfur in the form of sulfide and sulfate can be discriminated, but differences in the chemical environment of, for example, a sulfide, do not register.

In contrast, UPS causes electrons to be ejected from energy levels that change significantly with changes in molecular bonding. As a result, UPS is used more to investigate the types of chemical bonding present than for analyzing chemical makeup.

16.8.1.2 **Auger Spectrometry**

Strictly defined, the Auger electrons are also photoelectrons. However, it is convenient to separate them into their own category for a number of reasons. Irradiation with X-rays is comparatively inefficient in producing Auger electrons. Irradiation with accelerated electrons (a few hundred electron volts or so) is far more effective, and, thus, greater analytical sensitivity for the Auger method is obtained with electron excitation.

Auger spectrometry is primarily used to measure the elemental composition of a surface. Auger lines can be obtained for all elements except H and He. Its sensitivity is such that in the most favorable cases, less than 0.01 monolayer can be detected. Also, Auger spectra are characterized by a relatively high intensity, since the emitted electrons are able to be detected with excellent sensitivity. Further, the probability of Auger electron emission is high for elements lighter than zinc, making Auger spectrometry a complementary method to XPS. Auger spectrometry has become the method of choice for surface composition studies while XPS (ESCA) is used more for characterizing the chemical form of surface species.

If the electron beam which excites the Auger emission is focused onto a small (fraction of a micrometer) spot, the Auger electrons will be emitted only from the position of the spot on the surface. An Auger electron energy analysis can be used to quantify the elements at the one small part of the surface. The name given to this type of assay is electron microprobe analysis.

If the exciting electrons are focused to a small point and are scanned in a raster pattern over the sample (as done by the electron beam of a non-digital television screen to produce a picture), then a spatial distribution of the emissions can be made. For instance, either the characteristic X-rays or Auger electrons of a specific element can be monitored to obtain a picture of the spatial distribution of an element. An enlarged picture of the surface is generated on a television screen by simultaneously scanning the larger screen and a small area of the sample in synchrony. If the intensities of the Auger emissions for a specific element are recorded as a range of light intensity to produce a picture with good contrast, it will show the locations of the analyte. The Auger electrons without energy discrimination provide the basis for imaging in scanning electron microscopy (SEM).

16.8.2 **Secondary-Ion Mass Spectrometry**

If an accelerated charged particle impinges on a surface, it is stopped within a few atomic layers. One effect of this impact is to cause particles to be ejected from the surface. This is the process of sputtering. The emitted particles can be atoms, molecules, and molecule fragments. These may be analyzed with a mass spectrometer to determine information about the composition of the surface. The technique is called *secondary-ion mass spectrometry–SIMS*.

Of the sputtered material, only those that are ionized by the sputtering process can be accelerated directly into the mass spectrometer. As a result, complete analysis is problematic. However, quantitation and characterization can be more complete if all species are ionized. The neutral sputtered species can be ionized with UV light after their escape from the surface; this is photoionization. SIMS with photoionization has the potential to be far more reliable for quantitation.

Layer analysis can be carried out by obtaining a SIMS spectrum followed by ablation of the surface and another SIMS measurement. Alternate measurement and ablation can be reconstructed to view the chemical analysis of adjacent layers. Calibration of the ablation can provide an absolute depth scale.

16.8.3 Rutherford Backscattering

Rutherford backscattering involves, in effect, bouncing high-energy particles (usually helium ions with a few million electron volts of kinetic energy) off the nuclei of the surface atoms. Some of the energy of the incoming particles is transferred to the stationary atoms, and the amount of the energy transferred depends on the masses of the target atoms. As a result, the surface elemental composition can be probed by measuring the kinetic energy remaining with the probe particles after the collision. The technique is independent of the chemical bonding of the surface atoms, which results in this being a quantitative assay method. Absolute values of the coverage can be obtained without calibration of each sample since the scattering properties of each type of nucleus remain constant over the different samples. Less than a monolayer at the surface can be quantified if the surface layer is composed of heavier atoms on a solid composed of much lower weight atoms.

The mass resolution is not linear with the atomic mass of the analyte, and depends on the quality of the instrument. For instance, with 2-MeV ^4He probe particles, unit atomic masses can be resolved for atoms lighter than about chlorine, and, at the high end of atomic mass – near 200 amu – only masses differing by about 20 amu can be resolved.

16.9 Thermal Methods of Analysis

Energy added in the form of heat can cause changes in the structure or chemical content of materials. Examples of these changes are boiling (vaporization), melting (fusion), changes in crystalline form, transformation to and from a glass (glass transition), loss of water (dehydration), and chemical decomposition. These changes of structure occur at or above a characteristic temperature or range of temperature. Quantitation of the temperatures at which these changes occur and the amount of heat needed to cause these changes can be used to characterize or identify numerous materials. Among these are cements, ceramics, polymers, and plastics.

A large number of different techniques of thermal analysis have been devised including measurements of changes of electrical and magnetic characteristics, changes of elastic force, as well as changes in mass while a sample is heated. However, for chemical analysis, only a few are commonly used, and these are described briefly below.

16.9.1 Thermogravimetric and Evolved-Gas Analyses

Thermogravimetric analysis (TGA), or TG, involves measuring the change in mass of a substance as it is heated. TGAs involve recording the mass of a sample as its temperature is increased; mass is lost as decomposition occurs and volatile products escape, each process beginning at a characteristic temperature. Usually the temperature is raised at a steady rate, and the sample is kept in an atmosphere of flowing inert gas. Changes that occur in the sample that do not involve changes in mass, such as melting, cannot be studied by TG.

Characterizing the gaseous products of the decomposition as a function of sample temperature, a method called *evolved-gas analysis* is complementary to TG. The losses in mass from the sample may be due to loss of solvent or to breakdown of molecules in the material. Relatively detailed analyses can be done on sub-milligram samples. The analysis generally involves GC and/or MS of the volatiles.

16.9.2 Differential Scanning Calorimetry; Differential Thermal Analysis

The basis for calorimetric analyses is the measurement of the rise in temperature of a material for a given amount of heat added. The ratio of the temperature change to the heat added (T/Q) per unit weight is called the *heat capacity*. The heat capacity depends on the identity of the material and on the temperature at which the heat is added. Differential scanning calorimetry (DSC) works on the principle of keeping a constant rate of temperature change in a sample by regulating the amount of heat that is added. The rate of heat added is

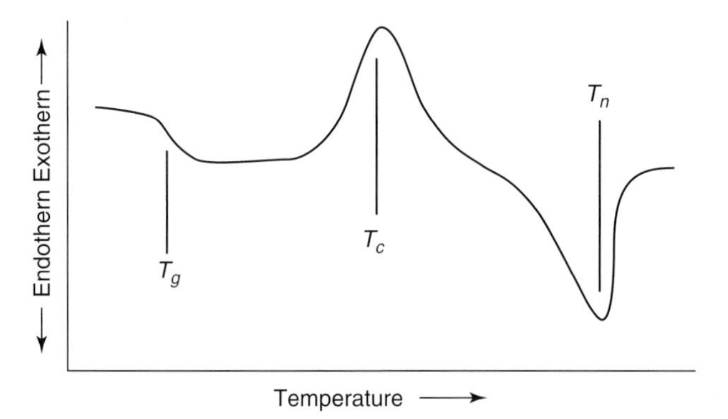

Fig. 16.24 Schematic differential scanning calorimetry thermogram of a polymeric material. The horizontal axis is a linear temperature scale, and vertical axis indicates the derivative of the amount of heat needed to obtain that temperature increase. The line would be flat if a material used the same amount of heat per degree to raise the temperature over the whole temperature range. The three deviations from the baseline (a flat line) are, from left to right, the glass transition, an exothermic (produces heat) chemical reaction, and endothermic (requires heat be added) melting.

plotted versus temperature as shown in Figure 16.24. This is called a *thermogram*.

An older method is differential thermal analysis (DTA) for which the heat is added at a constant rate, and the rate of temperature change is measured. The difference in temperature between a sample and standard is plotted versus temperature. The information obtained in DTA is similar to DSC, but they differ in experimental technique.

Acronyms of Analytical Techniques and Instruments

AA: Atomic absorption (spectrometry).
AAS: Atomic absorption spectrometry.
AE: Atomic emission (spectrometry).
AED: Atomic emission detector.
AES: Atomic emission spectrometry.
AES: Auger electron spectrometry.
AF: Atomic fluorescence.
ASV: Anodic stripping voltammetry.
ATR: Attenuated total reflection.
CARS: Coherent anti-Stokes Raman scattering (spectrometry).
CD: Circular dichroism spectrometry.
CI: Chemical ionization.
CI/MS: Chemical ionization mass spectrometry.
CP/MAS-NMR: Cross polarization magic-angle spinning nuclear magnetic resonance.
CW-NMR: Continuous-wave nuclear magnetic resonance.
DCP: Direct-current plasma.
DME: Dropping-mercury electrode.
DPP: Differential pulse polarography.
DSC: Differential scanning calorimetry.
DTA: Differential thermal analysis.
DTG: Differential thermogravimetry.
EAA: Electrothermal atomic absorption.
ECD: Electron-capture detector.
EDX: Energy-dispersive X-ray (spectrometers).
EDXF: Energy-dispersive X-ray fluorescence.
EGA: Evolved-gas analysis.
EI: Electron impact, electron ionization.
ELCD: Electrolytic conductivity detector.
EMP: Electron microprobe analysis.
ENDOR: Electron nuclear double resonance spectrometry.
EPR: Electron paramagnetic resonance (spectrometry).
ES: Emission spectrometry.
ESR: Electron spin resonance (spectrometry) (EPR).
ESCA: Electron spectrometry for chemical analysis.
ETA: Electrothermal atomization (electric furnace).
EXAFS: X-ray absorption fine structure.
FAA: Flame atomic absorption.
FAB: Fast atom bombardment.
FAE: Flame atomic emission.
FD: Field desorption.
FFF: Field-flow fractionation.
FFT: Fast Fourier transform.
FIA: Flow injection analysis.
FID: Flame ionization detector.
FT-IR: Fourier transform infrared (spectrometry).
FT-MS: Fourier transform mass spectrometry.
FT-NMR: Fourier transform nuclear magnetic resonance.
GC: Gas chromatography.
GC-IR: Gas chromatography with infrared detection.
GC-MS: Gas chromatography with mass spectrometry.
GDMS: Glow-discharge mass spectrometry.
GFAA: Graphite-furnace atomic absorption.
GPC: Gel permeation chromatography.

HPLC: High-performance liquid chromatography.

HPTLC: High-performance thin-layer chromatography.

IC: Ion chromatography.

ICAP: Inductively coupled argon plasma.

ICP: Inductively coupled plasma.

ICP/MS: Inductively coupled plasma/mass spectrometry.

IC: Ion chromatography.

INAA: Instrumental neutron activation analysis.

IR: Infrared (spectrometry).

ISE: Ion-selective electrode.

LAMMS: Laser microprobe mass spectrometry.

LC: Liquid chromatography.

LC/EC: Liquid chromatography/electrochemical detection.

LC/MS: Liquid chromatography/mass spectrometry.

LEED: Low-energy electron diffraction.

MAS-NMR: Magic-angle spinning nuclear magnetic resonance.

MCD: Magnetic circular dichroism (spectrometry).

MIP: Microwave induced plasma.

MS: Mass spectrometry.

NAA: Neutron activation analysis.

NHE: Normal hydrogen electrode.

NIR: Near infrared.

NMR: Nuclear magnetic resonance.

NPD: Nitrogen phosphorus detector.

NQR: Nuclear quadrupole resonance (spectrometry).

ORD: Optical rotatory dispersion (spectrometry).

PAS: Photoacoustic spectrometry.

PD: Plasma desorption (in mass spectrometry).

PID: Photoionization detector.

PIXE: Particle-induced X-ray emission.

PMT: Photomultiplier tube.

QCM: Quartz crystal microbalanceRBS: Rutherford backscattering (spectrometry).

RI: Refractive index.

RIS: Resonance ionization spectrometry.

RP: Reversed phase (in liquid chromatography).

SAM: Scanning Auger microscopy (spectrometry).

SAW: Surface acoustic wave (sensor)SCE: Saturated calomel electrode.

SEM: Scanning electron microscopy.

SERS: Surface-enhanced Raman scattering.

SFC: Supercritical fluid chromatography.

SHE: Standard hydrogen electrode.

SIMS: Secondary-ion mass spectrometry.

SNMS: Sputtered neutral mass spectrometry.

STM: Scanning tunneling microscopy.

TCD: Thermal conductivity detector.

TG: Thermogravimetry (see TGA).

TGA: Thermogravimetric analysis.

TLC: Thin-layer chromatography.

TLC: Thick-layer chromatography.

TMS: Tetramethylsilane.

TOF/MS: Time-of-flight mass spectrometry.

TS: Thermospray.

UV: Ultraviolet (light) spectrometry.

vis: Visible (light) spectrometry.

WDX: Wavelength dispersive X-ray (spectrometer).

XRF: X-ray fluorescence.

XPS: X-ray photoelectron spectrometry (see ESCA).

XPES: X-ray photoelectron spectrometry (see ESCA).

XANES: X-ray absorption near-edge spectrometry.

Glossary

Accuracy: A qualitative measure of the closeness with which an analytical result corresponds to the true content of a material.

Analyte: The species for which one is analyzing.

Assay: As a noun, a chemical analysis to determine the level of a specific analyte. As a verb, to carry out such an analysis.

Bias: The quantitative amount that a result with high precision varies from the "true content" as a result of the analysis shortcomings.

Calibrate: To determine the bias and precision of the methodology used for an analysis or to determine an instrument's response to some known quantity of analyte.

Contaminant: Any material that is picked up during an analytical procedure that may cause the sample not to be characteristic of the original material from which the sample was taken.

Destructive Method: An analytical method that requires some obvious portion of a material to be removed and destroyed in the process of analysis. The opposite of a nondestructive method.

Detection Limit (Limit of Detection): The limit below which a particular instrument or analytical method cannot be used to observe the presence or absence of some analyte in a sample.

Electromagnetic Radiation: Depending on the frequency of oscillation, also called radio waves, infrared, visible, and ultraviolet light (or, simply, light), X-rays, and γ *rays*. Electromagnetic radiation is also described as oscillating orthogonal electric and magnetic fields that travel through space.

Interferent(s): A component or components (not the analyte) in a sample that cause a measure of the analyte content to be greater or less than it would be in the absence of the interfering components.

Matrix: The part of the material to be analyzed that is not of interest in an analysis.

Micro Analysis: An analysis of a sample that weighs less than approximately 10 mg.

Nondestructive Method: An analytical method that, after analysis, leaves the material being analyzed with no apparent changes.

Potential: Electric potential difference. Voltage. The difference in electric potential energy per unit charge.

Parts per Million: A relative measure of content. One ppm is one part in 10^6 of the sample such as grams per gram.

Parts per Billion: A relative measure of content. In the United States, one ppb is one part in 10^9 of the sample such as nanograms per gram. In Europe, one ppb is one part in 10^{12} such as nanograms per kilogram.

Parts per Trillion: A relative measure of content. In the United States, one pptr is one part in 10^{12} of the sample. In Europe, one pptr is one part in 10^{18}.

Precision, Imprecision: A measure of the scatter of experimentally determined values of content. The highest precision corresponds to the least scatter. Not the same as accuracy.

Sample, Laboratory Sample: The part of the original material that is used to produce an analytical result. It should accurately represent the contents of the original material.

Spectrum: A graph of the changes in the amount (more precisely, power) of electromagnetic radiation that is absorbed or emitted by a sample versus the wavelength, frequency, or energy of the radiation. Also, instrument response versus mass in mass spectrometry.

Sensitivity: The ability of an analytical method to measure changes in content or concentration. Greater sensitivity means that smaller changes give a measurable difference in response.

Standard: Substances that are used to ascertain the accuracies and precisions of analytical methods.

Standard Addition: A method of running standards in which a measurement is made on a sample and then repeated after known quantities of the analyte are added to the sample. Used for complicated mixtures.

Trace Analysis: Analyses for components that are less than 0.01% (100 ppm) of a material but more than ~1 part in 10^9.

Ultra-microanalysis: An analysis of samples of total mass in the range below about 10μg.

Ultratrace Analysis: Analyses for components of a sample that are at levels below ~1 part in 10^9.

Verification: Demonstration that an analytical procedure measures what it is claimed to measure.

Voltage: See Potential. Electric potential difference.

References

Ahmad, S. *et al.* (**1979**) *J. Radioanal. Chem.*, **54**, 331–341.

Commission on Microchemical Techniques and Trace Analysis (**1982**) *Pure Appl. Chem.*, **54**, 1565–1577.

Derome, A.E. (**1987**) *Modern NMR Techniques For Chemistry Research*, Pergamon, Oxford.

Harris, R.K. (**1983**) *Nuclear Magnetic Resonance Spectroscopy*, Pitman, London.

Horwitz, W. (**1982**) *Anal. Chem.*, **54**, 67A–76A.

Morrison, G.H. (**1979**) *CRC Crit. Rev. Anal. Chem.*, **33**, 287–320.

Rubinson, K.A. and Rubinson, J.F. (**2000**) *Contemporary Instrumental Analysis*, Prentice Hall, Upper Saddle River.

Sandell, E.B. (**1959**) *Colorimetric Determination of Trace Metals*, 3rd edn, Interscience, New York.

Wawros, A., Talik, E., and Pastuszka, J.S. (**2003**) *Microsc. Microanal.*, **9**, 349–358.

Further Reading

Worsfold, P.J., Townshend, A., and Poole, C.F. (**2004**) *Compendia Encyclopedia of Analytical Science*, 2nd edn, Academic Press, New York.

Further Reading: Textbooks

Rubinson, K.A. and Rubinson, J.F. (**2000**) *Contemporary Instrumental Analysis*, Prentice Hall, Upper Saddle River.

Skoog, D.A., West, D.M., Holler, F.J., and Crouch, S.R. (**2006**) *Principles of Instrumental Analysis*, Brooks/Cole, Florence.

17
Photoacoustic Spectroscopy

Frans J. M. Harren, Stefan Persijn, and Simona M. Cristescu

Handbook of Metrology. Edited by Michael Gläser and Manfred Kochsiek

ISBN: 978-3-527-40666-1

17.1 Introduction

Photoacoustic (PA) spectroscopy is an indirect technique in the sense that an effect of light absorption is measured rather than light absorption itself. Light that is absorbed by a sample is, completely or partly, converted into heat, giving rise to pressure waves that can be detected. Using PA spectroscopy (PAs), regular absorption spectra can be obtained. PA spectroscopy is complementary to regular absorption/transmission spectroscopy, because only absorbed light contributes to the PA signal and drawbacks of regular spectroscopy (e.g., background absorption, scattered light, opaque media) do not play a role in PA spectroscopy. In general, when matter absorbs light the transmitted power follows Beer's law: $P = P_0 e^{-\alpha l}$ in which P_0 the incident laser power, P the transmitted laser power after distance l in centimeters through the sample, and α in cm^{-1} the absorption strength at that frequency. For low absorptions, this exponential expression can be linearized:

$$\Delta P = P_0 - P = P_0 \alpha l \tag{17.1}$$

It shows that the generated PA effect (i.e., decrease in power) is proportional to the incident power. On the other hand, other factors, such as the optical, thermal, and acoustical properties of the sample and media involved, are now more important and can even be the direct objective of study in many applications of PAs.

Like in all spectroscopy, the spectral dependence of absorption permits selectivity and identification of the absorbing components. The PA response itself can be fast, highly sensitive, and in principle, background-free. Thanks to these qualifications, no lengthy accumulation periods (e.g., in trace gas detection) or long absorption pathways are needed. Because transparency and scattering play no role, solids, whether crystalline, powdered, or amorphous, can be studied. Using the dependency of the PA signal on thermal and acoustic properties, surfaces and solids can be tested noninvasively for structural inhomogeneity. For these latter applications, the spectroscopical aspect plays a minor role. PA spectroscopy is especially suited to monitor local effects. Remote sensing, conversely, is alien to PAs.

A medium that absorbs electromagnetic radiation experiences a local temperature increase by relaxation processes that transfer electronic, vibrational, or rotational energy into translation energy (phonon energy in case of solids). This nonradiative relaxation process occurs if its relaxation time can compete with the radiative

Handbook of Metrology. Edited by Michael Gläser and Manfred Kochsiek

ISBN: 978-3-527-40666-1

lifetime of the excited energy levels. If the absorption is modulated, at an acoustic frequency, for example, by modulating the intensity of the radiation source, these temperature changes occur periodically and give rise to acoustic phenomena. Hence the name "photoacoustic"; light absorption is detected through its accompanying acoustic effect.

A sensitive microphone can detect the effect, especially, when it occurs in the gas phase. In the case of solids or fluids, a piezoelectric transducer is often used. Thus, the minimum requirement is a suitable source of modulated electromagnetic radiation for which the wavelength can be selected, such that the absorption spectrum of the substance investigated can be accessed. In addition, a detection device such as the mentioned microphone or piezo transducer must indicate that absorption takes place (Figure 17.1). This short description already clarifies that PA spectroscopy is based on a two-step process; first, electromagnetic radiation energy must be absorbed, this being the spectroscopically selective step; second, the absorbed energy must be transferred to sound-producing degrees of freedom, by relaxation processes. If this relaxation is too slow with respect to the employed modulation times, PA spectroscopy will not work. Note that the relaxation rate depends on the properties of the participating molecules.

The instrumentation required for PA spectroscopy can be relatively inexpensive and simple; the powerful and tunable laser that may be used forms the cost-determining part, but for some applications a simple mercury lamp and an inexpensive microphone suffice. However, the instrumentation has to be well designed and cleverly chosen, and its operation requires skillful people. Especially, the PA cell has to be designed with care, fitting with the chosen application.

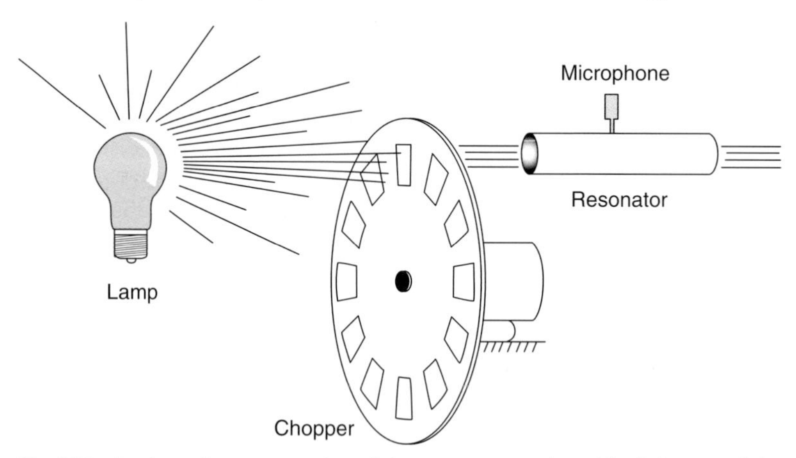

Fig. 17.1 A schematic representation of the photoacoustic setup. The lamp represents the light source, which was the sun in the original experiments by Bell. Nowadays lasers are used because of their tunability and brilliance. When, for example, a gas absorbs the light, the temperature of the gas increases, the gas expands adiabatically, generating a pressure increase. By switching off the light, temperature and pressure drop. The light is modulated, in this case mechanically chopped to generate an AC temperature and/or pressure wave (i.e., sound). If the modulation frequency corresponds to the resonance frequency of the acoustical resonator, the sound is amplified. A microphone is inserted into the resonator at the point where the resulting pressure standing wave has maximum amplitude.

The PA effect has been discovered and clarified almost simultaneously by Bell (1880), Röntgen (1881), and Tyndall (1881). In those days, neither laser sources (which are ideally suited for PA spectroscopy, owing to their brilliance) nor sensitive small microphones were available. Therefore, being well understood, but rather impractical, the effect remained slumbering in the cabinet of curiosities. In 1938, Veingerov refined the PA technique for the first spectroscopic gas analysis (Veingerov, 1938); and in 1943, Luft measured trace gas absorption spectra with an infrared broadband light source down to the part per million level (Luft, 1943). By the end of the 1960s, after the invention of the laser, the scientific interest expanded again. Kerr and Atwood (1968) utilized the laser PA detection to obtain the absorption spectra of small gaseous molecules. Owing to the high spectral brightness of lasers and improved phase-sensitive lock-in techniques to amplify the acoustic signal, they were able to determine low absorption strengths. Since then, a broad field of pure and applied spectroscopic methods has emerged using the PA effect. The last decades have seen an ever-increasing use of PA techniques in a very wide field of applications. These activities range from fundamental research (e.g., the investigation of vibrationally highly excited molecules) to applications (e.g., testing of solids, surfaces, biomedical applications). A number of useful reviews have appeared over this period (see Further Reading), covering most of these areas.

Firstly, the equipment is discussed. Various types of PA cells are introduced, which are resonant and nonresonant with respect to the modulation frequency of the light. In addition to the sound detection by microphones, related techniques such as thermal lensing and photothermal deflection are discussed, which are often applied if liquid- or solid-state samples are investigated. In Section 17.3, applications to solids, liquids, and surfaces and to gas spectroscopy are dealt with. As one of the more recent applications, trace gas detection is also considered with environmental and biological studies. Besides these key references, books and review articles are given for further reading.

17.2 Devices and Equipment

17.2.1 Light Sources

In principle, there is no need for lasers to operate a PA detection system. Light from lamps can be filtered by a monochromatic system to obtain radiation needed for spectroscopic discrimination to distinguish between components. Other highly sophisticated instruments use, for the selection of wavelength, the gaseous species under investigation itself (Figure 17.2); all wavelengths at which absorption takes place are simultaneously active. With the latter instrument (URAS, ABB Germany) detection limits in the sub-ppmv (1 ppmv = 1 part per million volume = $1 : 10^6$) range can be achieved for more than 25 gaseous compounds.

However, more often PA detection goes together with the application of laser light sources, ranging from the infrared to the visible and UV, including both continuous wave (CW) and pulsed lasers. Since the PA signal is proportional to the laser power, strong lasers are advantageously applied.

In the mid-infrared wavelength region, CW CO_2 (wavelength region from 9 to 11 μm) and CO lasers (4.6–8.2 μm) were

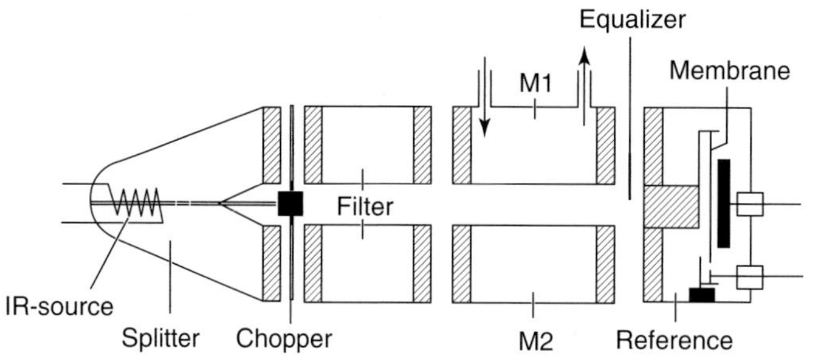

Fig. 17.2 The infrared absorption spectrometer setup (URAS). Light from the thermal IR source is divided into two paths. The chopper modulates the intensity for both paths. The filters serve to filter out light of wavelengths not needed for the detection process; they can be filled with gases whose absorption spectra do not overlap with those of the species under scrutiny. M1 represents the sampling cell through which the sampled gas is flowed. M2 serves as a reference cell with a calibrated gas mixture in which the traversing light is attenuated to equalize the light intensities at the entrance of the last cell at the far right. The light is equalized before the cells M1 and M2 are filled. The last cell consists of two compartments with a membrane in between. Both compartments are filled with the gas under scrutiny so that all wavelength characteristics for this gas contribute to the signal. If in M1 the attenuation differs from that in the reference cell, the membrane starts to oscillate with the frequency of the chopper. This oscillation (typically hertz) is detected capacitively.

among the most used. At present, developments in solid state physics and nonlinear optics have expanded the potential for PA spectroscopy to other wavelength regions. The recent availability of high power, pulsed, and CW quantum cascade lasers (QCL)s, operating at room temperature with power levels up to 1 W, has boosted PA spectroscopy in the mid-infrared wavelength region. Even more, single-mode tuning ranges, approaching 300 wavenumbers, are now achievable from commercially available external cavity QCL products. QCLs operate in the mid-infrared wavelength region from 3.5 to 24 μm (Kosterev and Tittel, 2002). For the 2.5–5 μm region CW periodically poled lithium niobate optical parametric oscillators (PPLN–OPOs) became available at high (several watts) power level and narrow linewidth. Nowadays, OPOs are pumped by high power, near-infrared lasers with excellent spectroscopic properties (narrow linewidth, Gaussian beam profile, fast wavelength scanning), which they can convert into the mid-infrared wavelength region due to their instantaneous nonlinear effects (Ebrahim-Zadeh and Sorokina, 2005).

In the near infrared wavelength region, CW titanium–sapphire ring lasers (650–1000 nm) are intensively used for high-resolution studies of weak overtone absorption spectra of molecular gases. High-overtone spectroscopy profits from sensitive PA detection, since one deals with high vibrational excitation, where fluorescence-based techniques do not work. Electronic spectroscopy with light sources in the visible or in the ultraviolet region does not immediately ask for PA detection because other techniques provide more powerful means (e.g., laser-induced fluorescence (LIF) or resonance-enhanced multiphoton ionization).

In addition to this, near-infrared "telecom" diode lasers have the advantage of small size, reliability, and low costs. Because of their low power the combination with PA spectroscopy is not always very favorable. However, the development of fiber amplifiers in this wavelength region could overcome this disadvantage. Telecom lasers are well suited to be used in gas sensing instruments in which the sensitivity is limited to the ppmv range. These lasers are tunable only over a limited wavelength region (1 cm^{-1}), but larger tuning ranges can be obtained by external cavity setups (external cavity diode laser). The near-infrared diode lasers have the main disadvantage that they operate at the weak overtone absorption bands instead of the strong fundamental vibrational absorption bands in the mid-infrared wavelength region.

By using pulsed lasers and gated detection, the pulse character of the signal allows suppression of a possible window signal by setting the boxcar delay such that the window signal has not yet arrived at the piezo transducer (or microphone) when the trace signal is detected. In general, however, the highest sensitivity of PA detection has been achieved for modulated CW lasers; if no other arguments prevail, the use of CW lasers is recommended.

17.2.2 Gas-Phase Photoacoustic Cells

The proper choice of the PA cell geometry depends on the intended application. For gas-phase measurements, mainly resonant cells are combined with frequency- or amplitude-modulated CW lasers. For condensed-phase spectroscopy (e.g., surfaces, thin surface layers, or depth profiling) with CW lasers, gas-coupled microphones are used within a small nonresonant cell. Utilizing pulsed lasers, the response is monitored with a piezoelectrical crystal directly coupled to the material. In competition with this is the noncontact detection by the photothermal effect, in which a laser beam is deflected or dispersed by the heat gradient from the sample.

17.2.2.1 Nonresonant Cells

If the modulation frequency of the incident light is much smaller than the lowest resonance frequency of the cell, it is to be considered a nonresonant cell. The performance of a cell is related to its efficiency in converting absorbed photon energy into acoustic energy by a cell constant F; that is, $p_{gas} = F\Delta P$ with p_{gas} being the generated acoustic pressure in the gas and ΔP derived from Equation 17.1 The cell constant F in (Pa m W^{-1}) can be derived by

$$F_{nonresonant} = \frac{G(\gamma - 1)l_{nr}}{2\pi \nu V_{nr}} \qquad (17.2)$$

with l_{nr} and V_{nr} being the length and the volume of the cell, respectively; γ the specific heat constant, ν the modulation frequency, and G a geometrical factor, typically of the order of 1. From the above formula, one can derive that the F is independent of the cell length. The diameter should be as small as possible to generate high signals, just large enough to allow proper alignment of the incident light beam. To minimize the detection of absorption by the cell windows, one has to maximize the cell length; the energy absorbed by the windows is then distributed over a larger volume. The PA signal decreases with higher modulation frequencies; therefore, it is advantageous to use low modulation frequencies.

For the investigation of solids, liquids, and surfaces, another type of nonresonant PA cell has to be designed. The sample has to be sealed in a gas cell having a window

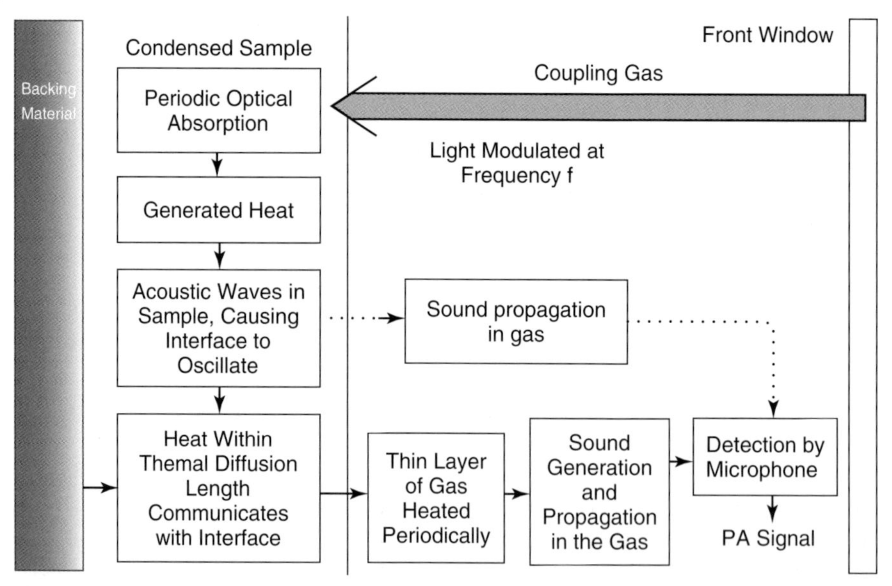

Fig. 17.3 Block diagram to explain photoacoustic detection of absorption in condensed matter using gas coupling. (Figure adapted from Tam, 1983.)

and a microphone. In the extreme case of an open cell, the wall opposite the window can be the material under investigation. The gas buffer volume in this cell should be designed as small as possible; in some cases, the heat generated by the absorption of (modulated) light diffuses to the surface and into the buffer gas (e.g., helium). The probing depth into the surface is determined by the modulation frequency: at low frequencies deeper layers of the material are investigated as compared to high frequencies. This probing depth, the thermal diffusion length, μ_s is given by

$$\mu_s = \left\{\frac{D}{\pi \nu}\right\}^{1/2} \tag{17.3}$$

where D is the thermal diffusivity of the sample and ν the modulation frequency of the light. To characterize the complete sample, the optical absorption length μ_α (the reciprocal of the optical absorption coefficient α) and the thickness of the sample L are also important (Figure 17.3). Because the values of all three properties (μ_s, μ_α, and L) can vary relative to each other and within orders of magnitude, six possible limiting cases can be distinguished, as elaborated by Tam (1983).

For pulsed operation, the bandwidth of the microphone is too small (typically 20 kHz) as compared to the fast nanosecond processes in materials. The microphone must then be replaced by a piezoelectric element, either ceramic or a polyvinylidene difluoride (PVDF) foil, which is directly attached to the material. These detectors can show an acoustic response up to tens of megahertz. High-output pulsed lasers combined with gated extraction of the acoustic signals yield a highly sensitive

($\alpha = 1 \times 10^{-7}$ cm^{-1}) technique that can discriminate against low-frequency noise and window absorption.

17.2.2.2 Resonant Cells

In a resonant cell setting, the modulation frequency is chosen to match one of the resonant frequencies of the cell. This produces a standing, amplified, sound wave in the resonator. For a gas-type resonant cell, two varieties of acoustical resonators have found widespread use: the Helmholtz resonator and the cylindrical resonator excited in a longitudinal, radial, and/or azimuthal mode (Figure 17.4).

A Helmholtz resonator consists of a closed volume (cavity) connected via a long narrow tube to the open air. It is the acoustic equivalent of a mechanical oscillator composed of a mass (long tube) and a spring (cavity). The oscillation frequency (ν_H) of the system depends on the length (l_H) and cross section (A_H) of the tube and the volume (V_H) of the cavity via the following relation:

$$\nu_H = V_s \left\{ \frac{A_H}{V_H l_H} \right\}^{1/2} \quad (17.4)$$

with V_s being the velocity of sound (Kapitanov *et al.*, 2001). Although the acoustical response of such a resonator is somewhat lower as compared to the response of the cylindrical resonators described below, the type of resonator has some advantages when low or high temperatures are needed. Microphones have a limited operating temperature range, that is, around room temperature. When gases are investigated at low (e.g., 80 K) or high (350 K) temperatures, it is advantageous to have a homogeneous temperature distribution across the gas volume while the microphone operates at room temperature. The temperature gradient is over the long narrow tube of the resonator at the end of the microphone is placed.

For cylindrical resonant cells, one has to multiply the cell constant F (Equation 17.2) with the quality factor Q of the generated acoustic resonance:

$$F_{\text{resonant}} = \frac{G(\gamma - 1) l_r Q}{2\pi \nu V_r} \quad (17.5)$$

with l_r and V_r being the length and the volume of the acoustic resonator, respectively; Q equals the ratio of the energy stored in the acoustical standing wave to the energy losses per cycle. For resonant cylindrical cells, one can derive for the cell constant: $F_{\text{resonant}} \propto \sqrt{l_r}/R$, with R being the radius of the resonator (Equation 17.5.) (Bijnen, Reuss, and Harren, 1996). Cell geometries with large diameter-to-length ratios, to excite the resonance in the radial or azimuthal acoustic mode, possess high Q-values and high resonance frequencies; they have, however, low F-values. PA cells with high Q-values are sensitive to long-term drifts, for example, owing to thermal expansion if the temperature is not carefully controlled. Such cells require active locking of the modulation frequency on the resonance frequency of the cell. In a longitudinally excited resonator, a smaller acoustic gain, as a consequence of a relatively low Q-value, is compensated for by the signal gain due to the smaller diameter.

The resonant acoustical amplification is limited by various dissipation processes, which can be divided into surface and volume effects. Main surface losses are viscous and thermal losses at the resonator surface; in addition there are microphone losses, and acoustic wave scattering losses at obstacles inside the cell. The less-important volumetric losses are radiation losses, free-space viscous and thermal losses, and dissipative relaxation losses within the absorbing gases. For a cylindrical resonator, the resonant frequencies are given by Hess (1983)

$$\nu_{\mathrm{nmp}} = V_{\mathrm{s}}\left[\left\{\frac{\alpha_{\mathrm{mn}}}{d}\right\}^2 + \left\{\frac{p}{2l}\right\}^2\right]^{1/2} \quad (17.6)$$

with V_s being the velocity of sound of the gas inside the cavity, d the cell diameter, l the axial length, and $p = 0, 1, 2, 3\ldots$ axial mode numbers; α_{mn} is the nth root of the derivative of the Bessel function, $dJ_m/dr = 0$ at $r = R$ (e.g., $\alpha_{00} = 0, \alpha_{01} = 1.2197$, $\alpha_{10} = 0.5861$, and $\alpha_{11} = 1.6970$).

To choose the proper frequency for an optimal signal-to-noise ratio, one has to take into account the noise spectrum. The theoretical minimum acoustical noise results from random pressure fluctuations and their frequency distribution due to Brownian motion. The total power of these pressure fluctuations is constant, but the noise spectrum depends on the Q-value of the acoustical resonance of the PA cell; a high Q-value will significantly amplify noise power at these resonances. However, often this Brownian noise level is below what is produced by other noise sources such as electronic amplifier noise and external acoustical noise from the surroundings; their power decreases with increasing frequency. It is then advantageous to utilize a resonant cell to improve the signal-to-noise ratio. The resonance frequency might be chosen such that contributions from Brownian noise and amplifier/external noise become comparable. At resonance, external acoustical noise within the cell will be amplified at a high Q-value. Here acoustical shielding helps, that is, a proper cell wall construction (massive), material choice (brass), and good design of in- and outlet ports are required. For the latter, $\lambda/4$ notch filters (λ the acoustical wavelength) are positioned in line with the gas in- and outlet (Figure 17.5). In practice, for the B&K 4179 condenser microphone mounted on a well-shielded resonant ($Q = 40$) PA cell, a noise level has been observed that is an order of magnitude above the noise of the complete system (microphone and pre-amplifier) quoted by the manufacturer (Bijnen, Reuss, and Harren, 1996).

By definition, noise components do not possess a fixed phase relation with the periodical modulation of the light intensity. External perturbing sources may have a fixed phase relation, for example, the directly generated acoustical sound caused by chopping the light beam. They must be minimized in the same way as the external acoustical noise. In the case of a mechanical chopper, a good choice of the chopper position and removal of objects in the neighborhood of the chopper wheel are advantageous. A more serious problem

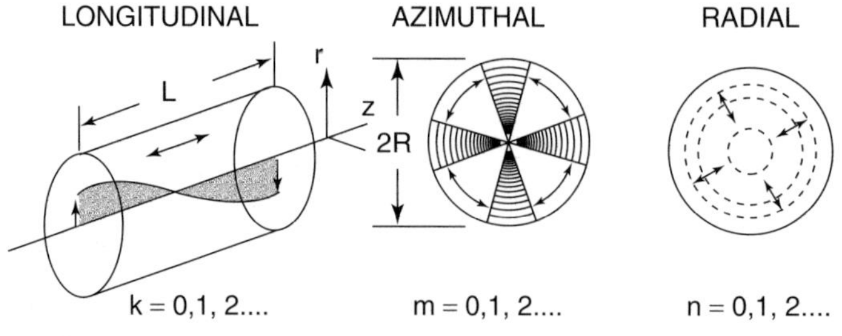

Fig. 17.4 Resonant acoustic modes of a cylindrical closed chamber, the fundamental longitudinal, azimuthal, and radial modes.

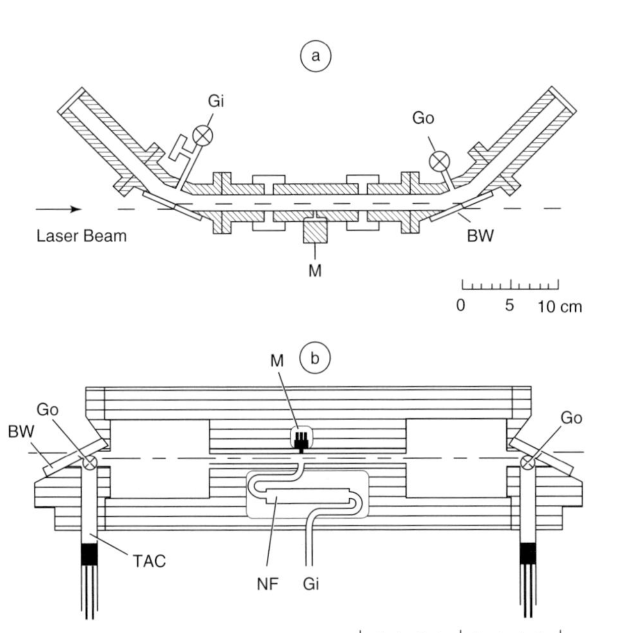

Fig. 17.5 Two photoacoustic cell types successfully used in trace gas detection. (a) The Zurich-banana cell, where the laser beam enters from the left through a Brewster window (BW) to leave the central part ($l = \lambda/2$) of the resonator at the right Brewster window. The position of the microphone M is indicated. The two pieces of length $\lambda/4$ before and after the bends serve to suppress the window signal. The total length of the banana cell amounts to λ. In the bends, the gas inlet (Gi) and gas outlet (Go) are shown. (b) The Harvard–Nijmegen open-organ-pipe cell; the main part is manufactured of a block of massive brass. The central organ pipe acts as an open resonator with length $\lambda/2$; the copper resonator tube has a highly polished, gold-coated inner surface to minimize wall heating by stray light. The central position of the gas inlet (Gi) is essential to obtain short measuring times; only the resonator volume of 2.8 ml must be replenished before an independent concentration measurement can be performed. TACs: $\lambda/4$ tunable air columns to suppress window signal. NF: $\lambda/4$ notch filter to suppress acoustic in coupling of noise via gas inlet.

is formed by the PA background signal; absorption of the light beam in the window material or light scattered/reflected from the resonator wall causes a signal at the frequency of the modulated light beam. In resonant cells, using $\lambda/4$-tubes next to the windows can diminish window signals. These tubes, placed perpendicular to the resonator axis close to the windows, are tuned to the resonator frequency and act as interference filters for the window signals (see Figure 17.5). Influence of scattered light on the PA background signal can be minimized by the use of highly reflecting polished material with a good thermal conductivity. In the case of infrared light a polished gold-coated copper wall of the acoustical resonator has been found to work satisfactorily.

We describe two successfully applied cells, the Zurich-banana-type cell and the Harvard–Nijmegen open-organ-pipe cell (Figure 17.5). In both cases, the modulation frequency ν is chosen in resonance with a longitudinal acoustic mode of the cell (typical length 150 mm and typical diameter

Tab. 17.1 Photoacoustic cell parameters; in the case of the open-organ-pipe cells in which the window signal is so far suppressed as to no longer limit the sensitivity, α_{min} is given for 1 W of laser power, that is, for 10 W of laser power α_{min} becomes smaller by a factor 10.

Cell type	Banana	Organ pipe
Microphone	Bruel and Kjaer	Knowless
Type number	4179	EK 3024
Microphone sensitivity (mV Pa^{-1})	1000	20
Microphone noise (μV $Hz^{-1/2}$)	5	0.5
Number of microphones	1	3
Cell constant F (Pa m W^{-1})	16.4	45.0
Q-value	43	32
Resonance frequency (Hz)	555	1600
Volume diameter ratio (ml mm^{-1})	80/15	3/6
α_{min} (cm^{-1} W^{-1})	5×10^{-8}	2×10^{-8}

10 mm). The banana-type cell has as an important advantage its small volume; it forms an acoustically closed system with a moderately high Q ($Q = 53$ for its utilized second longitudinal mode); the "bent" geometry allows Brewster windows (BWs) to be positioned at pressure nodes, thereby minimizing the window signals (similarly, the gas in- and outlet with their external acoustical noise). The cell is operated at 555 Hz. The end pieces beyond the bends are optimized to suppress the window signal by interference. The open-organ-pipe cell may have a somewhat lower Q-value ($Q = 32$, inner radius $R = 3$ mm) but a higher F-value due to the smaller radius of the resonator tube ($F_{\text{resonant}} \propto \sqrt{l_r}/R$). Table 17.1 shows the properties of the cells.

17.2.2.3 **Cantilever-Type Optical Microphone**

Microphones, where the flexible membrane is strained over a frame or the pressure sensor is a thinner portion in a metal plate, are not very sensitive and the response is not linear because of the fact that the material has to stretch out radially under the pressure variations. A cantilever-type pressure sensor has been proposed as shown in Figures 17.6 and 17.7 (Uotila, Koskinen, and Kauppinen, 2005). The cantilever size is 2 mm × 4 mm and made out of silicon. Because of its extreme thinness (5 μm) the cantilever moves like a flexible door owing to the pressure variations in the surrounding gas. The frame around the silicon cantilever is thick (380 μm) and there is a narrow (30 μm) gap between the frame and the cantilever on three sides. As the pressure varies, the cantilever only bends but it does not stretch. Therefore, the movement of the free end of the cantilever can be about 2 orders of magnitude greater than the movement of the middle point of the tightened membrane under the same pressure variation. With this setup a detection limit of 0.9 ppmv for ethylene, 0.3 ppmv for methane, and 0.1 ppmv for carbon dioxide was achieved with a mechanically chopped blackbody radiator.

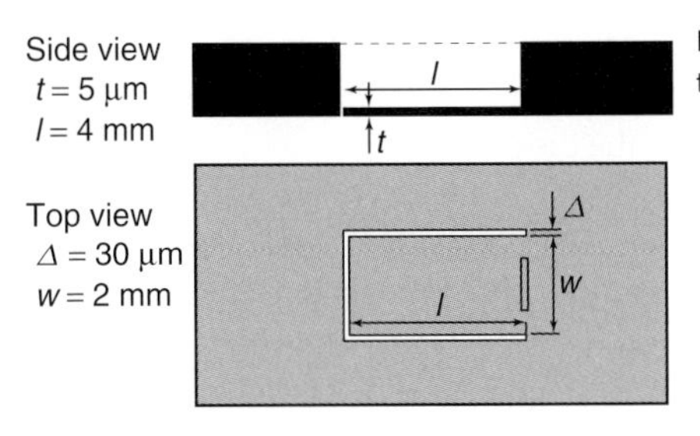

Fig. 17.6 The dimensions of the cantilever and the gap between the cantilever and the frame.

17.2.2.4 **Quartz Tuning Fork**

In the gas-phase spectroscopy, the common approach is to use a resonant PA cell and accumulate the absorbed energy in the gas. However, a recent developed approach is to store the absorbed energy in a sensitive element. A well-suited material for this is a high-Q crystal quartz element with piezoelectric properties. These elements are mass produced and inexpensive; every electronic watch or clock is built around a high-Q quartz crystal frequency standard. Usually, it is a quartz tuning fork (TF) with a resonant frequency close to 32 768 Hz. The mode at this frequency corresponds to a symmetric vibration (the prongs move in opposite directions). The antisymmetric vibration is piezoelectrically inactive. These quartz TFs have recently become widely used materials for atomic-force and optical near-field microscopy, and therefore their properties have been carefully analyzed (Kosterev *et al.*, 2005). A typical watch TF has a Q value of 20 000 in vacuum; at normal atmospheric pressure this is lower (~8000) because of the viscous properties of air. The typical energy accumulation time at atmospheric pressure is 250 ms, which is quite longer than any practical gas-filled resonator (Figure 17.8).

The special features of using a TF are as follows: (i) Ambient acoustic noise is very low above 10 kHz. (ii) External sound will always excite both prongs of the TF and, as a result, will not excite the piezoelectrically active mode, in which the two prongs move

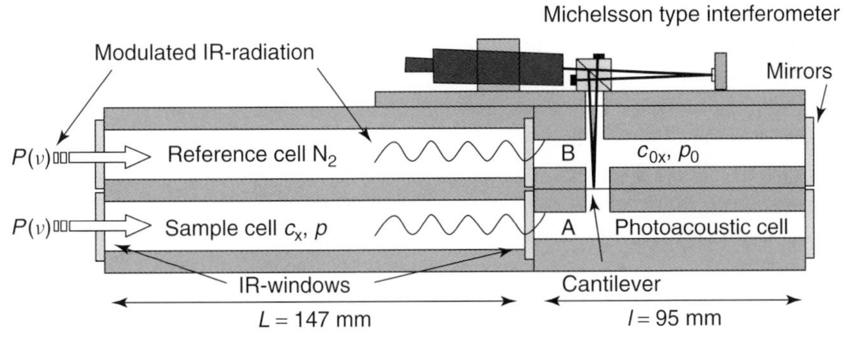

Fig. 17.7 The setup containing the cantilever is in principle the same setup as the nondispersive setup from Figure 17.2, but now with a cantilever detection setup. It consists of three cells: the reference cell, sample cell, and photoacoustic cell. The photoacoustic cell acts as a detector and includes high concentration of the gas to be detected. The cell is divided into two parts A and B which are separated by the cantilever and its frame. The position of the cantilever is proportional to the pressure difference between the cell parts A and B. The displacement of the cantilever end is measured continuously with a Michelson type interferometer.

in opposite directions. (iii) The width of the TF resonance at normal pressure is 4 Hz, and only frequency components in this narrow spectral band can produce efficient excitation of the TF vibration. (iv) The gas sampling volume is extremely small (0.15 mm^3).

17.2.3 Condensed-Matter Photoacoustic Cells

Most spectroscopy on solid samples is done with gas-coupled microphones. Little sample preparation is needed and PA cells are commercially available. Modulation frequencies are generally low, 2–500 Hz, and coupling volumes are as small as possible. One increasingly popular design is the open photoacoustic cell (OPC) in which the sample makes up one wall of the cell.

The investigation of condensed matter using gas coupling becomes more complex when the sample absorption coefficient is small and the modulation frequency is high (e.g., pulsed lasers). The sample vibration itself will start to contribute in a significant way to the microphone signal. In addition, gas microphone detection of acoustic waves inside a condensed sample is not an efficient way of detection, because the acoustic impedance mismatch at the sample–surface interface causes a low transmission across the interface (Tam, 1983). A better, direct coupling, method involves the insertion or attachment of a transducer (piezoelectric) into or onto the sample.

When CW lasers are used in combination with condensed-matter PA cells, wavelength tuning allows a considerable reduction of the background signal generated by light scattering and window absorption. Factors that limit sensitive detection are as follows: besides light scattering and window absorption, mechanical noise at low frequencies, heating of the sample (self-defocusing), and convection currents in the case of liquid.

These limitations are avoided with pulsed lasers. Pulsed lasers allow time gating to discriminate against window absorption (arrives later) and light scattering (is instantaneous). Heating of the sample is minimized by the low average power (pulses are at nanosecond timescale) in combination with low repetition rates. Mechanical noise and the electrical pickup can be eliminated because the pulsed PA signals are at high acoustic frequencies (>100 kHz).

Piezoelectric ceramics are (much more) suitable and many types are commercially available, for example, lead zirconate titanate (PZT), lead metaniobate, lithium niobate, crystalline quartz, and so on (Mason and Thurston, 1979). Thin, highly insulating polymeric films poled at elevated temperatures in strong electric fields also show a piezoelectric character: PVDF, Teflon, and Mylar are most commonly used. They combine a fast rise time (2 ns) with good flexibility and good acoustic impedance to liquids such as water (Tam, 1986).

17.2.4 Photothermal Detection

17.2.4.1 Photothermal Deflection Spectroscopy

Schemes different to PA detection are considered by Sigrist (1994). His conclusion is that for many applications, the PA technique utilizing a sensitive microphone appears to yield the simplest and the most sensitive solution. There are instances, however, in which PA fails. For the investigation of condensed matter using a microphone, a gas buffer is needed that results in low signals due to the acoustical mismatch between gas and solids. In addition, for fast (nanoseconds) experiments microphones are too limited due to their bandwidth. Piezoelectric transducers do have wider bandwidths but need to be attached to the condensed sample. When noncontact methods are preferred, techniques such as photothermal deflection and thermal lensing can be used. An additional advantage is that the bandwidth of detection is very wide and only limited by the bandwidth of the photodiodes.

In photothermal deflection spectroscopy, the basic principle rests on the change of refraction index n caused by a local temperature increase and a density decrease upon absorption of radiation. This change Δn results in deflection of a probe laser (in many cases a Helium–Neon laser with a good pointing stability), observed with the help of a position-sensitive detector. The basic idea can be implemented by two different geometries; the two lasers (absorption laser and probe laser) can either be aligned collinearly or perpendicularly (Figure 17.9). The advantages of the perpendicular arrangement are ease of reproducible adjustment and local probing since it defines a probing volume of some cubic millimeters. This small size is important in many applications such as local gas emission or local heating of the solid-state material for (sub)surface imaging. In the latter application, the excitation light is directed perpendicular to the surface, while the probe laser is parallel, and very close, to the surface.

17.2.4.2 **Thermal Lensing**

Like other photothermal phenomena, thermal lensing is based on the temperature increase after illumination of the sample and the nonradiative relaxation of the absorbed energy. In this case, a laser beam is needed because its radially symmetric Gaussian intensity distribution generates more heat at the center of the

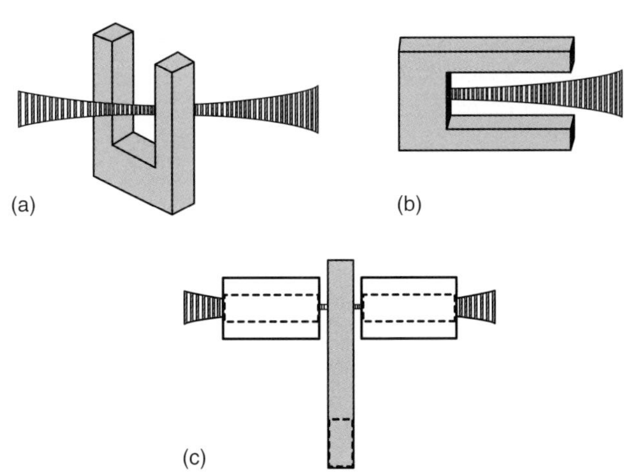

Fig. 17.8 Optical configurations for photoacoustic signal detection with a TF. (a) The laser beam is perpendicular to the TF plane. (b) The laser beam is in the TF plane. (c) An acoustic resonator (sound tube) is added to enhance the signal. The laser beam is directed through the tube. The pressure anti-node is at the center at the location of the tuning fork between the two pieces of the tube.

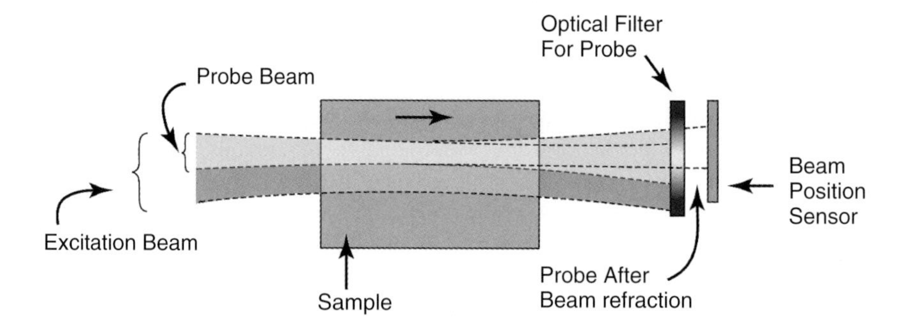

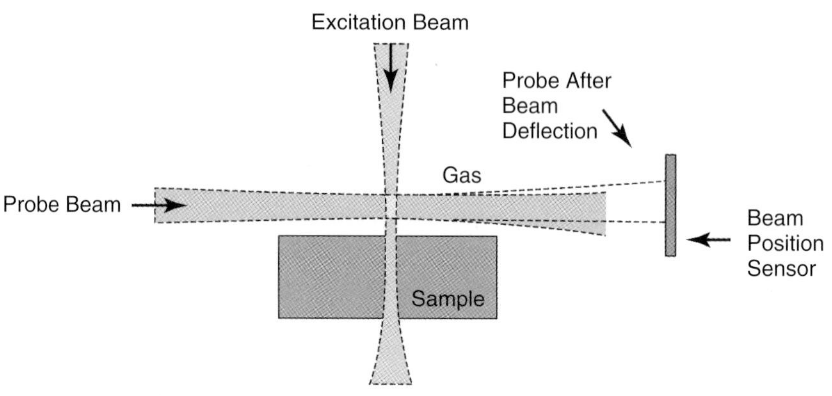

Fig. 17.9 Two schematic configurations of the photothermal beam deflection in which the sample absorbs the radiation of the pump laser and the probe laser (in many cases a He–Ne laser) is deflected by the refractive index gradient induced by the generated heat. (a) Parallel setup and (b) transverse setup.

beam than in the wings (Franko and Tran, 1996). Owing to this temperature gradient across the beam, a lens-like optical element is formed in the material. This thermal lens effect can be observed by a second laser beam or by the initial laser beam itself (single-beam thermal lensing). Single-beam instruments are relatively frequently used because of their simplicity and ease of operation compared to dual-beam instruments (see Figure 17.10). In such an instrument, the laser beam is focused with a lens and modulated by a chopper. After passing through the sample, the beam intensity is usually measured in the far field with a photodiode placed behind a pinhole. The thermal lens signal is linearly dependent on the energy of the pulsed laser or the power of the CW laser. The signal is also strongly dependent on the size of the pump beam; it is therefore necessary to focus the laser beam to obtain higher power densities and thus higher sensitivities. For a two-beam setup, a collinear configuration (two parallel laser beams) is more favorable as compared to a transverse configuration (beams crossing at a 90° angle) because of its higher sensitivity, although the second setup is sometimes preferable because of its small sample volume and independence of the sample length. Sampling volumes as small as a few microliters and absorbances as low as 10^{-7} have been detected and characterized by this technique.

17.2.5 Fourier-Transformed Infrared Photoacoustic Spectroscopy

Infrared lasers have mostly limited wavelength coverage. This restriction is circumvented if Fourier-transform techniques are combined with PA detection. As compared to dispersive techniques using filters or monochromators, Fourier-transformed infrared (FTIR) spectroscopy has the advantages of a relatively high spectral throughput (important since the PA signal is proportional to the light intensity), high spectral resolution, wide spectral range, and speed (Jiang, 2002).

The advantage of combining PA detection with a scanning FTIR derives from the signal being background-free; there is no contribution to the FTIR interferogram from IR frequencies where the sample does not absorb. In FTIR, the modulation frequency ν in Hz depends on the wavelength of the light λ and the velocity of the mirror V in the interferometer: $\nu = 2\,V/\lambda$. A drawback in using PAs is that now the generated PA signal depends on the wavelength of the light: it weakens at shorter wavelengths as the modulation frequency increases. An improvement is introduced by stepwise scanning of the interferometer. The PA signal is generated by an external intensity modulator (e.g., chopper) while the interferometer is scanned stepwise rather than continuously. The advantage of stepscan FTIR is that the strength of the PA signal only depends on the frequency of the external modulator. In addition, now relevant phase information can also be collected. In normal FTIR, short wavelengths will generate high Fourier frequencies and thus shorter penetration depth in solid-state material. Assuming equal thermal properties at both wavelengths, phase information will be different, with stepscan FTIR-PA proper phase information making spectral depth profiling possible for, for example, multilayer polymeric materials with a depth resolution of 1 μm. Initially, the stepscan interferometers were limited to only a few laboratories because they were very sensitive to environmental vibrations and had a poor mirror position control. Nowadays, they are commercially available, and use the computational power for interferometer function control, such as dynamic alignment, step and hold servo control, and modulation–demodulation signal generation.

17.3 Application

In general, PA spectroscopy can be an excellent spectroscopic tool, but more factors than just light absorption play a role. This might seem to complicate interpretation, but this makes it an ideal technique to monitor all kinds of other processes and to determine not only concentrations but also various material properties and local distributions of certain spectroscopically well-characterized compounds. In the following section, a number of typical studies will be shortly described with emphasis on those in which spectroscopic selectivity plays a role.

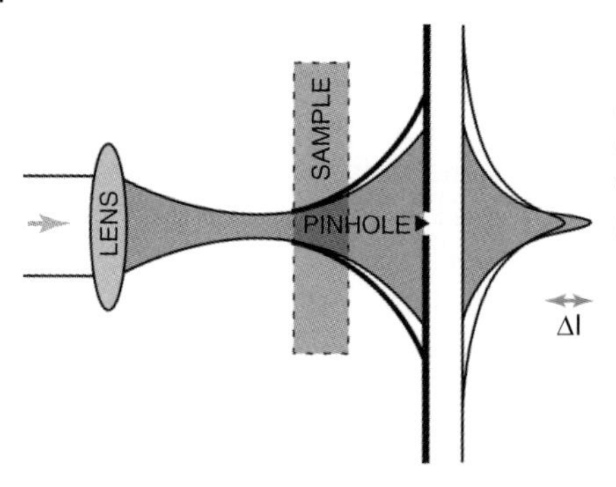

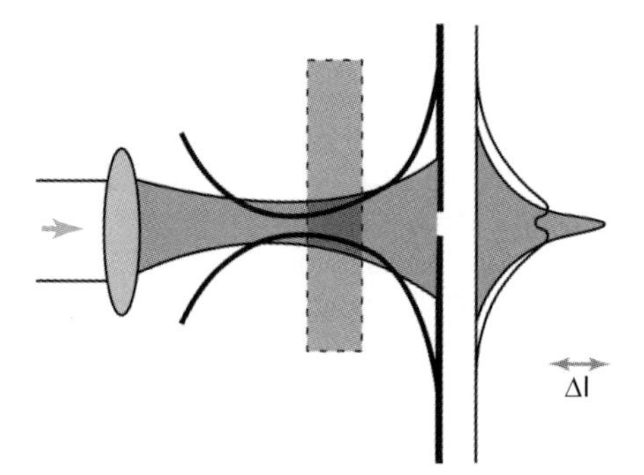

Fig. 17.10 Effect on the laser beam intensity behind a pinhole in a thermal lens effect. The upper panel shows a single-beam thermal lens effect. The lower panel shows a dual beam setup, with the effect of pump beam absorption on the probe beam intensity distribution.

17.3.1 Photoacoustic Spectroscopy in the Condensed Phase

The theory of PA spectroscopy in solids and semisolids was developed in the early seventies by Rosencwaig and Gersho at Bell labs (RG theory) not long after its revival in gaseous samples (Rosencwaig and Gersho, 1976). In those studies, chopped CW light and a gas-coupled microphone were used. In the late 1970s and early 1980s, Patel and Tam (1981) developed PA techniques on liquids and solids using pulsed laser light and a piezoelectric transducer as a contact device. Several aspects play an equally important role in the PA effect in the condensed phase: spectroscopic selectivity, optical, thermal, and acoustic properties of the sample, and the modulation frequency of the light, pulsed light being at the very high frequency end. This gives rise to a multitude of applications focusing on one or the other of these facets.

Despite their simple applicability, still, many of these have not found general use. This may very well be because of the sometimes complex nature of the PA signal in condensed samples. The amplitude and phase of the PA signal depend strongly on the modulation frequency, and both contain important information. There is a depth and absorption dependent phase lag in the signal, and if piezoelectric devices are used, the observed signal is a convolution of the PA signal and response properties of the detector itself.

For CW, light modulation frequencies between 1 and few 100 Hz are used. This can be by amplitude modulation (e.g., using a mechanical chopper), by frequency modulation, or by interferometry. Note that these frequencies are far lower than those applied in (resonance) gas spectroscopy. In

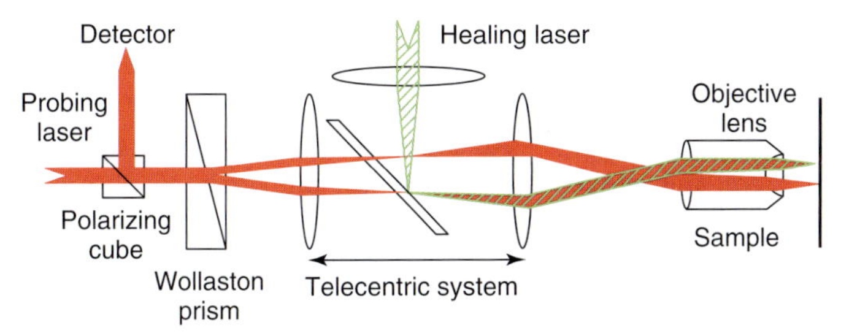

Fig. 17.11 The heating laser beam heats the sample containing the metallic nanoparticles behind the objective lens. The probing laser beam is modulated and split by a Wollaston prism into two probe interferometric arms, which are retro-reflected and again recombined by the Wollaston prism. After demodulation of the interferometer output, the modulated phase-shift induced by the heating can be detected when a nanoparticle is present. (Figure taken from Boyer *et al.*, 2002.)

pulsed experiments, the short light pulses (microseconds, nanoseconds) correspond to modulation frequencies several orders of magnitude higher (megahertz, gigahertz) and are comparable to the time domain of absorption and relaxation processes. There, chemists and biologists find important applications, since PA provides a method that is complementary to flash photolysis techniques. All the more, since calorimetric information is also contained in the PA signal, it represents the part of the absorbed light energy that is not used in luminescence or in chemical processes.

The production of a PA signal in a solid (and to some degree also in a liquid) can be characterized by three parameters as follows: sample thickness, optical absorption length (depth), and thermal (heat) diffusion length. The relative values of these determine what kind of information (spectroscopy, depth profile, mechanical) can be obtained from the signal. The theoretical analysis of these cases also shows that there is an optimum coupling gas thickness (Tam, 1983). The thermal diffusion length, usually denoted as μ, depends on the chopping frequency of the incident light: $\mu = (D/\pi\nu)^{1/2}$, with D being the thermal diffusivity of the sample and ν the chopping frequency. It represents the distance traveled by the generated heat wave during one modulation period of the light. For slow modulation, this is longer than that for fast modulation, which shows in the signal phase lag, and the signal originates from deeper inside the sample.

17.3.1.1 Solid-State Spectroscopy

Although in gases the PA signal is inherently stronger, application of the technique to solids has important advantages over regular solid-state spectroscopy. Since only absorbed light contributes to the PA signal, spectroscopy on scattering and opaque samples is very well possible. Numerous spectroscopic studies on powders, gels, crystals, multilayer solids, nanomaterials, and surfaces have been done. Applications range from determining spectra of organic/inorganic compounds or complexes, soil samples, powders, surfaces, or thin-layer compositions (e.g., paint layers), catalysis research, nanomaterials, to monitoring living biological tissue. PA cells can be designed for each application but they are commercially available as well. Some commercial FTIR instruments have a PA cell as a standard option. In principle, a black reference material is necessary to correct for the change in spectral intensity over the wavelength range of a scan. Black rubber or carbon is normally used.

As an example, an important application is the spectroscopy of silica and semiconductor ceramics, whether in powdered form or as a crystal or a mixture of both. The band gap within the material is usually observed as an increase of the PA signal together with a change (up to 30°) in the PA phase angle at the band edge. In powdered samples several other bands may be observed, depending on the particle size and caused by quantum effects in small particles. Spectroscopy on powdered or porous samples requires some modifications in the RG theory, becaue of the presence of interstitial gas. There is some debate on the exact role that is played by this.

There is good agreement between spectra obtained with a gas-coupled microphone and with a piezoelectric transducer, but, since the detection with a transducer is directly connected to the sample, and usually at the backside, (thermo-)acoustic properties of the material can be more prominently present in the signal.

17.3.1.2 Depth Profiling

Many studies combine PA spectroscopy with depth profiling. As shown above, $\mu = (D/\pi\nu)^{1/2}$ is the depth from which the detected PA signal originates, as determined by the chopping frequency ν and the thermal diffusivity D in $m^2\,s^{-1}$ of the sample. A plot of the PA signal or phase versus the chopping frequency can be read as a scan through the layers of material. With thermal diffusivities in the range of $10^{-5}-10^{-9}\,m^2\,s^{-1}$, penetration depths can be obtained varying from a few micrometers (kHz modulation) to a few hundred micrometers (~2 Hz). It is important to realize that the depth accessible with this technique depends heavily on the absorption and thermal properties of the sample.

An objective can be to study the thermal/mechanical properties of a sample, looking at defects, or to find a distribution of (different) compounds at different depths in the sample. Likewise, the oxidation/etching depths of materials have been followed in many studies. Other examples include multilayer studies, thickness analysis, and material defects. A nonspectroscopic method is to monitor material defects by detecting the acoustic waves traveling through the material originating from surface heating by a modulated light beam.

Recent examples of this are studies in which the penetration of cosmetics or drugs into the skin (or model layers) is monitored. Using modulation frequencies between 13 and 122, or 418 Hz, the diffusion of compounds into depths between 8 and 26 μm was followed over a time course of several hours. Diffusion coefficients could be calculated from these data for different parts of the penetration layer (Hanh *et al.*, 2001).

17.3.1.3 Thermal Diffusivity

As a related subject, the thermal diffusivity: $D = \kappa/\rho C_p$ in $m^2\,s^{-1}$ with κ in $W\,m^{-1}\,K^{-1}$, the thermal conductivity, C_p in $J\,kg^{-1}\,K^{-1}$, the heat capacity, and ρ in $kg\,m^{-3}$, the density of the material, can be determined from the same relationship between the chopping frequency and the PA signal, using differential equations for heat flow and applying the RG theory.

For the study of thermal properties such as thermal diffusivity in solid samples, the OPC is a popular device. The sample is mounted directly on top of the (gas-coupled) microphone and is illuminated from the other side, resulting in a heat transmission configuration. Advantages of this OPC are the minimal gas volume (as small as 9 mm^3) between sample and microphone, and little sample preparation. Of course, the sample has to close off the PA cell gastight. The thermal diffusivity can be determined from the relation between the modulation frequency and the PA signal or the signal phase. A double-log plot or a plot of the PA signal versus the square root of the frequency can be used to derive the diffusivity. The technique usually does not rely on spectroscopic selectivity.

Similar schemes are possible in liquid samples as well, but then a PA cell is used in a transmission mode. Detection takes place using a piezoelectric transducer. The group of Mandelis has presented various treatises on these techniques, proposing improved modulation and referencing methods (Balderas-López and Mandelis, 2001).

Thermal diffusivities are important parameters in the thermal characterization of materials (dental resins, epoxy resins, wood, metals, alloys) and fluids (heat exchange), and the technique has been applied to this wide range of subjects.

17.3.1.4 Phase Transitions

Thermal (and acoustic) material properties, such as conductivity, heat capacity, and thus also diffusivity, change abruptly at phase transitions. This implies that it should be simple to determine the exact phase-transition temperatures from changes in the amplitude and phase of the PA signal. This has been used to study solid–liquid and solid–solid phase transitions in solid and liquid crystals. The transition range can vary from several degrees Kelvin to a few tens of milliKelvin.

More often, however, a pyroelectric detector is used as it proves to be more sensitive in this application than a PA sensor. The photo-pyroelectric effect detects the periodic temperature variation in a sample that is illuminated with modulated light, rather than the acoustic signal. In general, the study of phase transitions is mostly used for liquid crystals, but also for the determination of the properties of natural oils and waxes.

17.3.1.5 Traces in Solution

Just as in trace gas detection, traces of solutes can be detected photoacoustically with great sensitivity and by a variety of related techniques. And just as in PA spectroscopy on solids, detection in opaque samples is possible. The most straightforward way is to detect photothermal changes directly by a piezo transducer immersed in the solvent or acoustically coupled to the cell wall, or using a gas-coupled microphone. The latter option has the obvious disadvantage of acoustic mismatch between liquid and coupling gas. Another possibility is to use thermal lensing. This technique has been used to follow chemical reactions in solution in real time, and, for example, the disappearance of reactants or the formation of products can be monitored. It was proposed to use PA detection methods as detectors in high performance liquid chromatography (HPLC) systems, which has definite advantages, owing to their high sensitivity.

The sensitivity for weak absorption in liquids was compared for pulsed and CW laser excitation (Faubel *et al.*, 1994). In both cases, a piezoelectric detector has been employed. For pulsed dye lasers, the limiting sensitivity is estimated to be 1×10^{-7} cm^{-1} (the minimum detectable absorption). The corresponding value found for a CW dye laser of 55 mW, chopped at about 80 kHz, was 1×10^{-5} cm^{-1} (Manzanares *et al.*, 1995). Note that applying PA to liquids has the distinct advantage of enhanced density of the investigated species, either traces in solution or pure liquids versus their vapors.

17.3.1.6 **Chemical Reactions**

Calorimetry If during light absorption processes the absorbed light energy is "used," "stored," or "lost," PA techniques can be very helpful in resolving timescales and energy balances of processes. One excellent example is time-resolved PA calorimetry; it has developed a distinct use in (bio)-chemical research since the early 1980s. When light is absorbed in the UV–VIS range other de-excitation channels become more important, depending on the chemical structure of the compound under study and of its surroundings (solvent/reactants). Chemical reactions, luminescence, or intersystem crossing to a long-living (e.g., triplet) state take away energy from the direct thermal de-excitation that produces the PA signal. A simple calculation involving excitation energy and PA intensity then gives the amount of energy "stored" in products or "lost" in luminescence:

$$E_a = \varphi_f E_f + \alpha E_a + \varphi_{st} E_{st} \qquad (17.7)$$

where E_a is the absorbed light energy ($h\nu$), ϕ_f and E_f are the quantum yield and energy of fluorescence, respectively, α is the fraction of absorbed energy immediately released as heat, and ϕ_{st} and E_{st} are the quantum yield and energy involved in energy-storing processes (reaction, intersystem crossing, all on a molar basis). By comparing the PA signals (over a range of concentrations) of the process under study with those of a standard that returns all absorbed light energy immediately as heat, calorimetric information is obtained. Over the years, knowledge on energy balance and PA response has become more detailed and it is now fairly well understood how reaction enthalpies, reaction volumes, and the timescales of processes are incorporated into the PA signal. It was also shown on many occasions that changes in chemical volume, solvent interactions, and enthalpy contributions of phenomena such as hydrogen bonding need to be taken into account (Laarhoven, Mulder, and Wayner, 1999).

Volume Effect The acoustical signal is not generated only by thermal expansion due to heating. Many chemical reactions are accompanied by changes in volume between reactants and products (reaction stoichiometry), or between ground state and transition state (e.g., protein folding). One of the reactants or products may be gaseous (photosynthesis), which produces a significant volume (decrease) increase. Also, differences in solvation between reactants and products can produce a volume effect (and also affect the reaction enthalpy) that can become incorporated in the observed PA signals.

Time Resolution Since most chemical reactions take place in the submicrosecond time domain, usually pulsed techniques are used in combination with piezoelectric transducers such as PZT (lead titanate zirconate crystal) or PVDF/polyvinyl difluoride (PVF_2). The first has a relatively slow time response and the signal obtained is a convolution of the PA response and the response/ringing of the piezo element. Deconvolution methods and a reference compound must be used to obtain time constants of the processes involved. The

PVDF film has a faster time response, and the PA trace obtained from this device is in the nanosecond–microsecond domain. If reaction rates are faster than the time response of the transducer, all energy used/produced in the reaction is integrated in the PA signal. If reactions are slower (factor sometimes used: 5τ, where τ is the instrument response time), the energy is stored. In the intermediate time domain, deconvolution methods must resolve reaction rates.

Because enthalpy, volume change, and time resolution are contained in one signal, a wealth of information can be obtained using PA techniques, but disentangling the signal can be cumbersome and requires an insight into the processes under study.

As a more recent development, the thermal grating method was proposed as a means to disentangle thermal and volume effects more easily (Terazima, Hara, and Hirota, 1995; Nishioku *et al.*, 2002). In this method, the interference of two light waves causes a transient grating in the solution. Photoexcitation of the sample by this light creates a (sinusoidal) modulation in the refractive index that contains contributions from the released reaction heat, the reaction volume change, and the change in absorption spectrum. The grating is probed by a second laser, similar to the thermal lensing system.

17.3.1.7 Biological Processes in Solution

Along the same line as the chemical investigations, important research on biologically relevant molecules and *in vivo* processes is in progress (Herbert, Han, and Vogelmann, 2000). For instance, inhibition of photosynthesis systems I and II has been studied photothermally in leaves and in cyanobacteria. Inhibited photosynthesis leads to an increase in heat produced under illumination. Various schemes to attach a PA cell to a leaf have been designed. The main complication is to assure gas exchange to keep the leaf functioning, while closing the cell tight enough to obtain a valid PA signal. The signal from an *in vivo* experiment has to be disentangled both for time resolution and for gas evolution (photobaric signal).

A great deal of research has been done on *in vitro* photosystems. Time resolution and volume changes associated with the different steps in the primary photochemistry of various rhodopsins have been studied by various laser-induced PA spectroscopy experiments (Losi *et al.*, 1999). Photothermal deflection measurements on bacteriorhodopsin have been shown useful to investigate conformational changes in photorefractive compounds (Schulenberg, Gaertner, and Braslavsky, 1994).

Laser-induced PA spectroscopy in biological tissue has developed itself toward three-dimensional tomography, to such an advanced state that animal brain structures can noninvasively be imaged, *in vivo* with skin and skull intact (Wang *et al.*, 2003). The imaging modality combines the advantages of optical contrast and ultrasonic resolution. The intrinsic optical contrast could not only reveal blood vessels but also other detailed brain structures.

Photothermal interference combined with high-frequency modulation and polarization interference provides an efficient, reproducible, and promising contrast method to visualize low amounts of proteins in cells (Boyer *et al.*, 2002). Membrane proteins in cells labeled with 10-nm gold nanoparticles could be visualized (Figure 17.11). The high sensitivity of the method and the stability of the signals allow 3D imaging of individual nanoparticles without the drawbacks of photobleaching and blinking of fluorescent dyes normally used for optically marking molecules. Furthermore, the photothermal image is immune to the effects of scattering background, which limits particle imaging through Rayleigh scattering to diameters larger than 40 nm.

17.3.2 Photoacoustic Spectroscopy in the Gas Phase

The full potential of molecular selectivity of the PA method becomes apparent when it is applied to the gas phase. It is not a high-resolution method (being limited by Doppler and/or pressure broadening); its successful application derives from high sensitivity, easy handling, and fast response.

17.3.2.1 Overtone Studies

Intracavity laser PA spectroscopy is one of the most sensitive techniques for studying (highly forbidden) vibrational overtone transitions. First used by Stella, Gelfand, and Smith (1976), this type of spectroscopy, utilizing tunable dye lasers or titanium–sapphire lasers, can determine weak overtone absorptions up to frequencies in the visible region because of its inherent high sensitivity. Overtone excitation deals with vibrational transitions. Since fluorescence lifetime of these energy levels in the infrared is much longer, as compared to collision-induced rotational–vibrational relaxation lifetimes, the sensitive LIF detection is not applicable. Information about highly excited vibrational levels is most useful to study molecular structure, anharmonicity constants, and (non-)localized bonds. Without these types of measurements, the local mode model for R–H bonds would be poorly developed. Typical spectra are obtained with Doppler-limited linewidths (<100 mbar) and at room temperature; the latter to ensure sufficient vapor pressure of larger organic molecules. A typical example is the high-resolution study of the 4 ν_1 band of monofluoroacetylene with the Ar^+-pumped titanium–sapphire ring laser (laser bandwidth 20 MHz) (Vaittinen *et al.*, 1993). Within the tunability of this laser (9090–17 800 cm^{-1}), Doppler-limited rotational lines of 58 overtone bands in the wavelength region of 10 750–14 500 cm^{-1} have been observed with a very good signal-to-noise ratio. In this way, a study of Fermi resonance interactions at high overtone excitations has been performed. A resonant PA cell was used, excited in its first longitudinal mode (resonator length 180 mm, cross section 2.1 mm × 5.3 mm, overall length 200 mm) (Lehmann, Scherer, and Klemperer, 1982).

The group of Demtröder (Hornberger *et al.*, 1995) has utilized a 500-mW single-mode F–Center laser (6300–6400 cm^{-1}) and a radially excited resonant PA multipass cell (50 passes) to detect weak combination bands of acetylene. In spite of the relatively modest laser power, a minimum detectable absorption of $\alpha_{min} = 10^{-9}$ cm^{-1} has been achieved (Figure 17.12). Often, multiple passing is considered to be a nuisance with PA detection because each mirror reflection produces a mirror signal (analogous to the window signal discussed earlier). Here, however, a radial acoustic mode has been excited, to which the mirror signal barely contributes.

17.3.2.2 Relaxation

There have been many proposals and attempts to study gaseous relaxation processes by PAs. Phase and amplitude of the PA signal change in a characteristic way when the modulation frequency ν and the relaxation time τ obey $2\pi\nu\tau = 1$. However, before the group of Hess really started

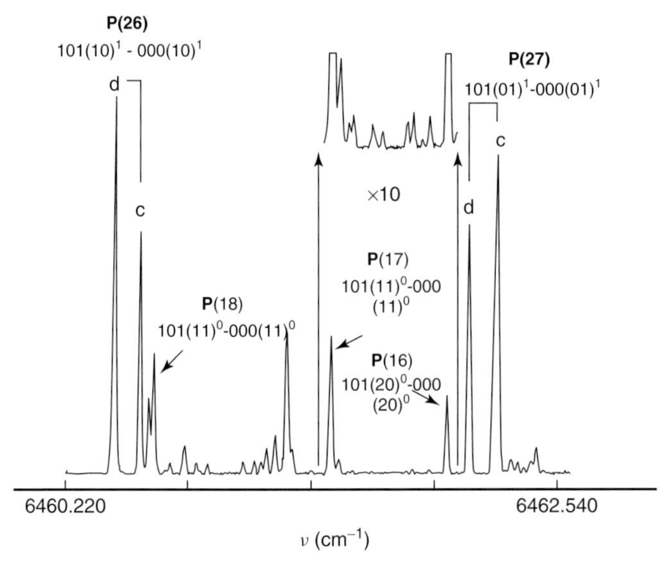

Fig. 17.12 Photoacoustic spectrum of C_2H_2 around 6461 cm^{-1}, the acetylene pressure is 16 mbar. The peaks belong to weak hotband transitions, that is, starting from vibrationally already excited levels with small populations at room temperature. The sensitivity of photoacoustic spectroscopy is illustrated by the inset, which has a 10-fold amplification (Hornberger *et al.*, 1995). (Reproduced with permission from Hornberger, C., Konig, M., Rai, S.B., and Demtroder, W. (1995) *Chem. Phys.*, **190**, 171–177).

an accurate analysis of laser excitation of acoustic modes in cylindrical and spherical high Q-resonators, the results suffered from ambiguities and systematic uncertainties (Karbach, Röper, and Hess, 1983). Essential in the treatment of Hess and coworkers was the theoretical and experimental study of the shift of the resonance frequency and the width of the acoustic resonance curve for various sample pressures. Consistent rate constants have been derived for several reactions, for example, $N_2O_4 \rightarrow 2NO_2$. Similarly, vibrational and rotational energy transfer has been studied for CH_4 excited to the (1,0,0) vibrational state at about 3000 cm^{-1}. At room temperature, rotational relaxation (equilibration) of 1.18 ns bar has been found to be very different from the value for vibrational relaxation (1.74 μs bar).

Vibrational relaxation occurring in the PA cell also has unintentional, practical consequences for other applications. Perhaps the most spectacular is the so-called kinetic cooling effect. PAs relies on (fast) relaxation into thermal heat. If the relaxation is too slow compared to the modulation frequency, the absorbed energy is not represented in the PA energy, that is, the energy is lost (the gas is kinetically cooled). The results are a lower signal and a phase shift. Well known for this is CO_2 in air; traces of CO_2 are excited by the CO_2 laser radiation (e.g., (02^00) $\rightarrow$ (001). Quasi-resonant, slightly endoergic vibrational energy transfer occurs to N_2 molecules: CO_2 (001) $\rightarrow$ N_2 $(v = 1)$. Some translational energy from the CO_2 molecules is thus transferred and stalled as internal vibrational energy of the N_2 molecules and, temporarily, cooling occurs. Collisional relaxation of the excited N_2 $(v = 1)$ molecules is too slow for PA

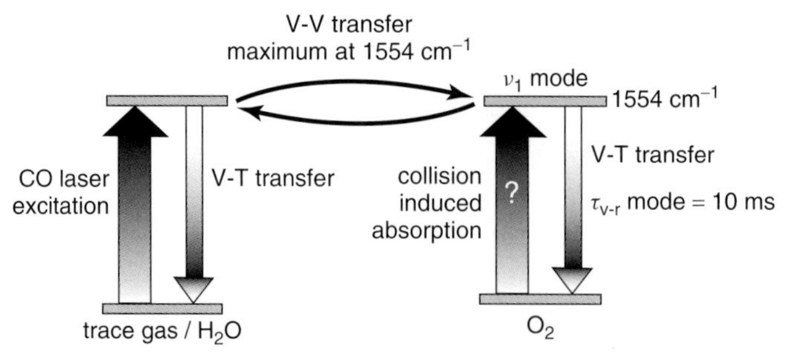

Fig. 17.13 Schematic overview of the excitation and energy transfer process in ethanol in the presence of O_2. Oxygen molecules can become vibrationally excited via vibrational energy transfer from a vibrationally excited trace gas, or directly via collision-induced absorption. Vibrational–translational transfer of H_2O and most trace gases takes place at a nanosecond timescale, while for O_2 this process takes about 10 ms.

measurements at high modulation frequencies and will therefore not contribute to the PA signal. A phase shift of the signal, however, can be observed at typical circumstances as a consequence of the temporary cooling. Addition of H_2O or SF_6 molecules (notorious relaxers) can reverse the situation.

An example is shown in Figure 17.13. Collisional relaxation allows observation of collision-induced absorption phenomena within oxygen. Oxygen molecules, whose ν_1 vibrational mode is centered at 1554 cm^{-1}, can be vibrationally excited via collisional energy transfer. Through collision-induced absorption, that is, a collision between oxygen and a trace molecule or a collision between two oxygen molecules, a dipole moment is induced, which creates an infrared absorption band. Because the induced dipole only exists for times that are on the order of the collision rate, the spectra exhibit a large bandwidth. The induced absorption strength of oxygen is relatively low but still significant due to the high abundance of O_2. The number of collisions required for a vibrationally excited O_2 molecule to relax to the ground state is 8.3×10^7 in pure O_2, while at standard temperature and pressure, 8×10^9 collisions per second take place (Lambert, 1977). Thus, relaxation of vibrationally excited O_2 to the ground state takes about 10 ms, which is long compared to a modulation period of 0.9 ms.

Figure 17.14 presents phase measurements in a mixture of water vapor and ethanol diluted in air and nitrogen. In addition, it shows the collision-induced absorption strength as taken from Thibault and coworkers (Thibault *et al.*, 1997). The observed phase has a pattern similar to the collision-induced absorption strength, however, with a long tail at higher frequencies. If collision-induced absorption would be the main mechanism, such a tail is not expected. Molecules excited at frequencies above 1554 cm^{-1} probably lose some of their energy via rotational–translational energy transfer because in polyatomic molecules the rotational–translational energy transfer is very rapid. As a result, they are closer in resonance to O_2 and vibrational energy transfer is likely to occur. A molecule excited at a frequency below 1554 cm^{-1} can lose some of its energy making vibrational energy transfer to O_2 less probable and hence no tail is expected.

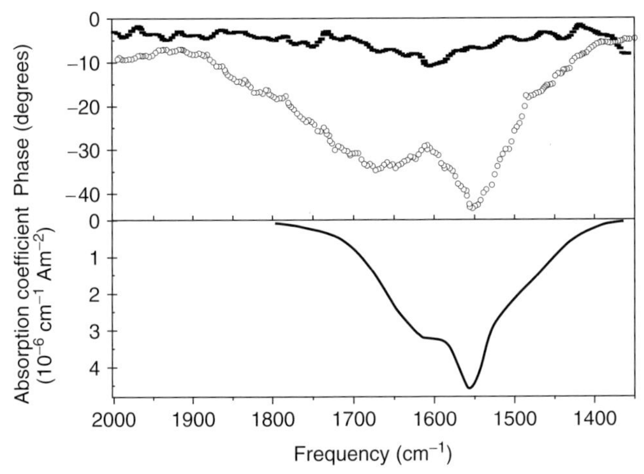

Fig. 17.14 The upper panel presents the phase angle for water vapor and ethanol, both at ppmv level recorded in air (◦) or nitrogen (■). In nitrogen, the remaining structure in the phase angle is partly due to out-of-phase contributions of the background signal. The lower panel presents the O_2-air normalized absorption coefficients for collision-induced absorption as taken from Thibault *et al.* (1997)). The main mechanism responsible for the observed phase lag is vibrational energy transfer from excited water/trace gas molecules to the slowly relaxing O_2 molecules.

17.3.3 Trace Gas Detection

PA spectroscopy shows its full flavor in trace gas detection for three main reasons. Firstly, the spectral selectivity becomes exploited; many gases have a clear spectrum of identifiable absorptions in the infrared. Secondly, PA and related techniques are fast and thirdly, they are sensitive. Trace gas detection techniques are important in view of environmental applications, but also for their possibilities in basic science (biology, agriculture, human health) (Cristescu *et al.*, 2008). With their help, it is possible to discover and control mechanisms in plant physiology such as: germination, blossoming, water household, stress reaction, respiration, fermentation, senescence, ripening, injury, and so on. In the following sections, examples of these important applications will be discussed. In the context of human health, an example of analyses of exhaled air and its interpretation toward the physiological status of the human body is given. For PA trace gas detection, a few, usually less sensitive, instruments are commercially available.

Fast and sensitive trace gas detection is not exclusively reserved for PAs. In Table 17.2, various spectroscopic methods are compared, all utilizing modern laser equipment. The first one, direct absorption (DA) spectroscopy, relies on long effective absorption path lengths, since in general the minimum observable laser power difference $\Delta P/P$ is in the order of 10^{-4}. This, combined with Beers' law: $P = P_0 e^{-\alpha l}$, in which P_0 is the incident laser power, P the transmitted laser power after distance l in centimeters through the sample, and α in cm^{-1} the absorption strength at that frequency, shows that sensitivity increases with increasing path length. Even better detection limits can be obtained by combining the multipass cell with a modulation technique, such as wavelength

Tab. 17.2 Feature comparison of the different spectroscopic detection schemes.

	DA	WMS	CRDS	PAS	QEPAS
Calibration required	Yes	No	Yes	No	No
Detection time constant	ms	ms	ms	s	s
Low pressure response (~10 mbar)	Good	Good	Good	Some loss	Good
Alignment, complexity	Average, multipass alignment	Average, multipass alignment	Difficult. Mode-matching	Easy	Average, tuning fork alignment
Spectral coverage, limiting factor	Wide, photodetector	Wide, photodetector	Limited, mirror coatings	Very wide, windows	Very wide, windows
Cell volume	300 ml	300 ml	500 ml	30 ml	1 μl
Ability to detect broad absorption features	Yes	No, only species till HWHM modulation amplitude	Yes	Yes	No, only species till HWHM modulation amplitude

[a] DA: Direct Absorption, WMS: Wavelength Modulation Spectroscopy, CRDS: Cavity Ring-Down Spectroscopy, PAS: Photoacoustic Spectroscopy, QEPAS: Quartz-Enhanced Photoacoustic Spectroscopy, Half-Width Half Maximum (HWHM) of the absorption profile.

modulation spectroscopy (WMS) (Bomse, Stanton, and Silver, 1992). Originally, WMS was designed for microwave spectroscopy where it is a standard method. However, now it is also commonly used in laser spectroscopy for fast and ultrasensitive trace gas detection applications. The cavity ring down (CRD) method combines accurate decay time measurements of light pulses in an optical cavity filled with weakly absorbing gas ($\Delta\tau/\tau = 10^{-4}$ is observable) with extremely long effective absorption lengths (Berden, Peeters, and Meijer, 2000). The last method, mainly advocated in this contribution, rests upon very sensitive detection of acoustical signals generated by infrared laser absorption in an intracavity-placed resonant PA cell and with quartz enhanced photoacoustic spectroscopy (QEPAS). Each of these spectroscopic detection methods can offer the sub-ppbv (1 ppbv = 1 part per billion volume = 1: 10^9, 1 pptv = part per trillion volume = 1 : 10^{12}) sensitivities required for trace gas detection (Kosterev and Tittel, 2002). However, they also have their own advantages and disadvantages, which should be considered carefully for each specific application. When the experiment, for example, requires determining absolute absorption values, cavity ring-down spectroscopy or DA with multipass enhancement is preferred. These two methods and WMS would also be good choices if highly dynamic processes are studied, since they require a short measurement time per point. However, the required gas volume of these methods is quite large (~liter) which can be an obstruction for small gas samples. In that case it will be better to use normal PA- or quartz-enhanced PA spectroscopy, which have volumes of

30 ml and 1 μl, respectively. The PA detection schemes also have the advantage of having a very broad spectral usefulness, covering a wide wavelength range.

17.3.3.1 Environmental Applications

At low absorption signal levels uncertainties in the measurements are determined by the microphone noise and/or background signal fluctuations. For stronger absorption signals, uncertainties can be either limited by the used calibration mixture, presence of spectral interfering compounds, or uncertainties in the used absorption coefficients. Under good controlled conditions a 1% uncertainty has been achieved (Morgando, 2008).

Measurement of practical samples can pose a real challenge as besides the compound(s) of interest other absorbing gases are present in often much higher levels. For instance, exhaled human breath contains typically 4–5% of both water vapor and carbon dioxide, both of which absorb in large part of the infrared spectrum. When no special precautions are taken, the measurement uncertainties will be considerably higher than for the ideal case of measurement in pure nitrogen or clean air. In most of our measurements, we lower the level of compounds showing spectral interference by using a cold trap and/or chemical scrubbers such as $CaCl_2$ for H_2O removal or KOH for CO_2 removal. Consequently, uncertainties in the measurements are only slightly higher as compared to the ideal case. For molecules like ammonia or nitric oxide these tricks cannot be played since they readily adsorb to most scrubber materials. As these molecules have a permanent electric (ammonia) or magnetic dipole (nitric oxide) moment they can be selectively detected with Stark or Zeeman modulation PA spectroscopy.

Early attempts have been made to measure stack gas emissions from power plants with laser PAs. Owing to their high amounts in the emissions, nitrogen oxide (NO and NO_2) compounds contribute significantly to photochemical smog formation and acidification of the soil. To reduce the total amount of NO_x in the stack gas, NH_3 is added in the exhaust gas toward the chimney. Using a voluminous catalyst, NO_x reduction (within the stack gas) takes place. To check the performance and to avoid an excess of ammonia injection, the ammonia concentration in the chimney is monitored. For this, a CO_2 laser-based PA detection system was successfully applied for *in situ* monitoring of ammonia concentrations (Olafsson *et al.*, 1989). Owing to the difficult, hostile environment halfway up the chimney (vibrations, temperature fluctuations, etc.), concessions have to be made as to the sensitivity of the apparatus. With a nonresonant PA cell at 125 °C a detection limit of 1 ppmv NH_3 was achieved in a multicomponent gas mixture containing 10–15% CO_2.

Another example of a mobile system was developed and used in a field campaign by Sigrist and coworkers (Moeckli, Fierz, and Sigrist, 1996). Installed in a small trailer, the stress on the equipment is less severe. Thanks to this approach, the system has been operational for years and has been applied for several field campaigns in urban and rural environments. It can operate, for example, over a consecutive five-day period continuously to analyze multicomponent gas mixtures during sunny days in the summer, using 12 laser lines; three of them representative for ammonia absorption, three for water vapor, two for ethene, and the others for CO_2, benzene, and toluene. The spectrum of these 12 selected lines could be measured within about 10 minutes, thereby determining the time resolution of the derived pollutant concentrations (Moeckli, Fierz, and Sigrist, 1996). In literature, extremely low detection limits are mentioned on the basis of a signal-to-noise ratio of one, extrapolated from larger quantities of trace gases in a clean buffer gas. In reality, the detection limits are higher due to multicomponent gas mixtures, which induce cross sensitivities in the absorption coefficients. A mathematical analysis of the PA spectra of multicomponent mixtures is based on the weighted least squares fit of the measured spectra with additional iteration steps. By choosing the best set of laser lines with known absorption constants of all the relevant gases, the error in the calculated concentrations can be reduced.

A disadvantage of lab-based systems is that they are still rather bulky. Nowadays, complete PA systems are integrated into durable, man portable, computer size boxes, completely protected from dust, precipitation, and damage, making them applicable in field measurements. Two examples are given, one for measuring ambient ammonia concentration in the lower ppbv (1 ppbv = 1 part per billion volume = $1 : 10^9$) concentration range (Huszar *et al.*, 2008). Within the environment, ammonia is the third most abundant nitrogen compound in the atmosphere and plays an important role in cloud and rainwater chemistry. Typical ammonia sources include livestock, fertilizers, soils, forest fires, humans, animals, oceans, industry, and traffic. The main emission source is the use of animal manure and fertilizers for agricultural soils (responsible for more than 80% of the total ammonia emission). The PA instrument is based on a wavelength-modulated, near-infrared, room temperature-operated diode laser, and a PA cell in its longitudinal resonance mode, with compact electronics for fully automated, long-term operation. The PA cell was made of polyvinylidene fluoride to reduce ammonia adsorption effects. Cross sensitivities (from atmospheric water vapor) were suppressed by using a multiwavelength approach. The system featured highly reliable, automatic operation achieving a detection limit of 50 ppbv for ammonia.

Another laser-based PA instrument was developed and used for aircraft measurements of ethene from industrial sources near Houston. This instrument provided 20-second measurements with a detection limit of ~0.7 ppbv (de Gouw *et al.*, 2009). Data collected from this instrument in flight were compared with gas chromatographic measurements and they agreed within 15% on average. Ethene fluxes from a chemical complex near Houston could be quantified during 10 different flights. The average measured fluxes were an order of magnitude higher than the values regulatory emission inventories indicate.

17.3.3.2 Plant Physiology

It is a fortunate coincidence that the ubiquitously active plant hormone ethylene, C_2H_4, is especially easy to detect with CO_2 laser-based PAs. Ethylene is the only known gaseous plant hormone; it is involved in virtually all aspects of the plant life starting from seed germination through growth and flowering, until fruit ripening and plant senescence. The biosynthesis in plants is well understood, leading from the amino acid methionine over two intermediates, S-adenosyl-methionine and amino-cyclo-propane-carboxylic acid, to C_2H_4. Ethylene is produced by the plant tissue as a response to many external factors such as wounding, pathogen attack, light/temperature stress, chilling, drought, and so on. Modern gas chromatographic systems can be used in many instances but still lack the time resolution and extreme sensitivity of PA spectroscopy. With a detection limit of 10 pptv and a short time constant of about 10 seconds, PA detection is ideally suited to obtaining an insight into C_2H_4-triggered plant physiological properties. Examples of recent investigations are listed in Table 17.3 (Ramina *et al.*, 2007).

Numerous studies of the cellular and molecular biology of plants use *Arabidopsis thaliana* as model system. This is an organism about which much is already known, that is, easily manipulated, genetically tractable and offers the ability of fast and efficient testing of hypotheses. By using Arabidopsis as a reference system, we gain knowledge to move forward with molecular biological and genetic analyses on many physiological processes involving ethylene. With PA system dynamics of many processes in plants become easily visible. For example, ethylene released by Arabidopsis displays a circadian rhythm (high emission during the light and low emission during dark) with a peak in the mid-subjective day. When the plants were grown for six days in a succession of light (16 hours) and dark (8 hours) and afterwards were kept in continuous light for several days, they showed an ethylene pattern as would follow the light–dark periods (Figure 17.15).

17.3.3.3 **Oxidative Stress**

In humans, animals, and plants, many diseases are caused by the imbalance between the formation of oxygen radicals and other reactive oxygen species (ROS), which might ultimately result in cell injury. This is a condition of oxidative stress. These reactive molecules can start a chain of reactions known as lipid peroxidation, which leads to degradation of the outer and inner membranes of cells (Halliwell and Gutterridge, 1989).

Ethane (C_2H_6) is a product of lipid peroxidation as a result of oxidative stress and can serve as a marker for membrane degradation. Ethane is detected very sensitively using a laser operating around 3 μm, such as in the OPO-based system. As a practical example, one has investigated the damage of the chilled cucumber leaves under light and dark conditions by monitoring the ethane emission. Under light, the damage of leaf tissues was associated with the presence of free radicals. Additionally, low temperatures slow down the electron transport chains, and inhibit alternative energy dissipation paths. Both factors enhance the formation of free radicals at low temperatures. Furthermore, low temperatures also decrease the activity of antioxidants. In consequence, light-chilling leaves show elevated ethane levels in correlation with a damage leaf area (yellow coloring) of 15–30% of the total area after one-day chilling and up to 65% in a two-day treatment. The dark-chilling leaves did not produce measurable ethane and present no damage.

Plant recognition of pathogen infection leads to the so-called hypersensitive response (HR) indicated by a fast, localized cell death at the site of infection. Upon pathogen invasion an oxidative burst can be observed in plant; however, ROS alone cannot stop the pathogen invasion. A synergistic mechanism has been proposed between ROS and NO during HR in plants. Within the HR, the plant initiates a suicidal strategy, signaled by the production of NO that helps to kill the infected cells and activate the defense mechanism to isolate the infection (Figure 17.16). Using laser-based detection, the first in planta and direct measurements of NO emission from plants as response to *Pseudomonas syringae* challenge has been obtained (Mur *et al.*, 2005).

Tab. 17.3 Recent biological photoacoustic investigations monitoring ethylene emission (Ramina *et al.*, 2007).

Studied process	Subject	Remarks
Germination	Pea, bean, rice seeds	Ethylene accompanies starting of germination, growth of shoots and roots
Root development	Arabidopsis seedlings	Ethylene emission during the root development
Flowering	Orchids, Carnations, tulips, Alstromeria, tobacco, transgenic tomatoes	Dynamics of ethylene emission during flower development and wilting
Pollination	Tobacco flower	Ethylene response to pollen tube growth in tobacco flower
Ripening	Fruit	Ethylene and CO_2 emission rates during ripening and post harvest of fruits
Cell death	Algae, tomato cells	Ethylene production during inducing cell death
Circadian rhythms	Arabidopsis plants	Circadian rhythms of ethylene emission in Arabidopsis
Wounding	Tomatoes, endive, cut fresh herbs	Wound-induced ethylene emission in tomatoes, endive, and cut fresh herbs
Water stress	Rumex plants	Ethylene production in relation with shoot elongation during submergence
Drought stress	Wheat, sun flowers	Ethylene rate in sensitive vs. tolerant wheat species and in sunflowers grown under potassium starvation and subjected to drought
Heavy metal stress	Aquatic plants	Ethylene from aquatic plants under heavy metals stress
Plant–insect interaction	Pine-sawfly *Diprion pini*	Plant responses induced by insect egg deposition
Plant–pathogen interaction	Tomato, Arabidopsis – *B. cinerea;* barley-powdery mildew; apples-bacteria	Ethylene production by *B. cinerea in vitro* and in tomato fruit; ethylene response from Powdery mildew infected barley plants; apples infected with *Erwinia amylovora*
Nitrogenase activity	Cyanobacteria	By adding acetylene, ethylene is produced functioning as a sensitive and fast indicator for nitrogen fixation
Hormone interaction	Mutated Arabidopsis plants, transgenic tomato	Cross talk between ethylene, abscisic acid, and auxin in mutated Arabidopsis and transgenic tomatoes

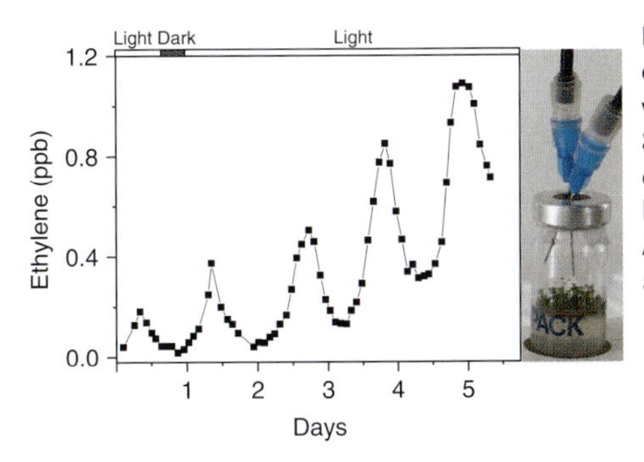

Fig. 17.15 Influence of day length on ethylene production in Arabidopsis. Seedlings were grown for six days in 16 hours light and 8 hours dark and then tested for ethylene emission during one light–dark cycle followed by constant light. Ethylene released by Arabidopsis displays a circadian rhythm with a peak in the mid-subjective day.

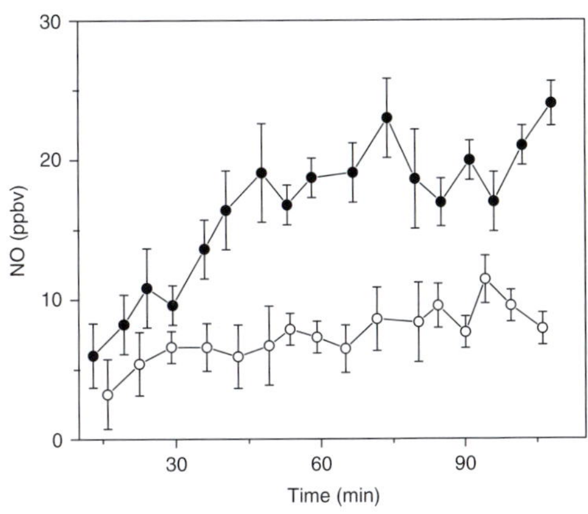

Fig. 17.16 NO formation after infection in tobacco leaves by virulent *P.Syringe pv. tabaci* (dark circles) and avirulent *P.Syringe pv. phaseolicola* (open circles). A suicidal strategy is applied by plant during the avirulent bacteria infection; plant localized fast the lesion due to the HR mechanism (a) compared to the one that did not recognize the bacteria and develops a typical disease response (b).

The quality of agricultural products at the time they arrive at the consumers strongly depends on the developmental stage at harvest, shipping, and storage conditions. Therefore it is commercially advantageous that fruits and vegetables have a long shelf life and do not deteriorate immediately after harvest. Fungal pathogens are known to be responsible for many post-harvest deteriorative processes. Enhanced production of the plant hormone ethylene is one of the earliest responses of plants to the perception of a pathogen.

Using a CO_2 laser-based PA detector the ethylene production by fungi was studied under *in vitro* conditions and then correlated with the ethylene emission from the infected host.

Infection of tomato fruits with *Botrytis cinerea,* a worldwide abundant fungus that attacks more than 200 plant species, resulted in enhanced ethylene release which started to rise before visible decay development (Figure 17.17). This demonstrates that ethylene can be considered a sensitive marker for early infection in harvested fresh products.

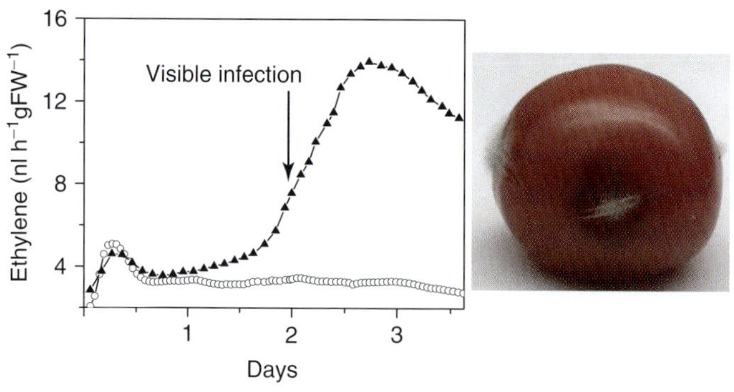

Fig. 17.17 Early detection of *B. cinerea* infection in tomato fruits. Ethylene production from artificially inoculated tomato fruits (triangles) starts to increase one day before visible infection. Production rate is given in nanoliter per hour per gram Fresh Weight (nl h^{-1} gFW^{-1}). For comparison ethylene emission of the uninfected fruit (circles) is shown (Cristescu *et al.*, 2002).

Besides ethylene, other compounds related to the quality control especially during storage can be monitored in real-time with the laser-based PA detectors. Currently, a trend in fruit storage is to reduce the oxygen level in order to slow down ripening and senescence. By lowering the oxygen level, aerobic respiration is gradually replaced by alcoholic fermentation that produces acetaldehyde and ethanol. These two compounds can thus serve as markers for suboptimal storage conditions; a timely observation is required in order to extend the storage period. Since in the storage room the concentration of these (and other trace gases) is normally below the detection limit of conventional gas analyzers, long accumulation periods are needed to obtain detectable concentrations. By avoiding the necessity of accumulation, online detectors such as laser-based PA trace gas detectors have proven to be capable of sensitive detection of ethanol and acetaldehyde at and below the part per billion levels.

Methane is an important greenhouse gas that originates both from anthropogenic and natural sources. In 2006 it has been reported that terrestrial plants may also emit methane under aerobic conditions by an unknown physiological process, and in this way may substantially contribute to the annual global methane budget with 10 Tg up to 260 Tg per year (Keppler *et al.*, 2006). The reported data have generated debates among the scientific community and also the general public. So far these findings were obtained in one single laboratory which was criticized for the experimental setup. Using an OPO-based system we have re-examined whether plants are producing methane in aerobic conditions by monitoring the $^{12}CH_4$ and $^{13}CH_4$ emissions of plants. Toward this aim ^{13}C-labeled plants grown under controlled conditions were used. Since about 99% of the carbon found in these plants was in the form of ^{13}C, one can expect that nearly 99% of the methane emitted by these plants is in the form of ^{13}C-methane. It is preferable to use these ^{13}C-labeled plants since the natural background concentration of ^{13}C-methane is 20 ppbv compared to 1.7 ppmv of ^{12}C-methane. An overall average production of 21 ng g^{-1} h^{-1} ^{13}C-methane was measured online which is 6–18 times lower than the average methane emission rates reported for the same plant species under "sunlight" and "no sun" conditions, respectively (Dueck *et al.*, 2007). Furthermore, less than 1 ppbv ^{13}C-methane was produced over a six-day period by the ^{13}C-labeled plants while the total plant biomass increased by 30%.

These tests indicated that plants produce insignificant amounts of methane in aerobic conditions, and the overestimated methane previously reported can be partly due to the diffusion of atmospheric methane from the plant tissue. Our results have been later confirmed by several other groups (Nisbet *et al.*, 2009; Beerling *et al.*, 2008).

17.3.3.4 **Microbiology**

The growth of all organisms depends on the availability of mineral nutrients, and nitrogen is one the most required nutrients in many biological reactions. Although the earth's atmosphere contains nearly 80% of nitrogen gas, this is unavailable for use by most organisms, being an inert molecule. In order for nitrogen to be used for growth it must be "fixed" (combined) in the form of ammonium or nitrate ions. This conversion process is known as nitrogen fixation and is exclusively performed by prokaryotes (bacteria) using an enzyme complex called nitrogenase. The acetylene reduction is the most widely used method for measuring the nitrogenase activity. It is based on the nitrogenase property to reduce compounds with a triple bond. Thus, by reducing nitrogen to ammonia, the nitrogenase reduces acetylene to ethylene. Both of these gases can be measured with high sensitivity by gas chromatography with flame ionization detection. However, due to its higher sensitivity and fast time response (about 20 seconds), laser PA spectroscopy offers the advantage of monitoring in real time small changes in the nitrogenase activity of low amounts of biomass.

The PA technique was applied for the first time to investigate the nitrogenase activity in a sample of the heterocystous cyanobacterium *Nodularia spumigena*. The

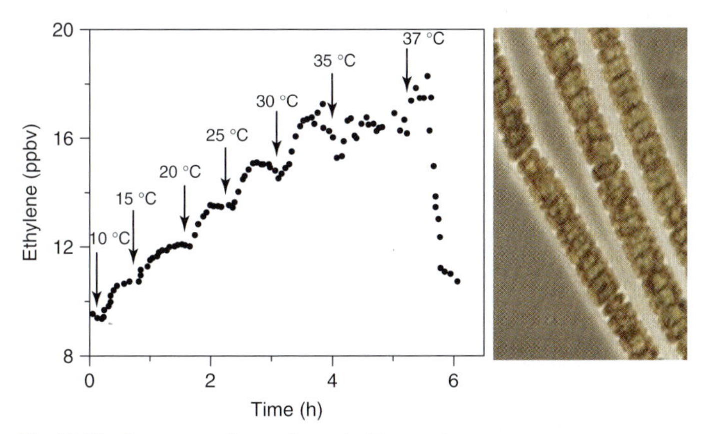

Fig. 17.18 Response of cyanobacteria *N. spumigena* to change in the temperature from 10 to 37 °C under light conditions. Significant instability in the nitrogenase activity is observed at 35 °C, leading to a short inhibition tendency at 37 °C, followed by a drastic decrease.

online monitoring allows measurements under constant conditions, which eliminates artifacts caused by changes in concentrations of oxygen or carbon dioxide due to respiration or photosynthesis during the incubations. Changes in light intensity and temperature, respectively, induce fast response in the nitrogenase activity. Knowing the level of saturation of the enzyme at different light intensities is important when rates of acetylene reduction are converted to rates of nitrogen fixation. At 35 °C and above this temperature, the nitrogenase activity is inhibited, most probably because nitrogenase, or because of the other substrate-generating enzyme systems (ATP and e^-) like the electron transport chain, are breaking down irreversibly (Figure 17.18).

17.3.3.5 Entomology

Laser PA trace gas detection is able to investigate the (respiration) behavior of the small insects like the Western Flower thrips, *Frankliniella occidentalis* (the smallest winged insects, size 1–2 mm, weight 50 μg) or single ant (weight of few milligrams) and *Drosophila melanogaster*, fruit fly (weight of 1–2 mg). In Figure 17.19 the respiration behavior of a single ant of 3.7 mg is shown. The ant is free moving in a 2-cm long tube with 4.5 mm internal diameter, and the CO_2 released is directed to the OPO-based system using air as the carrier gas. The respiration pattern periodically shows the release of CO_2 over several minutes that have been associated with the opening and closing, respectively, of the ant spiracles (the respiratory apertures that control the gas exchange of the insect).

Since the capability of the OPO-based system for monitoring CO_2 release of small insects has been demonstrated, one may think of investigating the metabolic changes in a single cell. In the future, this is in reach with the developments of more powerful lasers. For example, with a powerful pump laser which can supply more than 20 W of radiation between 1020 and 1050 nm, 10 times more power at the maximum of the CO_2 absorption can be

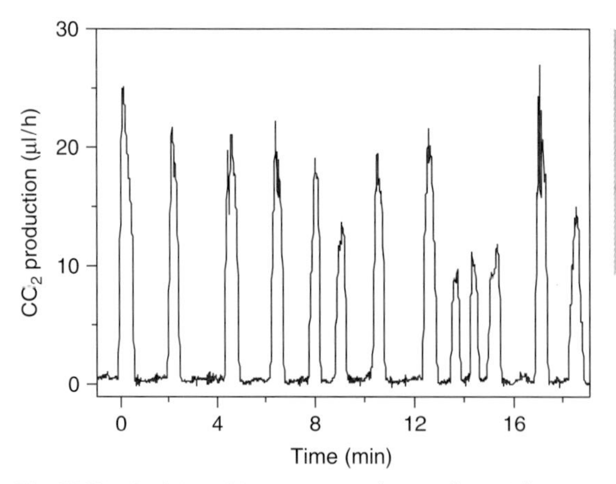

Fig. 17.19 Real-time CO_2 emission of a small ant. The respiration pattern displays periodic peaks of CO_2, which repeat on average once every 1.5 minutes. During the experiment the ant lives inside a small tube (diameter 4.5 mm) through which air is flowed.

Tab. 17.4 Molecules present in exhaled human breath with their physiological pathway and related health condition (Dueck *et al.*, 2007; Nisbet *et al.*, 2009; Beerling *et al.*, 2008).

Compound	Concentration in breath	Physiological origin	Biological-pathology Indication
Acetaldehyde	ppb	Ethanol metabolism	Metabolism of lung cancer cells
Acetone	ppm	Decarboxylation of acetoacetate	Diabetes, fasting response, glucose metabolism
Ammonia	ppb	Protein metabolism	Liver and renal diseases, fasting response
Carbon dioxide	%	Respiration product	$13CO_2/12CO_2$ test for detection of ulcer bacteria *Helicobacter pylori*
Carbon monoxide	ppm	Degradation of hemoglobin by the heme oxigenase enzyme	Blood flow promotion
Carbonyl sulfide	ppb	Gut bacteria	Acute lung rejection in transplantations, liver and bile duct diseases
Ethane	ppb	Oxidative stress Lipid peroxidation	Chronic obstructive pulmonary disease, Rheumatoid arthritis, diabetes mellitus, lung cancer
Ethanol	ppb	Alcoholic fermentation of glucose by gut bacteria and yeast	Glucose metabolism
Ethylene	ppb	Oxidative stress Lipid peroxidation	Free radicals-mediated damage in the human body
Isoprene	ppb	Cholesterol biosynthesis	Hepatocellular injury
Methane	ppm	Gut methanogenic bacteria	Bacterial overgrowth in the small intestine
Methanol	ppb	the degradation of pectin by bacteria in the colon	Carbohydrate malabsorption, pancreatin insufficiency, renal failure
Nitric Oxide	ppb	Oxidative stress, endogenously produced by nitric oxide synthase enzyme	Lung diseases (asthma, COPD, cancer, airways inflammation, etc.), vascular smooth muscle response
Pentane	ppb	Oxidative stress Lipid peroxidation	Inflammatory diseases, transplant rejection, breast and lung cancer

generated. This will lead to an increase in the sensitivity by the same factor and pushes the CO_2 detection limit under 1 ppbv.

17.3.3.6 Human Health Research

Since ancient times, it has been known that the smell of exhaled air can be used as an indicator for several processes taking place in the human body; uncontrolled diabetes produces a sweet, fruity odor; advanced liver diseases entail a musty, fishy reek; failing kidneys bring about a urinelike smell and a lung abscess can be brought to light by its putrid stench. In the past, several attempts to use trace gas detection of exhaled air have been performed with varying success because of the insufficient detection limits of the detectors available. Gases produced in the body are transported to the lungs and are diluted (at rest, healthy persons exhale approximately 1000 l per h) before being exhaled. To measure them, it is usually necessary to accumulate the air samples by adsorbing the gases on a carrier column and releasing the concentrated gas at a certain moment, which decreases accuracy and time resolution.

Under stress conditions (e.g., ionizing radiation, toxic chemical substances, diseases, etc.), the production of free radical in the body is significantly increased. Subsequently, the capacity of the free radical scavengers in the body is overloaded and a chain of chemical reactions is activated. This leads ultimately to cell membrane damage that plays an important role in the aging processes and pathogenesis of some diseases such as cancer, Alzheimer, atherosclerosis, and so on. The cell damage is accompanied by the production of small hydrocarbons such as ethane, pentane, and ethylene that can be easily and quickly measured using PAs.

Other exhaled compounds were found to be by-products of endogenous biological processes. Ethanol, for example, is produced via alcoholic fermentation of glucose by intestinal bacteria and yeast. Acetone, a volatile ketone of fruity aroma, in breath results from the oxidation of free fatty acids.

Laser PA spectroscopy offers several advantages over the GC/MS analysis: (i) allows single breath collection since small sample volumes (few hundred milliliters) are required, (ii) no preconcentration steps are needed, (iii) excellent selectivity, and (iv) the detection and analysis procedure is fast (seconds–minutes) and computer-automated.

A summary of several compounds out of more than 300 present in the breath with their physiological origin and biological–pathological indication is given in Table 17.4 (McCurdy *et al.*, 2007; Kharitonov and Barnes, 2001; Risby and Solga, 2006). Many of these compounds have well-defined biochemical pathway and can be monitored in real time by PAs.

Acknowledgment

The authors wish to thank Frans Bijnen, Huug de Vries, Jos Oomens, Tim Groot, Edi Santosa, Iulia Boamfa, Bas Moeskops, Anthony Ngai, Sacco te Lintel Hekkert, Luc-Jan Laarhoven, Marc Staal and Jörg Reuss for their valuable contributions to this article.

Glossary

Absorption Coefficient: The absorption coefficient α in per centimeter is the attenuation coefficient of the light transmitting through a material. The transmitted light intensity I through a material with thickness l in meters is related to the incident intensity I_0 according to Beers law: $P = P_0 e^{-\alpha l}$, since for small absorptions this equation can be linearized: $\Delta P = P_0 - P = P_0 \alpha l$; the minimum detectable absorption from a device can be denoted as $\alpha_{min}/\mathrm{cm}^{-1}\,\mathrm{W}^{-1}$

Acoustic Resonator: Here, part of the photoacoustic cell allows a standing wave pattern to be formed (resonance). The modulation frequency of the exciting light beam should be tuned to the resonance frequency of the resonator.

Band Gap: Energy difference between the valence band and the conduction band in a semiconductor.

Boxcar Amplifier: Amplifier in which signals that occur within a preset (subsecond) time gate are selectively sampled.

Brewster(s) Angle: The angle of incidence on a window at which plane-polarized light suffers no reflection loss.

Brownian Motion: Random motion of small particles due to statistical variations in molecular pressure around the particles.

Cavity Ring Down (CRD): Sensitive direct absorption method. A laser pulse is stored in a high finesse optical cavity containing the sample. The decay time of the light intensity indicates the absorption strength.

Cell Constant: Parameter that describes the efficiency of conversion of absorbed modulated light energy into acoustical pressure amplitude, in Pa m W^{-1}.

Continuous Wave (CW): Constant, steady state delivery of laser power, as opposed to pulsed.

Depth Profile: Dependence of the photoacoustic signal (phase of amplitude) on sample depth as a consequence of a concentration gradient or changes in the thermal properties of the substrate.

Ethylene: Trivial name of ethene, C_2H_4, mostly used in biology; a gaseous plant hormone, involved in many processes. It can be detected very sensitively by photoacoustic spectroscopy.

HPLC: High performance liquid chromatography. A technique that is widely used to separate compounds in the liquid phase on a packed, (particulate) solid phase column. Often equipped with a UV-VIS detector.

Intracavity Photoacoustic (IPA) Detection: Sensitive method for trace gas detection; the photoacoustic cell is placed inside the laser cavity; due to the higher laser power better sensitivities can be achieved.

Laser-induced Fluorescence (LIF): Sensitive detection of fluorescence – radiative relaxation of electronically excited molecules – is used to detect specific laser-excited molecules. A technique used, for instance, for monitoring molecules in a flame.

Line Tunability: Indicates the restriction for gas lasers to emit light only for a discrete set of wavelengths.

Lock-in Technique: Only signals with a well-defined AC frequency are sampled. A lock-in amplifier locks to and measures the photoacoustic signal using the resonance frequency, ignoring all other signals (noise).

Multiple Pass Setup: A method to increase the absorption pathlength x-fold by reflecting a light beam many times between two mirrors.

Notch Filter: A simple acoustical interference filter that suppresses specific acoustical wavelengths, for example, perturbations from outside entering the acoustical resonator of a photoacoustic detector.

Open "Organ Pipe" Resonator: The central part of an often-employed resonant photoacoustic cell; it consists of an open-ended straight $\lambda/2$ tube, with the microphone mounted at its center where the pressure amplitude has its maximum.

OPC, Open Photoacoustic Cell: Simple photoacoustic cell, in which the sample serves as one wall, usually opposite the (gas-coupled) microphone.

Optical Parametric Oscillator (OPO): A nonlinear optical device where incoming laser radiation of frequency ν_0 is transformed into two light beams of frequencies ν_1 and ν_2 with $\nu_1 + \nu_2 = \nu_0$.

Overtone: Vibrational transitions (due to infrared light absorption) to a higher quantum level than the first excited level. Usually observed as weak absorption lines at shorter wavelengths.

Phase (angle): Position on a periodic wave (sinus) relative to a reference point on that wave (usually a node). In a photoacoustic signal the phase carries temporal information.

Piezo (electric) Element: A material, mostly a ceramic crystal or a polymer foil, that undergoes charge separation when experiencing pressure; piezo = pressure.

Quality Factor: Number that characterizes the quality of a cavity. It is given by the ratio between the stored energy and the energy loss during one cycle under stationary operation.

Quantum Cascade Laser: Laser action based on photon emission by a well-defined cascade of electrons through the layered semiconductor material.

Relaxation: De-excitation of a species after absorption of (light) energy. Relaxation can be radiative (fluorescence, phosphorescence) or nonradiative and within a wide time-window.

Rhodopsin: Protein that efficiently absorbs light in the middle of the visual spectrum with a maximum at 500 nm. Part of the light receptor in the eye and in, for example, halobacterial photoreceptors.

Spectral Coincidence: (Near) resonance between an employed laser transition and one or more transitions of the investigated molecule.

Spectral Selectivity, Spectroscopic Selection: The structural and vibrational properties of molecules allow selective excitation (and detection) of one component in a gas mixture while the other species do not respond to the chosen wavelength of the laser.

Thermal Diffusivity: Thermal diffusivity: $D = \kappa/\rho C_p$ in $m^2\ s^{-1}$, with κ in $W\ m^{-1}\ K^{-1}$, the thermal conductivity, C_p in $J\ kg^{-1}\ K^{-1}$, the heat capacity, and ρ in $kg\ m^{-3}$, the density of the material; the thermal diffusivity determines the response of the material to heat produced by, for example, absorption of electromagnetic radiation.

Thermal Lensing: Thermal nonequilibrium as a consequence of local heating changes the refractive index of a (transparent) material in such a way that light rays become deflected as by the action of a lens.

Trace Gas Detection: Techniques to determine concentrations of species present below about 1 ppmv in the atmosphere.

Window Absorption: Loss of energy of a light beam traversing an optical window, for example, at the entrance or exit of a photoacoustic cell; the resulting heating produces an unwanted photoacoustic signal that must be suppressed.

References

Balderas-López, J.A. and Mandelis, A. (**2001**) *J. Appl. Phys.*, **90**, 2273–2279.

Beerling, D.J., Gardiner, T., Leggett, G., McLeod, A., and Quick, W.P. (**2008**) *Global Change Biol.*, **14**, 1821–1826.

Bell, A.G. (**1880**) *Am. J. Sci.*, **XX**, 305–324.

Berden, G., Peeters, R., and Meijer, G. (**2000**) *Int. Rev. Phys. Chem.*, **19**, 565–607.

Bijnen, F.G.C., Reuss, J., and Harren, F.J.M. (**1996**) *Rev. Sci. Instrum.*, **67**, 2914–2923.

Bomse, D.S., Stanton, A.C., and Silver, J.A. (**1992**) *App. Opt.*, **31**, 718–731.

Boyer, D., Tamarat, P., Maali, A., Lounis, B., and Orrit, M. (**2002**) *Science*, **297**, 1160–1163.

Cristescu, S.M., De Martinis, D., Te Lintel Hekkert, S., Parker, D.H., and Harren, F.J.M. (**2002**) *Appl. Environ. Microbiol.*, **68**, 5342–5350.

Cristescu, S.M., Persijn, S.T., te Lintel Hekkert, S., and Harren, F.J.M. (**2008**) *Appl. Phys. B*, **92**, 343–349.

Dueck, T., De Visser, R., Poorter, H., Persijn, S., Gorissen, A., De Visser, W., Schapendonk, A., Verhagen, J., Snel, J., Harren, F.J.M., Ngai, A.K.Y., Verstappen, F., Bouwmeester, H., Voesenek, L.A.C.J., and Van der Werf, A. (**2007**) *New Phytol.*, **175**, 29–35.

Ebrahim-Zadeh, M. and Sorokina, I.T. (eds) (**2005**) *Mid-infrared Coherent Sources and Applications*, Springer, Dordrecht.

Faubel, W., Schulz, T., Seidel, B.S., Steinle, E., and Ache, H.J. (**1994**) *J. Phys. IV*, **C7**, 531–534.

Franko, M. and Tran, C.D. (**1996**) *Rev. Sci. Instrum.*, **67**, 1–18.

de Gouw, J.A., te Lintel Hekkert, S., Mellqvist, J., Warneke, C., Atlas, E.L., Fehsenfeld, F.C., Fried, A., Frost, G.J., Harren, F.J.M., Holloway, J.S., Lefer, B., Lueb, R., Meagher, J.F., Parrish, D.D., Patel, M., Pope, L., Richter, D., Rivera, C., Ryerson, T.B., Samuelsson, J., Walega, J., Washenfelder, R.A., Weibring, P., and Zhu, X. (**2009**) *Environ. Sci. Technol.*, **3**, 2437–2442.

Halliwell, B. and Gutterridge, J.M.C. (eds) (**1989**) *Free Radicals in Biology and Medicine*, Oxford University Press, Oxford.

Hanh, B.D., Neubert, R.H.H., Wartewig, S., and Lasch, J. (**2001**) *J. Controlled Release*, **70**, 393–398.

Herbert, S.K., Han, T., and Vogelmann, T.C. (**2000**) *Photosynth. Res.*, **66**, 13–31.

Hess, P. (**1983**) *Top. Curr. Chem.*, **111**, 1–32.

Hornberger, C., König, M., Rai, S.B., and Demtröder, W. (**1995**) *Chem. Phys.*, **190**, 171–177.

Huszar, H., Pogany, A., Bozoki, Z., Mohacsi, A., Horvath, L., and Szabo, G. (**2008**) *Sens. Actuators B, Chem.*, **134**, 1027–1033.

Jiang, E.Y. (**2002**) *Spectroscopy*, **17**, 23–34.

Kapitanov, V.A., Ponomarev, Y.N., Song, K., Cha, H.K., and Lee, J. (**2001**) *Appl. Phys. B*, **73**, 745–750.

Karbach, A., Röper, J., and Hess, P. (**1983**) *Chem. Phys.*, **82**, 427–434.

Keppler, F., Hamilton, J.T.G., Brass, M., and Rockmann, T. (**2006**) *Nature*, **439**, 187–191.

Kerr, E.L. and Atwood, J.G. (**1968**) *Appl. Opt.*, **7**, 915–921.

Kharitonov, S.A. and Barnes, P.J. (**2001**) *Am. J. Respir. Crit. Care Med.*, **163**, 1693–1722.

Kosterev, A.A. and Tittel, F.K. (**2002**) *IEEE J. Quant. Electron.*, **38**, 582–591.

Kosterev, A.A., Tittel, F.K., Serebryakov, D.V., Malinovsky, A.L., and Morozov, I.V. (**2005**) *Rev. Sci. Instrum.*, **76**, 043105.

Laarhoven, L.J.J., Mulder, P., and Wayner, D.D.M. (**1999**) *Acc. Chem. Res.*, **32**, 342–349.

Lambert, J.D. (ed.) (**1977**) *Vibrational and Rotational Relaxation in Gases*, Clarendon Press, Oxford.

Lehmann, K.K., Scherer, G.J., and Klemperer, W. (**1982**) *J. Chem. Phys.*, **77**, 2853–2861.

Losi, A., Wegener, A.A., Engelhard, M., Gärtner, W., and Braslavsky, S.E. (**1999**) *Biophys. J.*, **77**, 3277–3286.

Luft, K.F. (**1943**) *Z. Tech. Phys.*, **5**, 97–104.

Manzanares, I.C., Mina-Camilde, N., Brock, A., Peng, J., and Blunt, V.M. (**1995**) *Rev. Sci. Instrum.*, **66**, 2644–2651.

Mason, W.P. and Thurston, R.N. (eds) (**1979**) *Physical Acoustics*, vol. **XIV**, Academic Press, New York.

McCurdy, M.R., Bakhirkin, Y., Wysocki, G., Lewicki, R., and Tittel, F.K. (**2007**) *J. Breath Res.*, **1**, 014001.

Moeckli, M.A., Fierz, M., and Sigrist, M.W. (**1996**) *Environ. Sci. Technol.*, **30**, 2864–2867.

Morgando, I. (**2008**) Design and realization of a primary and secondary leak standards for the measurements of leak flow rates of refrigerants, Thesis, L'Ecole Nationale Superieure des Mines de Paris, Paris.

Mur, L.A.J., Santosa, I.E., Laarhoven, L.J.J., Holton, N.J., Harren, F.J.M., and Smith, A.R. (**2005**) *Plant Physiol.*, **138**, 1247–1258.

Nisbet, R.E.R., Fisher, R., Nimmo, R.H., Bendall, D.S., Crill, P.M., Gallego-Sala, A.V., Hornibrook, E.R.C., López-Juez, E., Lowry, D., Nisbet, P.B.R., Shuckburgh, E.F.,

Sriskantharajah, S., Howe, C.J., and Nisbet, E.G. (**2009**) *Proc. R. Soc. B*, **276**, 1347–1354.
Nishioku, Y., Nakagawa, M., Tsuda, M., and Terazima, M. (**2002**) *Biophys. J.*, **83**, 1136–1146.
Olafsson, A., Hammerich, M., Bülow, J., and Henningsen, J. (**1989**) *Appl. Phys. B*, **49**, 91–97.
Patel, C.K.N. and Tam, A.C. (**1981**) *Rev. Mod. Phys.*, **53**, 517–550.
Ramina, A., Chang, C., Giovannoni, J., Klee, H., Terata, P., and Woltering, E. (eds) (**2007**) *Advances in Plant Ethylene Research*, Springer, Dordrecht.
Risby, T.H. and Solga, S.F. (**2006**) *Appl. Phys. B*, **85**, 421–426.
Röntgen, W.C. (**1881**) *Ann. Phys. Chem.*, **1**, 155–159.
Rosencwaig, A. and Gersho, A. (**1976**) *J. Appl. Phys.*, **47**, 64–69.
Schulenberg, P.J., Gaertner, W., and Braslavsky, S.E. (**1994**) *Biochim. Biophys. Acta*, **1185**, 92–96.
Sigrist, M.W. (ed.) (**1994**) *Air Monitoring by Spectroscopic Techniques*, John Wiley & Sons, Inc., New York.
Stella, G., Gelfand, J., and Smith, W.H. (**1976**) *Chem. Phys. Lett.*, **39**, 146–149.
Tam, A.C. (**1983**) Photoacoustics: spectroscopy and other applications, in *Ultrasensitive Laser Spectroscopy* (ed. D. Kliger), Academic Press, New York, pp. 1–108.
Tam, A.C. (**1986**) *Rev. Mod. Phys.*, **58**, 381–431.
Terazima, M., Hara, T., and Hirota, N. (**1995**) *Chem. Phys. Lett.*, **246**, 577–582.
Thibault, F., Le Doucen, R., Rosenmann, L., Hartmann, J.M., and Boulet, C. (**1997**) *Appl. Opt.*, **36**, 563–567.
Tyndall, J. (**1881**) *Proc. R. Soc. London*, **31**, 307–317.
Uotila, J., Koskinen, V., and Kauppinen, J. (**2005**) *Vib. Spectrosc.*, **38**, 3–9.
Vaittinen, O., Saarinen, M., Halonen, L., and Mills, I.M. (**1993**) *J. Chem. Phys.*, **99**, 3277–3287.
Veingerov, M.L. (**1938**) *Compt. Rend. Acad. Sci. USSR*, **19**, 687–688.
Wang, X., Pang, Y., Ku, G., Stoica, G., and Wang, L.V. (**2003**) *Opt. Lett.*, **28**, 1739–1741.

Further Reading

Kinsley, L.E., Fey, A.R., Coppens, A.B., and Sanders, J.V. (**2000**) *Fundamentals of Acoustics*, 4th edn, John Wiley & Sons, Inc., New York.
Mandelis, A. (ed.) (**1992**) *Progress in Photothermal and Photoacoustic Science and Technology*, vols. **1, 2 and 3**, Elsevier.
Miklos, A., Hess, P., and Bozoki, Z. (**2001**) *Rev. Sci. Instrum.*, **72**, 1937–1955.
Morse, P.M. and Ingard, U.K. (**1986**) *Theoretical Acoustics*, Princeton University Press, Princeton.
Tam, A.C. (**1983**) in *Ultra Sensitive Laser Spectroscopy*, Chapter 2 (ed. D. Kliger), Academic Press, New York.
Zharov, V.P. and Letokhov, V.S. (**1986**) *Laser Optoacoustic Spectroscopy*, Springer Series in Optical Science, Vol. 37, Springer-Verlag, Berlin, Heidelberg.

18
Radiation Detectors

Claus Grupen

Handbook of Metrology. Edited by Michael Gläser and Manfred Kochsiek

ISBN: 978-3-527-40666-1

18.1 Introduction

The development of particle detectors practically starts with the discovery of radioactivity by Henri Becquerel in the year 1896. He noticed that the radiation emanating from uranium salts could blacken photosensitive paper. Almost at the same time (1895), X rays, which originated from materials after the bombardment by energetic electrons, were discovered by Wilhelm Conrad Röntgen.

The first nuclear-particle detectors were thus extremely simple. Furthermore, the zinc-sulfide scintillators in use at the beginning of the twentieth century were very primitive. They were used, for example, for the study of scattering of alpha particles from nuclei. This technique required tedious and tiresome optical registration of scintillation light with the human eye.

Scintillations in the form of "northern lights" (aurora borealis) had already been observed since long. This fluorescence is due to charged solar particles entering the Earth's atmosphere along the magnetic field lines at the geomagnetic poles causing excitations of nitrogen and oxygen atoms. Also, already about 50 years before the discovery of Cherenkov radiation, Heaviside (1892) showed that charged particles, moving faster than light in a transparent medium with refractive index n, emit an electromagnetic radiation at a certain angle with respect to the particle direction. Madame Curie, too, noticed a faint light emitted from concentrated solutions of radium in water thereby operating unknowingly the first Cherenkov detector. The human eye can also act as Cherenkov detector. This has been shown by light flashes experienced by astronauts with eyes closed during their space missions. These light emissions are caused by energetic primary cosmic rays passing through the vitreous body of the eye.

In the course of time, the measurement methods were greatly refined. Nowadays, it is generally insufficient only to detect particles and radiation. One wants to identify their nature, that is, one would like to know whether one is dealing, for example, with electrons, muons, pions, or energetic γ rays. In addition, accurate energy and momentum measurements are often required. For most applications, an exact knowledge of the spatial coordinates of particle trajectories is of interest.

The trend of particle detection has shifted in the course of time from optical measurement to purely electronic means. In this development, ever higher resolutions, for example, of time, spatial reconstruction, and energy resolutions have been achieved. Early optical detectors,

Handbook of Metrology. Edited by Michael Gläser and Manfred Kochsiek

ISBN: 978-3-527-40666-1

like cloud chambers, only allowed rates of one event per minute, while modern devices, like fast organic scintillators, can process data rates in the gigahertz regime. Even with gigahertz rates, new problems arise and questions of radiation hardness and aging of detectors become issues.

Originally, particle detectors were used for the study of cosmic rays and in nuclear and particle physics. These devices have since found applications in medicine, biology, environmental science, metrology, radiation protection, oil exploration, civil engineering, archaeology, and arts, to name a few. While the most sophisticated detectors are still developed for particle physics and astroparticles, practical applications often require robust devices that also function in harsh environments.

Radiation detectors have contributed significantly to the advancement of science. New detection techniques, for example, cloud chambers, bubble chambers, multiwire proportional and drift chambers, and micropattern detectors allowed essential discoveries. The development of new techniques in this field was also recognized by a number of Nobel prizes (Charles T.R. Wilson, cloud chamber, 1927; Pavel A. Cherenkov, Ilja M. Frank, Igor J. Tamm, Cherenkov effect, 1958; Donald A. Glaser, bubble chamber, 1960; Luis W. Alvarez, bubble-chamber analysis, 1968; Georges Charpak, multiwire proportional chamber (MWPC), 1992; Raymond Davis Jr., Masatoshi Koshiba, Riccardo Giacconi, novel detection techniques for neutrinos and X rays, 2002).

Only a brief overview over the main radiation detectors can be given in this article. Within the scope of this article, the basic interaction mechanisms, which form the basis of radiation detectors, can only be sketched.

18.2 Historical Detectors

Historical radiation detectors are mainly optical devices that have been used in the early days of cosmic rays and particle physics (P. Galison, 1997). Even though some of these detectors have been "recycled" for recent elementary particle physics experiments, like nuclear emulsions for the discovery of the tau neutrino (2000) or bubble chambers with holographic readout for the measurement of short-lived hadrons, these optical devices are nowadays mainly integrated into demonstration experiments in exhibitions or employed as eye-catchers in lobbies of physics institutes.

18.2.1 Cloud Chambers

The expansion cloud chamber ("Wilson chamber") is one of the oldest detectors for track and ionization measurements. The positron (1932) and the muon (1937) were discovered in a cloud-chamber experiment. A cloud chamber is a container filled with a gas–vapor mixture (e.g., air–water vapor or argon–alcohol) at the vapor saturation pressure. If a charged particle traverses the cloud chamber, it produces an ionization trail. The lifetime of positive ions produced in the ionization process in the chamber gas is relatively long ($\approx$ms). Therefore, after the passage of the particle, a trigger signal, for example, can be derived from a coincidence of scintillation counters, which initiates a fast expansion of the chamber. By means of adiabatic expansion, the temperature of the gas mixture is lowered and the vapor gets supersaturated. It condenses on nuclei, which are represented by the positive ions yielding droplets marking the particle trajectory. The track consisting of droplets is illuminated and photographed (Figure 18.1).

In contrast to the expansion cloud chamber, a diffusion cloud chamber is permanently sensitive. The chamber is, like the expansion cloud chamber, filled with a gas–vapor mixture. A constant temperature gradient provides a region, where the vapor is in a permanently supersaturated state. Charged particles entering this region produce a trail automatically without any additional trigger requirement. Zone widths (i.e., regions in which trails can form) of 5 to 10 cm can be obtained with supersaturated vapor. A clearing field removes the positive ions from the chamber.

18.2.2 Bubble Chambers

The bubble chamber allows the recording and reconstruction of events of high complexity with high spatial resolution. Therefore, it is perfectly suited to study rare events (e.g., neutrino interactions). In a bubble chamber, the liquid gas (e.g., hydrogen, deuterium, neon, or Freon) is held in a pressure container close to the boiling point. Before the expected event, the chamber volume is expanded by retracting

Fig. 18.1 Example of an electromagnetic cascade in a multiplate cloud chamber; (G.D. Rochester, K.E. Turver (1981) Cosmic rays of ultra-high energy. Contemp. Phys., **22**, 425, and private communication by G.D. Rochester 1971; photo credit G.D. Rochester.)

a piston. The expansion of the chamber leads to a reduction in pressure, causing the temperature to exceed the boiling point of the bubble-chamber liquid. If in this superheated liquid state, a charged particle enters the chamber, bubble formation sets in along the particle track. The positive ions produced by the incident particles act as nuclei for bubble formation. The lifetime of these nuclei is too short to trigger the expansion of the chamber by the incoming particles. For this reason, the superheated state has to be reached before the arrival time of the particles. Therefore, bubble chambers can only be used at accelerators, where the arrival time of particles in the detector is known and, therefore, the chamber can be expanded in time.

In the superheated state, the bubbles grow until the growth is stopped by a termination of the expansion. At this moment, the bubbles are illuminated by light flashes and photographed (Figure 18.2). The inner walls of the container have to be extremely smooth so that the liquid "boils" only in those places where bubble formation should occur, namely, along the particle trajectory, and not on the chamber walls. To be able to measure short lifetimes in bubble chambers, the size of the bubbles must be limited. This means that the event under investigation must be photographed relatively soon after the onset of bubble formation when the bubble size is relatively small, thereby guaranteeing a good spatial resolution and, as a consequence, also good lifetime resolution for unstable particles. In any case, the bubble size must be small compared to the decay length of the particle.

18.2.3
Spark Chambers

In a spark chamber a number of parallel plates are mounted in a gas-filled volume. Typically, a mixture of helium and neon is used as counting gas. Alternatively, the plates are either grounded or connected to a high-voltage supply. The high-voltage pulse is normally triggered to every second electrode by a coincidence between two scintillation counters placed above and below the spark chamber. The gas

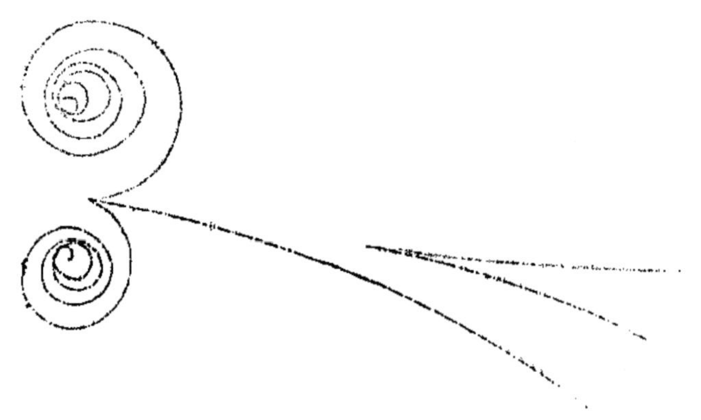

Fig. 18.2 This event shows the creation of an e^+e^- pair by a photon in the Coulomb field of an electron and an e^+e^- pair created in the field of a nucleus. In the first case, the target electron gets a substantial recoil and leaves a high-momentum track; F. Close and M. Marten 1987.

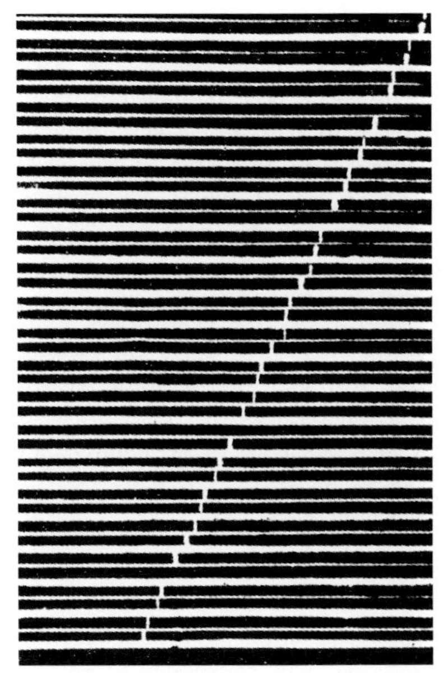

Fig. 18.3 Track of a cosmic-ray muon in a multiplate spark chamber; (V.S. Kaftanov, V.A. Liubimov (1963) Spark chamber use in high energy physics. Nucl. Instrum. Methods, **20**, 195, and private communication by V.S. Kraftanov 1995; photo credit V.S. Kraftanov.)

amplification is chosen in such a way that a spark discharge occurs at the point of the passage of the particle (Figure 18.3). This is obtained for gas amplifications between 10^8 and 10^9. For lower gas amplifications, sparks will not develop, while for larger gas amplifications sparking at unwanted positions (e.g., at spacers that separate the plates) can occur. The discharge channel essentially follows the electric field. Up to an angle of 30°, the conducting plasma channel can, however, follow the particle trajectory (track spark chamber). The ions produced between two discharges are removed from the detector volume by means of a clearing field. If the time delay between the passage of the particle and the high-voltage signal is less than the memory time of about 10 μs, the efficiency of the spark chamber is close to 100%. A clearing field, of course, removes also the primary ionization from the detector volume. For this reason, the time delay between the passage of the particle and the application of the high-voltage signal has to be chosen as short as possible to reach full efficiency. Furthermore, the rise time of the high-voltage pulse must be short because otherwise the leading edge acts as a clearing field before the critical field strength for spark formation is reached.

18.2.4 Nuclear Emulsions

Tracks of charged particles in nuclear emulsions can be recorded by the photographic method. Nuclear emulsions consist of fine-grained silver-halide crystals (AgBr and AgCl), which are embedded in a gelatine substrate. A charged particle produces a latent image in the emulsion. Owing to the free charge carriers liberated in the ionization process some halide molecules are reduced to metallic silver in the emulsion. In the subsequent development process, the silver-halide crystals are chemically reduced. This preferentially affects those microcrystals (nuclei) that are already disturbed and partly reduced. These are transformed into elemental silver. The process of fixation dissolves the remaining silver halide and removes it. Thereby the charge image, which has been transformed into elemental silver particles, remains stable.

The evaluation of the emulsion is usually done under a microscope by eye, but it can also be performed by using a charged-coupled-device (CCD) camera and a semiautomatic pattern-recognition device. Fully automated emulsion-analysis systems have also been developed. The sensitivity of the nuclear emulsion must be high enough so that the energy loss of minimum-ionizing particles is sufficient to produce individual silver-halide microcrystals along the track of a particle. Commercially available photoemulsions do not have this property. Furthermore, the silver grains that form the track and also the silver-halide microcrystals must be sufficiently small to enable a high spatial resolution. The requirements of high sensitivity and low grain size are in conflict and, therefore, demand a compromise. In most nuclear emulsions, the silver grains have a size of 0.1 to 0.2 μm and so are much smaller than in commercial films (1–10 μm). The mass fraction of the silver halide (mostly AgBr) in the emulsion amounts to approximately 80%. Nuclear emulsions allow high spatial resolution of complex events with large multiplicities. They are permanently sensitive and cannot be triggered.

18.2.5 Track-Etch Detectors

Particles of high electric charge destroy the local structure in a solid along their tracks. This local destruction can be intensified by etching and thereby made visible. Solids such as inorganic crystals, glasses, plastics, minerals, or even metals can be used for this purpose. The damaged parts of the material react with the etching agent more intensively than the undamaged material, and characteristic etch cones will be formed. If the etching process is not interrupted, the etch cones starting from the surface will merge and form a hole at the point of the particle track. The etching procedure will also remove some part of the surface material.

Track-etch detectors show a threshold effect: the minimum radiation damage caused by protons and α-particles is frequently insufficient to produce etchable tracks. The detection and measurement of heavy ions, for example, in primary cosmic rays ($Z \geq 3$), will consequently not be disturbed by a high background of protons and α-particles. The size of the etch cones (for a fixed etching time) is a measure of the energy loss of the particles. It allows, therefore, if the velocity of the particles is known, a determination of the charge of the nuclei. A stack of track-etch detectors, flown in a balloon at a residual atmosphere of several grams per square centimeter, thus permits a determination of the elemental abundance in primary cosmic rays. Specially prepared plastic materials can also be made sensitive to α-particles and neutrons. Such detectors are used in the field of radiation protection for radon monitoring and neutron measurement.

18.3 Gas Detectors

18.3.1 Bethe–Bloch Energy-Loss Formula

Particles and radiation cannot be detected directly, but rather only through their interactions with matter. There are specific interactions for charged particles, which are different from those of neutral particles, for example, of photons. The variety of interactions is quite rich and, as a consequence, a large number of detection devices for particles and radiation exist. The most important energy-loss mechanism for charged particles in gas detectors is the ionization loss, which can be described by the Bethe–Bloch formula, which represents the energy loss dE per length dx. It is given by (Particle Data Group W.-M. Yao *et al.* 2006)

$$-\frac{dE}{dx} = 4\pi N_A r_e^2 m_e c^2 z^2 \frac{Z}{A}\frac{1}{\beta^2} \times \left(\ln\frac{2m_e c^2\gamma^2\beta^2}{I} - \beta^2 - \frac{\delta}{2}\right) \quad (18.1)$$

where

z – charge of the incident particle in units of the elementary charge;
Z, A – atomic number and atomic weight of the absorber;
m_e – electron mass;
r_e – classical electron radius;
N_A – Avogadro number (= number of atoms per mole);
I – ionization constant, characteristic of the absorber material;
β – velocity of the particle v normalized to the velocity of light c, $\beta = v/c$;
γ – Lorentz factor $\gamma = 1/\sqrt{1-\beta^2}$;
δ – a parameter that describes how much the extended transverse electric field of incident relativistic particles is screened by the charge density of the atomic electrons. In this way, the energy loss is reduced ("density effect", "Fermi plateau" of the energy loss).

A useful constant appearing in the Bethe–Bloch equation is

$$4\pi N_A r_e^2 m_e c^2 = 0.3071\ \text{MeV}\,\text{cm}^2\text{g}^{-1}.$$

In the low-energy domain the energy loss decreases like $1/\beta^2$ and reaches a broad minimum of ionization near $\beta\gamma \approx 4$. Relativistic particles ($\beta \approx 1$), which have an energy loss corresponding to this minimum, are called *minimum-ionizing particles*. In light absorber materials, where the ratio $Z/A \approx 0.5$, the energy loss of minimum-ionizing particles can be roughly represented by

$$-\left.\frac{dE}{dx}\right|_{\min} \approx 2\ \text{MeV}\,\text{cm}^2\text{g}^{-1} \quad (18.2)$$

The energy loss increases again for $\beta\gamma > 4$ ("logarithmic rise" or "relativistic rise"). The increase follows approximately a dependence like $\ln\gamma$ (see Figure 18.4). For thin absorbers the ionization energy loss is subject to asymmetrical Landau fluctuations.

18.3.2 Ionization Yield

An interesting value for gas detectors is the ionization yield. One must distinguish between primary ionization, that is, the number of produced primary electron–ion pairs, and the total ionization. A sufficiently large amount of energy can be transferred to some primary electrons so that they also can ionize (knock-on electrons or δ rays). This secondary ionization together with the primary ionization forms the total ionization. The average energy required to form an electron–ion pair (W value) exceeds the ionization potential of the gas because inner shells of the gas atoms can also be involved in the ionization process, and a fraction of the energy of the incident particle can be dissipated by excitation processes that do not lead to free electrons. The W value of a material is constant for relativistic particles and

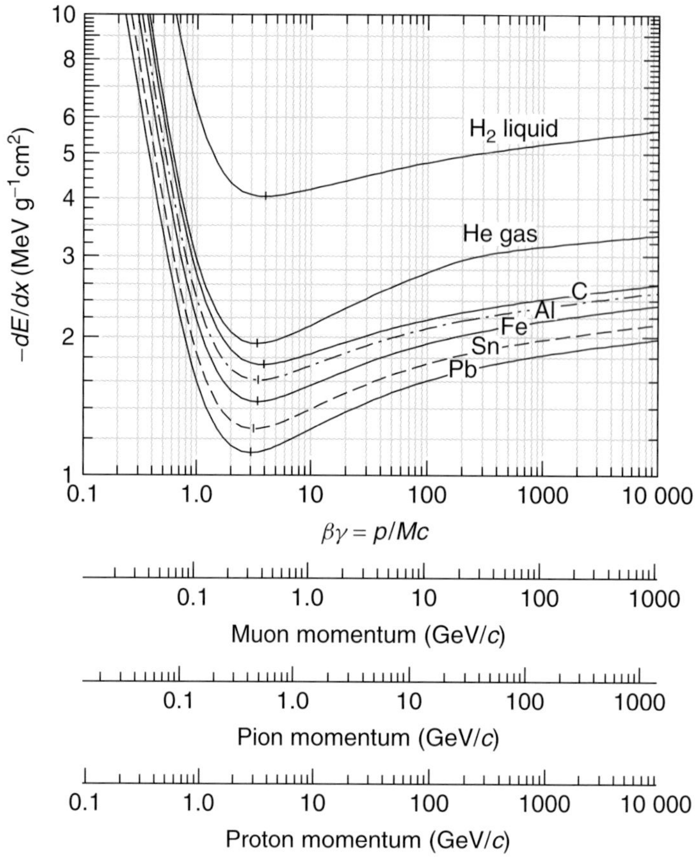

Fig. 18.4 Mean energy loss in liquid hydrogen, gaseous helium, carbon, aluminum, iron, tin, and lead. p is the momentum of the particle, M its mass, and c the velocity of light; Particle Data Group, W.-M. Yao *et al.* 2006.

increases only slightly for low velocities of incident particles.

For gases the W values are around 30 eV. They can, however, strongly depend on impurities in the gas. This number has to be compared to the average energy required for the production of an electron–hole pair in solid-state radiation detectors, which is only around 3.5 eV. Therefore, the statistical fluctuations in the number of produced charge carriers for a given energy loss are much smaller in solid-state detectors than in gaseous detectors.

18.3.3 Ionization Chambers

An ionization chamber is a gaseous detector that measures the amount of ionization produced by a charged particle passing through the gas volume. Neutral particles can also be detected by this device via secondary charged particles resulting from the interaction of the primary ones with electrons or nuclei. Charged particles are measured by separating the charge-carrier pairs produced by their ionization in an electric field and guiding the ionization products to the anode or cathode, respectively, where corresponding signals can be recorded. If a particle is totally absorbed in an ionization chamber, such a detector type measures its energy.

In the simplest case, an ionization chamber consists of a pair of parallel electrodes mounted in a gas-tight container that is filled with a gas mixture, which allows electron and ion drift. Charged particles produce electron–ion pairs along their track. The number of created charge carriers depends on the particle type and energy, and on the gas.

The parallel electrodes of the ionization chamber producing a homogeneous electric field act as a capacitor, which is initially charged to a certain voltage. The drifting charge carriers induce an electric charge on the electrodes, which leads to a certain change of the voltage, ΔU. The signal amplitude ΔU has contributions from fast-moving electrons and the slowly drifting ions. The disadvantage is that the output signal depends on where the particles enter the ionization chamber. This can be solved by introducing a Frisch grid, permitting the particles to enter the chamber only in a well-defined region.

Ionization chambers are mainly used in the spectroscopy of particles of higher charge, because in this case the deposited energies are, in general, larger (typically 5 MeV for α-particles from radioactive sources) compared to those of singly charged minimum-ionizing particles. A minimum-ionizing particle passing a layer of 4 cm of argon will deposit only about 11 keV, which provides about 400 electron-ion pairs. To detect such a small signal is a very difficult task.

Apart from planar ionization counters, cylindrical ionization counters are also in use. Because of the cylindrical arrangement of the electrodes, the electric field in this case is no longer constant, but increases inversely proportional to the distance to the anode, which is usually a thin wire. The important point with cylindrical ionization chambers is that no multiplication of charge carriers occurs in the gas. Only the originally produced number of electrons and ions is collected.

18.3.4 Proportional Counters

In contrast to ionization chambers, the high voltage in proportional chambers is increased so that the initially produced charge carriers gain enough energy between two collisions to produce secondary electrons. In this way, an avalanche of secondary and tertiary charge carriers is started.

If N_0 primary electrons are produced, the number of particles $N(x)$ at the point x is calculated to be

$$N(x) = N_0 \, e^{\alpha x} \tag{18.3}$$

The first Townsend coefficient α depends on the field strength $\boldsymbol{E}$ and the pressure in the gas counter. The proportional range of a counter is characterized by the fact that the gas-amplification factor takes a constant value. As a consequence, the measured signal is proportional to the produced ionization. Gas-amplification factors of up to 10^6 are possible in the proportional mode. Typical gas amplifications are usually in the range 10^4–10^5. At high field, collisions of electrons with atoms or molecules can cause not only ionization but also excitation. De-excitation is often followed by photon emission. However, photons produced in the course of the avalanche development are of no importance in the proportional regime.

Since the output signal in a proportional counter is proportional to the energy loss of the particle (or to the energy, if the particle stops in the counter), these radiation detectors can be used for spectroscopy, for example, of X rays or α-particles.

If many anode wires are stretched in a planar chamber in parallel, one arrives at a multi-wire proportional chamber (MWPC). The working principle of an MWPC is the same as the one for a proportional counter. In addition to energy-proportional signals, MWPCs provide accurate track information if the wires are stretched with relatively small separations. They can also be used in the field of radiation protection as contamination monitors to measure possible contamination over larger areas.

18.3.5 Geiger–Müller Counters

The increase of the field strength in a proportional counter leads to a copious production of photons during the avalanche formation. As a consequence, the probability to produce further new electrons by the photoelectric effect increases. This photoelectric effect can also occur at points distant from the region of production of the primary avalanche. These electrons liberated by the photoelectric effect will initiate new avalanches whereby the discharge will propagate along the anode wire, in contrast to the proportional chamber, where the discharge is localized to the point of particle passage.

The discharge is normally stopped by "quenching". In self-quenching counters a quench gas is admixed to the counting gas, which, in most cases, is a noble gas. Hydrocarbons like methane (CH_4), ethane (C_2H_6), isobutane (iC_4H_{10}), alcohols like ethyl alcohol (C_2H_5OH) or methylal ($CH_2(OCH_3)_2$), or halides, like ethylbromide, are suitable as quenchers. These additions will absorb photons in the ultraviolet range (wavelength 100–200 nm) thereby reducing their range to a few wire radii ($\approx$100 μm). The transverse propagation of the discharge proceeds only along and in the vicinity of the anode wire because of the short range of the photons. The photons have no chance to liberate electrons from the cathode by the photoelectric effect because they will be absorbed before they can reach the cathode. After a flux tube of positive ions has been formed along the anode wire, the external field is reduced by this space charge by such an amount that the avalanche development comes to an end. The positive ions drifting in the direction of the cathode will collide on their way with quench-gas molecules, thereby becoming neutralized. The molecule ions, however, have insufficient energy to liberate electrons from the cathode upon impact. Consequently, the discharge stops by itself.

18.3.6 Streamer Tubes

In Geiger counters the fraction of counting gas to quenching gas is typically 90 : 10. The anode wires have diameters of 30 μm and the anode voltage is around 1 kV. If the fraction of the quenching gas is considerably increased, the lateral propagation of the discharge along the anode wire can be completely suppressed. One again obtains, as in the proportional counter, a localized discharge with the advantage of large signals (gas amplification $\geq 10^{10}$ for sufficiently high anode voltages), which can be processed without any additional preamplifiers. These streamer tubes, sometimes also called *Iarocci tubes*, are operated with "thick" anode wires between 50 μm and 100 μm. Gas mixtures with $\leq$60% argon and $\geq$40% isobutane can be used. Streamer tubes operated with pure isobutane also proved to function well.

18.3.7
Drift Chambers

MWPCs allow only modest spatial resolutions. For a wire spacing of only 2 mm resolutions around 500 μm are achievable. The number of anode wires to cover large areas with MWPCs soon becomes prohibitive. A way out represents the drift chamber. The time Δt between the moment of the particle passage through such a chamber and the arrival time of the charge cloud at the anode wire depends on the point of passage of the particle through the chamber. If v^- is the constant drift velocity of the electrons, a linear relation between the drift time and the distance to the anode wire holds. The precise measurement of this drift time allows to determine the spatial coordinate with high accuracy. The resolution is only limited by longitudinal diffusion and by primary ion statistics. Another advantage of drift chambers is that the anode-wire spacing can be much larger than in MWPCs, typically several centimeters. With an appropriate time readout, spatial resolutions under 100 μm are achievable. The drift velocity depends on the gas mixture, with typical values around several centimeters per microsecond. The consequence is that this detector does not permit high repetition rates. The drift chamber cannot distinguish between particles passing to the right or to the left of the readout wire. This left–right ambiguity can easily be resolved by a double layer of chambers with staggered anode wires. To produce a suitable drift field, potential wires are introduced between neighboring anode wires.

Drift chambers can be made planar to cover large areas, such as muon chambers for fixed-target and collider experiments, but cylindrical arrangements are often in use as vertex detectors in e^+e^- colliders.

18.3.8
Time-Projection Chambers (TPCs)

The *crème de la crème* of track recording in cylindrical and planar gas detectors at the moment is realized with the time-projection chamber (TPC). Apart from the counting gas, this detector contains no other constructional elements and thereby represents the optimum as far as minimizing multiple scattering and photon conversions are concerned. The TPC is usually divided into two halves by means of a central electrode. A typical counting gas is a mixture of argon and methane (90 : 10).

The ionization produced by charged particles drifts in the electric field, in the direction of the endfaces of the chamber, which in most cases consist of MWPCs. For cylindrical chambers, the electric field is usually supplemented by a parallel magnetic field. The magnetic field suppresses the diffusion perpendicular to the field. This is achieved by the action of the magnetic forces on the drifting electrons which, as a consequence, spiral around the direction of the magnetic field. For typical values of electric and magnetic field strengths, Larmor radii below 1 μm are obtained. The MWPCs at the endfaces determine the projected coordinates x and y. The arrival time of the ionization electrons at the endplates supplies the z coordinate along the cylinder axis. Therefore the TPC is a truly three-dimensional imaging device (see Figure 18.5).

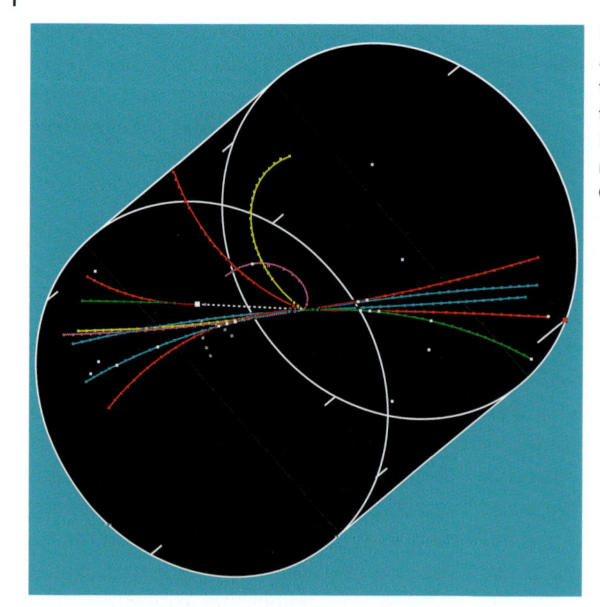

Fig. 18.5 Two-jet event from an electron–positron annihilation into hadrons as seen in the ALEPH TPC at the Large Electron–Positron storage ring (LEP); at CERN; ALEPH Collaboration, 2000.

The gas amplification of the drifting ionization takes place at the anode wires of the MWPC, which are mostly stretched in azimuthal direction. To obtain accurate three-dimensional coordinates the cathodes of the endcap MWPC segments are usually structured as pads. The number of the fired wire and pad provides the coordinates x and y.

In addition, the analog signals on the anode wires give information on the specific energy loss and can consequently be used for particle identification. Typical values of the magnetic field are around 1.5 Tesla and around 20 kV/m for the electric field. Since in most constructions electric and magnetic field are parallel, thc Lorentz angle is zero and the electrons drift parallel to $\boldsymbol{E}$ and $\boldsymbol{B}$ (there is no "$\boldsymbol{E} \times \boldsymbol{B}$ effect").

A problem, however, is caused by the large number of positive ions, which are produced in the gas-amplification process at the endplates and which have to drift a long way back to the central electrode. The strong space charge of the drifting positive ions would cause the field quality to deteriorate. This can be overcome by introducing an additional grid ("gate") between the drift volume and the endcap MWPC. The gate is normally closed. It is only opened for a short period of time if an external trigger signals an interesting event. In the closed state, the gate prevents ions from drifting back into the drift volume. Thereby the quality of the electric field in the sensitive detector volume remains unchanged. This means that the gate serves a dual purpose. On the one hand, electrons from the drift volume can be prevented from entering the gas-amplification region of the endcap MWPC if there is no trigger that would signal an interesting event. On the other hand – for interesting gas-amplified events – the positive ions are prevented from drifting back into the detector volume. TPCs can be made very large (diameter ≥ 3 m, length ≥ 5 m). They contain a large number of analog readout channels (number of anode wires ≈ 5000 and cathode pads $\approx 50\,000$). Several hundred energy-loss samples can be obtained per track, which ensures an excellent determination of the radius of curvature and allows an accurate measurement of the energy loss, which is essential for particle identification. The drawback of the TPC is the fact that high particle rates cannot be handled, because the drift time of the electrons in the detector volume amounts to 40 μs (for a drift path of 2 m) and the readout of the analog information also requires several μs.

18.3.9 Micropattern Gaseous Detectors

The construction of MWPCs would be simplified and their stability and flexibility would be greatly enhanced if anodes were made in the form of strips or dots on insulating or semiconducting surfaces instead of stretching anode wires in the counter volume. The rate capability would improve by more than one order of magnitude for these devices because of the short drift times. At present, the class of micropattern gaseous detectors is already rather wide and many new promising devices are under study.

These microstrip gas detectors (MSGC) are miniaturized MWPCs, in which the dimensions are reduced by about a factor of 10 in comparison to conventional chambers. The typical pitch is 100–200 μm and the gas gap ranges from 2 to 10 mm. This has been made possible because the electrode structures can be reduced with the help of electron lithography. The wires are replaced by strips that are evaporated onto a thin substrate. Cathode strips arranged between the anode strips allow for an improved field quality and a fast removal of positive ions. The segmentation of the otherwise planar cathodes in the form of strips or pixels also permits two-dimensional readout. The electrode structures can be mounted on ceramic substrates or, alternatively, they can also be arranged on thin plastic foils. In this way, even light, flexible detectors that exhibit a high spatial resolution can be constructed. Possible disadvantages lie in the electrostatic charging-up of the insulating plastic structures, which can lead to time-dependent amplification properties because of the modified electric fields. The obvious advantages of these microstrip detectors – apart from their excellent spatial resolution – are the low dead time (the positive ions being produced in the avalanche will drift the very short distance to the cathode strips in the vicinity of the anodes), the reduced radiation damage (because of the smaller sensitive area per readout element), the high-rate capability, and the reduced occupancy.

Another structure providing charge multiplication is the gas electron multiplier (GEM). This is a thin (≈50 μm) insulating kapton foil coated with a metal film on both sides. It contains chemically produced holes 50–100 μm in diameter with 100–200 μm pitch. The metal films have different potential to allow gas multiplication in the holes. A GEM-based detector contains a drift cathode separated from one or several GEM layers and an anode readout structure to read out the ionization information. The readout is then done in a similar way as in most other micropattern detectors.

18.3.10 Transition-Radiation Detectors (TRD)

Transition radiation is emitted when a charged particle traverses a medium with discontinuous dielectric constant. A charged particle moving toward a boundary, where the dielectric constant changes, can be considered to form together with its mirror charge an electric dipole whose field strength varies in time. The time-dependent dipole field causes the emission of electromagnetic radiation. This emission can be understood in such a way that although the normal component of the dielectric displacement $\boldsymbol{D} = \varepsilon\varepsilon_0 \boldsymbol{E}$ varies continuously in passing through a boundary, the electric field does not. The energy radiated from a single boundary (transition from vacuum to a medium with dielectric constant ε) is proportional to the Lorentz factor of the incident charged particle. This provides an excellent possibility to identify the nature of this particle.

Since the number of transition-radiation photons produced per boundary is rather low, usually a stack of transparent foils is employed. The generated photons – preferentially in the X-ray domain – are detected in a standard MWPC or drift chamber. It is important that the generator foils are made of low-Z material, so that they do not absorb the X-ray transition-radiation photons, and the counting gas in the MWPC is of high Z (e.g., xenon) to efficiently detect them. Transition-radiation detectors (TRDs) are mostly used for particle-identification purposes. This is related to the fact that the yield of transition-radiation photons is proportional to the Lorentz factor (i.e., energy) of the particle rather than the velocity, which saturates at high energies.

18.3.11 Neutron Counters

Neutrons, just as photons, can only be detected indirectly. Depending on the neutron energy, various reactions can be considered, which produce charged particles that are then detected via their ionization or scintillation in standard radiation detectors.

1. Low kinetic energies (<20 MeV)

$$\begin{array}{rcl} n + {}^{6}Li & \rightarrow & \alpha + {}^{3}H \\ n + {}^{10}B & \rightarrow & \alpha + {}^{7}Li \\ n + {}^{3}He & \rightarrow & p + {}^{3}H \\ n + p & \rightarrow & n + p \end{array} \tag{18.4}$$

   The conversion can be performed in a proportional counter with BF_3 or ^{3}He filling or a mixture with an addition of hydrocarbons, like CH_4. Also a scintillator consisting of suitable conversion material like LiI(Tl) can be used.
2. Medium energies (20 MeV $\leq E_{\text{kin}} \leq$ 1 GeV)
   The (n, p) recoil reaction can be used for neutron measurements in detectors that contain many quasi-free protons in their sensitive volume (e.g., hydrocarbons, like CH_4).
3. High energies ($E_{\text{kin}} > 1$ GeV)
   Neutrons of high energy initiate hadron cascades in inelastic interactions, which are easy to identify in hadron calorimeters.

Neutrons are detected with relatively high efficiency at very low energies. Therefore, it is often useful to slow down neutrons with substances containing many protons ("moderation") because neutrons can transfer a large amount of energy to collision partners of the same mass. In the field of radiation protection, it is also important to measure the neutron energy (e.g., by threshold counters) because the relative biological effectiveness depends on it.

18.4 Scintillators

Scintillator materials can be inorganic crystals, organic liquids, or plastics, and gases.

The scintillation mechanism in inorganic crystals is an effect of the lattice. Incident particles can transfer energy to the lattice by creating electron–hole pairs or taking electrons to higher energy levels below the conduction band. Recombination of electron–hole pairs may lead to the emission of light. Also electron–hole bound states (excitons) moving through the lattice can emit light when hitting an activator center and transferrring their binding energy to activator levels, which subsequently de-excite. Important parameters for inorganic scintillators are the light yield, the emission spectrum, and the decay time of the light signal. In thallium-doped NaI crystals about 25 eV are required to produce one scintillation photon. The decay time in inorganic scintillators can be quite long (e.g., 1 μs in CsI(Tl)), but there are also fast scintillators like BaF_2 with a decay component of 0.7 ns. BaF_2 has its emission maximum in the ultraviolet and therefore requires quartz windows for the readout with photomultipliers. If scintillators are used as sampling elements in electromagnetic calorimeters, materials with a short radiation length X_0, like bismuth germanate (BGO: $Bi_4Ge_3O_{12}$), cadmium or lead tungstate ($CdWO_4$, $PbWO_4$) or cerium-activated lutetium oxyorthosilicate (LSO: $Lu_2SiO_5(Ce)$) are required.

The scintillation mechanism is different in organic substances. Certain types of molecules will release a small fraction (≈3%) of the absorbed energy as optical photons. This process is especially marked in organic substances containing aromatic rings, such as polystyrene, polyvinyltoluene, and naphthalene. Liquids that scintillate include toluene and xylene. The primary scintillation light is preferentially emitted in the UV range. The absorption length for UV photons in the scintillation material is rather short: the scintillator is not transparent for its own scintillation light. Therefore, this light is transferred to a wavelength shifter, which absorbs the UV light and re-emits it at longer wavelengths (e.g., in the green). Owing to the lower concentration of the wavelength-shifter material the re-emitted light can get out of the scintillator and be detected by a photosensitive device (see Figure 18.6). The technique of wavelength

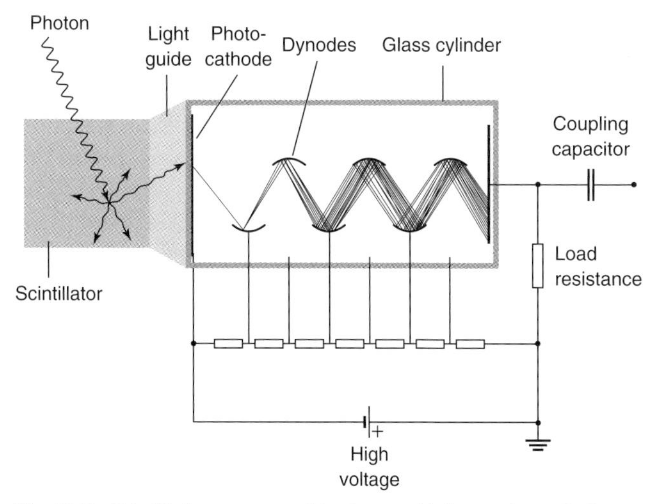

Fig. 18.6 Scintillation counter with photomultiplier readout; C. Grupen 2008/2010.

shifting is also used to match the emitted light to the spectral sensitivity of the photomultiplier. For plastic scintillators, the primary scintillator and wavelength shifter are mixed with an organic material to form a polymerizing structure. In liquid scintillators, the two active components are mixed with an organic base.

About 100 eV are required to produce one photon in an organic scintillator. The decay time of the light signal in plastic scintillators is, in most cases, substantially shorter compared to inorganic substances (e.g., 30 ns in naphthalene and around 2 to 3 ns in polystyrene). Also polystyrene-based scintillators containing soluted organic compounds of heavy elements with ultrashort decay times (0.5 ns) for fast-timing purposes are available.

Gas scintillators are used in cosmic rays for extensive air-shower experiments, where the atmospheric nitrogen is used as a fluorescence agent. Because of the low light absorption in gases there is no need for wavelength shifting.

Plastic scintillators do not respond linearly to the energy-loss density. The number of photons produced by charged particles is described by Birk's semi-empirical formula

$$N = N_0 \frac{dE/dx}{1 + k_B dE/dx} \tag{18.5}$$

where N_0 is the photon yield at low specific ionization density and k_B is Birk's density parameter. For low energy losses Equation (18.5) leads to a linear dependence

$$N = N_0 \cdot dE/dx \tag{18.6}$$

while for very high dE/dx saturation occurs at

$$N = N_0/k_B \tag{18.7}$$

18.5 Solid-State Detectors

Solid-state detectors are essentially ionization chambers with solids as a counting medium. Because of their high density compared to gaseous detectors, they can absorb particles of correspondingly higher energy. Charged particles or photons produce electron–hole pairs in a crystal. An electric field applied across the crystal allows the produced charge carriers to be collected.

The operating principle of solid-state detectors can be understood from the band model of solids. According to this theory, the electron energy levels of individual atoms or ions within a crystal are unified forming energy bands. According to the Pauli principle, each band can contain a finite number of electrons. So, some low-energy bands are fully filled with electrons while the high-energy bands are empty, at least at low temperature. The lowest partially filled or empty band is called the *conduction band* while the highest fully filled band is referred to as the *valence band*. The gap between these bands is called the *forbidden band* or *band gap*, E_g.

When the "conduction band" is partially filled, electrons can move easily under the influence of an electric field, hence, this solid is a conductor. Such a material cannot be used as an ionization counter. The solids that have basically empty conduction bands are divided conventionally into insulators (specific resistivity 10^{14}–$10^{22}\ \Omega \cdot \text{cm}$ at room temperature) and semiconductors (10^{9}–$10^{-2}\ \Omega \cdot \text{cm}$). The electric charge in these materials is carried by electrons, which have been excited to the conduction band from the valence band. The corresponding vacancies in the valence band are called *holes* and they are able to drift in the electric field as well.

The common solid-state ionization counters are based on semiconductors. The specific resistivity of the material is determined as

$$\varrho = \frac{1}{e(n\mu_e + p\mu_p)} \qquad (18.8)$$

where n and p are electron and hole concentrations, respectively, while μ_e, μ_p are their mobilities and e is the elementary charge. In a pure semiconductor the electron concentration, n, is equal to the hole concentration, p. Electron and hole concentrations in solid-state detectors are intentionally changed by special doping. A germanium or silicon crystal becomes n-conducting (i.e., conducting for negative charge carriers, electrons), when electron-donor impurities are introduced into the lattice. Germanium and silicon have four electrons in the outer shell. Atoms with five electrons in the outer shell act as electron-donor impurities. Similarly, germanium and silicon become p-conducting (i.e. conducting for holes) if trivalent atoms, which act as electron-acceptor impurities, are introduced into the crystal lattice. Phosphor and arsenic act as electron donors. The neighboring silicon (germanium) atoms can only bind four electrons. The fifth electron of the electron-donor impurity is only weakly bound and can easily reach the conduction band. The donor levels are situated approximately 0.05 eV below the edge of the conduction band. If trivalent electron-acceptor impurities like boron or indium are added to the lattice, one of the silicon bonds remains incomplete. This acceptor level, which is about 0.05 eV above the edge of the valence band, tries to attract one electron from a neighboring silicon atom. As a consequence, the state of the missing electron (the hole) migrates through the crystal. The addition of small amounts of energy can cause the holes to be transferred to the valence band, thereby causing a hole current. Lithium acts as an electron donor because it has only one weakly bound electron in the outer shell.

If a charged particle traverses an n- or p-conducting crystal, it will produce electron–hole pairs along its track. The primary electrons can produce further secondary electron–hole pairs or excite lattice vibrations. A plasma channel is produced along the particle track with charge-carrier concentrations from 10^{15} up to $10^{17}\,\mathrm{cm}^{-3}$. The working principle of a solid-state detector now consists of collecting the free charge carriers in an external drift field before they can recombine with the holes. If this is successful, the measured charge signal is proportional to the energy loss of the particle or, if the particle deposits its total energy in the sensitive volume of the detector, it is proportional to the particle energy.

Solid-state counters can be used in nuclear physics for α, β, and γ spectroscopy. High-purity germanium detectors provide excellent energy resolution at the level of $\Delta E/E \approx 10^{-3}$ in the mega-electronvolt range (see Figures 18.7 and 18.8). Outstanding spatial resolutions of about $10\,\mu\mathrm{m}$ can also be achieved if the readout planes of silicon detectors are segmented into strips or pixels. This, however, requires a large number of readout channels equipped with low-noise charge-sensitive amplifiers. The number of readout elements can be reduced if, in addition, the drift time from the moment of charge generation to the arrival time of the drifting charge at the readout element is measured (silicon drift chamber).

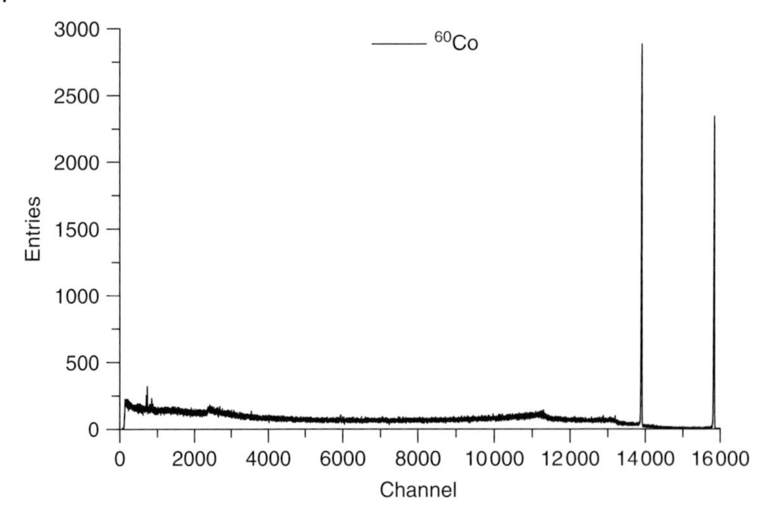

Fig. 18.7 γ-ray spectrum of ^{60}Co measured in a high-purity Ge detector. The energy resolution of the 1.17 MeV and 1.32 MeV γ-ray lines is rather impressive; (J. Hausmann (2002); University of Siegen, private communication photo credit Joachim Hausmann.)

For the operation of silicon pixel counters in harsh radiation environments aging effects also become an issue.

Even better energy resolutions can be achieved if quantum transitions, which require less energy than ionization or electron–hole pair creation, are used. For example, phonons in solid-state materials have energies around 10^{-5} eV for temperatures of 100 mK. Other types of quasiparticles at low temperature are Cooper pairs in a superconductor, which are bound states of two electrons with opposite spin that behave like bosons. Cooper pairs in superconductors have binding energies in the range between $4 \cdot 10^{-5}$ eV (Ir) and $3 \cdot 10^{-3}$ eV (Nb). Thus, even extremely low-energy depositions would produce or break up a large number of phonons or Cooper pairs, respectively. To avoid thermal excitations of these quantum processes, such radiation detectors, however, would have to be operated at extremely low temperatures, typically in the millikelvin range. For this reason, such devices are called *cryogenic detectors*. Cryogenic detectors can be subdivided into two main categories: detectors for quasiparticles in superconducting crystals, and phonon detectors in insulators. One detection method is based on the fact that the superconductivity of a substance is destroyed by the energy deposition if the detector element is sufficiently small. This is the working principle of superheated superconducting granules. The transition from the superconducting to the normal-conducting state can be detected by pick-up coils coupled to very sensitive preamplifiers or by superconducting quantum interference devices (SQUIDs). These quantum interferometers are extremely sensitive radiation detectors.

In contrast to Cooper pairs, phonons, which can be excited by energy deposition in insulators, can be detected with the methods of classical calorimetry. If ΔE is the absorbed energy, this results in a

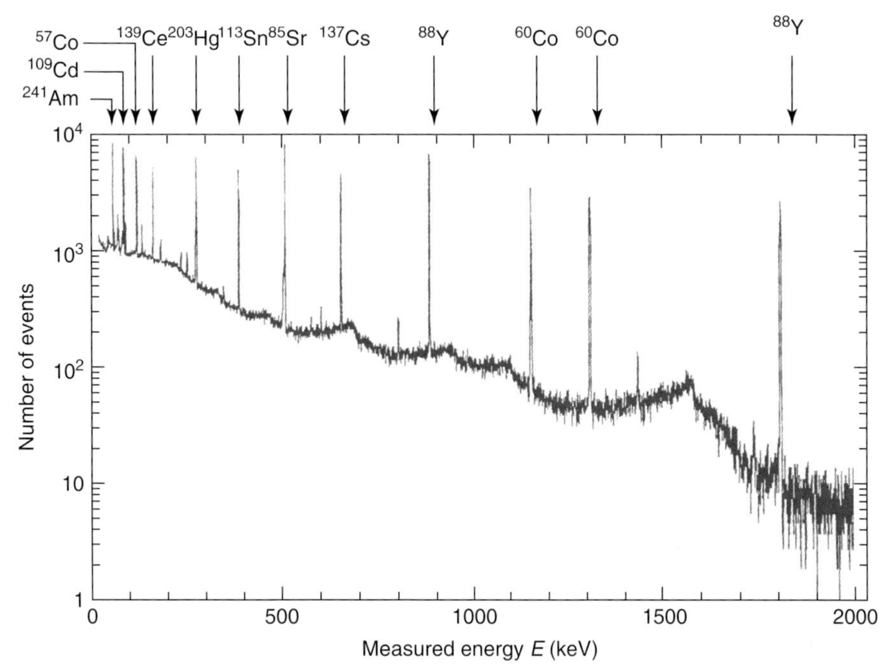

Fig. 18.8 Photopeak identification in a radioisotope sample using a high-purity germanium detector; C. Grupen 2008 and O. Kalthoff 1996.

temperature rise of

$$\Delta T = \Delta E / mc \tag{18.9}$$

where c is the specific heat capacity and m the mass of the detector. If these measurements are performed at very low temperatures, where c can be very small (the lattice contribution to the specific heat is proportional to T^3 at low temperatures), this method can be used to detect individual particles of extremely low energy.

18.6 Cherenkov Detectors

A charged particle, traversing a medium with refractive index n with a velocity v exceeding the velocity of light c/n in that medium, emits a characteristic electromagnetic radiation, called *Cherenkov radiation*. Cherenkov radiation is emitted because the charged particle polarizes atoms along its track so that they become electric dipoles. The time variation of the dipole field leads to the emission of electromagnetic radiation. As long as $v < c/n$, the dipoles are symmetrically arranged around the particle path, so that the dipole field integrated over all dipoles vanishes and no radiation occurs. If, however, the particle moves with $v > c/n$, the symmetry is broken resulting in a nonvanishing dipole moment, which leads to the radiation. The contribution of Cherenkov radiation to the energy loss is small compared to that from ionization and excitation even for minimum-ionizing particles. For gases with $Z \geq 7$ the energy loss by Cherenkov radiation amounts to less than 1% of the ionization loss of minimum-ionizing particles.

The angle between the emitted Cherenkov photons and the track of the charged particle can be worked out to be

$$\cos\theta_{\rm c} = \frac{1}{n\beta} \tag{18.10}$$

There is a threshold effect for the emission of Cherenkov radiation. Cherenkov radiation is emitted only if $\beta > \beta_{\rm c} = \frac{1}{n}$. At the threshold, Cherenkov radiation is emitted in the forward direction. The Cherenkov angle increases until it reaches a maximum for $\beta = 1$, namely,

$$\theta_{\rm c}^{\rm max} = \arccos\frac{1}{n} \tag{18.11}$$

Consequently, Cherenkov radiation is emitted only if the medium and the frequencies ν are such that $n(\nu) > 1$. The maximum emission angle, $\theta_{\rm c}^{\rm max}$, is small for gases ($\theta_{\rm c}^{\rm max} \approx 1.5^\circ$for air) and becomes large for condensed media (about 45° for usual glass).

For fixed energy, the threshold Lorentz factor depends on the mass of the particle. Therefore, the measurement of Cherenkov radiation is suited for particle-identification purposes.

The number of Cherenkov photons emitted per path length with wavelengths between λ_1 and λ_2 is given by

$$\frac{dN}{dx} = 2\pi\alpha z^2 \int_{\lambda_1}^{\lambda_2} \left(1 - \frac{1}{n^2\beta^2}\right)\frac{d\lambda}{\lambda^2} \tag{18.12}$$

for $n(\lambda) > 1$, where z is the electric charge of the particle producing Cherenkov radiation and α is the fine-structure constant.

Neglecting the dispersion of the medium (i.e., n independent of λ) leads to

$$\frac{dN}{dx} = 2\pi\alpha z^2 \cdot \sin^2\theta_{\rm c} \cdot \left(\frac{1}{\lambda_1} - \frac{1}{\lambda_2}\right) \tag{18.13}$$

For the optical range ($\lambda_1 = 400$ nm and $\lambda_2 = 700$ nm) one obtains for singly charged particles ($z = 1$)

$$\frac{dN}{dx} = 490\sin^2\theta_{\rm c}\ {\rm cm}^{-1} \tag{18.14}$$

All transparent materials are candidates for Cherenkov radiators. In particular, Cherenkov radiation is emitted in all scintillators and in the light guides, which are used for the readout. The scintillation light, however, is approximately 100 times more intense than the Cherenkov light. A large range of indices of refraction can be covered by the use of solid, liquid, or gaseous radiators. Even indices of refraction intermediate between solids and gas can be achieved by using transparent silica aerogels.

Cherenkov detectors are mainly used for particle identification in beams of particles of fixed momentum. Lighter particles of fixed momentum are faster than heavier ones. Therefore threshold Cherenkov counters, being sensitive to the velocity, can be adjusted to distinguish between these particles. Special imaging devices, like the ring imaging Cherenkov detector (RICH) or other differential Cherenkov counters, can also provide a direct measurement of the particle velocity.

18.7 Calorimeters

18.7.1 Electron–Photon Calorimeters

The development of cascades induced by electrons, positrons, or photons is governed by bremsstrahlung of electrons and pair production of photons. Secondary particle production continues until photons fall below the pair-production threshold, and energy losses of electrons other than bremsstrahlung start to dominate: the number of shower particles decays exponentially.

A very simple model can describe the main features of particle multiplication in electromagnetic cascades: a photon of energy E_0 starts the cascade by producing an e^+e^- pair after one radiation length. Assuming that the energy is shared symmetrically between the particles at each multiplication step, one gets at the depth t

$$N(t) = 2^t \tag{18.15}$$

particles with energy

$$E(t) = E_0 \cdot 2^{-t} \tag{18.16}$$

The multiplication continues until the electrons fall below the critical energy E_c,

$$E_c = E_0 \cdot 2^{-t_{max}} \tag{18.17}$$

From then on ($t > t_{max}$) the shower particles are only absorbed. The position of the shower maximum is obtained from Equation (18.17),

$$t_{max} = \frac{\ln(E_0/E_c)}{\ln 2} \approx \ln E_0 \tag{18.18}$$

The total number of shower particles is

$$\begin{aligned} S &= \sum_{t=0}^{t_{max}} N(t) = \sum 2^t \\ &= 2^{t_{max}+1} - 1 \approx 2^{t_{max}+1} \\ S &= 2 \cdot 2^{t_{max}} = 2 \cdot \frac{E_0}{E_c} \propto E_0 \end{aligned} \tag{18.19}$$

If the shower particles are sampled in steps t measured in units of the radiation length X_0, the total number of track segments is obtained as

$$S^* = \frac{S}{t} = 2\frac{E_0}{E_c} \cdot \frac{1}{t}, \tag{18.20}$$

which leads to an energy resolution of

$$\frac{\sigma}{E_0} = \frac{\sqrt{S^*}}{S^*} = \frac{\sqrt{t}}{\sqrt{2E_0/E_c}} \propto \frac{\sqrt{t}}{\sqrt{E_0}} \tag{18.21}$$

In a more realistic description the longitudinal development of the electron shower can be approximated by

$$\frac{dE}{dt} = \text{const} \cdot t^a \cdot e^{-bt} \tag{18.22}$$

where a, b are fit parameters.

With homogeneous electron–photon calorimeters, energy resolutions of

$$\frac{\sigma_E}{E} = \frac{3\%}{\sqrt{E/\text{GeV}}} \tag{18.23}$$

can be achieved.

The lateral spread of an electromagnetic shower is mainly caused by multiple scattering. It is described by the Molière radius

$$R_m = \frac{21\text{MeV}}{E_c} X_0 \,(\text{g cm}^{-2}) \tag{18.24}$$

Ninety-five percent of the shower energy in a homogeneous calorimeter is contained in a cylinder of radius $2R_m$ around the shower axis.

Since the position of the shower maximum depends only logarithmically on the energy, electron–photon calorimeters can cover quite a large range of energies from some hundred mega electronvolts up to several hundred giga electronvolts even at moderate detector sizes. Electron–photon calorimeters are usually total-absorption devices. Such calorimeters for electrons and photons are mostly made of scintillating crystals. However, Cherenkov media, like lead glass, can also be used as converter materials.

18.7.2
Hadron Calorimeters

The longitudinal development of electromagnetic cascades is characterized by the radiation length X_0 and their lateral width is determined by multiple scattering. In contrast to this, hadron showers are governed in their longitudinal structure by the nuclear interaction length λ and by transverse momenta of secondary particles as far as lateral width is concerned. Since, for most materials, $\lambda \gg X_0$ and $\langle p_T^{\text{interaction}} \rangle \gg \langle p_T^{\text{multiple scattering}} \rangle$, hadron showers are longer and wider (see Figure 18.9).

Part of the energy of the incident hadron is spent to break up nuclear bonds. This fraction of the energy is invisible in hadron calorimeters. Further energy is lost by escaping particles like neutrinos and muons as a result of hadron decays. Since the fraction of lost binding energy and escaping particles fluctuates considerably, the energy resolution of hadron calorimeters is systematically inferior to electron calorimeters.

Hadron calorimeters are usually sampling devices where active detector layers interspersed with passive targets collect the energy of shower particles. The active detector parts can be scintillators or liquid-argon or gaseous detectors. Lead, uranium, copper, or iron mostly are used as passive targets. With a high-performance uranium–liquid-argon calorimeter, energy resolutions of $\sigma_E/E = 35\%/\sqrt{E/\text{GeV}}$ can be achieved.

Acknowledgments

It is a pleasure to thank Dr. Tilo Stroh for a very careful reading of the manuscript

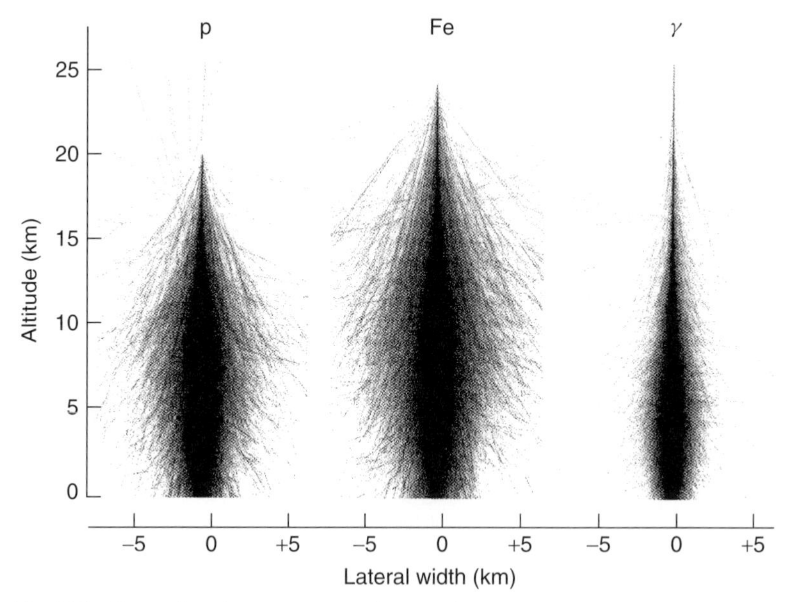

Fig. 18.9 Comparison between proton-, iron- and photon-induced cascades in the atmosphere. The primary energy in each case is 10^{14} eV; J. Knapp and D. Heck 1998; photo credit J. Knapp 2007.

and his efficient professional help with the layout of this article.

Glossary

Acceptor: Substance added as impurity to a semiconductor to create hole conductivity.

Activator: Dopant used in scintillators to de-excite excitons, which then provide photons.

Aging: Deterioration in the performance of detectors in harsh radiation environments.

Avalanche Formation: When drifting electrons gain enough energy in an electric field on their mean free path, they themselves can ionize other atoms and initiate avalanche formation.

Bethe–Bloch Formula: The formula describes the energy loss of charged particles by ionization and excitation.

Charge-Coupled Device (CCD): pixelized silicon detector, which digitizes the incident photon or particle pattern, which is then electronically read out.

Cherenkov Effect: The characteristic electromagnetic radiation created by particles with velocities v faster than light in media ($v \geq c/n$; n – index of refraction).

Critical Energy: Energy at which the energy loss of electrons due to ionization and bremstrahlung are the same.

Decay Length: Track length of a short-lived particle before it decays.

Donor: Substance added as impurity to a semiconductor, which can donate electrons to the conduction band.

Exciton: Electron–hole bound state.

Fermi Plateau: Saturation of the ionization loss at high energies due to polarization effects.

First Townsend Coefficient: Coefficient that characterizes the gas-amplification factor in its electric-field and pressure dependence.

Frisch Grid: Grid electrode in an ionization chamber, which provides signals that do not depend on the position of impact of the particles into the chamber.

Gating: Interesting events are selected by a grid in a TPC by applying suitable voltages to it.

Ionization Loss: Energy loss of charged particles by transferring part of their energy to atomic electrons.

Ionization Yield: Number of electron–ion pairs created by charged particles.

Knock-on Electron: Atomic electron that receives a rather large energy transfer in a collision with a charged particle (also called *δ ray*).

Landau Fluctuations: Energy-loss fluctuations due to the stochastic nature of the ionization process that lead to asymmetric energy-loss distributions especially for thin targets.

Larmor Radius: Bending radius of a charged particle gyrating in a magnetic field.

Lattice Vibrations: Quantized oscillations of a lattice: phonons.

Light Yield: Amount of light created in scintillators by charged particles or γ rays.

Lorentz Angle: In a combined electric and magnetic field, electrons drift in the direction of $\boldsymbol{E} \times \boldsymbol{B}$.

Lorentz Factor: Given by $\gamma = 1/\sqrt{1-\beta^2} = E/mc^2$, where $\beta = v/c$ and c is the velocity of light.

Minimum-Ionizing Particles (mips): Relativistic charged particles with $\beta \cdot \gamma \approx 4$ exhibiting the lowest energy loss by ionization and excitation.

Moderation: Slowing down of neutrons by materials containing many quasi-free protons.

Molière Radius: Characteristic size of an electromagnetic cascade describing its lateral spread.

Nuclear Interaction Length: Characteristic distance between two collisions of a hadron.

Occupancy: Probability that a readout element of a detector is occupied by a particle so that no further particles can be recorded on this element. High occupancy leads to dead times and inefficiency.

Pad: Cathode-structure element that enables the determination of the position of a charged particle along the wire in MWPCs.

Pitch: Distance between anodes or readout structures.

Pixel: Picture element as readout structure, mostly in silicon detectors.

Quenching: Vapor additions to noble gases in Geiger–Müller counters that limit the range of avalanche photons and cause the gas discharge to stop by itself.

Radiation Damage: Degradation of the performance of detectors in a harsh radiation environment. Detectors that can be operated under such conditions are called *radiation hard* or *radiation tolerant*.

Radiation Length X_0: Characteristic length describing the energy loss of electrons due to bremsstrahlung, $dE/dx = E/X_0$.

Relativistic Rise: Rising energy loss at intermediate energies owing to the relativistic deformation of the transverse electric field of fast charged particles.

RICH: Ring imaging Cherenkov counter for the measurement of the velocity of charged particles.

Silica Aerogel: Artificially produced Cherenkov medium of very low index of refraction consisting of a porous structure of silica and sub-micrometer air bubbles.

SQUID: Superconducting quantum interference device used for the measurement of small magnetic flux changes.

Wavelength Shifter: Agent that absorbs UV light and re-emits visible light at lower frequency.

***W* Value:** Energy required to create an electron–ion pair in a gas or an electron–hole pair in a semiconductor.

References

ALEPH Collaboration (**2000**) *http://aleph.web.cern.ch/aleph/*; photo credit ALEPH Collaboration; Roberto Tenchini.

Close, F. and Marten, M. (**1987**) *The Particle Explosion*, Oxford University Press, Oxford, photo credit Frank Close 2007.

Galison, P. (**1997**) *Image and Logic: A Material Culture of Microphysics*, The University of Chicago Press, Chicago.

(a) Grupen, C. (**2008**) *Grundkurs Strahlenschutz*, Springer, Heidelberg; (b) *Introduction to Radiation Protection*, Springer, Heidelberg, (**2010**); (c) Kalthoff, O. (**1996**) Berechnung der Photopeakeffizienz für koaxiale Reinstgermaniumdetektoren, Diploma Thesis, Siegen.

Hausmann, J. (**2002**) *Development of a low noise integrated readout electronic for pixel detectors in CMOS technology for a Compton camera*, PhD Thesis, Siegen.

Kaftanov, V.S. and Liubimov, V.A. (**1963**) Spark chamber use in high energy physics, *Nucl. Instr. Meth.*, **20**, 195.

Knapp, J. and Heck, D. (**1998**) Luftschauer-Simulationsrechnungen mit dem CORSIKA Programm. *Forschungszentrum Karlsruhe Nachr.*, **30**, 27; and *www.ast.leeds.ac.uk/~fs/*; photo credit Johannes Knapp 2007

Rochester, G.D. and Turver, K.E. (**1981**) Cosmic rays of ultra-high energy, *Contemp. Phys.*, **22**, 425.

Yao, W.-M. *et al.*, Particle Data Group (**2006**) Review of particle physics. *J. Phys. G*, **33**, 1–1232. *http://pdg.lbl.gov/*

Further Reading

There are many conferences that specialize in radiation detectors, like the annual "Nuclear Science Symposium" or the "Instrumentation and Measurement Technology Conference" organized by the IEEE (Institute of Electrical and Electronic Engineers), and the "Vienna Conference on Instrumentation". Results on radiation detectors are mostly published in the journal "Nuclear Instruments and Methods in Physics Research Section A". There are also a number of excellent monographs on radiation detectors, like K. Kleinknecht, *Detectors for Particle Radiation*, Cambridge University Press (1986); F. Sauli (ed.), *Instrumentation in High Energy Physics*, World Scientific, Singapore (1992); G.F. Knoll, *Radiation Detection and Measurement*, 3rd edition, John Wiley & Sons Inc., New York (1999/2000); D. Green, *The Physics of Particle Detectors*, Cambridge University Press (2000); C. Grupen & B. Shwartz, *Particle Detectors*, Cambridge University Press (2008).

19
X-ray Spectrometry

Thomas H. Markert and Eckhart Förster

Handbook of Metrology. Edited by Michael Gläser and Manfred Kochsiek

ISBN: 978-3-527-40666-1

19.1 Introduction

X-ray spectrometers are devices designed to study X-ray beams with the goals of determining the wavelengths (or equivalently the energies) and the fluxes of the incident X-rays. While spectroscopy of a sort was being carried out from within a few years after Röntgen's discovery of X-radiation (1895), an enormous breakthrough was achieved when crystal diffraction revealed the first detailed X-ray spectrum (Bragg and Bragg, 1913). The Bragg technique has been refined over the years but is still used extensively and, for many purposes, gives the most useful results (see Section 19.3.1). Since that time, X-ray spectroscopy has been applied with great success in many branches of pure and applied science (Agarwal, 1991). X-ray spectrometers have improved significantly, and new classes of such instrumentation are evolving rapidly (Tsuji, Injuk, and van Grieken, 2004; Janssens, Adams, and Rindby 2000). In this article, the tools of X-ray spectroscopy are briefly described and the most common spectrometer devices are discussed. The strengths and limitations of the various spectrometers are indicated, and brief examples of applications and the kinds of devices that might be appropriate for real-world experiments are provided.

To some degree, all X-ray detectors are spectrometers, since all have some ability to determine wavelengths and measure fluxes. However, here we concentrate on those detectors (and spectroscopic devices) that have moderate spectral resolving powers (roughly, the resolving power $E/\Delta E$ must be better than about 20 over at least some of the spectral range). This criterion excludes, in general, proportional counters, microchannel plate devices, and scintillation detectors. Useful discussions of these classes of detectors (including their spectroscopic capabilities) can be found in the books by Fraser (1989), Knoll (1989), and Michette and Buckley (1993).

The survey here is further restricted to the energy (wavelength) range of 100 eV (124 Å) to 100 keV (0.124 Å). Lower and higher energy spectrometers are discussed in articles and books on ultraviolet and γ-ray spectroscopy. Photons with energies above the creation threshold for electron–positron pairs ($E \geq 1.02$ MeV) interact with matter by initiating a cascade shower. Measurements on these hard photons allow their total energy, arrival position, and arrival time to be determined (Sauli, 1992; Ramana Murthy and Wolfendale, 1993). Although this article is mainly concerned with the energy range up to 30 keV, we briefly consider, at this

Handbook of Metrology. Edited by Michael Gläser and Manfred Kochsiek

ISBN: 978-3-527-40666-1

point, the range beyond this up to 511 keV, that is, half the electron–positron creation threshold. The resolving power of high-purity germanium (HPGe) detectors (see Section 19.4.1.2) is typically several hundreds, which is excellent for resolving lines of the highest energy atomic transitions and the lower energy nuclear transitions. The disadvantage of germanium detectors, which progressively worsens as energies increase, is that the percentage of photons that convert via the photoelectric effect becomes smaller with higher energies (about 75% at 100 keV to about 10% at 511 keV). Consequently, the full energy efficiency becomes small.

Sophisticated X-ray spectrometers have been recently designed to study X-ray emission of new types of hot plasma flashes in order to learn about space- and time-dependent ion distributions as well as electron density and temperatures (see Section 19.3.1.3). Commercially available X-ray spectrometers register X-ray fluorescence of a wide range of samples being either irradiated by an X-ray source or directly by an electron beam in order to obtain their elemental composition. Well-developed X-ray spectrometers with powerful, water-cooled X-ray tubes of many kilowatts power now compete with small spectrometers equipped by air-cooled X-ray mini-tubes of up to 50 W power and adapted focusing optics (e.g., polycapillaries).

Finally, X-ray spectrometers have been constructed to study X-ray absorption of species in experiments with bright synchrotron and hot plasma sources. The X-ray absorption fine structure provides information about the geometrical distribution of electrons in the vicinity of the absorbing atom. Some recent practical applications are outlined in Section 19.5.

19.2 General Concepts

The primary figures of merit of most spectrometers are the resolving power

$$R = \frac{\lambda}{\Delta\lambda} = \frac{E}{\Delta E'} \tag{19.1}$$

where $\Delta\lambda$ is the full width at half maximum (FWHM) (in wavelength units) of the spectrometer response to a monochromatic X-ray of wavelength λ. The energy E is, of course, related to λ by $E = hc/\lambda$, where h is Planck's constant and c is the speed of light. The relationship $E\lambda \sim 1.24$ (E in keV, λ in nanometer units) is often useful in X-ray spectroscopy. Note that X-ray spectroscopists also rely on the older units (angstroms and electronvolts or kiloelectronvolts) for wavelength and energy. In this article, we typically use these as well, since the interested reader will find them in most of the standard references. To convert to SI units, note that 1 Å = 0.1 nm and $1\,\text{eV} = 1.60218 \times 10^{-19}$ J. Besides the resolving power (Equation 19.1), other performance parameters of X-ray spectrometers are also relevant and may, in fact, be decisive in the selection of an instrument for a particular experiment. The sensitivity (or instrumental efficiency) is particularly important. Other parameters include linearity of (or at least information on) spectral response, maximum counting rate of the detector, breadth of spectral coverage, field of view, physical size, simplicity of operation, and cost. The strengths of various spectrometer types are are discussed below.

X-ray spectrometers can be divided into two general categories. The first are the dispersive spectrometers that use X-ray diffraction to spread out the incident spectrum (a prism is an optical analogy). A detector can then read off the dispersed

X-rays and associate a wavelength to the position of the detected photons. Dispersive spectrometers are sometimes called *wavelength-dispersive spectrometers.*

Nondispersive spectrometers, on the other hand, convert the energy of an incident X-ray to a number of "particles" (electron–ion pairs, electron–hole pairs, phonons, superconductor quasiparticles, etc.), and the energy resolution of the spectrometer is determined by the statistical fluctuations of the number of particles (this statement holds for ideal detectors only; no actual device is free of systematic uncertainties). All designs of nondispersive spectrometers suffer from the problem of "dead time," the time required to process an X-ray absorption event. Note that some authors refer to nondispersive spectrometers as *energy-dispersive spectrometers,* generalizing the word "dispersive" beyond simple spatial separation.

Broadly speaking, dispersive spectrometers work by taking advantage of the wave nature of the X-rays (interference produces the diffraction) and nondispersive (energy-dispersive) spectrometers work by taking advantage of the particle nature of the X-rays (photons ionize or otherwise interact with the detectors). Of course, this comment is overly simplistic, since dispersive spectrometers always use photon detectors to read out the dispersed spectrum. In fact, it is often useful to combine dispersive and nondispersive spectrometers, such as diffraction gratings and solid-state detectors (Sections 19.3.2 and 19.4.1), so as to combine the strengths of the different devices.

In Section 19.3, the two types of dispersive X-ray spectrometers, Bragg crystal diffractors and diffraction gratings, are discussed. In Section 19.4, the nondispersive detectors (those with reasonably high resolving powers) are considered.

19.3
Dispersive Spectrometers

19.3.1
Bragg Crystal Devices

19.3.1.1 Bragg's Law

Figure 19.1 illustrates the elementary derivation of the fundamental equation of crystal diffraction, that is, Bragg's law,

$$n\lambda = 2d \sin\theta \tag{19.2}$$

where λ is the wavelength of the incident X-ray, d is the spacing between the planes of the diffracting atoms, often called *net planes,* θ is the angle of incidence (and reflection) of the X-ray beam onto the crystal surface, and n is a positive integer,

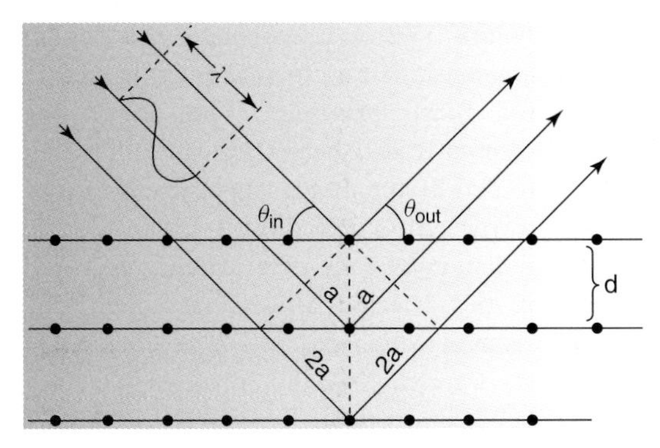

Fig. 19.1 Derivation of Bragg's law.

denoting the order of diffraction. Bragg's law states that only X-rays with wavelengths that satisfy Equation 19.2 will be efficiently reflected from the crystal. Figure 19.1 shows how Bragg's law is derived: X-rays of wavelength λ are incident on a crystal surface with a glancing angle θ_{in}. While each scattering site scatters in all directions, only those rays that satisfy $\theta_{in} = \theta_{out}$ will constructively interfere and result in a significant reflected flux (this is, of course, consistent with the general law of reflection in geometrical optics). Furthermore, a second constraint on the reflected ray arises from the requirement that X-rays diffracted from each crystal plane must also interfere constructively (since the X-rays have relatively high energies, they can typically penetrate through hundreds to thousands of crystal planes). The X-rays penetrate the crystal surfaces and are scattered by the various crystal molecular planes (the planes are separated by d). The differences in the path lengths between the X-ray waves reflected off the various planes are $2a$, $4a$, $6a$, and so on. Clearly, there will be constructive interference between the various waves only if the path-length differences are integral multiples of the wavelength λ, that is, if $n\lambda = 2a$, where n is a positive integer. Simple trigonometry shows that $a = d \sin\theta$, thereby giving Equation 19.2.

The crystal lattice parameter of silicon was linked to the SI meter base unit by combined optical and X-ray interferometry experiments with an experimental uncertainty error of about 10^{-8} (reviewed in Becker (2001)). For practical purposes, the lattice parameters (or net plane distances d) of any crystal can be derived at lower accuracy from a high-resolution X-ray diffractometer experiment. Here, Bragg angles θ of the silicon gauge crystal are compared with those of the crystals being measured (Klöpfel *et al.*, 1997). In such thorough evaluations of crystal parameters, several correction factors have to be considered, for example, refraction effects on the boundary between air and crystal. This leads to a correction of Bragg's law:

$$n\lambda = 2d \sin\theta \left(1 - \frac{\delta}{\sin^2\theta}\right) \tag{19.3}$$

where $1 - \delta$ is the real part of the index of refraction of the crystal at the wavelength of interest. Other effects such as the following must also be considered for a rigorous treatment of Bragg diffraction:

1. extinction (the primary beam cannot reach deeply into the crystal since it is reflected by intervening crystal layers);
2. absorption (again, the primary beam is absorbed by intervening crystal layers); and
3. multiple reflection (crystal layers reflect both "up" (toward the surface of the crystal) and "down" (away from the surface)).

Thorough discussions of these topics are beyond the scope of this article but may be found in some detail in the books by Compton and Allison (1935), Pinsker (1978), James (1982), and Authier (2001).

19.3.1.2 Flat Crystal Bragg Spectrometer

Figure 19.2 is a schematic drawing of a generic X-ray spectrometer based on the Bragg principle. Here, X-rays are collimated so that they strike the diffracting crystal at the Bragg angle for X-rays of wavelength λ. Only those X-rays that satisfy Equation 19.2 are reflected and registered in the detector. Since (presumably) the spacing d and incident angle θ are known, the wavelength of the diffracted X-ray can be determined to be within a factor of

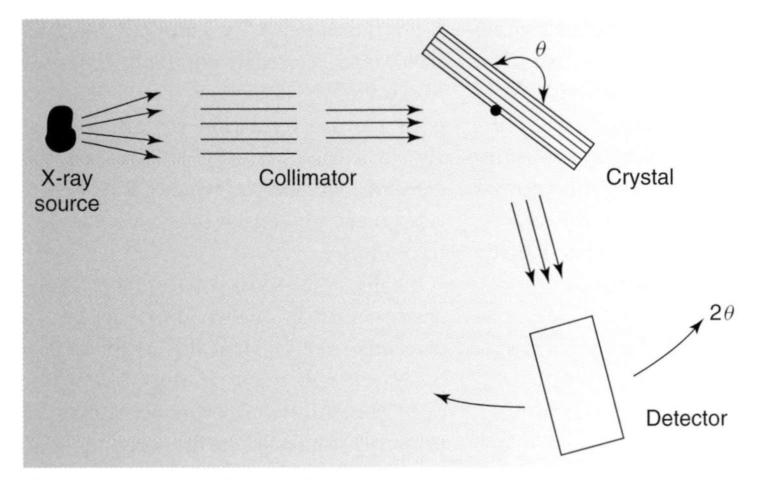

Fig. 19.2 Schematic drawing of a generic crystal spectrometer.

$1/n$ (n is the diffraction order). Removing order ambiguity is an issue for all dispersive spectrometers. Techniques for determining the actual X-ray wavelength (i.e., removing the $1/n$ uncertainty) are discussed below.

To be useful as a spectrometer, of course, it is necessary for a device to cover a range of wavelengths. This is achieved for the generic device in Figure 19.2 by mounting the crystal and the detector on coaxial rotary stages, so that the crystal can present a range of incident angles to the X-ray beam. Any rotation of the crystal by an angle $\Delta\theta$ is accompanied by a rotation of the detector by $2\Delta\theta$, so that the diffracted X-rays fall on the detector. To study a single X-ray emission line, for example, the crystal must be rotated over a range of angles that cover the response of the crystal to a monochromatic line. It should be apparent, from this discussion, that a Bragg device can only measure X-rays with wavelengths in the range $0 \leq \lambda \leq 2d$ (i.e., when $0^\circ \leq \theta \leq 90^\circ$); in practice, however, both the short- and long-wavelength limits are restricted by geometrical constraints. In order to measure longer wavelengths, one requires crystals (or multilayer diffractors discussed below) with larger $2d$ spacings.

From Bragg's law, the angular dispersion ($d\lambda/d\theta$) can be easily derived: after division by the reflection order n, the wavelength λ is a product of a constant factor ($2d/n$) and $\sin\theta$. If this factor is expressed by $\lambda/\sin\theta$ in the derivative, we obtain

$$\frac{d\lambda}{d\theta} = \frac{\lambda}{\tan\theta} \tag{19.4}$$

Thus, the angular dispersion is inversely proportional to $\tan\theta$, which means that the resolving power R is directly proportional to $\tan\theta$. If we have a set of crystals with different net plane distances d, we should choose the plane with $2d$ slightly larger than λ to obtain the highest resolving power R.

The FWHM of the response function of an ideal crystal spectrometer is dominated by the natural resolving power of the crystal (the angular dispersion of the collimated X-rays also contributes, however, and in some cases will dominate). The response function of the crystal alone is called the *rocking curve,* because the complete response of the crystal to a monochromatic

X-ray is measured by rotating (rocking) the crystal over a small angular range. The FWHM is the rocking curve width. Very early on, Compton and Allison (1935) had derived a general theory of X-ray spectrometers. In the case of Figure 19.2, the relation between the X-ray emissivity J, the rocking curve C, and the reflected intensity P holds with some minor simplifications:

$$P(\lambda) = \int d\lambda' J(\lambda - \lambda') \int_{\alpha_{\min}}^{\alpha_{\max}} d\alpha \int_{\phi_{\min}}^{\phi_{\max}} d\phi \cdot C\left(\alpha - \frac{\phi^2}{2}\tan\theta_0 - \frac{\lambda' - \lambda_0}{\lambda_0}\tan\theta_0\right) \quad (19.5)$$

where the divergence angles α and ϕ describe the angular source emission in and out of the reflection plane of Figure 19.2. To minimize the broadening of X-ray emissivity J by convolution with the rocking curve C, the FWHM of C should be much smaller than the divergence $(\alpha_{\max} - \alpha_{\min})$ in order not to lose X-ray photons in the Bragg reflection. The meridional divergence (multiplied by the angular dispersion of Equation 19.4) should, in adding, be much smaller than the FWHM of X-ray emissivity J. This is normally realized in wavelength-dispersive spectrometers, if the resolving power is high and the collimator is well adapted (cf. Figure 19.2). For small sagittal divergences $(\phi_{\max} - \phi_{\min})$, this broadening influence in Equation (19.5) can be neglected since it depends on the second order of ϕ.

If the absorption in the spectrometer crystal is very small (which holds for hard X-rays), the rocking curve C has the form of a cylinder head (Darwin curve) with a flat top of almost unity reflection (i.e., in the range of interference total reflection). For the general case of structurally perfect crystals, the rocking curves can be computed using the dynamical theory of X-ray interferences (see book by Authier (2001) and references therein). A narrow FWHM of the rocking curves leads to a good spectral resolution but also eventually to a loss of X-ray photons on the detector, whereas a wide FWHM provides many photons at the cost of spectral resolution.

Finally, the sensitivity of the Bragg devices must be known if one wishes to determine the incident flux as a function of wavelength (this is not always the case; sometimes, experimenters may be primarily interested in the locations (i.e., wavelengths) of spectral features or in the relative intensities (as opposed to the absolute fluxes)). To determine the flux in an X-ray line, the crystal must be rocked over the nominal Bragg angle of the line, covering a $\Delta\theta$ range that includes the response of the crystal (most of the width of the rocking curve), as well as contributions due to the dispersion of the incident beam. With the knowledge of the crystal reflectivity as a function of the angle (and other parameters such as the detector quantum efficiency), the incident flux in the X-ray line can be computed. The parameters of interest for computing the flux are not always easy to determine, however, particularly at all energies.

As noted above, Bragg's law permits a residual uncertainty in the X-ray wavelength because of the reflectivity of various diffraction orders of n. (This uncertainty is also present for grating spectrometers.) This ambiguity may be minimized or eliminated in a number of ways. The simplest way of eliminating it is to use a detector that has a sufficient inherent spectral resolution that the orders may be distinguished by the detector. Even low-resolution devices such as proportional counters can separate orders sufficiently in a wide energy range. Other techniques are to use filters and/or X-ray optics (which do not reflect efficiently at higher energies) to minimize the contribution from higher orders.

Collection and diffraction efficiency of a flat crystal spectrometer can be improved if parallel beam X-ray optics is incorporated. A diverging beam originating from the X-ray source is parallelized by polycapillary and grazing-incidence optics. The now parallel beam impinges on the spectrometer crystal (Love, 2002). This results in a much larger solid angle of X-rays that can be collected. The wavelength-dispersive unit can, for example, be incorporated in a scanning electron microscope.

19.3.1.3 **Bent-Crystal Spectrometer**

Besides industrial applications of X-ray spectrometers where a high X-ray flux is often available, there are tasks in which the detector flux is rather limited. This is the case for X-ray diagnostics of high-temperature plasmas and for electron-beam X-ray microanalysis (see Section 19.3.1.4). There is, therefore, a need for additional X-ray optic elements to collect and focus X-rays to the detector. Depending on the required resolving power and luminosity as well as the geometry of spectrometer, different X-ray optics, grazing-incidence mirrors, capillaries, bent gratings, bent crystals, and so on, must be selected.

Bent-crystal spectrometers employ, in the simplest form, a cylindrically bent crystal (instead of a flat one) as the dispersing element and a detector that is positioned on the circumference of the Rowland circle (see Figure 19.3a,b). The curvature of the reflecting net planes is twice that of the radius of the Rowland circle in this (Johann-type) spectrometer. Equation 19.5 is still applicable when the rocking curve is corrected for crystal bending (Uschmann, Malgrange, and Förster, 1997) and modified relations are used between the divergence angles α, ϕ, and the wavelength λ'. Since rocking curves are usually wider and bent crystals accept a diverging beam of X-rays, photon numbers on the detector can be increased by one or two orders of magnitude in comparison to the flat crystal spectrometer. Figure 19.3c shows the dominant resonance lines of hydrogen-like and helium-like aluminum ions together with intercombination and satellite lines on a relatively low continuum emission. The electron density and temperature can be obtained from plasma physics models from ratios of selected spectral lines (Attwood, 1999).

The highly excited aluminum ions were produced by focusing high-power Nd : glass laser pulses (200 J, 1 ns) on solid aluminum target (Renner *et al.*, 1994).

If the cylindrical crystal is rotated by 90° around its central surface normal, then X-rays are collected in the ϕ angle (instead of the α angle) direction. Figure 19.4a shows the von Hamos type spectrometer in which the distances from the crystal to the source and the detector are equal. By bending a thin crystalline wafer in two dimensions, spherical, toroidal, or ellipsoidal crystals can be produced with curvature radii down to 10 cm. As shown in Figure 19.4b,c, luminous X-ray spectrometry and quasi-monochromatic X-ray imaging can be performed. Mißalla *et al.* (1999) have discussed reflection and focusing properties of these schemes when short X-ray pulses in the kiloelectronvolt range are considered.

A cylindrically bent analyzer crystal can be combined with a flat position-sensitive detector in the von Hamos geometry (see Section 19.4.1.4). This makes parallel data collection possible for light elements in wavelength-dispersive spectrometers

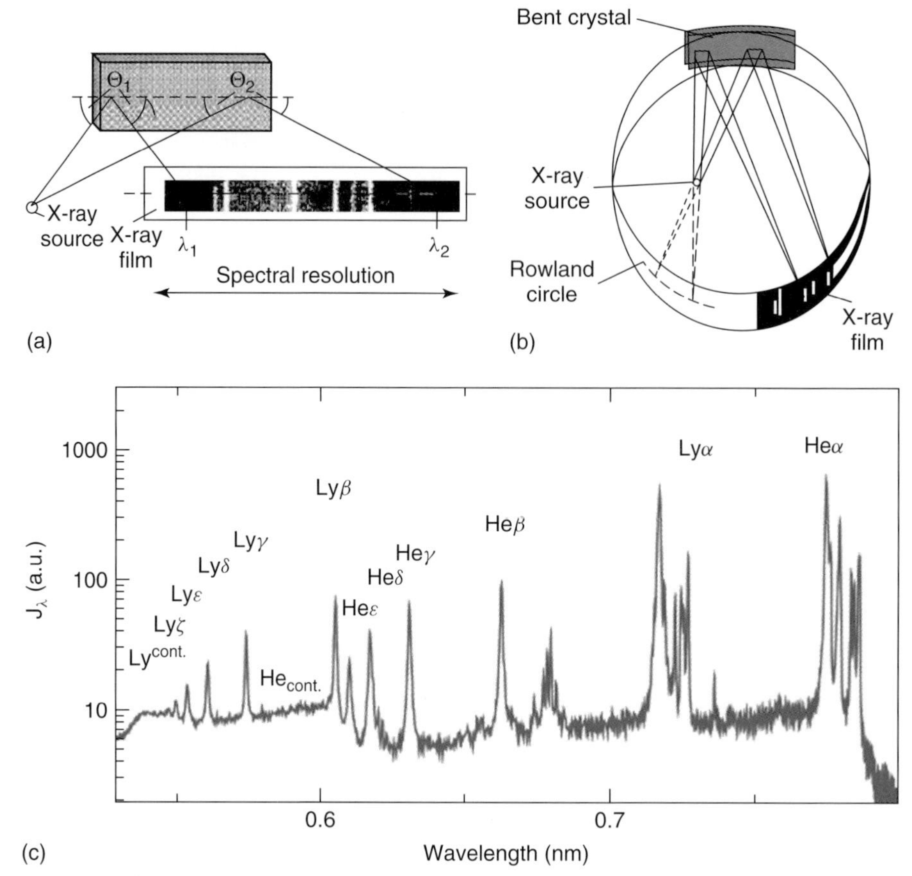

Fig. 19.3 Schematic drawings of (a) a flat crystal spectrometer; (b) a Johann-type bent-crystal spectrometer; and (c) X-ray spectrum of a laser-produced Al plasma.

(Love, 2002). Toroidal crystals combine focusing in both sagittal and meridional directions if the ratio of the respective radii is $\sin^2\theta$. They are characterized by high resolving power and efficiency but require much careful preparation and testing.

An even higher spectrometer efficiency can be achieved by applying ellipsoidal highly oriented pyrolithic graphite (HOPG) crystals if a wider reflection curve (a tenth of a degree) is tolerated.

All the spectrometer types discussed so far are based on the Bragg equation and its corresponding angular dispersion ($d\lambda/d\theta = \lambda/\tan\theta$). There is, however, a special variant of a Johann spectrometer in which a point source and film are both located on the Rowland circle (Figure 19.5b).

Assuming that the X-ray wavelength λ_0 is reflected by the cylindrical crystal in the Rowland circle plane, X-rays with nonzero sagittal angles ϕ impinge on the crystal under slightly smaller Bragg angles. Geometrical considerations lead to the following relation between wavelength λ

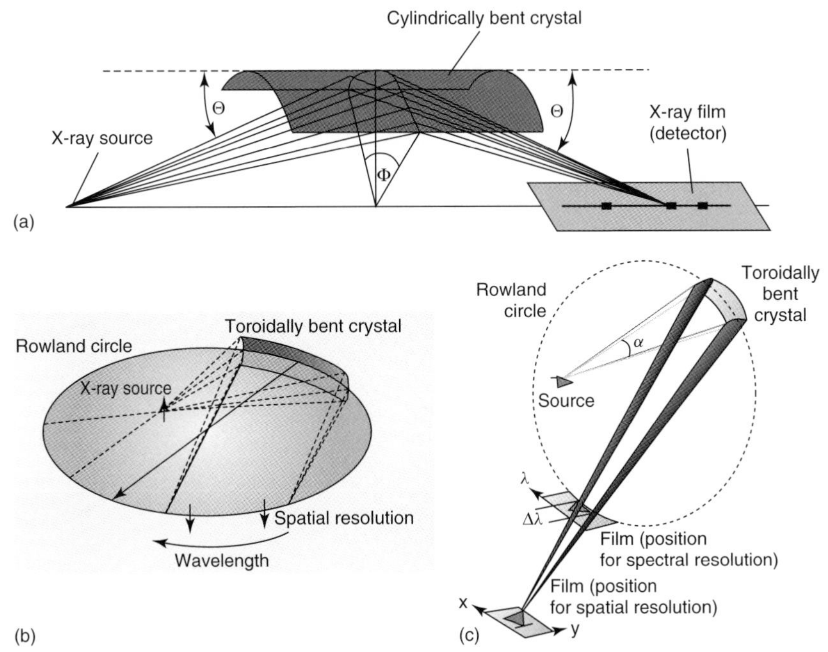

Fig. 19.4 Schematic drawings of bent-crystal spectrometers: (a) von Hamos type; (b) toroidally bent-crystal spectrometer; and (c) quasi-monochromatic X-ray imaging setting.

and the divergence angle ϕ:

$$\lambda = \lambda_0 \cos\phi \tag{19.6}$$

where the maximum wavelength λ_0 can be simply changed by moving the X-ray source and the film along the Rowland circle. This leads to a high luminosity and a high spectral resolving power (Renner *et al.*, 1994), which is needed for sophisticated X-ray plasma diagnostics, but is limited to a small spectral coverage. There is also a simple extension of the flat crystal spectrometer to a double-crystal spectrometer in which wavelengths are determined by Equation 19.6. It suffers, however, from a very low luminosity and is therefore applicable only for strong X-ray sources. Table 19.1 summarizes some key parameters of the X-ray spectrometers of Figures 19.3a,b and 19.5a,b, respectively.

19.3.1.4 Bragg Spectroscopy with Synthetic Crystals (Multilayers)

Most Bragg diffractors are either natural crystals or crystals grown in the laboratory according to certain chemical and thermal specifications. Crystals of this kind have $2d$ spacings in the range 0.2–2.7 nm and can be used to cover energies between ~500 eV and 35 keV (Attwood, 1999). For lower energies, synthetic multilayer diffractors sometimes called *layered synthetic microstructures* (*LSM*s) or simply multilayers are frequently used replacing, in most cases, the Langmuir–Blodgett soap films used for many years. Multilayer diffractors are made by depositing layers (using the sputtering or evaporation techniques of semiconductor technology) of alternating high-Z and low-Z materials (carbon and tungsten, for example) with precise thicknesses so as to perform the same

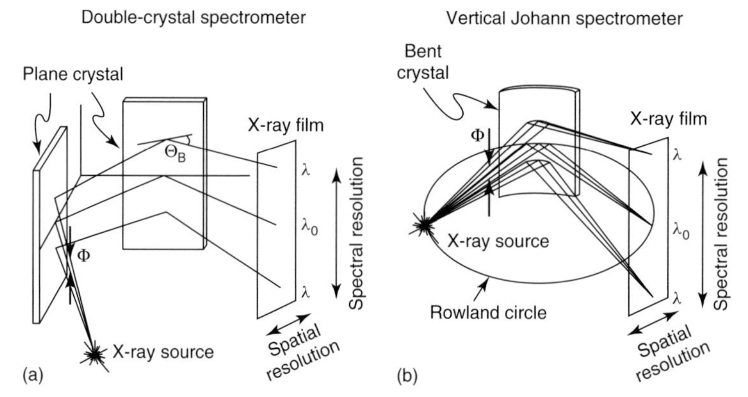

Fig. 19.5 Schematic drawings of high-resolution X-ray spectrometer with vertical (sagittal) dispersion: (a) double flat crystal and (b) Johann-type cylindrically bent-crystal setting.

Tab. 19.1 Key parameters for different X-ray crystal spectrometers: Quartz crystals of dimension $20 \times 30\,mm^2$ either flat or with curvature radius of 76.8 mm were used. Source-to-detector distance was 130 mm; source of size $1\,mm^2$ emits a certain number of photons/$mrad^2$/mm^2 into a solid angle 1 $mrad^2$ (Renner *et al.*, 1994).

Type of spectrometer	Luminosity ($mrad^2$)	Spectral resolution	Range (Å)
Flat single crystal	1.74×10^{-2}	4 000	1.14
Johann classical scheme	1.88×10^{-1}	12 800	1.34
Double-crystal vertical dispersion	4.36×10^{-3}	13 000	0.18
Johann vertical dispersion	2.72	10 400	0.18

functions as one-dimensional crystals. In the multilayer case, the effective spacing d is the sum of the thicknesses of the high-Z and low-Z materials. Up to several hundred layers can be laid down in this way (more layers are of little value since the relatively low-energy X-rays cannot penetrate deeper). Multilayers can now be deposited on curved substrates (up to 30-cm lateral size), even with gradients in depth and lateral thickness in order to optimize imaging performance. A comparison of focusing optics for femtosecond X-ray diffraction with Cu K alpha line radiation shows differences in the key parameters, focus size, flux, and photons per second (Bargheer *et al.*, 2005). The authors considered a toroidally bent Ge Crystal, multilayer optics, ellipsoidal monocapillary, and polycapillary X-ray optics.

A clear advantage of using multilayer films is that the spacing d can be selected to be any desired value (although d spacings smaller than about 25 Å are difficult to fabricate). Generally, the resolving power of multilayer diffractors is limited to about 50, which is superior to many other kinds of spectrometers but will not be adequate for some applications. As shown in Figure 19.6 effective methods

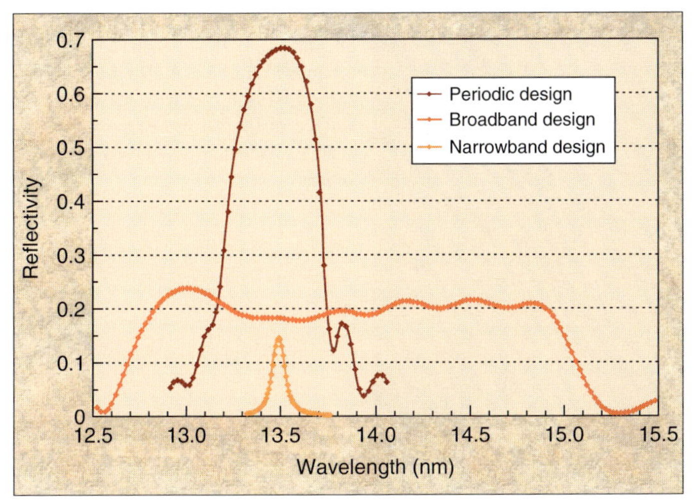

Fig. 19.6 Change in Mo/Si multilayer reflection curves to obtain higher resolution and higher sensitivity, respectively. (Courtesy of Torsten Feigl and Sergij Yulin.)

are available to change the form of the multilayer reflection curve. By using higher reflection orders n the FWHM of the reflection curve can be reduced to obtain higher resolution. In contrast, the FWHM can be broadened by using depth gradients in layer thicknesses to obtain higher spectrometer sensitivity on the expense of peak reflectivity.

In recent years, multilayers have been used as the reflecting surfaces of many different optical devices, providing enhanced reflection at certain soft X-ray wavelengths (i.e., those wavelengths in which the Bragg condition is met). For example, traditional X-ray telescopes utilize grazing incidence ($\theta <$ a few degrees) on polished metallic surfaces (Fraser, 1989). Such telescopes are difficult to fabricate and typically have a relatively small collecting area. More traditional telescope geometries (i.e., those with nearly normal incidence) can be used, however, if the reflecting surfaces are coated with multilayers. The resulting telescope forms sharp images over a narrow wavelength band. Figure 19.7 shows an image of the sun obtained with an X-ray telescope coated with layers of cobalt and carbon with a period of 31.8 Å. At nearly normal incidence, the mirrors reflect efficiently at a wavelength of 63.5 Å. The bandpass of 1.4 Å encompasses the wavelengths of two bright X-ray emission lines from ions of highly ionized Mg and Fe, which are sensitive diagnostics of the plasma temperature in the range $1 - 3 \times 10^6$ K, in which these ions are abundant. The figure shows the high-temperature material in the solar corona, in contrast to the much cooler photosphere which appears relatively dark at this wavelength (Golub, Zirin, and Wang, 1994).

19.3.2 Diffraction Gratings

Diffraction gratings are similar to crystals in that a periodic structure sets up an interference pattern that disperses the X-rays according to wavelength. They differ from crystals in that the structure is periodic only on the surface of the

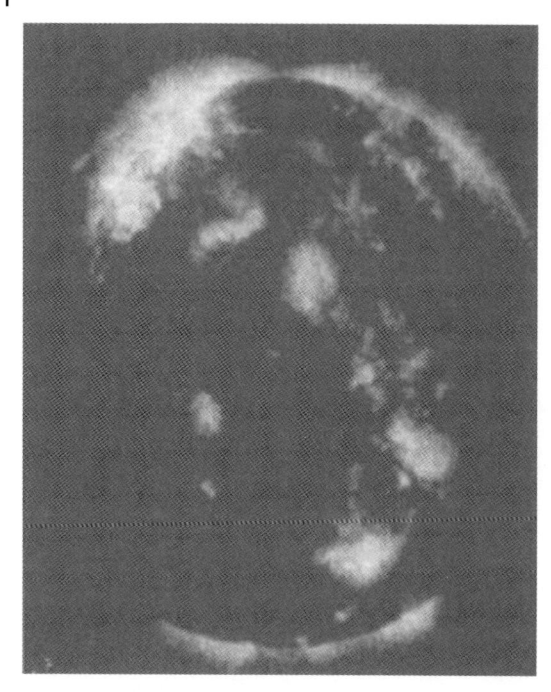

Fig. 19.7 Image of the sun taken on July 11, 1991 at $\lambda = 63.5$ Å with a bandwidth of 1.4 Å. The hot solar corona is seen clearly, while the cooler surface of the sun is relatively dark at this wavelength. (Courtesy of Leon Golub.)

diffractor instead of both on the surface and within the body of the diffractor, as for crystals. As a result, diffraction gratings reflect (or transmit) incident X-rays at all wavelengths, regardless of the incidence angle.

Gratings disperse light according to the grating equation

$$n\lambda = p(\cos\beta - \cos\alpha) \tag{19.7}$$

where the incident and diffracted angles (α and β) are defined in Figure 19.8 for the two types of gratings: reflection and transmission gratings. As was the case for crystal spectrometers, n is the integer order of diffraction (except that for gratings, n can be positive, negative, or zero). For gratings, p is the period of the grating lines or grooves. X-ray reflection gratings operate at small incidence angles ($\alpha < 2 - 3$) because surfaces reflect X-rays efficiently only at grazing angles (unless enhanced with multilayer coats; see Section 19.3.1.4 and Spiller (1994)). Transmission gratings generally operate at nearly normal incidence ($\alpha \sim 90^\circ$) where the transmitted efficiency is greatest, although they can be used with moderate efficiency over a large range of angles.

Ordinary ruled reflection gratings were among the first X-ray spectroscopic instruments, although they have been limited, until recent years, to the longer wavelengths ($\lambda > \sim 20$ Å) since it is difficult to rule gratings with sufficiently small periods. In recent years, maturing technology (essentially, replacement of the mechanical ruling method by lithography or holography) has led to reflection gratings that can be used at wavelengths as short as ~ 5 Å. As is the case for reflection gratings used at optical wavelengths, X-ray gratings can be blazed (i.e., can have asymmetric reflecting surfaces) so as to enhance reflections into particular orders.

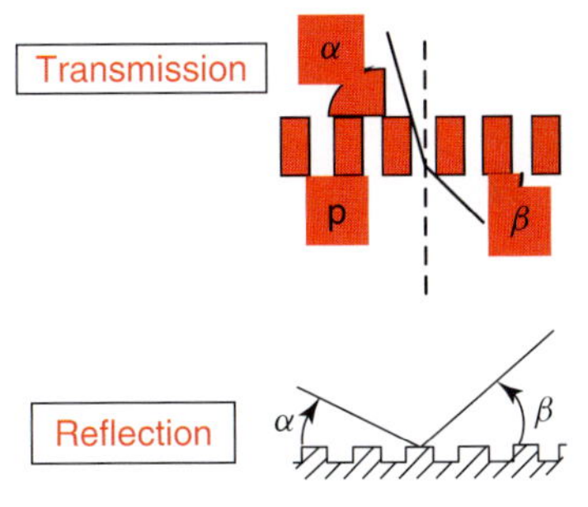

Fig. 19.8 Schematic illustration of the two kinds of X-ray gratings, the transmission and the reflection grating. Although not indicated in the diagram, both kinds of gratings can be blazed to enhance the efficiency in some of the orders.

X-ray transmission gratings are a relatively recent development. Originally conceived for applications in X-ray astronomy (Gursky and Zehnpfennig, 1966), they have also been applied as spectrometers in synchrotrons and in X-ray lasers, and as interferometers in atomic diffraction experiments. Advantages of transmission gratings are their light weight and that they are typically used at normal incidence so that the spectrometer can be more compact than for a reflection grating. Effective transmission gratings must thick enough for the grating bars to be at least partially opaque to the X-rays and must have small enough periods so that the dispersion angle is large enough to be useful. Figure 19.9 shows a sketch of a transmission grating developed for X-ray astronomy.

Transmission gratings are made with the techniques of semiconductor technology (microfabrication and nanofabrication). Periodic patterns are generated from holographic interference techniques or from mechanical ruling for larger period gratings. The periodic pattern is transferred to thicker metallic grating bars (gold is most often used). The first-generation transmission gratings were used primarily at lower energies ($E < 3\,\text{keV}$) because the grating-bar thicknesses were too small (gratings became transparent at higher energies) and/or the bar periods were too large (dispersion angles were small). Recent developments have extended the effective energy range of transmission gratings to $\sim 10\,\text{keV}$ (Schattenburg *et al.*, 1991).

Transmission gratings are usually fairly small (a few square centimeters, Figure 19.9), but individual gratings can be combined into a larger assembly. For X-ray astronomy applications, for example, grating areas of over $1000\,\text{cm}^2$ have been used at the objective of X-ray telescopes to provide a large collecting area for spectroscopy of cosmic sources. For laboratory applications, free-standing gold bars with up to 5000 lines/mm have been produced where the gold bars were fixed on a periodic gold support grid with $17\,\mu\text{m} \times 17\,\mu\text{m}$ active areas. In combination with a toroidal grazing-incidence mirror, these transmission gratings were used in X-ray spectroscopy of both laser-produced plasmas and scattered radiation from a soft X-ray-free electron laser (Jasny *et al.*, 1994; Höll *et al.*, 2007).

Like Bragg spectrometers, grating spectrometers can, in theory, achieve extremely large resolving powers. The theoretical maximum resolution for diffraction order n is given by

$$R = \frac{E}{\Delta E} = nN \tag{19.8}$$

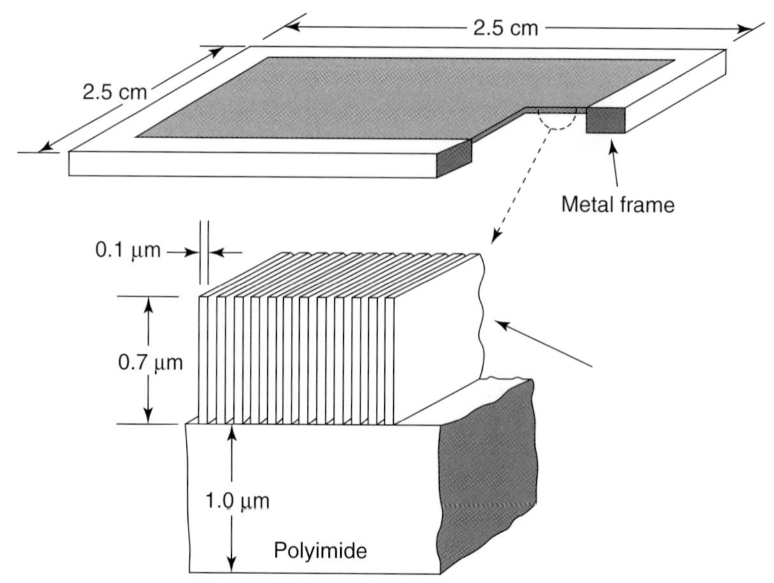

Fig. 19.9 Schematic of a high-energy grating facet being built for the US observatory Advanced X-ray Astrophysics Facility (AXAF). The gold grating bars (period = 2000 Å) are supported by a thin (1-μm) plastic film.

where N is the number of grating grooves or lines illuminated by the X-ray beam. For a grating with 1000 lines/mm (routine for transmission gratings), a 5-mm-wide X-ray beam can achieve a resolving power of 5000. It is, however, difficult to fabricate perfect periodic gratings, so that the variation in the grating period usually limits the resolving power to $p/\Delta p$, where Δp is the variation in the grating period over the illuminated surface of the grating.

Note that both classes of dispersive spectrometers (crystals and gratings) can, with relatively minor modifications, be used as monochromators as well as spectrometers; that is, they can select a narrow energy band out of a broad spectrum and transfer this band for further processing and analysis. Both grating and crystal monochromators are employed (generally, the crystals at the shorter wavelengths, where the gratings are less effective) in this way.

As noted in the discussion of crystal spectrometers, there are potential benefits to curving the diffractors. In such a design, the incident X-rays can diverge from a point (parallel or nearly parallel beams have been tacitly assumed for the flat diffractors discussed above), and the curved gratings perform both energy dispersion and focusing. In the Rowland configuration (which is the most commonly used for curved diffractors), a point source at the entrance slit will focus to a (slightly curved) line at the exit slit, each wavelength focusing to a different position as shown in Figure 19.4b,c. More complex curvatures (spheres, ellipsoids, toroids) can improve the focusing properties at some wavelengths, although it can be more difficult to fabricate the curved diffractor elements (Michette and Buckley, 1993).

19.4 Nondispersive (Energy-Dispersive) Spectrometers

19.4.1 Semiconductor Devices

Semiconductor-based detectors, sometimes simply called *solid-state detectors*, have been used as X-ray spectrometers for more than 30 years. A basic description of how such devices work can be found in Fraser (1989), Knoll (1989), and Tsuji, Injuk, and van Grieken (2004). The fundamental principle is that an X-ray will liberate electron–hole pairs as it interacts within the semiconductor. The number of such pairs is given on the average by $N = E/w$, where E is the X-ray energy and w the ionization energy. For silicon and germanium, the ionization energies w are approximately 3.62 and 2.96 eV respectively, and are slightly temperature dependent. An X-ray photon with energy 1 keV, therefore, will liberate on average several hundred electron–hole pairs. This free charge can then be gathered, as in a gas ionization detector, by applying an electric field to the ionized region. The amount of charge gathered is proportional to the energy of the incident X-ray.

Figure 19.10 shows a schematic diagram of a generic semiconductor detector. Two thin surfaces are heavily doped so as to be n-type (where electrons are the majority charge carriers) and p-type (where holes are the charge carriers). A central region, considerably thicker than the n- and p-types, is relatively free of charge carriers and is called the *depletion region*. The key feature of the depletion region is that it has a nonzero electric field, so that electrons and holes that are produced by photon interactions are easily separated. For X-ray detectors, the depletion region can consist of either an extremely pure semiconductor (typically germanium or silicon) or a volume of a semiconductor into which an element of opposite polarity (n- or p-type) has been diffused to compensate for the excess of electrons or holes (lithium-drifted into p-type silicon or germanium is the most common). This compensated region (often called the *intrinsic region* since the layer has physical properties similar to pure, intrinsic semiconductors) is where the X-ray interaction takes place. In order to be effective as an X-ray detector over a broad energy range, the depletion region (whether naturally free of charge carriers or compensated) should be adequately thick (so that high-energy X-rays will be stopped), and any dead or absorbing layers (the n- or p-type regions or other nondetecting layers) should be thin.

When an X-ray enters the depletion layer (assuming the layer is thick enough), it interacts with the semiconductor to produce electron–hole pairs. A voltage is applied to the surfaces to attract the electrons and holes, which migrate rapidly to the electrodes.

The heavily doped n- and p-type regions serve as barriers to any leakage current (the current that would flow in an undoped semiconductor as a result of the finite resistivity of the material). The leakage current must be suppressed for any prospect of detecting the small signal arising from an X-ray interaction. The signal goes to a preamplifier as shown in Figure 19.10 and is then processed electronically.

The collected charge following an X-ray photoelectric interaction is $N = E/w$ electrons (and holes). The deviation in the collected charge is related to X-ray energy

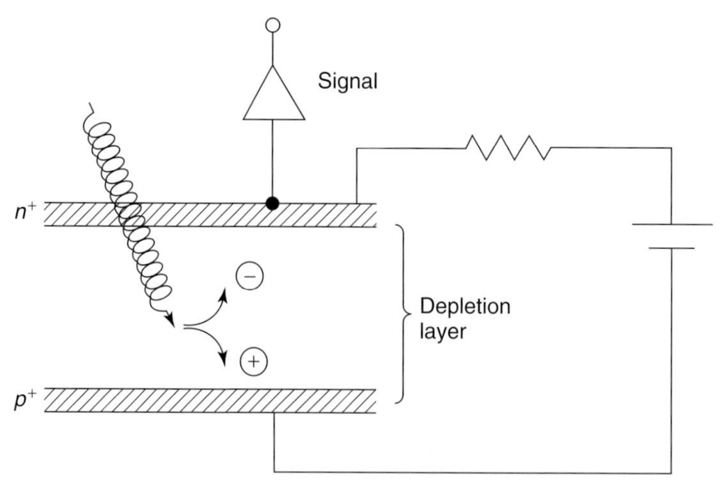

Fig. 19.10 Drawing of a (very) generic semiconductor spectrometer detector.

according to

$$\Delta E = 2.354w\left(\frac{FE}{w} + K^2 + D^2\right)^{0.5} \quad (19.9)$$

where ΔE is the FWHM response to a monochromatic X-ray line (see Equation 19.1), w is the ionization energy, E is the energy of the incident X-ray, K arises from the system electronic noise, and D is the dark current (the flow of electrons and holes that is present, owing to various impurities and imperfections in the device, even when there are no X-rays incident on the detector). Both K and D are measured in equivalent free electrons. F is the so-called Fano factor (Fano, 1947), which relates the statistical fluctuations in the number of electrons liberated to the expected number (E/w). (Fano factors for semiconductors have values of about 0.1, although there is an uncertainty in F of at least 50% due to the difficulties in precisely measuring this parameter.) Most semiconductor devices are operated at low temperatures (liquid nitrogen is typically used in the laboratory with $T = 77$ K), which helps to limit the dark current. The resolving power $E/\Delta E$ for a lithium-drifted silicon (Si(Li)) detector with system noise $\sqrt{K^2 + D^2}$ of 10 electrons is about 20 for 2 keV and grows to about 60 for 10 keV. At higher energies, the resolution is limited by the statistical fluctuations in the electron number; at low energies, the system noise dominates. Figure 19.11 shows a spectrum obtained with an Si(Li) device (see below), indicating the usefulness of a semiconductor device. The spectrum is the result of X-ray fluorescence of a sample consisting of thin films of gold, nickel, and chromium deposited on a silicon substrate. The intensities of the various fluorescent lines can be used to determine the thicknesses of the films. Note that the resolving power of the semiconductor device is adequate for the quantitative analysis required.

There are a number of semiconductor devices that are used for X-ray spectroscopy. Here, four general types are listed, giving some of the strengths and weaknesses of the various classes.

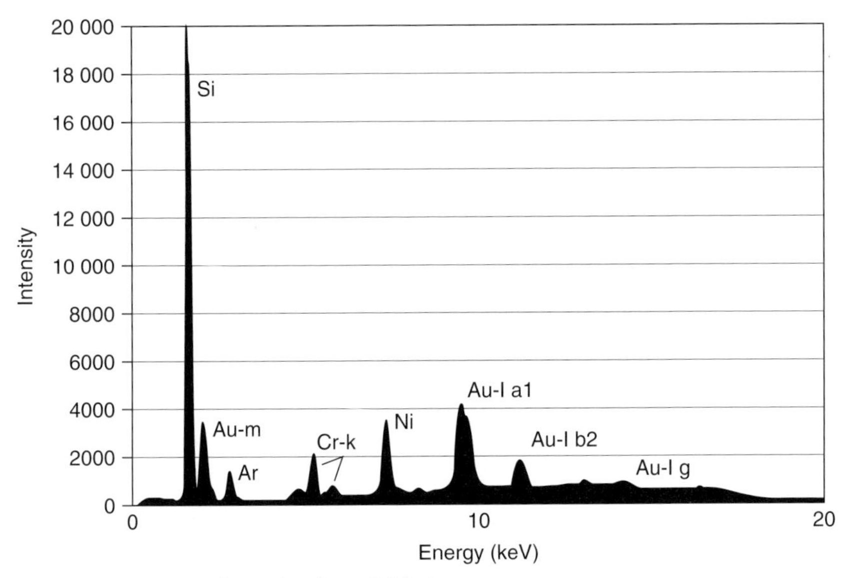

Fig. 19.11 Spectrum obtained with an Si(Li) detector.

All of these devices are available commercially. See Knoll (1989) and Fraser (1989) for good introductory discussions of these devices and more detailed references.

19.4.1.1 Lithium-Drifted Silicon [Si(Li)] Detectors

Si(Li) detectors have depletion layers of thickness of a few millimeters to $\sim$1 cm, manufactured by introducing ("drifting") a lithium dopant into slightly p-type silicon. The net result of this drifting process (which can take several weeks) is to produce a compensated depletion region that is essentially free of charge carriers. Si(Li) detectors can have significant efficiencies for energies up to about 30 keV. At lower energies, the dead layers limit the performance to energies greater than about 0.1 keV. For many commercial detectors, furthermore, a vacuum-tight window of 1-μm polymer with a cover of 30-nm aluminum is now used.

19.4.1.2 Germanium Detectors

Lithium-drifted germanium detectors (Ge (Li)) were used for many years, particularly at higher energies. In more recent years, extremely pure germanium crystals have been grown (HPGe) that have relatively few free charges and hence can be used directly as a depletion layer without the introduction of a compensating dopant. HPGe detectors complement Si(Li) detectors by having efficiencies at significantly higher energies (upto tens of mega electronvolts; see Section 19.5), but are usually not preferred at lower energies ($E < 20$ keV) since the Si(Li) detectors have better resolution and less-prominent escape peaks.

19.4.1.3 Silicon Drift Detectors (SDDs)

A drift chamber detector is a large semiconductor wafer of, for example, high-resistivity n-type silicon; it is fully depleted from a small n+ ohmic contact, which is positively biased with respect to the p+ contacts covering both wafer surfaces. By implanting a parallel p+ strip pattern and applying an electrical field on both wafer sides, electrons are forced to move to the central n+ readout anode (Strüder *et al.*, 1998). The silicon drift detectors (SDDs) incorporates an on-chip amplifier and has a very low capacitive loading; energy resolution is 127 eV for 6 kcal at -20°. A particular advantage is that commercially available SDDs (Röntec) can be employed for high X-ray counting rates in the range of 400 000 counts per second. The idea for this new detector scheme originated from Gatti and Rehak (1984); the SDDs are fabricated for a wide field of applications, for example, for several synchrotron radiation experiments (Strüder *et al.*, 1998).

19.4.1.4 Charge-Coupled Devices (CCDs)

Charge-coupled devices (CCDs) (see Imaging Detectors) are silicon-based detectors used extensively in television technology but are also effective (with some modifications) in the X-ray regime. While having spectral resolving powers similar (or even superior) to the best Si(Li) and HPGe devices, they are also excellent imaging detectors, with pixel sizes typically $<20\,\mu m$ and array dimensions 1024×1024 pixels or greater. The most common CCD design is a two-dimensional array of MOS capacitors on the surface of a semiconductor. (An alternative, first proposed by Gatti and Rehak (1984), is to use an array of p–n junctions on the surface of n-type silicon.) The charge from an X-ray interaction in the depletion layer of the semiconductor is stored, momentarily, in the nearest capacitor. By varying the surface voltages, the charge can then be passed onto adjacent capacitors until it reaches the edge of the detector, at which point it can be read out (in practice, the charge packet is read into a perpendicular register once it reaches the edge of the array, and is further transferred in this perpendicular direction; in this way, only a single amplifier circuit is required to read out an entire CCD chip). This charge transfer, or coupling, gives the device its name.

The depletion regions of CCDs are significantly shallower than that of the other semiconductor devices discussed here (of order of tens of microns), primarily because of the lower voltages required for the charge transfer. Consequently, currently available X-ray-sensitive CCDs are used most generally at energies $E <$ 10 keV since higher energy X-rays simply pass through the device. The capacitor structure on the front surface, furthermore, acts as a dead layer that can absorb photons with $E < 0.5$ keV.

19.4.1.5 Room-Temperature Devices

The solid-state devices described above operate most efficiently when cooled to temperatures <77 K (liquid-nitrogen temperatures). A class of detectors composed of semiconductor compounds (HgI_2, CdS, CdTe, and GaAs) is actually more effective at room temperatures than when cooled significantly. These materials have larger band gaps (than Si or Ge), so that thermal fluctuations result in a much lower leakage current. Furthermore, because all of these compounds have a much higher effective atomic number, a viable detector can be much thinner (have a thinner depletion layer) and still detect higher energy photons. Unfortunately, the room-temperature devices are difficult to work with, have poor mobility (particularly for holes), and have a generally inferior resolving power. In spite of some clear benefits, the various room-temperature detectors are not as commonly used as the silicon- and germanium-based devices.

19.4.2 Cryogenic Devices

A relatively recent advance in X-ray spectroscopy is the development of what are often called *cryogenic detectors.* These devices convert incident X-rays into thermal–acoustic quanta (phonons) or superconductor quasiparticles (broken Cooper pairs). The energy w required to create such a particle is quite small ($10^{-5} - 10^{-3}$ eV) so that extremely high resolving powers are possible in principle (the number of particles is large and so the relative uncertainty in their number is small). In order to achieve the optimal resolution of such devices (in theory ΔE can be as small as $\sim$1 eV), various noise sources must be suppressed by operating at extremely low temperatures ($T < 100$ mK typically), hence the generic name for such detectors – cryogenic devices. While cryogenic detectors are not yet available commercially, they hold significant promise for future applications.

Two basic classes of cryogenic detectors have been developed in recent years: calorimeters and quasiparticle detectors. The calorimeters measure the temperature in a low-temperature substrate due to the absorption of a single X-ray photon (hence, they are phonon detectors). The quasiparticle detectors most often discussed for X-ray spectroscopy are superconducting tunnel junctions (STJs) for which the mean particle energies are 0.001 eV. The quasiparticles are detected by their quantum-mechanical current tunneling through a thin oxide layer. While the STJ devices hold promise, they have not achieved, to date, the resolving power of the calorimeters and so are be discussed further (see Labov and Young (1993) for some recent articles on this topic).

The microcalorimeter has undergone extensive development over the past decade. Resolutions as high as $\Delta E \sim 2$ eV at 1.5 keV (Newbury *et al.*, 2002) have been achieved. Thus the microcalorimeter has a resolving power that is 10–50 times better than that of the best semiconductor devices. A schematic diagram of a generic calorimeter is shown in Figure 19.12. Basically, a calorimeter absorbs an incoming X-ray and converts the X-ray energy to heat (phonons), which raises the temperature of a cryogenically cooled substrate. The temperature increase is measured by a thermometer connected to the absorber. After a period of time, the equilibrium temperature is restored through the contact of the substrate with a cold bath. The temperature

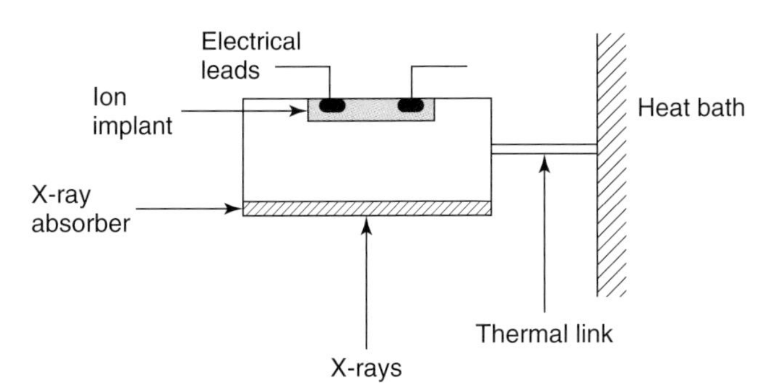

Fig. 19.12 Schematic diagram indicating the operation of an X-ray calorimeter (Holt, 1987).

increase is proportional to the energy of the X-ray.

The energy resolution of an ideal X-ray calorimeter is given by Moseley, Mather, and McCammon (1984)

$$\Delta E = 2.35\xi\sqrt{kT^2C} \qquad (19.10)$$

where T is the equilibrium temperature, C is the heat capacity of the device, k is Boltzmann's constant, and ξ is a nearly constant parameter with magnitude ~1–3. The parameter ξ is derived from the temperature variations of the heat capacity and thermal conductivity of the device, and the properties of the thermometer (Kelley *et al.*, 1988). For small detectors ($0.5 \times 0.5 \times 0.1\,\mathrm{mm}^3$ are typical dimensions), cooled to ~0.1 K, Equation 19.10 yields $\Delta E \sim 1\,\mathrm{eV}$ but, in practice, additional noise terms in the thermometer limit the current actual performance of calorimeters to ~7 eV. Note that the size of the detector is an important issue since ΔE scales as $C^{1/2}$. Note also that heat capacity often scales as T^3, so that the operating temperature is also of great importance; the best calorimeters built to date operate at 0.1 K.

The clear advantages of an X-ray calorimeter are

1. its grasp of a wide energy range (since it is nondispersive) and
2. its extremely high resolving power.

The resolution ΔE is essentially a constant; thus the resolving power ($E/\Delta E$) varies as E. Figure 19.13 shows the resolution E_{res} as a function of the X-ray energy E_ν for several types of X-ray spectrometers. If digital processing of microcalorimeter signals is performed, energy resolutions comparable to those of Bragg crystal devices in some spectral ranges (Newbury *et al.*, 2002) can be obtained.

The practical disadvantages of the X-ray calorimeter are

1. its small size (although mosaic detectors have been designed) and
2. the need to have a complex cryogenic system for the detector to function optimally.

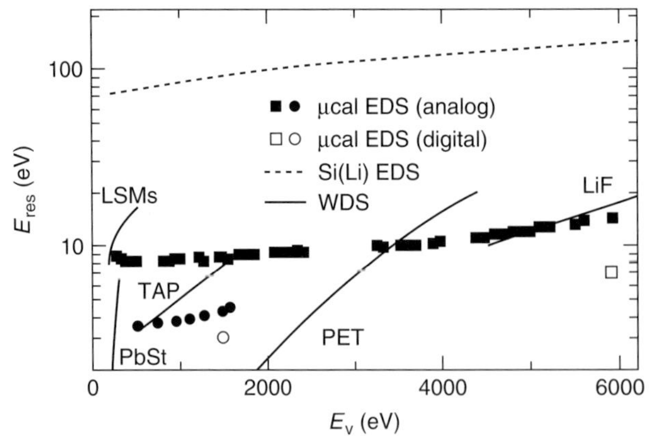

Fig. 19.13 Energy resolution (FWHM) of different types of X-ray spectrometers: microcalorimeter (µcal (energy-dispersive spectrometer (EDS)) with analog and digital data processing, lithium-drifted silicon detector (Si(Li) EDS), and wavelength-dispersive spectrometer (WDS). Dispersive elements for WDS are LSMs (layered synthetic microstructures), thallium acid phthalate (TAP), pentaerythrol (PET), lithium fluoride (LiF), and PbSt (lead stearate) (Newbury *et al.*, 2002).

The latter is a particular problem for space applications in which weight and risk are always issues. For typical designs, a large Dewar flask of liquid helium as well as an additional refrigeration stage are required.

X-ray microcalorimeters may not be the optimal choice of spectrometer in an environment in which an extremely high counting rate is expected, because of the relatively long time required to process each event (several milliseconds).

19.5 Applications

X-ray spectrometers are used in virtually every area of pure and applied science. Extensive discussion of specific applications can be found in the references following this article. In a general sense, however, X-ray spectrometers are employed for the following:

- to measure the properties (such as the temperatures, ionization states, and elemental compositions) of hot ($> 10^6$ K) plasmas, both cosmic and laboratory Silver and Kahn, 1993, where in extreme conditions time evolution and spatial distribution of laboratory plasma parameters have been obtained (Golovkin *et al.*, 2002);
- to determine compositions of various materials (qualitative and quantitative analysis via the monitoring of characteristic lines from X-ray fluorescence or direct bombardment with electrons or other charged particles; see, e.g., Jenkins (1988));
- to measure and interpret the absorption spectra of various solids at high resolution in order to determine the distribution of electrons. The study of X-ray absorption fine structure (XAFS), has blossomed in recent years with the advent of bright, monochromatic X-ray beams available from synchrotrons (Koningsberger and Prins, 1988). Ultrafast time-resolved X-ray absorption spectroscopy is a nascent technique that permits study of photo-induced processes on many systems down to the sub-femtosecond time domain (Bressler and Chergui, 2004).

Depending on the application, some of the various spectrometers discussed above may be more appropriate than others. For example, semiconductor devices, with their moderate resolving powers and relatively broadband energy response, are often used for X-ray fluorescence analysis (see Figure 19.11), since the fine structure of the spectral features is not generally an issue. When extremely high resolution is required, at least at the lower energies, crystal spectrometers are still the instruments of choice (the spectrum in Figure 19.3, e.g., would be extremely difficult to unravel with a semiconductor spectrometer). When low flux (or short exposure times) is an issue and/or a significant energy band is desired, the narrow bandpass of the crystal spectrometers argues in favor of a nondispersive device (or a grating spectrometer in which the entire dispersed spectrum can be read off simultaneously with an imaging detector). The desire for a broad bandpass and high resolution may lead the user to a calorimeter, in spite of its size, weight, and complexity. For some applications, the expected counting rate may be so high as to rule out some of the energy-dispersive detectors. For example, in many synchrotron applications, the detector of choice may be a photodiode (which measures current and not individual photons), and a dispersive device will be used as the spectrometer. Besides these complex and high-resolution experiment at synchrotron radiation centers, portable equipment for X-ray fluorescence analysis has been developed. The technological progress is based upon on miniature and air-cooled X-ray tubes, often polycapillary optics, thermoelectrically cooled X-ray detectors, multichannel analyzers and dedicated software (Tsuji, Injuk, and van Grieken, 2004). All components have small weight and are incorporated in a pistol-like housing. This portable energy-dispersive X-ray fluorescence equipment has been employed in many applications, such as archaeometry, analysis of lead in paint, environmental analysis, analysis of industrial alloys, soil analysis even on Mars, and so on.

Another factor that affects the choice of spectrometer is the nature of the optical system of which it is a component. For example, the small size of a calorimeter detector requires a tightly focused beam. Flat dispersive spectrometers (as in Figure 19.4) must receive all the radiation at the same incident angle, or the dispersed image will be blurred (thus effectively reducing the resolving power). As noted in Section 19.3.1.3, for some applications, crystals and gratings are curved so that the incident radiation will strike the crystal at nearly the same angle.

Acknowledgments

Thomas H. Markert benefited considerably from the helpful discussions with Richard Aucoin, Daniel Dewey, Keith Gendreau, George Clark, and Una Hwang.

Eckhart Förster wishes to thank Ingo Uschmann, Oldrich Renner, Konrad Goetz, and Michael Wendt for their critical remarks. Furthermore, Eckhart Förster is grateful to Jana Brusberg and Richard Hutcheon for their help in the preparation of the manuscript.

Glossary

Blazed Gratings: Gratings for which the bar, groove, or line shape is structured so as to enhance reflection into a particular order.

Cryogenic Devices: A relatively new class of X-ray spectrometers that operates at very low temperatures ($T < 2$ K typically). Such devices are capable of extremely high resolving powers ($\Delta E \sim 7$ eV has been demonstrated at 5.9 keV for at least one such device).

Diffraction: Modification of the intensity and/or phase of an electromagnetic wave by the presence of an object (such as a slit, hole, edge, etc.) in the path of the wave.

Dispersive Spectrometer: An X-ray spectrometer that employs wavelength dispersion. X-rays are dispersed (i.e., spatially separated) by the effects of diffraction. Dispersive spectrometers are sometimes called *wavelength-dispersive spectrometers.* The two classes of dispersive spectrometers are Bragg devices and diffraction gratings.

Escape Peak: For monochromatic incident X-ray photons with an energy E_x smaller than that of the absorption edge of detector atoms, the pulse height output is proportional to E_x. However, when the energy E_x of the monochromatic incident X-ray photons exceeds the absorption edge of the detector atoms, the output may contain two pulse height distributions. The additional or escape peak has a mean pulse height proportional to the difference between the energies of the incident photons E_x and the escaping (from active detector volume) photons $E_{K\alpha}$.

Monochromators: Devices used for producing radiation at a single wavelength (or, more realistically, in a narrow range of wavelengths). The elements of dispersive spectrometers (crystals, gratings) can also be used as X-ray monochromators.

Multilayers: Synthetic crystals formed by sputtering or evaporating alternating layers of high-Z and low-Z materials onto a substrate. Multilayers can be used as spectrometers or monochromators and can be a part of an X-ray optics system to enhance the performance of the system over a narrow band. Also called *layered synthetic microstructures* or *LSMs.*

Nondispersive Spectrometer: An X-ray spectrometer that does not employ wavelength dispersion. Such devices generally operate by converting the photon energy into some other sort of particle or quasiparticle (electron–ion pairs or phonons, for example). Nondispersive spectrometers are often called energy-dispersive spectrometers.

Resolving Power: The resolving power of a spectrometer (at an energy $E = hc/\lambda$) is usually defined as the ratio of the energy (or wavelength) of interest to the width of the response function of the spectrometer to a monochromatic X-ray line, that is, $R(E) = E/\Delta E = \lambda/\Delta\lambda$. The width of the line is usually chosen to be the FWHM of the spectral response function. Note that some authors invert this definition of resolving power so that it equals $\Delta E/E = \Delta\lambda/\lambda$ and often quote it as a percentage.

Rocking Curve: In crystal spectrometers, the rocking curve is a measure of the spectral resolving power of the crystal, independent of any geometric effects introduced by the geometry of the spectrometer. Specifically, it is the FWHM in degrees of the response of the crystal (as it is rotated or rocked) to a monochromatic X-ray beam.

Rowland Circle Configuration: This configuration is common in focusing X-ray spectrometers. The diverging X-ray source, the diffraction grating, and the detector all lie on a circle (the Rowland circle, after Henry Rowland) that has a diameter equal to the radius of curvature of the curved diffractor. Such a configuration gives one-dimensional focusing and disperses the spectrum along the circle. A variation, using a curved crystal rather than a grating, allows for imaging of small fields along the circle.

References

Agarwal, B.K. (**1991**) *X-ray Spectroscopy*, Springer, Berlin.

Attwood, D. (**1999**) *Soft X-rays and Extreme Ultraviolet Radiation: Principles and Applications*, Cambridge University Press, Cambridge.

Authier, A. (**2001**) *Dynamical Theory of X-ray Diffraction*, Oxford University Press, Oxford.

Bargheer, M., Zhavoronkov, N., Bruch, R., Legall, H., Stiel, H., Woerner, M., and Elsaesser, T. (**2005**) *Appl. Phys. B*, **80**, 715–719.

Becker, P. (**2001**) *Rep. Prog. Phys.*, **64**, 1945.

Bragg, W.H. and Bragg, W.L. (**1913**) *Proc. Phys. Soc. Lond. A*, **88**, 428–438.

Bressler, C. and Chergui, M. (**2004**) *Chem. Rev.*, **104**, 1781–1812.

Compton, A.H. and Allison, S.K. (**1935**) *X-rays in Theory and Experiment*, 2nd edn, Van Nostrand, Princeton.

Fano, U. (**1947**) *Phys. Rev.*, **72**, 26.

Fraser, G.W. (**1989**) *X-ray Detectors in Astronomy*, Cambridge University Press, Cambridge.

Gatti, E. and Rehak, P. (**1984**) *Nucl. Instrum. Methods A*, **225**, 608–614.

Golovkin, I., Mancini, R., Louis, S., Ochi, Y., Fujita, K., Nishimura, H., Shirga, H., Miyanaga, N., Azechi, H., Butzbach, R., Uschmann, I., Förster, E., Delettrez, J., Koch, J., Lee, R.W., and Klein, L. (**2002**) *Phys. Rev. Lett.*, **88**, 045002.1–045002.4.

Golub, L., Zirin, H., and Wang, H. (**1994**) *Solar Phys.*, **153**, 179–198.

Gursky, H. and Zehnpfennig, T. (**1966**) *Appl. Opt.*, **5**, 875–876.

Höll, A. *et al.* (**2007**) *High Energy Density Phys.*, **3**, 120–130.

Holt, S.S. (**1987**) *Astrophys. Lett. Commun.*, **26**, 35.

James, R.W. (**1982**) *The Optical Principles of the Diffraction of X-rays*, Ox Bow Press, Woodbridge.

Janssens, K., Adams, F., and Rindby, A. (**2000**) *Microscopic X-ray Fluorescence Analysis*, John Wiley & Sons, Inc.

Jasny, J. *et al.* (**1994**) *Rev. Sci. Instrum.*, **65**, 1631.

Jenkins, R. (**1988**) *X-ray Fluorescence Spectrometry*, John Wiley & Sons, Inc., New York.

Kelley, R.L., Holt, S.S., Madejski, G.M., Moseley, S.H., Schoelkopf, R.J., Szymkowiak, A.E., McCammon, D., Edwards, B., Juda, M., Skinner, M., and Zhang, J. (**1988**) in *X-ray Instrumentation in Astronomy II*, SPIE Proceedings No. 982 (ed. L.Golub), SPIE, Bellingham, p. 219.

Klöpfel, D., Hölzer, G., Förster, E., and Beiersdörfer, P. (**1997**) *Rev. Sci. Instrum.*, **68**, 3669.

Knoll, G.F. (**1989**) *Radiation Detection and Measurement*, 2nd edn, John Wiley & Sons, Inc., New York.

Koningsberger, D.C. and Prins, R. (eds) (**1988**) *X-ray Absorption: Principles, Applications, Techniques of EXAFS, SEXAFS, and XANES*, John Wiley & Sons, Inc., New York.

Labov, S.E. and Young, B.A. (eds) (**1993**) *J. Low Temp. Phys.*, **93**, 185–858.

Love, G. (**2002**) *Microchim. Acta*, **138**, 115.

Mißalla, T., Uschmann, I., Förster, E., Jenke, G., and von der Linde, D. (**1999**) *Rev. Sci. Instrum.*, **70**, 1288–1299.

Michette, A.G. and Buckley, C.J. (eds) (**1993**) *X-ray Science and Technology*, Institute of Physics Publishing, Bristol.

Moseley, S.H., Mather, J.C., and Mc Cammon, D. (**1984**) *J. Appl. Phys.*, **56**, 1257.

Newbury, D., Wollman, D., Nam, S.W., Hilton, G., Irwin, K., Small, J., and Martinis, J. (**2002**) *Microchim. Acta*, **138**, 265.

Pinsker, Z.G. (**1978**) *Dynamical Scattering of X-rays in Crystals*, Springer, Berlin.

Ramana Murthy, P.V. and Wolfendale, A.W. (**1993**) *Gamma – Ray Astronomy*, Cambridge University Press, Cambridge.

Renner, O., Kopecky, M., Krousky, E., Förster, E., Mißalla, T., and Wark, J.S. (**1994**) *Laser Particle Beams*, **12** (3), 539.

Sauli, F. (ed.) (**1992**) *Instrumentation in High Energy Physics*, World Scientific, Singapore.

Schattenburg, M.L., Canizares, C.R., Dewey, D., Flanagan, K.A., Levine, A.M., Lum, K.S., Manikkalingam, R., and Markert, T.H. (**1991**) *Opt. Eng.*, **30**, 1590–1600.

Silver, E. and Kahn, S. (eds) (**1993**) *UV and X-ray Spectroscopy of Laboratory and Astrophysical Plasmas*, Cambridge University Press, Cambridge.

Spiller, E. (**1994**) *Soft X-ray Optics*, SPIE Optical Engineering Press, Bellingham.

Strüder, L., Fiorini, C., Gatti, E., Hartmann, R., Holl, P., Krause, N., Lechner, P., Longoni, A., Lutz, G., Kemmer, J., Meidinger, M., Popp, M., Soltau, H., Weber, U., and von Zanthier, C. (**1998**) *J. Synchrotron Radiat.*, **5**, 268.

Tsuji, K., Injuk, J., and van Grieken, R. (**2004**) *X-ray Spectrometry: Recent Technological Advances*, John Wiley & Sons, Inc.

Uschmann, I., Malgrange, C., and Förster, E. (**1997**) *J. Appl. Crystallogr.*, **30**, 4554.

Further Reading

Bertin, E.P. (**1975**) *Principles and Practice of X-ray Spectrometric Analysis*, 2nd edn, Plenum, New York.

Dyson, N.A. (**1990**) *X-rays in Atomic and Nuclear Physics*, 2nd edn, Cambridge University Press, Cambridge.

Hows, M.J. and Morgan, D.V. (eds) (**1979**) *Charge-coupled Devices and Systems*, John Wiley & Sons, Ltd, Chichester.

Russ, J.C. (**1984**) *Fundamentals of Energy Dispersive X-ray Analysis*, Butterworths, London.

Van Grieken, R.E. and Markowicz, A.A. (eds) (**1993**) *Handbook of X-ray Spectroscopy: Methods and Techniques*, Marcel Dekker, New York.

20
Calorimetric Particle Detectors

Christian Fabjan

Handbook of Metrology. Edited by Michael Gläser and Manfred Kochsiek

ISBN: 978-3-527-40666-1

20.1
Introduction

20.1.1
The Task of Particle Detectors

In many areas of experimental physics (solid-state, nuclear, and particle physics) kinematical properties of particles or quanta of radiation need to be determined. The momentum **p** and the energy E, or equivalently the particle's velocity **v** and its mass m ($\mathbf{p} = m\beta\gamma$, $E = m\gamma c^2$, $\boldsymbol{\beta} = \mathbf{v}/c$, and $\gamma = (1 - \beta^2)^{-1/2}$), or momentum **p** and velocity **v** (Fernow, 1986) have to be measured.

A host of different particle detectors has been developed, frequently with very specialized functions. The momentum of charged particles is determined in magnetic spectrometers: particles with momentum p (GeV/c) in a field B (in tesla) describe a circular trajectory with radius r (in meters), such that $p = 0.2998Br$. Position detectors trace the trajectories. The methods used for the velocity measurement depend on the β-range to be covered. The flight time of particles traversing a known distance ("time-of-flight" (TOF) method) is clocked with sensors sensitive to the arrival time; accuracies of better than 100 ps (10^{-10} seconds) are achieved quite routinely. This method is applicable for particles velocities up to $\beta \leq 0.98$. For even more relativistic particles, the velocity dependence of the electromagnetic (EM) interaction of the charged particle in the medium is exploited. If the particle velocity v is larger than the velocity of light c/n in the medium, n being the index of refraction, the medium radiates photons at a characteristic angle $\theta = \arccos(1/\beta n)$, called *Cherenkov radiation*. Measuring the angle of the emitted photons provides useful velocity determination up to $\beta \leq 0.9999$ and hence particle identification up to $\approx$100 GeV/c.

Momentum and velocity measurements have found numerous and diverse applications despite two major shortcomings: (i) the method is limited to charged particles only; (ii) the relative accuracy scales as $\Delta p/p \sim p$; huge engineering efforts are required to employ very large (superconducting) magnets and position detectors comprising typically 10^6 or more position sensors, to achieve adequate momentum resolution at the 100 GeV/c scale.

For the measurement of neutral and very energetic particles a further technique is required: calorimetry.

20.1.2
Energy Measurement with Calorimeters

The measurement of the particle's energy E relies on "total absorption" in the detector

Handbook of Metrology. Edited by Michael Gläser and Manfred Kochsiek

ISBN: 978-3-527-40666-1

occurring within a few tens of nanoseconds through a cascade of interactions producing increasingly lower-energy particles. The kinetic energy of the particle is initially transformed into excitation or ionization energy of the absorber atoms. After approximately 10 μs, all energy appears in thermal form, hence the name "calorimeter." Calorimeters are instrumented blocks of mostly dense absorbers with sensors embedded to detect these atomic excitations that reflect the strength of the particle cascade and provide a measure related to the energy of the absorbed particle.

Calorimeters are widely used in experimental particle physics for several reasons:

1. They are sensitive to both charged (e.g., protons, pions, kaons, electrons) and neutral (e.g., neutrons, neutral pions, photons) particles.
2. The absorption process is statistical, so that in well-designed instruments the relative accuracy of energy measurements improves with the energy E, in principle, as $\sigma(E)/E \sim E^{-\frac{1}{2}}$.
3. The total depth D necessary to fully absorb (or "contain") the particle increases slowly with energy as $D \sim \ln E$, an important practical advantage.
4. The fundamental constituents of matter (called *quarks*) manifest themselves as tightly collimated groups of charged and neutral particles, whose total energy can only be measured with calorimeters.

In nuclear physics, the energy of particles (protons, nuclear fragments, electrons) or gamma quanta is typically in the mega electronvolt range; relatively small volumes of absorber (a few tens of cubic centimeters) are sufficient (Section 20.2).

In particle or high-energy physics (Sections 20.3–20.5), energies range from 100 MeV (0.1 GeV) to several 10^6 MeV (1 TeV). Total absorption proceeds, depending on the type of particle, via high-energy EM, strong, or even weak interactions.

Astroparticle physics (Section 20.6) has spawned the most original and diverse developments covering the widest spectrum of particles – known and unknown – as well as the largest energy range, from mega electronvolts to exa electronvolts (10^{19} eV).

The layout of a generic particle physics experiment is shown in Figure 20.1. The momentum of charged particles is derived from the track curvature in a magnetic field using position detectors that barely affect the particle momentum. Calorimeters measure the energy of particles, charged or neutral. Photons, electrons, and positrons require relatively little mass to be absorbed. Hadrons, interacting predominantly through the strong interaction, require considerably more material for total absorption. The destructive nature of the calorimetric energy measurement imposes the spatial sequence of the measurement tasks and hence fixes the layout of experiments.

20.2 Calorimetry Methods in Nuclear Physics

20.2.1 Total Absorption and Range of Particles

In nuclear physics, the energy measurement of charged (protons, alphas, nuclei, fission fragments, electrons, positrons) and neutral (neutral pions and, in particular, neutrons) particles is frequently required. Precision energy measurements of photons emitted from excited nuclear states are also important. The energies range typically from 100 keV to 100 MeV. In this energy range the absorption of charged

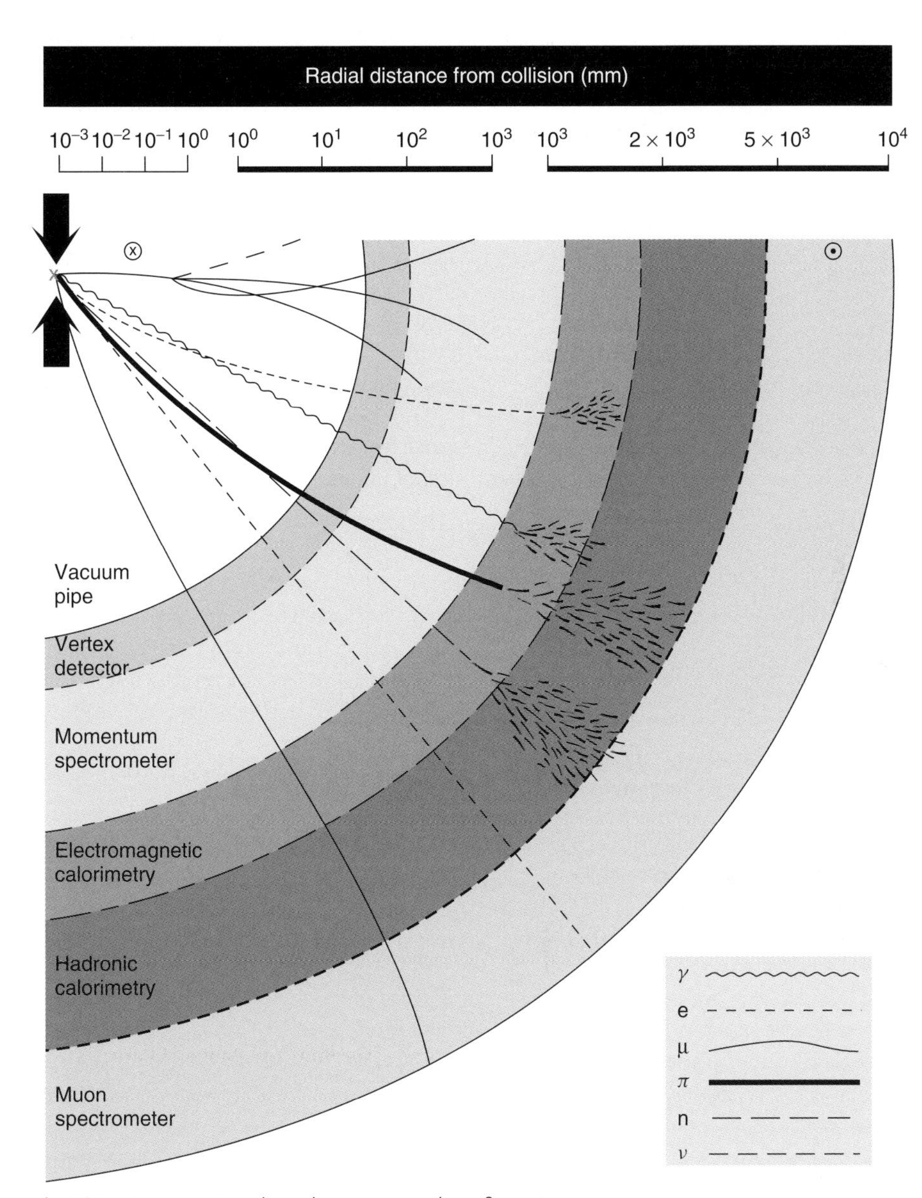

Fig. 20.1 Cross section through a conceptual configuration of a collider detector: Like "Russian dolls," a sequence of detectors enclose the collision point. In each detector layer, the particle is subjected to a specific measurement, shedding layer by layer, its physics information. Note the three different logarithmic scales of the radial dimensions.

particles proceeds essentially through the EM interaction.

The "mean energy loss per unit path length x" or "stopping power" is given to a good approximation by the Bethe–Bloch equation (Yao *et al.*, 2006):

$$\left[\frac{dE}{dx}\right]_{\text{inc}} = \frac{CZ_{\text{med}}\rho_{\text{med}}}{A_{\text{med}}}\left[\frac{Z_{\text{inc}}}{\beta}\right]^2 \times \left[\ln\left(\frac{2m_e\gamma^2\beta^2c^2}{I}\right) - \beta^2 + \text{correction terms}\right] \tag{20.1}$$

where $C = 0.307\,\text{MeV}\,\text{cm}^2\ \text{g}^{-1}$, Z_{med}, Z_{inc} denote the charge of the absorbing medium and incident particle respectively, m_e is the electron mass, and I the effective ionization potential of the absorber. Owing to the $1/\beta^2$ dependence on the particle velocity, most of the energy loss occurs near the end of the path. The energy loss, when expressed in units of length × density (grams per cm squared), that is, the number of atomic electrons or scattering centers encountered, is approximately independent of the material (Figure 20.2).

The depth at which half the initial particles are stopped or "ranged out" is called the *mean range* $R(E)$. It is related to

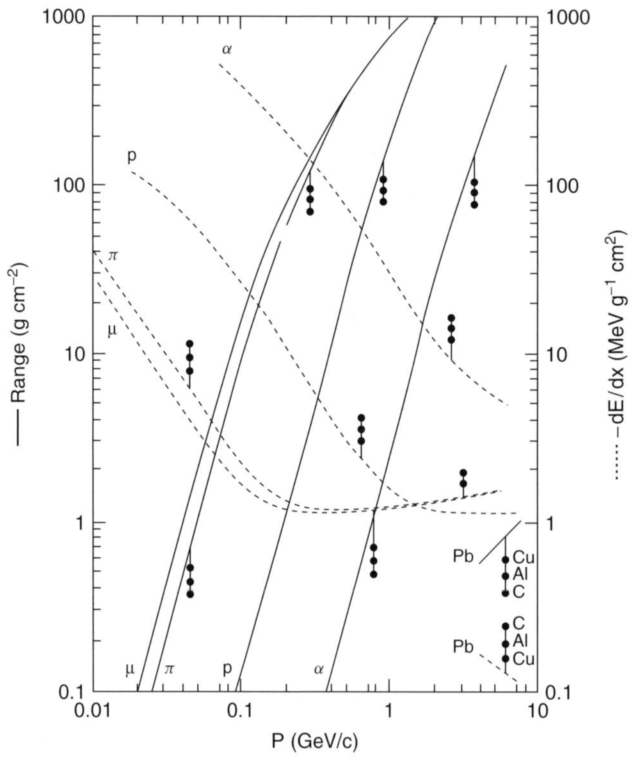

Fig. 20.2 Mean range (left ordinate, dashed lines) and energy loss (right ordinate, dashed lines) due to ionization for the indicated particles in Pb, with scaling to Cu, Al, and C indicated, using the Bethe–Bloch equation, including corrections (Yao *et al.*, 2006).

the energy loss by

$$R(E) = \int_{E}^{0} -dE/\left(dE/dx\right) \qquad (20.2)$$

Typical values of the mean energy loss and the range of various particles (Yao *et al.*, 2006) are given in Figure 20.2 and can be used to estimate the required absorber depth to range out a charged particles in this energy domain. As an example, a pion with a momentum of 100 MeV/c has a range of approximately 10 g cm^{-2} in lead, that is, 0.9 cm; a lead layer of $\approx$1 cm will absorb or range out such a particle.

Neutrons with mega electronvolt energies are detected through elastic scattering (via the strong force) off the detector nuclei. The scattering process transfers a fraction of the neutron kinetic energy to the target nucleus, usually hydrogen, deuterium, or helium, resulting in a recoil nucleus. The recoil nuclei in turn lose their energy through ionization of the detector medium. This process is repeated until essentially all of the kinetic energy of the neutron has been transferred to the detector nuclei.

Low-energy quanta in the electronvolts to kiloelectronvolts range will ionize the atoms of the detector medium and produce detectable electrons (atomic photo effect). At higher energies, pair production, the conversion into electron–positron pairs in the EM field of the nucleus, dominates. These processes are described in Section 20.3.

20.2.2 Principal Detection Techniques: Scintillators and Solid-State Detectors

Having described the absorption processes relevant for particles in the mega electronvolt region, in this section we address the two principal techniques, scintillators and semiconductor detectors, used for measuring particles in this energy range.

In scintillators, the molecules are excited by the interaction of the charged particles, converting a small fraction into visible photon emission. Time constants are typically 10^{-9}–10^{-6} seconds. Both inorganic (crystals) and organic scintillators are used. Inorganic scintillators are usually grown with an admixture of dopants acting as recombination centers for ionization charge with subsequent excitation, followed by light emission. Certain inorganic scintillation crystals are compounds of heavy elements and hence have a large absorption coefficient and an excellent energy resolution for gamma rays (see Section 20.4). More recently, the scintillating properties of noble liquids, in particular liquid xenon, have been investigated. These liquids combine high gamma absorption cross sections with a very fast time response (Doke, 2005).

Only a relatively small fraction of the absorbed energy, typically a few percent, is converted into visible light that can be detected with photon detectors (e.g., photomultipliers). Nevertheless, the light output is quite sufficient; in "good" scintillators there may be as many as 10^4 photons produced for 1 MeV of absorbed energy. This method is suitable, because the average fraction of energy converted into light is constant and the light output is proportional, apart from saturation effects (Section 20.4), to the energy deposited.

The second category of detectors, also pioneered by nuclear physicists, uses the measurement of ionization in semiconductors (predominantly silicon and germanium) (see ***Detectors, Semiconductor***). In these materials, the energy ε required to create the basic measurable excitations, electron–hole pairs, is approximately $\varepsilon \approx 3$ eV, compared to typically 100 eV/photon needed in a scintillator. Therefore, for a given energy deposit, the statistical signal fluctuations σ are smaller and the energy resolution can be better. Experimentally, one finds $\sigma^2(E) = (\mathrm{FE}/\varepsilon)\varepsilon^2$, and a "Fano" factor $F \approx 0.1$ for Si and Ge; the limit of energy resolution is therefore $\sigma \sim 500\,\mathrm{eV}$ for 1-MeV energy deposit (Fabjan and Gianotti, 2003).

Semiconductor detectors work as ionization chambers in which the charge created by the energy deposit is collected under the influence of an externally applied electric field. Registration of this ionization charge is, however, only possible if it is not masked by free charges, generated thermally or because of the presence of impurities. Adequate suppression is possible through purification of the material or implementing a diode structure in the detector material, operated in reversed bias mode. For Ge, impurity levels as low as 10^9 atoms/cm^3 have been achieved. Such ultrapure crystals are fabricated in sizes of hundreds of cubic centimeters, making them the detector of choice for γ-ray spectroscopy in the energy range of 10 keV to a few megaelectronvolts. Silicon cannot yet be purified to this level, limiting the useful depth of Si detectors to a few millimeters. Such detectors are still well suited for the energy measurement of X-rays and heavy charged particles, such as protons, alphas, and fission fragments.

20.3 Calorimetry Methods in Particle Physics: Photons, Leptons, and Hadrons

20.3.1 Absorption Mechanisms

In the "high-energy" domain ($E \geq 100$ MeV) of calorimetry the incident particle is sufficiently energetic to initiate particle production. This provides a further mechanism for absorption in addition to those previously discussed (ionization, excitation, etc.). With increasing energy of the incident particle, the energy of the secondary particles may be sufficient for subsequent, that is, tertiary production, and so forth. The image of an "avalanche" or shower of particles appropriately describes this process of energy degradation. The particle multiplication continues until the energy of the last generation drops below the energy threshold for further production; at this point, the shower has reached the maximum number of particles – the shower maximum – after which further degradation continues via the low-energy processes described in Section 20.3.2.

There are two distinct types of particle showers. The first type is initiated by energetic electrons (positrons) and photons and propagates through the EM interaction with the atoms of the absorber. The second type of cascade is produced by particles strongly interacting with the absorber, that is, hadrons, such as pions, kaons, and protons. In both types of cascade, the relevant interaction cross sections reach values that are rather energy independent for particle energies $E \geq 1\,\mathrm{GeV}$. Therefore, the cascade dimensions are described by a mean free path λ between collisions with a specific value of λ for a given material, in general, different for the EM and hadronic cascades. The geometrical dimensions of a shower of given energy in different absorber materials will be similar, when expressed in the appropriate physics units of the mean free path (scaling in λ).

The principle of the calorimetric energy measurement of energetic particles may be formulated as follows: particle cascading (through the EM or strong interactions) degrades the energy over rather small geometrical distances. The charged component in the cascade is measured through its ionizing or excitation effect on the medium, which provides a signal approximately proportional to the energy of the incident particle (Fabjan and Gianotti, 2003).

20.3.2 Physics of the Electromagnetic Cascade

In spite of the apparently complex phenomenology of shower development in a material, electrons and photons interact with matter via a few well-understood quantum electrodynamics (QED) processes, and the main shower features can be parameterized with simple empirical functions.

The average energy lost by electrons and the photon interaction cross section is shown in Figure 20.3 for a frequently used absorber material, lead, as a function of energy. Two main regimes can be identified. For energies larger than ~10 MeV, the main source of electron energy loss is bremsstrahlung. In this energy range, photon interactions mainly produce electron–positron pairs. Above 1 GeV, both processes become roughly energy independent. At low energies, electrons lose their energy mainly through collisions with the atoms and molecules of the material, giving rise to ionization and thermal excitation; photons lose their energy through Compton scattering and the photoelectric effect.

Electrons and photons of sufficiently high energy ($\geq$1 GeV) incident on a block of material produce secondary photons by bremsstrahlung, or secondary electrons and positrons by pair production. These secondary particles in turn produce other particles by the same mechanisms, thus giving rise to a cascade (shower) of particles with progressively degraded energies. The number of particles in the shower increases until the energy of the electron component falls below a critical energy ε, where energy is mainly dissipated by ionization and excitation and not in the generation of other particles.

The main features of EM showers (e.g., their longitudinal and lateral sizes) can be described in terms of one parameter, the radiation length X_0, which depends on the characteristics of the material (Yao *et al.*, 2006)

$$X_0\ (g/cm^2) \approx \frac{716\ gcm^{-2}A}{Z(Z+1)\ln(287/\sqrt{Z})} \tag{20.3}$$

where Z and A are the atomic number and weight of the material, respectively. The radiation length X_0 represents the average distance x an electron needs to travel in a material to reduce its energy to $1/e$ of its original energy E_0

$$\langle E(x)\rangle = E_0 \exp\left(-\frac{x}{X_0}\right) \tag{20.4}$$

Similarly, a photon beam of initial intensity I_0 traversing a block of material is absorbed mainly through pair production. After traveling a distance $x = 9/7\ X_0$ its intensity is reduced to $1/e$ of the original intensity

$$\langle I(x)\rangle = I_0 \exp\left(-\frac{7}{9}\frac{x}{X_0}\right) \tag{20.5}$$

These relations show that the physical scale over which a shower develops is similar for incident electrons and photons, and is independent of the material type if expressed in terms of X_0. Therefore, EM showers can be described in a universal way by using simple functions of the radiation length.

The shower maximum, that is, the depth at which the largest number of secondary particles is produced, is approximately

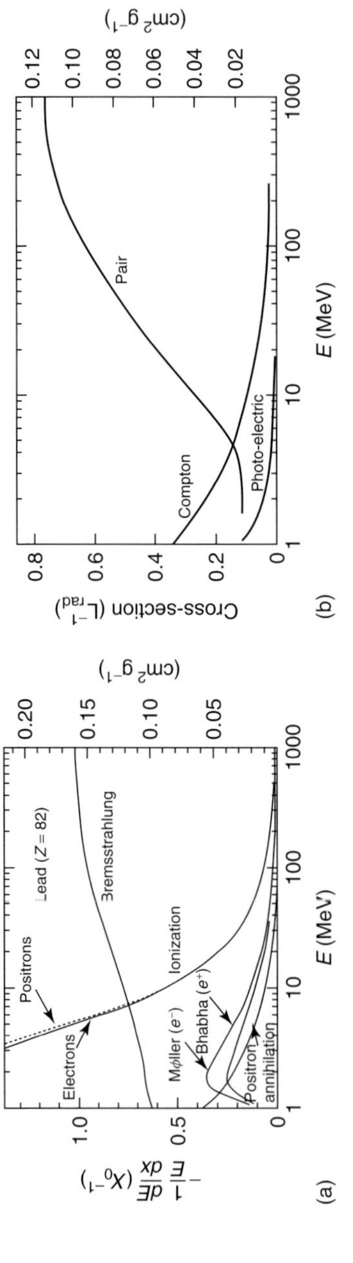

Fig. 20.3 (a) Fractional energy lost in lead by electron and positrons as a function of energy. (b) Photon interaction cross section in lead as a function of energy (Yao *et al.*, 2006).

located at $t_{max} \approx \ln(E/\varepsilon) + t_0$, where t_{max} is measured in radiation lengths, E is the incident particle energy, and $t_0 = -0.5(+0.5)$ for electrons (photons). This formula shows the logarithmic dependence of the shower length, and therefore of the detector thickness needed to absorb a shower, on the incident particle energy. Longitudinal shower profiles for different energies of the incident particles are shown in Figure 20.4a and for electrons at 6 GeV in different materials (Figure 20.4b). The calorimeter thickness, containing 95% of the shower energy, is approximately given by $t_{95\%} \approx t_{max} + 0.08Z + 9.6$, where t_{max} and $t_{95\%}$ are measured in radiation lengths. As an example, for a calorimeter with thickness $\approx 25\ X_0$, the longitudinal shower leakage beyond the end of the active detector is less than 1% up to incident electron energies of ~300 GeV. Therefore, even at the particle energies expected at the CERN Large Hadron Collider (LHC), of order ~teraelectronvolts, EM calorimeters are very compact devices: the ATLAS lead–liquid argon calorimeter (ATLAS Collaboration, 1996) and the CMS crystal calorimeter (CMS Collaboration, 1997) have thicknesses of ≈45 and ≈23 cm, respectively (the radiation lengths are ≈1.8 and ≈0.9 cm, respectively).

20.3.3 Physics of the Hadronic Cascade

Understanding the physics of the hadronic cascade is the basis for the construction of powerful, widely used instruments reaching performance limits imposed by the physics of the hadronic cascade.

By analogy with EM showers, the energy degradation of hadrons proceeds through

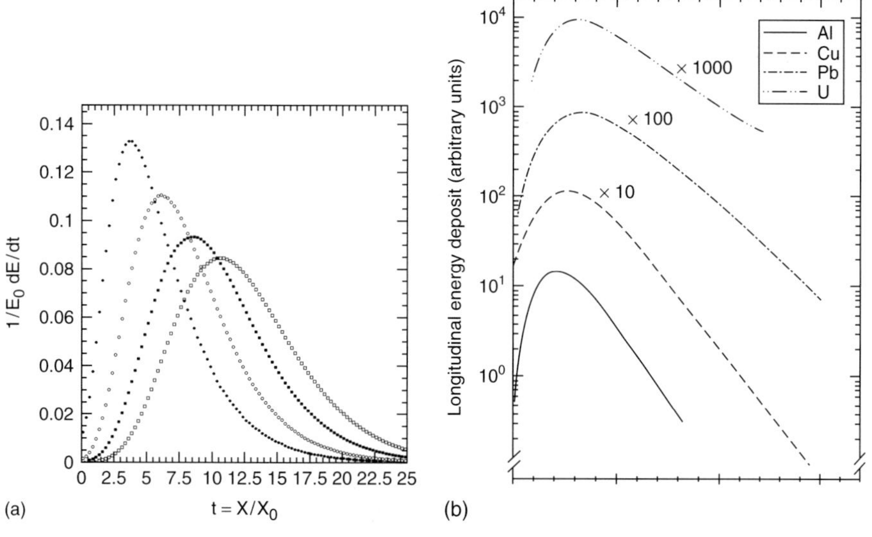

Fig. 20.4 (a) Simulated shower longitudinal profiles in $PbWO_4$. as a function of material thickness (expressed in radiation lengths), for incident electrons of energy (from left to right) 1, 10, 100, 1000 GeV. (b) Longitudinal shower development of 6 GeV/c electrons in four very different materials, showing the approximate scaling in units of radiation lengths X_0.

an increasing number of mostly strong interactions with the calorimeter material.

In analogy to EM cascades, a characteristic scaling length describes the longitudinal shower shape. This is the interaction length λ, defined as

$$\lambda = A/N_{\mathrm{Av}}\sigma_{\mathrm{i}}\left[\mathrm{g\,cm^{-2}}\right] \approx 35\ \mathrm{A}^{1/3}\ \mathrm{g\,cm^{-2}} \tag{20.6}$$

A being the mass number of the material, N_{Av} the Avogadro number, and σ_i the inelastic cross section. It describes the absorption of N hadrons in matter through the equation

$$dN/dx = -N/\lambda \tag{20.7}$$

For incident hadrons with sufficient energy to produce secondary hadrons ($E \geq 1$ GeV) a hadron shower is initiated. Typical longitudinal shower profiles in different absorbers are shown in Figure 20.5 and some useful features can be parameterized in approximate form:

1. shower maximum $S_{\max}[\lambda] \approx 0.2 \ln E$ (GeV) $+ 0.7$;
2. longitudinal depth for 95% containment $L(0.95)[\lambda] \approx S_{\max} + 2.5E^{0.13}$ (GeV); and
3. transverse radius for 95% containment $R[\lambda] \approx 1[\lambda]$.

The hadronic interaction produces two classes of effects. Energetic secondary hadrons are produced with a mean free path (interaction length) λ between interactions; their momenta typically add up to a fair fraction of the primary hadron momentum, that is, at the gigaelectronvolt scale. Second, in the hadronic collisions with the absorber nuclei, a significant part of the hadron energy is consumed in nuclear processes such as excitation, nucleon evaporation, spallation, and so on, producing particles with characteristic nuclear energies at the megaelectronvolt scale.

The fast, energetic hadronic component contains protons, neutrons, and charged and neutral pions. Owing to the charge independence of hadronic interactions, in each high-energy collision, on average, one third of the pions produced will

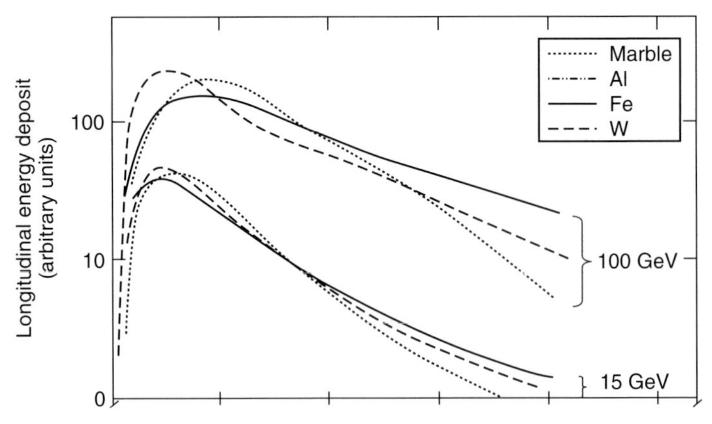

Fig. 20.5 Hadron-induced longitudinal shower development in different materials, showing approximate scaling in interaction length λ. The shower distributions are measured with respect to the face of the calorimeter (Fabjan and Ludlam, 1982; reproduced with permission of Annual Review of Nuclear and Particle Science.)

be π^0's, $F_{\pi 0} = 1/3$. This is the "odd man" in this group of energetic hadrons because these neutral pions will decay to two photons, $\pi^0 \rightarrow \gamma\gamma$, before having a chance to reinteract hadronically. These energetic photons will induce an EM cascade, proceeding along the laws of EM interactions (Section 20.3.2). This physics process acts like a "one-way diode," transferring energy from the hadronic part to the EM component, which will not contribute further to hadronic processes.

As the number of shower generations increases with increasing incident energy, so will the fraction of the EM cascade: there is an "identity change" of a predominantly hadron-rich cascade at low incident energy to a cascade developing into a mostly EM cascade at very high energies.

Second, contributions from nuclear interactions produce photons and neutrons in the few megaelectronvolt range. The total energy carried by photons from nuclear reactions is substantial: only a fraction, however, will be recorded in practical instruments, as most of these photons are emitted with a considerable time delay ($\geq 1\,\mu$s). These nuclear effects, represented by delayed photons and soft neutrons produce signals that are poorly or not at all detected in practical instruments. One part of the cascade particles is truly "invisible" or "missing," namely energy carried by escaping muons, neutrinos or backscattered (albedo) particles. Let η_e be the efficiency for observing a signal E^e_{meas} (measurable energy) from an EM shower, that is, $E^e_{\text{meas}} = \eta_e\, E$ (em); let η_h be the corresponding efficiency for purely hadronic energy to provide visible energy in an instrument. Therefore, for a pion-induced shower the visible energy E^{π}_{meas} is

$$E^{\pi}_{\text{meas}} = \eta_e F^0_{\pi} E + \eta_h F_h E \tag{20.8}$$

where E is the incident pion energy. In Figure 20.6, the dependence of the various components as a function of the energy of the incident hadron is shown.

The ratio of observable signals induced by EM and hadronic showers, usually denoted by e/π, is therefore

$$\frac{E^{\pi}_{\text{meas}}}{E^{e}_{\text{meas}}} = \left(\frac{e}{\pi}\right)^{-1} = 1 - \left(1 - \frac{\eta_h}{\eta_e}\right) F_h \tag{20.9}$$

In general, $\eta_h \neq \eta_e$; therefore, the average response of a hadron calorimeter as a function of energy will not be linear because F_h decreases with incident energy. More subtly, for $\eta_h \neq \eta_e$, we have to expect that event-by-event fluctuations in the F_h and F^0_{π} components will have an impact on the energy resolution of such instruments. The relative response e/π turns out to be the most important yardstick for gauging the performance of hadronic calorimeters (HCs) (Wigmans, 1991).

20.4 Performance of Calorimeters

Suitably instrumented calorimeters measure not only the energy but also the time, direction, and impact point of the incident particle. The quality of the measurement (resolution) is driven by fluctuations in the measurement process, both intrinsic due to the physics of particle absorption as well as due to a range of instrumental contributions.

With increasing energy the particle content of the shower increases and therefore intrinsic fluctuations will scale as $1/\sqrt{}$ (number of particles) $\sim 1/\sqrt{E}$. Instrumental contributions may scale differently with energy. The conventional parameterization for the energy resolution is $\sigma(E)/E = a/\sqrt{E} \oplus b/E \oplus c$, a denoting the stochastic,

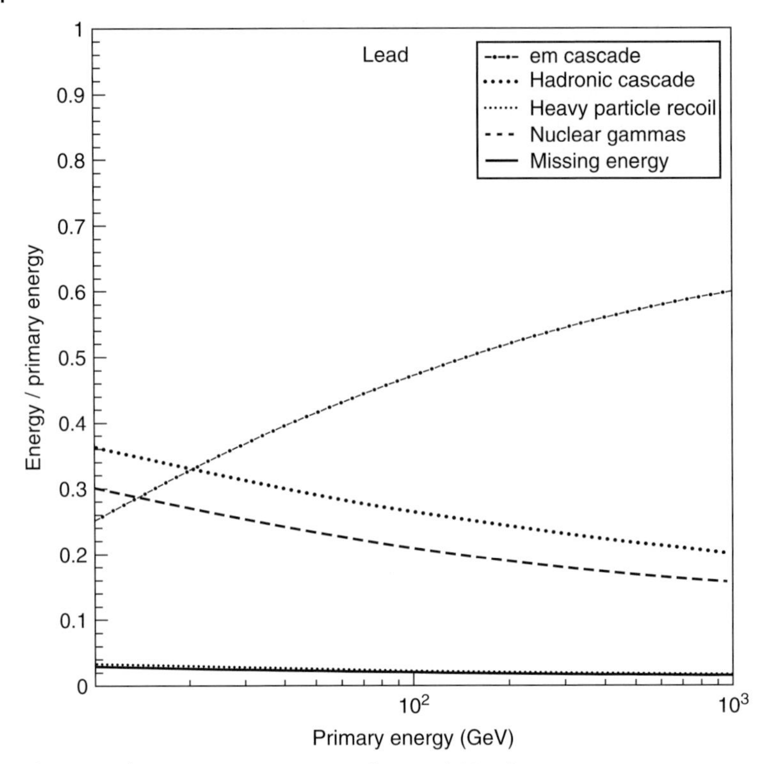

Fig. 20.6 Characteristic components of proton-initiated cascades in lead. With increasing primary energy the em (π^0) component increases. The component due to nuclear photons is in general time delayed and is detected with very low efficiency in practical instruments (A. Ferrari, 2001, private communication).

b the "noise term," and c a constant term. Measurement of the shower shape provides important information on the particle impact, its direction and time of arrival. Furthermore, the shower shape encodes information about the particle type.

20.4.1 **Energy Resolution in Homogeneous Electromagnetic Calorimeters**

Homogeneous calorimeters are instruments in which the absorber serves also as the detecting medium. These fully sensitive instruments provide the most complete information about the shower cascade and hence the potentially best measurement performance. This type of instrument is particularly suitable for precision EM calorimeters, for which crystals (Table 20.1) and liquefied noble gases (Table 20.2) are the most frequently used materials. At low (~ 1 MeV) energies the achieved energy resolutions are at the 1–2% level. At these low energies, the measured energy resolution for crystals is quite consistent with counting statistics, based on the number of registered photons. For noble liquids, however, the measured low-energy resolution is a large factor (5–10) worse compared to

Tab. 20.1 Characteristics of commonly used crystals for electromagnetic calorimeters.

Crystal	Density ($g\,cm^{-3}$)	Radiation length (cm)	Measured energy resolution (σ at 1 MeV)	Reference
CsI	4.53	1.85	2%	Moses (2002) and Lecoq *et al.* (2006)
BGO	7.13	1.12	3.1%	Moses (2002) and Lecoq *et al.* (2006)
PbW04	8.28	0.89	n.a	Moses (2002) and Lecoq *et al.* (2006)

Tab. 20.2 Characteristics of noble liquids used in electromagnetic calorimeters.

Liquid	Density ($g\,cm^{-3}$) at boiling point	Radiation length (cm)	Measured energy resolution (sigma at 1 MeV)	Reference
Argon	1.40	14	1.2%	Doke (2005)
Krypton	2.41	4.7		Doke (2005)
Xenon	2.94	2.8	1.6%	Doke (2005)

the theoretical one based on the number of created ion pairs. Recombination and incomplete collection of the generated ion pairs are thought to be the origin of this discrepancy.

For particle physics applications at high energy – above 1 GeV – naïve $1/\sqrt{E}$ scaling from the 1-MeV energy domain is not observed. Such scaling would give resolutions at the 0.03% level at 1 GeV, far better than the measured resolutions, observed at the 1% level. This fact is generally attributed to the nonlinear response of the detector medium to the very low energy component (below 1 MeV) of the EM cascade, frequently parameterized with a cutoff value below which the detector medium effectively becomes insensitive. Neither detailed simulations with state-of-the-art Monte Carlo codes nor the low-energy results support this explanation. Plausibly, several effects conspire – a certain fraction ($\leq 0.5\%$) of the shower energy escapes through the front, resulting in $\sigma(E)/E \sim 0.3\%/\sqrt{E}$. In practical crystal calorimeters inhomogeneous light production and transport also contribute at a comparable level as do counting statistics and electronics noise. Above 1 GeV this $1/\sqrt{E}$ scaling is observed to hold to a very good approximation over a very wide energy range from one to several hundred gigaelectronvolt (CMS Collaboration, 1997; Lecoq *et al.*, 2006).

20.4.2 Instrumental Contributions to the Electromagnetic Energy Resolution

Sampling fluctuations arise in calorimeters that are not uniformly (homogeneously) sensitive, that is, in instruments having their signal sensors interspersed in the absorber (see discussion in Section 20.5.2). Such constructions are very common for electromagnetic calorimeters (EMCs), and almost exclusively used HCs. The shower profile is sampled, leading to incomplete signal information with increased signal fluctuations. The magnitude can be estimated by considering the variations in the number of charged particles N_{ch} that cross the active layers and produce the detectable signal. This number is proportional to

$$N_{ch} \propto \frac{E_0}{\Delta E} \qquad (20.10)$$

where ΔE is the energy loss in one "unit cell" (absorber plate + active layer). Assuming statistically independent crossings of the active layers, the "sampling" contribution to the energy resolution due to the fluctuation of N_{ch} are

$$\frac{\sigma}{E} \div \frac{1}{\sqrt{N_{ch}}} \div \sqrt{\frac{\Delta E}{E_0(\text{GeV})}} = c\sqrt{\frac{\Delta E(\text{MeV})}{E_0(\text{GeV})}} \qquad (20.11)$$

where $c = 0.032$.

In reality, the particles are not statistically independent and cross the gap on average under an angle different from normal, increasing ΔE. A coefficient $c \approx 0.05$ is consistent with measurements in many different sampling EMCs (Fabjan and Gianotti, 2003; Wigmans, 2000).

In practical sampling, EMCs the energy resolution is in the range of 5–20%/$\sqrt{E(\text{GeV})}$. In well-designed instruments it is below 1% for 100-GeV photons or electrons, consistent with the detection requirements for the physics discovery at the CERN LHC (ATLAS Collaboration, 1996).

There are several other instrumental contributions to the energy resolution that can be kept at a level below the intrinsic resolution of homogeneous calorimeters, well below the sampling contribution. Incomplete shower containment (transverse or longitudinal shower leakage) affects the performance, as do mechanical or signal-collection nonuniformities. These instrumental contributions imply a lower limit to the resolution, usually parameterized with a constant term c. At low energies, electronic readout noise may contribute to the resolution, decreasing with energy as $1/E$. Achieving 1% energy resolution implies control of all these instrumental effects at a level well below 1%: a major engineering challenge.

20.4.3 Time, Space, and Angular Resolution in EMCs

While energy measurement is the principal task of calorimeters, such instruments also provide additional powerful information about the detected photons or electrons (positrons).

Pulse-shape analysis of the detected signal can provide information on the arrival time of the particle. The performance depends – as always – on the signal/noise ratio and therefore scales – at least in principle – with $1/\sqrt{E}$. In modern instruments, the timing resolution is typically at a level of a few nanoseconds/$\sqrt{E}$ (in gigaelectronvolts).

Spatial and angular resolution is achieved by appropriate subdivision of the detector in transverse and longitudinal directions. The scale for the subdivision is the radiation length X_0. Longitudinal cells typically have dimensions of a few X_0, while transverse subdivisions are at the scale of 1 X_0. Such granularity in the readout allows a spatial resolution of the impact point of the particle at the level of 1 cm $/\sqrt{E(\text{GeV})}$ and an angular measurement of ~50 mrad$/\sqrt{E(\text{GeV})}$, (ATLAS Collaboration, 1996).

20.4.4 Energy Resolution of Hadron Calorimeters

The general analysis presented for EMCs is conceptually also applicable to HCs, albeit with significant quantitative differences due to the physics of the much more complex hadronic cascade.

Inevitably, hadronic cascades imply nuclear interactions with their correlated production of low-energy neutrons and photons, only partially detectable (invisible energy) (Chapter 3). On average, for particles with the same incident energy, the signal response of a calorimeter to hadrons (e.g., pions, π) will be lower than that to electrons, that is, $e/\pi > 1$. For hadrons, the visible energy will fluctuate, event by event, between two extremes: from fully EM, yielding the same signal as an electron to fully hadronic with a maximum of invisible energy, as shown conceptually in Figure 20.7.

This simple analysis already provides the following qualitative conclusions for HCs with $e/\pi \neq 1$:

- fluctuations in $F_{\pi 0}$ are a major contribution to the energy resolution;
- the average value $< F_{\pi 0} >$ is energy dependent and therefore calorimeters have a response to hadrons, which is nonlinear with energy;
- the above-mentioned fluctuations are non-Gaussian and therefore the energy resolution scales weaker than $1/\sqrt{E}$.

Empirically one finds that such instruments show typically a $e/\pi \approx 1.4$ and an energy resolution of $\sigma/E \geq 0.4/\sqrt{E(\text{GeV})}$.

Understanding these effects suggests how to design instruments in which the response to this nuclear component is boosted or "compensated for," thereby achieving on average an equal response to electrons and pions, $e/\pi = 1$. Such instruments are referred to as *compensated calorimeters*.

Remarkably, it is possible to "tune" the e/π response of a calorimeter in the quest

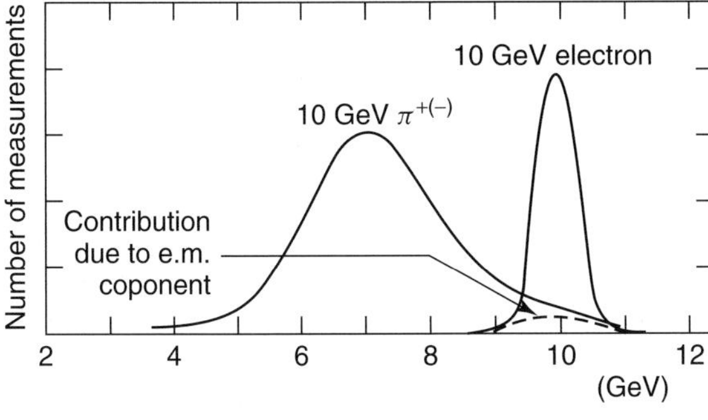

Fig. 20.7 Response function of calorimeters for electrons and hadrons. The curves are for an EMC with $\sigma/E = 0.1/\sqrt{E}$ and for a HC with $\sigma/E = 0.5$ and $e/\pi = 1.4$. A fraction of the hadron-induced cascade propagates almost completely electromagnetically, producing non-Gaussian tails and broadening of the response function.

to achieve $e/\pi = 1$, and thus optimize the performance. This "tuning" is based on the fact that practical hadron calorimeters are almost always built as sampling devices; the energy sampled in the active layers, E_{samp}, is typically a small fraction, a few percent or less, of the total incident energy.

The energetic hadrons lose relatively little energy ($\leq$10%) through ionization before being degraded to such low energies that nuclear processes dominate. Therefore the response of the calorimeter will be strongly influenced by the response to neutrons (n/mip) and photons (γ/mip) relative to the signal produced by the fast (mip) hadrons (in both the absorber and the readout materials).

As an example, it is possible to influence the response to the abundant neutrons by choosing a readout material rich in hydrogen. The n–p elastic cross section is large and on average half of the neutron kinetic energy is transferred to the recoil proton produced in the active material, contributing directly to the calorimeter signal: a 1 MeV proton has a range of $\sim$20 μm in a scintillator and is not sampled unlike a fast hadron (mip). This difference in response between high-Z absorbers and hydrogen-containing readout materials is exploited in the "tuning."

The mip signal will be inversely proportional to the thickness of the absorber plates: increasing the thickness of the absorber plates will decrease E_{samp}, whereas the signal from proton recoils will not be affected and hence neutrons will contribute more significantly to e/π. Changing the sampling fraction changes e/π (Wigmans, 1991).

Having tuned the response to $e/\pi \sim 1$ results also in much improved energy resolution, as conceptually suggested in Figure 20.6, and qualitatively borne out by shower and signal simulations, confirmed by measurements (Fabjan and Gianotti, 2003). One observes a significant reduction in the fluctuations and an intrinsic hadronic energy resolution of $\sigma/E \sim 0.2\sqrt{E(\text{GeV})}$. In principle, tuning of e/π can be applied to all sampling calorimeters, opening the way to better performance. Note that this tuning can only be achieved in sampling calorimeters: a homogeneous calorimeter, for example, a bismuth germanate (BGO) crystal calorimeter would show the typical response of an uncompensated device with a resolution of $\sim 0.4/\sqrt{\text{E(GeV)}}$.

Sampling fluctuations contribute of course to the energy resolution for which a similar empirical parameterization can be given as for EMCs:

$$\sigma_{\text{samp}}/E = c \times (\Delta E(\text{MeV})/E(\text{GeV}))^{1/2} \tag{20.12}$$

where ΔE is the energy lost in one sampling cell, with c(hadron) ≈ 0.09 for HCs (Fabjan and Gianotti, 2003). For high-performance hadron calorimeters, sampling fluctuations can usually not be neglected.

We can summarize the principles of modern, optimized hadron calorimetry as follows:

- The key performance parameter is $e/\pi = 1$, which guarantees linearity, $E^{-1/2}$ scaling of the energy resolution, and excellent resolution.
- With a proper choice of type and thickness of active and passive materials, the response can be tuned to obtain $e/\pi \approx 1$.
- The intrinsic resolution in practical hadron calorimeters can be as good as $(\sigma/E)\sqrt{E} \leq 0.2$;
- Sampling fluctuations contribute at the level of $\sigma/E \approx 0.09(\Delta E(\text{MeV})/E(\text{GeV}))^{1/2}$.

20.5 Calorimeter Systems

The attractive detection properties of calorimeters – energy measurement of charged and neutral particles, improved performance with increasing energy – have made them essential components in modern nuclear and particle physics experiments. In the following, we discuss the most important construction principles, frequently used instrumentation techniques, and calorimeter facilities at the LHC.

20.5.1 Calorimeter Construction Principles

A host of calorimeter techniques have been developed to optimize one or several of the following performance parameters:

1. linear response as a function of energy, covering a range from 100 MeV to >1000 GeV;
2. low instrumental noise of the sensors to avoid a degradation of the energy resolution;
3. high degree of readout segmentation, as required for position resolution and discrimination between electrons (photons) and hadrons based on the differences in shower shape;
4. insensitivity to magnetic fields, temperature variations, or ionizing radiation on the sensors.

Given the very different properties of EM and hadronic showers, different optimization criteria apply to EMCs and HCs.

EMCs may be homogeneous or of the sampling type; HCs are essentially always sampling instruments. The calorimeters must also measure the topology (position, direction) of complex particle collisions requiring calorimeters of high lateral and longitudinal subdivision. Many experiments require precision at the 1% level over the full area of calorimeter facilities.

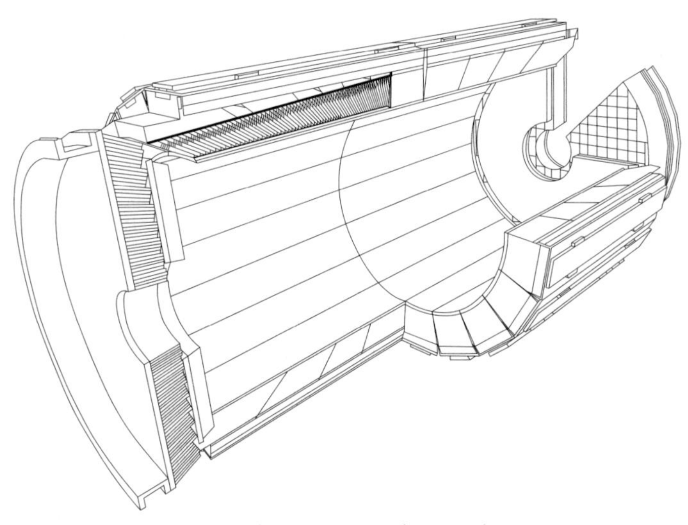

Fig. 20.8 Cutaway view of the CMS crystal EMC. The crystals point toward the collision point and are tapered for crackles coverage of the detection volume. The detector, consisting of 76 000 crystals, has a length of almost 8 m and a diameter of 3.6 m.

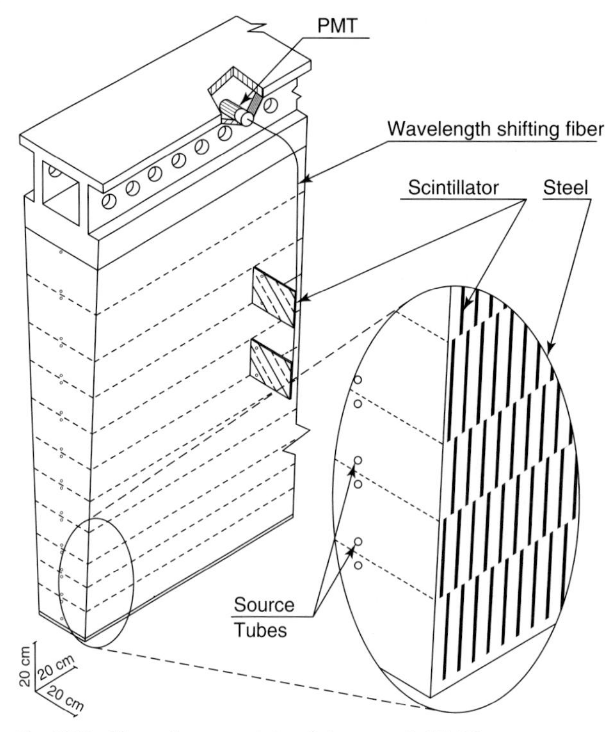

Fig. 20.9 View of one module of the central ATLAS hadronic calorimeter. Thirty-six such modules complete the cylindrical detector. Each of the longitudinally oriented scintillating tiles is read with two wavelength-shifting fibers.

20.5.2
Instrumentation Techniques

All detectors, sensitive to charged particles in the shower cascade, can and are being used to obtain a recordable signal. Two categories are predominant: scintillators (crystals, plastic scintillators, liquefied noble gases) and charge-collecting detectors (liquefied noble gases, semiconductor detectors). A modern example of a homogeneous EMC, using crystals, is the CMS calorimeter, which is installed at the CERN LHC, Figure 20.8. It uses approximately 76 000 $PbW0_4$ crystals (Table 20.1), surrounding the collision point almost "hermetically." This device has been developed to discover the putative Higgs particle via the decay channel $H \rightarrow \gamma\gamma$, requiring a precision of better than 1% at 100 GeV. $PbW0_4$ is the most recent example of a crystal being developed for particle physics by an international, 10-year long R&D effort with the aim to achieve the needed transparency and radiation hardness.

An example of a modern version of plastic scintillator readout is the ATLAS HC, installed at the LHC, Figure 20.9. It is doubly remarkable: the absorber plates, made from steel, are oriented longitudinally, along the direction of the incident hadron. This geometry works as well in the conventional, transverse orientation, because hadrons, after having traversed the preceding EMC, have developed a sufficiently broad shower to be insensitive to the precise orientation of the absorber plates. This geometry was chosen for ease of construction. Scintillation "tiles" are nested in slots; the light is transmitted to and absorbed in "wavelength-shifting" (WLS) fibers, running closely along two edges of the tile. The WLS fibers reemit the absorbed scintillation light at the larger wavelength, to which the fibers are transparent; a group of fibers from tiles covering a certain volume, a "cell," are routed to and read by a photomultiplier. The necessary subdivision of the calorimeter is naturally achieved.

Another novel use of scintillators is in the form of scintillating fibers, one to a few millimeters in diameter. Geometries with fibers perpendicular or along the direction of the incident particle have been used successfully (Fabjan and Gianotti, 2003).

Plastic scintillators are chosen for their potential to tune the e/π response (Section 20.4.4) and their relative facility in calorimeter construction. Systematic effects connected with light collection and material aging need, however, have to be very carefully handled, often requiring complex *in situ* calibration methods.

The second important instrumentation family is based on charge collection from liquid noble gases (Table 20.2). One recent example of a quasi-homogeneous liquid krypton EMC is shown in Figure 20.10. This instrument was built for a fixed target experiment at the CERN SPS to study CP violation, measuring the reaction $K^{\circ}_{s,,l} \rightarrow \pi^0\pi^0$ and $K^{\circ}_{s,,l} \rightarrow \pi^+\pi^-$. For this measurement, an EMC was needed with an energy resolution of $\sim 5\%/\sqrt{E}$ (GeV) and a position resolution of ~1 mm for 25 GeV photons (Fanti *et al.*, 2007). Liquid krypton was chosen as a compromise between the commonly used LAr, which has a large radiation length that makes the homogeneous detector quite bulky and the "ideal" noble liquid, LXe, which, for the volume needed, is prohibitively expensive. The use of LXe for precision photon and electron calorimetry is extremely attractive, the cost (and the necessary purity) however

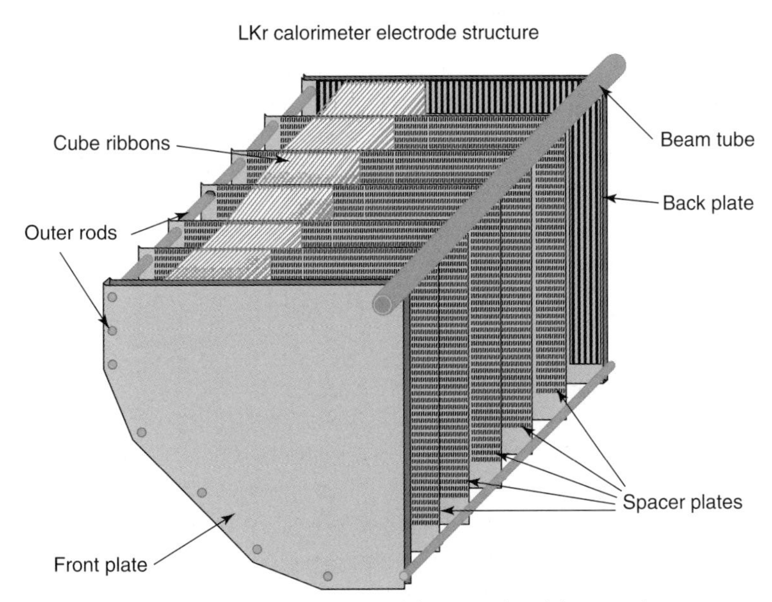

Fig. 20.10 Schematic view of a quarter of the NA48 liquid krypton electromagnetic calorimeter.

being a formidable obstacle. The largest LXe device to date is being prepared for a high-sensitivity search for the forbidden decay mode $\mu \to e\gamma$, in which a 53-MeV photon would be emitted. The MEG Collaboration has constructed a 900-l homogeneous LXe EMC with remarkable performance on energy resolution of $\sigma/E \leq 1\%$ at 100 MeV (a world record!) and a position resolution of $\sigma < 5$ mm is reported (Sawada, 2007). It should be noted that only the scintillation light is recorded.

Liquid argon has been used for more than three decades as a readout medium: it is cheap, can be easily purchased at sufficient purity, and can be conveniently refrigerated with LN_2. The most advanced incarnation is the ATLAS EMC (Figure 20.11). It is a sampling calorimeter with again an unusual absorber structure in the form of "accordions" made of folded Pb plates, covered with a stainless steel skin. This geometry was chosen to achieve the LHC requirements:

- uniform sensitivity to <1%
- relatively fast readout
- good transverse and longitudinal segmentation.

The geometry has allowed the construction of a cylindrical EMC without cracks or insensitive regions. Noble liquids are chosen for the advantages associated with charge collection and its superior control of systematic effects. Above 50 GeV, the ATLAS sampling EMC compares very well

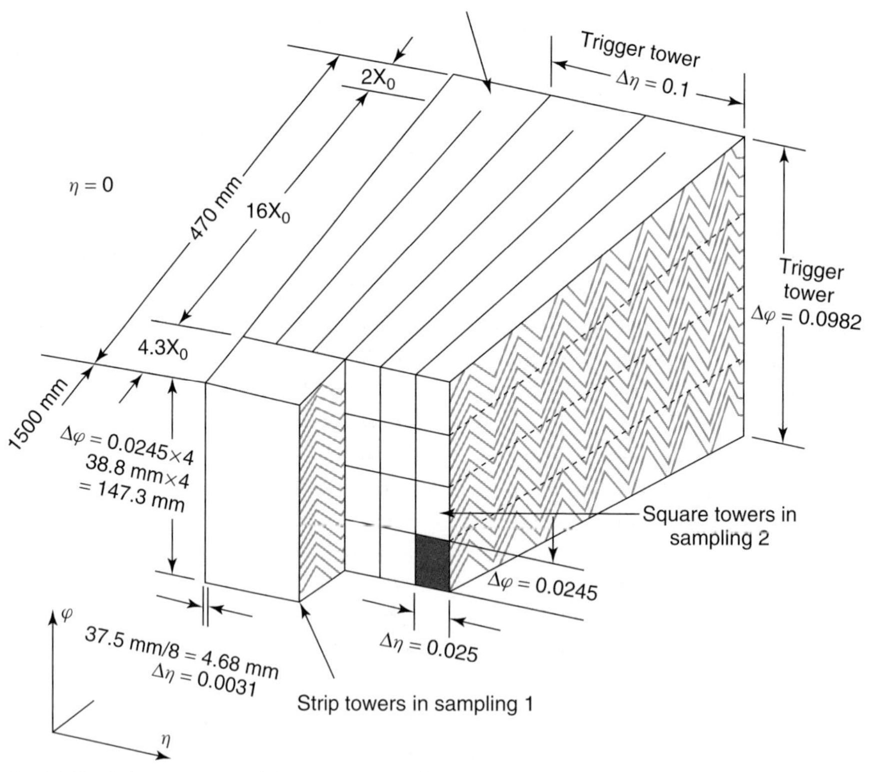

Fig. 20.11 Schematic view of the geometry and segmentation of the ATLAS electromagnetic calorimeter.

in performance with the crystal calorimeter of CMS.

20.5.3 Calorimeter Facilities

The physics and detector requirements for the LHC program have stimulated a major, worldwide R&D effort to meet the long list of conditions:

- **Fast response**: At the LHC design intensity, on average, 30 collisions will take place every 25 ns. Most of these collisions will reflect "conventional" physics with relatively low-energy particles. With fast calorimeter response – 50–100 ns – the integration over $\leq$100 events produces a "low-energy noise" in which the "new" physics signals are still well detectable.
- **Radiation hardness**: This is one of the severest technical requirements. At small angles to the collision axis radiation levels of $\sim 10^{17}$ neutrons/cm^2 and $\sim 10^7$ Gy will be accumulated after 10 years of operation.
- **Construction methods that minimize nonsensitive areas**: Neutrinos, a frequent component of "new" physics, are recognized through an apparent "missing" energy. This methods works, provided cracks do not fake this signal
- **Excellent electromagnetic energy resolution**: The "Holy Grail" of LHC physics is the search for the Higgs particle, which has an important decay channel of $H \rightarrow \gamma\gamma$. A resolution of 1% at 100 GeV is required, combined with an angular measurement of $\sim$5 mrad for 100-GeV photons.
- Large dynamic energy range for electrons, photons, and hadrons from a few gigaelectronvolts to few teraelectronvolts.

In Figure 20.12 the calorimeter facility developed for the ATLAS experiment is shown. The very stringent requirements, combined with more mundane considerations, such as cost, have led to different choices for the EMC and HC; the radiation levels in the small-angle regions are so high that the very robust LAr technique was also chosen for the HC.

20.6 Calorimeters in Astroparticle Physics

This relatively new field addresses questions in cosmology and particle physics that cannot be addressed with accelerator-based experimentation. At the energy frontier of 10^{20} eV are ultra-high-energy cosmic rays, impinging on the Earth at a rate of one every few seconds. Another high-energy "cosmic" phenomenon is the study of sources of very high (hundreds mega electronvolts to few tera electronvolts) photons. Not surprisingly, these studies are mostly based on extensions of the calorimetric principle. Given the rarity of these events, the calorimeter has to be gigantic. It turns out that the Earth's atmosphere is a most precious gift: at sea level it represents approximately 28 radiation and 17 collisions length – an ideal, homogeneously sensitive calorimeter.

Ingenious approaches have been developed to instrument this homogeneous absorber. The relevant performance parameters are as follows (Hoffman, 1999; Pretzl, 2005):

- The energy threshold for particle detection
- The systematic error on the energy scale
- The energy resolution
- Their potential to discriminate between different particles.

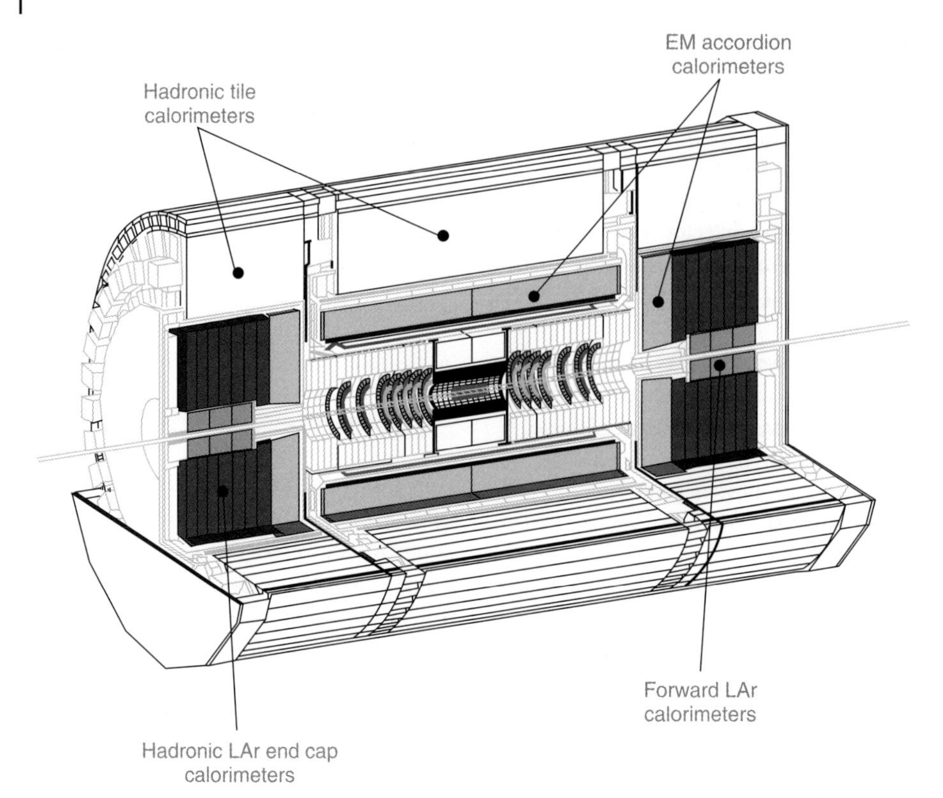

Fig. 20.12 Partial cutaway view of the ATLAS calorimeter system (and the inner charged-particle tracker). The diameter is 8.5 m and the total length 13 m. The outer cylinder and the external end caps measure the hadrons; the inner (light gray) cylinders and end caps are the "Accordion" LAr EMCs. In principle, all particles are absorbed in this calorimeter system with the exception of neutrinos and muons, the latter being measured in the large muon spectrometer surrounding the calorimeters (not shown).

One approach explores the Cherenkov light induced in the atmosphere by the relativistic component of a very high energy shower. This technique is tailored to explore the energy window between space-based detectors, rate limited to $E \leqslant 10\,\text{GeV}$, and ground-based systems, limited by signal-to-noise to $E \geqslant 300\,\text{GeV}$. The lateral distribution of the "Cherenkov shower" is narrowly concentrated around the shower axis and can be observed in a dark night sky. The Cherenkov light is collected with large mirrors and focused onto an array of light detectors, providing the energy, direction, and discrimination between photon- and hadron-induced showers. This technique allowed the recent observation of γ-point sources (Hoffman, 1999).

In another strategy, the fluorescence excited by the passage of the shower particle in the atmosphere provides the measure of the particle's energy. An array of telescopes is used to focus the light onto arrays of photo detectors. One observes a very large volume of the atmosphere, close to $10\,000\,\text{km}^2$ sr in a fine $1^0 \times 1^0$ grid. This technique measures the longitudinal shower profile and the shower maximum, sensitive to the composition of the cosmic rays. With its very high sensitivity, it has recorded some of the highest-energy cosmic rays to date.

The oldest of the three techniques has recently witnessed a remarkable renaissance. It uses ground-based detectors to record the tails of particle cascades initiated by sufficiently energetic primaries. The development of this technique (Cronin et al., 1993) aims at

- Increased sensitivity towards higher energies
- Better understanding of the absolute energy scale
- Identification of the primary composition.

The most ambitious project is the Auger observatory, covering an area of $3000\,\mathrm{km}^2$ with a variety of detectors: ground-based devices to measure the ratio between electromagnetic showers and muons, combined with detectors measuring the atmospheric fluorescence, providing complementary energy and particle-type information. A spectacular proof of the power of this method has been the recent observation that ultra-high-energy (close to $10^{20}\,\mathrm{eV}$) cosmic rays are emitted from extragalactic sources (Auger Collaboration, 2007).

Neutrino astronomy is another recent frontier requiring the most massive detectors: the water of the oceans and the Antarctic icecap have been instrumented to provide prototypes on the way to instrumenting one cubic kilometer (Learned, 2000).

While high-energy astrophysical applications use known calorimetric techniques in innovative ways, the detection of very low energy deposits in the milli electronvolt to electronvolt range requires totally new detection principles, opening the exploration of other fundamental physics questions, such as

1. improved lower limits on the rest mass of neutrinos;
2. the search for weakly interacting massive particles (WIMPs), which may be responsible for the component of dark (invisible and not yet accounted for) matter in the Universe;
3. the search for magnetic monopoles, which may be relics of the Big Bang; and
4. the search for relic neutrinos from the Big Bang, which may have energies of $\sim$1 MeV.

The basic detection principle, registration of the interaction of the particle with the detector during the absorption process, is again calorimetry. However, the energies involved are below the atomic excitation levels, and different detection processes have to be invoked.

These calorimeters are characterized by operation at cryogenic temperatures, in the few millikelvin to $\sim$1 K range. They exploit either some specific properties of the superconducting phase of matter or the reduction in the thermal noise of the detecting medium. Given the very low thresholds of the basic detection process, the number of elementary excitations can be considerably higher; these detectors, therefore, hold the promise of much improved energy resolution.

Operation at cryogenic temperatures opens a number of detection channels:

1. Cooper pairs in superconductors have binding energies in the 10^{-6} to 10^{-3} eV range and may be broken into quasiparticles by phonon (lattice vibration) absorption.
2. The specific heat c for dielectric crystals decreases as $\sim\exp(-1/kT)$, resulting in a measurable, relatively fast temperature rise for very low-energy absorption.

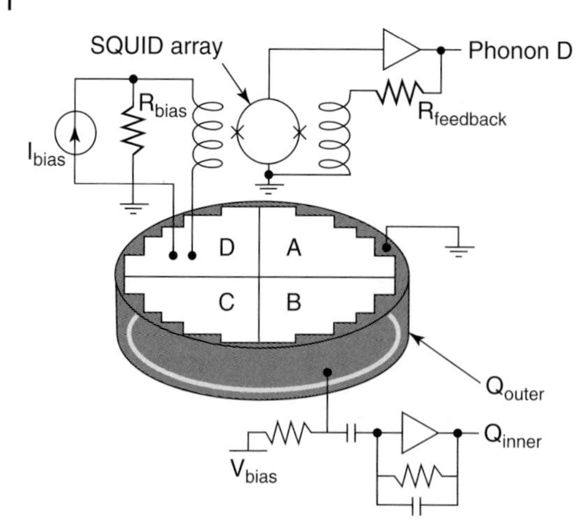

Fig. 20.13 Schematic diagram of a massive (a few hundred grams) Ge detector for dark-matter searches. Simultaneous measurements of phonons with transition-edge sensors and of the ionization charge provide the required background rejection (Cabrera, 2000).

3. Energy absorption may induce a change in magnetization, latent heat release, or quasiparticle multiplication.

A powerful application of the detection in superconductors is being developed for dispersive spectroscopy in the wavelength region from 0.5 nm to 2 μm (Peacock, 1999). Arrays of superconducting tunneling junctions serve the dual function of absorber and detector. The number of excess carriers N produced by the photon absorption as a function of the wavelength λ is

$$N(\lambda) \approx \frac{7 \times 10^5}{\lambda(\mathrm{nm})\Delta(T/T_c)(\mathrm{meV})} \tag{20.13}$$

where Δ is the bandgap. Typical values of N are 10^6 at $\lambda \approx 1$ nm and the spectroscopic resolution is at the level of a few percent.

"Bolometric" detectors are an example of an application of concept 2, the true calorimeters, in which the temperature rise owing to the interaction of particles in an insulating crystal maintained at very low temperatures may be measured with a resistive thermometer. The energy resolution (neglecting noise in the readout electronics) is evaluated to be $\sigma(E) \sim T^{5/2}M^{1/2}$ (M being the mass of the bolometer); X-rays have been measured with 0.1% energy resolution (at 6 keV), superior to that obtained with conventional semiconductor detectors. An example is shown in Figure 20.13: using Ge allows simultaneous measurement of temperature and ionization, providing powerful (and needed) discrimination between the putative WIMP and conventional background (Booth, 1996). Detectors with a mass between 100 and 1000 kg are being considered for sensitive dark-matter searches (Enns, 2005).

Glossary

Calorimetry: In its most general definition, a technique to measure the energy of a reaction or a process. Sometimes, the energy measurement is related to a temperature measurement, hence the name "calorimetry." In particle physics, it is the name given to techniques that measure the energy of particles or quanta.

Gamma Quanta: Energetic quanta of electromagnetic radiation. They may be emitted from the de-excitation of nuclei or created in very energetic interactions of elementary particles with matter.

Hadrons: Generic name for those elementary particles that are carriers of the "strong interaction" (e.g., protons, neutrons, and pions).

Leptons: Generic name for those elementary particles that do not participate in the strong interaction (e.g., electrons, muons, and their corresponding neutrinos).

Nucleon: Generic name for neutrons and protons, the constituents of the atomic nucleus.

Particle Detectors: Instruments to detect or measure properties of particles or radiation quanta. Here, particles designate the microscopic constituents of matter or elementary particles created in energetic collisions of matter. Typically, the trajectory, velocity, energy, or mass of such particles is detected.

Particle Physics: Branch of physics devoted to the study of elementary particles and forces between them. Frequently, very energetic accelerators are used to create or study such particles; this field is therefore often also called *high-energy physics*.

Scintillator: A group of materials that are widely used for particle detection; the material is excited by the passage of charged particles and emits photons (in the UV to visible range) in the de-excitation process. It has important applications outside particle physics.

Solid-state Detectors: Common class of particle detectors. Most frequently used are detectors based on very pure silicon, germanium, or gallium arsenide. The passage of a charged particle creates free charge carriers ("electron–hole" pairs), which may be collected on suitably arranged electrodes. They are used for the precision measurement of the energy loss or the position of charged particles.

References

ATLAS Collaboration (**1996**) CERN/LHCC/96-41.

Auger Collaboration (**2007**) *Science*, **318**, 938.

Booth, N.E. *et al.* (**1996**) *Annu. Rev. Nucl. Part. Sci.*, **46**, 471.

Cabrera, B. *et al.* (**2000**) *Phys. B*, **280**, 509.

CMS Collaboration (**1997**) CERN/LHCC/97-33.

Cronin, J.W. *et al.* (**1993**) *Annu. Rev. Nucl. Part. Sci.*, **43**, 883.

Doke, T. (**2005**) *Nucl. Instrum. Methods B*, **234**, 203.

Enns, C. (**2005**) *Cryogenic Particle Detection*, Springer.

Fabjan, C.W. and Gianotti, F. (**2003**) *Rev. Mod. Phys.*, **75**, 1243.

Fabjan, C.W. and Ludlam, T. (**1982**) *Annu. Rev. Nucl. Part. Sci.*, **32**, 335.

Fanti, V. *et al.* (**2007**) *Nucl. Instrum. Methods*, **A574**, 433.

Fernow, R.C. (**1986**) *Introduction to Experimental Particle Physics*, Cambridge University Press, Cambridge.

Hoffman, C.M. *et al.* (**1999**) *Rev. Mod. Phys.*, **71**, 897.

Learned, J.G. and Mannheim, K. (**2000**) *Annu. Rev. Nucl. Part. Sci.*, **50**, 679.

Lecoq, P. *et al.* (**2006**) *Inorganic Scintillators for Detector Systems*, Springer.

Moses, W.W. (**2002**) *Nucl. Instrum. Methods*, **A487**, 123.

Pretzl, K. (**2005**) *J. Phys.*, **G31**, 133.

Sawada, R. (**2007**) *Nucl. Instrum. Methods*, **A581**, 522.

Wigmans, R. (**1991**) *Annu. Rev. Nucl. Part. Sci.*, **41**, 133.

Wigmans, R. (**2000**) *Calorimetry*, Clarendon Press, Oxford.

Yao, W.M. *et al.*, PDG (Particle Data Group) (**2006**) *J. Phys.*, **G33**, 1.

Further Reading

Ahlen, S.P. (**1980**) *Rev. Mod. Phys.*, **52**, 121.

Froidevaux, D. and Sphicas, P. (**2006**) *Annu. Rev. Nucl. Part. Sci*, **56**, 378.

Grupen, C. (**1996**) *Particle Detectors*, Cambridge University Press.

Radeka, V. (**1988**) *Annu. Rev. Nucl. Part. Sci.*, **38**, 217.

Index

Handbook of Metrology. Edited by Michael Gläser and Manfred Kochsiek

ISBN: 978-3-527-40666-1

f

j

p

t

u

V